Ultraclean Technology Handbook

Volume 1

Ultrapure Water

edited by

Tadahiro Ohmi
Tohoku University
Sendai, Japan

CRC Press
Taylor & Francis Group
Boca Raton London New York

CRC Press is an imprint of the
Taylor & Francis Group, an **informa** business

CRC Press
Taylor & Francis Group
6000 Broken Sound Parkway NW, Suite 300
Boca Raton, FL 33487-2742

First issued in paperback 2019

© 1993 by Taylor & Francis Group, LLC
CRC Press is an imprint of Taylor & Francis Group, an Informa business

No claim to original U.S. Government works

ISBN-13: 978-0-8247-8753-0 (hbk)
ISBN-13: 978-0-367-40234-1 (pbk)

**Visit the Taylor & Francis Web site at
http://www.taylorandfrancis.com**

**and the CRC Press Web site at
http://www.crcpress.com**

Library of Congress Cataloging-in-Publication Data

Ultraclean technology handbook / edited by Tadahiro Ohmi.
    p. cm.
    Includes bibliographical references and index.
    Contents: v. 1. Ultrapure water
    ISBN 0-8247-8753-6 (v. 1)
    1. Manufacturing processes—Cleaning.   I. Ohmi, Tadahiro.
TS183.U46   1993
    670.42—dc20   93-19413
                                              CIP

Preface

The development of ultralarge-scale integration (ULSI) manufacturing has led to the need for ultraclean technology, and nowhere has this need been more prevalent than in semiconductor manufacturing. This technology now is moving into other areas of electronic manufacturing, as well as the pharmaceutical and medical fields.

Ultraclean technology is a total concept of advanced scientific semiconductor manufacturing establishing completely reproducible production of highly valuable ULSIs. This is achieved by creating an environment free of pollutants and contaminants—a substrate surface free from particles, impurities, native oxide and surface microroughness, and allowing perfect process parameter control. As the trend toward developing smaller components, devices, and systems continues, manufacturing lines supported by clean rooms and various utilities become increasingly important.

Ultrapure Water includes detailed chapters on ultrapure water technologies, such as production, component, construction/operation, and quality analysis techniques as well as wet chemical processings. The book has been prepared by over 75 leading experts in the field and includes many of the latest results, such as requirements of ultrapure water quality as it relates to the cleanliness of silicon wafer surfaces.

The book begins with a discussion of the history of structural studies of water which forms an essential basis for the field. It then goes on to detail the processing technologies necessary to build an ultrapure water production system. Following chapters discuss ultrapure water piping systems which are needed to connect production subsystems and use points. The manufacture of hot ultrapure water, which has recently been developed to clean wafers, is then discussed. Methodologies covered include deaeration, which is used exclusively in dissolved oxygen and static electricity removal methods. Specifications for high-purity monitors are also discussed.

These elements are brought together to form the most comprehensive and practical work on this subject to date. The emphasis in presenting the data has been to serve as a working guide for engineers and surface scientists. Therefore, the book will be useful to electrical and electronics, semiconductor process, semiconductor facility, liquid chemical manufacturing, clean gas manufacturing, clean components and manufacturing, assembly, and semiconductor manufacturing machine engineers, as well as all personnel involved in the design, construction, validation, operation, and monitoring of pharmaceutical and medical device clean rooms. It is hoped that this volume will be a valuable resource and a helpful guide in further developing quality manufacturing around the world.

The editor would like to thank the contributors for sharing their expertise. Thanks also to the staff of Marcel Dekker, Inc., for their interest, patience, and cooperation throughout the preparation of this book.

Tadahiro Ohmi

Contents

2. Cleaning Methods

Contributors

Mitsugu Abe Research and Development Laboratory, Nomura Micro Science Co., Ltd., Kanagawa, Japan

Mitsuo Abe Semiconductor Equipment Group, Tomco Manufacturing Ltd., Tokyo, Japan

Seiichiro Aigo Alpha Science Laboratory Co., Ltd., Chiba, Japan

Shin'ichi Akazawa Manufacturing Division, DKK Corporation, Tokyo, Japan

Hiroto Fujii Production and Engineering Department, Kubota George Fischer Ltd., Osaka, Japan

Takaaki Fukumoto Kumamoto Works, Mitsubishi Electronic Corporation, Kumamoto, Japan

Masaharu Hama LSI Laboratory, Mitsubishi Electronic Corporation, Hyogo, Japan

Yukio Hamano Plastics Pipe Engineering Research and Development Center, Sekisui Chemical Co., Ltd., Shiga, Japan

Hiroyuki Harada Technology Affairs Department, Mitsubishi Corporation, Tokyo, Japan

Nobuko Hashimoto Matsudo Research Laboratory, Hitachi Plant Engineering & Construction Co., Ltd., Chiba, Japan

Yoshiaki Hashimoto Design Department, Tokyo Keiso Co., Ltd., Kanagawa, Japan

Nobuyuki Hirose Designing Section, Environmental Systems Division, Shinko Pantec Co., Ltd., Kobe, Japan

Akihiko Hogetsu Planning Department, Shinko Pantec Co., Ltd., Kobe, Japan

Shigenori Hokari Technical Section, Nagano Keiki Seisakusho Co., Ltd., Ueda, Japan

Hiroyuki Horiki Sales Department, Nisso Engineering Co., Ltd., Tokyo, Japan

Riichi Ikegami Engineering Department, Semiconductor Plant Division, Japan Organo Co., Ltd., Tokyo, Japan

Takashi Imaoka Department of Electronics, Faculty of Engineering, Tohoku University, Sendai, Japan

Seiichi Inagaki Sales Department, Nomura Micro Science Co., Ltd., Kanagawa, Japan

Hiroaki Ishikawa Engineering Department, Ultrapure Water System Plant System Division, Kurita Water Industries, Ltd., Tokyo, Japan

Katsuhiko Ito TPI Development Department, Mitsui Toatsu Chemicals, Inc., Tokyo, Japan

Toshihiko Kaneko Taura Works Engineering Department, Hitachi Machinery & Engineering, Ltd., Kanagawa, Japan

Yoh'ichi Kanno UC Division, Motoyama Engineering Works, Ltd., Miyagi, Japan

Hisanao Kano Semiconductor Group, Nippon Rensui Co., Yokohama, Japan

Kotaro Karita Engineering Department, Shingu Factory, Teikoku Electric Manufacturing Co., Ltd., Hyogo, Japan

Michiya Kawakami Tokyo Central Research Laboratory, Mitsubishi Gas Chemical Company, Inc., Tokyo, Japan

Ichiro Kawanabe Analytical Section, Hashimoto Chemical Corporation, Sakai, Japan

Frederick W. Kern, Jr. General Technology Division, IBM Corporation, Essex Junction, Vermont

Hirohisa Kikuyama Research and Development Section, Hashimoto Chemical Corporation, Osaka, Japan

Hiromi Kohmoto Engineering Department, Engineering Division, Nomura Micro Science Co., Ltd., Kanagawa, Japan

Katsumi Koike Engineering Department, Semiconductor Plant Division, Japan Organo Co., Ltd., Tokyo, Japan

Shoji Kubota Hitachi Research Laboratory, Hitachi, Ltd., Ibaraki, Japan

Toshio Kumagai Research and Development Division, Kurita Water Industries, Ltd., Kanagawa, Japan

Junsuke Kyomen Plastic Pipe Research and Development Department, Kubota Corporation, Osaka, Japan

Takeo Makabe Product Development and Sales Department, Business Development Headquarters, Nomura Micro Science Co., Ltd., Kanagawa, Japan

Toshiki Manabe Research and Development Section, Semiconductor Plant Division, Japan Organo Co., Ltd., Saitama, Japan

Yoshiaki Matsushita Semiconductor Materials Engineering Department, Toshiba Corporation, Kawasaki, Japan

Nobuhiro Miki Hashimoto Chemical Corporation, Osaka, Japan

Hiroyuki Mishima Special Equipment and Chemicals Department, Tokuyama Soda Co., Ltd., Yamaguchi, Japan

Masami Miura Semiconductor Equipment Division, Hakuto Co., Ltd., Tokyo, Japan

Masayuki Miyashita   Research and Development Section, Hashimoto Chemical Corporation, Sakai, Japan

Tetsuo Mizuniwa   Research and Development Division, Kurita Water Industries, Ltd., Kanagawa, Japan

Yoshito Motomura   Ultra Pure Water Division, Kurita Water Industries, Ltd., Tokyo, Japan

Shigeharu Nakamura   Technical Section of Chofu-kita Plant, Kobe Steel Ltd., Shimonoseki, Japan

Takao Nakazawa   Sales Department, Haruna Incorporated, Tokyo, Japan

Kenji Oda   Research and Development Laboratory, Nippon Rensui Co., Yokohama, Japan

Shosuke Ohba   Production and Engineering Department, Kubota George Fischer Ltd., Osaka, Japan

Tadahiro Ohmi   Department of Electronics, Faculty of Engineering, Tohoku University, Sendai, Japan

Yoshiharu Ohta   Research and Development Laboratory, Nomura Micro Science Co., Ltd., Kanagawa, Japan

Makoto Ohwada   Taira Plant Engineering Section, Alps Electric Co., Ltd., Fukushima, Japan

Motohiro Okazaki   Membranes Laboratory, Polymers Research Laboratories, Toray Industries, Inc., Shiga, Japan

Makoto Saito   Manufacturing Division, DKK Corporation, Tokyo, Japan

Masao Saito   Water Treatment Department, Shinko Pantec Co., Ltd., Kobe, Japan

Takashi Sasaki   Technical Development Division, Shinko Pantec Co., Ltd., Kobe, Japan

Hitoshi Sato   Matsudo Research Laboratory, Hitachi Plant Engineering & Construction Co., Ltd., Chiba, Japan

Kenichi Sato   Scientific and Laboratory Services Division, Nihon Pall Ltd., Tokyo, Japan

Yukinobu Sato   Research and Development Division, Kurita Water Industries, Ltd., Atsugi, Japan

Makoto Satoda   Process Instruments Department, DKK Corporation, Tokyo, Japan

Rokuheiji Satoh   Design Section Pump Engineering Department, Nikuni Machinery Industry Co., Ltd., Kanagawa, Japan

Sachio Satoh   Engineering Department, Semiconductor Plant Division, Japan Organo Co., Ltd., Tokyo, Japan

Koichi Sawada   Research and Development Laboratory, Nomura Micro Science Co., Ltd., Kanagawa, Japan

Yoshio Senoo   Research Laboratory, Tokico Ltd., Kawasaki, Japan

Yoshiki Shibata   Electronic Products Factory, Toray Engineering Co., Ltd., Shiga, Japan

Hirotake Shigemi   Research and Development Division, Kurita Water Industries, Ltd., Kanagawa, Japan

Ikuo Shindo Research and Development Section, Semiconductor Plant Division, Japan Organo Co., Ltd., Saitama, Japan

Takeshi Shinoda Water Treatment Research and Development Department, Hitachi Plant Engineering & Construction Co., Ltd., Tokyo, Japan

Isamu Sugiyama Research and Development Laboratory, Nomura Micro Science Co., Ltd., Kanagawa, Japan

Sankichi Takahashi Energy Engineering Department, Hachinohe Institute of Technology, Aomori, Japan

Kazuhiko Takino Water Treatment Division, Hitachi Plant Engineering & Construction Co., Ltd., Tokyo, Japan

Mamoru Torii Osaka-Technoport Laboratory, Fujikin Incorporated, Osaka, Japan

Minami Tsuchizaki Japan Microbiological Clinic Co., Ltd., Kanagawa, Japan

Tomoyuki Ueda Molding Department, Asahi Yukizai Kogyo Co., Ltd., Miyazaki, Japan

Norihisa Urai Engineering Department, Semiconductor Plant Division, Japan Organo Co., Ltd., Tokyo, Japan

Kenichi Ushikoshi Environmental Systems Division, Technical Department, Shinko Pantec Co., Ltd., Kobe, Japan

Koichi Wada Manufacturing Department, Enameling Section, Chemical Process Equipment Division, Shinko Pantec Co., Ltd., Kobe, Japan

Koichi Yabe Technical Engineering Department, Kurita Water Industries, Ltd., Tokyo, Japan

Norikuni Yabumoto Center for Analytical Technology and Characterization, NTT, Kanagawa, Japan

Yasuyuki Yagi Matsudo Research Laboratory, Hitachi Plant Engineering & Construction Co., Ltd., Chiba, Japan

Akira Yamada System Department, Shinko Pantec Co., Ltd., Kobe, Japan

Yoshitaka Yamaki Research and Development Laboratory, Nomura Micro Science Co., Ltd., Kanagawa, Japan

Katsumi Yamazoe Fundamental Technology Section, Technical Development Division, Shinko Pantec Co., Ltd., Kobe, Japan

Motonori Yanagi LSI Laboratory, Mitsubishi Electronic Corporation, Hyogo, Japan

I
Quality of Ultrapure Water

1
Requirements for Ultrapure Water

Takaaki Fukumoto
Mitsubishi Electronic Corporation, Kumamoto, Japan

Current emphasis is focused on a semiconductor technology capable of handling fine processing, for 0.1–1.0 μm pattern formation, for example, or the formation of thin films of 10 nm or less, and of improving the yield of such fine-geometry devices. To achieve this goal, it is essential to find ways to minimize impurities and particle contamination in the production process. Impurities and particles should not be generated during the process or allowed to adhere to any surface. In other words, the entire environment, including materials, jigs (e.g., cassettes and chucks), operators, and robots, to which wafers are exposed during production—wafer processing, transport, and storage—demands rigorous standards of cleanliness and purity.

This is not effective enough, however, to remove contaminants on the wafer surface, and this is where the cleaning process plays an important role. Although the dry process has increasingly been introduced as devices become more complex and integrated, the cleaning process remains vital to remove metal contamination from the wafer surface and the surface excitation defect layer.

Pure water was first produced around 1945 using the ion-exchange resin commercialized in the United States. The quality of pure water has been greatly improved ever since, and accordingly the variables of quality to be controlled have been diversified. Recently the resistivity has neared the theoretical limit. It is necessary to lower the amount of fine particles and bacteria in pure water to the level of the detection limit. Moreover, to improve the cleanliness of the interface of ultrathin films formed on devices, concentration of harmful dissolved impurities should be lowered to the parts per billion to trillion level.

This section is a basic discussion of water, including its structure and the properties of ultrapure water.

2
The Structure of Water

Shoji Kubota
Hitachi, Ltd., Ibaraki, Japan

I. INTRODUCTION

Water is a vital material resource not only for all living creatures but also for industry.

Although water is generally expressed as H_2O, natural water is a mixture of nine isotopic compounds, as shown in Table 1. Since $H_2^{16}O$ predominates, water is usually represented by $H_2^{16}O$ [1].

Any substance can exist in three states—gas, liquid, and solid. These three states are defined by the interactions of a substance's constituent molecules, ions, and atoms. Liquid water is intermediate between ice, whose component particles are closest and have the strongest interaction, and steam, whose component particles have no interaction.

Table 1 Composition of Natural Water, H_2O

$H_2^{16}O$	99.76 (%)
$H_2^{18}O$	0.17
$H_2^{17}O$	0.037
$HD^{16}O$	0.032
$HD^{18}O$	0.00006
$HD^{17}O$	0.00001
$D_2^{16}O$	0.000003
$D_2^{18}O$ $D_2^{17}O$	0.000001

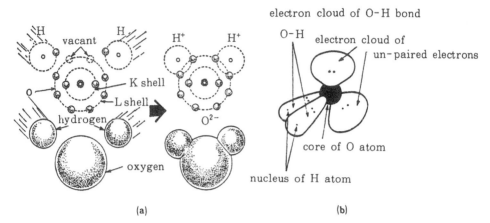

Figure 1   Formation of the water molecule and its electron cloud.

Unlike ice, liquid water does not have a regular structure, although various models have been suggested.

II. STRUCTURE OF THE WATER MOLECULE

The water molecule consists of a $1s$ electron on the K shell of a hydrogen atom bonded with the sp^3 composite orbit formed by the $2s2p_x2p_y2p_z$ orbit on the L shell of the oxygen atom. In the water molecule, oxygen has eight electrons on the outermost shell and hydrogen has two. This forms a closed shell structure for both oxygen and hydrogen, stabilizing the molecule. Figure 1a [2] shows the structure of the water molecule, and Fig. 1b [3] illustrates the electron cloud.

    Figure 2a illustrates the dimensions and bonding angle of the water molecule. ∠HOH, the bonding angle between oxygen and the hydrogen atoms, is 104.52°. This

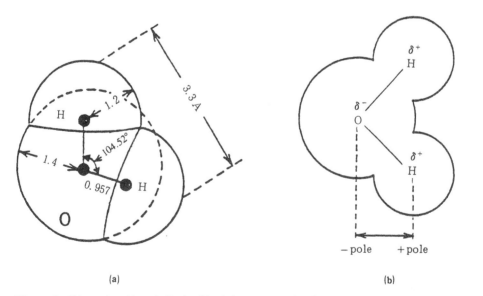

Figure 2   Dimension (a) and dipole (b) of the water molecule.

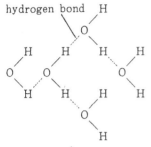

Figure 3   Hydrogen bonds of water molecules.

figure is rather large compared to the angles in the homologous series H_2S (92.16°), H_2Se (90.53°), H_2Te (90.25°), and so on. The angle 104.52° is close to the angle of a regular tetrahedron, 109.5°.

The center of gravity of the bonded electrons nears the oxygen atom slightly as the electrical negativity increases. As shown in Fig. 2b, the oxygen atom becomes more negative as the hydrogen atoms become more positive. Water is a polar molecule. Moreover, since the center of gravity of the positive electrical charge does not match that of the negative electrical charge, the water molecule has a dipole moment. Therefore, in the ice and liquid states shown in Fig. 3, each H_2O molecule forms hydrogen bonds, an association that can be loosely represented as $(H_2O)_n$.

III.   UNIQUE PROPERTIES OF WATER

Figure 4 [4–6] presents the relation between the temperature and density of pure water. When the temperature of water is lowered, the density rises to peak at 4°C. As the

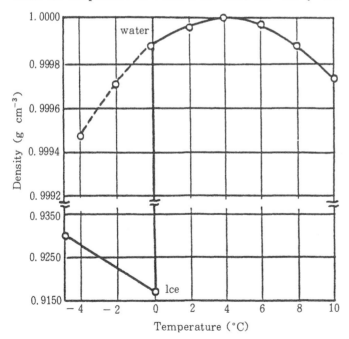

Figure 4   Relation between temperature and density of pure water (broken line shows supercooling condition).

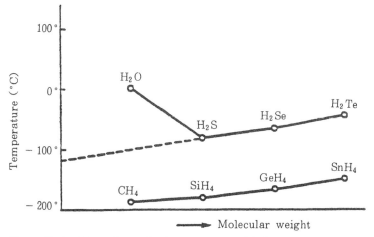

Figure 5 Melting points of hydrogen compound belonging to the homologous series.

temperature is lowered further, however, the density decreases, and when water turns to ice at 0°C, the density suddenly drops. Below zero, the density rises as the temperature decreases. In most liquids, the density increases as the temperature increases, and usually when a liquid becomes a solid, its density increases further.

This is one of the characteristics of water that demonstrate its difference from other liquids. This unique behavior of water helps fish to survive in ponds during winter. When the temperature of the air decreases in winter, the density of the upper layer of water rises, and this upper layer sinks toward the bottom of the pond. To replace this descending upper layer of water, a warmer bottom layer of water rises to the surface. Since the temperature of the bottom layer of water is thus maintained at 4°C, the bottom layer is never frozen, even when the pond surface is covered with ice, which enables fish to survive during winter.

Figures 5 and 6 show the melting points and boiling points of hydrogen compounds

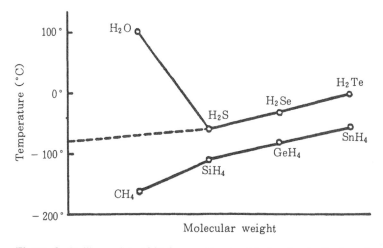

Figure 6 Boiling points of hydrogen compound belonging to the homologous series.

Table 2 Heat of Melting and Vaporization of Typical Liquids

	Heat of melting		Heat of vaporization	
	kJ/mol	cal/g	kJ/mol	cal/g
Water (H_2O)	6.01	80	40.7	540
Acetone (CH_3COCH_3)	5.69	23	29.0	125
Ethyl alcohol (C_2H_5OH)	5.02	23	38.6	204
Formic acid (HCOOH)	12.7	66	22.3	116
Benzene (C_6H_6)	9.84	30	31.7	94

belonging to the homologous series as a function of molecular weight [7]. The bottom curves in Figs. 5 and 6 present series IVB of the periodic table. In general, in a homologous series, both melting points and boiling points decrease as the molecular weight decreases. In series VIB, however, which includes water, the melting point and boiling point of water are much higher than the values that can be estimated from the values recorded for other numbers of the series. This is thought to be a result of the hydrogen bonds, which increase the size of the molecule.

Table 2 lists the heat of melting and the heat of vaporization of typical liquids. When values per gram are compared, the heat of melting and the heat of vaporization of water are the highest. It is also noted that, for water, the heat of vaporization is seven times as high as the heat of melting. This is thought to be because the conversion from liquid water to steam requires cutting seven times as many hydrogen bonds as the conversion from ice to liquid water.

Table 3 lists the surface tension of various liquids. The surface tension of water is the greatest of all the liquids shown here. A liquid always appears to shrink from its surface. Surface tension γ is defined as the energy required to increase the liquid surface by 1 cm^2. γ can be expressed as

$$\gamma = \frac{erg}{cm^2} = \frac{dyn \times cm}{cm^2} = \frac{dyn}{cm} = 10^{-5} \ N \cdot cm^{-1} \tag{1}$$

Although γ is measured in dynes per centimeter, it measures the amount of energy, not the degree of dynamics. The stronger the hydrogen bond or van der Waals force that attracts

Table 3 Surface Tension of Various Liquids ($\times 10^5$ N·cm^{-1}, 20°C)

Water (H_2O)	72.75	Chloroform ($CHCl_3$)	27.1
Carbon disulfide (CS_2)	32.4	Ethyl alcohol (C_2H_5OH)	22.3
Acetic acid (CH_3COOH)	27.6	Cyclohexane (C_6H_{12})	25.3
Acetone (CH_3COCH_3)	23.7	Ethyl acetate ($CH_3COC_2H_5$)	23.9
Carbon tetrachloride (CCl_4)	26.8	Diethyl ether ($C_2H_5OC_2H_5$)	17.0
Benzene (C_6H_6)	28.9	n-Hexane ($CH_3(CH_2)_4CH_3$)	18.4
Nitrobenzene ($C_6H_5NO_2$)	43.6		

molecules to each other, the greater is the energy required. Therefore, the surface tension of a liquid is in proportion to the cohesion among component molecules of the liquid. The high surface tension of water can be attributed mainly to its hydrogen bonds.

The high surface tension of water enables the human body to send blood to every part and tall trees to send water to their tops. Surface tension is also very important in connection with the efficiency of cleaning.

Table 4 shows the molar heat capacity at constant pressure and the specific heat capacity of typical liquids. The specific heat capacity of water is greater than that of other liquids. Along with the large heat of vaporization of water, the large specific heat capacity of water significantly contributes to the temperature control of the environment and of living creatures.

Table 5 summarizes the various properties of different kinds of molecular liquids [8]. The size of a molecule, that is, the molecular weight, is closely related to intermolecular actions. When the molecular weight increases, the polarizability rises because the number of electrons in the molecule increases. Consequently, the intermolecular action increases. Generally, when the molecular weight rises, the boiling point rises. Even with small molecular weights, if a special interaction is provided to molecules, such as hydrogen bonds, the melting point and boiling point increase. The melting point and boiling point of water are higher than those of other similar liquids, such as methanol and ethanol. For alcohol this is because the methyl group is not able to form hydrogen bonds. Consequently, the intermolecular action of alcohol is weaker than that of water although its molecular weight is larger.

Viscosity is defined as the energy per unit area and per unit time required to transfer a molecule to a neighboring position. Therefore, viscosity depends on the size of the molecule and the intensity of intermolecular action. In general, the larger the molecular weight, the higher is the viscosity. Water has strong hydrogen bonds, and thus the viscosity is relatively high despite its small molecular weight. Methanol and ethanol have hydrogen bonds similar to those of water. The viscosity of methanol is lower than that of water, although its molecular weight is greater. On the other hand, the viscosity of ethanol is higher than that of water. This is because water has stronger bonds than methanol and the molecular weight of ethanol effectively works to raise the viscosity.

The dipole moment of a molecule is one of the indicators of the intensity of intermolecular action. The intermolecular action of polar molecules is generally stronger

Table 4 Molar Heat Capacity at Constant Pressure and Specific Heat Capacity

Substance	Molecular weight	Molar heat capacity at constant pressure $JK^{-1}mol^{-1}$	Specific heat capacity $JK^{-1}g^{-1}$
Water (H_2O)	18.01	75.23	4.179
Acetone (CH_3COCH_3)	58.08	125	2.152
Ethyl alcohol (C_2H_5OH)	46.07	111.4	2.418
Formic acid (HCOOH)	46.03	99.04	2.152
Benzene (C_6H_6)	78.12	136.1	1.742

Table 5 Various Propertes of Different Kinds of Molecular Liquids [8]

Substance	M.W.	b.p. (°C)	m.p. (°C)	d g cm^{-3} (25°C)	η millipoise (25°C)	ε_r (25°C)	μ (10^{-30} cm)	δ (cal.cm^{-3})$^{1/2}$	D_N	A_N	pK_i
Acetic acid	60.05	117.8	16.64	1.0492*	12.2*	6.15	5.67	13.01	–	52.9	14.5
Acetone	58.08	56.2	−95.4	0.7845	3.02	20.7	9.76	9.62	17.0	12.5	–
Acetonitrile (AN)	41.05	81.6	−45.7	0.7766	3.39	35.95	13.06	12.11	14.1	18.9	28.5
Benzene	78.12	80.122	5.493	0.87903	6.49	2.28	0.00	9.16	0.1	8.2	–
Carbon tetrachloride	153.82	76.7	−22.6	1.595*	9.75*	2.23	0.00	8.55	–	8.6	–
Chloroform	119.38	61.27	−63.49	1.4891*	5.55**	4.724	3.40	9.16	–	23.1	–
N-dimethylformamide (DMA)	87.12	165.0	20	0.9366	9.19	37.78	–	–	27.8	13.6	–
Dimethylformamide (DMF)	73.10	158	−61	0.9443	7.96	36.71	12.86	11.79	26.6	16.0	–
Dimethyl sulfoxide (DMSO)	78.14	189.0	18.55	1.096	19.6	46.6	–	12.97	29.8	19.3	~32
Ethyl alcohol	46.07	78.32	−114.15	0.7851	10.78	24.3	5.63	12.78	20	37.1	18.9
Formamido (FA)	45.04	210.5	2.55	1.12918	33.02	111.0	11.2	–	24	39.8	–
Hexamethyl phosphoramide (HMPA)	179.20	235	7.20	1.024***	–	29.6	–	11.35	38.8	10.6	–
n-Hexane	86.18	68.7	−94.3	0.6594*	3.258*	1.90	0.00	7.27	–	0.0	–
Methyl alcohol	32.04	64.75	−97.68	0.7866	5.42	32.6	5.63	14.50	19.0	41.3	16.7
N-methyl formamide (NMF)	59.07	180−185	−3.8	0.9988	16.5	182.4	12.9	–	–	–	–
Nitrobenzene (NB)	123.11	210.80	5.76	1.1986	18.11	34.82	14.03	11.06	4.4	14.8	–
Nitromethane (NM)	61.04	101.2	−28.6	1.1312	6.27	35.94	11.53	12.90	2.7	20.5	19.5
Propylene carbonate (PC)	102.09	241	−49	1.19	25.3	64.4	–	13.34	15.1	18.3	–
Pyridine (Py)	79.10	115	−41.5	0.9779	8.824	12.01	7.17	10.62	33.1	14.2	–
Tetrahydrofuran (THF)	72.11	65.0	−108.5	0.880	4.6	7.39	5.67	9.52	20.0	–	–
Water	18.01	100.00	0.00	0.9971	8.903	78.54	6.47	23.53	18.0	54.8	14.0

* at 20°C.　** at 22.8°C.　*** at 30°C.

Notes　M.W. : Molecular weight　　　　d　: Density　　　　　　　　　　μ　: Dipole moment

b.p. : Boiling point　　　　　　η　: Viscosity　　　　　　　　　δ　: Solubility parameter

m.p. : Melting point　　　　　　ε_r　: Relative permittivity　　D_N : Donor number

A_N : Accepter number

pK_i : Autoprotolysis constant

11

than that of nonpolar molecules as the electrostatic action between the dipoles is applied. Although the electrostatic interaction is not the only factor that determines intermolecular action, there is a tendency for a substance composed of molecules with a larger dipole moment to have a higher melting point, boiling point, and viscosity.

It is widely considered that water is a good solvent because it is a polar compound with a large relative dielectric constant. It is true that the relative dielectric constant of water is large, but as shown in Table 5, formamide and N-methylformamide show a much higher relative dielectric constant than water. Many compounds, including nitrobenzene and acetonitrile, feature a much larger dipole moment than water. Nevertheless, water is superior to any other compound as a solvent.

It is clear from this discussion that any single property of water cannot determine all its various characteristics. The various properties of water enable living creatures to survive. Moreover, but for the various properties of water, the environment of the earth itself, which enables living creatures to survive, could not be maintained.

IV. STRUCTURE OF ICE

Ice [9] presents various modifications under different conditions of temperature and pressure. At present at least nine phases of ice have been confirmed. Under ambient pressures, however, there are only three kinds of ice: I_h, I_c, and glassy ice, where h is a hexagonal crystal and c is a cubic crystal.

I_c is formed when steam is sublimated and condensed on a metal surface that is cooled to $-100°C$ or lower. Glassy ice is reported to grow when steam is rapidly condensed at $-160°C$ or less. Both I_c and glassy ice are amorphous ice, and their details have not yet been revealed. Because they are only formed under special conditions, however, only I_c can be observed under ambient pressures.

Figures 7 and 8 show the structure of the I_c crystal. As shown in Fig. 7, four H_2O molecules are connected to each other by hydrogen bonds, forming the tetrahedral

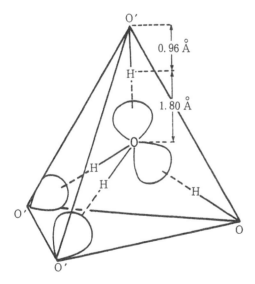

Figure 7   Tetrahedrally coordinated H_2O model in ice.

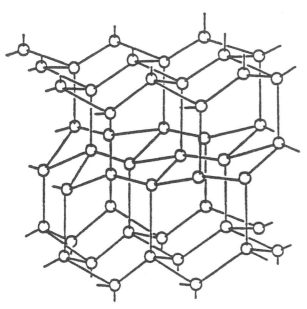

Figure 8 Structure of I_h. Circle denotes oxygen atom (hydrogen atom not indicated).

structure. At the same time, a hexagonal crystal structure similar to a diamond is formed, as shown in Fig. 8. The distance of the nearest neighbors, that is, the distance between the two closest oxygen atoms, is 2.76 Å (276 pm). This ice structure contains a large interior hollow space. The size of the hollow space corresponds to the size of a sphere with radius of about 3.50 Å (350 pm). The bonding angle $\angle O' \ldots O \ldots O'$ is 109.5°. This is the angle of a tetrahedral structure, which is 5° wider than the bonding angle of the water molecule (104.5°). The distance between the two closest molecules not attached by hydrogen bonds is 4.50 Å (450 pm). The number of O-H vibrations as determined by infrared spectroscopy is 3220 cm^{-1}.

Figure 9 is a phase diagram for H_2O [10]. When the pressure is raised by maintaining the equilibrium between ice and water, the temperature of ice decreases, as shown in Fig. 9. When the pressure is raised further, the ice-water equilibrium curve changes along with *T-E*. Even if the pressure is increased further, however, a critical point, such as point *C* in the gas-liquid equilibrium, does not appear. The *T-E* curve for ice has a negative slope, which is quite different from the usual gas-liquid equilibrium system. The slope of the equilibrium curve between ice and water can be expressed as

$$\frac{dT}{dP} = \frac{T(V_L - V_s)}{\Delta H_f} = -7.51 \times 10^{-3} \text{ K atm}^{-1} \quad 0°C \tag{2}$$

where: V_L = molar volume of water
V_S = molar volume of ice
ΔH_f = heat of melting of ice

Equation (2) shows that the melting point of ice falls by 0.0075 K when the pressure is raised by 1 atm. This means ice melts when the pressure is raised. This phenomenon explains how a skater can glide on the ice. Water has a smaller molar volume than ice. When ice is melted at 0°C, the volume is decreased:

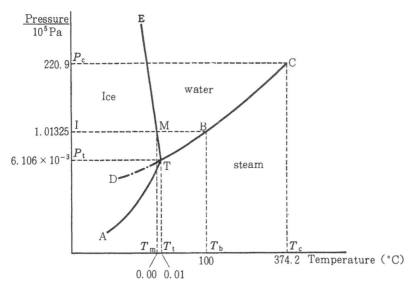

Figure 9   Phase diagram of water (both axes are arbitrary scale).

$$\Delta V = V_L - V_S = 18.0128 - 19.6463 = -1.6335 \text{ cm}^3 \text{ mol}^{-1} \tag{3}$$

1 mol per ice. When ice is melted to water, the volume decreases by about 8.3%. On the other hand, when water turns into ice, the volume increases by about 9.1%.

V.   STRUCTURE OF WATER

Water is a liquid [11]. Liquids do not have a clear structure like solids. Therefore, unlike ice, it is not easy to determine the structure of water. Various models have been suggested, but a final conclusion has not yet been presented. This section reviews studies of the structure of water and discusses the latest study conducted by Arakawa [11], which comprehensively examined the liquid structure of water.

A.   Three Major Models

The first study of the liquid structure of water was conducted by applying x-ray diffraction. The results were presented by Bernal and Fowler [12] in 1933 and by Morgan and Warren [13] in 1938. These studies were followed by the studies of Hall [14] in 1948 and Pople [15] in 1951. The study by Bernal and Fowler and Pople's study presented a starting point for the continuum model, one of the three major models of the liquid structure of water. This model is also called the *pseudocrystal structure model*. The study of Morgan and Warren first presented the *interstitial model*. The studies of Eucken and Hall presented the *mixture model* for the first time.

Bernal and Fowler presented an irregular four-coordination structure as the structure of water. The x-ray diffraction intensity of water and ice was measured, and then the density maximum, the shrinkage of volume in the process of melting, dielectric characteristics, and other properties were studied by comparing radial distribution curves, for example.

Warren conducted an x-ray diffraction test of water at 1.5–85°C using a unique tungsten (W) ray diffraction camera. Figures 10 and 11 show the results. In the radial distribution curve shown in Fig. 11, apart from 2.76 Å (276 pm), the second peak appears at about 4.50 Å (450 pm). The radial distribution curve expresses the density of molecules that usually exist around a molecule.

If it is assumed that each water molecule is surrounded by four water molecules in a tetrahedral manner, the second neighbor distance calculated based on the first neighbor distance of 2.76 Å (276 pm) is 4.51 Å (451 pm), which corresponds to the second peak. In Fig. 11, the location of the first peak indicates a greater distance as the temperature rises: 2.90 Å (290 pm) at 1.5°C and 3.05 Å (305 pm) at 83°C. Table 6 presents the number of nearest neighbors calculated from the area of the first peak. The number of nearest neighbors increased as the temperature rose: 4.4 at 1.5°C to 4.9 at 83°C. At room temperature, this was within the range 4.4–4.6. Bernal and Fowler's study indicated that for ice, the number of nearest neighbors was 4. For this deviation, Morgan and Warren calculated an icelike distribution curve *B* using the structure of ice as a reference. As shown in Fig. 12, Morgan and Warren found that when the calculated icelike distribution curve *B* was deducted from the observed radial distribution curve *A*, the result corresponded to curve *C* [13]. This means that for liquid water, apart from the icelike four-coordination structure, interstitial molecule exists between the first and second neighbor distances. The proportion of interstitial molecule increases as the temperature increases. Based on this finding, Morgan and Warren considered the liquid structure of water as a broken icelike structure.

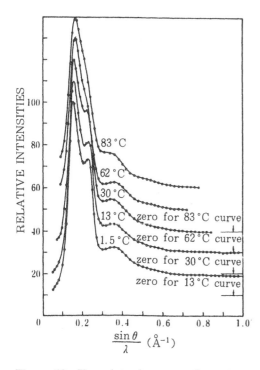

Figure 10 X-ray intensity curves for water, corrected for absorption and polarization. The intensity is in arbitrary units [13].

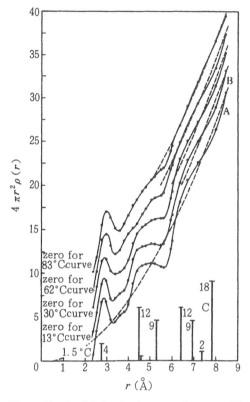

Figure 11   Radial distribution curve for water. The vertical lines at the bottom give the number and position of the neighbors in ice [13].

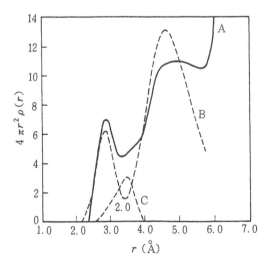

Figure 12   Comparison of the 1.5°C water distribution curve with a smoothed ice-like distribution [13]: (A) water distribution, (B) calculated distribution, and (C) difference.

Morgan and Warren presented the first interstitial model in this way, and the idea was further developed by Samoilov and by Narten et al. [16].

In 1946, Eucken considered water an equilibrium mixture of monomolecule, dimer, tetramer, and octamer. This was the beginning of the mixture model. The idea of a mixture model was first developed by Hall, who looked into the ultrasonic properties of water [14]. This study indicated that in ordinary water there were an icy water state I and a water state II. In terms of density and energy, water state I showed lower values and water state II high values. Actual water was considered a mixture of the two.

Meanwhile, Pople [15] suggested a continuum model, which is different from the idea of Bernal and Fowler [12]. Bernal and Fowler attributed the less perfect characteristics of the structure of water compared with the structure of ice to the breaking of hydrogen bonds. Pople explained this by pointing out the deformation from linearity.

In 1967, about 30 years after the Bernal and Fowler study, Narten et al. conducted an x-ray diffraction test of water at a temperature of 4–200°C, applying the latest technology [16]. Figure 13 shows the radial distribution curve based on the test results. Assuming the

Table 6 Number of Nearest Neighbors in Water [13]

Temperature (°C)	1.5	13	30	62	83
Number	4.4	4.4	4.6	4.9	4.9

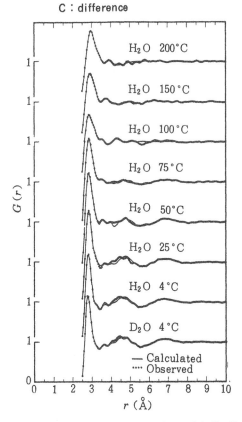

Figure 13 Observed and model radial distribution functions of water [16].

framework of the liquid structure of water is a structure like ice I, Narten et al. arranged interstitial molecules in the hollow space on the tertiary symmetrical axis in the direction of the C axis. The values observed are a good match with the calculated values in the radial distribution curve shown in Fig. 13. In the study by Narten et al., the number of first neighbors was stable at 4.4 in the temperature range 4–100°C and 4.5 at 200°C. Although the results are slightly different from those of the Morgan and Warren study in terms of temperature dependence, the results of the two studies are in good agreement at around room temperature. This determined the water structure observed with the x-ray diffraction method: water has the structure of interstitial model in which interstitial molecules exist in the hollow space of the ice I-like structure.

The x-ray diffraction method was able to observe the oxygen atom in large part but was not able to analyze the hydrogen atoms. Neutron diffraction was required, and it was not until the 1970s that this method was available.

During the 1960s, Nemethy and Scheraga [17] presented important results in terms of water structure by applying a method different from x-ray diffraction. From the standpoint of liquid theory, in 1962 Nemethy and Scheraga investigated the equilibrium mixture model by comparing the cluster structure formed by hydrogen bonds and monomolecular water using statistical dynamics [17]. They indicated that this cluster structure repeated formation and extinction within a short interval. Figure 14 shows the model. Since the cluster alternates formation and extinction, it is called a flickering cluster. Its lifetime was determined by Frank and Wen as approximately 10^{-11}–10^{-10} s. This is the time during which water molecules are able to maintain the translation movement 100–1000 times. At present, however, the relaxation time is converging to around 10^{-12} s.

Nemethy and Scheraga calculated various values in terms of thermodynamics and the proportions of each hydrogen bond state that forms monomolecule and cluster. They set the energy level, the partition function, and other criteria for a mixed system of five states: monomolecule and 1-, 2-, 3-, and 4-hydrogen bond states. They then conducted calculations by statistical dynamics. The calculated values in terms of thermodynamics, such as free energy, internal energy, and entropy, were almost the same as the observed values. The cluster size was determined as 50–60 water molecules at room temperature. This was

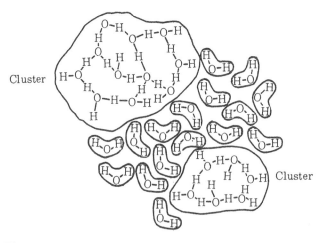

Figure 14   The model of liquid water. Nemethy-Scheraga [17].

modified, however, to fewer than 9 molecules by subsequent studies. It is now considered that a cluster is usually comprised of 5 or 6 molecules.

B. Basic Structural Model of Water

Arakawa reviewed liquid structure studies from that by Bernal and Fowler in 1933 to studies conducted in 1960s. Based on this review, Arakawa suggested that the structural model of water should satisfy the following conditions at minimum [18].

The model should be able to explain the temperature at maximum density. At 1 atm, H_2O has a density maximum temperature of about 4°C and D_2O, about 11.6°C. This characteristic disappears at about 2000 kg cm^{-2} or higher.

Second, the model should be able to explain the results of x-ray diffraction and neutron diffraction studies. In particular, under x-ray diffraction, the first peak of the radial distribution curve is located around 2.65 Å (265 pm) and the second peak around 4.5 Å (450 pm). The number of nearest neighbors calculated on the basis of the first peak is 4.4 at room temperature. For an ordinary liquid, the number of nearest neighbors is 8–10. In a close-packed structure, it is 12. This comparison indicates that water has much more vacant space than other liquids, which is its unique characteristic.

Third, the model should be able to explain the IR and Raman spectra. In the intermolecular mode, wide-band spectra are observed at vibrations of around 60, 100–200, and 400–800 cm^{-1}. There is as yet no conclusive explanation of the spectrum at 60 cm^{-1}, but it has been confirmed that the spectrum at 100–200 cm^{-1} corresponds to translation and the spectrum at 400–800 cm^{-1} corresponds to rotation.

Fourth, the model should be able to explain relaxation phenomena at the molecular level, such as dielectric, ultrasonic, and magnetic relaxation and neutron scattering. The conventional relaxation time for these relaxation phenomena was of the order of 10^{-12} s. The structure of water should support translation movement of a few times to 10 times at least. In other words, too large a cluster does not fit the practical conditions. Finally, the model should be described by a liquid study because water is a liquid.

All these conditions are appropriate. Although it may not be possible to satisfy all of them, the structural model of water should satisfy most of them.

C. Study of the Liquid Structure of Water by Computer Simulation

Rahman and Stillinger first presented the results of a computer simulation of water structure in 1974 [19]. This was the beginning of the third era of the study of the structure of water, following the first era from the 1930s to the 1950s and the second era of the 1960s.

Water has been structurally modeled as a continuum, an interstitial lattice, and a mixture. Although these models are very useful in visualizing water physically and geometrically, they can express only some of the physical characteristics of water because they are merely models. These models therefore have some weak points: the same experimental data were also explainable by theoretical calculations based on a completely different model, and the experimental results depend entirely on the choice of parameters even when the same model is used.

In the molecular dynamics (MD) method, one of the major computer simulation methods, no specific model is assumed but a potential function is assumed to solve the motion equations one after another for a limited number of molecules in a box. As a result, the position and energy of a molecule at a certain moment can be determined. The thermodynamic parameters can be calculated based on the results, which provide the information about the development over time of a certain system. In connection with the structure of water, computer simulation has mainly been used to count hydrogen bonds and to analyze the network structure.

Rahman and Stillinger placed $216 = 6^3$ hard water molecules in a cube to give a density of 1 g cm^{-3}. Therefore, one side of the cube was 18.02 Å (182 pm). MD calculation was carried out for this cube using an intermolecular action potential model that combined a Lennard-Jones potential and a four-point electrical charge potential [19]. Although six molecules on one side of the cube is rather a small number for water, interactions are thought to interact along the long range. The computer simulation, however, presented good results that matched well the observed values in terms of x-ray scattering intensity, neutron scattering intensity, self-diffusion coefficient, and thermodynamic parameters. The Rahman-Stillinger study was followed by many different computer simulations applying various potential functions.

D.   Study of Water Structure by Neutron Diffraction

The first study of water structure on the basis of neutron diffraction was conducted by Page and Powles in 1971 [20]. Narten also presented results in 1972 [21]. Neutron diffraction was later applied by many researchers.

As mentioned earlier, it is difficult to determine the location of a water molecule by x-ray diffraction. The neutron diffraction method, however, can determine the location with good accuracy because both light and heavy atoms show similar scattering amplitudes.

In 1982, Arakawa et al. determined the S_m (Q) of water over a wide range of Q and at a temperature of 25–95°C. Figure 15 shows the result [22]. Here Q is diffraction scattering area, a structural factor obtained by neutron diffraction. Q is $4\pi/\lambda$ sin θ, where 2θ is the scattering angle and λ is the wavelength of the neutron ray. For a molecular liquid like water, the contribution from the molecular structure is observed in the liquid over a wide range of $S_m(Q)$. In the middle Q range, much of the information gathered about liquid structure, such as intermolecular position and coordination, is determined by intermolecular action. In the range where Q is close to 0, thermodynamic characteristics appear. It is therefore necessary to obtain $S_m(Q)$ in the wide Q range for the study of the structure of a molecular liquid. In general, $S_m(Q)$ has a large peak around $Q = 2\pi/r_0$ (r_0 is the distance between the centers of the molecules).

In Fig. 15, the 2 $Å^{-1}$ (200 pm^{-1}) peak height barely fluctuates even as the temperature rises but the 4 $Å^{-1}$ (400 pm^{-1}) peaks decrease as the temperature rises.

Arakawa et al. considered water an equilibrium mixture of tetrahedral pentamers that has hydrogen bonds and a free water molecule monomer that does not have hydrogen bonds, as shown in Fig. 16. Arakawa et al. obtained a structural factor for neutron diffraction, $S_m(Q)$, taking the results of x-ray diffraction conducted by Narten into consideration. Figure 15 indicates the calculated values match the observed values well. A small peak around 4 $Å^{-1}$ (400 pm^{-1}) is caused by a pentamer.

Figure 17 shows the temperature dependence of the molar fraction of the pentamer (X_5). X_5 shows a polygonal line around 25°C. Even at 95°C, 55% of X_5 is in the pentamer state. This means that the water molecules are extremely strongly associated.

Furthermore, Arakawa et al. calculated the temperature fluctuation of each entropy of the degree of translational movement (S_{trans}), the degree of rotational movement (S_{rot}),

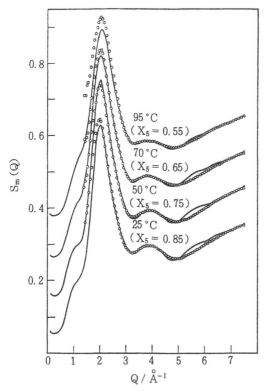

Figure 15 Comparison between the calculated $S_m(Q)$ for the pentamer-monomer mixture model and the observed neutron structure factors $S_m(Q)$ [12]: (circles) observed; (solid line) calculated.

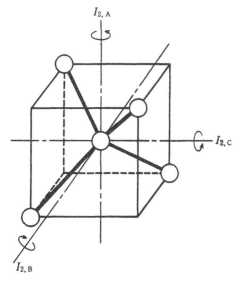

Figure 16 Tetrahedrally coordinated pentamer model.

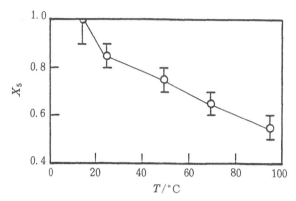

Figure 17 Pentamer fraction X_5 determined by fitting the $S_m(Q)_{calc}$ with the $S_m(Q)_{obs}$ at each temperature. The vertical line indicates the range of deviation in X [22].

and the internal capacity of the pentamer (S_{pint}) for the pentamer-monomer mixture. This calculation was conducted in a purely theoretical manner by statistical dynamics without using adjustment parameters. The results are shown in Fig. 18 [23], where p is pentamer and m is monomer. It is obvious that the contribution of the pentamer drops as the temperature rises and that the contribution is solely provided by the monomer at 60°C or higher.

In Fig. 18, $S_{ptrans} = S_{mtrans}$ at 15°C, $S_{prot} = S_{mrot}$ at 20°C, $S_{pint} = S_{mrot}$ and $S_{pint} = S_{mtrans}$ at 30–35°C, and $S_{mtrans} = S_{mrot}$ at 45°C. It is known that as 15, 30, and 45°C are harmful to living creatures, a physiologic phenomenon changes discontinuously [24]. It

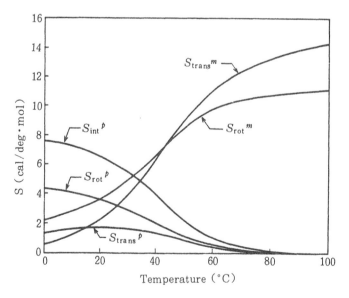

Figure 18 Contribution of pentamer and monomer to entropy S in the pentamer-monomer mixture model [23]: p, pentamer; m, monomer.

will be interesting to discover the relationship between the calculated results and this observation.

E. Comprehensive Discussion

In the 1980s, Arakawa et al. conducted a computer simulation of water using the MD method [25]. It was assumed that $1000 = 10^3$ molecules were placed in a cubic box with side 31.03 Å (3103 pm) to give a density of 1 g cm^{-3}. In this assumption, each molecule interacts on the basis of ST2 potential. The radial distribution relationship at 24 and 93°C, $g_{OO}(r)$, $g_{OH}(r)$, and $g_{HH}(r)$, and all structure factors $S_m(Q)$ were calculated by computer simulation. Figure 19 shows the results. The values measured by neutron diffraction, the values calculated by statistical dynamics using the pentamer-monomer model, and the values calculated by the MD method using the ST2 potential match each other well in the range up to $Q = 15$ Å^{-1} (1500 pm^{-1}).

The water we actually use has 10^{23} molecules. The assumed number of molecules—1000—is rather small compared with the actual figure, but it is close to the limit of the processing capability of conventional computers. In this sense, a discussion based on the molecule number of 1000 should be at the highest level. The reliability of the pentamer-monomer model is considered very high.

Another discussion was carried out to confirm whether there was a difference between the part near the center and the part near the interface in terms of the formation of hydrogen bonds. A cube with one side $L = 31.03$ Å was divided into two parts: $0.2L$,

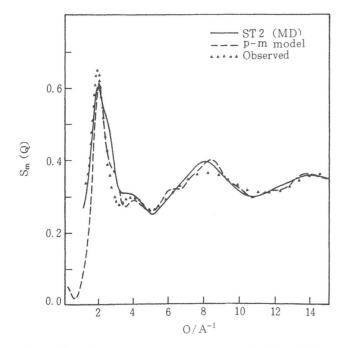

Figure 19 Coherent neutron structure factor $S_m(Q)$ at 24°C compared with the observed and calculated data for the pentamer-monomer mixture model [25].

which is close to the interface, and the part close to the center of cube. The first group was called group I and the second group was called group II. Because the molecule moves around, it is not possible to determine accurately the number of molecules belonging to each group. Roughly speaking, the number of molecules in group I is 210 and that in group II is 790. It was found that the proportion of hydrogen bonds in group I was smaller than that in group III by 10%. In group I, water molecules are strongly affected by the interface and the degree of freedom is strictly limited.

VI. CONCLUSIONS

This chapter describes the history of structural studies of water and the results of the study conducted by Arakawa et al. There are three reasons this is mentioned. The first is that the author wants to visualize the liquid structure of water in the simplest possible way. The second is there is a definite philosophy in the study of the liquid structure of water. The third is that these studies have been systematically conducted based on this philosophy. The results of past studies will be evaluated in the future in a real manner. Because water is a liquid, the determination of its structure is more difficult than for a solid. Furthermore, the structure of water continuously changes in 10^{-12} order, which makes the study much harder.

The x-ray diffraction method is useful for finding out the average conditions of the liquid structure of water. To describe the dynamic behavior of water, however, which is continuously changing, the computer simulation method based on molecular dynamics is superior. On the other hand, the capability of a conventional computer is not sufficient to handle actual water, with as many as 10^{23} molecules. It is expected further developments in computer capabilities will make it possible to conduct more accurate studies of the liquid structure of water.

REFERENCES

1. K. Suzuki, Mizu oyobi Suiyoeki, Kyoritsu Shuppan, p. 25 (1980).
2. Y. Kitano, Chikyu Kankyo no Kagaku, Shokabo, p. 41 (1983).
3. Denki Kagaku Kyokai, Wakai Gijutsusha no tameno Denki Kagaku, Maruzen, p. 3 (1983).
4. S. Kubota, DJIT, *222*, 23 (1986).
5. S. Kubota, Haguruma, *391*, 10 (1989).
6. K. Suzuki, Mizu oyobi Suiyoeki, Kyoritsu Shuppan, p. 14 (1980).
7. Y. Kitano, Chikyu Kankyo no Kagaku, Shokabo, pp. 39–40 (1983).
8. H. Otaki, Yoeki Kagaku, Shokabo, pp. 33–36 (1985).
9. K. Maeno, Kori no Kagaku, Hokkaido Daigaku Tosho Kankokai, p. 163 (1981).
10. H. Otaki, Yoeki Kagaku, Shokabo, pp. 8–9 (1985).
11. K. Arakawa, Suiyoeki Kei no Kozo to Butsusei, Hokkaido Daigaku Tosho Kankokai (1989).
12. J. D. Bernal and R. H. Fowler, J. Chem. Phys., *1*, 515 (1933).
13. J. Morgan and B. E. Warren, J. Chem. Phys., *6*, 666 (1938).
14. L. Hall, Phys. Rev., *73*, 775 (1948).
15. J. A. Pople, Proc. R. Soc., *A205*, 163 (1951).
16. A. H. Narten, M. D. Danford, and H. A. Levy, Discussions Faraday Soc., *43*, 97 (1967).
17. G. Nemethy and H. A. Scheraga, J. Chem. Phys., *36*, 3382 (1962).
18. K. Arakawa, Kagaku Sosetsu, *11*, 35 (1976); K. Arakawa, Suiyoeki Kei no Kozo to Butsusei, Hokkaido Daigaku Tosho Kankokai, p. 147 (1989).
19. A. Rahman and F. H. Stillinger, J. Chem. Phys., *55*, 3336 (1971); A. Rahman and F. H. Stillinger, J. Chem. Phys., *60*, 1545 (1974).

20. D. I. Page and J. G. Powles, Mol. Phys., *21*, 901 (1971).
21. A. H. Narten, J. Chem. Phys., *56*, 5681 (1972).
22. N. Ohtomo, K. Tokiwano, and K. Arakawa, Bull. Chem. Soc. Jpn. *55*, 2788 (1982).
23. K. Arakawa, Water and Metal Cations in Biological Systems, B. Pullman and K. Yagi, eds., Japan Sci. Soc. Press, Tokyo, (1980), p. 13.
24. H. Uedaira and A. Osaka, Seitaikei no Mizu, Kodansha, pp. 18–20 (1989).
25. K. Tokiwano and K. Arakawa, Bull. Chem. Soc. Jpn., *60*, 475 (1987).

3
Wafer Cleanliness and Requirements for Ultrapure Water Quality

Yoshiaki Matsushita

Toshiba Corporation, Kawasaki, Japan

I. EFFECT OF WAFER SURFACE DEFECTS ON VLSI

As devices are further integrated, with finer patterns, minute defects and trace impurities on the wafer surface affect performance more seriously. They affect the characteristics and reliability of devices, which may lower the yield of VLSI (very large scale integrated) circuit production. Wafer surface contamination not only damages the quality of substrate wafer and devices but also leads to dislocations at the edge of fine patterns, oxidation-induced stacking faults (OSF), and oxide breakdown defects. Devices are frequently affected by these defects, as well as by contamination. Damage to devices is often rather serious because of the combined effects of defects and contamination. For example, researchers present different values for leakage current per OSF, as shown in Table 1. This can be attributed to the difference in the degree of contamination at the OSF. In short, the degree of contamination constitutes a major factor in leakage of current. Table 2 shows the relation between wafer surface contamination and the characteristics of a device. The characteristics affected depend on the contaminants, but the major effects

Table 1 Leakage Current per OSF

Author		Leakage Current/OSF
Ravi et al.	(1973)	15 ~ 500 μA
Tanaka et al.	(1974)	0.06 ~ 6 μA
Rozgonyi and Kushner	(1976)	7 ~ 33 pA
Tanikawa et al.	(1976)	~ 10 pA
Ogden and Wilkinson	(1977)	~ 590 pA

Table 2   Correlation Between Wafer Surface Contamination and Device Failure

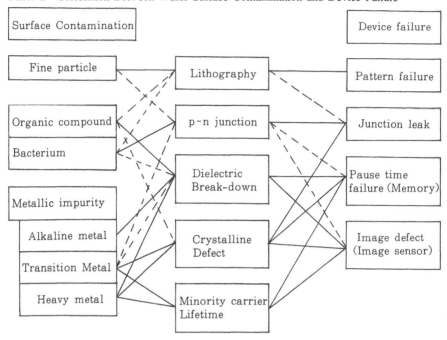

are leakage from *p-n* junctions, the pause time defect of memory, and the image defect in a charge-coupled device (CCD) image sensor.

As the geometry of a device increases in complexity, a smaller electrical charge is applied. As a result, defects induced by minute contamination, even if they are extremely low in density, can have fatal effects on devices. If we assume that defects and contamination introduced in the active area of device can lead to a defective device, the yield of a device Y can be estimated by Poisson distribution as

$$Y = e^{(-DA)} \tag{1}$$

where: D = defect density
       A = active area per chip

Figure 1 shows chip size and cell size in each generation of DRAM (dynamic random access memory). As shown, the active area increases along with the chip size. Figure 2 indicates the defect density dependence of device yield. It is necessary to limit the defect density significantly to maintain the same device yield: for 4M, the defect density should be lowered to half or less of that of 256 K and, for 64M, to half or less of that of 4M. Furthermore, as mentioned earlier, as devices pursue much finer geometry, extremely minute defects and contamination not formerly regarded as a problem will affect the performance of devices. When we consider that the critical defect size and critical contamination level will be lower, further stringent control will be required in the production process.

As VLSI becomes more integrated, the production process becomes more complex: for 4–16M, more than 200 process steps are required. Each process has cleaning steps, and devices are repeatedly rinsed. Water of ultrapure quality is therefore critical to keep wafers from contamination. If the quality of the ultrapure water is not sufficient, the wafer

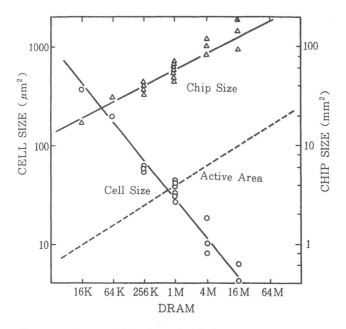

Figure 1 Trend of chip size and cell size in DRAM.

surface is contaminated by the water itself, which makes the cleaning process meaningless. Various methods have been investigated as ways to evaluate the ultrapurity of water, including resistivity, bacteria count, total organic carbon (TOC), fine particles, silica, metal, and concentration of ionic impurities. For the ultrapure water actually used in a production line, the level of impurity concentration has been lowered to below the detection limit in many cases. What should be eliminated, however, are the particles and impurities adsorbed on the wafer surface. The quality of ultrapure water should be

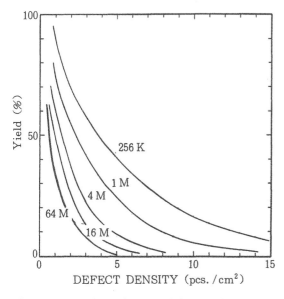

Figure 2 Correlation between defect density and device yield.

discussed based on the effect of impurities absorbed on the wafer surface of devices. In this sense, the method of evaluating the impurities on the wafer surface plays a critical role. The remainder of this chapter discusses conventional evaluation methods.

II. EVALUATION OF SURFACE CONTAMINATION

Since VLSI is produced in an ultraclean environment, the contamination caused by the production process is at an extremely low level. Because this minute contamination affects the performance of a device, however, it should be controlled to extremely low levels and evaluation technology is very important. In the VLSI production process, the relevant level of surface contamination is 10^{-10} atoms cm^{-2} or less. A technology to evaluate trace impurities of at least 10^{-9} atoms cm^{-2} is therefore required.

Table 3 lists the major methods actually used at present to evaluate trace impurities on a wafer surface. Vapor-phase decomposition (VPD) chemically decomposes impurities on the wafer surface, collects them, and then analyzes them with a high-sensitivity analysis method. Total reflection energy-dispersive x-ray fluorescence spectrometry (TREX), total reflection x-ray fluorescence spectrometry (TRXRF), and secondary ion mass spectrometry (SIMS) are methods of directly analyzing impurities on the wafer surface by x-ray or ion beam analysis. Deep-level transient spectroscopy (DLTS) and microwave

Table 3 Analysis of Trace Impurities on the Wafer Surface

Analysis Methods	Characteristics	Problems
(1) Chemical Analysis		
(a) VPD	Decompose by HF vapor	Destructive
Vapor Phase	Resolved solution is analyzed	Impossible to get impurity
Decomposition	by AA or ICP-MS.	distribution
	DL.~10^9 atoms/cm^2	DL. depends on solution
(2) Beam Analysis		
(a) SIMS	Small area analysis	Destructive
Secondary Ion Mass	In-depth profile measurement	Impossible to measure full
Spectroscopy	DL.~10^9 atoms/cm^2	surface area
		DL. depends on kinds of atoms
(b) TREX or TRXRF	Non-destructive, Non-contact	Sensitivity
Total Reflection	Surface distribution analysis	High power X-ray source
Energy dispersive	In-depth distribution	Difficult for light element
X-ray fluorescence	DL.~6×10^9 atoms/cm^2	
(3) Electrical Analysis		
(a) DLTS	Measure the density of deep	Heat process is necessary
Deep Level	level formed by metals	(Diode formation)
Transition	DL.~10^{12} atoms/cm^3	Quantitative analysis
Spectroscopy		
(b) μ-PCD	Measure the recombination	Heat oxidation is necessary
μ-wave Photo	lifetime.	Impossible to identify the
Conductive Decay	High sensitivity for Fe and	kinds of impurities
method	Ni.	

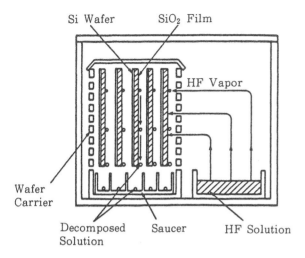

Figure 3 Instrument used for the VPD Method.

photoconductive decay (μ-PCD) are methods of diffusing impurities and electrically measuring the effect of the deep levels formed. The details of each method are as follows.

A. Chemical Method: VPD

Figure 3 is a schematic drawing of the VPD method [1,2]. In this method, oxide on the silicon wafer surface (native oxide or thermal oxide) is decomposed with HF vapor and then the solution is analyzed by a high-sensitivity analysis method. The solution is usually analyzed with frameless atomic absorption spectrometry or inductively coupled plasma mass spectrometry (ICP-MS).

HF solution is not adequate to decompose oxide on the wafer surface because the impurity concentration in the solution becomes too high to be evaluated by the high-sensitivity analysis method. It is comparatively easy to lower the background level when HF vapor is used, and a detection limit of surface contamination around -10^9 atoms cm^{-2} can be obtained. In this method, it is critical to completely recover impurities in the solution, but the recovery ratio varies from one element to another. In particular, the recovery ratio of Cu varies greatly among different solutions: it is only 2% when pure water is used, but it is 90% when a mixture of oxidant and HF, such as H_2O_2/HF, is used (see Table 4) [3].

This method is not adequate to evaluate local contamination and the distribution of contamination, although it is able to evaluate contamination over the entire surface.

Table 4 Recovery Ratio for Cu [3]

Collection solution	Analyzed value ($\times 10^{10}$ atoms/cm^2)	Recovery ratio (%)
H_2O	5	2
HF/H_2O_2	300	90

Cu contaminated wafer is used.
Contamination level ; 10^{12} atoms/cm^2

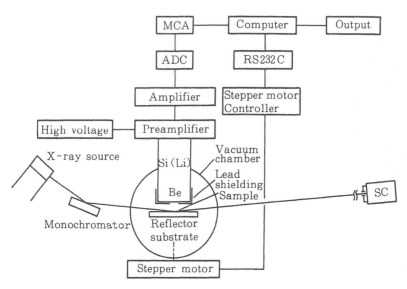

Figure 4   TREX.

B.   Beam Analysis

There are two methods that apply beam analysis: TRXRF or TREX, which applies x-rays, and SIMS, which uses an ion beam. SIMS is widely used as a high-sensitivity analysis method, and it is useful in analyzing the depth profile of impurities introduced by ion implantation, diffusion, and so on. For trace impurities on the wafer surface, however, SIMS is not adequate because it can cover only a limited area, shows different sensitivities for different elements, is less quantitative, and is a destructive inspection.

TREX or TRXRF is a method of directing x-rays toward the wafer surface with an incidence angle less than the critical angle of total reflection and then detecting the

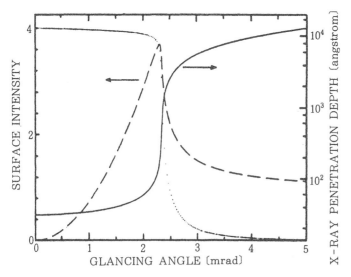

Figure 5   Surface x-ray intensity and x-ray penetration depth as a function of glancing angle. (From Ref. 5.)

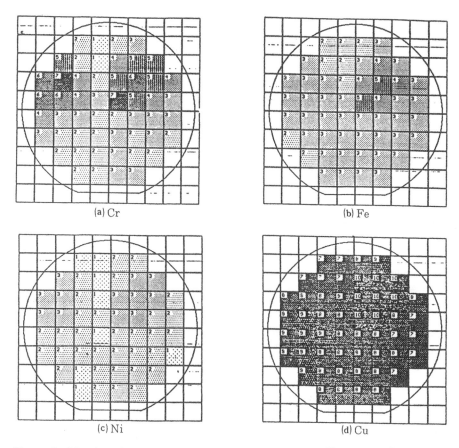

(a) Cr (b) Fe (c) Ni (d) Cu

Figure 6 Mapping of surface metallic impurities (unit, 1×10^{13} atoms cm^{-2}).

fluorescence x-ray generated by impurities on the surface by a solid-state detector (SSD). A schematic diagram is provided in Fig. 4. Figure 5 [4] shows the intensity of fluorescence x-rays generated by impurities on the wafer surface and the x-ray penetration depth as a function of the x-ray incidence angle [5]. The intensity of the fluorescence x-ray hits its peak around the critical angle of total reflection. The x-ray penetration depth is around 3 nm, and impurities on the surface can be detected with high sensitivity. The detection area is around 1 cm^2. It is possible to conduct the mapping evaluation of a wafer measuring 1 cm^2. Figure 6 shows examples of impurity mapping on a wafer surface [5].

When the x-ray incidence angle is altered to obtain different x-ray penetration depths, it is possible to evaluate the depth profile of impurities nondestructively [6], which makes it possible to conduct a three-dimensional evaluation of trace impurities on the wafer.

The detection limit of TREX depends on the ability of the impurities to generate x-ray fluorescence and on the detection sensitivity, which means that it varies slightly according to the atom number of the element to be analyzed. When a W target at 30 kV and 200 mA ($L\beta_1$) is used, the detection limit for Fe is 1×10^{10} cm^{-2} and that for Cu is 6×10^9 cm^{-2}. A strong x-ray source is required, such as SOR, to increase the sensitivity.

C. Electrical Evaluation

Electrical evaluation introduces impurities into the surface layer of a wafer by annealing and then the deep level affected by the impurities are detected by DLTS or recombination

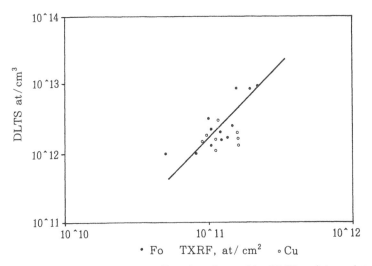

Figure 7   Correlation between Fe values detected by DLTS and those detected by TREX. (From Ref. 7.)

lifetime analysis. For Fe, DLTS is comparatively less quantitative. As shown in Fig. 7, a good correlation is observed between DLTS and TREX for a sample processed at 1100°C for 5 minutes [7]. Cu and Ni, however, are less quantitative as a result of fast diffusion.

Lifetime evaluation is based on the fact that the defect generated by impurities is a recombination center. The method is capable of evaluating Fe with high sensitivity (-10^9 atoms cm^{-2}) [3]. However, this method is not able to identify elements. Therefore, when the contaminant has been identified, this method is able to conduct quantitative evaluation to some extent, but when various contaminants exist together, which is the usual case, this method is not able to evaluate individual contaminants one by one. The method is effective as a line monitor since it can be easily applied to the VLSI production line.

III.   EFFECT OF SURFACE IMPURITIES ON DEFECT GENERATION

Wafer surface impurities and particles lead to crystalline defect generation, the thin oxide defect, and the deterioration of minority carrier lifetime.

The most critical surface impurity is the metallic impurity. As shown in Fig. 8, metallic impurities are diffused into silicon crystal within a short time [8,9] and are easily introduced to a crystal even at low temperatures. Also, the metallic impurities introduced into a silicon crystal form deep levels in the crystal, as shown in Fig. 9 [10]. This acts as a generation-recombination (G-R) center for minority carrier, which could lead to leakage from the *p-n* junction and the pause time degradation of memory. Metallic impurities also lead to OSF and dislocation, inducing a crystalline defect. The effect of surface metal impurities on OSF generation is shown in Fig. 10 [11]. In this study, the wafer surface was intentionally contaminated with a standard solution of metals. The correlation between metal impurity content and the OSF formed during thermal oxidation at 1000°C was studied. It is found that Fe and Ni have a strong effect on OSF generation. The critical level of OSF generation is around 10^{-12} atoms cm^{-2} for both Fe and Ni. When the wafer was oxidized at 1000°C after preprocessing at 1150°C, many OSF were formed when the

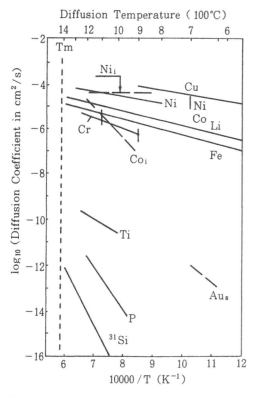

Diffusion Temperature (100°C)

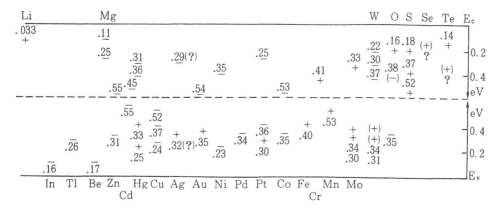

Figure 8   Diffusion coefficient of metallic impurities in silicon. (From Ref. 9.)

wafer was contaminated by Ni or Cu at a level of 10^{11} cm^{-2}, but when the wafer was contaminated with Fe, not as many OSF were formed [12]. When the wafer was preprocessed at 650–850°C before oxidation at 1000°C and was contaminated with Fe at a level of 10^{12} cm^{-2}, OSF formed rapidly (see Figure 11) [12]. In any case, surface contamination with Fe, Ni, or Cu at a level of 10^{11}–10^{12} atoms cm^{-2} seriously affects OSF generation. These tests were conducted on wafers free of surface damage. In actual VLSI

Figure 9   Deep levels reached by metallic impurities in silicon. (From Ref. 10.)

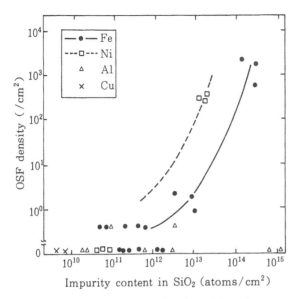

Figure 10   OSF density as a function of impurity content in the oxide film. (From Ref. 11.)

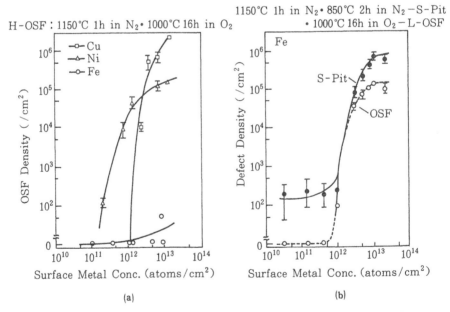

Figure 11   Dependence of OSF density upon surface metal concentration: (a) heat treatment, 1150°C for 1 h in N_2 + 1000°C for 16 h in O_2; (b) heat treatment, 1150°C for 2 h in N_2 + 1000°C for 16 h in O_2. (From Ref. 12.)

production, however, the wafer surface is often damaged during such processes as ion implantation, RIE, and CDE, and the contamination on the damaged surface could have a more serious effect on defect generation. Figure 12 shows the relationship between surface contamination and OSF generation when a wafer implanted with boron was heated [13]. In this test, the wafer was contaminated with Fe. When the wafer was free of damage, the OSF increased with an Fe contamination of 10^{12} atoms cm^{-2}, but when the wafer was damaged, the OSF were formed even with an Fe contamination of 10^{10} atoms cm^{-2}.

Surface contamination also affects dislocations at the edge of the pattern. Figure 13 shows the LOCOS dislocation defect generation ratio as a function of impurity content on the wafer surface [14]. LOCOS dislocation is observed at the oxide edge when the wafer is locally oxidized by so-called field oxidation. The width of the pattern, the structure of the upper film, and thermal stress are the major factors affecting LOCOS dislocation defect generation. Figure 13 indicates that the ratio surges when the wafer is contaminated at a level of 10^{10}–10^{11} atoms cm^{-2}.

As mentioned earlier, surface metal impurities have a serious effect on the generation of crystalline defects. In particular, in the VLSI production process, when the wafer surface is damaged, metal impurities adsorbed on the wafer surface should be suppressed at a level of 10^9 atoms cm^{-2} or less.

Surface contamination also has a significant effect on oxide breakdown. Figure 14 shows the correlation between Na concentration and the yield of oxide retention field of a 10 mm^2 capacitor, where the fraction of capacitors that showed a breakdown voltage

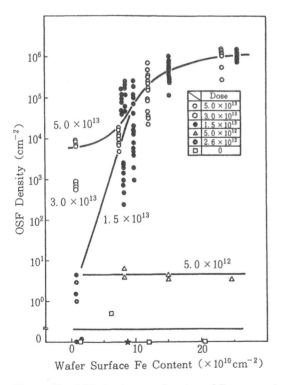

Figure 12 OSF density as a function of Fe content in the implanted wafer.

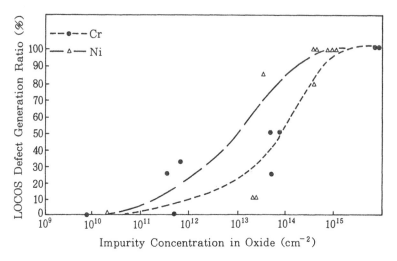

Figure 13   LOCOS dislocation generation ratio as a function of impurity concentration in the oxide film.

larger than 8 MV/cm (so-called C mode) is defined as *yield*. When the Na concentration at the surface exceeds 10^{11} atoms cm^{-2}, the yield obviously drops. Figure 15 indicates the effect of Cu concentration on the wafer surface on oxide breakdown. When the Cu concentration is low, only C mode (over 8 MV/cm) is observed, but as the surface is contaminated with Cu, the yield of C mode decreases and eventually disappears.

   Besides metallic contaminants absorbed on the wafer surface, when a crystal is

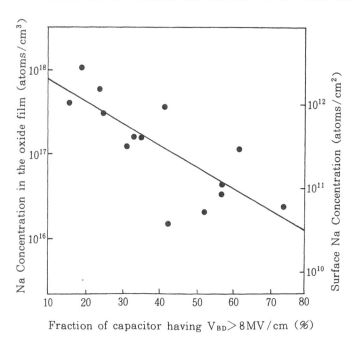

Figure 14   Correlation between yield of oxide retention field and Na concentration in the oxide film.

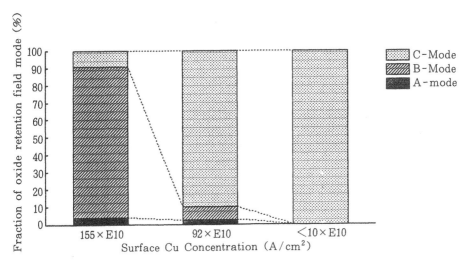

Figure 15 Effect of surface Cu concentration on oxide breakdown failure A mode, $V_{BD} \leqq 1$ MV/cm; B mode, 1 MV/cm $< V_{BD} \leqq 8$ MV/cm; C mode, 8 MV/cm $< V_{BD}$.

treated at high temperature a deterioration in lifetime in the active area of the device is caused. Figure 16 indicates recombination lifetime as a function of metallic impurities when a wafer is immersed in dilute HF (DHF) cleaning solution (HF/H_2O = 1:99) with different metals added for 60 minutes and then oxidized at 1000°C [15]. The lifetime deteriorates in each case. For Cu and Au, the lifetime deterioration is significant. This is because metals whose ionization tendency is smaller than that of Si are adsorbed on the Si surface (Table 5). Figure 17 shows the recombination lifetime as a function of the

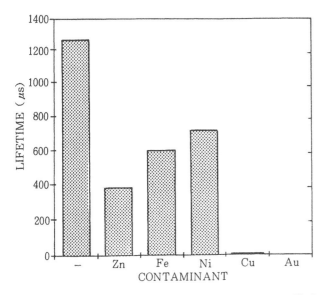

Figure 16 Dependence of recombination lifetime on metallic impurities introduced by using DHF (dilute HF) solution containing 50 ppb metals. (From Ref. 15.)

Table 5   Ionization Tendency and Electronegativity

Ionization Tendency	Zn > Fe > Ni > Si > Cu > Au
Electronegativity	1. 6 1. 8 1. 8 1. 8 1. 9 2. 4

concentration of metals adsorbed on the wafer surface [16]. When the surface concentration increases beyond 10^{11} atoms cm^{-2}, the lifetime drops sharply.

As mentioned, metallic impurities adsorbed on the wafer surface induce crystalline defects or cause oxide breakdown and recombination lifetime deterioration. These effects appear when the surface concentration of metals reaches 10^{10} atoms cm^{-2}, and they become significant at a level of 10^{11} atoms cm^{-2}. It is therefore required that the metallic impurities adsorbed on the wafer surface (including transition metals and alkali metals) be kept at a level of 10^{9} atoms cm^{-2} or less.

IV.   IMPURITIES IN ULTRAPURE WATER AND SURFACE CONTAMINATION

It is necessary to consider the adsorption of impurities in ultrapure water onto the wafer surface to work out the requirements for ultrapure water to be used as a cleaning solution. The adsorption level of metallic impurities on a wafer surface depends on the conditions of preprocessing the wafer surface: whether native oxide is formed and what kind of

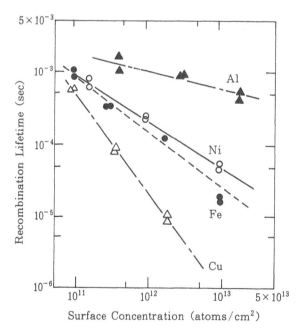

Figure 17   Dependence of recombination lifetime on the surface metal concentration. (From Ref. 16.)

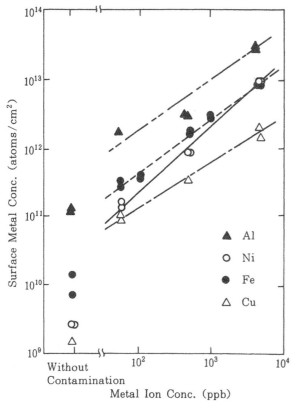

Figure 18 Correlation between metal ion concentration in solution and adsorbed surface metal concentration. (From Ref. 16.)

cleaning chemicals contaminated the surface. A high level of accuracy is also required in the detection method. This is why the data for adsorption levels vary among researchers [15–17]. Figure 18 shows the correlation between the metal ion concentration in solution and the concentration of metals adsorbed on the wafer surface when a standard solution for an atomic adsorption photometry is adsorbed on the wafer surface by the spin method [16]. It is clear that the concentration of metals adsorbed on the wafer surface increases as the metal ion concentration in solution rises. Kern [17] studied the concentration of metals adsorbed on the wafer surface with tracers and found that Cu at 1.9×10^{12} atoms cm^{-2} was adsorbed on a bare Si surface even when the Cu concentration in pure water was limited to 2 ppb. This means the Cu concentration in pure water should be lowered to less than 1 ppt to suppress the concentration of Cu adsorbed on the wafer surface to a level of 10^9 atoms cm^{-2} or less. The conditions for suppressing the surface concentration of Fe and Ni at 10^9 atoms cm^{-2} or less was studied based on Fig. 17. It was found that the Fe and Ni concentration in pure water should be limited to 0.1–1 ppb or less. All the tests were conducted using wafers with a flat mirrored surface. Because the wafers actually treated in the VLSI production process do not have a flat surface, however, it should be taken into consideration that the conditions of the wafer surface are quite different between tests and the actual case.

V. REQUIREMENTS FOR ULTRAPURE WATER QUALITY

When the quality of ultrapure water used in the VLSI production process is discussed, account should be taken of wafer contamination during the cleaning process. There are two kinds of contamination on the wafer surface. One is visible: a particle is a major example. The other is invisible contamination (or barely observed): organic materials, inorganic materials, and heavy metals are adsorbed on the wafer surface as atoms or molecules.

For particles adhered to a wafer, the patterning process in lithography is most seriously affected. The yield of LSI (Y) can be expressed by Poisson's equation as

$$Y \propto e^{(-D)}$$

where D = defect density of a pattern on the wafer.

If there are n critical patterning steps in lithography, this can be expressed as

$$Y \propto e^{(-nD)}$$

which means the yield is lowered exponentially as the number of adsorbed particles and the number of steps increase.

The largest acceptable particle size that does not affect a device even if it is adsorbed on the wafer is determined by the critical dimensions of the device. The alignment accuracy of dimension W is about $W/5$ [18]. If a particle as large as the alignment accuracy is adsorbed on an aligned part, the alignment accuracy is damaged. The particle size should therefore be smaller than half the alignment accuracy. The diameter of the largest acceptable particle should be $W/10$ or smaller. Table 6 shows the estimated diameters of the largest acceptable particles in each generation of DRAM. If the number of critical patterning steps is 10, the number of adsorbed particles larger than the largest acceptable particle size per process should be limited to less than 10 per wafer to guarantee the yield of VLSI.

In cleaning, the acceptable number of particles in pure water is determined by the amount of pure water contacting the wafer surface and the probability that particles will be adsorbed on the wafer surface. The details have not yet been revealed, however, in regard to microparticles of 0.1 μm or smaller. It is considered that the probability of adsorption

Table 6 Acceptable Particle Size for Individual Device Generation

Device (DRAM)	Min. line width (μm)	Max. acceptable particle size (μm)
64 K	3	0.3
256 K	2	0.2
1 M	1	0.1
4 M	0.8	0.08
16 M	0.6	0.06
64 M	0.4	0.04

Table 7 Example of Pure Water Quality Required in Semiconductor Engineering

Device	64 K	256 K	1 M	4 M	16 M
Resistivity (M$\Omega \cdot$cm, 25°C)	> 15	> 17	> 18	> 18	> 18
Particle (ps/cc) > 0.2 μm	100 – 200	< 50			
Particle (ps/cc) > 0.1 μm		< 100	< 20		
Particle (ps/cc) > 0.08 μm				< 1	
Particle (ps/cc) > 0.06 μm					< 0.5
TOC (μgC/l)	< 500	< 100	< 50	< 20	< 10
Bacterium (ps/cc)	< 1	< 0.1	< 0.05	< 0.01	< 0.005
Resolved oxygen (μgO/l)		< 100	< 100	< 50	< 50
Silica (μgSiO$_2$/l)		< 10	< 10	< 5	< 5

rises as the particle size decreases. If it is assumed that 1 L ultrapure water contacts each wafer surface, it is sufficient to have 10 particles or less per liter measuring 0.08 μm or greater. Practically, however, only one particle 0.08 μm or larger is allowed per cubic centimeter.

In terms of microscopic contamination, metallic impurities have the most significant effect. As mentioned earlier, metallic impurity leads to the crystalline defect and the oxide defect. It is fatal if these defects are introduced into the active area of a device. The surface absorption concentration of metallic impurities must be limited to 10^9 atoms cm^{-2} or less. For the most critical case, it is assumed that all the metals in ultrapure water held in the trench after cleaning are adsorbed. The trench groove is assumed to be 1 μm^2 and 10 μm in depth. To maintain the adsorbed metal concentration on the bottom of the trench at 10^9 atoms cm^{-2} or less, the metallic impurity concentration in the ultrapure water must be held at 30 ppt or less. In actuality, however, not all the metals are adsorbed and the conditions are not as strict as this assumption.

The requirements in terms of resistivity, TOC, and dissolved oxygen are shown in Table 7 [19,20].

REFERENCES

1. A. Shimazaki, H. Hratsuka, Y. Matsushita, and S. Yoshii, Extended Abstracts, 16th Conference Solids State Device and Materials, Tokyo, 1984, p. 281.
2. Y. Tanizoe, S. Sumita, M. Sano, N. Fujino, and T. Shiraiwa, Bunseki Kagaku, *38*, 177 (1989).
3. T. Shimono and M. Tsuji, Proceeding 1st Workshop on Ultra Clean Technology, Tokyo, 1989, p. 51.
4. A. Iida, K. Sakurai, A. Yoshida, and Y. Gohshi, Nucl. Inst. Phys. Res., *A246*, 736 (1986).
5. N. Tsuchiya and Y. Matsushita, Abstracts 50th Ouyoubutsuri-Gakkai, Fukuoka, 1989, p. 298.
6. A. Iida, Extended Abstracts, 12th Conference Solid State Device and Materials, Tokyo, 1989, p. 501.
7. D. Huber, P. Eichinger, and E. Englmueller, Proc. 9th Int. Conference Crystal Growth, Sendai, 1989.
8. E. R. Weber, Appl. Phys, *A30*, 1 (1983).

9. M. Yoshida, Kesshou Kakou to Hyouka Gijutsu 145 Iinkai, 47th Kenkyuukai, Karatsu, 1989, p. 40.
10. A. G. Milness, Deep Impurities in Semiconductors, John Wiley and Sons, New York, 1973.
11. A. S. Maeda and M. Ogino, Extended Abstracts, Electrochemical Society, 1986 Spring Meeting, 1986, No. 254.
12. N. Fujino, S. Sumita, K. Murakami, and K. Hiramoto, Kesshou Kakou to Hyouka Gijutsu 145 Iinkai, 47th Kennkyuukai, Karatsu, 1989, p. 61.
13. Y. Matsushita, Proceedings 1st Workshop Ultra Clean Technology, Tokyo, 1989, p. 1.
14. Y. Matsushita, Extended Abstracts, Symposium VLSI Technology, Kyoto, 1989, p. 5.
15. J. Atsumi, S. Ohtsuka, S. Munehira, and K. Kajiyama, Extended Abstracts, Electrochemical Society, 1989 Fall Meeting, 1989, No. 385.
16. M. Hourai, T. Naridomi, Y. Oka, K. Murakami, S. Sumita, N. Fujino, and T. Shiraiwa, Jpn. J. Appl. Phys., *27L*, 2361 (1988).
17. W. Kern, RCA Rev., *234* (1970-6).
18. H. Kamata, Semiconductor World, 123 (1988-7).
19. ASTM D-19 Proposal, p. 172.
20. Y. Satoh, Densi Zairyou, *27*, 52 (1988-8).

4
Properties of Ultrapure Water

Shoji Kubota
Hitachi, Ltd., Ibaraki, Japan

I. INTRODUCTION

Water is an excellent solvent and cleaning agent in which many substances are well dissolved. In the semiconductor industry, a leading high-technology industry, large amounts of ultrapure water are used as a cleaning agent. In the past, when semiconductors were not highly integrated, the quality of conventional pure water was sufficient. As the integration level improves, however, pure water must have higher purity to further upgrade the yield and reliability of semiconductors. The quality of ultrapure water required in the semiconductor industry has been improved along with advancements in technology to evaluate the quality of the water.

It is very important to understand the properties of ultrapure water to make the best use of it. This chapter describes the properties of ultrapure water in connection with the cleaning of the Si wafers used in semiconductor production.

II. WHAT IS ULTRAPURE WATER, AND HOW SHOULD IT BE USED?

The concept of ultrapure water or super ultrapure water has developed along with the semiconductor industry [1]. The conventional concept of pure water was based on an electrolyte in solution. Therefore mainly electrical conductivity was studied. In considering ultrapure water, not only the electrolyte but also substances that are dissolved or diffused in water, such as organic materials, bacteria, and particles, should be covered. The final goal of the semiconductor industry is theoretically pure, 100% water. Theoretical pure water is an ideal, however, and is not available in practice. This does not simply mean that it is impossible to produce such theoretical pure water, but even if it could be produced, it would not be possible to maintain its purity because of contamination and elution from manufacturing equipment and containers. Therefore the ultrapure water or super ultrapure water available in practice is that closest to theoretical pure water.

In using such high-purity ultrapure water, the following points should be noted.

First, theoretical pure water features the highest available purity of water. The resistivity is 18.24 MΩ cm^{-1} and its electrical conductivity is 0.05482 μS cm^{-1} at 25°C. This conductivity value is based on the dissociation of water itself: $H_2O = H^+ + OH^-$ (more precisely, $2H_2O = H_3O^+ + OH^-$).

Ultrapure water or super ultrapure water is very close to theoretical pure water in terms of resistivity. Therefore, no electrolytes other than hydrogen ion and hydroxide ion, which are created by the dissociation of water, exist in ultrapure water. In addition to particles, organic materials, bacteria, silica, and other impurities are removed as much as possible. In this sense, ultrapure or super ultrapure water features very high purity. The purer water becomes, the less saturated it becomes. Substances are better dissolved in unsaturated water. This means that such high-purity water is easily contaminated by elution from its containers.

Second, a production line making highly integrated semiconductors needs to have high-quality water with a resistivity of 18 MΩ/cm or higher. When Fe^{2+} at 1 ppb (μg/L) is dissolved in pure water of 18 MΩ/cm, the resistivity drops to about 17 MΩ/cm. The critical level of concentration in this ultrapure water is measured in parts per billion (10^{-9}). For example, assume that the world population is 5 billion and 5 individuals have rare beauty: this proportion is 1 ppb.

Even if the same sample is measured by the same instrument using the same method, the measured value depends on who conducts the measurement. Even when the same person conducts the measurement, a value measured in the morning can be different from a value measured in the afternoon. If specific parameters are properly controlled in the measuring process, however, it is possible to obtain data with good reproducibility. It is still difficult, however, to determine the reliability of the data in ppb level measurement. When discussing ppm (parts per million, or 10^{-6}), elution from the materials of containers, pumps, and piping systems, for example, can be ignored. These factors cannot be ignored however, when considering ppb. This is the difficulty in handling extremely low levels of concentration. In short, elution at the ppb level is inevitable, which affects the purity of ultrapure water.

As semiconductors are further integrated, the water quality required is rising from ppb to ppt (parts per trillion, or 10^{-12}). With such an upgrading trend, the problem of elution should be handled more carefully.

Third, water, even potable water, is a so-called semiconductor with a resistivity of a few kiloohms per centimeter. Ultrapure water or super ultrapure water features a resistivity of 18 MΩ cm^{-1} or higher. As water becomes purer, it becomes more of an insulator. For a solid semiconductor like silicon, the electrical conductivity is determined by its electrons and positive holes. For water, however, the electrical conductivity is determined by the ions. Static electricity is therefore easily generated in ultrapure water or super ultrapure water. When city water is treated to improve its purity, static electricity begins to be generated at a resistivity level of 10 to several hundreds of kΩ cm^{-1}, and the electrical potential generated increases as water is further purified. The generation of static electricity depends on such conditions as pipe materials and the flow rate of water. For example, when pure water flows through a pipe of acrylic resin, the pipe is charged positive and pure water is charged negative. The charge condition can be reversed, however, depending on the pipe materials.

III. CHANGE IN PURE WATER PROPERTIES AS A RESULT OF TEMPERATURE CHANGE

For the quality of ultrapure water required by the semiconductor industry, electrical conductivity is the only specific parameter that has been investigated. Not even measurement technology has been developed for other parameters. This is why this chapter mainly uses electrical conductivity data to evaluate the purity of water.

A. Electrical Conductivity

The electrical conductivity of an aqueous solution can be expressed by the equation (Kohlraush's additive ratio)

$$k = \Sigma C \alpha l 10^{-3} \tag{1}$$

where: C = ionic strength
α = degree of ionization
l = equivalent electrical conductivity

Water dissociates, as expressed in Eq. (2), maintaining the equilibrium:

$$H_2O = H^+ + OH^- \tag{2}$$

$$\frac{[H^+][OH^-]}{[H_2O]} = K \tag{3}$$

Equation (3) is converted to Eq. (4)

$$[H^+][OH^-] = K[H_2O] \equiv K_w \tag{4}$$

where K_w = dissociation constant of water or ion product of water.

As the dissociation of water fluctuates with temperature, K_w changes as a function of temperature. Table 1 lists the dissociation constants for water [2].

Because no dissolved substances are contained in pure water, the only ions existing in

Table 1 Dissociation Constants for Water[a]

Temperature (°C)	−log Kw	Temperature (°C)	−log Kw	Temperature (°C)	−log Kw
0	14.943	25	13.996	50	13.261
5	14.733	30	13.833	55	13.136
10	14.534	35	13.680	60	13.017
15	14.346	40	13.534	18[b]	14.23932
20	14.166	45	13.396	100[c]	12.295

(a) H. S. Harned, R. A. Robinson : Trans Faraday Soc., **36**, 973 (1940).
(b) H. C. Dueker *et al.:J.Phys.Chem* ., **66**, 225 (1962).
(c) G. N. Lewis *et al.*:Thermodynamics and the Free Energy of Chemical Substances New York (1923).

aqueous solution are H^+ and OH^-. Therefore, if K_w, the ion product of H^+ and OH^-, can be obtained, it is possible to calculate k in Eq. (5):

$$k = (lH^+ + lOH^-)\rho\sqrt{K_w}\ 10^{-3} \tag{5}$$

where ρ = density of pure water.

The terms in Eq. (5), l, ρ, and K_w, fluctuate in accordance with temperature. If the following relationships are obtained, it is possible to calculate k based on Eq. (5) by substituting the figures at each temperature for l, ρ, and K_w:

Relationship between the temperature of a solution and the limit equivalent ion electrical conductivity
Relationship between the temperature of a solution and the density
Relationship between the temperature of a solution and K_w

Figure 1 shows the results of the calculation [3]. As shown in Fig. 1, as the temperature rises, the values of k increase but the resistivity decreases. For most conductors, as temperature rises, resistivity rises and electrical conductivity falls. The results for pure water, however, show the opposite phenomena, with the same tendency as most semicon-

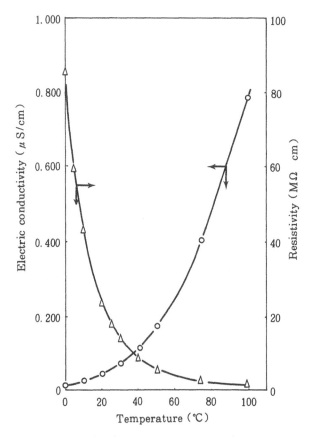

Figure 1   Relation between temperature and the electrical conductivity and resistivity of pure water.

ductors. At 25°C, k is 0.0548 μS cm^{-1}. This becomes 0.7370 μS cm^{-1}, about 13 times larger, at 100°C and 0.0101 μS/cm, falling about ⅓, at 0°C. The temperature coefficient of resistivity at 25°C is about -0.84 MΩ · cm °C^{-1}. Therefore, at a temperature of around 25°C, the resistivity decreases by 0.84 MΩ · cm when temperature rises by 1°C. To change the first decimal place, the fluctuation in water temperature needs to be controlled within 0.1°C, and to change the second decimal place, it needs to be controlled within 0.01°C.

B.   pH

pH is defined in Eq. (6):

$$pH = -\log[H^+] \tag{6}$$

Although [H$^+$] is not ionic strength but activity, at extremely low ionic strength, [H$^+$] can be expressed as ionic strength. Because the only ions in pure water are hydrogen ion and hydroxide ion and their strength is extremely low, they can be expressed by strength. Furthermore, the principle of charge neutralization proves [H$^+$] = [OH$^-$] and Eq. (6) is converted to Eq. (7):

$$pH = -\log \sqrt{K_w} = -\frac{1}{2} - \log K_w \tag{7}$$

As K_w increases in accordance with the rise in temperature, the pH decreases. Figure 2 illustrates this relationship.

As shown in Fig. 2, although pure water is at pH 7 at 25°C, it becomes acid at 100°C, that is, pH 6.14. In the usual aqueous solution, however, the equilibrium shown in Eq. (3)

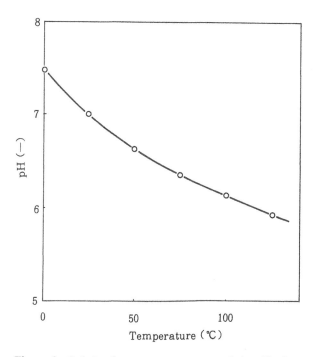

Figure 2   Relation between temperature and the pH of pure water.

is observed. When the temperature is fixed, the concentration of hydroxide ions decreases as the concentration of hydrogen ions increases. The solution therefore becomes acid because the number of hydrogen ions is greater than that of hydroxide ions. When pure water becomes acid as the temperature rises, the number of hydrogen ions is the same as the number of hydroxide ions. In this sense, the phenomenon is unique.

C. Vapor Pressure

Figure 3 indicates the relation between the temperature of pure water and the vapor pressure [4]. Vapor pressure rapidly increases as temperature increases. Compared with the amount of vapor pressure at 25°C (25 mmHg), that at 100°C (760 mmHg) is about 30 times greater and that at 0°C (4.6 mmHg) is about ⅕.

The relationship between temperature and vapor pressure is well known as the Clausius-Clapeyron equation:

$$\frac{d \ln P}{dT} = \frac{L}{RT^2} \tag{8}$$

where: P = vapor pressure
T = temperature
L = heat of vaporization
R = gas constant

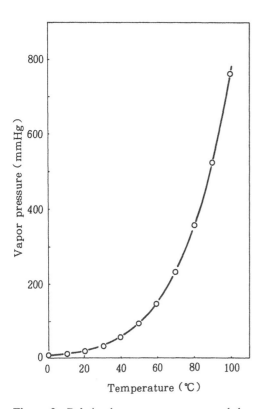

Figure 3   Relation between temperature and the vapor pressure of pure water.

When Eq. (8) is integrated, with R and L assumed constant,

$$\ln P = -\frac{L}{R} - \frac{1}{T} + C \qquad (9)$$

where C = integration constant. Equation (9) indicates that the logarithm of the vapor pressure of a liquid is linear with the reciprocal of temperature.

D. Surface Tension

Figure 4 shows the relation between surface tension and temperature [5]. As temperature rises, surface tension gradually decreases. The change is not as drastic as that of vapor pressure, however: compared with the surface tension at 25°C (72.8 dyn cm^{-1}), that at 100°C (58.9 dyn cm^{-1}) is smaller by only about 18% and that at 0°C (75.6 dyn cm^{-1}) is greater by 4%.

In general, boundary tension always exists where two phases contact each other. When one of the two phases is gas, the boundary tension is called surface tension. Surface tension is the amount of work required to push back against the tendency of a surface to shrink. The tendency to shrink is the cohesive force of the molecules comprising the liquid and has a close relationship with the interaction of molecules. Generally, surface tension tends to increase as the molecular weight increases. The surface tension of water is strong, although its molecular weight is low. This is because hydrogen bonds make water molecules appear larger than their actual size. Since such water clusters become smaller clusters or water molecules as the temperature rises or falls, the surface tension weakens

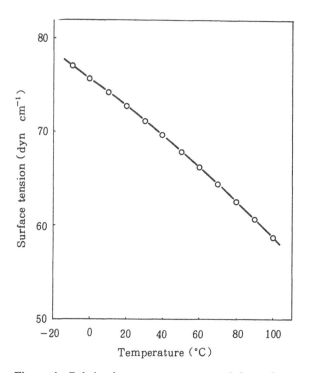

Figure 4   Relation between temperature and the surface tension of pure water.

and the evaporation of water is facilitated. As a result, the vapor pressure increases. The effect of temperature on the vapor pressure is much greater than that on surface tension. This is because even if the hydrogen bonds are cut because of the temperature rise, they are still able to help molecules to evaporate.

It is very interesting to study how many water molecules should be gathered for the concept of surface tension to have physical significance. The details have not yet been determined. Taking the capillary condensation phenomenon in pores into consideration, it can be presumed that at least four water molecules should be placed in a line. If the diameter of a water molecule is assumed to be 3.3 Å, the area will be around 137 Å^2.

Considering cleaning with ultrapure water, the cleaning efficiency is affected not only by the surface tension of water but also by the boundary tension between the water and the Si surface and between the Si surface and substances adhered to it. It is not easy to measure such boundary tension, however, so surface tension is used to explain cleaning efficiency.

Cleaning efficiency improves as the temperature of water rises. In accordance with the temperature rise, the thermal motion of water is more active, which cuts hydrogen bonds to reduce the size of water clusters and to weaken surface tension. Consequently it becomes easier for water to penetrate the smaller space between contaminants and wafer. The thermal motion of contaminants is also activated as a result of the temperature rise and they are easily removed.

E.   Viscosity

Figure 5 shows the relationship between the temperature of pure water and viscosity [6]. As the temperature rises, the viscosity decreases together with surface tension and resistivity. When the temperature increases from 25°C to 100°C, the viscosity decreases about 68%, from 0.00894 to 0.00284 P. The viscosity at 0°C (0.01792) is about twice that at 25°C. The effect of temperature on viscosity is smaller than that on vapor pressure but it is greater by far than the effect on surface tension. The mechanism of viscosity change in accordance with temperature is considered the same as the mechanism of surface tension change.

F.   Dielectric Constant

Figure 6 shows the relationship between temperature and dielectric constant [7]. Dielectric constant is defined as the C/C_0 ratio, where C is electrostatic capacity for an insulator inserted without play between parallel electrodes and C_0 is electrostatic capacity when an insulator is replaced by air. The C/C_0 ratio always indicates values greater than 1. Figure 6 shows that the dielectric constant decreases as the temperature of water increases. It is commonly believed that substances are well dissolved in water because water is a polar compound with a high dielectric constant. However, substances are dissolved even better when the temperature of the water is higher although its dielectric constant is decreased.

In 1939, Kirkwood considered that the dielectric constant ϵ_r of a polar compound is related to the bipolar moment of the molecule and its orientation and suggested the equation [8]:

$$\epsilon_r = 2\pi N \frac{\mu^2 g}{kT} \tag{10}$$

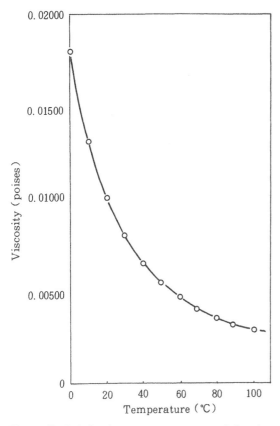

Figure 5 Relation between temperature and the viscosity of pure water.

where: N = number of molecules per unit area

μ = bipolar moment of central molecule surrounded by neighbors

When the inductive bipolar moment can be ignored, μ is equal to the eternal bipolar moment of the molecule. The value of g is determined by the relative orientation of its neighbors. When bipolar molecules are aligned in the same direction, the value of g increases.

G. Solubility of Gases

Figures 7 through 9 show the relationship between the temperature and solubility of air, oxygen, and carbon dioxide, respectively [9]. In every case, the solubility decreases as the temperature rises.

Figure 7 shows that 2.83×10^{-2} ml oxygen is dissolved at a partial pressure of oxygen of 760 mmHg at 25°C and that 1.70×10^{-2} ml oxygen is dissolved at the same partial pressure at 100°C. Therefore, since the oxygen concentration in the air is 21%, in accordance with Henry's law we know that a maximum of 8.5 mg L^{-1} of oxygen can be dissolved at a temperature of 25°C, and that a maximum of 5.1 mg L^{-1} of oxygen can be dissolved at 100°C. (These figures are converted at 0°C.)

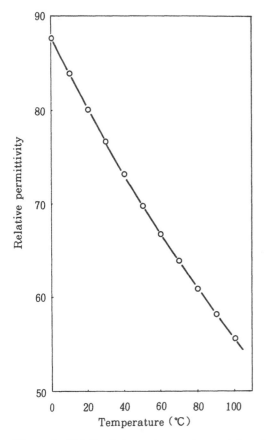

Figure 6   Relation between temperature and the relative permittivity of pure water.

H.   SiO₂ Solubility

Figure 10 shows the solubility of SiO_2 [10]: 120 mg SiO_2 is dissolved into 1 kg water at 25°C, 430 mg is dissolved at 100°C, and 30 mg is dissolved even at 0°C. Therefore, when Si, water, and O_2 exist together, it should be checked whether SiO_2 $[Si(OH)_4 = SiO_2 \cdot 2H_2O]$ is eluted.

IV.   SI WAFER CLEANING WITH ULTRAPURE WATER

A.   What Is Cleaning?

Cleaning means to remove unwanted substances that adhere to or are generated on a surface and to recover the cleanliness of the surface. Although ordinarily a very simple process, at the ultrapure water level, cleaning is very difficult and complex. This is because there is no way to guarantee the cleanliness of a surface after cleaning. In general, it is necessary to use pure water with a higher purity level than the surface being cleaned to improve its cleanliness. In some cases, however, the cleanliness level does not improve sufficiently even after the cleaning process. The difficulty of cleaning with ultrapure water is a result of its characteristics: the cleanliness level is extremely high and so it is not easy to confirm the cleanliness.

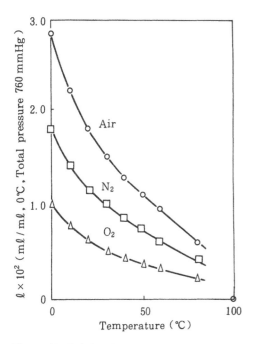

Figure 7   Relation between temperature and the solubility of air, N₂, and O₂ in pure water.

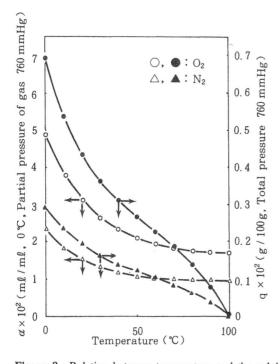

Figure 8   Relation between temperature and the solubility of O₂ and N₂ in pure water.

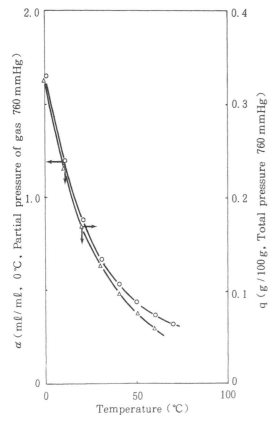

Figure 9   Relation between temperature and the solubility of CO_2 in pure water.

Generally, contaminants are considered adhered to a Si wafer by one or a combination of the following forces:

1. Physical adsorption by van der Waals force, for example
2. Electrical forces, such as Coulombic force
3. Chemical adsorption by means of chemical reactions
4. Mechanical forces

To efficiently remove contaminants, surfactant is added to reduce the surface tension of water or to electrically neutralize contaminants. On a semiconductor production line, however, surfactants are not permitted. Therefore, more importance is placed on physical cleaning than is usual.

B.   Mechanisms of Cleaning

Cleaning can be divided into two major categories:

1. Mechanically remove deposited contaminants
2. Dissolve deposited contaminants to remove them

The importance of mechanical cleaning was touched upon previously. Because water, in particular ultrapure water, has quite good cleaning ability even without using surfactants,

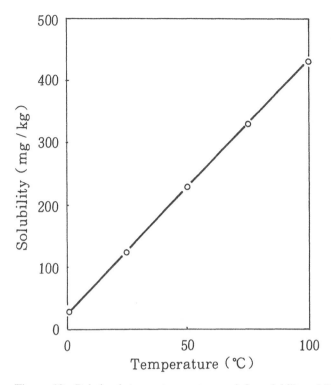

Figure 10   Relation between temperature and the solubility of SiO₂ in pure water.

mechanical cleaning with ultrapure water is an important cleaning method on the semicon-ductor production line. However, mechanical cleaning with ultrapure water is effective only with polar and ionic substances. It is not effective with nonpolar substances. As shown later in discussing the properties of various compounds (Table 5 on p. 11), closely related to the solvent characteristic of water are surface tension, dielectric constant, molecular weight (size and shape of molecules), bipolar moment, donor number, acceptor number, and other characteristics, but it is impossible to explain the wide-ranging solvent ability of water by a single factor. Hildebrand suggested the solubility parameter δ, which is thought to cover water solvent ability to some extent [11,12]:

$$\delta = \left(\Delta H^V - \frac{RT}{V} \right)^{1/2} \qquad cal \cdot cm^{-3} \tag{11}$$

where: ΔH^V = heat of vaporization
R = gas constant
T = absolute temperature
V = molar volume of solvent

ΔH^V is a parameter that strongly depends on intermolecular interaction. It also has a close relationship to the intensity of hydrogen bonds of molecules in a liquid, surface tension, and bipolar moment. V has a strong correlation with molecular weight. Later it can be seen (Table 5 on p. 11) that water has the largest δ value (23.53 cal cm^{-3}), which proves δ is well able to explain the good solvent ability of water.

C.   What It Means to Be Dissolved in Water

1.   *Nonpolar Substances*

Gases composed of such nonpolar molecules as O_2, N_2, Ar, CH_4, and C_2H_6 are not well dissolved in water [13]. Their solubility is in conformity with Henry's law. Liquids like CCl_4 and C_6H_6 and solids like $C_{10}H_8$ are also nonpolar substances that are barely dissolved in water. Table 2 lists the Gibbs standard free energy and standard enthalpy, both of which are thermodynamic parameters of the dissolution of substances in water. Table 2 shows that the values of ΔG°_{soln} are positive and that it is very difficult for the dissolution reaction to occur. On the other hand, the values of ΔH°_{soln} are negative and dissolution is an exothermic reaction. The relationship between ΔG°_{soln} and ΔH°_{soln} can be expressed as

$$\Delta G^{\circ}_{soln} = \Delta H^{\circ}16_{soln} - T \, \Delta S^{\circ}_{soln} \tag{12}$$

where ΔS°_{soln} = enthalpy change of dissolution. Since $\Delta G^{\circ}_{soln} > 0$ and $\Delta H^{\circ}_{soln} < 0$ in Eq. (12), it is found that ΔS°_{soln} is a large negative value. In short, the dissolution of nonpolar substances in water is an advantageous reaction in terms of enthalpy but very disadvantageous in terms of entropy.

These phenomena can be described schematically. Since nonpolar molecules are not electrostatically attracted to water molecules and have a very weak van der Waals force, it is difficult for nonpolar molecules to penetrate between water molecules that are held together by relatively strong hydrogen bonds. Because the intermolecular interaction between water molecules and nonpolar molecules is weak, the water molecules surrounding the nonpolar molecule are more strongly tied to each other than the water molecules that form bulk water. In short, the water molecules around nonpolar molecules are more organized. The structure of water nears that of ice. This reaction is therefore exothermic. Also, when water molecules are systematically arranged, the entropy decreases.

The icelike structure of water at temperatures that usually keep water liquid is not stable. In such a structure, water molecules have a tendency to recover the original structure of water by pushing out nonpolar molecules. This is why $G^{\circ}_{soln} > 0$ and also why nonpolar substances are barely dissolved in water. Nevertheless, nonpolar substances, such as O_2, have significant effects on the semiconductor production line although they are barely dissolved in water.

2.   *Dissolution of Polar Substances*

Substances composed of polar molecules, such as HCl, NH_3, C_2H_5OH, and CH_3COOH, are by far better dissolved in water than nonpolar substances in general. For gases, Henry's law does not apply. Moreover, some or most of the molecules with high polarity dissociate to ions.

Table 3 indicates thermodynamic parameters for major polar substances dissolved in water. All the ΔG°_{soln} show large negative values, although the values of ΔH°_{soln} are also negative, which is the same as for nonpolar substances. This proves that these substances

Table 2   The ΔH°_{soln} and ΔG°_{soln} of Nonpolar Gases dissolved in Pure Water

Gases	ΔH°_{soln} / kJ mol^{-1}	ΔG°_{soln} / kJ mol^{-1}	Gases	ΔH°_{soln} / kJ mol^{-1}	ΔG°_{soln} / kJ mol^{-1}
Ne	− 4.6	19	C_2H_6	− 17.4	15.8
Ar	− 12	16	H_2	− 4.2	18
CH_4	− 14.2	16.4	O_2	− 12	16

Table 3   The ΔH°_{soln} and ΔG°_{soln} of Polar Gases Dissolved in Pure Water

Gases	ΔH°_{soln} / kJ mol^{-1}	ΔG°_{soln} / kJ mol^{-1}	Gases	ΔH°_{soln} / kJ mol^{-1}	ΔG°_{soln} / kJ mol^{-1}
NH$_3$	− 34.2	− 10.1	HBr	− 85.1	− 50.5
HCl	− 74.9	− 36.0	HI	− 81.7	− 53.3

are easily dissolved. In the dissolution of polar substances, a very strong force of hydration, which is able to cut the covalent bonds between molecules, is generated by the interaction between water molecules and polar molecules. This kind of strong interaction makes the system extremely stable in terms of enthalpy, which facilitates dissolution with a great amount of exothermic energy. When strong hydration is generated, water molecules are systematically arranged around solute molecules and ions, creating a new order. Consequently, the entropy of the system decreases, which makes the $T \Delta S^\circ_{soln}$ terms in Eq. (12) negative. Because the contribution of ΔH°_{soln} is greater, however, ΔG°_{soln} becomes negative and dissolution occurs extremely easily.

3.   Ionic Substances

Table 4 indicates the change in Gibbs standard free energy and standard enthalpy when various ionic crystals are dissolved in water. In Table 4, both a and b group crystals are well dissolved in water; group a is dissolved exothermically but group b is dissolved endothermically. Group c is barely dissolved in water.

As shown in Eq. (13), ΔG°_{soln} is defined as the difference between the Gibbs standard hydration energy of an ion (ΔG°_{solv}) and the standard lattice energy (ΔG°_{lat}):

$$\Delta G^\circ_{soln} = \Delta G^\circ_{solv} - \Delta G_{lat} \tag{13}$$

Although ΔG°_{soln} can be calculated with Eq. (13), the accuracy is not good because various corrections are needed. Experimentally the value of ΔG°_{soln} can be obtained through the solubility product of a salt. When $A_p B_q$ completely dissociates into ion A and ion B in solution,

$$A_p B_q(S) + nH_2O \Leftrightarrow pA(H_2O)_j + qB(H_2O)_k \tag{14}$$

Table 4   The ΔH°_{soln} and ΔG°_{soln} of Ionic Crystals Dissolved in Pure Water

Ionic crystal	$\dfrac{\Delta H^\circ_{soln}}{\text{kJ mol}^{-1}}$	$\dfrac{\Delta G^\circ_{soln}}{\text{kJ mol}^{-1}}$	Ionic crystal	$\dfrac{\Delta H^\circ_{soln}}{\text{kJ mol}^{-1}}$	$\dfrac{\Delta G^\circ_{soln}}{\text{kJ mol}^{-1}}$
(a) AlF$_3$	− 326	− 134	(b) AgNO$_3$	+ 22.6	− 0.75
AlCl$_3$	− 331	− 251	NaCl	+ 3.88	—
CuSO$_4$	− 73.1	− 17	KCl	+ 17.22	—
FeCl$_2$	− 81.6	− 39.0	KI	+ 20.3	—
MnSO$_4$	− 64.9	− 15			
ZnCl$_2$	− 73.1	− 40.1	(c) AgCl	+ 65.5	+ 55.7
ZnSO$_4$	− 80.3	− 17	AgBr	+ 84.4	+ 70.0
			AgI	+ 112.2	+ 91.7
(b) NH$_4$Cl	+ 14.8	− 7.7			
NH$_4$Br	+ 16.8	− 7.9			

The equilibrium constant of dissolution K_s can be expressed as

$$K_S = \frac{a_A^p(H_2O)ja_B^q(H_2O)_k}{a_{ApBq(S)}a_{H_2O}^n} \tag{15}$$

where: S = solid
$\quad\quad A(H_2O)j$ = hydration of ion A
$\quad\quad B(H_2O)k$ = hydration of ion B
$\quad\quad a$ = activity

In a solution with low ionic strength, the activity can be regarded as equal to the ionic strength and the activity of a pure solid is defined as 1. Because water as solvent in a solution of low ionic strength is very close to pure water, $a_{H_2O} = 1$. The equilibrium constant of dissolution K_S is also called the solubility product, and in general it can be obtained as the product of ionic strength:

$$K_S = a_A^p a_B^q = [A]p[B]q \tag{16}$$

The relationship between ΔG_{soln}° and K_S can be expressed as

$$\Delta G_{soln}^\circ = -RT \ln K_S \tag{17}$$

Therefore, if the electrolyte K_S is given, ΔG_{soln}° can be easily calculated. The change in the standard enthalpy of dissolution can be calculated by extrapolating the heat of dissolution of a salt to 0 concentration.

Furthermore, there is a relationship in Eq. (12) between ΔG_{soln}°, ΔH_{soln}°, and ΔS_{soln}°. Some substances with $\Delta G_{soln}^\circ < 0$, which tend to be dissolved easily, are therefore dissolved endothermically, and substances with $\Delta G_{soln}^\circ > 0$, which tend to be barely dissolved, are dissolved exothermically.

D.   Cleaning the Si Wafer

The types of contaminants that adhere to a wafer surface vary from one cleaning process to another. Because there are several major adhering contaminants, such as ionic substances, polar substances, and nonpolar substances, however, in practice most contamination is mixed. It is most important in the cleaning process to understand the types and characteristics of adhering contaminants to select the most effective cleaning method. However, because the concentration of these contaminants is extremely low and the different types of contaminants adhere together, it is very difficult to identify each contaminant.

It is thought that contaminants reach the wafer surface by such actions as inertial force and diffusion and adhere to the wafer surface by means of van der Waals forces, static electricity, or chemical reactions.

There are two major methods of cleaning adhered contaminants: a physical method and a chemical method. As mentioned earlier, physical or mechanical cleaning is the basic method. In chemical cleaning, organic materials are dissolved and removed by organic solvents, and metals, metal oxides, and ions are removed by acid cleaning. Chemical cleaning generally finishes with overflow cleaning using ultrapure water. Physical cleaning includes ultrasonic cleaning, blow cleaning with high-pressure water, and blow cleaning with fine-particle ice. Because chemical cleaning requires a physical treatment to enhance its effects, both chemical and physical cleaning methods are applied together in most cases.

Contaminants often reach the Si wafer surface in aqueous solution by means of electrical and chemical actions [14,15]. Many substances are charged negative (or positive) in solution. Substances with the same charge maintain a balance between repulsion and attraction as a result of intermolecular interaction. The Si surface is exposed when a wafer is treated with HF, so contaminants easily adhere to the surface. The indicator of the charge is the Zeta (ζ) potential. ζ can be obtained using the Helmholtz-Smoluchowski equation,

$$\zeta = \frac{4\pi\eta v}{\epsilon E} \tag{18}$$

where: E = intensity of electrical field
$\quad\quad\;\; v$ = shift speed of particles in electrical field
$\quad\quad\;\; \eta$ = viscosity of liquid
$\quad\quad\;\; \epsilon$ = dielectric constant

Polystyrene particles adhere to a bare Si wafer surface treated with 99.99% HF, 10–100 times as many as to a hydrophilic Si wafer surface covered with native oxide. The ζ potential of polystyrene is negative. The ζ potential of a Si wafer surface covered with native oxide is also negative, but that of a bare Si wafer surface treated with diluted HF is positive. This means that the adhesion of polystyrene particles can be qualitatively explained by the ζ potential. It is therefore possible to enhance the cleaning effect by controlling the intensity of the ζ potential.

The watermark method is a way to investigate the purity of ultrapure water. In this method, a droplet of ultrapure water is placed on a clean wafer and dried to see whether the watermark remains. When ultrapure water is contaminated, a watermark is observed. For example, it is assumed that an impurity of 1 ppm (1 mg L^{-1}) is contained in water and 0.1 g water is dropped onto a wafer surface 1 cm^2 in area. When this droplet is dried, an impurity of 1 mg m^{-2} is left on wafer surface. If the molecular weight of the impurity is assumed to be 50, the number of molecules is 1.2×10^{19} pieces m^{-2}. If the diameter of the molecule is 5 Å, this is almost as thick as two molecule layers. Although this method must be used very carefully [16], it is simple and practical.

When a Si wafer is cleaned with ultrapure water in which various substances, including ions, are dissolved or diffused, it is possible that ions will cause the following exchange reaction. The Si wafer surface is covered with native oxide, and as shown in Eq. (19), it changes to silanol relatively easily in water.

$$
\begin{array}{c}
\mathrm{-Si} \\
\!\!>\!\mathrm{O} \\
\mathrm{-Si} \\
\mathrm{-Si} \\
\!\!>\!\mathrm{O} \\
\mathrm{-Si}
\end{array}
\quad \xrightarrow{\;\; H_2O \;\;} \quad
\begin{array}{c}
\mathrm{-Si-OH} \\
\mathrm{-Si-OH} \\
\mathrm{-Si-OH} \\
\mathrm{-Si-OH}
\end{array}
\tag{19}
$$

Therefore, if $Cu^{2+}Na^+$ and Cu^{2+} are present, the following ion-exchange reactions take place:

$$
\begin{array}{l}
-\overset{|}{\underset{|}{Si}}-OH \\[4pt]
-\overset{|}{\underset{|}{Si}}-OH \\[4pt]
-\overset{|}{\underset{|}{Si}}-OH \quad + Na^+ \longrightarrow \quad -\overset{|}{\underset{|}{Si}}-ONa + H^+ \\[4pt]
-\overset{|}{\underset{|}{Si}}-OH
\end{array}
\tag{20}
$$

$$
\begin{array}{l}
-\overset{|}{\underset{|}{Si}}-OH \\[4pt]
-\overset{|}{\underset{|}{Si}}-OH \\[4pt]
-\overset{|}{\underset{|}{Si}}-OH \quad + Cu^{2+} \longrightarrow \quad
\begin{array}{l} -\overset{|}{\underset{|}{Si}}-O \\ \qquad\qquad \diagdown Cu \ + 2\,H^+ \\ -\overset{|}{\underset{|}{Si}}-O \diagup \end{array} \\[4pt]
-\overset{|}{\underset{|}{Si}}-OH
\end{array}
\tag{21}
$$

If the electrically active ions are diffused into silicon, the characteristics of the semiconductor are adversely affected.

If a TOC (total organic carbon) like alcohol is present, the following reaction takes place:

$$
\begin{array}{l}
-\overset{|}{\underset{|}{Si}}-OH \\[4pt]
-\overset{|}{\underset{|}{Si}}-OH \\[4pt]
-\overset{|}{\underset{|}{Si}}-OH \quad + CH_3OH \longrightarrow \quad -\overset{|}{\underset{|}{Si}}-O-CH_3 + H_2O \\[4pt]
-\overset{|}{\underset{|}{Si}}-OH
\end{array}
\tag{22}
$$

The Si surface then becomes hydrophobic if the concentration of alcohol is high enough. When the temperature is raised under anaerobic conditions carbon is precipitated.

When colloidal silica $[Si(OH)_4 = SiO_2 \cdot 2H_2O]$ in water adheres to a Si wafer and the temperature is raised, the following dehydration reaction takes place, which can become the source of particle generation:

$$
Si(OH)_4 = SiO_2 + 2H_2O
\tag{23}
$$

The chemical adsorption accompanying chemical reactions is the main topic here, but on a Si surface, physical adsorption takes place more easily than chemical adsorption. Particularly when a bare Si surface is exposed by HF treatment, the characteristics of chemical adsorption become much stronger.

V. SUMMARY

This chapter describes the basic properties of ultrapure water and the basics of ultrapure water cleaning. Few articles have been published about the properties of water. A discussion of the properties of ultrapure water should be based on the liquid structure of water. At present, however, the details of the structure of water have not yet been confirmed and further study is required.

The new field of ultraclean technology has developed along with the semiconductor industry. The technology to produce, evaluate, and use ultrapure water is one of the most important pillars that support this new technology. Ultraclean technology is expected to become vital not only in the semiconductor industry but also in other areas of high technology, including biotechnology.

REFERENCES

1. S. Kubota, Haguruma, *393*, 7 (1989); S. Kubota, Haguruma, *395*, 12 (1989).
2. H. S. Harned and R. A. Robinson, Trans. Faraday Soc., *36*, 973 (1940).
3. T. Sakamoto and K. Koike, Kogyo Yosui, *305*, 8 (1984).
4. Nihon Kagaku Kai, Kagaku Binran Kisohen, *II*, 117 (1984).
5. Landolt-Börnstein Tabellen, 6 Aufl., II Band 3 Teil, Springer-Verlag, 1956.
6. T. R. Camp, Water and Its Impurities, Reinhold, New York, (1963), p. 15.
7. Nihon Kagaku Kai, Kagaku Binran Kisohen, *II*, 501 (1984).
8. H. Otaki, Yoeki Kagaku, Shokabo, 35–36 (1985).
9. Nihon Kagaku Kai, Kagaku Binran Kisohen, *II*, 158–159 (1984).
10. Nihon Kagaku Kai, Kagaku Binran Kisohen, *II*, 175 (1984).
11. M. Imoto, MOL, No. 2, 38 (1986).
12. H. Otaki, Yoeki Kagaku, Shokabo, 37 (1985).
13. Denkikagaku Kyokai, Wakai Gijitsusha no tameno Denki Kagaku, Maruzen, 6–12 (1983).
14. H. Tamura and A. Saeki, Denshi Jyoho Tsushin Gatsukai, Gijitsu Houkoku, SDM87-190, p. 45 (1988).
15. M. Watanabe, Hitachi Hyoron, No. 5, 39 (1989).
16. M. Hamamoto et al., Denshi Jyoho Tsushin Gatsukai, Gijitsu Hokoku, SDM87-188, p. 33 (1988).

5
Afterword

Takaaki Fukumoto
Mitsubishi Electronic Corporation, Kumamoto, Japan

This part has described the properties of water and ultrapure water and various requirements for the quality of ultrapure water employed in the production line of memory chips from megabit to gigabit. In future technology development, a good balance should be maintained by studying the mechanisms of surface cleaning as well as properties.

The remainder of this book touches upon complete ultrapure water production systems, installation methods, water quality evaluation methods, and wafer cleaning technology. These technologies are closely related, and comprehensive understanding is necessary. The quality of ultrapure water also needs to be determined based on the cumulative effect of these technologies. We suggest that the reader return to Part I after finishing the book.

II
Ultrapure Production Systems

1
Introduction

Koichi Yabe
Kurita Water Industries, Ltd., Tokyo, Japan

Natural water to be processed to ultrapure water contains various impurities that cannot be completely removed by any single method. To remove these impurities completely, various processing technologies are combined to build an ultrapure water production system. This system is composed of a pretreatment system, a primary treatment system, an ultrapure water system, a piping system, and a wastewater reclamation system.

The design of an ultrapure water production system requires a full understanding of the characteristics and behavior of natural water, as well as the characteristics and limitations of each technology applied to each component system. At the initial stage, existing technologies for the water supply to a boiler were merely adapted. Since it was revealed that these existing technologies were incapable of dealing with the high purity level demanded for ultrapure water, however, great efforts have been made to develop new technologies as well as to improve existing technologies. The present technologies of ultrapure water production in the semiconductor field are advanced enough to cope with the higher purity level and larger quantity required.

The primary pure water system is mainly composed of a reverse osmosis unit, a deaerator, and an ion-exchange unit. The functions of the reverse osmosis membrane are critical, and improvements in the reverse osmosis membrane have contributed markedly to progress in the ultrapure water production system. The ion-exchange unit is also vital to remove inorganic ions economically.

The ultrapure water system is mainly composed of an ultraviolet (UV) sterilization unit, an oxidation unit, an ion-exchange unit, and an ultrafiltration unit. The ultrafiltration unit plays an important role as the final filter for removing trace particles and bacteria. Ultrafiltration technology, the technology peculiar to ultrapure water, has been developed and upgraded.

A piping system is required to distribute ultrapure water with high purity to use points without degrading its quality. Specific technologies have been developed for piping materials, the design of piping networks, installation, cleaning, and sterilization.

The wastewater recycling system is composed of a reverse osmosis unit, an active carbon adsorption unit, an ion-exchange unit, and a UV oxidation unit. In Japan, the wastewater reclamation system is relatively prevalent as one countermeasure to deal with restrictions on industrial water consumption and cost increases. Based on 10 years of experience, the basic technology for wastewater reclamation has been established.

2
Pretreatment System

Kenichi Ushikoshi
Shinko Pantec Co., Ltd., Kobe, Japan

I. PURPOSE OF PRETREATMENT

Ultrapure water is produced by treating various kinds of source water in various processes. An effective ultrapure water production system must be designed for the stable production of ultrapure water of the required quantity and quality at a low cost. An ultrapure water production system is usually composed of a pretreatment system, a primary treatment system, a secondary treatment system, and a distribution piping system.

Several treatment methods can be applied in the primary treatment system, depending on the quality of the raw water and the process characteristics. As shown in Fig. 1, these methods can be divided into two major types; the RO + IX type where a reverse osmosis unit (RO) is placed upstream of an ion-exchange unit (IX); and the IX + RO type where the RO unit is placed downstream of the IX unit.

The primary treatment system, followed by the polishing system (the secondary treatment system), is an important system that removes 99–99.99% of the impurities contained in raw water. The pretreatment system must therefore be designed and installed to secure its safe and perfect performance.

The pretreatment process depends on the type of primary treatment system that follows it. Table 1 compares the difference of the inlet water quantity required by RO +

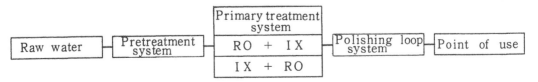

Figure 1 Process flow of ultrapure water treatment system.

Table 1   Inlet Water Quality for RO + IX Versus IX + RO Systems

Process Flow	Turbidity (Unit)	SDI (15min.)	Iron (mg/ℓ)	Silica (mg/ℓ)	Residual chlorine (mg/ℓ)	pH	TDS (mg/ℓ)
RO+IX	<0.5	<3−5	<0.05	<100 (RO brine)	0 or <1.0	4.5 − 7.5 (for scale control)	−
IX+RO	<2.0	−	<0.3	−	<0.2	−	<500

IX and IX + RO. The inlet water quality in IX + RO is not as crucial as that in RO + IX. City water in Japan well satisfies the requirements of the IX + RO. On the other hand, in the RO + IX, in which pretreated water is directly fed to the reverse osmosis unit, much attention should be paid to pretreatment to avoid fouling the reverse osmosis unit.

In current ultrapure water production systems, membrane filtration by reverse osmosis and ultrafiltration, for example, is employed at some point in the process. Therefore, one of the important challenges to successfully operate an ultrapure water production system is to secure the stable performance of the membrane filtration unit. In the IX + RO type, effective and stable reverse osmosis operation can be expected because the ion-exchange unit removes suspended matter and the dissolved salts that cause scaling, which leads to fouling of the subsequent reverse osmosis unit. The IX unit also reduces coagulation and adhesion in the membrane module by lowering the interface potential. On the other hand, the RO + IX type is widely employed because it can save cost, space, and energy as well as reduce wastewater by making the best use of the demineralization function that is a feature of reverse osmosis.

Table 2 lists typical types of fouling of the reverse osmosis system and countermeasures to prevent them. This chapter describes the major technologies of pretreatment: coagulation, precipitation, filtration, softening, silica removal, iron removal, manganese removal, and adsorption.

Table 2   Fouling of RO System Versus Pretreatment for Foulants

Type of Fouling	Foulants	Pretreatment for Foulants	Index for Safety Operation
Scaling	$CaCO_3$, $SrSO_4$ $CaSO_4$, $BaSO_4$ $Ca_3(PO_4)_2$, SiO_2 $Fe(OH)_3$, $Al(OH)_3$	Sodium cycle softening, lime softening, silica removal, ion removal, pH adjustment, injection of dispersant	LSI 0 (at brine) SiO_2 100 mg/ℓ (at brine) Fe, Al 0.05 mg/ℓ
Colloidal fouling	Organic colloid Inorganic colloid	Coagulation, sedimentation, filtration	SDI 3 − 5
Clogging	Suspended matter	Coagulation, sedimentation, filtration, cartridge filter	Put 5 − 10 micron filter
Biological fouling	Bacteria, slime, organic	Coagulation, sedimentation, filtration, sanitization, chlorine injection, carbon filter	Proper sanitization
Chemical degradation	Cl_2, oxidative reagent, detergents, etc.	Carbon filter, reducing reagent injection, pH adjustment etc.	Cl_2 0 (for PA, TFC membrane) Cl_2 0.2 − 1.0 mg/ℓ (for CA membrane)

II. WATER SOURCE AND WATER QUALITY

Table 3 reports water consumption and water sources in Japan issued by the National Land Agency [1]. Water is also consumed for agricultural use (585×10^8 m^3). Industrial use is almost the same as domestic use in volume. As for municipal water, surface water accounts for close to 70%.

A. Difference in Characteristics of Water Quality Among Various Water Sources

1. *Water from Rivers*

Water flows into rivers both through the surface and underground. Surface water contains much suspended matter, but inorganic salts are dissolved in underground water depending on the stratum. After rain, the concentration of suspended matter in river water increases because a large amount of surface water flows into rivers. On the other hand, when it does not rain a lot, river water becomes clean. One of the major characteristics of river water is the wide fluctuation in suspended matter concentration.

When rivers flow through forests or peat bogs, organic materials derived from humus can be added to the water. Sewage or industrial wastewater may be mixed into river water. Therefore, the design of a pretreatment unit should be based on a full study of river water quality when this is used as raw water.

2. *Water from Lakes*

Once river water flows into lakes or dams, its quality drastically changes. Since contaminants sediment when water is detained in lakes for long periods of time, the suspended matter concentration is lowered. On the other hand, water is significantly affected by the biological activity during the detention. When eutrophication progresses in lakes as a result of the influx of nutritive substances from upstream, algae propagate. Since the specific gravity of algae is low, coagulation and settling are interfered, and the filter bed is thus subject to clogging. The water pH also rises because dissolved CO_2 is consumed through carbonic acid assimilation.

When water stratifies within lakes, a shortage of oxygen occurs in the bottom layer, which leads to the redissolution of iron and manganese and the formation of hydrogen sulfide under the anaerobic conditions.

Table 3 Water Consumption and Water Sources in Japan, 1985 (Yearly Data) [11]

Category		x 10^8 m^3	(%)
Water consumption	Domestic use	152	49.7
	Industrial use	154	50.3
Water source of municipal water	Dam	52.9	35.5
	Lake	1.9	1.3
	River	48.7	32.7
	Well	40.5	27.2
	Other	4.9	3.3

(Annual Report of Water Source −1989 by National Land Agency of Japan)

When lakes and dams are considered to be intake sources, it is important to study the treatment method in advance and not to take in bottom water.

3.   *Underground Water*

Since underground water flows very slowly between strata, all contaminants precipitate or are adsorbed, and eventually the suspended matter concentration drops to very close to zero. Underground water is strongly affected by the layers through which it flows, however: underground water that travels through limestone layers is high in calcium hardness, underground water flowing through volcanic areas has a high silica concentration.

Since underground water does not come in contact with the surface, it contains limited oxygen and it is under a reducing atmosphere. Therefore it often contains ferrous compounds or manganese, and sometimes it contains hydrogen sulfide.

Although the suspended matter concentration is very low in underground water and its temperature is stable throughout the year, pretreatment is required to use underground water as raw water because it often contains hardness materials, silica, and heavy metal ions.

Table 4 lists analytic data for major rivers, lakes, and underground streams in Japan. Because raw water in Japan contains limited amounts of dissolved salts and hardness materials, in general, softening treatment by the cold lime-soda process is rarely employed. In designing a pretreatment system located in Japan, the high silica concentration, the necessity for iron and manganese removal, and the growth of algae must be considered depending on the water intake source.

When city water is introduced as raw water, there should be few problems: city water is processed to meet the municipal water quality standards set by the Ministry of Health and Welfare. It is still necessary, however, to trace back the source of the city water. SDI values and silica content should also be checked to protect the reverse osmosis unit.

Table 4   Typical Raw Water Analytic Data in Japan

Item	Unit	Water source		
		River	Lake	Well
Turbidity	STD unit	10	4	1
Color	STD unit	8	5	2
pH	STD unit	7.1	8.3	7.4
Conductivity	$\mu S/cm$	196	123	295
Ca−Hardness	mg/ℓ as $CaCO_3$	62.5	25	41
Mg−Hardness	mg/ℓ as $CaCO_3$	8.2	8.8	23
M−alkalinity	mg/ℓ as $CaCO_3$	30.6	29.1	71
Chloride	mg/ℓ as Cl	23.3	8.5	32.6
Sulfate	mg/ℓ as SO_4	28.5	9.0	7.4
Nitrate	mg/ℓ as NO_3	6.2	0.2	2.0
Silica	mg/ℓ as SiO_2	7.5	15	57.8
Iron	mg/ℓ as Fe	0.15	0.1	0.03
TOC	mg/ℓ as C	2.2	1.5	0.9

B. Water Sources and Water Quality in Silicon Valley, California and Silicon Plain, Texas [2]

In Silicon Valley, California, city water is always used as the raw water for ultrapure water production systems. City water satisfies the U.S. Environmental Protection Agency (EPA) standards of municipal water. There are two sources for city water in Silicon Valley: well water with high calcium and high alkalinity concentration and some silt and reservoir water which is characterized by limited calcium hardness and alkalinity, however er containing high TOC, and a high volume of suspended matter and bacteria. The city water source is alternated in accordance with water consumption. The water quality at points of use therefore fluctuates greatly over the course of a day.

City water is also used in Silicon Plain, Texas. In Austin, city water is taken from Lake Travis or Lake Austin, and in Dallas, the source is surface water. Since the water hardness is high in Silicon Plain, the cold lime-soda softening method is applied.

Table 5 lists analytic data for raw water in the United States.

III. PRETREATMENT PROCESS AND UNITS

The most appropriate choice of pretreatment process and units is based on a study of raw water quality. Major processes and pretreatment units are described here.

A. Coagulation, Settling, and Filtration

Most city water in Japan is treated with coagulation, settling, and filtration. When industrial water with a turbidity of more than 10 is used as raw water, these processes are recommended.

Table 5 U.S. Raw Water Analytic Data [2]

Item	Unit	Texas	Northern California		
			Hetch Hetchy Reservoir	Calaveras Reservoir	Pleasanton Well Field
Hardness	mg/ℓ as $CaCO_3$	110	4.0	96	340
Alkalinity	mg/ℓ as $CaCO_3$	68	6.2	92	232
TDS	STD units	200	12.5	141	465
Silica	mg/ℓ as SiO_2	7.0	4.4	11.7	19.7
pH	STD units	10.1	6.9	8.1	7.5
TOC	mg/ℓ Act.wt.	6—8	3—5	5—6	3—5

(Supplied by Arrowhead Industrial Water Inc.)

Table 6 Suspended Matter in Water

Item	Size (μm)
Colloid	0.001 – 1.0
Turbidity	0.1 – 100
Color	0.001 – 0.01
Virus	0.01 – 0.1
Bacteria	0.2 – 1.0
Algae	1.0 – 100

1. Coagulation

Table 6 shows the distribution of suspended matter in water. Generally, particles that can be removed through sand filtration are greater than 10–50 μm in diameter. Most of the particles in water must be treated with coagulation first and then removed by filtration.

The coagulant used for coagulation has the following functions:

1. Charge neutralization: Most suspended matter in water has a negative charge. Coagulant neutralizes the electrical charge of suspended particles so that they can be coagulated. The intensity of the electrical charge of the particles is called the ζ potential. Particles with a ζ potential from –10 to 10 mV can be coagulated by van der Waals forces [3].
2. Cross-linkage (bridge formation): Coagulated particles can be linked to make larger particles. The coagulant, then, must have larger molecules.

The salt of Al and Fe goes through hydrolysis in water to become a positively charged hydroxide polymer. PAC or $[Al_2(OH)_nCl_{6-n}]m$, alum $[Al_2(SO_4)_3 \cdot 18H_2O]$, and $Fe_2(SO_4)_3$ are often used as coagulants. PAC in particular, is very easy to handle because it contains polynuclear condensed ions that are hydrolyzed, such as $[Al_8(OH)_{20}]^{4+}$, and it does not consume as much alkalinity when it is reacted with raw water. Also, the coagulation pH range of PAC is 6–8. Table 7 [4] lists specifications for PAC. Figure 2 [4] shows the relationship between coagulation using PAC and the pH.

Alkali coagulants can be used to compensate for alkalinity, and polymer coagulant can be used for making large and stronger flocs. Flocs are generally formed in two stages:

1. Rapid mixing for 1–5 minutes to mix water and form small flocs
2. Slow mixing for 20–40 minutes to form large flocs

Rapid mixing is necessary to mix water and coagulant well. Slow mixing is required to grow large enough flocs to precipitate and be filtered. GCT is used as the indicator of slow mixing [5].

$$GCT = \sqrt{\frac{\epsilon_0}{\mu}}$$

where: ϵ_0 = energy consumption ratio of mixing blades, s^{-1}
   μ = viscosity coefficient of water, g cm·s^{-1}
   C = turbidity, unit
   T = time required to form flocs, s

Table 7 Specifications for PAC [4]

Item	PAC−250A	PAC−250AD
Appearance	pale amber clear liquid	hygroscopic powder
Al_2O_3 (%)	10.25 ± 0.25	over 30.0
Cl (%)	9.0 ± 0.5	26.5 ± 2.5
Fe (%)	under 0.01	under 0.03
As (mg/ℓ)	under 5	under 20
Heavy metal (mg/ℓ as Pb)	under 50	under 150
Basicity (%)	50.0 ± 5.0	50.0 ± 5.0
Specific gravity (25 ℃)	1.20 ± 0.01	0.85 ± 0.5
pH (25 ℃)	2.6 ± 0.3	−
pH (1/20 dilution 25 ℃)	4.4 ± 0.3	−
Viscosity (cp, 25 ℃)	4.0 ± 0.5	−
Freezing point (℃)	−12.0 ± 1.0	−
Stability	stable for years under normal conditions	

(Supplied by Taki Chemical Co., Ltd.)

When GCT is assumed fixed at 10^6 and G is set at 20 (G varies from 20 to 70), T is calculated as follows:

Turbidity	Mixing time T(s)
Low (5°)	10^4 (167 minutes)
Normal (50°)	10^3 (17 minutes)

Since a longer mixing time is required when the turbidity is low, it is more effective to apply the microfloc filtration method than the precipitation method. If G is set too

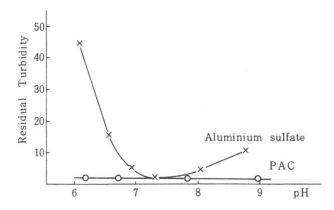

Testing water : Artificial-turbid water prepared by dispersing kaolin
pH 7.3, Turbidity 54, Alkalinity 47.6 mg/ℓ
Amount of coagulant (as Al_2O_3): 4 mg/ℓ

Figure 2 Residual turbidity and pH. (From Ref. 4.)

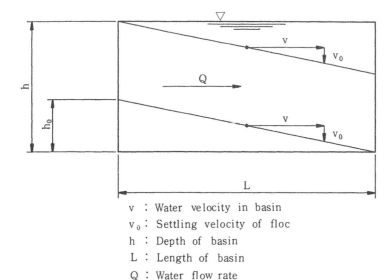

v : Water velocity in basin
v_0 : Settling velocity of floc
h : Depth of basin
L : Length of basin
Q : Water flow rate

Figure 3   Settling basin. (From Ref. 6.)

high to increase the mixing intensity, the flocs formed are easily broken. In this case a tapered flocculation method is applied in which several mixing steps are employed and mixing intensity is gradually decreased.

2.   Settling

The Stokes equation is used to predict settling velocity in general water treatment:

$$W = \frac{1}{18} g \frac{\rho_s - \rho}{\mu} d^2$$

where: W = particle settling velocity, cm s^{-1}
ρ_s = density of particles, g cm^{-3}
ρ = density of water, g cm^{-3}
μ = viscosity coefficient of water, g cm^{-1}s^{-1}
d = diameter of particle, cm
g = gravitational acceleration, cm s^{-2}

Because the settling velocity is proportional to the particle diameter squared, the greater the floc diameter, the higher the settling velocity becomes. Actually, however, when the floc diameter is increased, the density of the particles decreases. This means that the settling velocity does not actually increase to the theoretical value.

Figure 3 shows how flocs settle in a settling basin. If the settling efficiency is expressed as [6],

$$E = \frac{h_0}{h} = \frac{v_0}{Q/A}$$

where: E = settling efficiency, efficiency of removing settling flocs
A = settling area of basin
v_0 = settling velocity of floc
h_0 = distance that flocs with settling velocity v_0 settle in the settling basin
Q = flow rate
h = depth of settling basin

This equation indicates three ways to improve settling efficiency:

1. Increase A
2. Increase v_0
3. Decrease Q

If Q and v_0 are given, A should be increased. A can be increased by creating two to three layers in the settling basin. Based on this idea, a settling basin with a tube settler or inclined plates was developed. This structure improves the settling efficiency, and it is possible to set Q/A at 3–6 m/h. Figure 4 shows a typical settling basin with tube settler. Figure 5 [7] shows one kind of tube settler.

a. High-Speed Settling Basin. A high-speed settling basin was developed to improve coagulation efficiency by ensuring that flocs come into contact with high-concentration flocs that are kept in the basin in the course of floc formation. There are two types of high-speed settling basin, and the two may also be combined.

1. Slurry recirculation settler (see Fig. 6): Raw water is mixed with coagulant to form a slurry. The circulating flow then enters the treatment section, where particles settle.
2. Sludge blanket settler (see Fig. 7): Raw water is mixed with chemicals and flocs are formed. The raw water then passes through the upper sludge blanket with blanket concentration to be treated.

The high-speed settler is widely used for the treatment of industrial water. If the raw water turbidity is low, however, it is difficult to maintain the blanket concentration. On the other hand, if the raw water turbidity is high, the recovery ratio of water is not good because the concentration of discharged sludge cannot be set high.

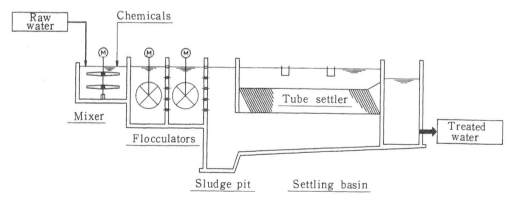

Figure 4 Typical settling basin.

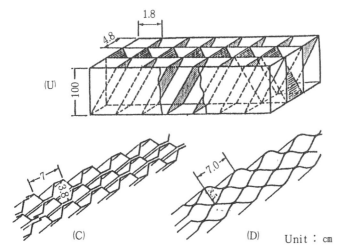

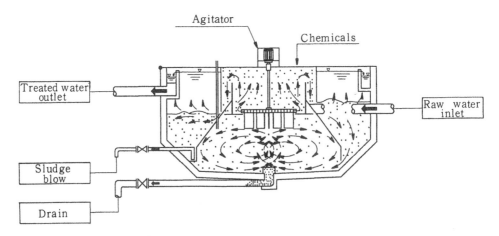

Figure 5   Typical tube settler. (From Ref. 7.)

If the sludge in the upper basin is not properly renewed, the coagulation capability is degraded as the sludge is used longer. Also, if captured flocs flow into the water, they pass through the filtration unit easily.

The high-speed settler needs to be properly and carefully operated based on full knowledge and skill.

3.   Filtration

There are two types of filtration: slow and rapid. In the slow filtration method, particles are filtered at a flow rate of 3–5 m/day. Particles do not settle in the basin with coagulant but are filtered through the biofiltration membrane formed on the surface of the filtration basin. Since this method employs biochemical treatment, it works well for city water but it is not suitable for high-turbidity raw water. This method is not employed in industrial water treatment because it requires a large space for installation.

The rapid filtration method is widely used for both city water and industrial water. Filter sand with a diameter of 0.3–0.7 mm is used, and the filtration velocity is 5–10

Figure 6   Slurry recirculation settler.

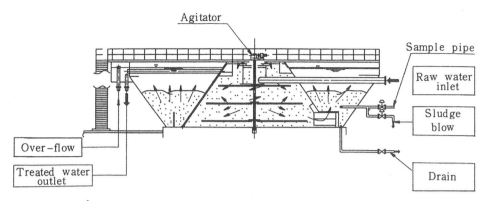

Figure 7 Sludge blanket settler.

m/h. Because the rapid filtration method does not employ biofiltration, it is usually combined with a coagulation settling unit.

Sand filtration can remove particles with a diameter of 10–50 μm. As raw water is filtered, more and more contaminants are trapped and deposited between the filter media, which increases the resistance and generates a head loss. In general, when the head loss reaches 1–5 mH$_2$O, the filter media should be recovered with backwash cleaning. Backwash cleaning is conducted for 5–10 minutes. In a bid to increase the cleaning effect, it is sometimes combined with surface wash and/or air backwash.

To further improve the filtration efficiency, the multimedia filtration method can be applied. In this method, coarse filtering media is placed in the upper layer and finer media in the lower layer to prevent contaminants from leaking. Usually, a combination of anthracite and sand, or anthracite, sand, and garnet is used.

Figure 8 illustrates three types of filters.

B. Microfloc and Filtration

The microfloc-filtration method is applied to raw water with a low turbidity. Flocculant is injected directly into the pipe to supply raw water to the filter, and microflocs are formed by mixing in the pipe and are filtered directly.

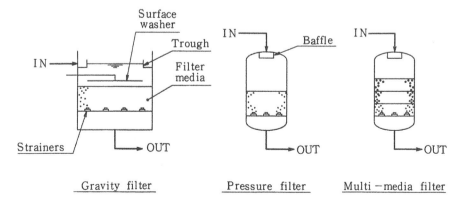

Figure 8 Filters.

The microfloc filtration is applied to underground water and water from lakes with low turbidity. This method is also used as the pretreatment for the reverse osmosis unit to further reduce the SDI value of city water. The volume of coagulant should be confirmed in each practical case. The coagulant volume needs to be determined carefully. If it is too small, raw water is not filtered properly; if it is too large, the water quality is degraded and the filter must be backwashed more frequently.

Because raw water travels from the flocculant injection point to the filtration layer in only 10–20 minutes, afterdeposit might be induced by excessive coagulant. (*Afterdeposit* is a phenomenon in which flocs are formed in a basin where the filtered water remains for any length of time.) Figure 9 shows experimental data for the degradation of water quality caused by the injection of excess coagulant. As shown in Fig. 9, the degradation of water quality due to recoagulation continues for 60 minutes after filtration, followed by saturation. PN in Fig. 9 is described in detail in Sec. IV.

Since the appropriate amount of coagulant injection must be maintained to ensure adequate operation of the microfloc filtration method, raw water with a limited fluctuation of quality is preferable. Afterdeposit flocs can be removed by polishing filtration. It is therefore preferable to install the RO inlet filter following the microfloc filtration units to securely conduct the pretreatment in the reverse osmosis unit. Figure 10 is a flow diagram of the microfloc filtration method.

In the microfloc-filtration method, all the contaminants in raw water and all the flocs formed as a result of flocculant injection should be trapped in the filtration layer. This requires the filter to capture many contaminants: generally a dual media filter with 0.4–0.6 mm sand and 0.7–1.5 mm anthracite is applied. The filter must be backwashed thoroughly with water, commonly with a combination of air and water.

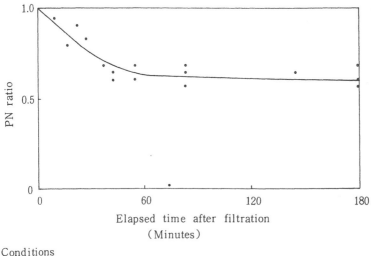

Conditions

Raw water	Lake water	Date of test	Dec. 19. 20 '74

Turbidity   4 − 6 mg/ℓ

PAC injection   37.5 mg/ℓ

$$\text{PN ratio} = \frac{\text{PN after elapsed time}}{\text{PN at filter outlet}}$$

Figure 9   Filtered water quality after microfloc filtration.

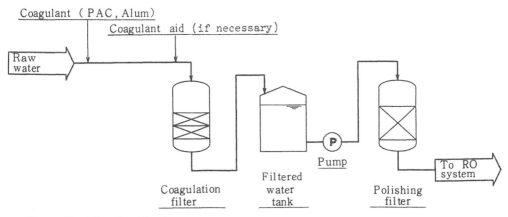

Figure 10   Microfloc filtration.

C.   Other Pretreatment Processes

1.   *Softening Method*

As mentioned earlier, because raw water in Japan contains a limited amount of hardness, softening treatment is not necessary. In most cases, Japanese raw water requires merely pH adjustment or dispersant injection. If the raw water needs to be treated by softening, the cold lime-soda method is effective. The sodium cycle ion-exchange softening method is also applied in many cases because it is easy to conduct and it prevents fouling of the reverse osmosis membrane.

a.   Cold Lime-Soda Method.   In the cold lime-soda method, hardness materials precipitate in such forms as $CaCO_3$ and $Mg(OH)_2$ at room temperature by injecting hydrated lime, soda ash, and alum to raise the pH to 10–11.
The major reaction formulas are as follows:

$$Ca(HCO_3)_2 + Ca(OH)_2 \rightarrow 2CaCO_3 \downarrow + 2H_2O$$
$$CaSO_4 + Na_2CO_3 \rightarrow CaCO_3 \downarrow + Na_2SO_4$$
$$Mg(HCO_3)_2 + 2Ca(OH)_2 \rightarrow Mg(OH)_2 \downarrow + 2CaCO_3 \downarrow + 2H_2O$$

The hardness is not totally removed: $CaCO_3$ at 35 mg L^{-1} is expected to remain. Figure 11 is a flow diagram of the cold lime-soda method.

b.   Ion-Exchange Softening Method.   This is a method of applying the sodium cycle cation-exchange method. The reaction formulas are as follows:

$$2R\text{-}SO_3Na + CaCl_2 \rightarrow (R\text{-}SO_3)_2Ca + 2NaCl \qquad \text{softening}$$
$$(R\text{-}SO_3)_2Ca + 2NaCl \rightarrow 2R\text{-}SO_3Na + CaCl_2 \qquad \text{regeneration}$$

The capacity of ion exchange per 1 L cation resin is 30–50 g $CaCO_3$. Sodium chloride solution or seawater is used for the regeneration. Water softened by the ion-exchange method can prevent semicolloidal fouling because it reduces scale formation and the ζ potential of suspended matter at the reverse osmosis membrane. In this sense, this is a very good method of pretreatment in the reverse osmosis unit. Figure 12 is a flow diagram of the ion-exchange softening method.

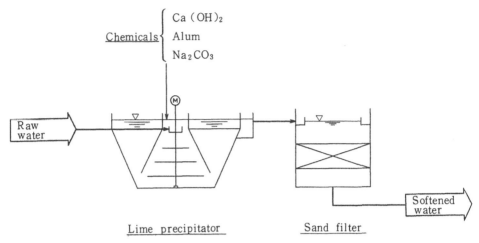

Figure 11   Cold lime-soda softening.

2.   *Silica Removal*

When the water source is located in volcanic areas, the silica concentration is sometimes high. The solubility of ionic silica is about 100 mg L^{-1} at room temperature (25°C). If ionic silica is concentrated in a reverse osmosis unit to a level higher than its solubility, silica compounds can be rapidly deposited if iron or aluminum is also present, eventually causing serious trouble to the reverse osmosis membrane. Unlike calcium carbonate or iron scaling, the silica deposit is hard to remove by a chemical cleaning process. The reverse osmosis unit must therefore be replaced with a new unit, although this is very expensive. It is critical to prevent silica scaling on reverse osmosis membranes if raw water with a high silica concentration is processed. A possible solution is the IX + RO method, which can be effective when the total dissolved salts is low. Another is to apply a silica removal process during pretreatment to supply raw water with a reduced silica concentration to the reverse osmosis unit. The silica removal process applied here is a coprecipitation method using a flocculant like alum.

    a.   Cold Lime-Soda Method.   This method of softening water hardness described earlier can also be applied to reduce silica in raw water. When magnesium precipitates

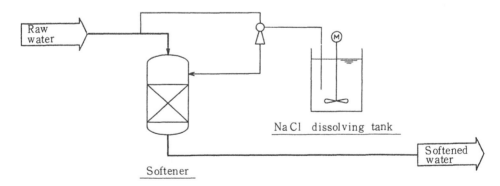

Figure 12   Sodium cycle ion-exchange softener.

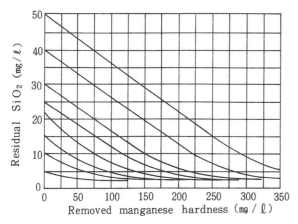

Figure 13 Removed magnesium hardness versus residual silica (at 21°C). (From Ref. 8.)

in the form of magnesium hydroxide at high pH, silica coprecipitates. Since the silica removal efficiency drops along with the decrease in magnesium hardness in raw water, magnesium oxide (MgO) or dolomite [MgCa(CO$_3$)$_2$] should be added to maintain the hardness. Figure 13 [8] shows the residual silica as a function of removed magnesium hardness. Figure 14 [9] shows the relationship between pH and the silica removal rate. The silica removal rate depends strongly on the pH; the best rate can be obtained in the pH range 10.3–10.5. The temperature dependence of the silica removal rate is also high, and the removal rate greatly increases when the temperature of the water is raised.

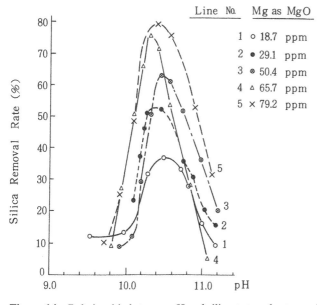

Figure 14 Relationship between pH and silica removal rate as a function of magnesium compound (at 30°C). (From Ref. 9.)

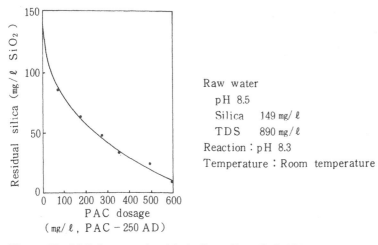

Raw water
  pH 8.5
  Silica   149 mg/ℓ
  TDS   890 mg/ℓ
Reaction : pH 8.3
Temperature : Room temperature

Figure 15   PAC dosage and residual silica. (From Ref. 10.)

b.   Method Using Aluminum Salt.   Silica can be removed by adding a large amount of such coagulants as PAC and alum. Figure 15 [10] shows the relationship between the residual silica and the amount of PAC-250AD (powder).

Figure 16 [10] indicates the relationship between the residual silica and the pH. The silica removal efficiency is highest in the pH range 8–8.5.

The method using aluminum salt is simple and easy to handle. To limit the residual aluminum to protect the following reverse osmosis unit, a pH level of 7–8 is appropriate if PAC is used. When the pH level is higher than this, the residual aluminum significantly increases. Compared with alum, PAC shows a higher silica removal efficiency at a low pH range. In this sense, PAC is preferable.

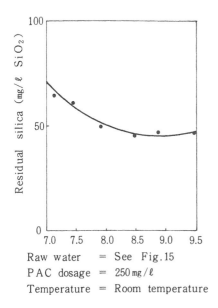

Raw water   = See Fig.15
PAC dosage = 250 mg/ℓ
Temperature = Room temperature

Figure 16   pH and residual silica. (From Ref. 10.)

Both the cold lime method and the aluminum salt method can be performed in a conventional high-speed settling basin. The sludge should be mixed well to facilitate the adsorption of silica. The blow volume of the sludge and the method of scale removal also require careful attention.

3. Iron and Manganese Removal

Iron and manganese removal is one of the most important steps in pretreatment. Iron in particular has a major role in fouling of the reverse osmosis unit. When iron is oxidized to become ferric, its solubility decreases and it exists as suspended matter. Such suspended iron exists in surface water from rivers and lakes. It can be easily removed by sedimentation with coagulation or filtration. In underground water and the bottom layer of water in lakes, however, which have a reducing atmosphere, iron is dissolved in the form of ferrous ion at a high concentration.

If the iron exists completely as dissolved iron in the reducing atmosphere of raw water, it need not be removed and the raw water can be supplied to a reverse osmosis unit without being processed. On the other hand, however, if the raw water is stored in a tank and exposed to the air, all the dissolved iron should be oxidized and removed. Such oxidative reagents as oxygen, chlorine, and potassium permanganate are used. The oxidative reagents take a long time to act in the neutral pH range, especially manganese oxidation, which requires a catalyst. Table 8 lists the oxidative reagents required to oxidize iron and manganese.

The oxidation process is usually followed by coagulation, settling, and filtration. For underground water with lower turbidity, however, contact filtration is employed using a filter coated with catalyst. FeOOH is an effective catalyst for iron and manganese sands coated with manganese oxide and are useful for manganese. Figure 17 shows typical processes for iron and manganese removal. When iron is oxidized by the aeration method, it may turn into a colloid, and be hardly coagulated, or be filtered when silica content is high in raw water.

4. Activated Carbon Filtration

The activated carbon filtration method has two major applications: the removal of organics from raw water and the removal of residual chlorine from raw water. A removal test should be carried out in advance because removal characteristics differ depending on organic materials in raw water. In general, the space velocity (SV) is set at $2–5$ h^{-1}. A tower, which has a structure similar to that used in the pressurized filtration method, is employed. In some cases, two to three towers are combined, one tower operates while the rest stand by. Exhausted carbon in the standby unit is replaced with new activated or reactivated carbon in series.

When a synthetic reverse osmosis membrane that does not feature chlorine resistance

Table 8 Required Oxidative Reagent for Fe and Mn Oxidation

Unit : mg/ ℓ

Item	Fe	Mn
Cl_2	0.63	1.29 *
$KMnO_4$	0.94	1.92

* Catalyzer required

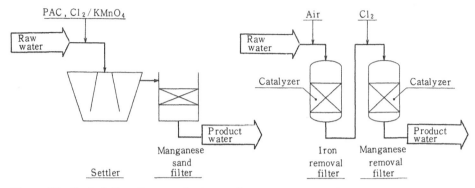

Figure 17 Typical flow diagrams for iron and manganese removal.

or an ion-exchange unit follows the filtration unit, the activated carbon filtration method is employed to remove residual chlorine. The activated carbon bed is 600–900 mm high, and its line velocity is 10–20 m h^{-1}. Activated carbon has a long life. In one case activated carbon was used for 2–5 years.

Fouling of the reverse osmosis membrane by bacteria should be noted when the residual chlorine is removed from this process. To suppress or eliminate the increased bacteria, the activated carbon filtration unit should be regularly sterilized with hot water or used carbon should be replaced with new carbon every 6–12 months. The reverse osmosis membrane also requires sterilization treatment—the injection of sterilizer or sterilization cleaning.

Activated carbon filtration is also effective to prevent the fouling caused by organic materials. In this case, an organic adsorption resin is sometimes employed.

D. Pretreatment in California and Texas

Because raw water is from municipal water supplies in Silicon Valley and Silicon Plain, it is well processed by the local authorities [2].

In Silicon Valley, California, the coagulation, settling, and filtration method is applied to surface water and underground water is treated with chlorine injection. In general, the cold lime-soda method is not employed because the hardness of the water is low. In Silicon Plain, Texas, the cold lime-soda method is employed for both the surface and underground water, which are high in hardness. This method maintains a high pH of the water, and thus sodium hexametaphosphate is added to prevent scaling in the pipes.

In existing plants, such pretreatments as filtration, activated carbon filtration, and Na cycle softening are employed, depending on necessity. In Silicon Plain, Texas, neutralizing agent is introduced upstream of the reverse osmosis process to lower the pH of the raw water. Figure 18 shows two U.S. water treatment systems.

IV. ANALYSIS AND EVALUATION

To properly design the pretreatment system and to evaluate the pretreated water, the analysis must be appropriately conducted.

A. Items to Be Analyzed and Analytic Method

Analysis items must include not only turbidity, pH, and electrical conductivity but also ions because the RO and IX processes follow. Items peculiar to water analysis such as SDI

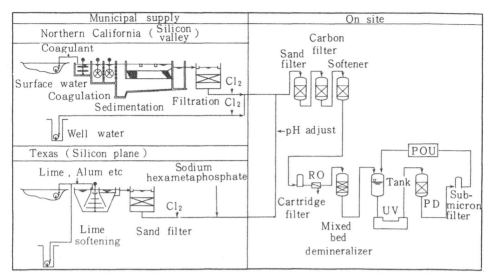

Figure 18   Water treatment systems in Silicon Valley, California, and Silicon Plain, Texas. (From Ref. 2.)

must be checked as well. The desirable items to analyze include pH, water temperature, turbidity, color, electrical conductivity, M-alkalinity, P-alkalinity, Ca-hardness, Mg-hardness, chloride ion, sulfate ion, phosphate ion, nitrate ion, ammonium ion, strontium ion, barium ion, COD, total organic carbon (TOC), carbon dioxide, colloidal silica, silica, total iron, iron ion, manganese ion, total aluminum, residual chlorine free and combined types, total residue, SDI, and bacteria.

The JIS K0101 industrial water examination method and standard methods (APHA, AWWA, and WPCF), are good sources for analytic methods.

B.   Trace Suspended Matter Measurement Method

Colloidal fouling is a serious problem in the reverse osmosis unit. Trace particles in colloidal form adhere to the membrane surface of a RO module in various ways. A rise in the salt concentration in the condensed water in a RO module leads to neutralization of the ζ potential, resulting in coagulation. In this way, large particles of tens of micrometers are formed and captured in the RO module. Pretreatment is critical to prevent colloidal fouling. There are several methods of measuring and evaluating the quality of the pretreated water.

1.   SDI Value

The SDI value was formerly applied only to the hollow fiber RO unit, but it is now applied to the spiral unit as well. Millipore filter HAWP (47 mm dia., 0.45 μm pore size) is used in the measurement, and the flow rate is measured for a certain amount of time (5 and 15 minutes) under a filtration pressure of 2.1 kg cm^{-2} (30 psi) at the inlet and outlet. The SDI value is the damping ratio of flow rate divided by filtration time:

$$\text{SDI (5 minutes)} = \frac{(1 - t_0/t_5)100}{5}$$

$$\text{SDI (15 minutes)} = \frac{(1 - t_0/t_{15})100}{15}$$

where: t_0 = time of taking sample of filtered water in a measuring cylinder of 100 or 500 ml at the inlet

t_5 = after 5 minutes

t_{15} = after 15 minutes

Although the SDI value of 15 minutes is usually measured, when the volume of suspended matter is large, the 5 minute value is measured. Figure 19 shows the method of measuring the SDI value.

The SDI value measured depends on the type of reverse osmosis membrane and the ζ potential of the particles. The following are general SDI values.

Type of reverse osmosis membrane	SDI
hollow fiber	<3–4
spiral	<4–5

Table 9 [11] lists typical SDI values for raw water taken from different water sources.

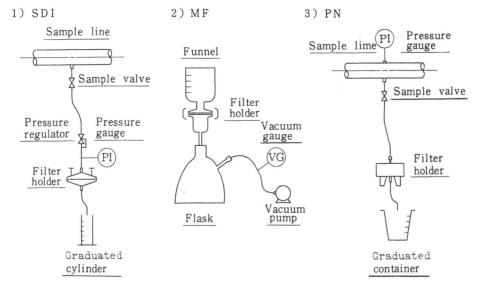

Figure 19   Method of measurement for colloidal matter.

Table 9   Raw Water Source and SDI

Raw Water Source	SDI
Well water	1 − 4
Municipal water	3 − 12
Surface water	6 −

Because the SDI value at 5 minutes does not match with that at 15 minutes, the measurement time must be indicated clearly. Figure 20 [11] shows the correlation between 5- and 15-minute measurement values.

2. MF Value

The MF value is the time (seconds) required to conduct the suction filtration of a 1 L sample of water under a negative pressure of –50 cmHg with Millipore filter HAWP (47 mm dia., 0.45 μm pore size) installed as shown in Fig. 19. The filtration time is affected by the temperature of water, and it should be converted to a value at a temperature of 25°C. Figure 21 [12] shows the relationship between SDI and MF. This method of measuring MF is very easy because only 1 L water is required.

3. PN Value

The PN value measurement was developed by Ushikoshi et al. This is a useful method for controlling the quality of pretreated water because the measurement is simple and various types of water can be tested with high accuracy.

Nuclepore filter (25 mm dia., 1 μm pore size) is used. The water filtered at 3.0 kgf cm^{-2} forms laminar flow at the outset. Over the course of filtration, however, the filtered water assumes a droplet shape as a result of clogging of the filter. The PN value is the volume (liters) of water collected before the filtered water attains the droplet shape. Figure 19 illustrates the method.

Since various sizes of filter pore can be used in this method, the PN value should be expressed as

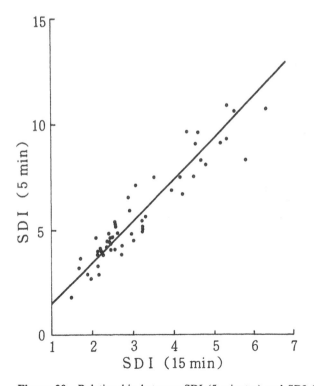

Figure 20   Relationship between SDI (5 minutes) and SDI (15 minutes). (From Ref. 11.)

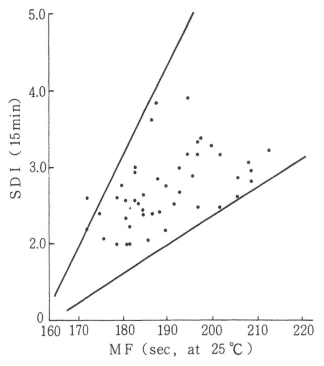

Figure 21   Relationship between SDI and MF. (From Ref. 11.)

PN[*ABC* (*D* − *E*) *F*] = *G*

where: *ABC* = type of filter: NPF, Nuclepore filter; MPF, Millipore filter
       D = diameter of filter (nominal): 13 = 13 mm dia. 25 = 25 mm dia. 47 = 47
              mm dia.
       E = pore size: 0.2 = 0.2 μm, 0.4 = 0.4 μm, 1.0 = 1.0 μm
       F = measurement pressure of 3.0 kg^f/cm^{-2}g (when compensation is com-
              pleted, S should be put down)
       G = PN value

For measurement of the quality of water pretreated upstream of the reverse osmosis unit,
Nuclepore filter (25 mm dia., 1 μm pore size) is employed. Figure 22 [11] shows the
pressure coefficient. Therefore, the standard pressure compensation PN value is expressed
as

$$PN = \frac{\text{measured PN value}}{\text{pressure coefficient}}$$

Figure 23 [11] shows the relationship between the PN value and the SDI value. The
correlation is very high:

PN value	SDI value
5	5
10	4
20	3

Thus the volume of filtered water doubles every time the SDI value increases by 1.0.

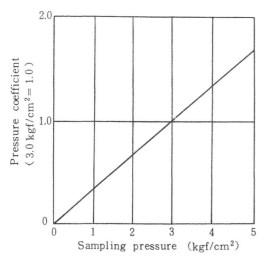

Figure 22 Sampling pressure versus pressure coefficient. (From Ref. 11.)

Figure 24 [11] indicates the relationship between the PN value and the volume of suspended matter. Water at PN 20 (SDI 3), which is usually allowed to be supplied to the reverse osmosis unit, contains 20–30 μg L^{-1} of suspended matter.

Figure 25 [11] shows the relationship between the PN value and the turbidity, which is measured by SEP-PL (Nippon Seimitsu Co., Ltd.). The turbidity should be maintained around 0.05 in the pretreatment process for the reverse osmosis unit.

V. SUMMARY

To operate the ultrapure water production system in a safe and stable manner, a certain level of water quality should be obtained in the pretreatment process. Full understanding

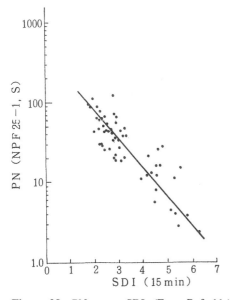

Figure 23 PN versus SDI. (From Ref. 11.)

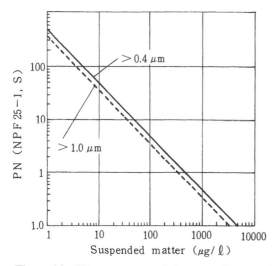

Figure 24   PN versus suspended matter. (From Ref. 11.)

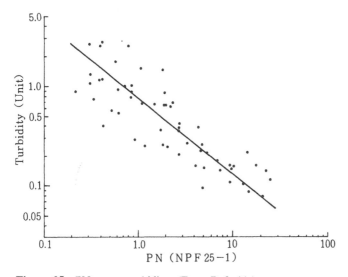

Figure 25   PN versus turbidity. (From Ref. 11.)

of the raw water quality is critical. Considering the subsequent primary treatment system, the pretreatment system must be adequately designed and well maintained.

REFERENCES

1. Annual Report of Water Resources, Water Resources Society, Reviewed by National Land Agency, 1989, pp. 31, 95.
2. Technical Data, Arrowhead Industrial Water, Inc.
3. N. Tanbo, J. Jpn. Water Works Assoc., *361*, 2–12 (1964).
4. Technical Report of Taki Chemical Co, Ltd.

5. N. Tanbo and K. Ogasawara, Jyosui no Gijutsu, Gihodo, p. 57.

6. Design Criteria for Water Works Facilities, JWWA, Reviewed by the Minstry of Health and Welfare, p. 160.

7. Design Criteria for Water Works Facilities, JWWA, Reviewed by the Minstry of Health and Welfare, p. 169.

8. S. T. Powell, Water Conditioning for Industry, McGraw-Hill, New York, 1954.

9. M. Akahane, Chem. Eng., *56*, 746 (1953); *58*, 402 (1955).

10. K. Ushikoshi, Shinko-Pfaudler News, *23*(3, 4), 30 (1979).

11. S. Koga and K. Ushikoshi, J. Water Reuse Technol., *4*(1), 60–65 (1978).

12. Report on Development for Energy Saving Sea Water Desalination Technology, Water Reuse Promotion Center, 1977.

3

The Primary Treatment System

Isamu Sugiyama and Yoshitaka Yamaki
Nomura Micro Science Co., Ltd., Kanagawa, Japan

I. OUTLINE

A. General

The primary treatment system is located between the pretreatment system and the subsystem and produces the primary pure water. The system is large, composed of a very tall recovery ion-exchange tower, a vacuum degasifier tower (VDG tower), a reverse osmosis unit that often causes noise and vibration, and a large tank. In many cases, tall and large units are installed outdoors. When the ultrapure water production system is very large, the subsystems are often separately installed close to each use point. Even in this case, however, the primary treatment system for ultrapure water is usually installed as is without being subdivided. The primary treatment system is required to maintain a continuous supply of water of stable quality to the subsequent subsystem despite the regular and irregular fluctuation of water quality at the inlet.

Figure 1 shows the different arrangements of the primary treatment system in the overall process. The primary treatment system had a very simple structure 10 years ago (Fig. 1, part 1). In response to various demands from users, including water quality improvement, system stability, and reduction in initial and running costs, the primary treatment system has become more and more complicated (see Fig. 1, parts 2–4).

Many plants now recycle ultrapure water. The mixture of recycled water and municipal or well water is supplied at the inlet of the primary treatment system. Recycled water of high purity is returned directly to the primary treatment system inlet without being processed (see Fig. 1, part 3). This means the primary treatment system must function as a recycling system as well. On the other hand, when a separate recycling system is installed to recycle a large volume of pure water, as shown in Fig. 1 (part 4), recycled pure water is sent to different stages of the primary treatment system, depending on the purity level.

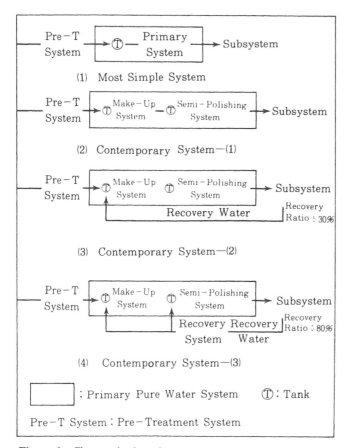

Figure 1   Changes in the primary pure water system.

Current primary treatment systems may include even more processing steps. Some parts of the conventional recycling system and subsystem are now used in the primary treatment system. Although the primary treatment system now handles an increasing number of processing steps, it can still be defined simply as a system for further treatment and purification of pretreated raw water and recycled water for delivery to the final tank.

B.   Targeted Quality of Water and Current Level of Water Quality

Table 1 summarizes the relationship between pure water quality and large-scale integration (LSI) [1]. Although the data in Table 1 are somewhat out of date and there may be slight differences from projected values for 1 and 4 Mbit devices, the basic idea is the same. (Actual figures are often more stringent than the data in Table 1.)

An ultrapure water production system, including primary treatment, must be designed to maintain the water quality required by the integration level of the circuitry that will be produced at the point of use. Since the water quality cannot be improved solely by the final subsystem, the quality at the outlet of the primary treatment system must be upgraded steadily to keep pace with advances in LSI. At present, the water quality at the primary treatment outlet is not usually guaranteed. In the future, however, in the 1 and 4

Table 1 Relation of Pure Water Quality and Integration Level of LSI Circuits [1]

Item	Unit	64K	256K	1 M	4 M	16M
Resistivity	$M\Omega \cdot cm$ at 25°C	>17	>17.5	>17.5	>18	>17.5
Number of Particles	pcs/mℓ	<50	<20, <50	<10, <30	<1, <10	
Counting Size	μm	(0.2)	(0.2)(0.1)	(0.2)(0.1)	(0.2)(0.1)	(0.05)
Number of Microorganisms	cfu/100mℓ	<25	<10	<1	<0.5	<0.1
TOC	μg C/ℓ	<200	<100	<50	<30	<30
SiO_2	μg SiO_2/ℓ	<20	<10	<5	<3	<3
Dissolved Oxygen	mg O/ℓ	<0.2	<0.1	<0.1	<0.05	<0.05
Na	μg Na/ℓ	<1	<1	<1	<0.1	<0.1
K	μg K/ℓ	<1	<1	<1	<0.1	<0.05
Cl^-	μg Cl^-/ℓ	<5	<5	<1	<1	<0.1
Cu	μg Cu/ℓ	<2	<2	<1	<1	<0.1
Fe	μg Fe/ℓ	—	—	<1	<1	<0.1
Zn	μg Zn/ℓ	<5	<2	<1	<1	<0.1
Cr	μg Cr/ℓ	—	—	<1	<0.1	<0.02
Mn	μg Mn/ℓ	—	—	<1	<0.5	<0.05

Mbit devices, the water quality at the primary treatment outlet will be the same as that needed for a 64 kbit production line.

To upgrade the water quality, the following improvements are being proposed for the primary treatment system:

1. Reverse osmosis module with new materials: low-pressure complex membrane reverse osmosis module with high particle removal efficiency and a large volume of product water
2. New two-stage reverse osmosis with a low-pressure complex membrane reverse osmosis module
3. New ion-exchange resins with limited elution exclusively designed for ultrapure water production
4. Combination of new ion-exchange resins to perform the ion exchange by combining an anion-exchange resin tower to prevent silica leakage and a mixed-bed ion-exchange tower (MB tower)
5. New catalytic and N_2 deaeration methods to further reduce the concentration of dissolved oxygen
6. New circular production line that includes the entire primary treatment system or the primary treatment system and the subsystem

The details are not touched upon here (refer to appropriate chapters). The primary treatment system is certain to become more and more complicated and advanced to meet more stringent demands for water quality at the outlet.

II. TYPICAL FLOW AND CHARACTERISTICS

A. Development

As described before, the primary treatment system has developed along with LSI technology. The primary treatment system is composed of the following units:

Membrane system
   Reverse osmosis (RO) unit
   Ultrafiltration (UF) unit
   Membrane filter (MF)
Ion-exchange system
   Two-bed ultrapure water unit (2B tower)
   Mixed-bed ion-exchange tower (MB tower)
Deaeration system
   Degasifier (DG tower)
   Vacuum degasifier (VDG tower)
Others
   Pump
   Tank
   Ultraviolet sterilizer (UV)
   Chemical supply unit

Figure 2 summarizes the developments in two major units: reverse osmosis and ion exchange. The data are derived from the literature [2, 3].

1.   Ion-Exchange Unit Only

When pure water production systems were first installed in plants during the 1960s, resistivity was regarded as an important indicator of water quality. The quality required at

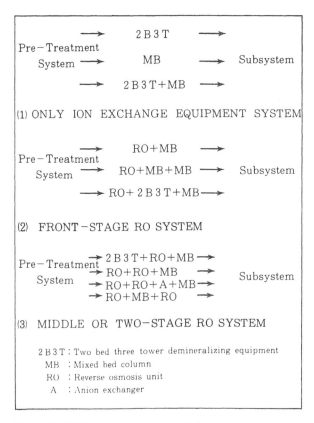

Figure 2   Different arrangements of the primary pure water system.

the point of use was 10–16 MΩ·cm at 25°C. Therefore, as for the primary treatment system, the two-bed, three-tower pure water system (2B3T) or the mixed-bed (MB) ion-exchange unit was used only to remove ions, as shown in Fig. 2 (part 1).

Later, in the 1970s, as requirements for the resistivity at use point were tightened, the 2B3T system was combined with the MB system to improve performance. Depending on the quality of raw water desired, a multiple-bed ion-exchange system, for example a four-bed, five-tower pure water system (4B5T), was sometimes used instead of the 2B3T system.

Because this system employed only the ion-exchange unit, however, it was plagued by organic contamination of the ion-exchange resin. The total organic carbon (TOC) and particle problems also remained unsolved.

2. Front-Stage RO System

In the late 1970s, ultrafiltration began to be used as the final filter, replacing the membrane filter in removing particles and bacteria at use point. Initially, either RO or UF was employed in a system, but gradually ultrapure water production came to employ RO in primary treatment and UF in the subsystem, as shown in Fig. 2 (part 2). A middle-pressure cellulose acetate spiral RO module was installed as the front-stage RO (RO placed upstream of the ion-exchange unit).

The front-stage RO unit removed almost all the ions in the supply water, which dramatically decreased the load on the ion-exchange unit and thus greatly increased the volume of water produced. Moreover, the water quality was significantly improved: RO removed almost all the particles, bacteria, silica, and TOC.

On the other hand, there was a problem with the RO unit: fouling and scaling of the module membrane surface. This can interfere with the stable operation of the system. Here it led to a reduction in the amount of product water, a rise in the differential pressure, and degradation of removal efficiency. This problem was dealt with by improving the pretreatment process and by chemical cleaning. The FI and SDI values were used as indicators to confirm the improvement.

In a bid to improve the recovery efficiency and to suppress ion leakage from the ion-exchange unit, various ion-exchange towers, including the multiple-bed type and counter-current regeneration type, were proposed. Ion-exchange resins with less TOC elution were also required. Ion-exchange resin was inspected at the inlet by production lot, and cleaning methods were improved. Moreover, to prevent the generation of colloidal silica, the regeneration cycle was controlled by the volume of supply water and regeneration with hot NaOH was studied.

Until this stage, the primary treatment system had not been operated on a continuous basis. Primary water treatment began to be operated continuously to prevent the increase in bacteria caused by a change in water quality or entrapment and subsequent stagnation of water when the ion-exchange or RO unit was started up. Thus the water quality at the primary treatment outlet was improved and stabilized.

3. Middle-Stage and Two-Stage RO Systems

In the 1980s, the midpressure and low-pressure complex membrane RO modules were developed and commercialized in Japan and North America. With regard to water quality at the use point, TOC and metallic ions, as well as particles and bacteria, were closely examined.

In the early 1980s, further advanced RO and UF modules were used in the same positions in the production line as before to cope with user demands. In primary treatment, the front-stage RO was replaced by a low-pressure complex membrane RO module, which was very effective in reducing energy costs and improving the removal efficiency of TOC and silica. Besides enabling the primary system to maintain a neutral pH, this new RO module also contributed to the prevention of colloidal silica generation downstream.

Compared with conventional cellulose acetate reverse osmosis, the complex membrane module was less efficient in terms of oxidative reagent and fouling resistance. In general, the material of which the complex membrane RO module is made has a low oxidative reagent resistance and thus cannot be sterilized using soda hypochlorite. This problem was addressed by adopting intermittent sterilization to reduce the frequency of reagent injection or by using sterilizers with less oxidizing action. Another problem was that the module electrically adsorbed foulants in supply water because of its strong charge characteristics, clogging the membrane. This problem was overcome by reducing the FI and SDI value or changing such conditions as the recovery ratio and the volume of product water.

In the second half of the 1980s, increasing numbers of low-pressure complex membrane RO modules were marketed (Fig. 2, part 3).

The first was middle-stage reverse osmosis. This is a method of supplying to the reverse osmosis unit pure water treated in a two-bed, three-tower unit and a mixed-bed unit. In the past, the RO module was sometimes placed in the middle stage if the recovery ratio was suppressed as a result of a high concentration of silica in the raw water. In the late 1980s, a RO module at low pressure and with high removal efficiency was placed in the middle stage for further stabilizing water quality and suppressing TOC.

The second two-stage RO system had two RO units in the primary treatment system. There were variations:

Arrange two low-pressure RO modules in series, and supply water product in the first RO module directly to the second RO module.
Store the water product in the first RO module in a tank, and then supply it to the second module.
Place the ion-exchange unit between the first and second RO modules.

The goals were to stabilize the TOC concentration at extremely low levels in water produced in the primary treatment system and to simplify the entire system.

When two-stage RO systems were employed in the front stage, the product water was rich in anions. The ion-exchange process was therefore often equipped with both an anion-exchange resin and a mixed-bed ion-exchange tower. The ion-exchange resin was specially developed for ultrapure water treatment at every processing step, including production, recovery, cleaning, and inspection.

In the circular production line, continuous primary treatment became more prevalent. A large circular line often included the subsystem as well as primary treatment, to stabilize TOC and dissolved oxygen at extremely low levels at the use point.

B.   Recent Primary Treatment Systems

1.   *Two-Bed + RO + MB System*

This primary treatment system is composed of a two-bed, three-tower pure water unit, a RO unit, a MB unit, a vacuum degasifier, an ultraviolet sterilization unit, a microfilter unit, and a heat exchanger. This arrangement is quite common. An example of this system is shown in Fig. 3, and Table 2 shows the water quality at the outlet of the system. Each unit is described here according to its position in the system.

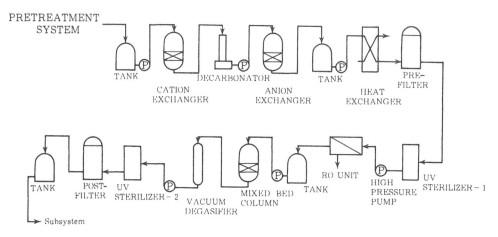

Figure 3 Typical arrangement of the 2B3T + RO + MB system.

Municipal water is processed in the pretreatment system by chemical injection before being supplied to the primary treatment system. The pretreated water is also sterilized with NaClO.

The two-bed, three-tower pure water unit produces deionized water. The residual chlorine in the water is reduced upstream with $NaHSO_3$. An on-line residual chlorine meter is mounted at this position to constantly monitor the chlorine concentration. The standards for water treated in the two-bed, three tower unit are conductivity $\leqq$ 10 μS/cm and silica concentration $\leqq$ 0.1 mg/L. The actual values, however, as shown in Table 2, are silica concentration $\leqq$ 0.10–0.03 mg/L and resistivity at a constant 2–7 MΩ·cm. Stable water quality can be maintained by complete backwash with double volume of agent. The cation-exchange resin tower uses a gel type of strongly acid cation-exchange resin; the anion-exchange resin tower uses a gel type of strongly basic anion-exchange resin (type II).

The next unit is the plate heat exchanger. The temperature of the water is adjusted to 25°C by this heat exchanger before the water is sent to the RO unit through the spool

Table 2 Water Quality of 2B3T + RO + MB System

Parameter	Resistivity(MΩ·cm)	SiO_2 (mg/ℓ)	TOC (mg C/ℓ)	Number of Bacteria (counts/mℓ)	Total Number of Particles >0.2 μm (particles/mℓ)
Pretreatment Water	1) 50 ~ 100	7 ~ 15	0.4 ~ 0.5	0	> 10⁵
Post-2B3T	2 ~ 7	<0.01~0.03	0.1 ~ 0.4	10 ~ 30	10³
Post–RO	2 ~ 6	< 0.01	< 0.05	0	< 10
Post–MB	> 17.5	< 0.01	< 0.05	2) 20 ~ 40	—
Post–Filter	> 17.5	< 0.01	< 0.05	2) 0	< 10

1) Conductivity (μS/cm)
2) counts/100mℓ

prefilter and the UV sterilizer. The prefilter not only decreases particles but also prevents resin leakage to the RO unit should the resin accidentally escape the 2B3T pure water system. To prevent water containing bacteria from entering the RO unit and multiplying in the module, the UV sterilizer is placed immediately upstream of the RO unit and regular sterilization using chemicals is also conducted.

The RO module employs a low-pressure complex membrane spiral RO unit. Because the conductivity of the water supplied to the RO unit is low enough that TOC, particles, bacteria, and colloidal materials can be removed, the major function of the unit. The operation pressure of this unit is 15 kg/cm^2, and the recovery ratio is 85%. The conductivity of the concentrated water drained from the RO unit is also monitored. When it reaches $\leqq 10$ μS/cm, the concentrated water is recycled to the pretreatment system or used as the regeneration water for the two-bed, three-tower unit. After reverse osmosis, the TOC concentration decreases to $\leqq 0.03$ μg/L and the number of particles is reduced to 0–1 particle per ml.

Next, the MB unit removes almost all the residual ions. Although the standard for water treated with this unit is set at $\geqq 10$ MΩ·cm, it is always maintained at $\geqq 17$ MΩ·cm. Since the water-processing cycle of a single MB unit is close to 1 week, there is concern about bacterial propagation while the water is stored before being fed to the unit. This is why two units are operated at the same time. When the water in one is being regenerated, the volume of water processed by the other is raised. In this MB unit, strongly acid cation-exchange and strongly basic anion-exchange resin gels (type I) are employed. The TOC concentration of the feedwater is very low, and thus the resin must be conditioned before being mounted in the unit.

Water processed in the mixed-bed unit is then sent to the vacuum degasifier, which removes dissolved oxygen. The vacuum degasifier is made of stainless steel, with a filling part and a storage part. The degree of vacuum is 20–30 mmHg. It is 8 m tall, so it is installed outdoors. After treatment in this unit, dissolved oxygen is lowered to 50–100 μg/L.

The processed water flows through the UV sterilizer and the postfilter and is stored in the primary pure water tank. To prevent the absorption of carbon dioxide, the tank is purged with N$_2$ gas.

The advantages of this system are as follows:

1.   The two-bed, three-tower unit is placed at the front stage so that the slight fluctuation in the raw water quality does not affect later stages as greatly.
2.   Materials eluted from the 2B3T unit can be removed during reverse osmosis.
3.   The water fed to the RO unit was already treated in the two-bed, three-tower unit, and thus it contains limited hardness materials and silica, which enables the RO unit to operate at a high recovery ratio.
4.   The regeneration cycle of the mixed-bed unit is long enough to guarantee stable water quality for long periods of time.

It is reported that the two-bed, three-tower unit is superior to reverse osmosis in coping with the wide and frequent fluctuation in raw water quality due to turbidity and ions.

2.   RO + 2B3T + MB + RO System

This system incorporates the elements of the preceding system. What is peculiar to this system is the placement of the RO units upstream and downstream of the ion-exchange unit (2B3T + MB), for the following reasons:

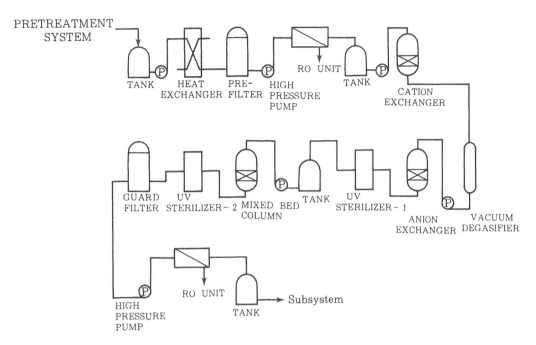

Figure 4   Typical RO + 2B3T + MB + RO system.

1. Even if the volume of water drained by regeneration from the two-bed, three-tower unit increases in the future should the system be expanded, the present waste water treatment system will be able to deal with the increased volume because the amount of ion load on the unit is reduced by reverse osmosis to set the regeneration cycle at a few days.
2. By removing organic materials and colloidal materials by reverse osmosis, organic contamination in the 2B3T unit can be prevented.

This system flow is shown in Fig. 4, and Table 3 shows the water quality at the outlet. Each component is described here in accordance with its position in the system.

Industrial water is fed to this system as raw water. The sludge contact clarifier and

Table 3   Water Quality of RO + 2B3T + MB + RO System

Parameter	Resistivity(MΩ·cm)	SiO₂ (mg/ℓ)	TOC (mgC/ℓ)	Number of Bacteria (counts/mℓ)	Total Number of Particles (particles/mℓ)
Pretreatment Water	1) 150 ~ 200	10 ~ 30	1.0 ~ 1.5	0	—
Post–RO	1) 15 ~ 35	0.4 ~ 0.5	0.1 ~ 0.3	0	30 ~ 50
Post–2B3T	1 ~ 5	0.01 ~ 0.03	0.1 ~ 0.2	0 ~ 10	10 ~ 30
Post–MB	> 18.0	< 0.01	0.05 ~ 0.1	0 ~ 10	10 ~ 20
Post–RO	> 17.5	< 0.01	∴ 0.05	2) 10 ~ 50	< 5

1) Conductivity (μS/cm)
2) counts/100mℓ

the gravity sand filtration unit are installed in the pretreatment section. Since the quality of raw water fluctuates greatly, the pH is constantly adjusted. The raw water after pretreatment is stored in the filtered water tank to be fed to the primary treatment system.

The temperature of the water is first adjusted to 25°C by a plate heat exchanger. The water flows through the spool prefilter to be sent to the reverse osmosis unit, which employs a low-pressure cellulose acetate spiral RO module. As shown in Table 4, although this is a cellulose acetate reverse osmosis membrane, it can be operated at low pressure. The removal efficiency of NaCl is not very good, but that of $MgSO_4$ is as high as 98%. This RO unit is capable of removing 90% TOC and 80% ions. The pH level of the feedwater is controlled by injecting H_2SO_4. NaClO injection kills bacteria.

The water processed in the RO unit is stored in the RO tank and then treated in the two-bed, three-tower pure water unit. To prevent oxidative reagent from entering the unit, $NaHSO_3$ as reducing agent is injected and an on-line residual chlorine meter is installed. The vacuum degasifier is mounted on the 2B3T unit, which removes both carbon dioxide and dissolved oxygen. The concentration of dissolved oxygen at the outlet of the vacuum degasifier is $\leqq100$ μg/L. The cation-exchange resin tower uses a strongly acid cation-exchange resin gel, and the anion-exchange resin tower uses a strongly basic anion-exchange resin gel (type II). Just like the system described in Sec. II.B.1, the actual quality of the water is much better than standard. The actual resistivity is 1–5 MΩ·cm, and the actual silica concentration is $\leqq0.01$–0.03 mg/L. The processed water is stored in the deionized water tank after treatment in the UV sterilizer.

Next the water is sent to the mixed-bed unit. This unit employs not only the resins available on the market but also a gel type of strongly acid cation-exchange resin and strongly basic anion-exchange resin (type I), both of which undergo conditioning by the resin manufacturer. As a result, even immediately after start-up, the TOC level is 0.23

Table 4   Specifications for SC-L200 RO Membrane

Type		
Element	Spiral Wound Type	
Membrane	Modified Cellulose Acetate	
Performance Data		
Rejection	NaCl	70%
	$MgSO_4$	98%
Product Flow Rate	25 ~ 30 m³/d	
Operating Range		
	Recommended	Allowable
Applied Pressure	15kg/cm²	30kg/cm²
Feedwater Temperature	<25°C	35°C
Feedwater pH	5 ~ 6	4 ~ 7
Feedwater Chlorine Concentration	0.2~0.5ppm	0.2~ 1 ppm

mgC/L at the inlet and 0.12 mgC/L at the outlet, which means no elution takes places in the unit and organic materials are removed from the supply water.

The water, now processed to $\geqq 17.0$ MΩ·cm, flows through the guard filter and the UV sterilizer to be fed to the RO unit.

Reverse osmosis with a low-pressure complex membrane spiral RO module is operated under almost the same conditions as those described in Sec. II.B.1. Since the resistivity of the water supplied to this RO unit is very high, however, the ions are hardly removed. The water is stored after being processed in the RO unit. The quality of this water is almost as high as that of ultrapure water.

Figure 5 summarizes data on the number of particles ($\geqq 0.2$ μm) in ultrapure water production systems other than that described in Sec. II.B.2 that employ the primary treatment system RO + 2B3T + MB. Although the number of particles remains the same in the pretreatment section (rapid filtration and cartridge filtration), it markedly drops in the RO unit. (In the RO unit, however, bacteria accounts for 60–70% of the total number of particles.) The particles increase in number in the cation-exchange resin tower and then decrease in the anion-exchange resin tower. It is thought that bacteria propagate and adhered or adsorbed particles are desorbed in the cation-exchange resin tower. On the other hand, in the anion-exchange resin tower, it is thought that the number of particles is lowered by the sterilization effect and adsorption of negatively charged particles. There is no change in the number of particles in the MB unit or the UV sterilizer. Particles decrease in the membrane filter unit. In the subsystem, the number of particles drops in the ultrafiltration unit.

As shown in Fig. 5, the primary treatment system contributes greatly to the reduction of particles in the ultrapure water production system. This is why the primary treatment system needs careful design, manufacture, and operation.

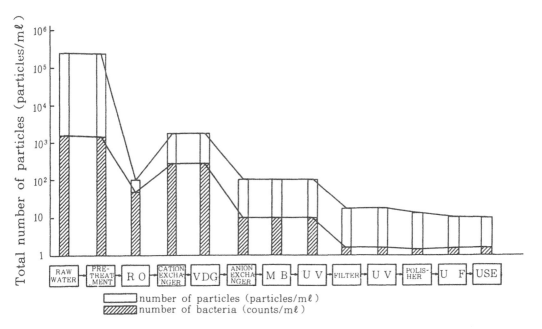

Figure 5 Change in number of particles in an ultrapure water process.

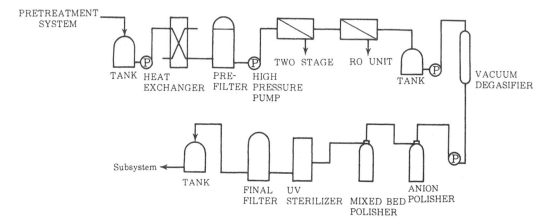

Figure 6 Typical two-stage RO + MB system.

3. Two-Stage RO + MB System

The two-stage RO unit is what makes this system unique. Figure 6 shows the system flow, and Table 5 shows the water quality at the outlet. Each component unit is described in sequence.

The raw water fed to the primary treatment system is well water, which has stable quality and low turbidity. The well water flows through the chemical-injected sand filtration unit and is stored in the tank to be fed to the primary treatment system.

In the primary treatment system, the water flows through the heat exchanger and the spool prefilter and is then treated in the two-stage RO unit. The two-stage RO unit is equipped with the following chemical injection systems:

1. Sterilization chemicals
2. pH adjustment chemicals (e.g., NaOH)
3. Scale prevention chemicals

For sterilization, an agent with limited oxidizing action on the reverse osmosis membrane is injected for 1 h once a day. The pH is adjusted using NaOH:

Table 5 Water Quality of Two-Stage RO + MB System

Parameter	Resistivity($M\Omega \cdot cm$)	SiO_2 (mg/ℓ)	TOC (mgC/ℓ)	Number of Bacteria (counts/100mℓ)	Total Number of Particles (particles/mℓ)
Pretreatment Water	*1) 50 ~ 100	10 ~ 30	1 ~ 3	0	—
Post- Two Stage RO	1.0 ~ 3.0	0.01~0.03	< 0.05	* 0 ~ 6	10 ~ 60
Post-MB	> 17.5	< 0.01	< 0.05	—	—
Post-Filter	> 17.5	< 0.01	< 0.05	—	—

* 1) Conductivity (μS/cm)

1. Silica becomes an ion under basic conditions, which can be easily removed by reverse osmosis.
2. TOC, including organic acid, can be removed more easily under alkali (basic) conditions.
3. Under basic conditions, carbon dioxide in the water becomes bicarbonate ion for removal in the RO unit.
4. The removal efficiency of the low-pressure complex membrane RO module employed in two-stage reverse osmosis is slightly higher at alkali pH.

Conversely, the alkalinity may induce scaling caused by hardness materials. To address this problem, acrylic acid polymer instead of phosphoric acid is used in this system because acrylic acid polymer is more effective in preventing scaling. In designing this system, the following experiment was conducted and it was confirmed that no scaling caused by hardness materials occurred. Figure 7 shows the experimental apparatus. Acrylic acid polymer (AF-600, B. F. Goodrich) at 4 ppm was injected in one side and nothing was injected in the other. City water with pH adjusted to with NaOH was then supplied to both sides. To eliminate the effect of particles, a 0.2 μm filter was installed upstream of the RO module. Figure 8 shows the results. Although the line without the scale inhibitor showed a decrease in product water in about 60 h, the line injected with the scale inhibitor kept working without problems for 1000 h. Based on these results, this two-stage RO unit is now operated under the conditions listed in Table 6. The water quality at the outlet of this system is generally very good, exhibiting excellent values for TOC, particles, and bacteria.

The water processed in the two-stage RO unit is stored in the RO water tank. The resistivity of this water is improved by a cartridge anion polisher and a cartridge

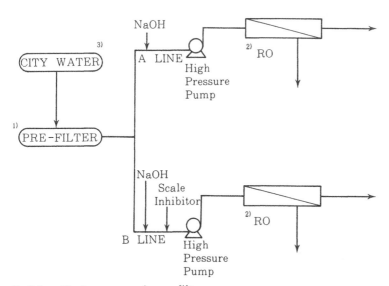

1) 0.2 μ Nuclepore membrane filter
2) Operation pressure : 15 kg/cm^2
3) Water temp. 25°C

Figure 7   Test unit for scale inhibitor.

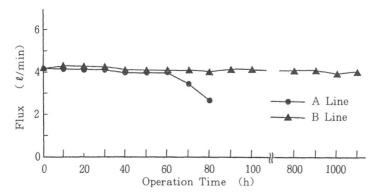

Figure 8   Results of scale inhibitor test.

mixed-bed polisher. The reasons for using the cartridge polishers at this point in the primary treatment system are as follows:

1. Water drained in regeneration from the regeneration type of unit cannot be processed further.
2. The cartridge type is relatively small.
3. Since the water processed in the two-stage RO unit has a high quality, the frequency of replacing the cartridge polisher is relatively low.

The anion polisher is installed here because the silica should be processed and the alkaline water must be fed to the mixed-bed polisher. (The details are described in Sec. III.B.)

   After the resistivity reaches the appropriate level, the water flows through the final filter and is stored in the primary pure water tank to be supplied to the subsystem.

C.   Other Primary Treatment Systems

Some of the new components of the primary treatment system are described here.

1. N_2 Gas Bubbling

Although a vacuum degasifier is mainly used to remove dissolved oxygen, the nitrogen gas bubbling method is also sometimes employed. In this method, nitrogen gas is dispersed in the tower through a diffuser in the form of bubbles, and it contacts water to remove dissolved oxygen. The details are explained in another section.

   Figure 9 shows the N_2 gas bubbling system jointly developed by Hitachi and the

Table 6   Operating Conditions for Two-Stage RO

	Feed	1st Product (Conc.)	2nd Product (Conc.)
Pressure(kg/cm²)	16.0	10.3 (14.2)	3.2 (9.9)
Flux (m³/h)	2.7	2.0	1.8
Recovery (%)	—	74	90
pH	8.4	7.0	6.5

Figure 9 N_2 gas bubbling system.

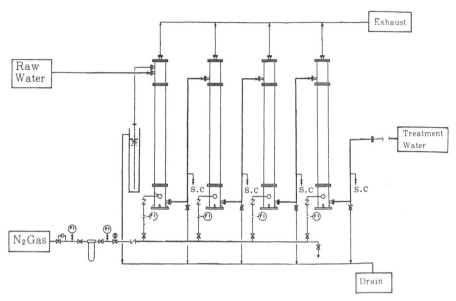

Figure 10   N₂ degasifier unit.

authors. Figure 10 illustrates its operation flow. Polyvinyl chloride (PVC) pipes are used for the main body. The diffuser is made of polymethyl methacrylate. The average diameter of the bubbles released from this diffuser is 30 μm. To reduce dissolved oxygen to 50 μg/L, four towers are arranged in series. As a result, the concentration of dissolved oxygen at the outlet was lowered to 30 μg/L with a liquid-gas ratio of 1:0.8.

Table 7 compares the performance of the vacuum degasifier and the N₂ gas bubbling system when they are applied to the primary treatment system. As shown in Table 7, the N₂ gas bubbling system consumes a large amount of nitrogen gas, resulting in high running costs. This is why only small N₂ bubbling systems are now manufactured.

2. Multiple-Bed Ion-Exchange Unit

The multiple-bed ion-exchange unit has strong and weak resins in a single tower in two layers. Ion exchange is conducted by utilizing the difference in the two resins in terms of specific gravity and ion selectivity. This unit used less regeneration agent than the conventional two-bed, three-tower unit (1:1.5) [4]. Figure 11 shows the multiple-bed ion-exchange unit applied to the cation-exchange resin tower. The details are described in another section.

Table 7   Comparison Between Degassing by Vacuum and Nitrogen

Parameter	Vacuum Degasifier	N₂ Degasifier
Initial Cost	High	Low
Running Cost	Low	High
Maintenance	Need	Almost No Need
Tower Height	7~10 m	3~4 m

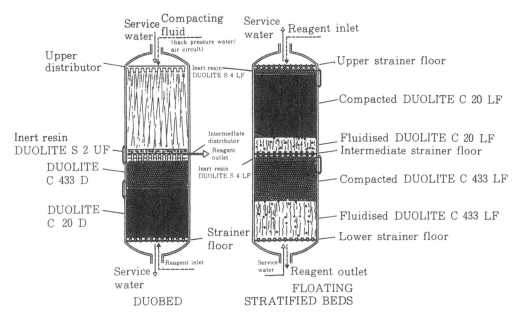

Figure 11   Stratified cation beds (Duolite INF. 80001 A November 1980).

When the multiple-bed ion-exchange unit is employed in the primary treatment system, the quality of raw water must be checked carefully. Raw water should contain hardness ions and mineral acid ions in large volume and silica in relatively small volume. The initial cost is high, and thus it is preferable to have a large amount of water to be processed. Thorough discussion with the engineering company is necessary when the multiple-bed ion-exchange unit is considered.

3.   *Anion Mixed-Bed Ion-Exchange Unit*

In the conventional mixed-bed ion-exchange tower, ionic silica is converted to colloidal silica and low-polymer silica [5]. Countermeasures are as follows:

1. Lower the ionic composition ratio (anion/cation and silica/anion) in the water fed to the MB unit.
2. Expand the load distribution in the ion-exchange tower by increasing the velocity of the supply water to prevent the conversion of SiO_2.
3. Regenerate the strongly basic anion-exchange resin at high temperatures.
4. Do not conduct overrun operations.
5. Place the anion-exchange resin tower upstream of the MB unit.

The anion mixed-bed unit (A-MB tower) is now being studied to develop a new unit to satisfy these five requirements [6]. Figure 12 illustrates the anion mixed-bed unit. The unit applies a unique merry-go-round (or roundabout) method to reduce not only silica but other ions as much as possible. Even after regeneration, trace amounts of regeneration agent remain in the resin layer, which eventually leak to degrade the processed water. To overcome this problem, two units are arranged in series. The residual trace impurities are adsorbed and removed in the second unit.

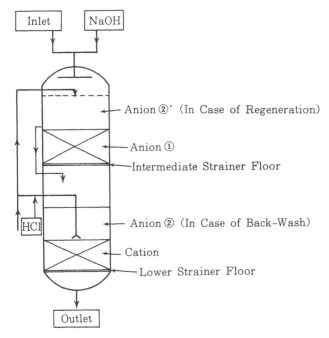

Figure 12   Mixed-bed anion-exchange equipment system.

III.   FUNCTIONS OF THE COMPONENTS OF PRIMARY TREATMENT

The major components in the primary treatment system are the reverse osmosis unit and the ion-exchange unit. The two units are described in details in this section.

A.   RO Unit

The RO unit is essential to primary treatment. Basically, the RO unit must remove almost all contaminants, including ions, particles, bacteria, TOC, silica, and colloid. The water quality can be greatly improved by placing the RO unit in the primary treatment system. Also, unlike the ion-exchange unit, the RO unit does not have fluctuating water quality because of tower switching and regeneration start-up.

The front-stage, middle-stage, and two-stage reverse osmosis units are described briefly here.

1.   *Comparison Between Front-Stage and Middle-Stage RO*

In a primary treatment system with one RO unit, the front-stage RO (placed upstream of the ion-exchange unit) and the middle-stage RO (placed downstream of the ion-exchange unit) are operated and used quite differently.Table 8 summarizes this difference. In the comparison, the front-stage RO unit uses a middle- or low-pressure cellulose acetate spiral RO module and a low-pressure complex membrane spiral RO module and the middle-stage RO unit uses a low-pressure complex membrane spiral RO module.

a.   Functions.   First, the greatest difference between the two is the targeted contaminants. As mentioned earlier, the first priority of the front-stage RO unit is to remove ions from the water to reduce the load on the ion-exchange unit that follows. The

Table 8 Comparison of Front-Stage and Middle-Stage RO

	Front−Stage RO	Middle−Stage RO
Main Purpose	Ion Removal	TOC, Particle Removal
Feed Water TDS	50~250ppm	0.1~5 ppm
Feed Water FI	3~5	0.5~3
RO Recovery	~75%	~90%
Cleaning period	2~12 times/Year	2~4 times/Year
Sterilization Method	NaClO, Chloramine T	H_2O_2
Measurement Method of change of Performance	Product Flowrate, Operating Pressure Ion Rejection	Product Flowrate, Operating Pressure TOC・Particle Rejection Inspection by Pull out Modules
Module Type	Low Pressure Thinfilm Membrane Medium Pressure, Cellulose Acetate Membrane	Low Pressure Thinfilm Membrane

front-stage unit is also capable of removing colloid, organic material, particles, and bacteria. The front-stage unit also maintains a stable processed water quality even when the quality of the raw water supply fluctuates.

For the middle-stage RO unit, TOC and particles are the most critical targets to be removed since the ion concentration in the water fed to the unit is already very low. The RO module for pure water, which is designed for limited TOC elution and particulation, is therefore often mounted on the middle-stage RO unit. When the spiral RO module is set on the vessel, operators are required to wear gloves. The silicon grease that is usually applied to an O ring to enhance the sealing effect should not be used.

b. Design conditions. The recovery ratio design and the volume of product water are also different, mainly because of the difference in the turbidity (FI and SDI values) and the silica concentration.

Table 9 shows the relationship between the design conditions and FI values for the low-pressure complex membrane RO module. It was found that the fluctuation in the turbidity in the water fed to the RO unit (FI value) greatly affects the clogging of the module. In the front-stage RO unit, the volume ratio between concentrated water and product water should be more than 5; in the middle-stage RO unit, it is 1–4. The middle-stage RO unit requires fewer modules.

The silica concentration also affects the recovery ratio of the front-stage RO unit. The solubility of silica depends on temperature and pH. At neutral pH and room temperature,

Table 9 Example of Design Conditions for RO Module

Feed Water FI	Product Flowrate Ratio	Concentrate/Product Flowrate Ratio
~1	1.8	>1
1~3	1.3	>4
3~4	1.0	>5
4~5	0.6	>6

Module Type : Low Pressure Thinfilm Membrane

the solubility is 100–120 ppm. A recovery ratio is required that does not exceed the silica solubility in the concentrated water in the RO unit.

Taking these two conditions into consideration, the recovery ratio of the front-stage RO unit should be set at 75% at maximum, and that of the middle-stage RO unit can be as high as 90%.

     c.   Operation Control (Cleaning and Sterilization).   Since the turbidity and the FI value are high in the front-stage RO unit, the module becomes clogged, which leads to a rapid rise in differential and inlet pressures. This means cleaning should be more frequent in the front-stage than the middle-stage unit.

Different materials adhere to the surface of the module membrane in the front-stage and middle-stage RO units. In the front-stage RO unit, the membrane is clogged with iron, aluminum, and suspended matter contained in the raw water. In some cases, the membrane is clogged with calcium scaling and silica scaling. The selection of cleaning chemicals depends on what clogs the membrane. Acid, alkali, and surfactant are often used to clean the membrane.

It was reported that the scaling on the RO membrane surface is caused by the following mechanisms [7, 8].

1.   Supersaturation is caused by materials with low solubility. In $CaSO_4$, for instance, Ca^{2+} and SO_4^{2-}, both of which are dissolved in water, collide with each other and form clusters. The frequency of collision rises as the temperature increases. Creation of the supersaturation state depends on solubility and temperature.
2.   Nucleus generation. The clusters also repeatedly collide against ions dissolved in water. Some clusters are dissolved in water again, but others grow in the course of collision to form stable nuclei. The nucleus generation process is affected by the number, size, and shape of the clusters.
3.   Growth to crystal. The stable nucleus grows further: ions and molecules move from bulk solution to the surface of the nucleus, ions and molecules are adsorbed on the nucleus surface, and finally nuclei are arranged in a lattice. Eventually this process yields a visible crystal.

On the other hand, clogging with such contaminants as ion and particles occurs seldom in the usual operation of a middle-stage RO unit. Clogging of the middle-stage RO unit is often caused by bacteria. Regular sterilization of the middle-stage unit is therefore critical.

There are several points to be noted in connection with sterilization. The oxidizing action of H_2O_2 is greatly intensified in the presence of heavy metals, such as iron. If there is a possibility that iron is adhering to the middle-stage RO module, the module should be cleaned with citric acid to remove iron or iron ion should be chelated by adding 100 ppm EDTA to H_2O_2. The second point is that the rise in differential pressure can be suppressed but it is difficult to decrease the differential pressure. When the differential pressure must be lowered, the unit must be cleaned with alkali or surfactant. The resistivity of the water fed to the middle-stage RO unit is usually very high, however, and thus it takes a long time to return the resistivity to the appropriate level after cleaning. Table 10 shows the time required to attain a sufficient resistivity level after cleaning (sterilization) of the middle-stage RO unit with H_2O_2 and surfactant.

In both case A and case B the middle-stage RO unit was placed at the outlet of the mixed-bed ion-exchange unit. Both the resistivity of the water fed to the RO unit and the resistivity of the water produced in the RO unit were 18 MΩ·cm at 25°C. In the RO unit

Table 10 Increasing Resistivity After Cleaning a Middle-Stage RO

	Case A	Case B
Module Size	8 ″	8 ″
Module Array	6 M× (6 − 2)	5 M× (4−2−1)
Module Quantity	48 pcs	35 pcs
Cleaning Solution	2 % Detergent	1 % H_2O_2
Feed Water	MB Outlet	MB Outlet
Feed Water Resistivity	18MΩ • cm at 25°C	18MΩ • cm at 25°C
Product Water Resistivity (Before Cleaning)	18MΩ • cm at 25°C	18MΩ • cm at 25°C
Product Water Resistivity (After Cleaning)		
1 Hr	—	17.5 MΩ • cm at 25°C
6 Hr	5 MΩ • cm at 25°C	—
15Hr	7	—
24Hr	13	—
40Hr	15	—
50Hr	16	—
75Hr	17	—

sterilized with 1% H_2O_2, the resistivity returned to 17.5 MΩ·cm at 25°C (almost the same as the level before cleaning) 1 h after cleaning. On the other hand, the resistivity reached 17 MΩ·cm at 25°C in 75 h after cleaning in the RO unit sterilized with 2% surfactant. Generally it is thought that the types of chemicals have nothing to do with the start-up time of the RO unit after cleaning. These data show quite different results, however. It must be noted that recovery of resistivity takes some time when the middle-stage RO unit is cleaned.

Another problem with the middle-stage RO unit is the initial elution from the ion-exchange resin (aromatic sulfonic acid or aromatic carboxylic acid from the cation resin and polystyrene 4-grade ammonium salt from the anion resin) [9] and the initial elution from pipes, fiber-reinforced plastic (FRP) tanks, and concrete pits. The module is sometimes clogged with eluted materials. When a new system is started up and operated on a trial basis or the ion-exchange resin is replaced, the system should be cleaned well before water is supplied to it.

d. Evaluation of Change in Performance. It is relatively easy to detect a change in performance of the front-stage RO unit by monitoring the volume of product water, differential pressure, and ion removal efficiency. In other words, RO module performance can be monitored and controlled through daily operation data.

The middle-stage RO unit can be also controlled quite easily for clogging and decreased product water, but it is difficult to detect a change in the ion removal efficiency. The meter for measuring ion removal efficiency is not mounted on the middle-stage RO unit because the ion intensity is very low there. Although it is possible to predict ion removal efficiency by measuring the resistivity and conductivity of feedwater and product water, the prediction is not always accurate. In many cases, the TOC concentration and the number of particles in the feedwater and product water are used as indicators of ion removal performance. Recently an on-line TOC meter and particle counter with high accuracy were developed that can be applied for this purpose.

Table 11 lists particle counts in the middle-stage RO unit under normal and abnormal conditions. Under abnormal conditions, some of the O-rings that connect modules are broken during operation, which causes partial leakage of feedwater to product water. The figures for feedwater and concentrated water pressures demonstrate some change, but the differential pressure and the resistivity of the product water remain almost the same. Since the number of particles in the product water increases 140 times from the normal condition, the UF modules downstream can be clogged. To prevent any dangerous conditions, regular analysis of the product water is critical.

There is a method of regularly removing the module from the RO unit to measure standard module performance more accurately.

Table 12 lists the performance data for a middle-stage RO module used for 6 years. In this measurement, the module was removed from the system. There was no major change in the volume of product water or the removal efficiency of this RO module. The water quality of the RO unit itself was also measured, and no great difference was observed from initial values. When the module can be removed from the RO unit for analysis, the evaluation data can be compared under fixed conditions. In this sense, this is quite an effective way to observe module performance.

If degradation of performance is observed in the RO module, it is often disassembled for further analysis. Recently analysis technology has improved greatly and an appropriate analytic method can be chosen [10].

For example, various different analytic methods can be applied to analyze materials that adhere to the membrane surface: optical microscopy, scanning electron microscopy

Table 11   Measurement of Particle of Middle-Stage RO

Item	Unit	Case A Abnormal Condition (Damage of O−ring Seal)	Case B Standard Condition
Feed Water Pressure	kg/cm^2	22.5	28.8
Conc. Water Pressure	kg/cm^2	20.4	25.3
$\triangle P$	kg/cm^2	2.1	3.3
Product Flowrate	m^3/Hr	33.0	33.5
Feed Water Resistivity	$M\Omega \cdot cm$ at 25°C	15.2	12.3
Product Water Resistivity	$M\Omega \cdot cm$ at 25°C	17.9	18.0
Feed Water Particle (0.2 μ)	pcs/mℓ	—	1.9×10^5
Product Water Particle (0.2 μ)	pcs/mℓ	7×10^3	5.2×10

Table 12 Example of Measurement of Middle-Stage RO Module Performance

	Initial	After 6 Years
Product Flowrate Ratio	1.0	1.0
Rejection	96.8%	93.3%
ΔP [1]	$0.05 \, kg/cm^2$	$0.05 \, kg/cm^2$

Module : Medium Pressure Cellulose Acetate Membrane
Feed Water : MB Outlet Water
Cleaning or Sterilization : None
Operating Period : 6 Years
Test Condition : Feed Pressure $30 \, kg/cm^2$
                Temp.   25 °C
                pH   6.5
                Conc. Flowrate   40 ℓ/min
                NaCl   1500 ppm
[1] $\Delta P =$ (Feed Pressure — Conc. Pressure)

(SEM), energy-dispersive x-ray analysis (EDXA), and others. Infrared spectroscopy measures hydrolysis of the cellulose acetate membrane.

2. *Operation Examples of Front-Stage and Middle-Stage RO*

Some more interesting examples of the primary treatment system are presented here. Table 13 shows an example of damage to the front-stage RO module. This system uses a midpressure cellulose acetate RO module. There are two wells in this plant. When well A is used, the removal efficiency does not degrade even if NaClO is added. When well B is operated in the same way, however, the removal efficiency drops sharply.
     The results of analysis of the degraded RO module are as follows:

Table 13 Example of Damage to Front-Stage RO from Well Water

Feed Water	Treatment System	Decreasing Ratio of RO Ion Rejection
Well A	NaClO	0% per month
Well B	NaClO	4.0% per month
Well B	NaClO + NaHSO₃	0.7% per month
Well B	AC + NaClO	0.6% per month

Module Type : Medium Pressure, Cellulose Acetate Membrane
Well A, B : Same Factory, Same site
NaClO : 0.5 ppm as Cl_2 Adding
$NaHSO_3$: 1.0 ppm as $NaHSO_3$ Adding
   AC   : Activated Carbon Filter

1.   Neither well A nor well B has problems with iron or manganese.
2.   Neither well A nor well B has problems with THM or TOX.
3.   No materials are deposited on the membrane surface, which is free of scaling and
     fouling.
4.   Observation with the differential interference microscope detects no attack by bacte-
     ria or related problems.
5.   Although the intrinsic viscosity of the membrane is slightly decreased, the decrease in
     viscosity is not as serious as the oxidation degradation.
6.   Observation by XSCA detects no problem with the membrane.
7.   Transmission electron microscopy (TEM) detects roughness and loss on the surface
     skin layer of the membrane profile. This is the cause of the drop in the removal
     efficiency.

To further study the causes of loss on the surface skin layer, another test was
conducted. The water from well B was treated in the active carbon tower, and NaClO and
then $NaHSO_3$ were added. As a result, the decrease in removal efficiency was reduced by
about one-sixth. The causes of this result have not yet been pinpointed. This is thought to
be a special oxidation degradation that is different from common chlorine degradation. It
is believed that the strong oxidation degradation phenomenon occurs on the RO membrane when free chlorine and some material contained only in the water from well B,
which is thought to be partly removed by active carbon, coexist. This strong oxidation
degradation leads to the loss of the surface skin layer, resulting in the drop in the removal
efficiency of the RO module. It was learned that the drop in the removal efficiency can be
prevented by using water only from well A or by intermittently adding NaClO when water
from well B is used.

Figure 13 and Tables 14 and 15 show the measurement results for concentrated water
in the middle-stage RO unit. As shown in Fig. 13, this RO unit is fed with water treated in
the two-bed, three-tower unit. The concentrated water is used as the regeneration water
for the 2B3T unit; it is recycled to the raw water tank and partially drained. Table 14 lists
the FI values measured at each component of the system when the volume of water

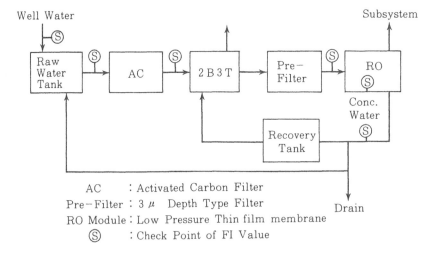

    AC   : Activated Carbon Filter
Pre-Filter : 3 μ Depth Type Filter
RO Module : Low Pressure Thin film membrane
    Ⓢ   : Check Point of FI Value

Figure 13   Middle-stage RO system.

Table 14 Change in FI When RO Concentrate Recovered

	Case A RO Conc. Recovered	Case B RO Conc. Drain
Recovery Ratio	90 %	0 %
Feed Water FI	0.47	—
AC Inlet FI	5.8	0.3
AC Outlet FI	6.0	0.8
RO Inlet FI	4.6	0.3

recycled to the raw water tank is changed. Table 15 lists the FI values for concentrated water at each bank of the RO unit. The measurement is conducted when the FI value in the supply water is high and when the FI value in the supply water is low. As shown in Table 15, the FI values rise considerably in the later stage in both cases. Therefore, when too much concentrated water is recycled back to the raw water tank, the FI value in the entire system rises, which may lead to clogging of the RO module. In actual operation, the volume of water recycled to the raw water tank is maintained at a level that keeps the FI value in the system lower than the set level. It must be noted, however, that the FI value does not always increase in the RO unit concentrated water. Sometimes the concentrated water FI values barely increase. This is thought to be because the FI value increase is affected by the turbidity and particle distribution in the supply water.

Figure 14 and Table 16 indicate the TOC removal efficiency of the middle-stage RO unit. As shown in Fig. 14, this system employs a low-pressure complex membrane RO and the supply water for this system is treated in the MB unit. The UV sterilizer is mounted at the inlet of the RO unit to suppress the increase in bacteria. As indicated in Table 16, the TOC removal efficiency of the RO unit is affected by the operation of the UV sterilizer. The ultraviolet wavelength normally used in the sterilization is 254 nm. This wavelength reportedly has no effect on TOC decomposition. Moreover, the TOC concentration at the outlet does not change whether the UV sterilizer is operated or not. This phenomenon results when some of the TOC materials contained in the water treated

Table 15 Change in FI in RO System

	Case−A Low FI Value of Feed Water	Case−B High FI Value of Feed Water
Feed Water	1.1	4.8
1st−Bank Conc.	2.0	6.2
2nd−Bank Conc.	3.6	6.4
3rd−Bank Conc.	5.4	6.4

Module : Low Pressure Thin film Membrane
Feed Water : 2B3T Outlet Water
Module Array : 3 M × (2−2−1)

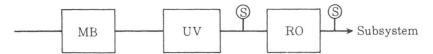

RO Module : Low Pressure Thinfilm membrane
UV   : Ultraviolet Sterilizer (254nm)
Ⓢ   : Sampling Point of TOC

Figure 14   Middle-stage RO used to test removal efficiency.

in the MB unit decompose especially easily, and here they are decomposed by ultraviolet from the sterilizer TOC materials of lower molecule weights.

3.   Two-Stage RO

a.   RO + RO System.   This is a front-stage RO unit in which the water produced in the first RO is sent directly to the second RO. This system was realized as a result of the recent development of the low-pressure complex RO module. Even when two RO modules are arranged in series, the pressure of the supply water at the inlet of the first RO unit can be maintained at the same level as when one midpressure RO unit is employed. This means that the two-stage RO unit can be operated with a conventional high-pressure pump. (It is also possible to place a booster pump between the first and second RO units.)

The final goal of this system is to reduce the ion and silica level to below that of a conventional one-stage RO unit to lighten the load on the degasifier and primary ion-exchange unit that follow. If the resistivity of the water produced in the second RO unit is kept at a steady 5–10 MΩ·cm at 25°C, the produced water can be sent directly to the subsystem. The current performance of the RO module has not yet reached this high level. This is why the ion-exchange tower is placed downstream of the two-stage RO unit.

To effectively operate the two-stage RO unit, various ways of injecting chemicals have been proposed [11, 14]. Figures 15 and 16 show the relationship between the pH of the water supplied to the two-stage RO and the quality of the water produced in the RO unit. As shown in Fig. 16, the pH is adjusted to the alkali side only at the inlet of the first RO unit. The same low-pressure complex membrane RO modules are mounted on both the first and second modules. Figure 15 indicates that the resistivity of the water at the outlet of the second RO module is greatly affected by the pH of the feedwater. This is because the removal efficiency of the mounted RO module is high and it exhibits a high dependence on pH. In the system shown in Fig. 15, the resistivity of the produced water

Table 16   Influence of UV on TOC Rejection of Middle-Stage RO

	UV − ON	UV − OFF
Feed Water TOC	80ppb	80ppb
Product Water TOC	50ppb	30ppb

RO Module : Low Pressure Thin film Membrane
Feed Water : MB Outlet Water
UV   : Ultraviolet Sterilizer (254nm)

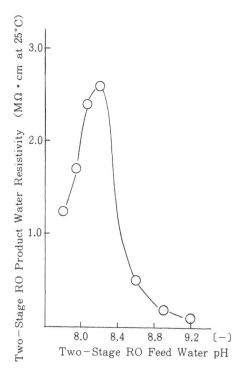

RO Module   : Low−Pressure Thin film Membrane
Feed Pressure : 1 st −Stage   19 kg/cm²
                2 nd−Stage   17 kg/cm²
Recovery   : 1 st −Stage   75 %
                2 nd−Stage   90 %

Figure 15 Comparison of feedwater pH with product water resistivity in two-stage RO system.

changes more than 1 MΩ·cm at 25°C as the pH of the feedwater changes by only 0.2. Since during the actual operation of the pH cannot be adjusted in such a precise manner, it is expected that a new module with a high removal efficiency and a smaller pH dependence will be developed.

Water produced in the two-stage RO unit contains quite a lot of sodium, silica, and bicarbonate ion. Because it is in a strongly anion-rich state, the ion-exchange unit after the two-stage RO unit is usually composed of an anion-exchange resin tower and a mixed-bed ion-exchange tower.

Data have been reported on the combination of modules mounted on the first and second RO units [15]. Figure 17 shows the removal efficiency when three modules A, B, and C are combined. The removal efficiencies of A and B differ. The removal efficiency of C is completely different from that of the A-C combination and the B-C combination. The conductivity of the water fed to C is less than 10 μS/cm. It was reported that the removal efficiency of C drops sharply as the ion intensity of the feedwater is lowered.

Table 17 shows the TOC removal efficiency of different combinations of three RO modules: midpressure and low-pressure cellulose acetate RO and low-pressure complex membrane RO. The removal efficiency of the entire system is changed by the different

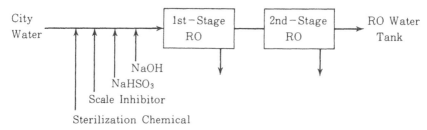

Figure 16   System flow of two-stage RO.

membrane combinations. In Table 17, the combination of membranes of different materials works better than the combination of membranes of the same material in terms of the reduction in TOC concentration. This is thought to be because the cellulose acetate and complex membranes are charged differently, and each membrane removes different types of TOC materials.

b.   RO + MB + RO System.   The RO + MB + RO and RO + tank + RO are variations of the two-stage RO unit. Basically the RO + tank + RO system is the same as the RO + RO system. Since the RO + tank + RO system requires high facility costs, it is employed only in special cases, for example when a two-stage RO system is employed for some portion of the water fed to the primary treatment system (when the water quality required is different at each use point).

When the RO + MB + RO system is employed, the front-stage and middle-stage RO units are installed in the primary treatment system to make the best use of the characteristics of the two units. This system was developed to cope with the recent tendency to pursue much lower TOC concentrations at use points.

Figure 17   Performance test of two-stage RO. (From Ref. 15.)

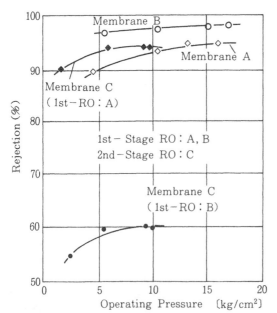

Table 17 Comparison of TOC Rejection in Combination RO Module

No	Feed Water TOC	1st-Stage RO	2nd-Stage RO
1)		MP-CA	LP-CA
	3000 ppb	100 ppb	75 ppb
2)		MP-CA	LP-TFM
	3000 ppb	90 ppb	60 ppb
3)		LP-TFM	LP-TFM
	1300 ppb	95 ppb	75 ppb
4)		LP-TFM	LP-CA
	1200 ppb	105 ppb	50 ppb

MP : Medium Pressure
LP : Low Pressure
CA : Cellulose Acetate Membrane
TFM : Thin film Membrane

It is obvious that the RO unit plays a very important role in the primary treatment system. Sometimes, how to use the RO unit is decided first and the structure of the other units is considered based on that decision. Using the RO unit effectively is a key to successful operation of the primary treatment system.

B. Ion-Exchange Unit

The functions of the ion-exchange unit in the primary treatment system are described here. The ion-exchange unit is installed in the primary treatment system to remove the ions that cannot be removed in the RO unit and to raise the resistivity to 10–15 MΩ·cm. Another function of the ion-exchange unit is to suppress organic materials, particles, and bacteria. In particular, when the RO unit is placed upstream of the ion-exchange unit, extra attention should be paid to the impurity increase since the impurity concentration was decreased to a very low level in the RO unit. The ion-exchange unit, the ion-exchange resin, and the operation of the unit should be carefully researched.

Regeneration is necessary for the ion-exchange unit. The timing of the regeneration is very critical, and the regeneration operation should be improved so that it is completed as soon as possible.

The ion-exchange units are as follows: mixed-bed; two-bed (two-bed, three tower) + mixed bed; three-bed, three-tower; four-bed, four-tower; four-bed, five-tower units. Refer to the books and articles on these units for details [16, 17]. Most ion-exchange units are regenerable.

1.   Requirements for the Unit

As shown in Fig. 18, the ion-exchange unit is a cylindrical structure made of steel plates. The materials for the inner surface should feature high acid and alkali resistance.

The inner wall of the tower usually has a hard lining. Synthetic resin (such as fluorocarbon resin) is much better because it has limited elution properties. At present, however, it is too expensive and hard to process. The inner pipes, supports, bolts, and nuts are made of SUS 304 and SUS 316. The distributors, collectors, and strainers are made of SUS, acrylonitrile-butadiene-styrene (ABS), and polypropylene (PP). Saran net is also used at many spots. All the materials should be improved to reduce elution.

To prevent the ion-exchange resin from leaking from the unit, a collector placed at the outlet employs a 0.3 mm strainer through which water can pass but resin cannot pass. To secure the prevention of resin leakage, a resin catcher and membrane filter are placed downstream of the ion-exchange unit.

HCl and H_2SO_4 are used for the regeneration of cation-exchange resin, and NaOH is used for the regeneration of anion-exchange resin. In the usual operation, 30–35% HCl and 20–25% NaOH are diluted 10% just before the inlet of the resin tower to be injected. As shown in Tables 18 and 19, there are various regeneration agents of different grades. It is desirable to use an agent with as small a concentration of impurity as possible.

Current semiconductor manufacturers demand completion of the regeneration opera-

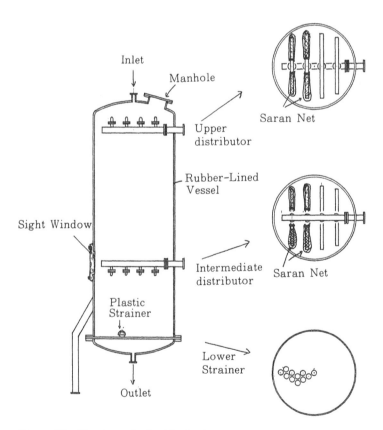

Figure 18   Structure of mixed-bed column.

Table 18 Quality of HCl

	No. 1	No. 2	No. 3
HCl (%)	≧ 37	≧ 35	≧ 35
Fe (%)	≦ 0.0005	≦ 0.002	—
Ignition Residue (%)	≦ 0.005	≦ 0.01	—

[JIS K 1310−1959]

tion within a short period of time. Table 20 shows the improved regeneration flow of a MB unit, which can be completed within a shorter time. The resin, size of pipes, chemical concentration, and chemical injection speed are modified. From the technical point of view, this modification may not be favorable. In this case, the quality of treated water and the cycle time remain unchanged before and after the modification.

2.   *Requirements for the Resin*

The resin for the ion-exchange unit in the primary treatment system requires the following features:

1.   Fast ion-exchange speed
2.   High physical and chemical strength
3.   High regeneration efficiency
4.   Large ion-exchange capacity
5.   Infrequent and mimimal oxidation degradation and organic contamination
6.   Limited elution from resin

Table 19 Quality of NaOH (%)

Component	Mercury Process	Diaphragm Process	Ion Exchange Membrane Process
NaOH	48 ± 1	49 ± 1	20
Na_2CO_3	0.04	0.1	
NaCl	0.003	1.0	0.002
Na_2SO_4	0.0013	0.02	No detect
$NaClO_3$	0.0001	0.07	≦ 0.001
CaO	0.0003	0.002	0.0001
MgO			
Al_2O_3	0.0001	0.001	≦ 0.0001
SiO_2	0.002	0.01	0.0004
Fe (ppm)	1	5	≦ 1
Ni (ppm)			
Cu (ppm)			
Mn (ppm)	0.5	0.4	0.01

Table 20   Comparison of Regeneration Times

	before change		after change	
Regeneration Time Table (min)	Back-Wash	10	Back-Wash	10
	Rest	5	Rest	5
	NaOH Injection	37	NaOH Injection	20
	HCl Injection	30	HCl Injection	20
	HCl Rinse	30	HCl Rinse	15
	1st Rinse	20	1st Rinse	10
	Drain	18	Drain	10
	Mix	15	Mix	6
	Fill Up	10	Fill Up	10
	2nd Rinse	30	2nd Rinse	40
Cation Resin Effective Size	0.45 ± 0.03 mm		0.60 ± 0.03 mm	
Uniformity Coefficient	$\leqq 1.8$		$\leqq 1.6$	
Anion Resin Effective Size	0.45 ± 0.03 mm		0.50 ± 0.03 mm	
Uniformity Coefficient	$\leqq 1.8$		$\leqq 1.6$	

The two-bed (2B3T) and mixed-bed units often use the gel type of strongly acid cation-exchange resin and strongly basic anion-exchange resin (types I and II). If necessary, the porous type, which features high physical strength and less organic contamination, is applied.

In the MB unit, the resin is separated by using the difference in specific gravity between the cation-exchange and anion-exchange resins. If the two resins are not properly separated at this point, water quality is damaged [18]. To completely separate the two resins, inert resin with a specific gravity between that of the cation-exchange resin and that of the anion-exchange resin is sometimes mixed into the MB unit. Since there is an extremely small possibility that the cation-exchange resin and NaOH or the anion-exchange resin and HCl may come into contact, a high water quality level can be obtained.

A gel resin for the MB unit with a very uniform particle size and excellent physical and chemical strength has recently become available on the market. This new resin can be separated perfectly without using inert resin of intermediate specific gravity [19].

Among the requirements for the ion-exchange resin, most closely studied is elution from the resin. Two materials are mainly eluted from the ion-exchange resin:

1. The water-soluble material formed when the resin is manufactured
2. Material cut from the carbon chain as a result of chemical shock or oxidation degradation

Several articles are available on the composition of eluted materials and the causes of elution [20–22]. The volume is gradually decreased as regeneration is repeated. Figure 19

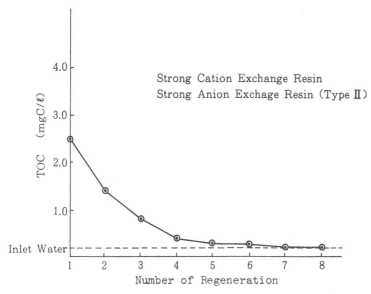

Figure 19 Variation in TOC levels after 30 minutes of operation.

shows the relationship between the number of regeneration operations and the TOC value immediately after start-up. After five regeneration operations, TOC elution from the resin is decreased.

Several methods have been proposed to suppress elution from the ion-exchange resin [23–26]. The most common method is to conduct conditioning before using the resin. Some ion-exchange resins are manufactured exclusively for ultrapure water production, and resin manufacturers pay careful attention to the manufacturing process and raw materials. These resins are shipped to market after conditioning at the production site. Table 21 compares these specially manufactured resins and other resins. In this test, 50 ml resin is immersed in 1 L ultrapure water and left for 24 h with vibration applied from time to time. After 24 h, the TOC concentration in the surface layer of the ultrapure water is measured. The specially manufactured resins give quite a good result. The specially manufactured anion-exchange resins are type I. A special type II resin should also be developed.

Ion-exchange resin is inevitably degraded with time. It undergoes regeneration at least once every few days or once a day in the most frequent case. The physical and chemical loads are quite large in this regeneration. Figure 20 shows the degradation of the ion-exchange resin. The strongly acid cation-exchange resin does not become degraded in 1 or 2 years, but the deterioration of the strongly basic anion-exchange resin proceeds faster, and type II deteriorates much faster than type I. Regular sampling is therefore necessary to analyze the total exchange capacity, the neutral salt decomposition capacity, the moisture content, and other factors. It is desirable to replace all of the resin with new resin every 1–3 years. In addition, over the course of operation, parts of the resin may be broken, generating fine resin. The fine resin is discharged outside the tower little by little at the time of backwashing of regeneration. The surface layer of the resin should be replaced once every 6–12 months.

More and more users want to cut the downtime due to resin exchange and obtain high-purity water immediately after start-up. Resin manufacturers are therefore expected

Table 21   Comparison of TOC of Normal Resin and Pure Resin

	TOC (mg C/ ℓ)
Normal Cation Resin (A Company)	0.27
Normal Anion Resin (A Company)	2.2
Normal Cation Resin (B Company)	2.2
Normal Anion Resin (B Company)	0.96
Pure Cation Resin (A Company)	0.06
Pure Anion Resin (A Company)	0.08
Pure Cation Resin (C Company)	0.12
Pure Anion Resin (C Company)	0.55

1)   All Resin is Gel Type.
2)   Cation Resin is Strong Acid Cation Exchange Resin.
3)   Anion Resin is Strong Base Anion Exchange Resin.

(Type I)

to deliver resin that has undergone regeneration conditioning. Note the following important points:

1.   The regeneration resin must be placed in the unit as soon as possible.
2.   The cation-exchange resin and the anion-exchange resin should not be transported as a mixture.
3.   After placement in the unit, the cation-exchange and anion-exchange resins should be rinsed as long as possible.

Point 1 is necessary because the regenerated resin easily decomposes outside the unit. Points 2 and 3 are required to prevent the cross-contamination that arises when the negatively charged materials eluted from the cation-exchange resin and the positively charged materials eluted from the anion-exchange resin are adsorded on each other.

3.   Requirements for Operation

The current two-bed, unit three-tower and the mixed-bed unit can process high-purity water immediately after startup. However, it is desirable to continue operation as long as possible because suspension of operation leads to elution and water entrapment in the tower.

As water is continuously supplied to the system, the H^+ and OH^- concentration in the resin decreases. At a certain point, the ions in raw water begin to be released to processed water. This point is called the breakpoint. The exchange capacity of resin up to this breakpoint is the *breakpoint exchange capacity*.

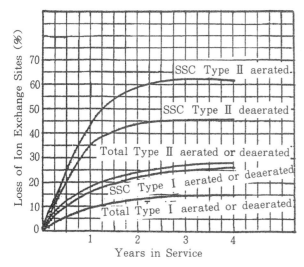

Figure 20   Typical loss of ion-exchange site (type I and II strongly basic resin versus time in service).

With a cation-exchange resin, monovalent ions, such as Na^+ and K^+ leak just after the breakpoint. On the other hand, with anion-exchange resin, SiO_2 leaks. Since Na^+ and K^+ can be detected easily using a conductivity or resistivity meter, the breakpoint can be detected immediately. SiO_2 does not affect the conductivity or resistivity, however, so it is hard to identify the breakpoint. If SiO_2 is allowed to leak continuously, processed water with high silica concentration is released to the subsequent process (Fig. 21). This problem is *overrun*, which should be prevented by noting the following points:

1. The resin volume and the regeneration level should be adjusted at the designing stage so that the cation leaks earlier than anion.
2. In addition to the conductivity and resistivity meters, an accumulative flow rate meter should be installed to control the volume of supply water.
3. An on-line silica monitor should be installed to constantly monitor fluctuation in the soluble silica concentration.

It has been reported when using mixed bed, particularly in the RO + MB system, that the ionic silica is polymerized because of the influx of acid water into the MB unit, eventually generating low-polymer silica and colloidal silica [17]. The low-polymer and colloidal silica reportedly affect the water quality at the use points even if they flow through the polisher and the ultrafiltration unit. Figure 22 shows the fluctuation in water quality at the outlet when silicic acid soda is supplied to a mixed-bed polisher. The processed water quality is evaluated using a residue after evaporation meter, a resistivity meter, a silica monitor, and silica analysis. Silica begins to leak before the resistivity drops, but before silica leakage, the total silica concentration rises. The breakpoint can be identified by the change in the total silica concentration. The residue after evaporation meter is the most appropriate unit to detect this [27, 28].

Figure 23 shows measurement data for particle concentration, TOC, SiO_2, and resistivity in an experiment in which city water is continuously supplied to the MB unit, which is adjusted to that the cation begins to leak before the anion. As a result, the

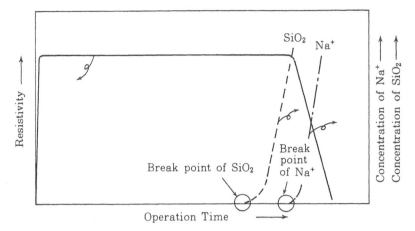

Figure 21   Relationship between ion and operation time.

resistivity begins to drop before silica leakage, but particle leakage takes place earlier than the resistivity drop. This is thought to be because the particles that are initally adsorbed to the ion-exchange resin by a weak adsorption force start to be desorbed as the anions are adsorbed to the ion-exchange resin. In any case, to conduct the regeneration operation properly, it is important to monitor the breakpoint accurately.

A reverse osmosis unit installed in the primary treatment system often uses a oxidative reagent, such as NaClO, for sterilization. Although the oxidative reagent is processed with the reducing agent or active carbon, if it flows into the ion-exchange unit by mistake, the ion-exchange resin begins oxidizing decomposition. This phenomenon

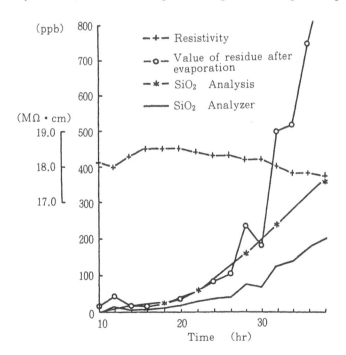

Figure 22   Variation in residue after evaporation, SiO_2 at anion break.

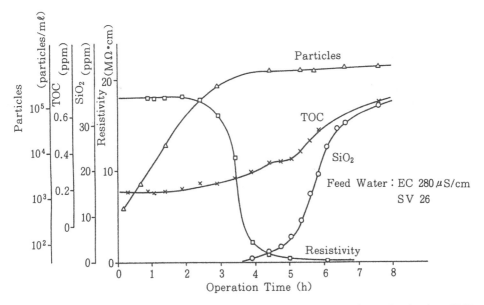

Figure 23   Variation in resistivity, SiO$_2$, TOC, and particle concentration at breakpoint of MB.

leads to an increase in the differential pressure of the subsequent UF unit and a drop in production. The oxidative reagent must be carefully handled in the primary treatment system.

IV.   FUTURE PRIMARY TREATMENT SYSTEM

A.   Easy Maintenance

On-line monitors for constant checking of water quality and a computerized operation monitoring system to compile and process the operation data have recently been introduced. By further upgrading these monitoring systems, future primary treatment systems should be equipped with an automatic operation controlling function for the automatic gathering and analyzing data, lifetime prediction of components that are not durable, and calculation of the optimum operating conditions.

Sterilization is an important regular maintenance operation. In the future, the bacteria level in the primary treatment system must be suppressed to even lower levels. To cope with this task, systems with longer sterilization cycles are expected. Improved sterilization methods to replace conventional H$_2$O$_2$ sterilization should be developed with an eye to decreased costs and handling requirements.

B.   Less Energy Consumption

The RO module must be modified to use lower pressures. Moreover, reverse osmosis at superlow pressure and pump inverter control should be adopted.

The volume of N$_2$ gas used for the N$_2$ purge should be reduced. At present there are problems with contamination and purity and with the cost of the N$_2$ gas. The introduction of pressure swing adsorption (PSA) and purification of N$_2$ gas by gas deaeration membrane are expected [29].

C.   Improvement of Components

The removal efficiency of TOC, silica, and sodium, fouling resistance, oxidative reagent resistance, and the pH dependence of removal capability must be further upgraded. In the two-stage RO unit, if the later-stage ion-exchange unit can be omitted, the entire system can be simplified to obtain significant advantages in terms of cost and space.

To improve the ion-exchange resin, TOC elution and the leakage of sodium and silica must be suppressed. One of the possible countermeasures is modification of the cleaning method applied before the operation [30].

The problem with the vacuum degasifier is its height. Now at 8–10 m, it requires a great deal of space for installation. Smaller deaeration systems, membrane deaeration, and the N_2 gas bubbling method are now being studied to reduce the size of this unit.

The materials and structure of tanks, pipes, pumps, and UV sterilizers should be improved to eliminate the elution of metallic ions and the increase in TOC, corrosion, and bacteria.

As mentioned in the introduction, the water quality at the outlet of the primary treatment system will be required to be much higher in the future. Components employed at present in the recycling system and subsystem will be used in the primary treatment system. For example, high-pressure UV oxidation, low-pressure UV oxidation, advanced piping materials (PVDF and stainless steel processed with electricochemical buffing [30], and loose UF at the end of the primary treatment system will be studied. It is certain that the primary treatment system will play a more important role in future ultrapure water production systems.

REFERENCES

1.   T. Ohmi, T. Nitta, ed., Ultra High-Purity Gas Supply System. Realize. (1986).
2.   Y. Hiratsuka, Semicon News, *11*, 58 (1985).
3.   Y. Haraguchi, T. Imaoka, S. Takano, A New Ultrapure Water System For Megabit Semiconductor Chip Production. Semiconductor Pure Water Conference (Sixth), (1987), p. 22.
4.   H. Shimizu, Industrial Water, *347*(8), 31 (1987).
5.   B. J. Hoffman, M. J. Gavaghan, A Chemical Approach to Understanding Ion Exchange Performance in Ultra Pure Water Systems, Semiconductor Pure Water Conference.
6.   Japanese Patent Application HEI 1-249651.
7.   Walter, D., Himelstein, Zahid Amjad : The Role of Water Analysis. Scale Control, and Cleaning Agents in Reverse Osmosis. Semiconductor Pure Water, March/April (1985).
8.   Zahid Amjad, Mechanistic Aspects of Reverse Osmosis Mineral Scale for Mation and Inhibition, Ultrapure Water, September (1988).
9.   M. Furukawa, Ultra Pure Water Supply (I) Component Ultra-Clean Technology (1989), p. 109.
10.   Zahid Amjad, The Role of Analytical Techniques in Solving Reverse Osmosis Fouling Problems, Ultrapure Water, July/August, p. 20 (1988).
11.   Japanese Patent Laid Open SHO 59-11890.
12.   Japanese Patent Laid Open SHO 61-4591.
13.   Japanese Patent Laid Open SHO 62-42787.
14.   Japanese Patent Laid Open SHO 62-110795.
15.   Y. Hiratsuka, H. Sato, N. Hashimoto, T. Shinoda, Senjo-Sekkei, p. 2 (1987).
16.   H. Ohya, Junsui-TYOjunsui Seizouhou, Saiwai, p. 36 (1985).
17.   F. M. Cutler, Evaluation and Selection of Ion Exchange Resin for Producing High Purity Water, Semiconductor Pure Water Conference, January (1987).

18. T. Ohmi, T. Nitta, ed., Ultra High-Purity Gas Supply System, Realize, Tokyo, (1986), p. 117.
19. Dowex Monosphere Resin Catalogue.
20. Ohashi, Mori, Kogyo-Kagaku Kai shi. *74*(9), 166 (1971).
21. Mizuno, Kogyo-Kagaku Kai shi. *69*(6). 170 (1966).
22. J. R. Stahibush et al, 48th Int. Water Conf. IWC-85, (1987) p. 22.
23. Japanese Patent Laid Open SHO 60-166040.
24. Japanese Patent Laid Open SHO 62-132585.
25. Japanese Patent Laid Open SHO 62-4447.
26. Japanese Patent Laid Open SHO 62-114662.
27. Y. Ohta, S. Inagaki, Denshi-Zairyo, *6*, 1 (1988).
28. Y. Ohta, I. Sugiyama, Senjo-Sekkei, Summer (1989), p. 21.
29. MOL. *11*, 79 (1989).
30. K. Ushikoshi, M. Saito, Senjo-Sekkei, Summer (1989), p. 13.

4
Ultrapure Water System

Koichi Yabe
Kurita Water Industries, Ltd., Tokyo, Japan

I. INTRODUCTION

The ultrapure water system itself is usually also called the subsystem or the secondary treatment system. It is part of the ultrapure water production system, which is composed of a primary treatment system and a secondary treatment system (subsystem). The role of the ultrapure water system is to upgrade the quality of water sent from the primary treatment system. As shown in Fig. 1, early ultrapure water production systems were simply an ion-exchange unit and a microfiltration unit. At this time, the treatment system was not divided into two stages—primary and secondary.

In the early 1980s, as shown in Figs. 2 and 3, the reverse osmosis and ultrafiltration membranes were introduced to replace the microfiltration unit. Around this time, the treatment system was divided into the primary treatment system and the ultrapure water system (secondary treatment system). In this method a larger volume of water than the actual consumption volume at the use point is supplied and circulated in the ultrapure water system. By introducing to the secondary treatment system a membrane with an advanced capability of removing microparticles, for example ultrafiltration, and by constantly circulating water in the secondary treatment system and the subsequent piping system, the quality of ultrapure water at the use point was markedly improved.

The ultrafiltration element has been modified several times in the past. At first, a spiral ultrafiltration element was employed. The main purpose of the ultrafiltration

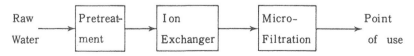

Figure 1 Ultrapure water system of the 1970s.

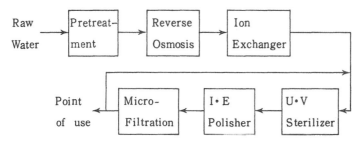

Figure 2   Ultrapure water system of the early 1980s.

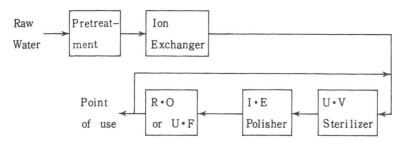

Figure 3   Ultrapure water system of the early 1980s.

element at that time was its superiority to microfiltration in removing polymeric organic materials that cannot be removed by microfiltration, as well as microparticles. Hollow fiber ultrafiltration was then developed. This new element has a simple structure, and it can be sterilized with hot water. The hollow fiber ultrafiltration element gradually replaced the spiral type.

In the mid-1980s, as shown in Fig. 4, the two-stage membrane system prevailed. The reverse osmosis unit was employed in the primary treatment system and the ultrafiltration unit was employed in the secondary treatment system. Since most of the particles in industrial water are removed in the reverse osmosis unit, the particle load on the ultrafiltration unit was greatly reduced. From this time, the hollow fiber type has been the method of choice for secondary treatment.

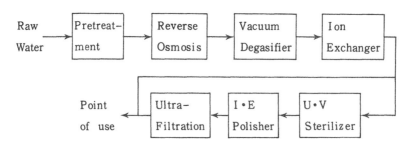

Figure 4   Ultrapure water system of the middle 1980s.

II. LATEST SECONDARY TREATMENT SYSTEM

A. Water Quality

Table 1 shows the quality of ultrapure water for 256 Kbit to 1 Mbit dynamic random access memory (DRAM) and for 4 Mbit DRAM [1]. The quality of water required in production of the 4 Mbit DRAM is by far higher than that required for the 256 Kbit to 1 Mbit DRAM generation. Because of the improved accuracy of the resistivity meter, a resistivity of 18 MΩ·cm can be guaranteed [2,3]. The size of particles to be removed is 1/10 the geometry of ultralarge scale integrated (ULSI) technology. In 4 Mbit DRAM production, the number of particles ≥ 0.1 μm should be suppressed to less than five per milliliter. The number of bacteria should be less than one per 100 ml. The total organic carbon (TOC) concentration has also been greatly improved. In the 256 Kbit to 1 Mbit DRAM range, it was more than 10 ppb. However, 4 Mbit DRAM production requires a TOC concentration of less than 10 ppb. The silica concentration must be less than 1 ppb. Trace ions, such as Na^- and Cl^-, are increasingly monitored and controlled. Generally, however, it is said that the trace ion concentration can be less than 0.1 ppb if the resistivity is maintained at higher than 18.0 MΩ·cm.

B. Structure of the System

Figure 5 shows the latest secondary treatment system. The structure of the primary treatment system differs depending on the quality of the raw water supply, but in any case, the water quality at the outlet of any primary treatment system can be considered the same. For water processed in the primary treatment system, the following are required: resistivity of more than 15 MΩ·cm, trace particles over 0.1 μm at 100 per ml or less, TOC of less than 50 ppb, and SiO_2 of less than 10 ppb. A secondary treatment system is therefore required to upgrade the quality of the water produced by primary treatment to the

Table 1 Ultrapure Water Quality for Conventional and Advanced Semiconductor Manufacturing in Japan

		CONVENTIONAL PLANT	ADVANCED PLANT
DESIGN RULE OF LSI		$1 \sim 1.5$ μm	$0.8 \sim 1.0$ μm
DRAM		256 K $\sim$ 1 M	4 M $\sim$
RESISTIVITY	(MΩ · cm)	17.7 $\sim$ 18.3	> 18.0
PARTICLE	(Counts / mℓ)	0.1μm, $7 \sim 0.3$	0.1μm <5
BACTERIA	(Counts / mℓ)	$0.1 \sim <0.01$	<0.01
TOC	(ppb)	$60 \sim 11$	<10
SiO$_2$	(ppb)	<5	<1
O$_2$	(ppb)	$50 \sim 80$	<50
Na	(ppb)	—	<0.1
Cl	(ppb)	—	<0.1

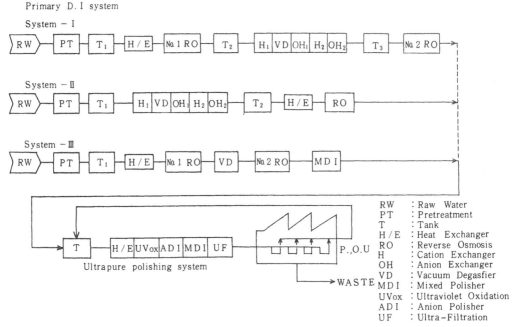

Figure 5   Advanced ultrapure water system.

level of ultrapure water. The secondary treatment system is mainly composed of a pure water tank, a pump, a heat exchanger, an ultraviolet (UV) oxidation unit, an ion-exchange polisher, and an ultrafiltration unit.

The ultrapure water manufactured in the secondary treatment system is supplied to use points through a piping system specially prepared for ultrapure water. If ultrapure water stays in this piping system for a long period of time, the water quality gradually deteriorates. To maintain the water quality, 20–30% more water than the maximum consumption volume at the use point is supplied and the excessive water is returned to the secondary treatment system. The quality of the ultrapure water is stabilized by having a greater volume of circulating return water than the water consumption volume at the use point. If more water is circulated back to the secondary treatment system, however, the system must process large amounts of water, which raises the initial and running costs. For design purposes in Japan, the circulating water volume is estimated at 120–130% of the maximum consumption volume at the use points. In the United States, this is estimated at 200%.

The secondary treatment system and the piping system that follows should be operated continuously even when the ultrapure water is not completely consumed at the use points. If operation is suspended, the secondary treatment system and the piping system should be sterilized before restarting. In this way, the system can be maintained free of the bacteria that might multiply in the system during suspension.

The major characteristics of the latest secondary treatment system are the low-pressure UV oxidation unit, the outer pressure hollow fiber ultrafiltration unit, and hot-water sterilization.

Introduction of the low-pressure UV oxidation unit to decompose organic materials has made it possible to stably suppress the TOC concentration in ultrapure water to less than 10 ppb. Although TOC still accounts for a large proportion of the impurities in ultrapure water, its concentration is being significantly reduced by the development of high-efficiency organic material removal technology.

The hollow fiber ultrafiltration element that has been used to date is an inner pressure type that has problems with adhered particles that are released to the ultrapure water by vibration. These problems have been overcome in the outer pressure hollow fiber ultrafiltration unit [4].

III. FUNCTIONS OF EACH UNIT

A. UV Sterilization Unit

In the UV sterilization unit, ultraviolet light of wavelength 253.7 nm is irradiated by mercury lamp to eliminate the bacteria in ultrapure water. The secondary treatment system generally employs a water flow type of UV sterilization unit. The basic structure is stainless steel tower with mirror finish, a sterilizing lamp, and outer pipes. UV sterilization is superior to chemical sterilization because it does not change the water quality and it is simple.

The sterilizing effect of ultraviolet light can be expressed as the product of the irradiation time and the irradiation intensity under the same conditions. It is usually expressed as irradiation intensity (μW/cm^{-2}) × irradiation time (s) = ultraviolet irradiation volume (μW/s·cm^{-2})

The secondary treatment system applies ultraviolet irradiation at 50,000–100,000 (μW/s·cm^{-2}). With this volume of irradiation, 99.9% of the bacteria in ultrapure water can be eliminated. More irradiation is required to eliminate the same percentages of some kinds of mold.

During continuous operation the sterilizing lamp deteriorates with time: in 7000–8000 h the output is reduced to 65–70% of normal. To maintain a stable sterilizing effect, the lamp should be replaced regularly.

B. UV Oxidation Unit

The UV oxidation unit conducts the oxidation decomposition of organic materials in ultrapure water by irradiation with ultraviolet light at a wavelength of 185 nm [4]. This is also a water flow unit. The organic materials decomposed by ultraviolet give rise to carbon dioxide via intermediate products, such as organic acids. Since organic acids and carbon dioxide lower the resistivity of ultrapure water, they should be removed with anion-exchange resin. The UV oxidation unit in the secondary treatment system should therefore be followed by the anion polisher. The ultraviolet lamp also irradiates at a sterilizing wavelength of 253.7 nm, so the UV oxidation unit features a sterilization function as well.

The organic materials to be removed in the secondary treatment system are of relatively low molecular weight since they were not removed in the primary treatment system [4]. The decomposition rate of these organic materials is correlated with the irradiation volume of ultraviolet: the TOC decreases as the UV irradiation volume increases. This means that TOC removal efficiency can be estimated by calculating the UV irradiation volume. By introducing this technology for quantitative control of the TOC concentration, the TOC level in ultrapure water can be guaranteed.

The UV lamp deteriorates with time during continuous operation, decreasing its output. The lamp must be replaced on a regular basis to obtain a stable oxidation effect. The lifetime of the UV lamp is the same as that of the UV sterilizing lamp.

The low-pressure UV oxidation unit is smaller than the high-pressure UV oxidation unit. It does not need oxidative reagents like H_2O_2 and O_3. In this sense, the low-pressure UV oxidation unit is very suitable for the secondary treatment system. To greatly reduce the TOC concentration in ultrapure water with this system, however, the initial and running costs are very high. The TOC removal system should therefore be designed by combining an UV oxidation unit and a reverse osmosis unit, which is another TOC removal technique.

C.   Ion-Exchange Polisher

The ion-exchange polisher is installed in the secondary treatment system to remove the remaining ions. The resistivity of water processed in the primary treatment system is over 10 MΩ·cm, and the ion concentration is a few ppb.

The ion-exchange polisher is a mixed-bed unit composed of cation-exchange and anion-exchange resins of high purity. Since the ion concentration to be removed is extremely low, the consumption of the exchange resin is limited even when water is supplied at a high SV for a long time.

The ion-exchange resin for the ion polisher is fully cleaned to eliminate in advance organic material and functional group elution from the resin. Because the exchange ratio into H and OH is higher, the ion-exchange resin lasts longer. What is more important is if the basic ion is greatly contained, the basic ions leak into ultrapure water. This is because the H^+ and OH^- formed in the course of the dissociation of H_2O repels the basic ions. The replacement frequency of the ion-exchange polisher depends on the quality of the water processed in the primary treatment system, but generally the unit should be replaced once a year.

When the UV oxidation unit is installed in the secondary treatment system, an anion-exchange resin polisher should be employed to remove the organic materials and carbon dioxide formed in the UV oxidation unit. The lifetime of the anion-exchange polisher is about 6–12 months.

D.   Ultrafiltration Unit

The ultrafiltration unit placed in the final stage of the secondary treatment system has an important function as a final filter to remove microparticles. The ultrafiltration unit removes particles that leak from the primary treatment system or are mixed into or generated in the secondary treatment system and the subsequent piping system. The ultrafiltration unit has another function—to completely eliminate bacteria to prevent the bacterial contamination of the piping system.

There are two types of ultrafiltration elements: spiral and hollow fiber. The spiral type has a complicated structure that allows particle leakage. On the other hand, the hollow fiber structure is simple, capable of removing particles completely. Both the membrane and the vessel are made of polysulfone and the adhesive is epoxy, materials that have enough heat resistance to be processed by hot-water sterilization. Because it can be processed by hot-water sterilization, the hollow fiber type is used more often than the spiral type.

Originally hollow fiber ultrafiltration used the inner pressure configuration in which raw water is filtered from the inside of the hollow fiber to the outside. A recent study has found that this inner pressure type cannot get rid of the particle contamination in the filtered water side. This is why outer pressure ultrafiltration has recently become more prevalent. In the outer pressure type, raw water is filtered from the outside of the hollow fiber to the inside.

In the inner pressure type, 90% of raw water is filtered, with the remaining 10% discharged as condensed water. This is based on the idea that the cross-flow of water inside the hollow fiber keeps the membrane from being contaminated because it reduces the number of particles that adhere to the membrane surface. This cross-flow cannot be created in the outer pressure type. If particle leakage from the primary treatment system is less than 100 per milliliter, however, the clogging due to particle does not constitute that much of a problem and the decrease in filtered water as a result of particle contamination is very limited even with the outer pressure type.

Replacement of the ultrafiltration element depends on the decrease in the volume of filtered water caused by contamination of the membrane. On average, the element should be renewed once every 2–3 years.

The sterilization method for the ultrafiltration unit is determined on the basis of the materials used. The hollow fiber ultrafiltration element can well withstand repeated sterilization with 1% H_2O_2 and hot water at 80–90°C. Sterilization with O_3 at a few ppm cannot be applied because the polysulfone membrane has no resistance against O_3. There are other chemicals for general sterilization, such as formalin and hypochlorous acid soda, but these chemicals cannot be rinsed off easily after sterilization. Formalin also interferes with the TOC concentration at start-up, and hypochlorous acid soda interferes with the resistivity. These disadvantages mean that such chemicals cannot be used for sterilization during operation of secondary treatment system.

E.   Measurement of Water Quality

Since control of the final water quality of the secondary treatment system is very important, it is constantly monitored by on-line measurements. Resistivity, number of particles, TOC concentration, and the temperature of the water are measured. The resistivity reflects the concentration of trace ions in the ultrapure water, so the resistivity meter is always placed on the secondary treatment system. Particle counters and TOC meters are mounted on an increasing number of secondary treatment systems. The concentration of dissolved oxygen and the silica concentration can be monitored on-line as well.

IV.   CHANGE IN WATER QUALITY IN THE SECONDARY TREATMENT SYSTEM

A.   Bacteria

At present, the secondary treatment system and the subsequent piping system are so advanced that an increase in bacteria is well suppressed. Present requirements call for measures to prevent bacteria from penetrating the secondary system and, in the case of bacteria penetration, to suppress their increase. To prevent bacteria from penetrating the piping system, an ultrafiltration unit that traps all bacteria is mounted as the final filter of the secondary treatment system, or the piping system is pressurized to avoid the penetration of outside air. Even when bacteria get into the piping system, they should not be allowed to increase but should be discharged immediately. Constant circulation in the piping and secondary treatment systems is employed to discharge bacteria quickly. Moreover, the UV sterilization and ultrafiltration units are mounted to remove the bacteria that do reach the inlet to the secondary system.

If bacteria are trapped in the dead legs in the piping system or adhere to the inner surface of pipes, they increase there and can contaminate the entire system. Bacterial entrapment can be eliminated by paying attention to the design and installation of the piping system, but there is no way to prevent the adhesion of bacteria to the inner surface of pipes. Even if the water velocity is raised, it cannot be guaranteed that no bacteria adhere. The bacteria effectively take the nutrients from ultrapure water and attach themselves to the inner surface. They produce adhesive materials that facilitate surface attachment or clumping together.

Figure 6 illustrates the experimental procedures used to determine the presence of live bacteria in the secondary treatment system. Figure 7 shows scanning electron microscope (SEM) images of bacteria adhered to the inner surface of a polyvinylidene fluoride (PVDF) piping system. Water was continuously supplied to the experimental apparatus for about 6 months, and after this period bacteria were injected at the inlet of ultrafiltration unit for 4 weeks to raise the bacteria content to 1–10 per milliliter. Many bacteria are observed on the inner surface of the water supply to the ultrafiltration unit. No bacteria are found in the water filtered in the ultrafiltration unit. This is because bacteria are removed during ultrafiltration. It has been proved that the piping system can be contaminated by only 1–10 bacteria in 1 ml water. Once the bacteria adhere to the surface and start to increase, the concentration in ultrapure water rises significantly.

Table 2 shows the result of hot-water sterilization. The number of bacteria on the inner surface was 6.1×105 mm^{-2}. The 1 h immersion in hot water at 80°C was not effective in removing the bacteria: 6.3×105 mm^{-2} remained on the surface. When a water flow of 0.5 m/s was applied, the bacteria decreased by half to 3.5×105 mm^{-2}. Once bacteria adhere to the surface, they remain there as dead bacteria even after sterilization. This means that even after bacteria are killed, they are released from the wall little by little, causing particle contamination. It is desirable to conduct regular measurements of bacteria in the system so that countermeasures can be instituted at an early stage.

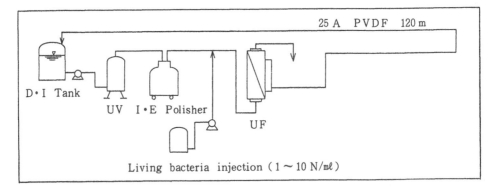

Figure 6   Living bacteria injection test flow in an ultrapure water system.

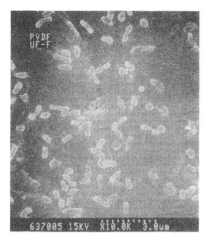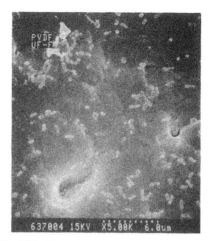

PVDF PIPE OF UF FEED

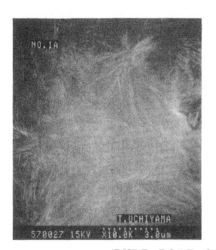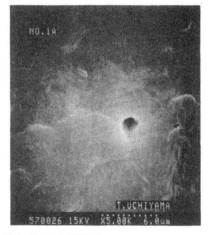

PVDF PIPE OF UF EFFLUENT

Figure 7 Bacteria adhesion to PVDF pipe surface.

Table 2 Adhesion Bacteria Rinse Test by Hot Water

	Adhesion bacteria($N/m\textrm{m}^2$)
1) Blank	6.1×10^5
2) After hot water immersion ($80°C \times 1$ hr)	6.3×10^5
3) After hot water rinsing ($80°C \times 1$ hr, 0.5 m/sec)	3.5×10^5

Sample , PVDF

SEM : $\times 8000$

Counting average number of sample

Figure 8 shows the result of comparing the frequency of hot-water sterilization and the bacteria level. The hollow fiber heat resistant ultrafiltration element is mounted on the final stage of the secondary treatment system, and the piping system employs PEEK. The ultrafiltration unit and the piping system are sterilized once a week. The bacteria levels at the outlet of the ultrafiltration unit and in the return water of the piping system are measured. As shown in Fig. 8, the bacteria are maintained at a very low level at both points.

Hot-water sterilization, H_2O_2 sterilization, and O_3 sterilization can be used in the piping system. Most prevalent in Japan is immersion into 0.5–1% H_2O_2 for 4–6 h. In North America the O_3 sterilization method is often employed.

B.   Resistivity

When water stays in the secondary treatment system and the subsequent piping system for a long time, trace ions are eluted from the inner surface and deteriorate the resistivity of the ultrapure water. No piping material is free of impurity elution. Constant water circulation in the secondary treatment and piping systems is essential to suppress the degradation of ultrapure water quality.

Figure 9 shows the change in resistivity in the secondary treatment system. In the pure water tank, the supplemental water for the primary treatment system and the return water from the piping system are mixed. If carbon dioxide in the air is dissolved in water in the pure water tank, the resistivity drops drastically and the ion load volume on the ion-exchange polisher mounted on the secondary treatment system is increased. To avoid

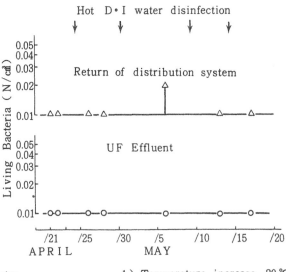

Figure 8   Hot-water disinfection of hollow fiber ultrafiltration and distribution piping system.

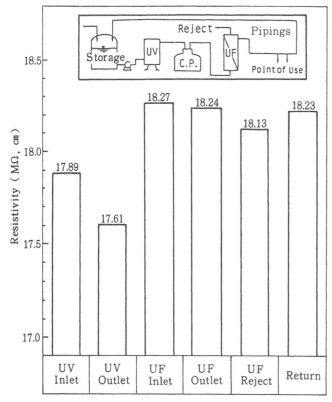

Figure 9 Actual plant resistivities.

this problem, the pure water tank is sealed with N_2 gas. By sealing the tank with high-purity N_2 gas, impurities like particles and bacteria can be shut out as well.

The UV sterilization unit in the secondary treatment system often slightly lowers the resistivity. This change is very subtle, 0.1–0.3 $M\Omega\cdot cm$, and it is frequently overlooked. The lowered resistivity is returned to a high level in the subsequent ion-exchange polisher, so it does not affect the final quality of ultrapure water. It is obvious, however, that if the UV sterilization unit is placed downstream of the ion-exchange polisher, the final resistivity will not be optimal.

In principle, the resistivity does not change in the ultrafiltration unit. In practice, however, there are examples of resistivity changes in the ultrafiltration unit. One reason is contamination of the unit. If the unit is contaminated by large numbers of bacteria, the resistivity decreases because of the carbon dioxide generated by the bacteria. This problem can be resolved by sterilization of the ultrafiltration unit.

The resistivity can rise when polymeric organic materials leak into ultrapure water. When polymeric organic materials with a molecular weight of over 100,000 are eluted from porous ion-exchange resin because of trouble in the primary treatment system, they can be removed by the ultrafiltration membrane. Since materials eluted from the ion-exchange resin contain functional groups, organic acid is eluted from the cation-exchange resin and organic base is eluted from the anion exchange resin. Although the resistivity is improved when these polymeric organic acids and bases are removed, they are adsorbed on the ultrafiltration membrane and contaminate it, which sharply reduces the volume of filtered water within a short time. This means it is not a good idea, from the viewpoint of the stable operation of the system, to use the ultrafiltration unit for the removal of polymeric organic materials.

Figure 10 shows the resistivity distribution of water processed with the relatively new secondary treatment systems. All the secondary systems have achieved a resistivity of over 18 MΩ·cm, and 70% have achieved the very high purity level of over 18.21 MΩ·cm.

C.   TOC

When ultrapure water stays in the secondary system or the subsequent piping system for a long time, the TOC concentration rises as a result of the elution of organic materials from piping materials, ion-exchange resin, and tank materials. A UV oxidation unit capable of removing TOC must be installed in the secondary treatment system, or the TOC concentration will increase even if water is circulated in the secondary treatment and piping systems.

Figure 11 shows the TOC concentration when ultrapure water is not consumed at the

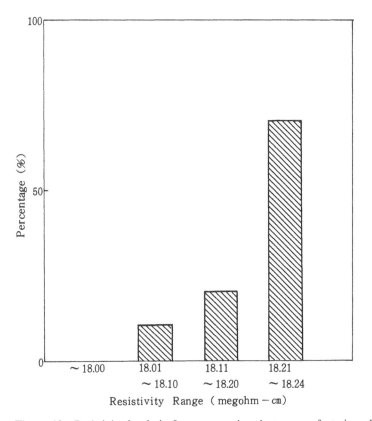

Figure 10   Resistivity levels in Japanese semiconductor manufacturing plants.

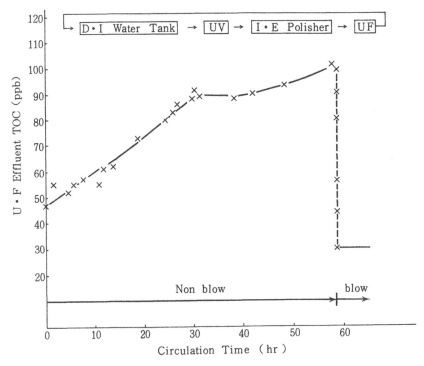

Figure 11 TOC increase in ultrapure water system.

Table 3 Ultraviolet Oxidation TOC Data for Advanced Ultrapure Water System

Plant	UV Oxidation Inlet (ppb)	UV Oxidation Outlet (ppb)	I E Polisher Outlet (ppb)	TOC Meter
A	6.8	4.4	3.4	ANATEL A − 100
B	15	−	3.0	"
C	5	−	< 1.0	"
D	20	2.3	1.4	"
E	3	−	< 1.0	TOKIKO 1000

use point but instead is circulated through the secondary treatment system and the subsequent piping system. It is obvious that the TOC concentration gradually increases. The TOC increase is attributed to elution from the tank of water processed in the primary treatment system, the ion-exchange resin polisher, and the PVDF piping system. Table 3 lists operation data for the latest secondary treatment system with a low-pressure UV oxidation unit. The TOC concentration is decreased by the UV oxidation unit to 10 ppb or less.

REFERENCES

1. K. Yabe, Technical Proceedings SEMICON EAST' 89, September 1989, pp. 16–25.
2. K. Yabe, T. Kumagai, H. Ishikawa, S. Akiyama, T. Mizuniwa, and T. Ohmi, Microcontamination, 25–30 (1989).
3. K. Yabe and T. Mizuniwa, Proceedings of 9th International Symposium on Contamination Control, Los Angeles, 1988, pp. 509–515.
4. K. Yabe, Y. Motomura, T. Ohmi, H. Ishikawa, and T. Mizuniwa, Microcontamination, 37–46 (February 1989).

5
Piping System

Takeshi Shinoda
Hitachi Plant Engineering & Construction Co., Ltd., Tokyo, Japan

I. INTRODUCTION

The ultrapure water piping system that connects the production subsystem and the use points plays an important role in maintaining water quality and pressure at optimum levels. To properly maintain high water purity, the piping materials, installation of the piping system, and sterilization methods, as well as piping system itself, are very important.

As shown in Table 1, the requirements for ultrapure water have become more stringent as the integration level of semiconductors has increased. With this in mind, the piping is considered one of the most important parts of the total system.

II. DESIGN OF THE PIPING SYSTEM

In principle, circulating water is employed in the piping to prevent the degradation of water quality caused by water entrapment. One approach to reduce entrapment is to decrease piping bends as much as possible in the design stage and to eliminate dead zone in instruments that measure pressure and mass flow, for example.

The inner surface of the piping system and tanks should be as smooth as possible to prevent the adhesion of bacteria and particles.

The circulating water method provides ultrapure water to several clean rooms through a main pipe a few hundreds meters long. Some portion, in general, 10–20% of the total volume of circulating water, of the ultrapure water must be returned constantly to the production system even though a certain amount of water is consumed at the use points. Ultrapure water supplied to each clean room is divided into many branch pipes to be supplied to hundreds of use points.

Since the entire piping system is so long and complicated, degradation in water quality is inevitable even if care is taken in the design stage. For example, the flow rate in the piping system affects the degradation in resistivity; this has something to do with the

Table 1 Example of Water Quality Requirements for Ultrapure Water

Degree of integration		256 K	1 M	4M(estimate)	16M(estimate)	Measurement method
Design rule(μm)		1.5 ~ 2.0	1.0 ~ 1.2	0.8	0.6	
Resistivity(M$\Omega \cdot$ cm at 25°C)		over 18	over 18	over 18	over 18.1	Specific resistance meter
Fine particles	diameter(μm)	0.1	0.1	0.1	0.05	SEM method
	number (number/mℓ)	less than 50	less than 10	less than 10	less than 10	
Live bacillus(number/mℓ)		less than 0.05	less than 0.01	less than 0.01	less than 0.005	M-TGE culture method
TOC(μg/ℓ)		less than 50	less than 30	less than 10	less than 5	Wet oxidation TOC meter
Dissolved oxygen meter(μg/ℓ)		less than 100	less than 100	less than 50	less than 10	Dissolved oxygen meter
SiO$_2$(μg/ℓ)		less than 10	less than 10	less than 5	less than 1	
Cr(μg/ℓ)				less than 0.007	less than 0.007	
Fe(μg/ℓ)		less than 1	less than 0.1	less than 0.003	less than 0.003	
Mn(μg/ℓ)				less than 0.05	less than 0.05	
Zn(μg/ℓ)				less than 0.02	less than 0.02	Ion chromatograph
Cu(μg/ℓ)		less then 1	less than 0.1	less than 0.002	less than 0.002	
Na(μg/ℓ)		less than 1	less than 0.1	less than 0.1	less than 0.1	
K (μg/ℓ)				less than 0.1	less than 0.1	
Cl(μg/ℓ)		less than 1	less than 0.1	less than 0.1	less than 0.1	

elution of impurities. In general, the flow rate should be set at 1.5–3.0 m/s to minimize the decrease in resistivity and the adhesion of particles and bacteria. The pipe diameter is critical to maintain this optimum flow rate. Table 2 shows the relationship between the flow rate in the piping system and the deterioration in the resistivity.

Another factor that degrades the water quality is the concentration of dissolved oxygen. Lately this factor has attracted more attention because it affects the SiO$_2$ thickness in the semiconductor production process. Figure 1 compares bacterial generation in degassed and nondegassed water trapped in the system.

It is important in piping design to eliminate water entrapment in the system and to minimize the total length of piping.

Table 2 Relation Between Water Quality and Water Velocity

	Velocity (m/sec)	Resistivity (M$\Omega \cdot$ cm at 25°C)		
		Outlet of ultra-pure water plant	50m outlet	100 m outlet
Continuous operation	1.6	18.0	17.9	17.9
	0.8	18.0	17.5	17.0
Intermittent operation	1.6	18.0	16.0	15.0

(NOTE) Piping material : PVC

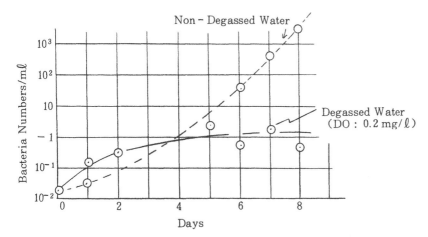

Figure 1 Effect of degassed water on bacterial growth.

1. The flow rate of ultrapure water in the piping should be set at 1.5–3.0 m/s.
2. The length of piping between production and use points should be minimized.
3. Every route should have almost the same length.
4. The consumption volume at each use point should be clearly articulated in the course of piping design.
5. In particular, for the reverse return system, the pressure in the water supply pipe should be different from that in the return pipe to prevent ultrapure water, once it flows through a use point, from entering the water supply pipe.

III. CHARACTERISTICS OF DIFFERENT PIPING SYSTEMS

Figure 2 shows the piping system most frequently employed in the initial stage. This is called the one-way system: a single loop of pipe from the ultrapure water production unit travels around to each use point and returns to the ultrapure water tank. At each use point, the branch pipes receive ultrapure water from the loop pipe for supply to the equipment at the use point, for example the clean bench. This system has been adapted to a small clean room supplied by a loop pipe of less than 2 in. and with no more than 10–15 use points.

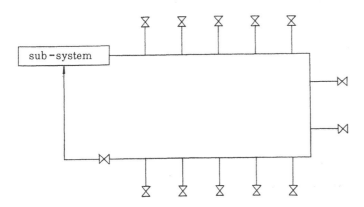

Figure 2 Piping system at point of use (one variation).

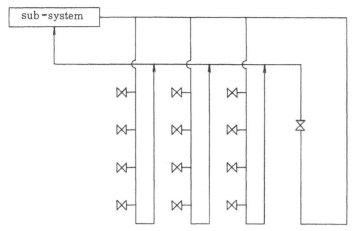

Figure 3   Piping system at point of use (two variations).

The problem with this system is that the branch pipes at the use points create water entrapment. In particular, branch pipes that are not frequently used have bacterial generation that results in a serious degradation in purity of the ultrapure water. This system is therefore not adequate to supply use points that require ultrapure water of high purity.

Figures 3 and 4 show improved versions of the one-way system. The length of the branch pipes is minimized as much as possible to suppress water entrapment. The piping network, however, is more complicated than the one-way system shown in Fig. 2.

Figure 5 shows the system that separates the water supply pipe from the water return pipe, in contrast to Figs. 3 and 4. In this system, the two parts are separated and the pressure in the two parts is different so that ultrapure water, once it flows through a use

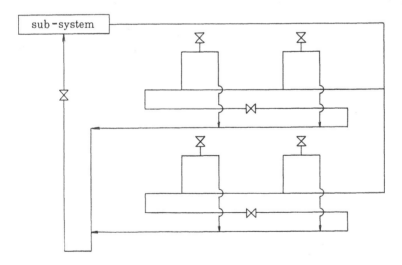

Figure 4   Piping system at point of use (three variations).

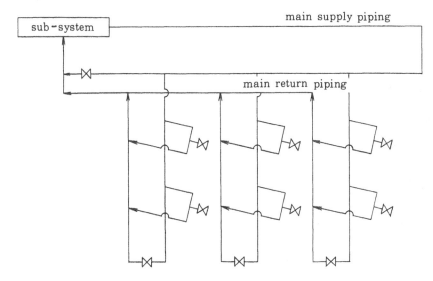

Figure 5   Piping system at point of use (four variations).

point, is prevented from flowing into another use point. The system shown in Fig. 5 is the full reverse return system and lately has been widely adopted in large clean rooms.

IV. IMPORTANT POINTS OF PIPING SYSTEM INSTALLATION

To maintain the high purity of the ultrapure water in a long piping system, the installation method, in particular the joint method, should be carefully selected. There are two joint methods: the adhesion joint and the fusion joint. The adhesion joint method was formerly widely employed, but the fusion joint method is now replacing it. This is because the adhesive used in the adhesion joint method is inevitably eluted into the ultrapure water, causing a increase in the total organic carbon (TOC) concentration.

A. Adhesion Joint

1. The pipe end caps to maintain the cleanliness of the pipes should be taken off at the very last minute.
2. Extra joints that are not used immediately should not be taken out of the package.
3. The pipes should be cut at a right angle. The cut edge should be planed off.
4. The reference line should be noted, and the zero point should be confirmed.
5. The glue should be specified by manufacturer.
6. The area to be adhered should be cleaned with absorbent cotton and acetone.
7. The glue should be carefully applied to the inner surface of joints in a thin and uniform manner. It should not be applied to the inside of the pipes. Somewhat more glue should be applied to the outside of pipes.
8. After joining the two parts, they should be supported for 1 minute in summer and more than 2 minutes in winter.
9. At room temperature, the joint part should be carefully treated for more than 24 h. When the temperature is low, however, it should be carefully watched for more than 72 h.

10. Joining work should not be done when the temperature outside is lower than 5°C (or a heater should be used).
11. The joint jigs that touch the area to adhere should be fully cleaned of oil.
12. A pair of clean gloves should be used.
13. Adhesive and acetone should be kept away from fire.

B.   Fusion Joint Method

1. The pipe end cap to maintain the cleanliness of the pipe should be taken off at the very last minute.
2. Extra joints should not be taken out of the package.
3. The pipe should be cut at a right angle. The cut edge should be planed off.
4. Conditions for the fusion joint, including heater temperature, fusion time, and length of pipe to be inserted, should be observed fastidiously.
5. Fusion joint work should be conducted where no wind effect is expected.
6. The parts should be inserted within 5 s after completion of the fusion.
7. Fusion joint work should be conducted by more than two operators.
8. The joint jigs to touch the area to adhere should be fully cleaned so that they are free of oil.
9. A pair of clean gloves should be used.
10. No people other than the operators should operate the fusion machine. The operator should be responsible for the key to the jet switch to start the oil hydraulic pump.
11. The fusion machine should be adjusted with the oil hydraulic pump turned off.

The method of supporting the piping system depends on the piping materials. Tables 3 through 5 show the average support distances for major piping materials.

V.   STERILIZATION

To maintain the water quality at the use points, the piping system should be sterilized on a regular basis. Generally the operation should be suspended twice a year to carry out H_2O_2 sterilization. Recently the hot-water sterilization method at 80–90°C has also been adopted as the water quality required at the use points has increased. To adopt these sterilization methods, the piping materials should be heat resistant. The sterilization effect and recovery of water quality after sterilization are discussed for hot-water and hot H_2O_2 sterilization.

Additional pipe was connected to the piping of an ultrapure water production unit for test purposes. Ultrapure water was supplied to the additional pipe for 2 months. This pipe was cut into pieces in the clean room to prepare samples. The samples were immersed in ultrapure water that was free of bacteria. Table 6 shows the results of experiments in which the samples were treated under different conditions of H_2O_2 concentration, sterilization temperature, and sterilization time.

H_2O_2 sterilization at room temperature is effective when the H_2O_2 concentration is higher than 0.5%. For hot H_2O_2 sterilization, however, the same sterilizing effect can be obtained with a H_2O_2 concentration of 0.25% for 1 h or 0.1% for 3 h when the H_2O_2 temperature is higher than 40°C.

H_2O_2 should be treated with sufficient blow down; otherwise, it remains an ion and directly affects the resistivity.

Table 3 Support Distance: PVDF

Unit : cm

Diameter(A) / Max Temp.	13	16	20	25	30	40	50	65	75
~ 40°C	80	90	95	100	115	130	140	155	165
~ 60°C	75	80	90	95	110	120	130	140	155
~ 80°C	70	75	85	90	100	115	120	130	145
~ 100°C	65	70	80	85	95	110	115	125	135
~ 120°C	60	65	75	80	90	100	105	115	125

Table 4 Support Distance: HT-PCV.

Unit : cm

Diameter(A) / Max Temp.	13	16	20	25	30	40	50	65	75	100	125	150
~ 60°C	60	65	70	75	80	90	95	105	115	135	140	150
~ 80°C	55	60	65	70	75	85	95	100	110	125	130	140
~ 90°C	55	60	60	70	75	80	90	100	110	125	130	135

Table 5 Support Distance: PVC

Unit : cm

Diameter(A) / Max Temp.	13	16	20	25	30	40	50	65	75	100	125	150
Ambient Temp.			100				150			200		

The hot H_2O_2 sterilization method is also effective at low H_2O_2 concentrations. In hot-water sterilization at 80°C, the sample is well sterilized in 2 h.

Recovery of water quality after sterilization was also studied. The sterilization conditions were changed as shown in Table 7. First, the pipe was sterilized as indicated and the sterilization solution was drained from the system. The pipe was then treated by continuous blow down ultrapure water. Figure 6 shows the resistivity as a function of blow down time. In H_2O_2 sterilization at room temperature, it takes 4 h to lower the resistivity of the blow down water to 16 MΩ·cm. On the other hand, it takes only 2.5 h in hot H_2O_2 sterilization. This is thought to be because the concentration of H_2O_2 giving rise to the residual ion is low in hot H_2O_2 sterilization.

In hot-water sterilization free of chemicals, the resistivity recovers more quickly than in the preceding examples.

The stability of the water quality (the TOC concentration) was also studied when normal operation started followed by ultrapure water blow down. Figure 7 shows the result. The vertical axis corresponds to the gap in the TOC concentration between the use point and the outlet of the ultrafiltration unit. In other words, the vertical axis indicates the amount of eluted TOC. In H_2O_2 sterilization at room temperature and hot H_2O_2 sterilization, TOC returns to the stable low level in 4 h after the operation resumes. In hot-water sterilization, however, the TOC concentration is raised, probably as a result of elution from piping materials triggered by the high temperature, resulting in deterioration in the stability of the water quality.

Table 6   Relation of Bacterial Growth and Disinfection Conditions

Disinfection condition		Culture days (d)	Disinfection duration(h)		
H_2O_2 (vol %)	Temp (°C)		1	2	3
0.75	25	3	O	O	O
		7	O	O	O
		10	+	O	O
0.5	25	3	+	O	O
		7	+	+	O
		10	+	+	O
0.25	40	3	O	O	O
		7	O	O	O
		10	O	O	O
	60	3	O	O	O
		7	O	O	O
		10	O	O	O
0.1	40	3	+	+	O
		7	+	+	O
		10	+	+	O
	60	3	+	+	O
		7	+	O	O
		10	+	O	O
0	60	3	+	+	O
		7	+	+	O
		10	+	+	O
	80	3	+	O	O
		7	+	O	O
		10	+	O	O

(Note) + : Detect, O : Non-detect
(Detect limit= 0.05 Numbers/cm^2)

Table 7   Disinfection Conditions

Disinfection Method	Temp(°C)	H_2O_2 conc.(%)	Disinfection Time(H)	Piping Materials
H_2O_2 + Hot water	40	0.1	3	C - PVC
H_2O_2 + water	25	0.75	3	C - PVC
Hot water	80	—	3	HT - PVC

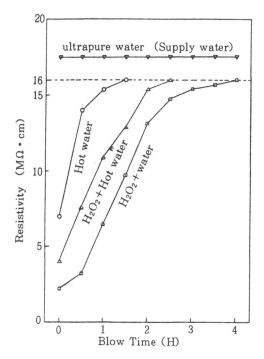

Figure 6 Relation of resistivity to blow down time.

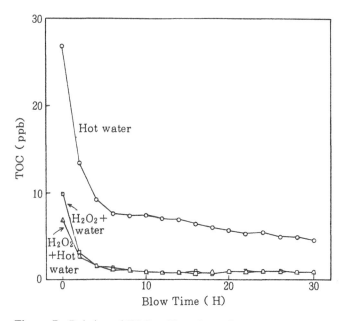

Figure 7 Relation of TOC to blow down time.

6
Reuse of Wastewater

Riichi Ikegami
Japan Organo Co., Ltd., Tokyo, Japan

I. INTRODUCTION

Current semiconductor technology is making incredibly rapid progress. Capital investment has also been active in a bid to expand production facilities and to improve the integration of chips. Each process in semiconductor production requires ultrapure water in large quantities to rinse off the impurities and chemicals that adhere to the wafer surface.

As the mainstream in VLSI (very large scale integrated) circuit production shifts from 1 to 4 Mbit, the requirements for ultrapure water quality are further toughened. The consumption of ultrapure water has also significantly increased as more processes are introduced in semiconductor production and as the volume increases.

The restrictions on the supply and discharge of water are becoming more stringent, including restraints on industrial and city water supply, the contamination level of wastewater, and the total amount of discharged water. Therefore, a secure water supply for plants is one of the major challenges to further expansion of production in the semiconductor industry.

This chapter describes ways of recycling and reusing ultrapure water from the rinsing process, which is certain to become more important as a water source in the future.

II. OUTLINE FOR RECYCLING WASTEWATER

A. Advantages

In the past, most of the ultrapure water used at use points was treated for discharge outside plants, but recently there have been an increasing number of factories in which the wastewater is recycled and reused. This is mainly because the reuse of water can cut costs and resolve problems caused by the limitations and restrictions on water.

Recycling can cut the costs of supply water. When city water is used to produce ultrapure water, it costs more than ¥150–200/m³. The cost of industrial water also exceeds ¥100/m³ in some cases. On the other hand, the cost of reclaiming waste rinsing water (water less contaminated) is only ¥50–70/m³. Even when the waste rinsing water is treated by ultraviolet (UV) oxidation to decompose organic materials, the cost remains at the ¥150–200/m³ level.

Recycling can cut the costs of treating wastewater for discharge. When wastewater is discharged to sewage, the restrictions on its contamination level are not very tough, which keeps the wastewater processing cost low, but factories are supposed to pay ¥150–200/m³ as the discharge cost.

On the other hand, when wastewater is discharged to rivers or oceans, the contamination level of the discharged water is set increasingly lower to deal with the problems of eutrophication and otherwise to protect the environment. Fluorine sometimes requires advanced treatment, and the restrictions on nitrogen and phosphorus are being introduced in an increasing number of places. This trend raises the initial cost of installing wastewater treatment facilities and the cost of treating, but reusing wastewater can lower these costs by decreasing the amount of wastewater discharged outside plants.

Recycling can cope with a limited supply of water. As more and more water is used, various restrictions are and will be imposed in terms of the water supply: restraints on the pumping of underground water and limits to the capability of facilities to supply city and industrial water. Whether wastewater can be reused to reduce the amount of fresh water, will greatly affect the production program of the entire factory.

For protecting the environment, restrictions are imposed on the total amount of wastewater discharged from a plant. When a new plant is built or a plant is expanded, approval to discharge additional wastewater sometimes cannot be obtained, which forces the plant to recycle the wastewater.

Also, when wastewater discharge is completely prohibited to protect the environment, a closed system must be introduced that does not discharge wastewater.

B. Chemicals Used in the VLSI Production

Table 1 lists some of the chemicals used in semiconductor production. It is likely that some of them find their way into the waste rinsing water to be reclaimed. In accordance with changes in the production process and the development of new chemicals triggered by the further integration of chips, many other chemicals are expected to be contained in the waste rinsing water in the future.

Inorganic acid and alkali chemicals can be adsorbed and removed by ion-exchange resins. Problems can arise when waste chemicals in high concentration are mixed in the wastewater to be reclaimed: even wastewater with a lower concentration is further contaminated by mixing with high-concentration waste chemicals, and the reclamation efficiency deteriorates since the frequency of regeneration of ion-exchange resins rises because of the large amount of ions. In the usual situation in which only the chemicals that adhere to wafers are mixed into the waste rinsing water, however, the inorganic ions can be easily removed merely by adjusting the design ionic load on the ion-exchange resins.

More critical chemicals are organic chemicals that are the components of TOC (total oxidizable carbon: *oxidizable* is used here instead of *organic* because TOC analyzers commonly used at present measure only those components that can be oxidized and decomposed to the level of CO_2 with UV). Low-molecular-weight organic solvents are

Table 1 Chemicals Used in Semiconductor Manufacturing

Inorganic Chemicals	Acids	Hydrofluoric Acid	HF
		Hydrochloric Acid	HCl
		Nitric Acid	HNO_3
		Sulfuric Acid	H_2SO_4
		Phosphoric Acid	H_3PO_4
	Alkalis	Ammonia	NH_4OH
		Potassium Hydroxide	KOH
		Sodium Hydroxide	NaOH
	Salts	Ammonium Fluoride	NH_4F
		Sodium Phosphate	Na_3PO_4
	Oxidizing Agent	Hydrogen Peroxide	H_2O_2
	Reducing Agents	Hydrazine	NH_2NH_2
		Ammonium Persulfate	$(NH_3)_2SO_5$
Organic Chemical	Organic Acid	Acetic Acid	CH_3COOH
	Organic Alkalis	TMAH	$N(CH_3)_4OH^+$
		Choline	$[HOCH_2CH_2N^+(CH_3)_3]OH$
	Alcohols	Methyl Alcohol	CH_3OH
		Ethyl Alcohol	C_2H_5OH
		IPA	$CH_3CH(OH)CH_3$
	Organic Solvents	Acetone	CH_3COCH_3
		Methyl Ethyl Ketone	$CH_3COCH_2CH_3$
		Ethyl Acetate	$CH_3COOCH_2CH_3$
		Butyl Acetate	$CH_3COO(CH_2)_3CH_3$
		Xylene	$C_6H_4(CH_3)_2$
		Phenol	C_6H_5OH
		Carbon Tetrachloride	CCl_4
		Trichloroethylene	$CHClCCl_2$
		Tetrachloroethylene	CCl_2CCl_2
		Butyl Cellosolve Acetate	$CH_2OHCH_2OHOC_3H_6COCH_3$
		Triethylene glycol	$CH_3OC_2H_4OC_2H_4OC_2H_4OH$
		Ethylenediamine	$NH_2CH_2CH_2NH_2$
	Surface Active Agents	Non-ionic Surface Active Agent	
		Amphoteric Surface Active Agent	

often used in semiconductor production. Some of them are surfactants that may plug the reverse osmosis membrane.

Other than these chemicals, residues of the etching process and reaction products that adhere to jigs may find their way into waste rinsing water in small quantities. To make the wastewater reclamation efficient, it is important to suppress the contamination by these impurities.

C. Sources of Wastewater and Methods to Reclaim It

The preceding chemicals are versatile in use and purpose. Usually it is possible to group certain chemicals used in each process. In the VLSI production line, the following four processes use and drain ultrapure water:

1. Photolithography: developers, such as tetramethylammonium hydride (TMAH) and choline
2. Pretreatment and wet etching: inorganic acids and alkalis and oxidizing agents, such as hydrogen peroxide
3. Subprocesses, such as maintenance and jig cleaning: various chemicals
4. Backgrind and dicing process: Si particles are generated

This chapter does not discuss the treatment or disposal of waste chemicals drained at the time of replacement of chemicals in chemical baths, which are usually handled by a contractor. Even in these four processes, the concentration and combination of chemicals vary in accordance with their use.

To successfully treat and reuse waste rinsing water, it is necessary to fully understand its characteristics and any variations to provide appropriate treatment. The following details the characteristics of wastewaters drained from the four main processes.

1.   Wastewater from Photolithography

In this process a wafer is coated with photoresist. Circuit patterns are printed with a stepper, the exposed photoresist is stabilized with developer, and the unwanted photoresist is removed from wafer surface to form circuit patterns. The wastewater drained from this process contains such organic developers as TMAH and choline at a few thousand ppm, as well as the photoresist.

These developers are organic materials, which are hard to decompose. Organic materials can be treated biologically, however, provided that the wastewater is not contaminated with any substances inimical to microorganisms, such as surface-active agents. Since the total amount of this wastewater is small, however, it is more economical to have a contractor handle them.

When this wastewater with low concentrations of chemicals is drained in large quantities, the chemicals can be adsorbed with a strongly acid cation-exchange resin. It is a good idea to adsorb and concentrate them on the ion-exchange resin and to have a contractor handle the concentrated wastewater resulting from regeneration of the ion-exchange resin.

2.   Wastewater from Pretreatment and Wet Etching

In this process photoresist and unnecessary by-products of etching in chemical baths are removed from the wafer. In the VLSI production line, this process uses the largest amount of ultrapure water, which means this is where a large amount of wastewater is available as source water for reclamation.

Inorganic acids and alkalis and oxidizing agents are used in reclamation. Although IPA (isopropyl alcohol) is also employed in some cases, it is not directly mixed into the wastewater because it is used in the drying step after rinsing. (Some IPA can still be mixed through the wafer-carrier transportation chuck.) Surfactants are also used to remove photoresist. Since the surfactants can plug the reverse osmosis membrane because they are not decomposed even under UV oxidation, it is better not to reuse wastewater containing such surfactants as source water for ultrapure water.

The chemicals and impurities that adhere to the wafer in the rinsing bath and are then mixed into the waste rinsing water are almost completely rinsed off during the initial rinsing. As rinsing progresses, the chemicals and impurities are significantly decreased. The wastewater therefore contains fewer impurities at the later stage of rinsing. When only the wastewater drained from the later rinsing stage is recycled, it can be easily treated without employing special methods, which greatly cuts costs.

When a large amount of wastewater must be recycled because of a limited water supply to plants, the wastewater drained from the earlier rinsing stage must be reused as well. This wastewater contains many chemicals and impurities and should not be recycled to the ultrapure water production unit to stabilize the ultrapure water quality. This wastewater should be used for general purposes, for example cooling water for air conditioners.

To reuse as much water as possible and to stabilize the wastewater recycling, the waste rinsing water should be separately collected in the earlier rinsing stage and the later rinsing stage, which requires separate drain piping systems. Counters should be used to count the numbers of rinsings in a batch rinsing method, such as the quick dump rinse, and timers should be used to measure the rinsing time in a continuous rinsing method, such as overflow rinse, to switch the paths of drain water flow by automatic valves. The timing of path switching should be fully studied because it should vary with the rinsing methods and chemicals used.

In treating the waste rinsing water, the following unit processes are combined in different ways:

Activated carbon adsorption: absorption and removal of organic matter, decomposition of oxidizing agents
Ion exchange: adsorption and removal of inorganic ions, organic acids, and organic matter
UV oxidation: decomposition of organic matter
Reverse osmosis: separation and removal of organic matter and fine particles

The details of combinations of these unit processes are described in Sec. 3. Table 2 indicates the performance of each unit process in removing contaminant chemicals.

Table 2   Removal of Contaminant Chemicals by Unit Processes

			Activated Carbon Adsorption	Ion Exchange	UV Oxidation	RO
Inorganic Chemicals	Acids	HF, HNO_3 and others		○		
	Alkalis	NH_4OH and others		○		
	Salts	NH_4F and others		○		
	Oxidizing Agent	H_2O_2	○			
Organic Chemical	Organic Acids	CH_3COOH and others		○		
	Organic Alkalis	TMAH and others		○		△
	Alcohols	CH_3OH IPA and others			○	△○
	Phenol		○			
	Ketones	Acetone and others			○	
	Chlorinated Compounds	Trichloroethylene and others	○			
	Cellosolves				△	○
	Surface Active Agents		○			

3. Wastewater from Subprocesses

In these processes, ultrapure water is used to clean wafer carriers and the components of semiconductor production equipment, such as the chemical vapor deposition (CVD) tube, to remove adhered particles and reaction products. The use of such organic chemicals as organic solvent requires extra attention. Although the proportion of drain water containing such organic materials is estimated as small compared to the total amount of drain water, it raises the TOC level of the composite wastewater for recycling when it is mixed with other wastewater. It is clear that a higher TOC level is undesirable because TOC leads to bacterial propagation and because the initial and running costs of facilities to reduce the TOC level are very high. Wastewater containing organic chemicals should therefore be processed and discharged or treated separately from the ultrapure water production line and used for general purposes.

To reduce the TOC level in wastewater containing organic chemicals, a new process has been developed in which the organic materials contained in wastewater are digested by bacteria grown on a packed bed.

There is no difficulty in treating wastewater containing small amounts of organic chemicals just like wastewater from the pretreatment and wet etching process (mentioned previously). Since these subprocesses are not carried out in any automatic, standardized, and continuous mode of operation, however, grouping wastewaters with different chemical concentrations is difficult. In this sense, it is better not to recycle these wastewaters to the ultrapure water production line.

4. Wastewater from Backgrinder and Dicer

The back of the wafer is precisely ground in the backgrinder process, and the wafer is cut to chip size in the dicing process. The ultrapure water drained from those processes contains ground Si particles. Most of the Si particles are less than 1 μm in size. The Si concentration is 200–250 ppm in the wastewater from the backgrinder process and 20–30 ppm in the wastewater from the dicing process.

It is rather difficult to remove fine Si particles with conventional technology. At present Si-bearing wastewater is conventionally treated after removing Si particles with a filter cloth using cationic and anionic coagulant aids. In some plants, the Si particles are concentrated and removed in the ultrafiltration membrane (UF unit), and then the permeate water is recycled.

When using the ultrafiltration membrane, it should not be contaminated with any other wastewater or chemicals. Otherwise, scaling by potassium silicate and the like forms on the membrane surface. The metallic silica particles are concentrated in the water rejected by the ultrafiltration membrane, and hydrogen is generated when the metallic silica and water react with each other. A tank that stores the reject water for a long time must be equipped with appropriate ventilation to prevent explosion. This reaction proceeds more rapidly on the alkali side.

Another method of removing fine Si particles is by coagulation and sedimentation and filtration, both units adopted in a closed system. Specifically, the Si-bearing wastewater is added to wastewater containing high concentrations of SS, such as filter backwash waste, which is coagulated by use of a coagulant, such as PAC, and filtered in the closed water and waste treatment system.

Figure 1 shows a typical example of a water supply and drain system in an integrated circuit (IC) plant, including scrubber blowdown and cooling tower blowdown, as well as the drains from respective points of use.

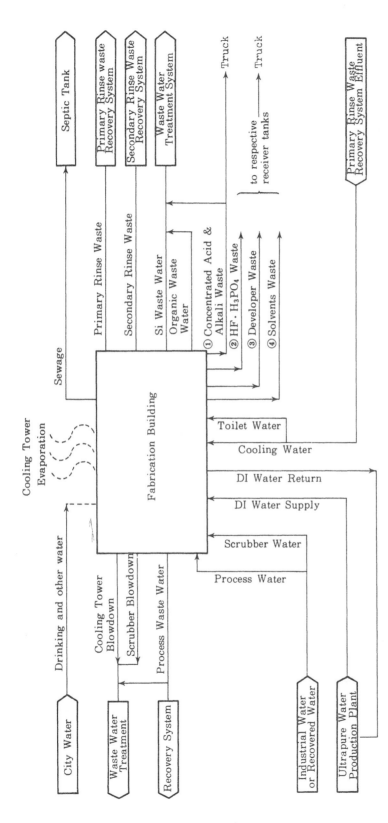

Figure 1 Classification of water and wastewater in an integrated circuit (IC) fabrication building.

D.   Uses for Recycled Water

There are mainly two ways to reuse reclaimed water: in the ultrapure water production unit for general plant purposes, such as makeup water for the cooling tower and scrubber. When the wastewater is recycled to the ultrapure water production unit or restraints on the water supply dictate a higher wastewater recovery ratio, the wastewater should be fractionated according to its nature and each fraction should separately be treated.

1.   Secondary Rinse Waste

This is wastewater drained from the latter half of the rinsing stage in the first rinsing bath of the wet etching process plus wastewater from the second bath of the wet etching process. It is virtually deionized, containing little chemical.

This secondary rinse waste is reused as source water in the ultrapure water production unit. Typically, when the recovery ratio of ultrapure water used in the wet etching process is kept below 50%, the secondary rinse waste has an electrical conductivity of 10 μS/cm or less and a TOC of 0.1 ppm or less. In this case, the secondary rinse waste can be recycled to the ultrapure water production unit without pretreatment.

2.   Primary Rinse Waste

This is wastewater drained from the first half of the rinsing stage in the first rinsing bath of the wet etching process plus wastewater drained from subprocesses, such as jig cleaning. This wastewater contains the residues from wafer etching and the reaction products that adhere to jigs, as well as chemicals, so that treating it to a quality level high enough for reuse as source water for the ultrapure water production unit is too expensive.

Although it is possible to process primary rinse waste with ion exchange, reverse osmosis, and UV oxidation to raise its quality level enough so that it can be recycled to the ultrapure water production unit, it is recommended to reuse this primary rinse waste for a cooling tower or scrubber, which consumes water in large volume. The only treatment then required for the wastewater is ion exchange, which removes inorganic acids and alkalis from it, thereby producing deionized water of low quality.

Because it contains organic materials, such as IPA, the deionized water obtained from the primary rinse waste may become a breeding ground for bacteria. To prevent bacterial propagation, organic slime control agents, which are relatively expensive, are usually used as sterilizers. Although it is less expensive, sodium hypochlorite should not be used since it corrodes piping and other materials in deionized water containing almost no ions, such as calcium.

When it is to be used as cooling water, deionized water reclaimed from the primary rinse waste should be used only as makeup for cooling towers or indirect-cooling systems. This deionized water, which may give rinse to bacterial contamination, should not be used for any direct-cooling systems.

3.   Si Wastewater

This is the wastewater drained from such processes as backgrinding and the dicing. Fine silica particles are filtered by cloth or concentrated and removed by ultrafilitration membrane. The resultant reclaimed water is reused for the backgrinding and dicing process or for miscellaneous purposes.

4.   Organic Wastewater

Organic wastewater is generated in such subprocesses as jig cleaning. It must be separately fractionated and treated for discharge or reuse as cooling water and other miscellaneous purposes.

5.   *Modification of Conventional Facilities to Recycle and Reuse Wastewater*

When conventional facilities to produce ultrapure water and treat wastewater are to be modified to recycle and reuse the plant wastewater, it is often the case that there is no space available at points of use where rinse wastewater can be fractionated. Treating and reusing the rinse wastewater as source water for ultrapure water production without fractionating it according to its contaminants requires such expensive unit processes as reverse osmosis and UV oxidation. If this is the case, the rinse wastewater is better used entirely for miscellaneous purposes after simple treatment, such as coagulation and sedimentation.

E.   Handling Chemical Wastes at the Time of Replacement

What follows is a description of the characteristics and disposal of the five groups of waste chemicals that are used repeatedly in the wet etching process and the like.

1.   *Developer Waste*

As mentioned in Sec. II.C.1, this waste contains such developers as TMAH and choline, which are hard to decompose. In most cases, it is disposed of by a contractor.

2.   *Acid Waste Containing Fluorine*

This waste contains fluorine at concentrations of a few thousands to a few tens of thousands ppm. To remove fluorine from this waste, a calcium compound, such as slaked lime, is added, precipitating fluorine in the form of CaF_2 by effecting coagulation and sedimentation at neutral pH. The CaF_2 precipitate is separated by dehydrator. The supernatant from the dehydrator is mixed with other wastewaters and treated for discharge. Separating and removing fluorine in this manner entails considerable chemical and equipment costs. Besides, the dehydrator, which is liable to scaling, requires a large amount of maintenance. Hence, it is more economical to have the fluorine-bearing acid waste disposed of by a contractor.

3.   *Acid Waste Not Containing Fluorine*

This is strong acid waste containing mainly sulfuric acid. This waste can be discharged after neutralization alone in Japan because there are no discharge limits for salt concentration.

When it is mixed with wastewater containing fluorine that is treated for fluorine removal, calcium sulfate is generated through reaction of the calcium with sulfuric acid. This calcium sulfate can become an added load on the dehydrator. Moreover, residual hydrogen peroxide remaining in the waste may pose a problem: it gives rinse to foaming, thereby curling the flocs formed in the coagulation and sedimentation step. Where COD discharge limits are more stringent, any such residual hydrogen peroxide must be removed before discharge. Where there are discharge limits for ammonia and nitrate nitrogen, the acid waste before discharge should be subjected to nitrification and denitrification in addition to neutralization.

In semiconductor plants where there is no closed water or wastewater treatment system, acid wastewater containing no fluorine is disposed of by a contractor. In some plants sulfuric acid is recovered at a 50% or higher solution and reused for acid cleaning of steel.

4. *Ammonia and Other Alkali Wastes*

In most cases, these alkali wastes are drained from the same sources as the acid wastes described earlier. Where there is no discharge limit for nitrogen, including a nitrogen discharge limit for sewage, these alkaline wastes may be discharged after neutralization just as the acid waste containing no fluorine mentioned earlier. If there is discharge limit for nitrogen, the alkaline wastes can be treated separately by the ammonia breakpoint method in which ammonia in the wastes is oxidized and decomposed by sodium hypochlorite.

5. *Organic Solvent Wastes*

These wastes mainly consist of alcoholic solvents and can best be disposed of by incineration. Since this incineration is costly, these wastes are usually handled by a contractor. When installing storage tanks for these wastes, due care should be used because these storage tanks are dangerous objects under the Fire Prevention Act.

III. SYSTEM COMPONENTS AND THEIR FUNCTION

A. Fractionation of Wastewaters

1. *Fractionation at the Source*

To treat wastewaters separately according to their characteristics, they must be fractionated at the source or as they are drained from the production line. Unless these wastewaters are fractionated in an orderly manner, the treatment of each fraction of wastewater does not go well.

   To fractionate wastewaters automatically at the wafer rinsing step after the wet etching process, automatic valves are installed at the drain pipes of rinsing baths, as shown in Fig. 2. In the multiple-bath rinsing system illustrated in Fig. 2, the fractionation of wastewaters is based on the movement of the water carrier: a rinsing counter is used with a batch rinsing system, such as the quick dump rinse, and a timer is used with a

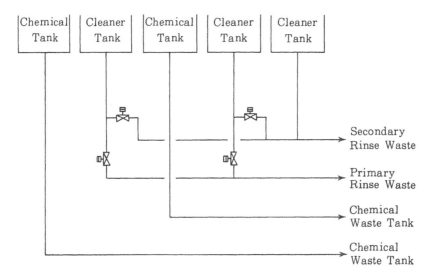

Figure 2 Routes for cleaning (rinse) wastes.

continuous rinsing system, such as overflow rinse. The direction of flow of each wastewater is switched by automatic valves that work in unison with the rinsing counter or timer. The same mechanism works in the one-bath automatic rinsing unit. The timing of switching wastewater flows needs to be studied in detail because it greatly depends on the rinsing method and the chemicals used. The automatic valve switching should be operated so as not simultaneously to close two valves on the same drain line.

To secure the optimum switch timing, the use of a conductivity meter has been proposed that dictates valve switching depending upon the conductivity of the wastewater. A conventional conductivity meter with platinum cells is not suitable for this purpose because platinum is attacked by fluorine and oxidizing agents. Cells resistant to oxidation have now been developed.

To establish a successful wastewater recycling system, careful consideration should be given to the fractionation of wastewaters during the planning and design of the semiconductor production line and machinery. Otherwise, even well-designed wastewater treatment systems cannot work as expected.

Another point to be noted is that wastewater fractionation facilities require a larger space than the conventional drain system because more drain pipes need to be installed. Furthermore, the piping space should be designed so that there is sufficient room for future modifications.

2. *Fractionation at the Recovery Inlet*

This fractionation is designed to guard against the possibility of wastewater of abnormal quality flowing into the wastewater recovery system in question. This fractionation is not required if the fractionation of wastewater at the source is effected appropriately. The electrical conductivity, pH, TOC, and other quality indices are monitored to effect this fractionation.

a. Electrical conductivity. This index is monitored to prevent wastewater of extraordinarily high concentrations from being fed to the wastewater recovery system. If an abnormal upswing in the electrical conductivity of the wastewater as a result of the inflow of acid or alkali is sensed by the conductivity meter, the wastewater is routed not to a recovery system but to a treatment system for discharge.

b. pH. The electrical conductivity of hydrogen ion (H^+) and hydroxyl ion (OH) is much greater than that of other ions. Thus the electrical conductivity of wastewater is largely affected by its pH. Assuming wastewaters with much the same concentrations of impurity ions, the ionic conductivity is greater at low or high pH than at neutral pH. Hence one cannot accurately assess the load of impurity ions in wastewater by conductivity alone. That is, the load of impurity ions in wastewater should be monitored by both electrical conductivity and pH. In practice, however, most chemicals used in the semiconductor production line are strong acids, which usually acidify the waste rinsing water. Therefore pH monitoring can be omitted once the wastewater pH is checked after the plant starts operating.

c. TOC. As mentioned earlier, utmost care should be paid to the TOC if the rinse wastewater is to be recycled to the ultrapure water production line. As long as the fractionation of wastewater at its source is carried out in an orderly manner, there is no possibility of wastewater rich in TOC flowing into the ultrapure water production line. To cope with unexpected trouble or operational failures, however, a TOC meter should be equipped to route wastewater rich in TOC to a wastewater treatment unit for discharge.

Currently available TOC meters with a measurement range of ppm order take 10

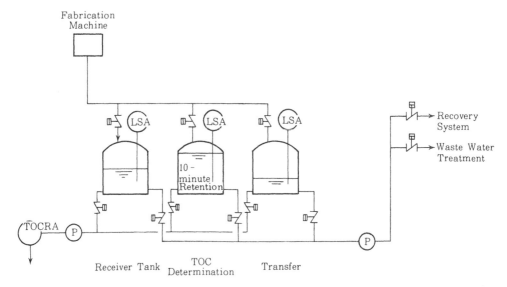

Figure 3   Fractionation of recovered wastewater by TOC.

minutes to determine a TOC value. In other words, a TOC value read now on a TOC meter is the TOC of the wastewater that passed the measuring point about 10 minutes ago. This time lag should be compensated for in one way or another. Exact compensation is effected by providing three tanks arranged as shown in Fig. 3: a first receiver tank, a second TOC determination tank, and a third transfer tank, each having a retention time of more than 10 minutes.

Such a three-tank system can become very complicated. The piping leading from points of use can be switched by automatic valves. Each tank can be equipped with a discharge pump. Tank outlets can be switched by automatic valves.

Malfunction of any of the automatic valves on the piping leading from points of use may have adverse effects on the equipment. As pointed out before, the mixture of high concentrations of TOC into wastewater to be recycled to the ultrapure water production line should in principle be avoided by assuring the exact fractionation of wastewater at the source.

Specifically, the fractionation should be planned after careful characterization of the wastewater drained from each piece of equipment at use point to prevent any significant mixture of TOC into the secondary rinse waste unless there is an accident or operational failure. Therefore, a one-tank system as illustrated in Fig. 4 will suffice.

In this simpler system, wastewater is retained in the tank for more than 10 minutes as long as its TOC value is low. If higher TOC values are detected, the wastewater retained in the tank is routed to a wastewater treatment system for discharge. Although this system cannot prevent wastewater of high TOC concentration from entering the wastewater recovery system by diffusion or short-circuit, it serves the purpose without serious problems.

B.   Activated Carbon Filter

This filter is installed in the front of the secondary rinse waste recovery system. It is designed to remove organics, surfactants, and other contaminants from the rinse waste. It

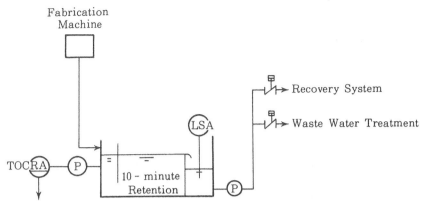

Figure 4   Fractionation of recovered wastewater by TOC.

is also designed to decompose oxidizing agents, such as hydrogen peroxide and sodium hypochlorite, which are dosed to the raw wastewater for bacterial control, thereby protecting ion-exchange resins used downstream in the ion-exchange unit.

It is known that organic compounds of higher molecular weight are better adsorbed on activated carbon than those of lower molecular weight, such as methanol and acetone. The field results of TOC removal by activated carbon are given in Table 3.

In this filter is charged granular and porous activated carbon with a large surface area and correspondingly high adsorption capacity. This activated carbon is disposable and is replaced by fresh carbon after a given period of working time. The frequency of this replacement depends upon the concentration of TOC and other adsorbates in the feed wastewater, but activated carbon is usually replaced once a year.

In operating this activated carbon filter, special attention should be paid to bacterial growth in the activated carbon filter bed. The higher the TOC concentration in the feed wastewater, the more proliferation of bacteria. If the worse comes to the worst,

Table 3   TOC Values of Treated Water

Measuring Points

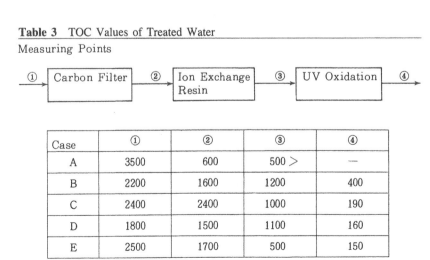

Case	①	②	③	④
A	3500	600	500 >	—
B	2200	1600	1200	400
C	2400	2400	1000	190
D	1800	1500	1100	160
E	2500	1700	500	150

TOC (μg / ℓ)

Case A ~ E are some semiconductor factories.

bacteria grow not only on carbon surfaces but also in the interiors of the carbon granules. To prevent such bacterial proliferation, sodium hypochlorite is injected on the feed wastewater so that it remains effective to the inlet point of the carbon filter. A more important countermeasure is again the appropriate fractionation of rinse wastewater at points of use so that the concentration of TOC on which bacteria feed is minimized.

C.   Ion-Exchange Equipment

Ion-exchange resins are used to remove inorganic ions from wastewater to be recycled. As shown in Table 4, wastewaters generated in semiconductor plants are acid and contain high concentrations of inorganic anions, such as sulfate and nitrate ions. These anions are well adsorbed on weak base anion-exchange resins. These weak base anion-exchange resins are incapable of splitting neutral salts, such as sodium chloride (NaCl), and of adsorbing their anionic moieties or other weak base anions, such as silica and carbonate ions. These resins are very effective in adsorbing acid anions, such as the Cl^- of hydrofluoric acid (HCl). Another feature of the weak base anion resins is that they are regenerated very effectively with almost an equivalent amount of alkali, such as caustic soda (NaOH). In general, anion-exchange resins can adsorb organics more effectively under acid conditions. Weak base anion resins are good at adsorbing organics. In treating wastewaters generated in a semiconductor plant, advantage is taken of these features of weak base anion resins, and a column charged with weak base anion resin is installed at the front of the ion-exchange equipment. This first ion-exchange column is followed by two columns charged with a strongly acid cation resin and a strongly basic anion resin, respectively, or a mixed-bed column charged with a mixture of these two resins. These two columns or one mixed-bed column is designed to remove the inorganic ions the forward weak base anion column cannot remove: cations, anionic moieties of neutral salts, and weak base anions, such as silica and carbonate ions.

Table 4   Wastewater Qualities from Different Points of Use

Ca^{2+}	$mgCaCO_3/\ell$	14. 4	16. 0	10. 6	8. 7	13. 8
Mg^{2+}	"	17. 2	10. 3	2. 6	3. 7	11. 6
Na^+	"	48. 5	65. 3	—	—	5. 7
NH_4^+	"	8. 8	38. 2	0. 6	18. 5	66. 3
SO_4^{2-}	"	310. 0	9. 2	16. 7	2. 0	28. 9
Cl^-	"	95. 3	83. 3	4. 7	1. 7	22. 7
PO_4^{3-}	"	76. 4	0. 9	—	—	4. 9
NO_3^-	"	85. 0	104. 9	129. 9	0. 8	28. 9
F^-	"	1. 3	421. 4	7. 9	9. 7	68. 5
TOC	mgC/ℓ	3. 5	3. 1	3. 4	0. 3	2. 7
Conductibility $\mu S/cm$		2340	1800	1140	33	388
pH		2. 3	2. 5	2. 6	5. 7	3. 6

D.   UV Oxidation Unit

This unit decomposes and removes organic materials in the wastewater by applying ultraviolet rays. UV has long been used for sterilization purposes in the food industry and wastewater treatment for organics removal.

In the semiconductor industry, UV was first used for the sterilization of ultrapure water. As the recycling of waste rinsing water was promoted in the semiconductor industry, UV began to be used to decompose organic materials in this wastewater. Recently the low-pressure UV oxidation method was introduced to further reduce TOC in ultrapure water.

The UV oxidation method is able to remove low-level alcohol and ketone, which cannot be easily removed with ion exchange or reverse osmosis. In this sense, UV oxidation plays a very important role in recycling the wastewater to be sent back to the ultrapure water production unit.

In the UV oxidation unit, organic materials are oxidized by oxidizing agents (radicals) excited by the UV energy emitted from the UV lamp. At the same time, the excited molecules (organic materials) adsorb the UV energy and decompose. These two reactions have synergistic effects in the UV oxidation unit. The intermolecular bonds of the organic materials are cut by the energy of hydroxyl radical (OH) and ultraviolet rays. They eventually decompose to carbon dioxide and water. The photon energy used in these reactions is in inverse proportion to the ultraviolet wavelength in accordance with Planck's equation.

There are several types of UV lamp emitting UV rays: the mercury lamp emits a certain amount of the excitation line, and the xenon lamp emits a countinuous spectrum. The mercury lamp is generally used, and there are two major types in terms of the vapor pressure of filled mercury: a high-pressure lamp (vapor pressure of filled mercury of 0.1–10 atm) and a low-pressure lamp (vapor pressure of filled mercury of 0.001–0.1 atm). Figure 5 and 6 show some examples of wavelength distribution.

Since the high-pressure lamp has its peak at 365 nm and the low-pressure lamp at 253.5 nm, the low-pressure lamp, which is able to emit short-wavelength ultraviolet, is slightly superior in terms of energy efficiency. In practice, however, it is rare that all the irradiation energy is used effectively. Therefore an irradiation amount of 1–5 kWh/m^3 is required in both lamps.

In the recycling system, the high-pressure lamp is used at present more often because of the lower initial and running costs. Because the low-pressure lamp is superior in terms of energy, a large-capacity low-pressure lamp is expected to be developed.

Wastewater that contains a large amount of TOC to be decomposed requires a large oxidation potential. A short-wavelength energy of less than 185 nm must be emitted to decompose H_2O to produce hydroxyl radical, but it is rare to obtain this short a wavelength even with the low-pressure lamp. (It is impossible with the high-pressure lamp.) Oxidizing agents, such as hydrogen peroxide, however, can generate hydroxyl radical at a wavelength of 400 nm or less. This is why oxidizing agent is added to the recycling system. Hydrogen peroxide is often used as the oxidizing agent because it is relatively stable and, unlike chlorine, it does not raise the Cl concentration and thus corrode the piping materials. Hydrogen peroxide must be added at 20–50 ppm at least, although the amount depends on the types and concentration of TOC to be decomposed.

Moreover, the UV irradiation intensity and the time of exposure to the UV ray are the factors that affect the TOC decomposition. A short pass of wastewater cuts the exposure

Planck's equation

$$Ee = h \cdot \frac{C}{\lambda} \cdot Na$$

  Ee ; Photon energy
  h ; Planck's constant
       (6.626 * 10^{-27} erg · sec)
  C ; Light velocity in vacuum
       (2.998 * 10^{10} cm / sec)
  λ ; Wavelength of UV ray
  Na ; Avogadro's number

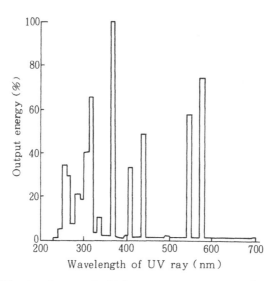

Figure 5   Wavelength range for high-pressure mercury vapor lamp.

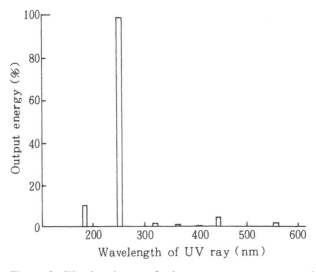

Figure 6   Wavelength range for low-pressure mercury vapor lamp.

time, resulting in the insufficient decomposition of TOC. To prevent a short pass, a baffle is placed in the system or the wastewater is stirred with air. When stirring with air, it is important not to generate bubbles because they interfere with the UV ray.

The UV oxidation unit is used together with the active carbon tower, which adsorbs the decomposed organic matter and decomposes the remaining hydrogen peroxide.

E. Reverse Osmosis Unit

Because the UV oxidation unit is very inefficient, it is possible to place the reverse osmosis (RO) unit ahead of the UV oxidation unit to reduce the load on the UV oxidation unit when wastewater containing a large amount of high-molecular-weight organic materials is treated. The latest synthetic complex membrane reverse osmosis has made remarkable progress in performance, removing TOC at a very high efficiency.

For the removal of ions, the RO system has a high capability of ion rejection at neutral pH, and therefore the ion rejection ratio decreases at low pH, such as pH 3–4.

F. Bacteria Filtration Unit

The conventional unit for recycling wastewater contining TOC in high concentration has a problem with bacterial propagation. This is because TOC is a nutrient for bacteria. The low-molecular-weight alcohol used in semiconductor production is also a good nutrient for bacteria: consider that methanol is used as the nutrient to cultivate bacteria in the BOD and COD processes (methods of processing wastewater with bacteria). This idea was used to develop a bacterial filtration unit containing bacteria to decompose TOC.

In this unit, the bacteria breed in a layer packed with a material like active carbon and digest and decompose TOC. Oxygen and nutrient should be supplied to cultivate the bacteria if necessary.

In principle, the separation of different wastewaters should be so efficiently operated that wastewater containing TOC is not recycled. The bacterial filtration unit is not the best solution for removing the TOC because it may lead to problems, including the metabolic products of bacteria and the penetration of bacteria to stages downstream of the recycling system. Nevertheless, when a large amount of wastewater needs to be recycled and a high concentration of TOC must be processed at low cost, the bacteria filtration unit should be introduced. It is a future challenge to come up with countermeasures to these problems as well as to develop a way of securely maintaining the cultivated bacteria.

IV. EXAMPLES

Initially wastewaters were not distinguished in the production equipment but were distinguished in the recycling unit by applying electrical conductivity. These days, planning of the recycling system begins with a study of how to distinguish wastewaters.

The original production equipment cannot distinguish wastewaters. Wastewaters are checked only by electrical conductivity meter. The recycled water is reused for general purposes in the plant (see Fig. 7).

The UV oxidation unit and the active carbon tower were added to the basic system. The recycled water is sent back to the ultrapure water production unit (see Fig. 8).

Functions were added to the production equipment to distinguish wastewaters. The water drained from the secondary rinsing bath is recycled directly to the ultrapure water

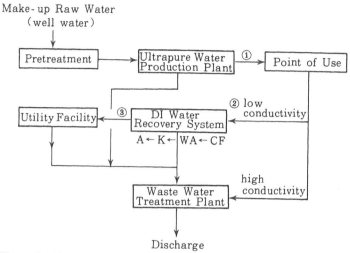

Water Quality

Point	Resistivity	TOC (mg/ℓ)	Particles (nos./mℓ > 0.2 μm)
①	17.5 ~ 18MΩ·cm	0.05 ~ 0.1	2 ~ 10
②	1000 ~ 3000 μS/cm	2.0 ~ 4.0	—
③	10 ~ 20 μS/cm	—	—

CF : Carbon Filter   K : Cation Resin
WA : Weak base Anion Resin   A : Anion Resin

Figure 7   Water and waste treatment.

production unit without processing. The water drained from the primary rinsing bath is processed in the same way as the basic system to be reused for general purposes (Fig. 9).

In the automatic rinsing unit, counters are attached to each rinsing bath in the quick dump rinse to distinguish wastewaters. Wastewater drained during the first half of the rinsing stage is processed and discharged. Water drained during the second half is recycled directly back to the ultrapure water production unit without processing (see Fig. 10).

Wastewaters drained from production equipment are now distinguished precisely. The water drained from the secondary rinsing is recycled to the ultrapure water production unit; the wastewater from the primary rinsing is reused for general purposes. More than 90% of the ultrapure water used in the production line is recycled (see Fig. 11).

V.   OPERATION AND MAINTENANCE

A.   Waste Rinse Water

Although it is very important to accurately distinguish wastewaters to increase the recycling rate, the types of wastewaters, which greatly depend on the chemicals used and

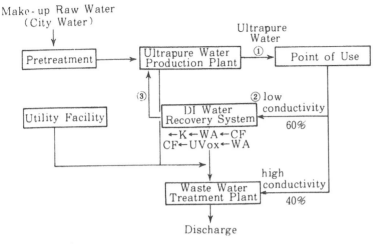

Water Quality

Point	Resistivity	TOC (mg/ℓ)	Particles (nos./mℓ > 0.2 μm)
①	17 ~ 17.5 MΩ·cm	0.05 ~ 0.07	2 ~ 10
②	500 ~ 4000 μS/cm	1.0 ~ 2.0	—
③	5 ~ 10 μS/cm	—	—

CF : Carbon Filter
WA : Weak base Anion Resin
K   : Cation Resin
WA : Weak base Anion Resin
UVox : UV Oxidation
CF : Carbon Filter

Figure 8 Water and waste treatment.

how long the rinsing lasts, cannot be determined at the planning stage. The recycling unit, the amount of ion-exchange resin, and other parameters should therefore be designed on the basis of estimated amounts of chemicals contained in wastewater. The reference values to be monitored, such as the electrical conductivity, pH, and TOC, are also set based on experience.

For efficient and stable operation of the recycling unit, it is important to understand the design concept well and to compare design assumptions with the actual conditions. The reference values for electrical conductivity, pH, and TOC should be revised to reflect the actual condition of the wastewater.

Wastewater conditions fluctuate along with changes in the production volume. Periodic checking of the conditions, as well as at start-up, is critical. In particular, changes in the semiconductor production process are sometimes accompanied by changes in the type and amount of chemicals, so communication with the production staff is also important.

Make-up Raw Water (Industry Water)

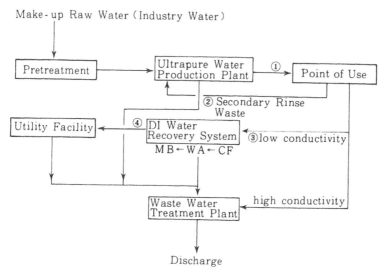

Discharge

Water Quality

Point	Resistivity	TOC (mg/ℓ)	Particles (nos./mℓ > 0.2 μm)
①	>18MΩ·cm	0.02 ~ 0.03	6 ~ 10
②	1 ~ 10 μS/cm	0.05 ~ 0.1	800 − 300
③	10 ~ 1000 μS/cm	—	—
④	0.2 ~ 1 μS/cm	—	—

CF : Carbon Filter
WA : Weak Base Anion Resin
MB : Mixed-bed Ion Exchange

Figure 9   Water and waste treatment.

B.   Countermeasures Against Bacteria

As mentioned earlier, TOC and the propagation of bacteria are the major problems in the recycling unit. Although the waste rinse water is acid water with low pH values and contains hydrogen peroxide, bacteria propagate rapidly in it.

In some cases, bacteria produce large amounts of slime in the active carbon tower. If no measures are taken, bacteria propagate at the distributor of the ion-exchange resin tower, which can lead to serious trouble, such as a reduction in the wastewater flow rate. Bacteria propagate rapidly in wastewater with a high TOC concentration, but even in water with little TOC, bacteria can breed. It is necessary to sterilize the waste rinse water with NaClO at the recycling unit inlet, to check the differential pressure in each tower, and to measure the number of bacteria at certain points, as well as to strictly prevent TOC from mixing in the wastewater.

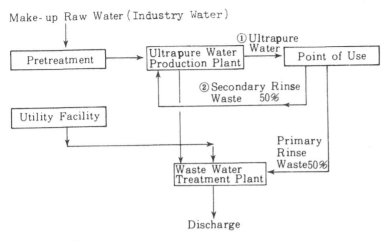

Make- up Raw Water (Industry Water)

Figure 10   Water and waste treatment.

Water Quality

Point	Resistivity	TOC (mg/ℓ)	Particles (nos./mℓ > 0.2 μm)
①	>18MΩ•cm	0.01 ~ 0.03	0 ~ 2
②	1 ~ 20 μS/cm	0.05 ~ 0.1	—

C. Routine Checkpoints

The condition of water in the recycling unit fluctuates all the time, which is quite different from the ultrapure water production unit. The volume of wastewater to be processed in the unit and the amount of the resin in the towers should be set in accordance with changes in the wastewater. Table 5 list routine checkpoints for a field example.

D. Periodical Checkpoints

The performance of the active carbon and ion-exchange resin deteriorates gradually as a result of the absorption of organic materials, oxidation, and hydrolysis. The reverse osmosis unit and the ultrafiltration unit also deteriorate. The irradiation efficiency of the UV lamp is also reduced. These basic items should therefore be periodically checked and replaced or refilled as necessary.

In periodic checks, the components inside the towers, the rotating parts, the automatic valves, and all the other parts that cannot be checked when the unit is operating should be closely examined.

Table 6 list periodic checks for a field example.

VI. CLOSED SYSTEM

The remainder of this chapter concerns the closed system, which is regarded as the ultimate recycling system. In a bid to secure the water supply, to cope with restrictions on

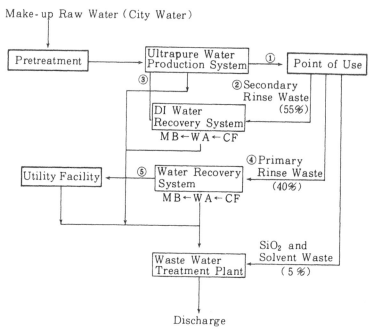

Make-up Raw Water (City Water)

Water Quality

Point	Resistivity	TOC (mg/ℓ)	Particles (nos./mℓ > 0.2 μm)
①	>18MΩ•cm	<0.01	0 ~ 2
②	1 ~ 100 μS/cm	0.02 ~ 0.2	—
③	2 ~ 5 MΩ•cm	—	—
④	200 ~ 600 μS/cm	0.5 ~ 1.5	—
⑤	2 ~ 5 MΩ•cm	—	—

CF : Carbon Filter
WA : Weak Base Anion Resin
MB : Mixed-bed Ion Exchange

Figure 11 Water and waste treatment.

water for discharge, and to find a site for plant construction, increasing numbers of newly built plants employ the closed system. From the viewpoint of protecting the environment, semiconductor plants, which use various chemicals and gases, are encouraged to introduce the closed system so that industrial wastewater is not discharged outside the plant. This attempt is also welcomed by local communities.

A completely closed system means that *everything,* including the solid wastes and discharged chemicals handled at present by contractors, is processed in a plant. At present it is not possible for companies to employ a completely closed system on a commercial basis. The current system is a closed system.

In a plant employing a closed system, all the water used on the premises makes one

Table 5 Routine Checkpoints (Field Example)

	Checkup Point	Remarks
Raw Water	Raw Waste Water Quality (μS/cm, pH, TOC, etc.)	Confirm the state and extent of change
	Raw Water Residual Chlorine	Check for an allowable range of $0.3 \sim 0.5$ mg
Activated Carbon Column	Pressure Differential	Check for any abnormal value
	Effluent Residual Chlorine	Confirm no detection
	Regeneration	Confirm that regeneration is carried out as required
Ion Exchange Column	Treated Water Quality (μS/cm, pH, etc.)	Check for any abnormal value
	Throughput Capacity	Check for any abnormal value
	Pressure Differential	Check for any abnormal value
	Regeneration	Confirm that regeneration is carried out as required
UV Oxidation Unit	Inlet and Outlet TOC	Confirm the effectiveness of UV oxidation
	Lamps	Check for any abnormality
RO Unit	Feed Water Residual Chlorine	Check for any residual chlorine depending upon whether or not the membrane is resistant to chlorine
	Feed Water Pressure and Pressure Differential	Check for any change in feed water pressure and any rise in pressure differential
	Permeate Flow Rate	Check for any change in permeate flow rate (as converted for the same feed water pressure)
	Feed and Permeate Water Quality (μS/cm and pH)	Check for any abnormal value
Common	Water Quality	Check for water quality at respective points in order to confirm the proper performance of respective pieces of equipment
	Calibration of Measuring Instruments	Calibrate each measuring instrument as specified
	Working Condition of Rotatory Machines	Check for any abnormal noise, vibration, oil quantity, ampere value, etc.
	Working Condition of Control Equipment	Check for any abnormal condition
	Working Condition of Valves	Check for any abnormal condition
	Chemicals	Check for dosage levels and quanties of remaining chemicals
	Replacement and Cleaning of Strainer and Filter Elements	Replace and cleanse as required (check for pressure differential, etc.)

Table 6   Periodic Checks (Field Example)

Activated Carbon Column	Measure the iodine adsorption capacity, density, etc. in order to determine the makeup or replacement ratio
Ion Exchange Column	Measure the total ion exchange capacity in order to determine the make-up or replacement ratio
UV Oxidation Unit	Replace lamps as necessary after checking inlet and out TOC values
RO	Replace membrane modules after checking the membrane performance, pressure differential, etc.
Interiors of Vessels, Tanks and the like	Check for any pinhole of lining (touch up if necessary)
Internals of Vessels and Columns	Disassemble and cleanse strainers, distributors, etc.
Rotatory Machinery	Disassemble and shake down
Valves	Disassemble and shake down (replace diaphragms if necessary)
Instruments and Controls	Carry out overall checkups, replace parts and calibrate

single flow cycle. Consequently, a defect or malfunction in one part of the cycle can affect the entire system, which might result in serious loss. This is why precise design is critical to securing the reliability of the system: grasp the characteristics of wastewaters in the plant, distinguish the wastewaters properly, build a backup system, provide the system with extra processing capability, come up with a way to deal with the extra flow of water, and introduce a central monitoring system. Central monitoring is introduced more and more often to many systems other than the closed system because it works well in saving labor during operation.

For chemical waste discharged during chemical replacement, as mentioned in Sec. II. E, technically it is possible to process this in the plant. This processing costs a great deal, however, so some plants hire contractors to handle the chemical waste. In this case the cost is around ¥10–20/kg. When acid waste is drained in large quantities from plants, it might be economical to store it in a tank without diluting it, although the initial cost for the tank material is high.

Figure 12 is a block diagram of a closed system, and Table 7 indicates the ultrapure water quality in a conventional closed system.

The closed system is mainly composed of three groups of equipment:

1. Water supply unit and ultrapure water production unit. Some water is evaporated through the cooling tower. To make up the water loss, water is newly supplied to the plant after flowing through the filtration unit. The newly supplied water together with the recycled water is processed in the ion-exchange unit, reverse osmosis unit,

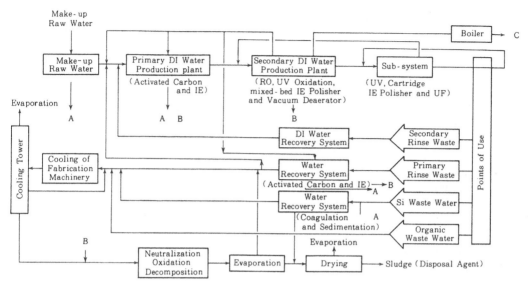

Figure 12 Closed system.

Table 7 UPW Qualities Obtained at Existing Closed Systems

Item of Analysis		Unit	Design Standard	Average	Range	Analysis Method
Resistivity		MΩ・cm	>18	18. 1	18. 0~18. 2	Foxboro
Particles	0. 1 μm<	count/ ml	10>	3	0 ~ 6	SEM or Particle counter < 0. 11 μm
	0. 2 μm<	count/ ml	1>	1	0 ~ 1	Direct microscopic observation or Particle counter
Bacteria		colonies/ 100 ml	1>	1>	0 ~ 1	Membrane filter (cultivation)
TOC		μg C / l	20>	10>	5 >~15	Wet UV oxidation TOC analyzers
Silica		μg SiO₂/ℓ	5 >	1	1 ~ 3	Molybdene blue
Na⁺		μg Na/ l	0. 1 >	0. 1 >	0. 1 >	Flameless atomic absorption
Zn		μg Zn / l	0. 1 >	0. 1 >	0. 1 >	Flameless atomic absorption
Cu		μg Cu / l	1 >	0. 1 >	0. 1 >	Flameless atomic absorption
Cl⁻		μg Cl / l	1 >	0. 5 >	0. 5 >	Ion chromatography
DO		μg O / l	50>	30	25~35	Color Comparison method

vacuum deaerator unit, and UV oxidation unit to be purified to the level of the ultrapure water unit. There is also a subsystem that polishes the ultrapure water in the ultrafiltration unit and transports the ultrapure water to the use points.

2. Wastewater recycling unit. Other than the units described earlier, a coagulation and sedimentation unit is also employed. This unit coagulates and sediments wastewater drained from the backgrinder and dicer together with wastewater from the filtration unit and the active carbon filtration unit. It also filters the upper layer of the wastewater to be recycled back to the ultrapure water production unit. The slug generated is condensed in a condensing bath and solidified in the drum drier or dehydrator and is them handled by the contractor.

3. High-concentration wastewater processing unit. The high-concentration wastewater processing unit is unique to the closed system. (It is not employed by a conventional recycling system.) This unit is composed of two processes: the condensation and evaporation process and the drying process. In the condensation process, water drained from the production equipment and the boiler drain and cooling tower blowdown water are condensed by the evaporator. The condensed liquid is solidified in the drying process.

The last two processes have the highest cost of all the units in the closed system. To reduce the volume of wastewater for treatment by the high-concentration wastewater treatment system, minimization of the amount of wastewater regenerated from the ion exchanger must be implemented.

A vapor-compressor evaporator that efficiently saves energy has been introduced as the condensation unit. This evaporator condenses the wastewater with a Cl concentration of 1% or less to the level of 12–15%. The distilled water is recycled to be reused for general purposes, such as in the cooling tower. To prevent the ammonia condensed in the ion-exchange unit from penetrating into the distilled water, the ammonia should be decomposed in the neutralization and breakpoint unit before the recycled water is transported to the evaporator.

The condensed liquid is evaporated and solidified in the drum drier to become solid waste with a water concentration of 15% or less. This solid waste is handled by a contractor. In the drum drier, vapor is sent to the rotating drum to solidify the wastewater, which is placed on the outer surface of the drum. The volume of vapor required is 1.3–1.4 times as much as the volume of the processed wastewater. Countermeasures should be worked out to take care of solidified powder scattering and the generation of chlorine gas.

VII. SUMMARY

The wastewater recycling system, if it is carefully designed together with the production staff, can achieve a level for practical use.

The future challenge is to upgrade the conventional system to as high a level as possible, a completely closed system. To achieve this goal securely, safely, and economically, more field data should be gathered and each unit should be improved further. The system should be improved in terms of water processing technology, so that the organic materials in the wastewater can be efficiently processed. Further study based on concrete field data should be carried out in the future.

7 Field Data

A. Operation Data for Actual Units

Ikuo Shindo
Japan Organo Co., Ltd., Saitama, Japan

I. DESCRIPTION OF THE SYSTEM

This chapter presents operation data from the ultrapure water production and pure water recycling systems delivered in 1986 by Japan Organo Co., Ltd. Figure 1 is a schematic of the system, and Figure 2 is a flow diagram. The circled numbers in Figure 2 indicate points of measurement.

A. Pretreatment System

This system is mainly used for makeup water.

The industrial water used as the raw water is first processed in the coagulation filtration unit, which employs PAC (polyammonium chloride) as flocculant. After the turbidity is lowered in this process, the industrial water is stored in the filtered water tank. It is then sent to the two-bed, three-tower pure water unit for salt removal. Following these treatments, the industrial water is stored in the desalted water tank.

B. Pure Water Recycling System

This system recycles that portion of the ultrapure water from the use points of sufficiently low electrical conductivity for the raw water supply to the ultrapure water production unit.

The recycled pure water is sterilized by ultraviolet (UV) and then filtered by a membrane filter. In UV oxidation tank 1, H_2O_2 is added to the processed pure water to decompose the total organic carbon (TOC). The recycled water is then sent to the desalted water tank, the same tank to which the water processed in the pretreatment system is sent.

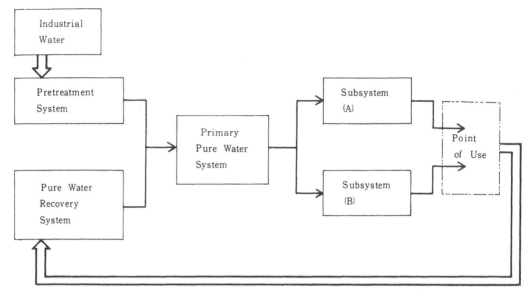

Figure 1   Schematic diagram of the system.

The lamp employed in the UV oxidation unit is a low-pressure UV lamp with main wavelength 254 nm.

C.   Primary Treatment System

In the UV oxidation unit 2, H_2O_2 is added to the desalted water to decompose TOC. The water is then sent to the active carbon tower to remove the remaining unreacted H_2O_2. After residual H_2O_2 removal, the water is sent to the primary pure water tank. UV oxidation unit 2 is equipped with the same low-pressure UV lamp as the UV oxidation unit 1.

The water flows through the heat exchanger, the UV sterilization unit, and membrane filter, and then trace amounts of TOC and particles are removed in the primary reverse osmosis unit. It is stored in the filtered water tank. The primary reverse osmosis unit employs a polyamide type of spiral composite membrane element whose removal efficiency is 90%.

In the vacuum degasifier, dissolved oxygen is removed from the pure water. The residual salts are eliminated in the mixed-bed ion-exchange tower. Following these treatments, pure water passes through the UV sterilization unit and the membrane filter to be stored in the secondary pure water tank, which is sealed with nitrogen gas.

D.   Subsystem

The subsystem is composed of two parts, A and B. The volume of water to be processed is 24 m^3h^{-1} in part A and 11 m^3h^{-1} in part B.

In part A, the pure water processed in the primary treatment system passes through the heat exchanger, the UV sterilization unit, and the secondary reverse osmosis unit. It is then treated in the UV sterilization unit again. The residual salt is removed by cartridge

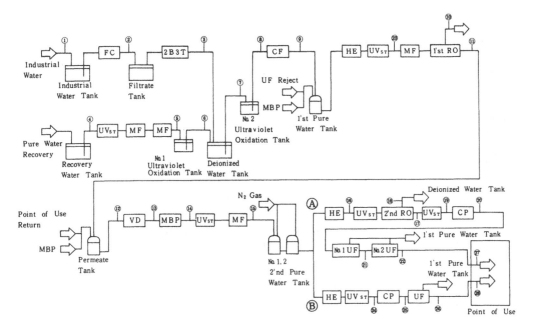

2B3T	two bed three tower	HE	heat exchanger	UVox	ultraviolet oxidation
	demineralizing equipment	MBP	mixed bed polisher	UVsт	ultraviolet sterilizer
CF	carbon filter	MF	membrane filter	VD	vacuum degasifier
CP	cartridge polisher	RO	reverse osmosis unit		
FC	filtration with coagulation system	UF	ultrafiltration equipment		

Figure 2 Flow-diagram of the system.

polisher. The secondary reverse osmosis unit employs a polyvinyl alcohol type of spiral composite membrane element whose removal efficiency is 90%. The pure water from the cartridge polisher is processed in the two-stage ultrafiltration unit to be supplied to the use points as ultrapure water.

The first stage of the ultrafiltration unit employs a polyacrylonitrile hollow-fiber membrane; the second stage employs a hollow-fiber polysulfone membrane. The second-stage ultrafiltration unit can be cleaned with hot water because it employs a heat-resistant membrane.

In part B, the water processed in the primary treatment system flows through the heat exchanger and the UV sterilization unit. Residual salt is then removed in the cartridge polisher. The pure water from the cartridge polisher is processed in a single-stage ultrafiltration unit to be supplied to the use points as ultrapure water. The ultrafiltration employs a polysulfone type of hollow-fiber membrane that is heat resistant and thus can be cleaned with hot water.

II.   MEASUREMENT ITEMS AND MEASUREMENT INSTRUMENTS

The operation was conducted in three phases.

Phase 1: operation before start-up of the pure water recycling system
Phase 2: operation after start-up of the pure water recycling system
Phase 3: operation with the UV lamp of UV oxidation unit 2 off and with no H_2O_2

The items to be measured and the measurement methods are as follows.

1.   TOC

XERTEX COA-1000: an oxygen bubbling and UV oxidation method to monitor the
    electrical conductivity and for in-line sampling
TORAY ASTRO 1800 LTO: an oxygen bubbling method following by oxidative reagent
    injection and UV oxidation for nondispersive infrared analysis and bottle sampling
ANATEL A-100: no bubbling agent or oxidative reagent added; a UV oxidation method
    to monitor resistivity and for in-line sampling

2.   TOX

The Mitsubishi Chemical TOX-10 is used for adsorption and desorption to and from the
active carbon. First, the inorganic chlorine compound is desorbed. High temperatures are
then applied to generate chlorine gas, and then the potential gap is measured by silver
nitrate solution absorption of the chlorine gas.

3.   Particles

The Fuji Electric PC-1000 detects side-direction scattering by He-Ne laser and is capable
of measuring microparticles of 0.11 μm or more.

4.   Bacteria

In the ASTM method, bacteria are captured by a 0.45 μm membrane filter and counted
after cultivation at 35°C for 48 h. The volume of sample was set at 50 ml or less.

   In the OT method, bacteria are captured by a 0.45 μm membrane filter and counted
after cultivation at 35°C for 48 h. This unit has an improved sampling kit for filtering a
few tens of liters when it is necessary to increase the volume of sample because the
number of bacteria is too small. In this test, the volume of sample was set at 20 L.

5.   Resistivity

The Organo MH-4 is used to measure resistivity.

III.   OPERATION DATA

A.   Analytic Results for Each Ion

Table 1 lists the analytic results for ions in the raw water (industrial and recycled water),
the water at the inlet of the active carbon tower, the water at the outlet of the mixed-bed
ion-exchange tower, the water at the outlet of the ultrafiltration unit part A, and the water
at the outlet of the ultrafiltration unit part B.

B.   Measurement Results

Each item to be measured was traced from one point to another. The results are as
follows.

Table 1   Analysis Results

Sampling Point	1 Industrial Water	4 Pure Water Recovery	9 Carbon Filter Outlet	14 MBP Outlet	22 A UF Outlet	26 B UF Outlet
COD Mn (mgO/1)	3.9	< 0.1	< 0.1	< 0.1	< 0.1	< 0.1
HCO₃⁻ (mgCaCO₃/1)	35.6	< 1	< 1	< 1	< 1	< 1
NO₃⁻ (μgNO₃/1)	7434	200	400	< 2	< 2	< 2
SO₄²⁻ (μgSO₄²⁻/1)	31090	14	< 4	< 4	< 4	< 4
Cl⁻ (μgCl⁻/1)	23030	4	170	< 1	< 1	< 1
Calcium (μgCa/1)	14090	5	70	< 0.5	< 0.5	< 0.5
Magnesium (μgMg/1)	2914	20	30	< 0.5	< 0.5	< 0.5
Silica (μgSiO₂/1)	6365	7	36	< 2	< 2	< 2
Iron (μgFe/1)	39	< 1	3	< 1	< 1	< 1

1.   Phase 1

Phase 1 is the state before start-up of the pure water recycling system. Figure 3 shows the change in TOC and TOX in phase 1. Figure 4 shows the change in particles, and Fig. 5 shows the change in bacteria.

The TOC concentration drops at UV oxidation unit 2 and decreases further at the outlet of the active carbon tower. It remains almost flat in the following treatments. By ANATEL measurement, however, the TOC concentration in both part A and part B is very low, at 2–3 mg CL^{-1} in the subsystem.

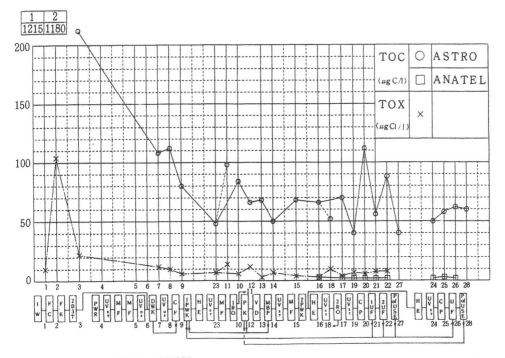

Figure 3   Phase 1 TOC and TOX.

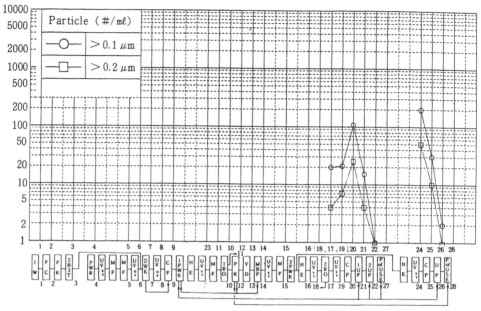

Figure 4 Phase 1 particle.

Since sodium hypochlorite is injected in the course of coagulation filtration, the TOX level shows a surge at the outlet of the filtration unit. TOX is removed in the 2B3T unit, however, and stabilizes at 10 ppb or less in the following process.

At the outlet of ultrafiltration unit, the final subsystem stage in both part A and part B, the number of particles ≥ 0.1 μm is less than 10 ml^{-1}. The number of bacteria at these outlets is less than 1 ml^{-1}.

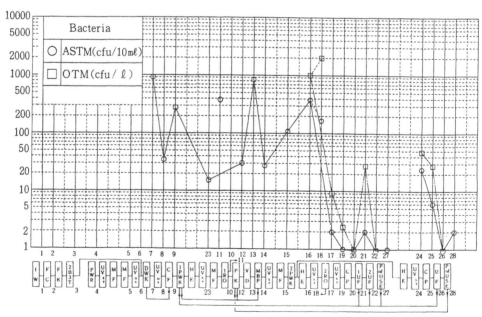

Figure 5 Phase 1 bacteria.

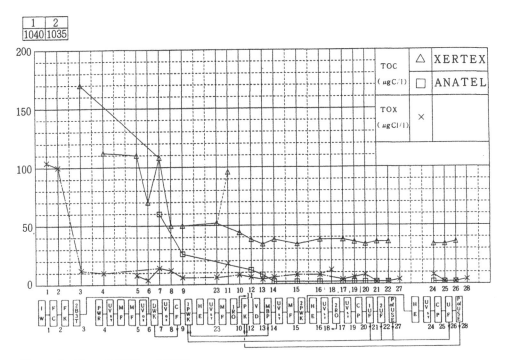

Figure 6 Phase 2 TOC and TOX.

2. *Phase 2*

Phase 2 is the state of the system after the pure water recycling system is started up. Figure 6 shows the change in TOC and TOX, Fig. 7 the particle change, and Fig. 8 the change in bacteria.

Although the TOC concentration drops at UV oxidation units 1 and 2, unlike phase 1, it does not decrease at the active carbon tower. It decreases further at the primary reverse osmosis unit but remains flat in the following process. By ANATEL measurement, however, it is very low, at 2–3 mg CL^{-1} in part A and part B subsystems.

Little TOX is carried over from the pure water recycling system. As in phase 1, in the pretreatment system the TOX level downstream of the desalting unit remains stable at 10 ppb or less.

Just as in phase 1, the number of particles ≥ 0.1 μm at the ultrafiltration outlet, at the subsystem final stage in part A and part B, is less than 10 ml^{-1}. Also as in phase 1, the number of bacteria here is less than 1 ml^{-1}.

3. *Phase 3*

Phase 3 is the state in which the UV lamp in UV oxidation unit 2 is turned off and H_2O_2 injection is suspended. Figure 9 shows the change in TOC and TOX, Fig. 10 the particle change, and Fig. 11 the change in bacteria.

The TOC concentration does fall in UV oxidation unit 2 but drops at the outlet of the active carbon tower. By ANATEL measurement, however, it is very low, at 2–3 mg CL^{-1} in both part A and part B subsystems.

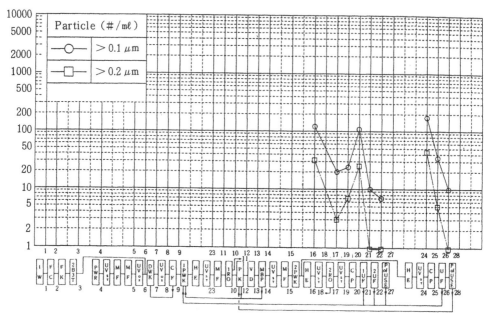

Figure 7   Phase 2 particle.

As in phase 2, the TOX level is almost stable at 10 ppb or lower.

At the outlet of the ultrafiltration unit at the final stage of the subsystem in part A, the number of particles ≥ 0.1 μm is 10 ml^{-1}. In part B, however, it is 18 ml^{-1}. Also as in phase 1 and phase 2, the number of bacteria is less than 1 ml^{-1} at the outlet of the ultrafiltration unit at the final stage of the subsystem in both part A and part B.

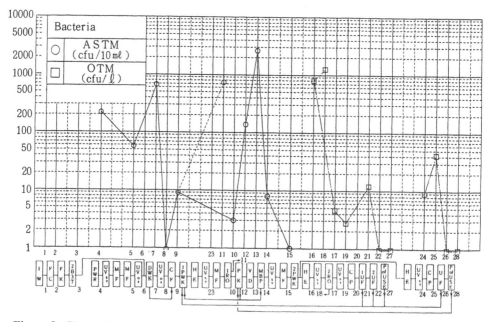

Figure 8   Phase 2 bacteria.

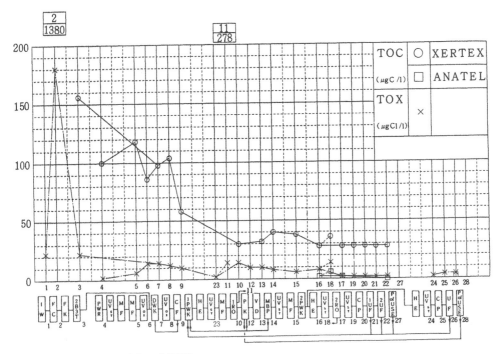

Figure 9 Phase 3 TOC and TOX.

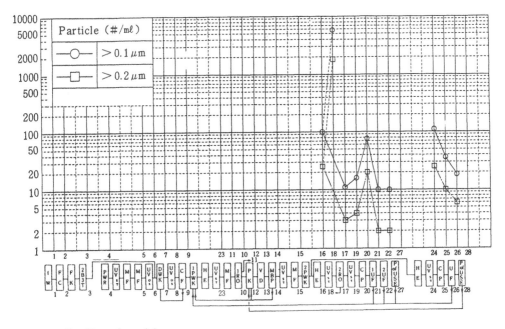

Figure 10 Phase 3 particle.

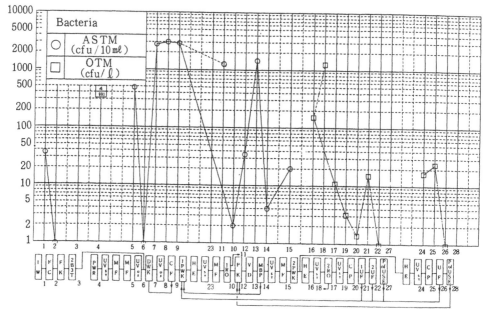

Figure 11 Phase 3 bacteria.

IV. SUMMARY OF OPERATION DATA

The operation data for phases 1–3 are summarized as follows (see Figs. 3 through 11 for details).

A. Coagulation Filtration Unit

The values at point 1 and point 2 are compared.

The TOC concentration is 1000–1700 μg CL^{-1} at both points. No removal effect is observed. The TOC concentration rises from point 1 (6–105 μg CL^{-1}) to point 2 (90–200 μg CL^{-1}). This is thought to be because of the injection of sodium hypochlorite.

B. 2B3T Unit

The values at points 2 and 3 are compared.

The TOC concentration is 1000–1700 μg CL^{-1} at point 2 and 100–300 μg CL^{-1} at point 3, which exhibits the removal effect. The removal efficiency is about 90%.

The TOX concentration is 90–200 μg ClL^{-1} at point 2 and 10–20 μg ClL^{-1} at point 3, which exhibits the removal effect. TOX is removed by bubbling in the decarboxylation tower.

Stabilization of water quality in the pretreatment system depends on the performance of these units. The water quality is carefully controlled at point 3.

C. Pure Water Recycling System

The effect of the UV sterilization unit and the membrane filter is indicated by comparing the values at points 4 and 5.

The TOC concentration is about 100 μg CL^{-1} at point 4 and 71–86 μgCL^{-1} at point 5, which exhibits a slight removal effect.

The electrical conductivity value is 1.2–1.5 μS $\cdot$ cm (25°C) at points 4 and 5.

The number of bacteria decreases from 200–1300 cfu/ml at point 4 to 0 cfu/ml at point 5. This is thought to be because of the effect of ultraviolet and H_2O_2.

D.   UV Oxidation Unit 2 and Active Carbon Tower

The TOC concentration at each point in phases 2 and 3 is as follows (μg CL^{-1}):

Phase	Point 7	Point 8	Point 9
2	108	51	50
3	97	105	57

In phase 2, the effect of UV oxidation can be observed. The data from phase 3, when the UV lamp is turned off, show that TOC is removed in the active carbon tower.

The TOX concentration at each point in phases 2 and 3 is as follows (μg CL^{-1}):

Phase	Point 7	Point 8	Point 9
2	4.5, 13.5	6.2, 12.2	4.8, 5.5
3	8.2, 15.1	3.5, 11.4	9.7, 4.9

No major change is observed in either phase 2 or phase 3.

The number of bacteria at each point in phase 2 and phase 3 is as follows (cfu/ml):

Phase	Point 7	Point 8	Point 9
2	700	0–1	9, 41
3	2790	2990	2830

Although the effect of the UV oxidation and H_2O_2 is recognized in phase 2, in phase 3, when neither the UV lamp or H_2O_2 is in use, most bacteria are not removed.

E.   Primary Reverse Osmosis Unit

The TOC concentration at each point in phases 2 and 3 is as follows (μg CL^{-1}):

Phase	Point 9	Point 10
2	50 (removal efficiency 14%)	43
3	57 (removal efficiency 42%)	29

A slight removal effect is recognized in both phase 2 and phase 3.

The number of bacteria at each point in phases 2 and 3 is as follows (cfu/ml):

Phase	Point 9	Point 10
2	9, 46	3, 1
3	2830	2

It is clear that almost all the bacteria are removed by the reverse osmosis membrane in phases 2 and 3.

F.   Vacuum Degasifier + Mixed-Bed Ion-Exchange Tower

The TOC concentration at each point in phase 2 and phase 3 is as follows (μg CL^{-1}):

Phase	Point 12	Point 14
2	38	38
3	—	39

No major change is observed in either phase 2 or phase 3.

The TOX concentration at each point in phase 2 and phase 3 is as follows (μg CL^{-1}):

Phase	Point 12	Point 14
2	5.0, 12.5	5.1, 6.7
3	10.8	9.2

No major change is observed either in phase 2 or phase 3.

The number of bacteria at each point in phase 2 and phase 3 is as follows (cfu/ml):

Phase	Point 12	Point 14
2	135, 51	8, 0
3	34	4

Bacteria are removed in the mixed-bed tower in both phases.

G.   Water Processed in the Primary Treatment System and the Subsystem

In phases 1–3, the quality of ultrapure water is compared at the outlet of the primary treatment system (point 15) and at the outlet of ultrafiltration unit at the final stage of the subsystem. There is no major difference in the water quality rated by TOX, bacteria, or resistivity between these two points.

Since three different instruments were used for the measurement, the values for each instrument are listed. The TOC concentration at each point in phases 1–3 is as follows (μg CL^{-1}):

Point		15		22	26
Device	Phase 1	Phase 2	Phase 3	Part A	Part B
ASTRO	55, 69	67, 45	38	50–90	31–62
XERTEX	—	34	38	29–35	36
ANATEL	—	2.3	—	1.7–2.1	1.3–2.5

Although the values vary among the instruments, no major change is observed in TOC concentration between the water processed in the primary treatment system and that treated in the subsystem ultrafiltration unit.

The number of particles ≥ 0.1 μm at each point in phases 1–3 is as follows (particles per ml):

Point	15			22	26
Phase 1	Phase 2	Phase 3		Part A	Part B
—	123, 185	149		4, 7	3, 10

There is a large difference in the number of particles between the outlet of the primary treatment system and the outlet of the ultrafiltration unit at the final stage of the subsystem. The effect of ultrafiltration can be recognized.

V.   SUMMARY

Some examples of operating an ultrapure water production system and pure water recycling system are described here. These measurements were made in 1986. The instruments available then had higher detection limits than current equipment. In this sense, some of the date must be reviewed. It would be a great pleasure, though, if these data indicate new directions for study. The system described here still works well.

7 Field Data

B. Examples of Operation Data

Yoshito Motomura
Kurita Water Industries, Ltd., Tokyo, Japan

I. EXAMPLE 1

The data presented in this chapter are from the two ultrapure water production systems installed in the mini-superclean room of the Engineering Faculty of Tohoku University in 1989.

A. System Flow

Figures 1 and 2 show the system flow of the two ultrapure water production systems. Both are small systems with a daily production of 6 m³/day. They are equipped with reverse osmosis, ion-exchange, ultraviolet (UV) oxidation, and ultrafiltration, units, among others, to remove ions, organic materials, and particles. A two-stage dissolved oxygen removal unit is also installed in this system, for lowering the dissolved oxygen concentration to 10 ppb or less.

The major characteristics of the two systems are as follows:

1. The systems use a two-stage dissolved oxygen removal method (deaeration membrane and catalytic resin deaeration) or a nitrogen bubbling method.
2. Organic materials are removed by combing UV oxidation (185 nm) and ion-exchange resin.
3. The piping materials are resistant to elution of impurities (polyetheretherketone, PEEK or polyvinylidenefluoride, PVDF)
4. A hot-water sterilization system is used.

For removing dissolved oxygen, system 1 employs a combination deaeration membrane and catalytic resin. System 2 combines a deaeration membrane and nitrogen bubbling.

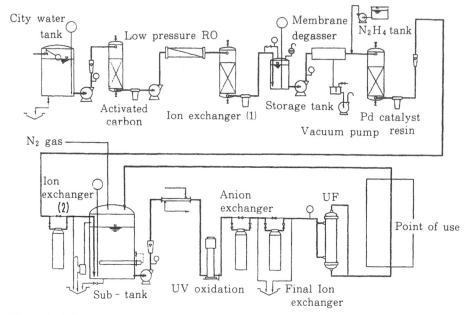

Figure 1   Ultrapure water production system 1.

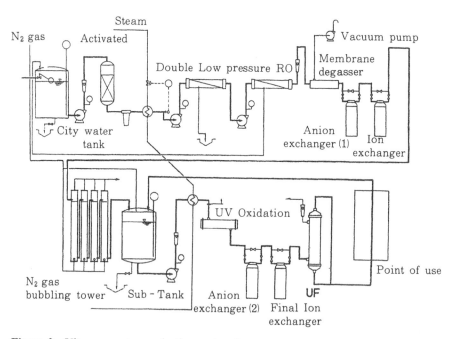

Figure 2   Ultrapure water production system 2.

Table 1   Analytic Monitors in Ultrapure Water

Items	Monitors
Resistivity	KURITA MX − 4
TOC	Tokico TOC − 1000
	Anatel A − 100
Dissolved Oxygen	DKK DOH − 2
Silica (SiO₂)	DKK SLC − 1605
Particle	Kurita K − LAMIC − 100
Total Residue	Nomuramicro HPM − 1000

B.   Analysis and Evaluation Units

Table 1 lists the analysis units used to evaluate these systems. As of 1989 they have the greatest accuracy. All measurements were carried out on-line.

C.   Operation Data

Figures 3 and 4 diagram water quality at the outlet of the two systems. In both cases, impurities are removed in each unit. At the outlet of the primary treatment system, the

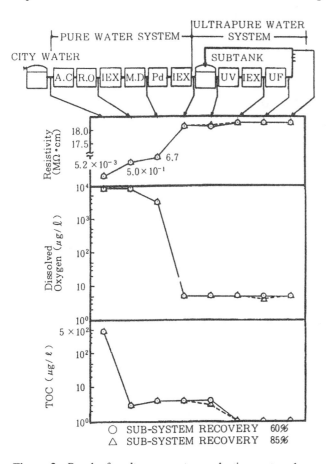

Figure 3   Results for ultrapure water production system 1.

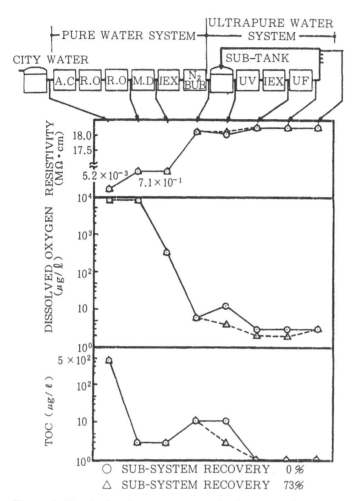

Figure 4 Results for ultrapure water production system 2.

resistivity is 18 MΩ·cm or higher, the concentration of dissolved oxygen is 10 ppb or less, and the total organic carbon (TOC) concentration is 10 ppb or less in systems 1 and 2.

In system 2, the TOC level rises at the outlet of the primary treatment system (the outlet of the nitrogen bubbling unit). This is thought to be because of elution from the ion-exchange resin placed upstream. Both system 1 (catalytic resin deaeration method) and system 2 (nitrogen bubbling method) succeed in achieving a level of 5 ppb of dissolved oxygen at the outlet of the primary treatment system.

The major function of the secondary treatment system is to remove the residual impurities the primary treatment system fails to remove. The trace amount of TOC is removed in the UV oxidation unit (ultraviolet wavelength of 185 nm) and the nonrecovery ion-exchange tower, which is filled with high-purity ion-exchange resin. Trace ions are removed in the nonrecovery ion-exchange tower, and microparticles are removed by an ultrafiltration membrane at the final stage. Both systems successfully achieve

Table 2   Results for Ultrapure Water in Systems 1 and 2

	Resistivity (MΩ·cm)	TOC ($\mu g/\ell$)	SiO$_2$ ($\mu g/\ell$)	Dissolved Oxygen ($\mu g/\ell$)	Total Residue ($\mu g/\ell$)	Particles (counts /mℓ, 0.07 μm)
System 1	18.25	<1	<1	3 ~ 5	<1	1 ~ 2
System 2	18.25	<1	<1	6 ~10	<1	1 ~ 2

a high purity of ultrapure water at the outlet, with resistivity 18.25 MΩ · cm and TOC less than 1 ppb.

The piping materials used in the two systems differ: system 1 uses PEEK, and system 2 uses PVDF. In both cases, however, no degradation in water quality is observed at the use points. The high level of water quality achieved at the outlet of the ultrafiltration unit is maintained and ultrapure water is supplied to use points.

Table 2 lists the other indicators of water quality, all of which are excellent.

II.   EXAMPLE 2

The data presented here are the analytic results for volatile organic materials at the outlet of each component unit of the ultrapure water production system.

A.   Analytic Method

Headspace gas chromatography is used in this analysis. Table 3 lists the detection limits of this measurement method.

B.   System Flow and Operation Data

1.   *Ultrapure Water Production Using Recycled Water*

Figures 5 and 6 and Tables 4 and 5 show the system flow and list measurement results. In Fig. 5 and Table 4, the concentrations of organic materials are too high to be fully processed by the ultrapure water recycling system. This results in residual organic materials in the ultrapure water. The volume of low-molecular-weight organic materials in

Table 3   Determination Limit with Distillation-Headspace Gas Chromatography

	Determination limit (ppb)	Coefficient of variation (%)
Methanol	10	6.2
Ethanol	5.0	9.8
Acetone	2.0	8.3
2 — Propanol	2.0	2.8
Diethyl Ether	5.0	36
Methyl Ethyl Ketone	2.0	14

Supply Water

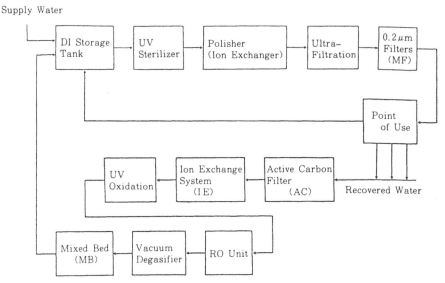

Figure 5   Ultrapure water production system with a recovery loop (case 1).

the ultrapure water (membrane filter, MF, product) accounts for about 70% of the total TOC concentration of 160 ppb. In Fig. 6 and Table 5, the concentration of organic materials in the recycled water is low enough to be fully processed in the ultrapure water recycling system. No organic materials are detected in the ultrapure water.

2.   Ultrapure Water Production Without Recycled Water

Figures 7 and 8 and Tables 6 and 7 show the system flow and list the measurement results. In Fig. 7 and Table 6, no low-molecular-weight organic materials are detected in the

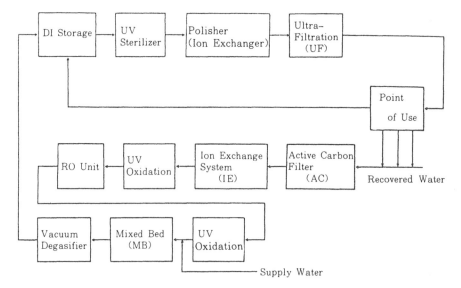

Figure 6   Ultrapure water production system with a recovery loop (case 2).

Table 4 Results of Measurements in an Ultrapure Water Production System with a Recovery Loop (Case 1)

Sample	Methanol (ppb)	Ethanol (ppb)	Acetone (ppb)	2-Propanol (ppb)	Diethyl Ether (ppb)	Methyl Ethyl Ketone (ppb)	Calculative * TOC (ppb)	TOC ** (ppb)	Ratio *** (%)
Recovered Water	4100	62	450	4600	< 5.0	2.5	4600	5800	79
AC Product	4200	66	450	4900	< 5.0	2.5	4800	5500	87
IE Product	4400	66	310	5000	< 5.0	2.0	4900	5200	94
UV Oxidation Product	44	< 5.0	750	4.5	< 5.0	< 2.0	480	1100	44
RO Product	56	< 5.0	720	5.5	< 5.0	< 2.0	470	520	90
DI Storage Tank	76	< 5.0	120	3.0	< 5.0	< 2.0	93	190	49
Polisher Product	110	< 5.0	110	3.0	< 5.0	< 2.0	110	170	65
MF Product	120	< 5.0	110	2.0	< 5.0	< 2.0	110	160	69

*   Calculative TOC : Concentration of the components in terms of TOC
**   Analysis by TOC analyzer (whole TOC)
***   Ratio of Calculative TOC/whole TOC

Table 5 Results of Measurements in an Ultrapure Water Production System with a Recovery Loop (Case 2)

Sample	Methanol (ppb)	Ethanol (ppb)	Acetone (ppb)	2-Propanol (ppb)	Diethyl Ether (ppb)	Methyl Ethyl Ketone (ppb)	Calculative * TOC (ppb)	TOC ** (ppb)	Ratio *** (%)
AC Product	110	< 5.0	17	7.5	< 5.0	< 2.0	57	220	26
IE Product	120	< 5.0	17	9.0	< 5.0	< 2.0	61	190	32
UV Oxidation Product	< 10	< 5.0	< 2.0	< 2.0	< 5.0	< 2.0	—	<50	—
RO Product	< 10	< 5.0	< 2.0	< 2.0	< 5.0	< 2.0	—	<50	—
MB Product	< 10	< 5.0	< 2.0	< 2.0	< 5.0	< 2.0	—	<50	—
Va Product	< 10	< 5.0	< 2.0	< 2.0	< 5.0	< 2.0	—	<50	—
UF Product	< 10	< 5.0	< 2.0	< 2.0	< 5.0	< 2.0	—	<50	—

*   Calculative TOC : Concentration of the components in terms of TOC
**   Analysis by TOC analyzer (whole TOC)
***   Ratio of Calculative TOC/whole TOC

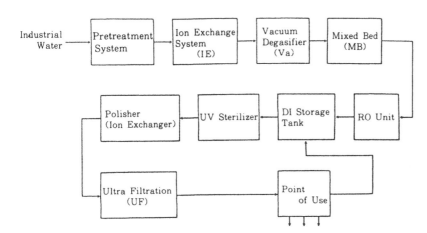

Figure 7 Ultrapure water production system 3.

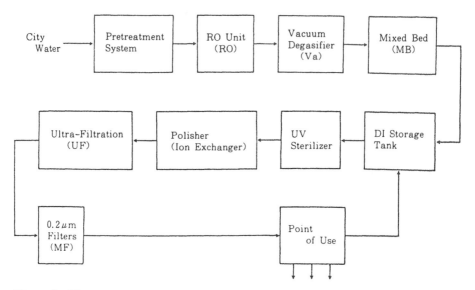

Figure 8   Ultrapure water production system 4.

Table 6   Results of Measurements in Ultrapure Water Production System 3

Sample	Methanol (ppb)	Ethanol (ppb)	Acetone (ppb)	2 - Pro-panol (ppb)	Diethyl Ether (ppb)	Methyl Ethyl Ketone (ppb)	Calculative * TOC (ppb)	TOC ** (ppb)	Ratio *** (%)
Industrial Water	< 10	< 5	< 2	< 2	< 5	< 2	—	600	—
MB Product	15	< 5	< 2	< 2	< 5	< 2	5.6	30	19
RO Product	15	< 5	< 2	< 2	< 5	< 2	5.6	20	28
UF Product	20	< 5	< 2	< 2	< 5	< 2	7.5	20	38

    *   Calculative TOC : Concentration of the components in terms of TOC
   **   Analysis by TOC analyzer (whole TOC)
 *** Ratio of Calculative TOC/whole TOC

Table 7   Results of Measurement in Ultrapure Water Production System 4

Sample	Methanol (ppb)	Ethanol (ppb)	Acetone (ppb)	2 -Pro-panol (ppb)	Diethyl Ether (ppb)	Methyl Ethyl Ketone (ppb)	TOC * (ppb)
RO Entrance	17	6.8	4.8	< 2.0	< 5.0	< 2.0	480
RO Product	< 10	< 5.0	5.1	< 2.0	< 5.0	< 2.0	90
Va Product	< 10	< 5.0	3.3	< 2.0	< 5.0	< 2.0	< 50
MB Product	< 10	< 5.0	2.3	< 2.0	< 5.0	< 2.0	< 50
MF Product	< 10	< 5.0	2.1	< 2.0	< 5.0	< 2.0	< 50

   * Analysis by TOC analyzer

Table 8 Results of Measurement of Ultrapure Water

Factory	Methanol (ppb)	Ethanol (ppb)	Acetone (ppb)	2-Propanol (ppb)	Diethyl Ether (ppb)	Methyl Ethyl Ketone (ppb)	TOC* (ppb)
A	120	< 5.0	110	2.0	< 5.0	< 2.0	160
B	< 10	< 5.0	< 2.0	< 2.0	< 5.0	< 2.0	< 50
C	20	< 5.0	< 2.0	< 2.0	< 5.0	< 2.0	< 50
D	< 10	< 5.0	2.1	< 2.0	< 5.0	< 2.0	< 50
E	< 10	< 5.0	< 2.0	< 2.0	< 5.0	< 2.0	< 50
F	< 10	< 5.0	< 2.0	< 2.0	< 5.0	< 2.0	< 50
G	< 10	< 5.0	< 2.0	< 2.0	< 5.0	< 2.0	—
H	< 10	< 5.0	2.5	< 2.0	< 5.0	< 2.0	< 50

* Analysis by TOC Analyzer

industrial water or the raw water, but methanol is detected in the mixed-bed (MB) product. The concentration of methanol is increased to 20 ppb in the ultrapure water (ultrafiltration, UF product). If this concentration of methanol is converted to TOC in the ultrapure water, it accounts for about 40% of the total volume of TOC in ultrapure water. This clearly indicates that methanol is generated in the ultrapure water production unit. It is thought that the mixed-bed unit and the polisher have something to do with the methanol generation. It has been confirmed that methanol is contained in materials eluted from the anion-exchange resin. Methanol is therefore believed to be formed in the anion-exchange resin.

In Fig. 8 and Table 7, methanol, ethanol, and acetone are detected in the water at the inlet of the reverse osmosis unit. Acetone remains in ultrapure water at 2 ppb, without being processed in the ultrapure water production system.

3. Various Types of Ultrapure Water Production Systems

Table 8 lists the results of analyzing low-molecular-weight organic materials in ultrapure water in eight different ultrapure water production systems, including results described in Refs. 1 and 2. As shown in Table 8, low-molecular-weight organic materials are not detected in most of these systems. Methanol is detected in two systems, acetone is detected in three, and 2-propanol is detected in one.

In summary, the TOC in ultrapure water contains a relatively large volume of low-molecular-weight organic materials.

REFERENCES

1. Y. Yagi, M. Kawakami, and T. Imaoka, Ultra Clean Technol., *1*(2), 67–74 (1990).
2. Y. Hirota and Y. Motomura, Ultra Pure Water Conference, 1987.

8
Operation Management and Cost

Hiromi Kohmoto
Nomura Micro Science Co., Ltd., Kanagawa, Japan

I. INTRODUCTION

The term "ultrapure water" was created more than 10 years ago. At present ultrapure water is widely applied, from biotechnology to ultra–large-scale integration (ULSI). The quality of ultrapure water has been greatly improved, from initial resistivities of 16 MΩ·cm and particle size 0.2 μm to, lately, resistivities over 18 MΩ·cm and particle size 0.05 μm. Ultrapure water production technology has been upgraded, triggered by the rapid development of ULSI. Recently ULSI production plants have increased greatly in size, and it is necessary to operate them without downtime to increase productivity. In this respect, ultrapure water should be provided with stable quality and in a constant manner. These two requirements for ultrapure water are somewhat contradictory. For example, to suppress the bacteria generated mainly in the piping system connected to the use point, regular sterilization with chemicals or hot water is necessary, but this sterilization requires operation of the system to be suspended. To overcome such contradictions, the characteristics of ultrapure water must be well understood. This chapter describes important points for a stable supply of ultrapure water: (1) the design of each processing system and (2) the characteristics of each component unit.

II. MANAGEMENT OF THE OPERATING SYSTEM

Generally, the ultrapure water production system is composed of pretreatment, primary treatment, and secondary treatment. The combination and the function of each component system have been and will be upgraded significantly along with improvements in water quality. The role of each component remains the same. Table 1 lists the reasons for installing each component system and the items to be checked.

Table 1   Purpose of Each System in Total System and Check Items

NAME	PURPOSE	CHECK ITEM	REMARKS
Pretreatment System	1. SS removal. RO membrane Protection, Ion exchanger resin protection. 2. Fe, Mn etc. removal. RO membrane & Ion exchanger resin protection. 3. Organic matter removal. Ion exchanger resin contaminant protection by Organic.	Fouling index (FI) TOC Fe Mn SiO_2	To moderate seasonable and / or any raw water quality variation, and reduce effect to down stream. Check items should be added depending on raw Water quality.
Make-up System	1. Stabilize primary water quality up to almost same quality of Ultra-pure Water.	Resistivity, TOC, SiO_2, DO, Particles, Living bacteria.	Check point should be decided according to make-up system specifications.
Polishing System	1. Final polishing.	Same as make-up system.	

A.   Pretreatment System

The major function of the pretreatment system is to protect the primary treatment system that follows it. Pretreatment should be carefully monitored to avoid rapid degradation of the processed water, which leads to clogging in the reverse osmosis unit.

In general, to remove major amounts of suspended material in the pretreatment system:

1.   Use chemical injection and filtration when the turbidity of the raw water is 5 or less.
2.   Use coagulation + sand filtration when the turbidity of the raw water is 5–10.
3.   Use coagulation and settling or pressurized floating + sand filtration when the turbidity of the raw water is 10 or more.

Sand filtration includes not only simple sand filtration but also filters with two or more layers. When the raw water contains large amounts of Fe and Mn, Fe and Mn removal units are necessary. When polarized organic materials, such as trihalomethane, are present, the raw water should be processed with active carbon. Since the pretreatment system includes many long physical processing steps, such as coagulation and settling, the following points are important:

1.   Keep operating the system without downtime as much as possible.
2.   During intermittent operation, note the time required to stabilize the treatment after start-up.
3.   Check changes in raw water quality so that appropriate action can be taken quickly.

With regard to point 3, when the raw water undergoes preliminary processing outside the system (for example, industrial water), the processing procedures used should be fully studied and grasped. Changes in water quality with the seasons should be also noted. Figure 1 shows the operation cycle and fouling index (FI) values for sand filtration with chemical injection. Figure 2 illustrates quality change in industrial water within a short span of time.

Both examples 1 and 2 in Fig. 1 indicate it takes considerable time to return to

Example—1
Raw Water = City Water

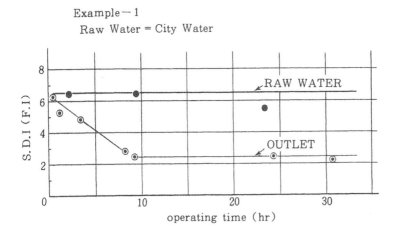

Example—2
Raw Water = Industrial Water

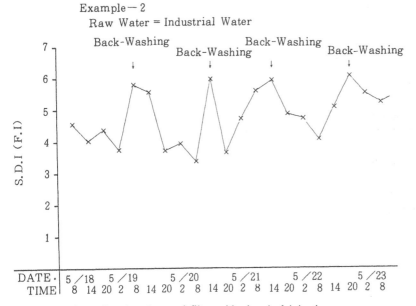

Figure 1 Operating data for sand filter with chemical injection.

optimum water quality after backwashing. It is necessary to understand the time required to return to acceptable levels of water quality.

Figure 2 diagrams changes in the raw water quality with weather. When it rains, the alkalinity falls. This is thought to be because the band sulfate and polyammonium chloride (PAC) injected into raw water to upgrade its quality to the level of industrial water lower the alkalinity of the original raw water. Because of this low alkalinity, the appropriate coagulation effect is not obtained in the pretreatment system even when a flocculant like PAC is injected into the water. To obtain sufficient coagulation, the characteristics of the raw water, including the pH, should be studied in advance to find out which flocculant should be injected.

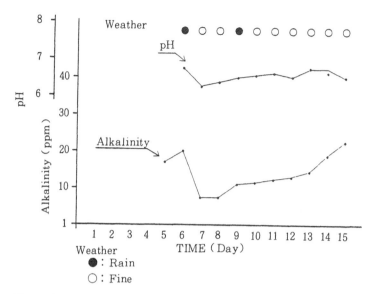

Figure 2   Example of industrial water quality fluctuation.

B.   Primary Treatment System

To operate the secondary treatment system in a stable manner for a long time, the primary treatment system needs to be managed in almost the same way as the secondary treatment system.

Depending on the quality of the raw water, the primary treatment system employs different components, such as front-stage reverse osmosis and end-stage reverse osmosis. Lately a two-stage reverse osmosis system has been developed that can achieve resistivities of 1–5 MΩ·cm by a reverse osmosis membrane.

Although different component units require different management, they share the advantages of coping with seasonal changes in raw water quality and ease of care in the event of the unexpected. The details are described in another chapter.

C.   Secondary Treatment System

Since this is the final stage of processing ultrapure water, which has a direct effect on ultrapure water quality, the real-time management of water quality is desirable.

The secondary treatment system is composed of the pure water tank, the circulation pump, the cartridge polisher, the UV sterilization unit, and the ultrafiltration unit. The following should be noted to maintain the system:

1.   Trend management of water quality based on regular water quality inspection
2.   Regular maintenance program
     a. Overhaul of pumps
     b. Calibration of instruments
     c. Replacement of nondurable parts (lifetime prediction)
     d. Sterilization cycle (sterilization method)

Table 2 Barometer of Treated Water Quality for Each System

ITEM	PRETREATMENT SYSTEM	MAKE–UP SYSTEM	POLISHING SYSTEM
F. I.	< 4	—	—
Residue (ppm)	—	< 0.02	< 0.005
TOC (ppm)	< 2	0.03 ~ 0.1	0.01 ~ 0.05
Particles (pcs/mℓ)	—	100 ~ 500 (0.2 μ)	5 ~ 20 (0.1 μ)
Bacteria (cfu/100 mℓ)	50 ~ 500	50 ~ 100	< 10
Dissolbed, SiO_2 (ppb)	< 25000	< 10	< 3
Resistivity (MΩ · cm)	—	10 ~ 16	> 18
Aℓ (ppb)	—	—	< 1.0
Cr (ppb)	—	—	< 0.1
Cu (ppb)	—	< 1.0	< 0.5
Fe (ppb)	—	< 1.0	< 0.5
Mn (ppb)	—	—	< 0.5
K (ppb)	—	< 0.5	< 0.1
Na (ppb)	—	< 0.5	< 0.1
Zn (ppb)	—	—	< 0.5
Cℓ (ppb)	—	< 10	< 0.5

Table 2 lists reference values for water quality processed in each system. Table 3 lists the functions of each component unit. Because the quality of the water processed in each system is determined by the performance of the component units, every component unit should be operated under optimum conditions. Important management points for each component unit are described here.

III. MANAGEMENT OF COMPONENT UNITS

A. Reverse Osmosis Unit

The reverse osmosis unit is capable of removing not only ions but also colloidal particles and TOC. This unit is always installed in the primary treatment system. Although the reverse osmosis unit is very functional, it should be treated properly to avoid serious trouble.

Lately the low-pressure complex membrane has been widely adopted. From the management point of view, although the complex membrane processes a large amount of water with a high desalting efficiency, it is easily contaminated and is subject to oxidation deterioration (it has a low free chlorine resistance).

Table 4 compares the cellulose acetate (CA) membrane and the complex membrane in terms of management. It is obvious that the complex membrane requires more stringent management of raw water quality than the cellulose acetate membrane.

Table 5 lists the causes of trouble in the reverse osmosis unit and countermeasures. The following should be properly managed for stable operation of the reverse osmosis unit.

Table 3   Function of Equipment for Ultrapure Water System [1]

Equipment \ Impurity to be removed	Suspended Solid	Ions	Particles	Micro-Organism	Organic Mater	CO	O	Colloidal Mater
Flocculator	+ +		+		+			+ +
Sand Filter	+ +		+					
Activated Carton Filter					+ +			
Cartridge Filter (1 ~ 5 μM)	+ + + +		+ +					
Reverse Osmosis		+ + +	+ + + +	+ + + +	+ + +			+ + + +
Degasifier						+ + +		
Vacuum Degasifier						+ + + +	+ + + +	
Deionizer (Ion Exchange resin)	+ + + +				+			+ +
UV Sterilizer				+ + + +				
UV Oxdizer					+ + + + (Decompose)			
MF Membrane filter			+ + + +	+ + + +				
Ultra-filter			+ + + +	+ + + +				+ + + +

Effectiveness :   + Fair   + + Good   + + + Better   + + + + Excellent

Table 4   Comparison of CA and Composite Elements

	C. A. Elements	Composite Elements
Material of membrane	C. A.	Polyamide or Polyimide
Product Water Flow Rate m^3/day 8″ Element (High rejection type)	17 ~ 23 at 30 kgf/cm^2	22 ~ 28 at 15 kgf/cm^2
Salt rejection (%) 1500 ppm NaCl	94 ~ 97	(96) ~ 99
Max. Operating Temperature	35℃	40 ~ 45℃
pH Range: Operating Cleaning	4 ~ 7 3 ~ 8	4 ~ 9 1 ~ 12
Feed chlorine Tolerance (Recommend)	< 1 ppm	< 0.1 ppm (0 ppm)
Max. Feed F. I (S. D. I) (Recommend)	< 5 (< 4)	< 4 (< 3)

Table 5 Trouble in RO (Reverse Osmosis) Plants and Countermeasures [2]

Troubles of R. O. could be guessed by checking the extraordinal
changes of flux, rejection and pressure drop of R. O. membrane modules.
Therefore it is advisable for clients to check daily and to take
prompt countermeasures to eliminate troubles in the diagram below.

CAUSES		PHENOMENA			CHECK POINTS	COUNTER-MEASURE
		flux	rejection	pressure drop		
MODULE	membrane degemeration	↗ increase	⬊ decrease (main)	↘ decrease	operating time, feed temp & pH, residued Cl₂ concentration	clean'g or REPLACE element
	leakage in R. O. element	↗ increase	⬊ decrease (main)	↘ decrease	vibrat'n, back-pressure, shock pressure	— " —
	compaction of membrane	⬊ decrease (main)	↗ increase	↗ increase	Temp / Press of feed, operation time	— " —
	leakage from O-ring	↗ increase	⬊ decrease (main)	↘ decrease	vibrat'n, shock pressure	change O-ring
	inproper brine seal	↘ decrease	⬊ decrease (main)	↘ decrease	element fitting, material withe-ring, direct'n	change brine seal
	inter-connector break	↗ increase	⬊ decrease (main)	↘ decrease	big pressure drop, High Temp	change connector
	center pipe break	↗ increase	↘ decrease	↘ decrease	— " —	change module
	deformation of element	↘ decrease	→ decrease	⬈ increase (main)	— " —	— " —
	membrane fouling (suspended solid)	↘ decrease	↘ decrease	⬈ increase (main)	pretreat't, feed water spec.	clean'g
	membrane fouling (scale)	↘ decrease	↘ decrease	↗ increase	pretreat't, feed water spec.	clean'g
	membrane fouling (organic oil)	⬊ decrease (main)	↘ decrease	↗ increase	— " —	— " —
FEED WATER · PRETREATMENT	temperature — high	⬈ increase (main)	↘ decrease	↘ decrease	seasonal change, pump efficiency	pressure adjust, cool'g
	temperature — low	⬊ decrease (main)	→	↗ increase	seasonal change, heater	pressure adjust, heating
	pressure — high	⬈ increase (main)	↗ increase	↘ decrease	pump, valve	pressure adjust
	pressure — low	⬊ decrease (main)	↘ decrease	↘ decrease	pump, valve filter	— " —
	brine flow rate — big	→ increase	→ increase	⬈ increase (main)	feed flow rate, press control valve	flow adjust't,
	brine flow rate — small	↘ decrease	↘ decrease	⬊ decrease (main)	feed flow rate, press control valve, press drop	— " —
	pH (membrane withering)	↗ increase	↘ decrease	↘ decrease	pH control	pH control
	salt concentration — high	⬊ decrease (main)	↘ decrease	↘ decrease	feed water	pressure cont'l
	salt concentration — low	⬈ increase (main)	↗ increase	↗ increase	— " —	— " —
	insoluble matters (scaling)	↘ decrease	↘ decrease	↗ increase	feed water quality, recovery	pressure cont'l
	residued Cl₂ (longtime) — high	↗ increase	⬈ increase (main)	↘ decrease	chlorinator, membrane withering	Cl₂ conc. adjust
	residued Cl₂ (longtime) — low	↘ decrease	↘ decrease	↗ increase	— " —	chemical clean'g

Note:
↗ : increase
↘ : decrease
⬈ ⬊ : main phenomenon

1. Salt Removal Efficiency

In general, the salt removal efficiency indicated in the reverse osmosis membrane catalogue is for a single membrane. It can be expressed as

$$\text{Salt removal ratio} = 1 - \frac{\text{ion concentration in filtered water}}{\text{average ion concentration of raw water}} \times 100$$

where the average ion concentration of raw water = the average ion concentration in raw water + the ion concentration of condensed water.

For convenience in the actual management of the reverse osmosis unit, the following equation is usually employed to determine salt removal efficiency:

$$\text{Salt removal ratio of unit} = 1 - \frac{\text{ion concentration of filtered water}}{\text{ion concentration of raw water}} \times 100$$

The salt removal efficiency of the unit is in general lower than that of the membrane.

Now let us discuss the operation model shown in Fig. 3. In case 1, the unit is operated under normal conditions and the removal efficiency gradually decreases. In case 2, on the other hand, the removal efficiency drops sharply. This sharp drop is thought to be caused by mechanical destruction of the membrane or sealing parts, and emergency countermeasures should be taken. In case 3, the removal efficiency is degraded in a short period of time. This is thought to be because of a chemical change in the membrane (hydrolysis or oxidation degradation) or sudden fouling or scaling on the membrane surface.

2. pH

Generally the reverse osmosis membrane is used on the acid side (pH 5.0–6.5) to suppress scaling on the membrane surface by increasing the solubility of Ca ion and Mg ion and to prevent hydrolysis of the cellulose acetate membrane.

The recently developed complex membrane, however, can be used on the alkali side. This is because the ion removal efficiency of the complex membrane is greatly affected by

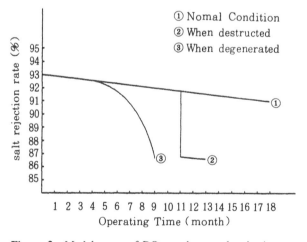

Figure 3   Model curve of RO membrane salt rejection rate. (From Ref. 3, p. 284.)

the pH, as shown in Fig. 4. In this case, measures should be taken to prevent the deposition of Ca ion and Mg ion: an effective scaling inhibitor is usually installed.

3. *FI, Volume of Filtered Water, and Differential Membrane Pressure*

The fouling index is an indicator of raw water turbidity. This index is often applied to manage the reverse osmosis unit. When raw water with a high FI value is used, more material adheres to the reverse osmosis membrane, resulting in degradation of removal efficiency and reducing the volume of filtrated water. This problem is treated by cleaning the membrane. Table 6 lists appropriate chemicals for different membrane contaminants. Cleaning efficiency is affected by the contaminants in the raw water, the materials and structure of the membrane, and the types of cleaning chemicals.

In general, the stability of the volume of filtered water is correlated with the filtration volume per membrane. Figure 5 shows the relation in a test operation. If the filtration volume per membrane is large, fouling occurs easily. In this case, the cleaning should be done frequently.

4. *Recovery Ratio*

The recovery ratio is the proportion of filtered water volume to the volume of raw water. It can be expressed as

$$\text{Recovery ratio} = \frac{\text{volume of filtered water}}{\text{volume of raw water}} \times 100$$

The recovery ratio should be based on the following points:

1. The concentration of materials with relatively low solubility, such as SiO_2, should not exceed the saturation solubility.
2. Sufficient condensed water is required to drain particles and other contaminants.

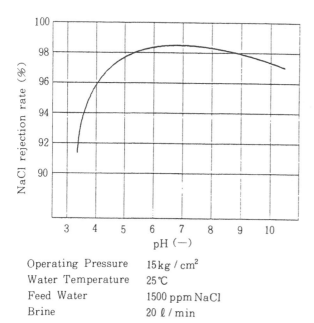

Operating Pressure   15 kg / cm²
Water Temperature   25 °C
Feed Water   1500 ppm NaCl
Brine   20 ℓ / min

Figure 4 Relation between pH and composite RO membrane rejection rate. (From Ref. 2.)

Table 6   Chemical Cleaning Reagents [2]

Fouling Substance	Chemical Reagent	Conditions
Calcium scale	Citric acid	1 — 2% solution, pH 3 — 4 adjusted with ammonia
	EDTA	1 — 2% solution, pH 7 — 8 adjusted with ammonia or sodium hydroxide
Metal hydroxide	Citric acid	1 — 2% solution, pH 3 — 4 adjusted with ammonia
Organic soil	Detergent	0.1 — 1.0% solution, pH 7 — 8 adjusted with sulfuric acid or sodium hydroxide
Inorganic colloid	Citric acid	1 — 2% solution, pH 3 — 4 adjusted with ammonia
Bacterial matter	Formaldehyde	0.1 — 1.0% solution
	Detergent	0.1 — 1.0% solution, pH 7 — 8 adjusted with sulfuric acid or sodium hydroxide

Figure 6 plots the allowable SiO_2 concentration for the reverse osmosis system. Table 7 lists operation specifications for the reverse osmosis module. As shown in Fig. 6, the recovery ratio for reverse osmosis is not determined only by the saturation solubility of silica but is also affected by the thickness of the concentration polarization layer.

A data sheet like that in Table 8 is very helpful in managing the reverse osmosis system.

B.   Ion-Exchange Unit and Ion-Exchange Resin

Since the ion-exchange unit (ion-exchange resin) is capable of removing almost all the electrolytes in raw water, it is always employed in the ultrapure water production system. It removes only electrolytes. It does not remove nonelectrolytes, including colloidal materials (such as colloidal silica), or organic materials. The nonelectrolytes should be removed by other methods.

It must be noted that the ion-exchange resin is made of organic materials. Organic materials are subject to oxidation decomposition and mechanical destruction, and organic materials that have not polymerized in the course of production may elute.

Important points in managing the ion-exchange resin installed in the ultrapure water production system are described here.

1.   Oxidation Degradation

The ion-exchange resin suffers from oxidation degradation as a result of the oxidative reagent in raw water. In particular, the cation-exchange resin is hygroscopic; it gradually absorbs water and eventually turns to powder. In the process of this change, a significant volume of organic material is released as the polymerization chains of the ion-exchange resin are cut.

In the mixed-bed method, mutual contamination takes place between the positive ion-exchange resin and the negative ion-exchange resin, resulting in decreased resistivity,

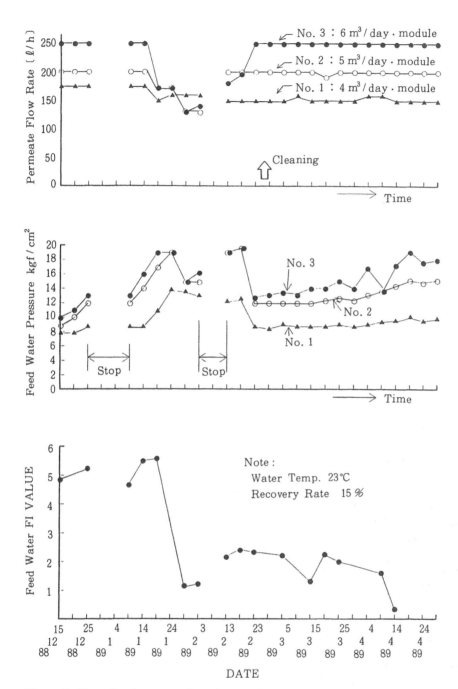

Figure 5 Example of test operation of composite RO membrane.

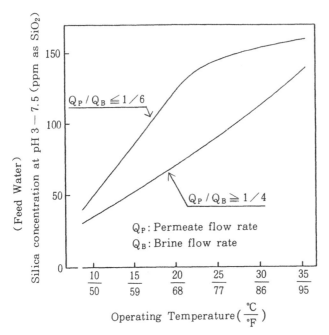

Figure 6   Allowable SiO$_2$ concentration for RO system (RO unit operating conditions). (From Ref. 2.)

silica leakage, and decreased volume of processed water. Furthermore, the units that follow, such as the reverse osmosis unit and the ultrafiltration unit, are clogged.

In the two-bed, three-tower method, the front-stage positive ion-exchange resin undergoes oxidation decomposition and the end-stage negative ion-exchange resin is contaminated, which leads to silica leakage and a decrease in the processed water volume. The mixed-bed ion-exchange unit, the reverse osmosis membrane, and the ultrafiltration membrane, all of which are located downstream, are also affected.

Figure 7 shows the variation in ultrafiltration with time as the membrane is clogged as a result of oxidation degradation of the ion-exchange resin. Oxidation degradation can be suppressed by injecting reducing agents, such as Na$_2$SO$_3$ and NaHSO$_3$, or by installing an active carbon tower. When reducing chemicals are injected, constant monitoring should

Table 7   Recommended Operating Conditions for RO Element

	CA Element		Composite Element	
	4″	8″	4″	8″
1. Feed Flow Rate	< 45 ℓ/min	< 180 ℓ/min	< 50 ℓ/min	< 200 ℓ/min
2. Brine Flow Rate	> 10 ℓ/min	> 40 ℓ/min	> 10 ℓ/min	> 40 ℓ/min
3. Brine/Permeate Flow Ratio	> 6	> 6	> 6	> 6
4. Pressure Drop (Per Element) (Per Vessel)	< 2 kgf/cm^2 < 3 kgf/cm^2	< 2 kgf/cm^2 < 3 kgf/cm^2	< 1 kgf/cm^2 < 2 kgf/cm^2	< 1 kgf/cm^2 < 2 kgf/cm^2

Table 8 Sample Date Sheet [2]

Parameters		Design Value	Daily Operation Data
1. Date & Time for data logging		Design	XX. YY
2. Total Operating Hours		Initial	ZZ
3. Number of Vessels in operation		6 / 3	6 / 3
4. Feedwater Conductivity	μ S / cm	3000	2000
5. Feedwater pH		6.5	6.5
6. Feedwater FI (SDI 15)		< 4	3
7. Feedwater Temperature	℃	20	25
8. Feedwater Pressure	kg / cm^2	14.4	12.3
9. Feedwater Chlorine Concentration	ppm	0	0
10. Feedwater each ion concentration		—	—
11. Brine Conductivity	μ S / cm	—	7300
12. Brine pH		7.0	6.9
13. Pressure Drop for each Bank	kg / cm^2	1.0 / 1.0	1.0 / 1.0
14. Brine Flow Rate	m^3 / d	325	325
15. Total Permeate Conductivity	μ S / cm	125	89
16. Permeate Conductivity for each Vessel		—	—
17. Permeate Pressure	kg / cm^2	0.3	0.3
18. Total Permeate Flow Rate	m^3 / d	975	975
19. Permeate Flow Rate for each Bank	m^3 / d	705 / 270	—
20. Permeate each ion concentration		—	—
21. Total recovery Ratio	%	75	75
22. Recovery Ratio for each Bank	%	54 / 46	—

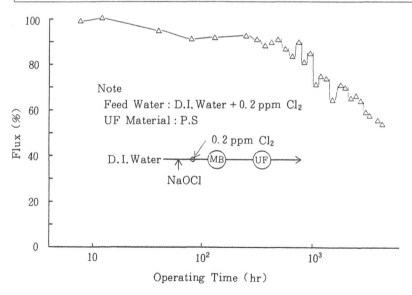

Figure 7 Example of UF flux decrease due to oxidized degradation of mixed-bed ion-exchange resin.

be introduced (e.g., a free chlorine meter and ORP meter) to prevent possible accident to the injection pump.

   H_2O_2 sterilization of the polisher should be conducted carefully so that no H_2O_2 remains in the unit.

2.   Organic Contaminants

The ion-exchange resin is usually recovered with chemicals (NaOH or HCl) and is used repeatedly. When it is contaminated with organic materials in raw water, however, these organic materials cannot be fully removed by normal recovery procedures. This is why organic materials accumulate on the ion-exchange resin. New ion-exchange resin for use in a polisher resin is compared with resin that has been recovered several times. The Na concentrations in water processed with the two resins are as follows:

Raw water: 0.1–0.2 μgL^{-1}
Water processed with the new resin: <0.01 μgL^{-1}
Water processed with the spent resin: 0.2–0.8 μgL^{-1}

The negative ion-exchange resin is contaminated with organic materials (mainly organic acid). NaOH is used to recover this resin. However, the Na in this NaOH is adsorbed on the resin and eventually desorbed into the processed water. Therefore new resin should always be used in a polisher.

   The resin used in the primary treatment system is always recovered several times. It is effective to use hot NaOH in the recovery to facilitate removal of organic materials during recovery procedures. When hot NaOH recovery does not work effectively, NaCl or NaCl + NaOH recovery can be done.

3.   Organic Material Elution at the Time of Resin Replacement

The resin used for the polisher is recovered in advance in many cases, but the resin used in the pretreatment system is not usually recovered. The pretreatment resin should be recovered several times in series to reduce the elution of organic materials. (See Primary Treatment System.)

   The manufacturing and recovery of resins have improved a great deal, and the elution of organic materials can be decreased rapidly to very low levels with current resins. Figure

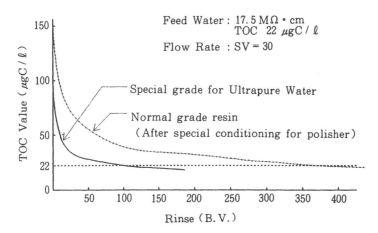

Figure 8   TOC extract curve from polisher ion-exchange resin.

8 compares improved resin and conventional resin in terms of organic material elution. With our goal of further decreases in the TOC concentration in ultrapure water the polisher at the final stage should employ the improved resin.

4.   Recovery of Negative Ion-Exchange Resin with Hot NaOH

Some recovery methods allow the SiO_2 in raw water to accumulate on the negative ion-exchange resin, which raises the SiO_2 concentration in the processed water. To prevent this, type I negative ion-exchange resin must be recovered with hot NaOH at 40–50°C. It is reported that even if the SiO_2 concentration in the raw water is low, the concentration rises in the resin to form the low-polymer silica. Since such silica cannot be desorbed by normal recovery procedures, physical cleaning with hot NaOH should be conducted. Besides, the colloidal silica formed in this way cannot be removed in the ion-exchange resin unit that follows. Therefore perfect control of temperature during negative ion-exchange resin recovery is necessary. SiO_2 leakage cannot be detected with the resistivity meter.)

It is important in properly managing the ion-exchange resin to detect the quality degradation in the processed water as soon as possible. For this purpose, constant monitoring of resistivity (or conductivity), SiO_2, and TOC is desirable. The High Pure Monitor is a meter that conducts comprehensive management of the processed water. This meter can measure the total residues on evaporation to the level of 1 ppb.

C.   UV Sterilization Unit

UV sterilization is widely adopted in ultrapure water production because it does not use chemicals and it can suppress running costs. The low-pressure mercury lamp should be replaced periodically to maintain the same level of irradiation intensity. Figure 9 shows the change in irradiation intensity of the low pressure mercury lamp over time. The lamp should be replaced once a year. Figure 9 shows that, after 1 year of use, the irradiation intensity is about 60% of that of a new lamp. Regular inspection for bacteria is required at the outlet of the sterilization unit.

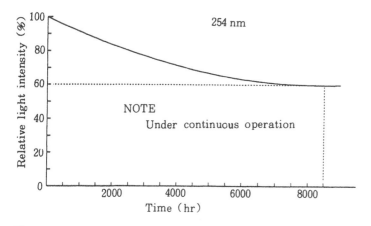

Figure 9   Relative energy curve. (From Ref. 6.)

D.   Ultrafiltration Unit

The ultrafiltration unit is installed as the final filter in almost all ultrapure water production systems. This is because the ultrafiltration membrane is superior to the
reverse osmosis membrane in the following areas: (1) it can be operated at lower pressures, (2) it has better chemical resistance and can be easily cleaned and sterilized, and
(3) its structure is simple and it is relatively impervious to secondary contamination
from the processed water. The water fed to the ultrafiltration membrane should be very
clean, but in a long run, the membrane becomes clogged and the membrane intensity
decreases.

Figure 10 shows the relationship between the number of particles in ultrapure water
and the volume of the filtered water.

Membrane area is determined based on prediction of the number of particles in the
feedwater. It is necessary to monitor the operation every day, however. Actual examples
of sudden clogging of a membrane are described here.

Oxidation decomposition of the ion-exchange resin upstream
Malfunction of the reducing agent injection pump
Backflow of oxidative reagent during a power failure
Leakage from the reverse osmosis membrane placed in the middle stage

Because the water quality at the outlet of the reverse osmosis membrane is monitored
only in terms of resistivity, the leakage from the reverse osmosis membrane was not
noticed.

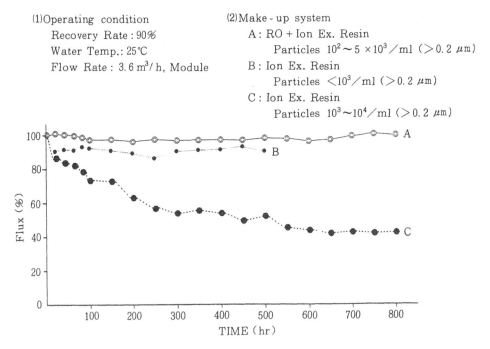

(1) Operating condition
   Recovery Rate : 90%
   Water Temp.: 25°C
   Flow Rate : 3.6 m³/h, Module

(2) Make-up system
   A : RO + Ion Ex. Resin
      Particles $10^2 \sim 5 \times 10^3$/ml (>0.2 μm)
   B : Ion Ex. Resin
      Particles $<10^3$/ml (>0.2 μm)
   C : Ion Ex. Resin
      Particles $10^3 \sim 10^4$/ml (>0.2 μm)

Figure 10   UF flux rate curve comparison of different make-up system configurations. (From Ref.
4.)

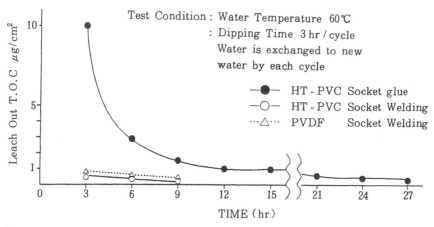

Figure 11   Elution test of piping material (TOC).

E.   Piping System

The pipes are cleaned well at the time of installation, and piping materials with limited elution properties are selected carefully. Nevertheless, elevated elution levels are detected a certain time after installation. Figure 11 shows the TOC elution from piping materials. In a piping system installed with the adhesion method, 27 h cleaning with hot pure water at 60°C is required. The generation of bacteria in the piping system should also be noted. Regular sterilization (once every 1–4 months) is effective.

Figures 12 through 14 plot the measurement results of particles and bacteria in a

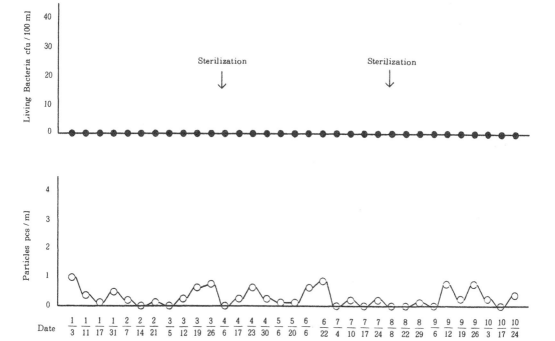

Figure 12   Particles and living bacteria at UF outlet.

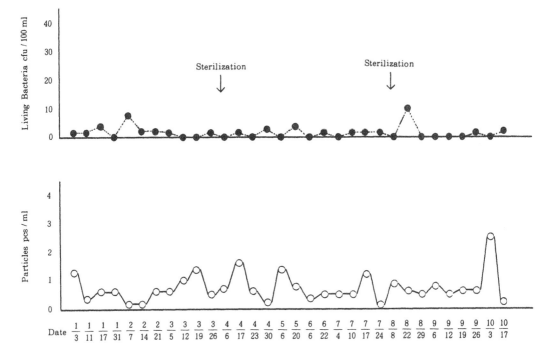

Figure 13 Point of use A.

piping system actually installed in a plant. Figure 12 shows the measurement results at the outlet of the ultrafiltration unit; Figs. 13 and 14 show the results at arbitrary use points.

At the outlet of the ultrafiltration unit, the number of bacteria is stable at 0 cfu per 100 ml and the number of particles is also well suppressed at 1 ml^{-1}. At use points A and B, however, both bacteria and particles increase. This is obviously because of the generation of bacteria in the piping system. Based on these findings, it is clear that regular sterilization is required in the ultrapure water piping system. To the best of our knowledge, no difference has been recognized between clean PVC pipe and PVDF pipe in terms of bacterial increase. Figure 15 shows bacterial increase in pure water over time. The data in Figure 15 are from a beaker-scale experiment. It is expected that a similar phenomenon takes place in microentrapment sites on the inner surface of pipes. Sterilization methods applied in the secondary treatment system and their effect are described here.

1. H_2O_2 Sterilization

Because H_2O_2 sterilization is very easy to handle, it is widely employed. In the usual H_2O_2 sterilization, 1–2% H_2O_2 solution is circulated in the system and left for 1–2 h. In this case, the purity of the H_2O_2 should be the same as that used in the semiconductor production process to avoid contamination of the entire piping system. Table 9 shows the effect of H_2O_2 sterilization.

2. Formalin Sterilization

If H_2O_2 is left in the system for a long time, it may cause oxidation degradation in the ultrafiltration and reverse osmosis membranes. Also, H_2O_2 itself is decomposed, losing its sterilizing capability. The formalin solution should therefore be confined should

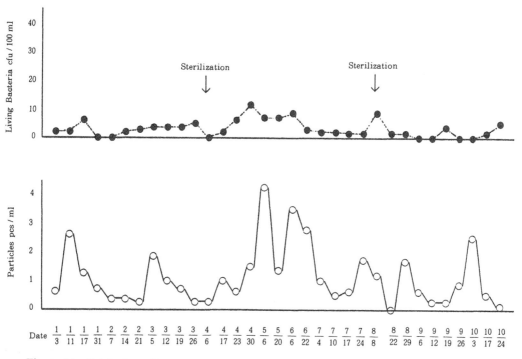

Figure 14 Point of use B.

operation of the system be suspended for long. The optimum concentration of formalin for long-term confinement is around 0.5%.

3. *Ozone Sterilization*

Ozone has excellent sterilizing effects because of its strong oxidizing action. Moreover, the ozone dissolved in water can be easily decomposed by ultraviolet at a wavelength of

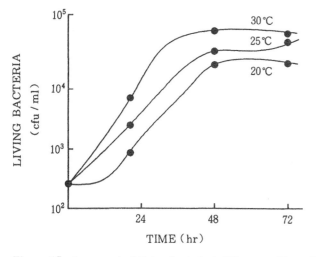

Figure 15 Increment of living bacteria in DI water. (From Ref. 3, p. 286.)

Table 9   Sterilization Effect of Hydrogen Peroxide

Agent	Method Concentration, Time	Bacillus subtilis	Escherichia coli	Bacteria in DI Water (Nonidentification)
H₂O₂	3% 30 min	—	—	
	non-sterilized area	⧣	⧣	
	1% 10 min			—
	30 min			—
	60 min			—
	3% 10 min			—
	non-sterilized area			⧣

NOTE. —: 0 cfu / mℓ,   ⧣: 10³ cfu / mℓ

Source: From Ref. 3, p. 288.

254 nm. These are the major relevant characteristics of ozone. Various sterilization experiments with ozone indicate that when bacteria in the pipes reach a level of 5–10 cfuml⁻¹, the appropriate conditions for sterilization are

1.   ozone concentration of 0.5 ppm or more
2.   sterilization time of 1–2 h

Since ozone has a short half-life, it must be noted that its sterilizing effect in an ultrapure water piping system may be considerably weakened. The dead space and portions with low flow rate further deteriorate the sterilizing effect. Another factor to be taken into consideration with regard to ozone sterilization is the increase in dissolved oxygen.

4.   Hot-Water Sterilization

Hot-water sterilization has been applied to pure water systems for the medical products industry, and its effect has been fully confirmed. In general, hot-water sterilization at 80°C for 30 minutes exhibits sufficient effect. Since this method is also free of chemicals, the system is not contaminated, and it takes only a short time to recover the original level of water quality after sterilization. In H_2O_2 sterilization, it takes 5–10 h to return the water quality to the sufficient level, but hot-water sterilization requires only 2–3 h. Figure 16 is a model of the hot-water sterilization method. The resistivity returns to 18 MΩ·cm within 30 minutes after the polisher is connected.

The secondary effect of hot-water sterilization is the capability of membrane cleaning. Figure 17 shows an example. Hot-water sterilization lowers the average filtration differential pressure. (This effect is greater at 90°C than at 60°C.) The temperature of the piping system in hot-water sterilization, is often kept at 80°C. Table 10 illustrates the heat resistance of bacteria living in pure water. It is said that the bacteria in pure water can be killed by hot-water sterilization at 60°C for 20 minutes.

An ultrafiltration membrane treated with hot-water cleaning is compared to an uncleaned ultrafiltration membrane. The volume of metal adhering to the two membranes is quite different, which means that hot-water sterilization is effective in removing materials that adhere to the membrane (Table 11).

There are disadvantages to hot-water sterilization, as well.

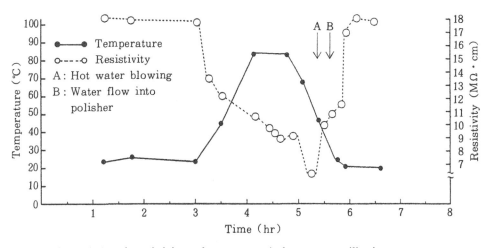

Figure 16   Variations in resistivity and temperature in hot-water sterilization.

It consumes a large amount of thermal energy. The heat is usually given by vapor. In a
   system of 50 m^3h^{-1}, for example, 1500–2000 kgh^{-1} of vapor is required.

The heat resistance and the elution resistance (at high temperatures) of the piping
   materials and the secondary treatment system must be considered. The equipment at
   use points and the drain pipes should be carefully selected for heat resistance. In
   practice, it is not easy to conduct hot-water sterilization in the entire system. If
   hot-water sterilization cannot be applied to the entire system, perfect results cannot be
   expected.

5.   HF Sterilization

It has been learned that HF injection at 10 ppm into ultrapure water exhibits the same
sterilizing effect as H_2O_2. Since HF sterilization is effective at even lower concentration

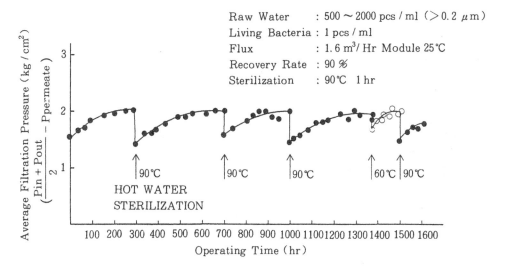

Figure 17   UF module sterilization effect by hot water. (From Ref. 5, p. 86.)

Table 10 Heat Durability of Microorganisms Under Hot and Wet Conditions

	Heat Durability	
Kinds of micro organisms	Temp (℃)	Time (min)
(1) Bacteria in Ultrapure Water		
Pseudomonas	60	20
Flavobacterium	60	10
Alcaligenes	60	20
Micrococcus	60	20
Aeromonas	60	20
(2) Molds in Ultrapure Water (Hypha)	60	20
(3) Yeasts in Ultrapure Water	60	20

Source: From Ref. 5, p. 82.

than H_2O_2 and the residual HF can be removed by the polisher, this method is very useful. Various experiments are now underway to study this method further.

IV. MAINTENANCE AND COST

A. Maintenance

The pure water system requires both regular maintenance and nonregular maintenance. The nonregular maintenance is required in the case of unexpected accidents, which should be avoided as much as possible.

On the other hand, regular maintenance means preventive maintenance, which is essential to proper operation of the system. To conduct regular maintenance effectively, every checkpoint should be fully understood and the appropriate methods should be employed. (Table 8 is a sample data sheet listing appropriate methods.)

In general,

1. Sterilize the system every 1–6 months.
2. Clean the reverse osmosis unit every 1–12 months. (depending on the installed part and the quality of the raw water).
3. Replace nondurable parts.
    a. Filters should be replaced when the rise in the differential pressure is detected. When filters are used in the primary and secondary treatment systems, however, it is desirable to replace them at least once a year to suppress contamination by bacteria.
    b. The reverse osmosis membrane should be replaced every 2–3 years. The salt removal efficiency is the indicator of the membrane condition.
    c. The ultrafiltration membrane should be replaced every 2–3 years. The membrane

Table 11 Effect on UF Module of Hot-Water Sterilization

	6 months after Start – up with sterilization	1 year after Start – up without sterilization
Na	9	73
K	3	25
Ca	7	20
Fe	21	96
Co	1	2
Cu	1	13
Zn	1	29
Pb	4	31
Al	3	28
Si	< 5	83

Heat Water Sterilization Condition : Cycle / 2 weeks
Unit : mg / dry membrane 1 kg
Source: From Ref. 5, p. 90.

differential pressure, the particle removal efficiency, and the physical strength of membrane are indicators of the membrane condition.

d. Depending on degree of degradation of the ion-exchange resin installed in the pretreatment system, it should be replaced every 1–3 years. The polisher should be renewed once every 6–12 months.

e. The UV lamp is replaced once a year.

f. Sliding parts of the rotating equipments are replaced once a year.

4. Each instrument that monitors conditions at the checkpoints should be calibrated once a year.

5. Water quality analysis is essential to understand the operational state of the entire system. Table 12 lists the items and the frequency with which they should be checked. The analytic results should be plotted (see Figs. 12–14).

B. Maintenance Cost and Initial Cost

Table 13 is a general example of maintenance cost and initial cost. These costs depend on the size of the system and other conditions. The costs of water quality analysis, personnel expenses, depreciation costs of equipment, and repair costs are excluded from Table 13. The cost of water quality analysis is estimated around ¥10–15 million for large systems if the items in Table 12 are fully covered.

Table 13 shows that the running cost decreases if the some of the drain water is recycled, for two reasons: (1) city water is the raw water, at a high cost of ¥310 m^{-3}, and (2) the cost required to heat the ultrapure water (usually 23–25°C) can be partly saved if recycled water is used. In Japan, raw water costs are likely to increase further, and the water supply is expected to be limited in the future. Therefore the large ultrapure water production systems being manufactured often adopt methods to recycle water.

Other methods can be adopted to reduce running costs: use a low-pressure reverse osmosis membrane, and consume electrical energy effectively by employing an inverter.

Table 12 Recommended Periodic Check Items

	Within 1 year after start up			Over 1 year after start up		
	Raw Water	Primary Water	Ultrapure Water	Raw Water	Primary Water	Ultrapure Water
pH	continuous	———	———	continuous	———	———
Turbidity	1 th / m	———	———	1 th / 3 m	———	———
TOC	1 th / m	continuous	continuous	1 th / 3 m	continuous	continuous
Fl Value	1 th / week	———	———	1 th / week	———	———
Living Bacteria	1 th / m	1 th / 2 weeks	1 th / 2 weeks	1 th / 3 m	1 th / m	1 th / m
Particle	———	1 th / 2 weeks	1 th / 2 weeks	———	1 th / m	1 th / m
Resistivity	1 th / m	continuous	continuous	1 th / 3 m	continuous	continuous
Ca	1 th / m	1 th / m	1 th / m	1 th / 3 m	1 th / m	1 th / m
Mg	1 th / m	1 th / m	1 th / m	1 th / 3 m	1 th / m	1 th / m
K	1 th / m	1 th / m	1 th / m	1 th / 3 m	1 th / m	1 th / m
Na	1 th / m	1 th / m	1 th / m	1 th / 3 m	1 th / m	1 th / m
Fe	1 th / m	1 th / m	1 th / m	1 th / 3 m	1 th / m	1 th / m
Al	1 th / m	1 th / m	1 th / m	1 th / 3 m	1 th / m	1 th / m
Mn	1 th / m	1 th / m	1 th / m	1 th / 3 m	1 th / m	1 th / m
Cu	1 th / m	1 th / m	1 th / m	1 th / 3 m	1 th / m	1 th / m
Cl	1 th / m	1 th / m	1 th / m	1 th / 3 m	1 th / m	1 th / m
SO_4	1 th / m	1 th / m	1 th / m	1 th / 3 m	1 th / m	1 th / m

V. MONITORING SYSTEM AND PREVENTIVE MAINTENANCE

In recent semiconductor plants, it is more and more often required that the system keep operating without downtime. Ultrapure water should be supplied continuously to the system. To maintain the best conditions in the ultrapure water production system, computers are introduced to collect, process, and log various data. An alarm system is also adopted. However, there is still great room for improvement. The on-line monitor for particles and bacteria is too expensive or in some cases has not been developed. AI computer systems should be introduced. The computer will be able to understand the problems and to send appropriate maintenance information. It is expected that such an advanced computer system will be developed in the future to enhance the preventive maintenance system.

REFERENCES

1. Y. Ohta, Ultra Pure Water System and High Pure Chemical Supply System, Ultra Clean Society, REALIZE, Inc., Tokyo, 1986, p. 157.
2. TORAY Industries, Inc., Technical Report.
3. A. Hoshio, Practical, Clean Technology of LSI Factory, Science Forum, Inc., Tokyo, 1985.
4. ASAHI Chemical Industry Co., Ltd., Technical Report.
5. Y. Fujii, Ultra Pure Water System and High Pure Chemical Supply System, Ultra Clean Society, REALIZE, Inc., Tokyo, 1986.
6. Nihon Photo Science Co., Ltd., Technical Report.

Table 13 Comparison of Initial and Running Costs

ITEM	Non recovery FRONT STAGE RO	Non recovery FINAL STAGE RO	30% recovery FINAL STAGE RO	50% recovery FINAL STAGE RO	75% recovery FINAL STAGE RO	REMARKS
1. Running Cost						Working hour is 24 h/d, 360 d/Y Other conditions are as per attached
1) Electricity @ 310 円/m³	65 円/m³	55 円/m³	55 円/m³	55 円/m³	55 円/m³	
2) City water @ 5 円/m³	480 円/m³	365 円/m³	300 円/m³	210 円/m³	135 円/m³	Remarks: Recovery rate formula is
3) Steam @ 20 円/m³	80 円/m³	105 円/m³	70 円/m³	50 円/m³	30 円/m³	
4) Chilled water	—	—	5 円/m³	10 円/m³	10 円/m³	$\text{Recovery rate} = \dfrac{\text{Recovered drain water}}{\text{Ultrapure water consumption}} \times 100$
5) TOC decomposer (Lamp+Electricity)				50 円/m³	70 円/m³	
6) Consumable { chemical, resin, RO, UF membrane, filter etc. }	100 円/m³	125 円/m³	125 円/m³	120 円/m³	120 円/m³	
Remarks: Maintenance cost is not included						
Running Cost Total	725 円/m³	650 円/m³	555 円/m³	495 円/m³	420 円/m³	
2. Waste water rate						Waste water rate
From use point	1.0	1.0	0.7	0.5	0.25	$= \dfrac{\text{Total waste water m}^3}{\text{Ultrapure water consumption m}^3}$
From system	0.4 ~ 0.5	0.1 ~ 0.2	0.1	0.1	0.1	
total	1.4 ~ 1.5	1.1 ~ 1.2	0.9	0.6	0.35	
3. Initial cost 2000 m³/d	650,000 千円	720,000 千円	750,000 千円	800,000 千円	850,000 千円	
4. Floor space						
Make-up	$7\ m^2/m^3h^{-1}$	$8\ m^2/m^3h^{-1}$	$10\ m^2/m^3h^{-1}$	$10\ m^2/m^3h^{-1}$	$10\ m^2/m^3h^{-1}$	
Polishing	$2.5 \sim 3\ m^2/m^3h^{-1}$	$2.5 \sim 3\ m^2/m^3h^{-1}$	$2.5 \sim 3\ m^2/m^3h^{-1}$	$2.5 \sim 3\ m^2/m^3h^{-1}$	$2.5 \sim 3\ m^2/m^3h^{-1}$	

Note: 円 = yen; 千円 = yen (thousands).

9
Hot Ultrapure Water System

Yoshito Motomura
Kurita Water Industries, Ltd., Tokyo, Japan

I. INTRODUCTION

As the patterns used in ultra–large-scale integration (ULSI) become further complex and finer, the wet process plays a more important role in the production line. Needless to say, the impurities in ultrapure water that have direct contact with wafers should be reduced. At the same time, the wet process is required for effective removal of trace amounts of contaminants on the water surface with high reproducibility to keep the wafer surface ultraclean. This chapter describes the manufacture of hot ultrapure water recently adopted to clean wafers.

II. HOT ULTRAPURE WATER PRODUCTION SYSTEM

A. Configuration of the System

Figure 1 shows the basic configuration of the hot ultrapure water production system [1]. The major difference from an ordinary ultrapure water production system is that this system must use heat-resistant materials. The system is composed of heat exchanger, heat-resistant ultrafiltration module, and heat-resistant piping system.

1. Heat Exchanger
Stainless steel, PFA, and titanium are the materials used at present in the heat exchanger. Even in an ultrapure water production system at room temperature, elution from the component materials is a serious problem. The quality of hot ultrapure water is more likely to be affected by elution. Table 1 lists the results of elution tests using different materials for the heat exchanger. Large amounts of iron, nickle, and manganese elute from stainless steel. Fluorine and carbon elute from PFA. Titanium has extremely low elution properties. Based on these results, titanium is considered the most appropriate

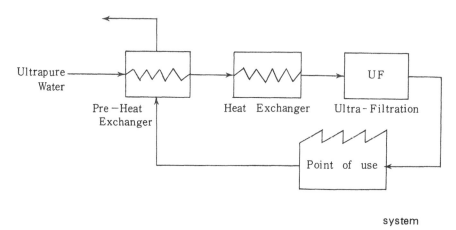

system

Figure 1   The hot ultrapure water production system.

material for the heat exchanger at present. Figure 2 shows the structure of the titanium heat exchanger that was actually adopted.

2.   *Ultrafiltration Module*

Since the ultrafiltration module is mounted at the final stage of the hot ultrapure water production system, no elution of total organic carbon (TOC) and heavy metals from its

Table 1   Confinement Test in SUS, Ti, and PFA Piping

	SUS 316 L	Ti	PFA
Fe	85	14	—
Cr	< 4	< 4	—
Ni	260	< 16	—
Mn	40	< 2	—
Al	< 4	< 4	—
Ti	—	< 60	—
F⁻	—	—	9600
TOC	—	—	12000

$(\mu g/m^2 \cdot 30\,Days)$

Condition

Temperature   80 ℃

Period   30 days

Piping   Size   25 mm ∅ × 1000mm L

Test piping: glued cap on

bottom and detachable cap on top

Imprisoned Liquid : Ultrapure water

S_1 : Inlet of Hi − Temp

S_2 : Outlet of Low − Temp

S_3 : Outlet of Hi − Temp

S_4 : Inlet of Low − Temp

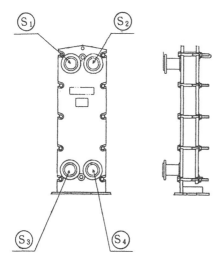

Figure 2 Heat exchanger.

material and no particulation should be allowed. Hot ultrapure water has excellent cleaning abilities, so contaminants on the module would have a serious effect on the water quality. In particular, the contaminants that adhere to the module during the production process are released from the module at the initial stage of operation, significantly increasing the numbers of particles. (Details are presented in Chap. 3, Sec. II.B.)

Figure 3 shows the structure of the ultrafiltration module (KU-1010-HS) used in a hot ultrapure water production system. Table 2 lists the major specifications for this module. The materials are polysulfone resin, heat-resistant epoxy resin, and heat-resistant polypropylene resin. The major characteristics are as follows:

Low particulation. The development of the hollow-fiber outer pressure and double skin structure, as well as the improved cleanliness on the membrane surface on the filtered water side, have enabled the ultrafiltration module to suppress the 0.1 μm particles, at less than $5 \cdot ml^{-1}$.

Quick initial start-up. Because the structure on the filtered water side is simple, particles are quickly removed to speed the initial start-up. The resistivity and the TOC concentration also start up very quickly.

Excellent heat resistance. The module can withstand temperatures of 90°C or higher. Hot ultrapure water can be fed in a stable manner.

Large volume of filtered water. The volume of filtered water per module is 3.5 $m^3 h^{-1}$ (1 kgf cm^{-2}, 25°C).

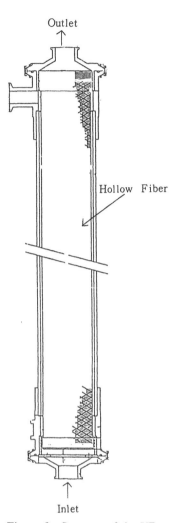

Outlet

Hollow Fiber

Inlet

Figure 3   Structure of the UF module.

3.   *Piping System*

The heat-resistant piping materials available on the market at present are SUS, PP, PVDF, PFA, and PEEK. Of these, PVDF and PEEK are often used in pipes for hot ultrapure water [2, 3]. Figure 4 shows the results of a TOC elution test of PEEK and PVDF pipe; Fig. 5 shows the results of an elution test for all other impurities than TOC (including heavy metals and ions).

   In these tests, ultrapure water was confined in a 1 m tube with a diameter of 25 mm (with one end capped by the same material as the pipe) at 25°C and at 80°C to analyze the eluted materials. The same pipe was used in a series of tests: the first sample was taken from the pipe after 1 day, and then ultrapure water was poured into the same pipe to continue the same test for 2–7 days and 8–30 days.

   As shown in Fig. 5, the amount of TOC eluted from PEEK pipe is smaller than that from PVDF pipe by one order of magnitude. The total ion elution from PEEK is very

Table 2 Specifications for the UF Module

Module No	KU-1010-HS
Fiber diameter	Inner 1.0 mm ∅
	Outside 1.6 mm ∅
Effective surface area	5 m²
Module diameter	106 mm
Module size	1150 mm
Separation molecular weight	80000
Product flow rate	3.5 m³/Hvas 1 atom, 25 ℃
Temperature of use (max)	95 ℃
Main materials	Polysulfone resin
	Epoxy resin
	Polypropylene
	Fluorine contained resin

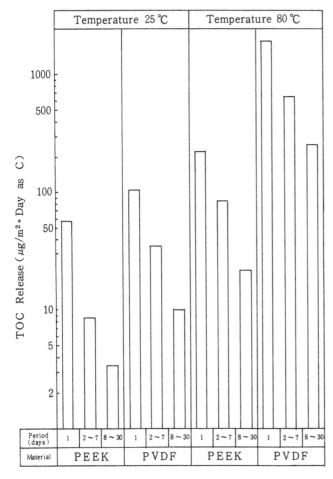

Figure 4 Confinement test in PEEK and PVDF pipe.

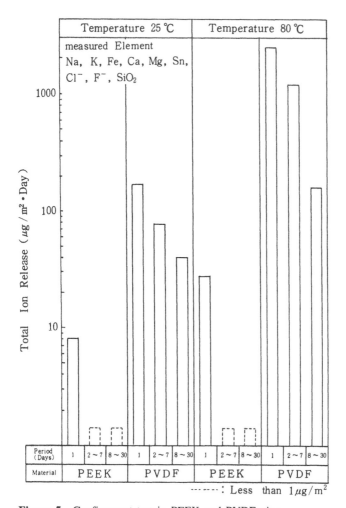

Figure 5   Confinement test in PEEK and PVDF pipe.

limited; a large amount is eluted from PVDF. Although elution from PEEK is detected on the first day, no elution is detected after that (less than 1 μg m^{-2} day^{-1}). The majority of the materials eluted from PVDF is fluoride ion. Considerable elution of TOC and total ion is detected at the initial stage from both PEEK and PVDF. The amount of elution decreases over time in both cases.

Based on these findings, PEEK is considered the most suitable material for the hot ultrapure water piping system.

B. Water Quality Analysis

It is only recently that hot ultrapure water production is actually employed in plants. Therefore, not many systems have been introduced and operated. A hot ultrapure water production system actually operated at present is introduced here, together with the analytic results of water quality.

Figure 6 shows a hot ultrapure water production system. Tables 3 and 4 list the analytic results for water quality. The system structure is almost the same in Tables 3 and

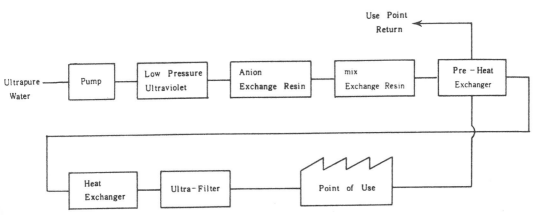

Figure 6 Hot ultrapure water system.

4, but the system in Table 3 adopts PEEK as piping material and that in Table 4 uses PVDF. The material for the preheater and the heater is titanium in both cases.

As shown in Table 3 (PEEK), the degradation in quality of the use point return water is indicated by a slight drop in resistivity of 0.4 M$\Omega \cdot$cm. No other items show degradation. For PVDF (in Table 4), a drop in resistivity of 2 M$\Omega \cdot$cm is detected in the use point return water. It is not possible simply to compare the measurement results in both cases because the length of the piping system and the water consumption at the use points are not the same. Nevertheless, the difference in the elution shown in Fig. 5 is thought to be reflected in the difference in the quality of the use point return water.

Table 3 Results of Measurements in Hot Ultrapure Production System 1

	Heat Exchanger Inlet	UF Product	Use Point Return
Resistivity (M$\Omega \cdot$cm)	18.3	18.1	17.7
TOC (ppb)	25	30	30
Na (ppb)	< 0.1	< 0.1	< 0.1
Fe (ppb)	< 0.5	< 0.5	< 0.5
Ti (ppb)	< 1.0	< 1.0	< 1.0
Cl$^-$ (ppb)	0.1	0.1	0.1
SiO$_2$ (ppb)	< 5	< 5	< 5

Product Water Temperature : 60 ℃
Heat Exchanger Materials : Ti
Piping Materials : PEEK

Table 4 Results of Measurements in Hot Ultrapure Production System 2

	Heat Exchanger Outlet	UF Product	Use Point Return
Resistivity (M$\Omega\cdot$cm)	18.0 <	18.0 <	16.3
TOC (ppb)	20	22	24
Na (ppb)	—	< 0.1	< 0.1
K (ppb)	—	< 0.1	< 0.1
Fe (ppb)	—	< 0.5	< 0.5
Al (ppb)	—	< 0.5	< 0.5
Mn (ppb)	—	< 0.1	< 0.1
Cu (ppb)	—	< 0.5	< 0.5
Mn (ppb)	—	< 0.1	< 0.1
Mg (ppb)	· ·	< 0.4	< 0.4
Cl$^-$ (ppb)	< 0.5	< 0.5	< 0.5
SO$_4^{2-}$ (ppb)	< 1.0	< 1.0	< 1.0
F$^-$ (ppb)	< 5	< 5	< 5

Product Water Temperature : 60 ℃
Heat Exchanger Materials : Ti
Piping Materials : PVDF

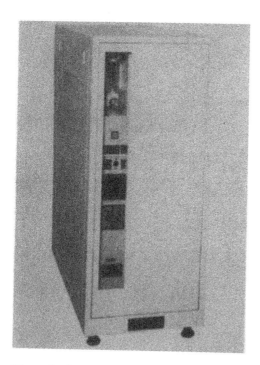

Figure 7 Hot ultrapure water production unit.

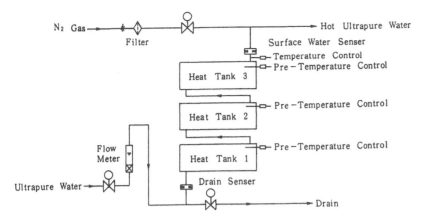

Figure 8 Hot ultrapure water production unit (Model K-HP9).

III. SMALL HEATER FOR ULTRAPURE WATER

This unit provides hot ultrapure water to a few cleaning baths. It is small and light enough to be mounted on the use points. Figure 7 is a picture of this unit. Figure 8 is the flow chart. Table 5 describes the basic specifications.

Since this unit is equipped with three quartz glass heating tanks with an infrared heater, it is capable of constantly supplying hot ultrapure water to the necessary points. The major characteristics are as follows:

1. It employes a high-purity transparent quartz glass heating tank. Impurity elution, such as metallic ion elution, is extremely low.
2. It is equipped with a high-accuracy temperature controller to supply hot ultrapure water with limited temperature fluctuation.
3. It is equipped with safety defices to protect against a drop in flow rate and overheating.
4. Maintenance and handling are very simple.

Table 5 Specifications for a Hot Ultrapure Water Production Unit

Flow limits	$1 \sim 10 \, \ell/\text{min}$
Number of heat tank	3
Consumption of electric power	$36 \, \text{kw} \cdot \text{Hv}$
Max product temperature	$85 \, ^\circ\text{C}$
Temperature accuracy	$\pm 2 \, ^\circ\text{C}$
Heat tank material	Hi − purity clear quartz
Electric power	AC 200V, 125 A
	Three phase current
Outward size	$520 \, \text{mm W} \times 1050 \, \text{mm D} \times 1270 \, \text{mm H}$
Weight	187 kg

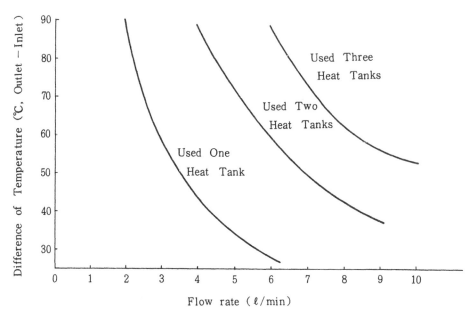

Figure 9   Results of measurements of product water temperature.

Figure 9 shows the temperature rise as a function of the flow rate. When ultrapure water at 20°C is fed, with three heating tanks, hot ultrapure water of 80°C can be processed at 8–9 L minute^{-1}.

IV.   AFTERWORD

Along with further developments in semiconductors, thin-film technology and fine-pattern processing technology are likely to become more and more advanced. The cleaning technology that supports such progress will be required to cope with higher cleanliness levels and further diversified states on the water surface. To meet these requirements, it is essential to upgrade and stabilize the entire wet cleaning process further, as well as to improve the purity of the ultrapure water. The adoption of hot ultrapure water is one possible way to achieve these goals. Hot ultrapure water technology needs to be studied further to improve purity and applications.

What is the most important in hot ultrapure water production technology is to suppress elution from system materials as much as possible. At present, PEEK and titanium are used. The development of the new materials with much lower elution properrties is expected.

REFERENCES

1.   Technical Information from Kurita Water Industries, Ltd., UF Module for Mega Bit Ultrapure Water, 1988.
2.   K. Yabe and M. Takano, Proceedings of 5th Symposium on ULSI Ultra Clean Technology, Tokyo, 1987.
3.   K. Yabe, Y. Motomura, and H. Harada, Proceeding for 7th Symposium on ULSI Ultra Clean Technology, Tokyo, 1988.

10
Afterword

Koichi Yabe
Kurita Water Industries, Ltd., Tokyo, Japan

The purity level required for ultrapure water is higher in the semiconductor industry than in any other industry. The required quality is also rising along with upgraded ULSI. Ultrapure water consumption by a semiconductor production line is very large, around 40–60 m^3 h^{-1}. Such stringent demands for ultrapure water have triggered rapid progress in ultrapure water technology. In the future, it should be further improved to meet the ULSI requirements.

Present ultrapure water production and analysis technology cannot further improve the purity of ultrapure water. Membrane separation technology has been adopted and will be upgraded for ultrapure water production. A possible new technology that might replace present technology is a more advanced and effective distillation. Many obstacles, however, including higher costs, must be overcome.

Membrane separation is very promising for the future. This technology has been widely adopted to water treatment, creating a huge market. Many leading chemical companies are involved in the development of new membrane separation technologies. Considerable advancement is therefore expected in this field. In particular, the desalting efficiency of the reverse osmosis membrane will be greatly improved by introducing a charged membrane.

A more important point in upgrading the purity of ultrapure water is how to suppress the elution of impurities from the component materials. This is the great challenge to further development.

Purity levels of parts per trillion to quadrillion will be achieved when the ultrapure water production system is built exclusively with ultrapure-water-grade components with high functionality and quality, including membranes, ion-exchange resins, tanks, piping materials, and pumps.

III
Elementary Technology

1
Introduction

Yoshiharu Ohta
Nomura Micro Science Co., Ltd., Kanagawa, Japan

There are a variety of elementary techniques of ultrapure water manufacturing. Membrane separation, ion exchange, dissolved gas removal, sterilization, reduction of total organic carbon (TOC), and removal of static electricity are among the elementary techniques described in this section.

The specific topics covered here include, for membrane separation (membrane filtration, MF), confirmation of completeness of membrane rejectivity, the materials of membranes, charged membranes, and heat-resistant membranes. For ultrafiltration (UF), we discuss the particle-reducing effect of hollow-fiber in outside-to-inside filtration. The characteristics, rejectivity, and application of reverse osmosis (RO) to recent ultrapure water systems using low-pressure composite membranes are also described.

We cover the characteristics and uses of various ion-exchange resins, the elution of organic compounds from the resin itself, and problems that may be encountered with the resin.

Degasification, vacuum degasification, and the catalytic resin and nitrogen purge methods are described in relation to removal of dissolved gas.

Sterilization by chemicals, hot water, ozone, and ultraviolet (UV) is described.

Reverse osmosis, low-pressure UV oxidation, ion exchange, and high-pressure UV oxidation are described for TOC reduction.

The removal of static electricity by direct injection of carbon dioxide and hydrophobic membranes are mentioned.

Such elementary techniques used in manufacturing ultrapure water differ from the techniques used in general industry in that contaminants in very small amounts must be removed. Because separations using a slight concentration gradient operate mainly at dilute ranges of concentration, this situation gives rise to several problems.

First, complete separation is required. Total rejectivities of at least 99.9–99.99999% must be reached. Second, contamination originating in the separation materials themselves cannot be ignored. Particles adhere to and ionic or organic contaminants elute from separation materials. Also, examination of the removal effect is very difficult. This means that the efficiency of unit processes in ultrapure water systems is difficult to examine. The analysis of microcontaminants is greatly affected. Thus it might even be said that the performance of the latest ultrapure water systems cannot be examined because of the limitations imposed by present analysis techniques!

Finally, there are problems of economics. Even if they are effective, costly unit processes cannot be adopted by industry, with current costs of \$4–5/m^3 for ultrapure water. Now that the merits of ultrapure water are being recognized as the only pollution-free chemical for cleaning, this restriction will presumably be gradually overcome.

2 Membranes

A. Membrane Technology

Norihisa Urai
Japan Organo Co., Ltd., Tokyo, Japan

I. REVERSE OSMOSIS

A. Current Reverse Osmosis Membrane Technology

1. Principles and History of Development

When a septum permeable to water (solvent) but not to salt (solute) is interposed between pure water and salt solution, the pure water passes through the septum (semipermeable membrane) into the solution, as shown in Fig. 1. The increment of the salt solution level corresponds to the osmotic pressure. If pressure exceeding the osmotic pressure is applied to the salt solution, then water in the salt solution moves toward the pure water. This phenomenon is called reverse osmosis (RO). RO separation is a technology that makes use of this phenomenon under a pressure difference.

Since the U.S. government adopted the RO method as one of the processes used in their seawater desalination project, the method has come to attract world attention. In 1960, Loeb and Sourirajan invented a process for manufacturing a membrane through which a large flux rate of water can pass, using cellulose acetate as basic material. Since then this field of study has made rapid progress.

For seawater desalination, which was the subject of early study, a membrane has been developed that is capable of, in a single stage, desalinating seawater and generating drinking water. The method has technically been exceeding the evaporation method. Today RO is applied to various other fields in addition to desalination of seawater or brackish water, including desalinating recycling of sewage and condensing valuable minerals in the food industry. Especially in the semiconductor and other ultrapure water industries, the method has been spread and developed rapidly. Its merit in manufacturing ultrapure water is that it can remove organic matter, colloidal substances, fine particles, and microorganisms contained in the water as well as removal of salt (removal of ions),

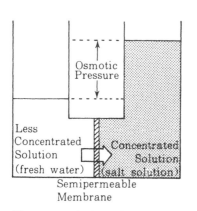
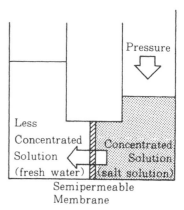

Figure 1   Principle of the RO process.

the original subject of desalination. The practical value of the method is therefore significant enough to say that there would be no ultrapure water today if reverse osmosis had not been developed.

The first RO boom in Japan was in the late 1970s, when the mainstream was an improved version of Loeb's acetylcellulose membrane, which required operating pressures of approximately 30 kg cm^{-2}.

In the 1980s, RO membranes were developed that require an operating pressure of around 15–20 kg cm^{-2}. Recently RO membranes serving at pressures of 10 kg cm^{-2} or less have been put into practical use. These RO membranes are usually made of synthetic polymer, such as polyamide. As a result of developments and improvements in these membranes, the number of fields applicable to the RO method has been steadily increasing.

2.   *Classification of RO Membranes*

There are several ways of classifying RO membranes, generally by operating pressure or material.

a.   Classification by Operating Pressure.   It is not academic to classify RO membranes by operating pressure, and therefore there are no specific definitions for this classification. However, the membranes are generally classified as follows:

1.   High-pressure: normal operating pressure 40 kg cm^{-2} or over
2.   Medium-pressure: normal operating pressure approximately 30 kg cm^{-2}
3.   Low-pressure: normal operating pressure approximately 15 kg cm^{-2}
4.   Ultra–low-pressure: normal operating pressure 10 kg cm^{-2} or under

As indicated, the membranes are classified by the normal operating pressure necessary for them to function in a satisfactory manner. Thus, under some service conditions, it is possible to operate a high-pressure membrane at a lower pressure, say 15–20 kg cm^{-2}, or a low-pressure membrane at a higher pressure.

b.   Classification by Material and Structure.   RO membranes are roughly classified into the following two groups according to material:

1.   Cellulose acetate and its derivative membrane
2.   Synthetic polymer membrane

Membranes can also be roughly classified by structure:

1. Asymmetric
2. Composite

Excluding some exceptions on the market, such as duPont's Permasep B-9 and B-10, RO membranes generally used today are

1. Cellulose acetate asymmetric
2. Synthetic polymer composite

Figure 2 shows a typical RO membrane manufacturing process. Figure 3 is an electron micrograph of a reverse osmosis membrane. The cellulose acetate asymmetric RO membrane has recently been improved on the basis of Loeb's original membrane.

The synthetic polymer composite RO membrane is based on a prototype invented by Cadotte and his associates in the 1970s. Since then this kind of membrane has been

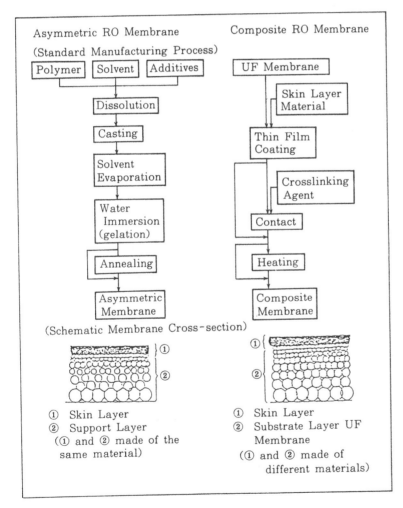

Figure 2 Manufacture of RO membranes. (From Ref. 1.)

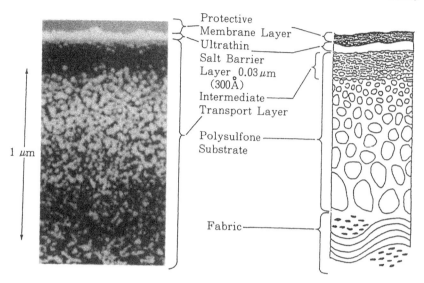

Protective
Membrane Layer
Ultrathin
Salt Barrier
Layer 0.03 μm
(300Å)
Intermediate
Transport Layer

Polysulfone
Substrate

1 μm

Fabric

Figure 3   Structure of a composite RO membrane. (From Ref. 2.)

developed industrywide. Current products are extremely diverse in material and ability. As illustrated in Fig. 3, the composite RO membrane is a "functional" membrane. It is manufactured by applying, over a substrate, a very thin film of a different material with desalinating ability. The substrate itself (the UF membrane) has no desalinating capability. In other words, this type of membrane is created by selecting the most suitable material for each layer on the basis of model analysis of the structure and function of the asymmetric membrane. Such configurations offer a wide range of choice of materials (for substrate and thin film). Most composite RO membranes currently available are constituted of polysulfone substrate and polyamide polymer thin film. This is the result of selection of specific materials by considering the various requirements for the RO membrane as summarized in Table 1.

Figure 4 shows the performance of representative RO membranes currently on the market. It should be noted, however, that these figures are based on values indicated in company catalogues. The exact capabilities of particular products may be somewhat different.

The main characteristics of RO membranes are discussed here. Manufacturers are indicated in parentheses.

High-pressure and medium-pressure membranes.   The PEC-1000 (Toray) is a polyether membrane offering the highest desalinating ability among the RO membranes currently on the market. It is used mainly in seawater desalination. Its chief drawback is extreme sensitivity to oxidizing agents, which degrade performance.

The NTR-7100 (Nitto Denko) is a polyamide membrane consisting of a aliphatic and an aromatic group. It is inferior to PEC-1000 in desalinating ability but somewhat more resistant to oxidizing agent. It is not remarkably damaged by dissolved oxygen.

Cellulose acetate membranes (UOP, Toray, and others) are marketed by many companies, including UOP, Toray, and Nitto Denko (Hydranautics). These are the most widely used medium-pressure membranes. They are not usable in the alkaline range in which hydrolysis is accelerated. They are easily attacked by bacteria but resistant to

Table 1 Conditions Required for RO Membranes

Reverse Osmosis Performance	High selectivity and rejection High water permeability
Resistance to Membrane Compaction	Little decline in flux as a function of operation time
Resistance to Oxidizing Agents	Subject to little change in performance characteristics due to oxidizing agents such as dissolved oxygen, chlorine, hydrogen peroxide and the like
Applicable pH Range	Subject to little change in performance characteristics when used in a wide range of pH
Thermostability	Subject to little change in performance characteristics when used at high temperatures
Resistance to Bacteria	Subject to little if any bacterial degradation
Resistance to Fouling	Subject to little if any fouling due to adsorption of foulants
Resistance to Chemicals	Subject to little change in performance characteristics due to membrane degradation by various chemicals

oxidizing agent and can withstand 0.5 mg L^{-1} chlorine solution. In addition, since contamination by impurities in water (metallic hydroxide, organic colloid, and others) is relatively less frequent, they permit relatively easy operation.

Low-pressure membrane and ultra-low-pressure membrane. The FT-30 (FilmTec) has generated much interest because of its high desalinating capability at low operating pressures, around 15 kg cm^{-2}. Each manufacturer has achieved a low-pressure membrane performance equivalent to that of the FT-30. It is made of polyamide constituted totally of aromatic groups and is not sensitive to pH change. It is poorer in oxidizing agent resistance than the cellulose acetate (CA) membrane. It is therefore not applicable in a system that contains residual chlorine.

The material of the SU-700 (Toray) is, as with the FT-30, polyamide polymer totally constituted of aromatic groups, called cross-linked aramide. It is equivalent or superior to the FT-30 in performance.

Like the FT-30 and SU-700, the NTR-759 (Nitto Denko) membrane is polyamide polymer entirely composed of aromatic groups. Its performance and behavior are also almost the same.

The NTR-729 (Nitto Denko) is an improved version of NTR-7250, the first low-pressure membrane marketed in Japan. The rate of monovalent ion desalination with the NTR-7250 was about 50%, whereas the improved version operates at 90% or more. The material may be polyamide-based, cross-linked polyvinylalcohol. It is characterized by a chlorine resistance almost equal to that of the CA membrane.

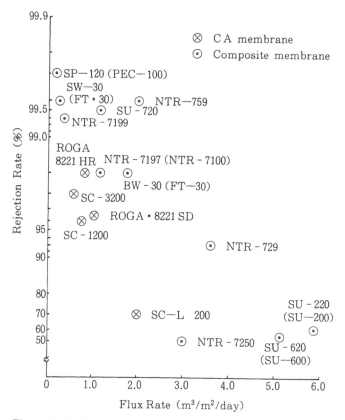

Figure 4   Performance of representative RO membranes (8 in. element).

The monovalent ion desalination efficiency of the SU-600 (Toray) is low, around 50%, but it can be used at extremely low pressures of 5 kg cm^{-2} or less. It has some chlorine resistance.

II. CLASSIFICATION AND PERFORMANCE OF RO MODULES

RO membranes are only about 100 μm thick. In operation 5–70 kg cm^{-2} pressure is applied to these thin films, and thus some consideration must be made for industrial use. The RO module, generally composed of membrane and supporting structure, works well for industrial purposes. Modules currently in practical use are classified into these four types:

1. Hollow fiber
2. Spiral wound
3. Plate and frame
4. Tubular

A. Hollow-Fiber Module

The hollow-fiber module embodies in a pressure vessel a RO membrane spun of hollow fibers of 100–200 μm in outer diameter and 20–50 μm in wall thickness. Its typical

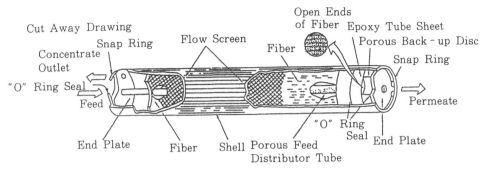

Figure 5   Structure of hollow-fiber RO module.

configuration is illustrated in Fig. 5. Pressurized raw water flows along the outer surface of the hollow fibers, part of which permeates through the membrane into the inside of the hollow fibers to be extracted as product.

Famous hollow-fiber modules are the Permasep B-9 and B-10 series from duPont, with a polyamide membrane. Modules with a cellulose acetate membrane (from Toyobo and others) are also available.

B.   Spiral-Wound Module

Two membranes are bonded together to form an envelope (one end is open and the other end closed) into which a porous material (nonwoven fabric) is placed. The open end is connected to the collecting tube. Sets of such a unit are wound together around the collecting tube, with spacers for allowing feedwater to pass. This assembly is generally called the element. In operation, one to six elements are connected to each other in series and installed in a pressure vessel made of fiber-reinforced plastics, for example. Figure 6 shows the typical configuration, and Figure 7 illustrates the structure of the CA membrane spiral-wound element from Toray. It is characterized by the forced passage of raw water in the direction normal to the central tube.

C.   Plate and Frame Module

Circular flat film, spacer for liquid passage, and porous support are alternated to form a pressed filter element from which permeating water and concentrate are taken. The module from DDS (De Danske Sukker Fabrikken) is famous. The chief merit of this type is easy disassembly for cleaning the membrane. A typical configuration is shown in Figure 8.

D.   Tubular Module

The modules of this type are classified into two systems, internal pressure and external pressure.

1.   Internal Pressure System

A tubular film is applied over the inner wall of porous pipe (generally half an inch in inner diameter). The raw water is directed into the pipe. The water permeating through the porous pipe is collected outside. The porous pipe itself serves as a pressure vessel. Thus

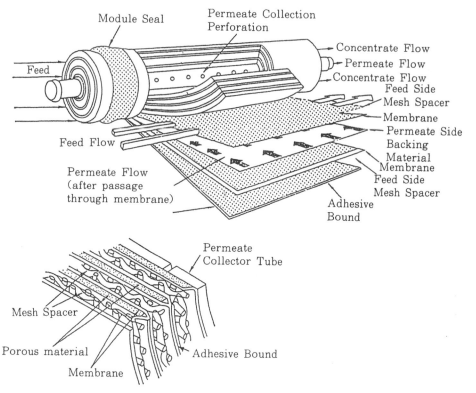

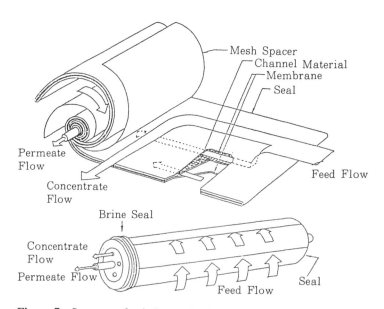

Figure 6 Structure of spiral-wound RO elements.

Figure 7 Structure of spiral-wound RO elements (Toray SC series).

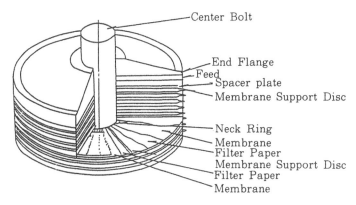

Figure 8 Structure of plate and frame RO modules (DDS type).

pressure resistance is not required for the outer jacket for collecting permeating flow, which can be made of PVC, for example. The typical configuration of the internal pressure system is shown in Fig. 9.

2. External Pressure System

Here the outer surface of the porous pipe is lined with a membrane. The raw water is directed outside the pipe, and the permeating water is collected inside the pipe. Therefore, a pressure vessel housing the entire assembly is required. However, the merit is that it can be disassembled for cleaning the membrane surface.

E. Summary

These four types are the most representative modules. Their characteristics are summarized in Table 2. The spiral-wound type is the most versatile, and it sees the most extensive use in current ultrapure water facilities. The membrane area of the spiral module varies with the diameter of the element assembly. A suitable diameter is selected according to the capacity of the plant. The 4 or 8 in. diameters are generally used (the length is 40 in. for both sizes), but recently a larger 12 in. diameter assembly has come on the market in the United States.

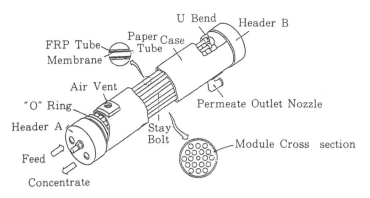

Figure 9 Structure of tubular RO modules.

Table 2 Comparison of Features of Various RO Membrane Module Configurations

Module Configuration	Membrane Pressure Vessel Packing Density (m^2/m^3)	Features	Major Applications
Hollow Fiber	9,000~10,000	This membrane configuration results in such a large membrane area per vessel volume that even membranes with relatively lower flux rates can be put to commercial use. As the rate of water transport across the membrane is very low, the membrane is more liable to fouling and scaling. Therefore, hollow fiber membranes demand attention to the pretreatment of feed water. This membrane configuration is less amenable to cleaning.	Seawater desalination, etc.
Spiral - Wound	650~1,000	This membrane configuration falls between the hollow fiber and tubular configurations in the membrane area per vessel volume, the degree of membrane fouling, etc.	Seawater and brackish water desalination, production of ultra pure water, concentration of valuable products, etc.
Tubular	33~330	This membrane configuration has a small membrane area per vessel volume. Therefore, the economics of this configuration is not favorable. As this membrane configuration permits large rates of water transport across the membrane, the membrane is less liable to fouling. This membrane configuration is amenable to mechanical sponge ball cleaning technique.	Recovery of wastewater and sewage, etc.
Plate and Flame	160~500	Like the tubular membrane configuration, the economics of the plate and flame membrane configuration is not favorable. It is easy to disassemble the plate and flame structure and cleanse the membrane.	Food processing, wastewater treatment, etc.

The performance of the typical 8 in. element is given in Table 3. In the 4 in. element, the membrane area is reduced about one-quarter. Hence the flux rate alone is a quarter of the values indicated in Table 3, but the other characteristics are equivalent.

III.   CONCEPTUALIZATION OF AN RO MODULE

A.   Basic Formulas

Reverse osmosis is a system of separating solute from solution through a membrane, always in the balance formulated by the following equations:

$$Q_f = Q_p + Q_r \tag{1}$$
$$C_f Q_f = C_p Q_p + C_r Q_r \tag{2}$$

where: Q_f = feed water flow rate
Q_p = permeating flow rate
Q_r = concentrate flow rate
C_f = feedwater salt concentration
C_p = permeating water salt concentration
C_r = concentrate salt concentration

The salt rejection (desalination ratio) R (%) and the water recovery Y (%) are given by

$$R(\%) = \left(1 - \frac{C_p}{C_f}\right) 100 \tag{3}$$

$$Y(\%) = \frac{Q_p}{Q_f} 100 \tag{4}$$

Note that in many cases, in technical documents or membrane manufacturers' catalogs, the salt rejection rate is defined as

$$R'(\%) = \left\{1 - \frac{C_p}{(C_f + C_r)/2}\right\} \times 100 \tag{3'}$$

In actual water treatment plants, the salt passage (SP) is frequently used in place of the rejection ratio R. SP is given by

$$SP(\%) = \frac{C_p}{C_f} 100 \tag{5}$$

That is, the relationship between the salt rejection R (%) and salt passage SP (%) is formulated by

$$SP\ (\%) = 100 - R\ (\%) \tag{6}$$

When we treat problems that may arise in relation to the water treatment plant, including ultrapure water plants, in many cases they are discussed using SP for the following reason. Let us assume an RO system running with a 98% desalination ratio. This value becomes 96% after a certain lapse of time. That is, the rejection ratio is decreased 2%, and this value seems not to be very important. However, the SP is doubled, from 2 to 4%. In other words, the water quality (salt concentration) becomes two times larger,

Table 3 Characteristics of Representative Japanese RO Membrane Modules (8 in. Diameter Spiral Modules)

		Medium Pressure Type		Low Pressure Type					Remarks
Model No.		SC-3200	NTR-7197	SU-220 S	SU-720	SU-620	NTR-729HF	NTR-759HR	
Manufacturer		Toray	Nitto	Toray	Toray	Toray	Nitto	Nitto	
Membrane Material		Cellulose Acetate	Polyamide Composite	Polyamide Composite	Polyamide Composite	Polyamide Composite	Polyamide Composite	Polyamide Composite	
Fundamental	Standard Operating Pressure(kg/cm²)	30	30 (25)	7.5	15	3.5	10	15	Figures in parenthesis are minimum values.
	Flux Rate (m³/day)	17.6 (16.0)	35 (25)	44 (38)	26 (22)	18	36 (29)	30 (25)	
Operating Conditions	Max. Operating Pressure (kg/cm²)	42 (25~30)	42 (25~30)	15 (3~10)	42 (10~20)	42 (3~6)	30 (7~15)	30 (7~20)	Figures in parenthesis are standard design values.
	PH	3~8.5 (4~6.5)	4~11 (5~9)	3~9 (6~8)	2~10 (5~8)	2~9 (5~8)	1~10 (5~8)	1~11 (5~8)	
	Max. Operating Temp.(°C)	40 (30>)	40 (30>)	40 (35>)	45 (35>)	45 (35>)	40 (30>)	40 (35>)	
	Resistance to Chlorine(ppm)	1 (0.1~0.7)	0	0.2 (0)	常用 0	1 (0~0.5)	1 (0~0.5)	1 (0~0.5)	
% Salt Rejection	NaCl	97 (96)	98 (97)	60 (50)	99.4 (98)	55	92 (90)	99.5 (98)	Figures in parenthesis are minimum values.
	MgSO₄	—	—	99 (98)	—	99	—	—	
(%)Organic Rejection	Methanol (32)	5	19	0	14	1	—	15	
	Ethanol (46)	9	51	12	54	10	25	42	
	IPA (60)	36	90	23	96	35	70	96	
	N-Butanol (74)	11	77	—	—	—	52	—	
	Ethylene Clycol (62)	42	81	37	—	—	—	—	
	Glycerine (92)	92	95	—	—	—	—	—	
	Phenol (94)	0	64	0	—	—	—	—	
	Glucose (180)	99<	99<	96	99<	95	97	99<	
	Saccharose (342)	99<	99<	99<	99<	99<	99<	99<	
	Raffinose (504)	99<	99<	—	—	—	—	—	
	Acetic Acid (60)	6	34	4	54	2	—	—	
	Oxalic Acid (90)	68	—	35	—	—	—	—	
	Citric Acid (192)	99	97	86	99<	86	—	—	
	Urea (60)	26	52	10	63	11	—	—	
	Formaldehyde (30)	33	33	—	32	9	—	—	
	Ethylenediamine (60)	76	—	67	95	18	—	—	

262

with a larger loading effect acting on the ion exchanger downstream of the ultrapure water plant.

Such a discussion is made from a particular viewpoint. The rejection ratio and the salt passage do not have the same function, but the operational problems of the water treatment plant are more easily tackled by using the salt passage SP.

B.   Basic Characteristics

As understood from the preceding section, the basic performance of the RO system is indicated by permeating flow rate and desalination ratio (salt passage), whose fundamental characteristics can be formulated by the simplified equations

$$J_v = A(P - \Delta\pi) \tag{7}$$
$$J_s = B\,(\Delta C) \tag{8}$$

where: J_v = water flux
   J_s = salt flux
   P = operational pressure
   $\Delta\pi$ = osmotic pressure difference between the two interfaces separated by the membrane (water feed side and permeating water side)
   ΔC = concentration difference of the two solute interfaces separated by the membrane
   A = water permeability coefficient of the membrane
   B = salt permeability coefficient of the membrane

Consequently, it can be said that a membrane with larger A is of higher capability in water permeation and a membrane with larger B has a higher desalination capability. Equations (7) and (8) have the following meanings, which are important to understand the basic characteristics of RO.

1.   Permeate Water Is Proportional to Effective Pressure

That is, the permeating water increases proportionally to the operational pressure, and permeation of water is enhanced accordingly. The osmotic pressure increases along with the salt concentration, and consequently the effective pressure decreases, reducing the permeating water.

The osmotic pressure π can be approximately formulated by the following equation. As easily understood from this equation, π is negligible, especially for normal industrial waters.

$$\pi \ (\text{kg/cm}^2) = 8 \times 10^{-4}(\text{TDS}) \tag{9}$$

where TDS (total dissolved solids) is in mg L^{-1}. TDS of industrial water being about 100–200 mg L^{-1}, the osmotic pressure π is about 0.08–0.16 kg cm^{-2}.

2.   Salt Rejection Ratio Increases with Effective Pressure

In other words, the permeating volume of the salt does not depend upon the pressure but on the volume of permeating water, and consequently the apparent salt rejection ratio increases with the pressure.

3.   Water Recovery and Salt Rejection Ratio Inversely Related

This means that the concentration of solute necessarily increases along with the water recovery, with higher ΔC. Consequently, permeation of solute is enhanced by the smaller apparent salt rejection ratio.

Furthermore, as one of the primary characteristics of RO, there is the effect of water temperature. In the preceding equation, the effect of water temperature may be considered to be included in A and B, but it should be noted that the water temperature exerts a large influence, especially on the volume of permeating water.

We know the volume of the permeating water is inversely proportional to the water viscosity, but in addition the permeating water volume decreases 2.5–3% for each 1°C temperature drop. For example, when the permeating water volume at 25°C is assumed as unity, that at 15°C is approximately 0.75. Therefore, the effect of water temperature is one of the primary parameters in the design of an RO module.

The basic characteristics of the RO module have been discussed. Problems remain to be solved, however, such as concentration polarization and membrane wall flow velocity, which have been treated from the viewpoint of chemical engineering. These problems are extremely specialized, and our discussion here is limited to their outlines.

Figure 10 shows the concept of concentration polarization. Water (solvent) passes through the membrane, but salt (solute) barely permeates. As a result, the concentration of salt (solute) near the membrane is extremely high compared with that in the incoming feedwater. This phenomenon is referred to as concentration polarization and is one of the most important problems in desalination performance and membrane scaling in chemical engineering. A membrane with excellent water permeation but with a lower salt passage value is most suitable for reverse osmosis, and these qualities lead to concentration polarization. To weaken the influence of concentration polarization, it is necessary to enhance the dispersion of salt (solute) near the membrane by raising the flow speed on the membrane surface. These parameters are taken into consideration when designing the structure of the elements and module, but for the design of an RO module, special attention should be paid so that water can flow as uniformly as possible at each element (module) while making the concentrated water flow constant. If for some reason enhanced concentration polarization occurs during operation, scaling may be produced at these locations and accelerate as a result of the polarization.

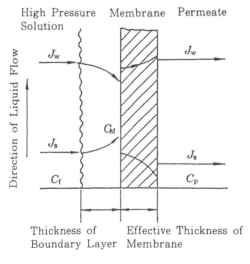

Figure 10   Concentration polarization.

C. Role of RO Modules in the Ultrapure Water Plant

Generally, the role of a RO module in the ultrapure water plant varies more or less along with its position; that is, the RO module is located before or after the ion exchanger. The former is referred to as the upstream RO system and the latter the downstream RO system. The upstream RO module is used for desalination and removal of organic matter, but the downstream RO module removes organic matter (TOC component), fine particles, and microbes without the desalination inherent to the RO system.

New types of reverse osmosis called two-staged systems, have been put into practice, which consist of recirculating water treated by RO into an RO system or forming a line of RO + ion exchange + RO by combining upstream and downstream RO modules.

Here we discuss the required function and capability of RO modules used in the ultrapure water plant.

1. Desalination Performance

As shown in Table 3, the desalination performance of the RO module is given generally as the rejection ratio of sodium chloride. The following trend is confirmed in terms of iron removal:

1. Removal of polyhydric ions is higher that of monohydric ions.
2. Free carbonic acid is not removed.
3. Removal of silica and ammonia is relatively low in comparison with other substances.

Recently, however, special synthetic polymer composite membranes have been developed that are capable of the highly efficient removal of silica. Some have a specific electrical charge by which monohydric ions are more easily removed than dihydric ions.

Furthermore, since a synthetic polymer composite membrane has little charge, negative or positive, ion performance removal may vary considerably along with the pH of the feedwater. The desalination performance listed in catalogs differs depending upon the manufacturer because of measurement conditions. Sufficient attention should be paid when using the desalination performance given in catalogs to avoid erroneous evaluation.

2. Removal of Organic Matter (TOC)

The goal of reverse osmosis in the ultrapure water plant is removing all organics. Removal of organic matter differs according to material quality and the texture density of the membrane.

Generally, the cellulose acetate-based membrane has a low removal of organic matters of light molecular weight (molecular weight 200–300 or less). For example, most low alcohols, such as methanol, propanol, and acetone, cannot be removed. Phenol, tends to be negatively removed, that is, concentrated on the permeating water side. On the other hand, compared to cellulose acetate membrane with the same desalination abilities, synthetic polymer membrane is relatively excellent in removing organic matter. The recently developed low-pressure membrane in particular is capable of efficiently removing organic matter of extremely low molecular weight, including 2-propanol (molecular weight 60) and ethanol (molecular weight 46). Figure 11 shows an example of 2-propanol removal.

Removal of organic matter varies with the pH. Removal of carboxylic acid in particular is highly influenced by pH, due to the dissociation of organic matter by pH.

Of course, the removal performance of RO in the ultrapure water system varies with

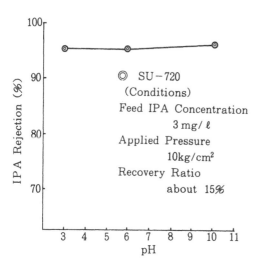

Figure 11 IPA removal performance of the RO membrane.

the TOC components. Generally speaking, removal of TOC is 70–90% in the upstream RO and about 40–70% in the downstream RO.

In any case, with the RO systems actually in place, it is difficult to remove a trace organic matter of low molecular weight effectively. For future treatment of ultrapure water in which a TOC of 10–20 μg L^{-1} or less is imposed, we must inevitably use some method of oxidant decomposition by ultraviolet radiation. Furthermore, when treating trace organic matter, it is necessary to study problems associated with TOC components eluting from the element. In the widely used spiral element, some problems to be solved arise because there are many parts that contact the liquid, such as the spacer. For this reason, development of a hollow-fiber RO is being studied, which is equivalent in material composition to the UF membrane.

3. *Removal of Fine Particles and Microorganisms*

Removal of colloidal substance, fine particles, and microorganisms is one of the most important functions of reverse osmosis. The size of colloids and fungi of various types is shown in Fig. 12. Fine particles are currently measured at 0.1–0.2 μm, and when considering the inherent structure of the RO membrane, 100% of such substances must theoretically be removed. However, in measuring the fine particles (0.2 μm or more) in actual RO devices, about 50–100 particles ml^{-1} for permeating water and $10^4 \times 10^6$ particles for feedwater are often detected at the upstream RO, and at the downstream RO, 10 particles ml^{-1} for permeating water and $10^2 \times 10^3$ particles for feedwater are detected. Several causes can be considered: membrane defects, defect in the bonded and sealed parts, and pollution from the external environment, which occurs when manufacturing membranes and modules.

That a considerable number of particles have adhered to the permeating water side of the RO membrane can be ascertained by checking the particle count in the permeating flow. The count can be changed by altering the flow rate during initial operation (Fig. 13).

Considering these factors, as was pointed out also for TOC, it is necessary to proceed with further study to improve the element and the manufacturing process.

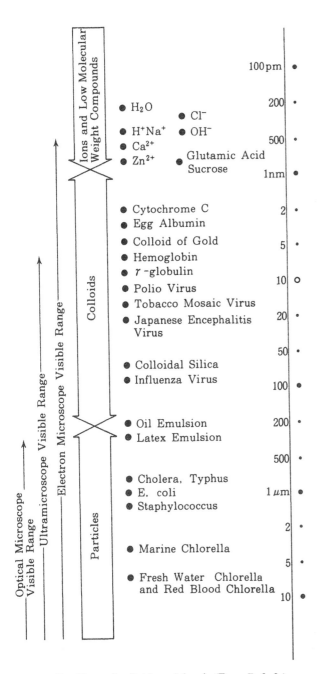

Figure 12 Sizes of colloids and fungi. (From Ref. 2.)

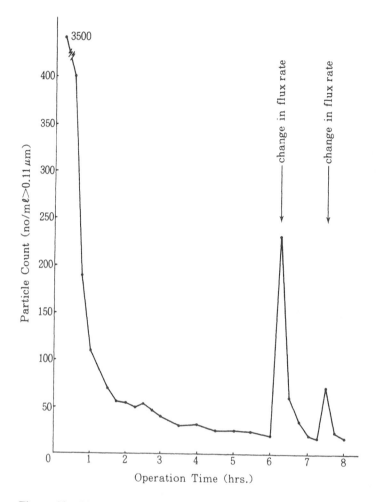

Figure 13   Change in permeate particle count during initial operation of the RO unit.

IV.   OPERATIONAL MANAGEMENT OF THE RO MODULE

A.   Reduction in Performance of the Membrane and Its Remedy

Because the RO module, in contrast to the filter and ion exchanger, requires no regeneration procedure, the water treatment once started can be continued without interruption. As time elapses in operation, however, the performance of the membrane degrades for various reasons. Consequently, to prevent such deterioration to the maximum extent and maximize the life of the membrane, operational control is the most important factor to be considered.

We can know the actual performance of the membrane by checking the rejection ratio and permeating flux rate directly. Decreased desalination may be indicated by a deterioration in the water quality and, in terms of the permeating flux rate, as an increase or decrease in volume, and these symptoms can appear separately or jointly. As an indirect index, an increase in the differential pressure (between the feedwater pressure and the concentrated water pressure) is an important item to check. The causes of deterioration in RO performance may be categorized as follows.

1.   Degradation

This type of performance drop arises when the membrane itself has deteriorated because of chemical and physical damage: pH effects, such as hydrolysis, oxidants, such as chlorine and hydrogen peroxide, water at high temperatures, and damage to the membrane surface by foreign substances and bacterial attack.

2.   Pollution

RO performance degradation in this category includes the following phenomena:

The surface of the membrane is covered with metal hydroxides, such as iron and aluminum (fouling), or bacterial slime or suffers from the formation of scales, such as silica and calcium carbonate.
Surfactant agent or macromolecular coagulant, for example, may adhere to the membrane surface.

Other physical damage may occur, although rarely, at the O-ring seal or the center pipe in the module.

Contamination of the RO membrane can be remedied in most cases by chemical cleaning (however, silica scaling is very difficult to remove). When the membrane is damaged, regeneration is frequently difficult. (The damaged membrane can be restored by coating the surface with a chemical product, but its life is limited.) Table 4 summarizes the main causes of degradation in RO membrane performance and the symptoms, and Table 5 lists the cleaning chemicals that are effective for contaminants of RO membrane.

Many of the synthetic polymer RO membranes that recently have found extensive use are negatively or positively charged, and minute attention is required when washing these membranes with surfactant. For example, if anionic surfactant is applied to a positively charged membrane, the surfactant adheres to the membrane surface, leading to extreme reductions in permeating flux volume (the same problem may occur when cationic surfactant is applied to a negatively charged membrane).

B.   Considerations for Operational Control

As mentioned previously, daily operational control is very important to minimize the performance degradation of the RO membrane and prevent trouble. In terms of the primary requirements of such control, however, there is a slight difference between the upstream and downstream RO modules.

1.   Upstream RO Membrane

Among the factors leading to performance degradation, some items of first importance, including adjustment of pH, restriction of bactericide concentration, and removal of oxidant, are directly related to membrane deterioration. However, the most frequently observed causes of trouble are fouling and scaling, which lead directly to membrane contamination.

An effective procedure to prevent fouling of the RO membrane is to establish higher pretreatment requirements and minimize the amount of suspended substances (hydroxides, such as iron and aluminum) flowing into the RO module. Pretreatment varies depending upon the characteristics of the raw water and the scale of the RO module.

Table 4 Causes and Symptoms of Drop in RO Membrane Performance

Change in Membrane	Cause	Symptom			Remark
		Rejection Rate	Flux Rate	Pressure Differential	
Chemical change	Hydrolysis	Decrease	Increase		Cellulose acetate membrane is liable to hydrolysis.
	Oxidation	Decrease	Increase		Synthetic membrane is liable to oxidation (depending upon pH).
Physical change	Damage to membrane surface	Decrease	Increase		
	Bacterial degradation	Decrease	Increase		Cellulose acetate membrane is liable to bacterial degradation.
No change	Fouling	Decrease	Decrease	Increase	This fouling occurs when the SDI of feed water is high.
	Fouling	Little change	Decrease	Little change	This fouling occurs when there is leakage of aluminum and the like from the pretreatment section.
	Adsorption	Little change	Decrease	Little change	Synthetic membrane tends to adsorb coagulants, surface active agents and the like.
	Scale formation	Decrease	Decrease	Increase	Silica scaling which is difficult to remove demands special attension.
	Bacterial slime	Decrease	Decrease	Increase	
	Mechanical faults	Decrease	Increase		Damage to 'O' ring seal and the like.

Table 5 RO Membrane Foulants and Cleaning Chemicals

	Organic		Inorganic	
	Microorganisms Colloids, etc.	Metal oxides (iron, aluminum, etc.)	Calcium scale ($CaSO_4$, $CaCO_3$, etc.)	Silica scale
Acids (citric, oxalic, etc.)		○	○	○
Acids (HF, NH_4F, etc.)		○	○	○
Chelating Agents (EDTA, etc.)	○			
Alkalis (NaOH, NH_4OH, etc.)	○			
Oxidizing Agents (NaClO and H_2O_2)	○			
Surface Active Agents	○			
Reducing Agents ($NaHSO_3$, etc.)	○	○		

note   This table indicates those chemicals which are expected to produce appreciable cleaning effects.
The optimum concentration, pH, etc. of each chemical cleaning solution should be selected depending
upon the nature of RO membrane in question such as its resistance to the chemical used.
(A particular chemical may not be used for a certain membrane.)

Coagulation sedimentation + filtration and coagulation filtration are widely used processes. It is important to conduct operational control of these pretreatment methods under appropriate conditions: continuous follow-up of variations in water quality and injection of coagulant and pH adjustment according to these variations.

The SDI (silt density index) is generally used to monitor RO feedwater quality. The SDI is given as the percentage of filter clogging per minute, which corresponds to the lowering of the filtering speed when water is passed through a 0.45 μm membrane filter at constant pressure.

For example, SDI 5 indicates clogging of 5% minute^{-1}: the larger the value of SDI, the more suspended substance and fine particles the water contains. Measurement of the SDI is outlined in Fig. 14. SDI 4–5 is required for the spiral RO module. If the water quality is stabilized and maintained at this degree of quality, troublesome fouling of the RO membrane will hardly ever occur. However, even if a low amount of suspended substance with a small SDI flows into the RO unit, it is inevitable that suspended substances will accumulate and gradually adhere to the membrane surface. Consequently, it is necessary to wash the RO membrane with cleaning chemicals one to three times a year.

Fouling may occur even at a small SDI, although this is rare. Fouling takes place easily with relatively pure raw water. It is assumed to take place when, during pretreatment, excessive polyaluminum chloride coagulant is added and the remaining inactive aluminum leaks from the pretreatment module and covers the RO membrane surface.

Once the RO membrane is fouled, the primary circuit (water feed) narrows as a result of clogging, increasing the differential pressure (pressure difference between feed side and concentrated water) in the RO module. Therefore, by continuous checking of the differential pressure, it is possible to confirm fouling. Fouling depends on the feedwater, so that determination of fouling should be made under the same flow rates.

If the RO membrane is covered with inactive aluminum, the permeating flux rate normally increases before the differential pressure, with a somewhat different behavior.

When an RO module at a higher differential pressure is left as is, not only is regeneration difficult, but the element is also subjected to deformation or deterioration. Such an RO module should be washed with cleaning chemicals at an earlier stage.

2. *Prevention of Scaling*

With natural water used as raw water, carbonate of lime and silica may precipitate as scales on the concentrated water side. Such scaling can be prevented without significant

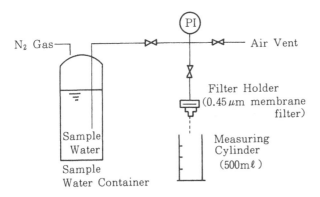

Figure 14 Method for determination of SDI.

alteration in raw water quality because the allowable recovery of each substance at the RO module can be determined using such parameters as raw water quality (concentration of the relevant substances), pH, and water temperature. Normally, attention should be paid to the following.

a. Lime carbonate scale. Formation of lime carbonate scale may be greatly influenced by variations in pH. Although there is almost no possibility of precipitation of scaled lime carbonate in the weak acid range of pH 6 or less, in the neutral and strong base range where such scale is easily formed, minute attention should be paid to the operation conditions.

In general, to determine the probability of formation of such substances in the concentrated water of the RO unit, the Langelier index I is commonly used; when I is in the negative range, there is no possibility of scaling.

$$I = pH - pH_s$$

where: pH $\ =\ $ pH of RO-concentrated water
$pH_s\ =\ pH_s = (9.3 + A + B) - (C + D)$

The values of A, B, C, and D are given in Table 6.

b. Silica. Special attention should be paid to the formation of silica scale, because precipitation then occurs in high probability. Effective cleaning chemicals are not available, so silica scales, once formed, cannot be removed. The solubility of silica around the neutral range, as shown in Fig. 15, varies greatly depending upon the water temperature. For example, the solubility is 100 mg L^{-1} at 25°C and decreases to 60 mg L^{-1} at 10°C. On the other hand, the concentration of silica in raw water is about 20 mg L^{-1} in Japan, except in the Kyusyu District, where the concentration is exceptionally high.

Assuming the RO module is operated with a recovery of 75% using water of 20 mg L^{-1} in silica concentration, silica may not precipitate in the form of scale at 25°C, but at 10°C scaling may occur. Consequently, if the RO unit is shut down for a long period in winter, scaling-preventive measures should be taken, such as replacement of concentrated water in the RO module.

Under normal operation conditions, however, silica precipitation does not occur at 25°C up to a concentration of 120 mg L^{-1}. As long as the water used circulates constantly, sedimentation of silica scale does not occur on the membrane surface even if the water is saturated with silica.

Formation of scales is accelerated further by polarized concentration on the membrane. If the membrane is fouled a little or suffers a trace of scaling, the place surrounding such spots is polarized more easily and scaling is accelerated. Consequently, when the RO module is in operation at the critical condition of scale formation, the measure of first urgency is to prevent fouling to avoid silica scaling on the membrane.

As confirmed for fouling, the formation of scales can be monitored by referring to the differential pressure. In general, in fouling, the differential pressure is generated from near the feedwater side and in scaling from near the concentrated water side.

3. Downstream RO Module

Unlike the upstream RO module, the downstream RO module suffers less frequently from performance degradation as a result of membrane contamination by fouling and scaling. Instead, because this unit is placed at the end of the system, the membrane is subject to bacterial contamination, and restriction of bacterial propagation is one of the important operational control items.

Table 6 Table for Calculation of the Langelier Index[a]

$$pH = (9.3 + A + B) - (C + D)$$

(A)

Total Solids (ppm)	A
50− 300	0.1
400−1,000	0.2

(B)

Temperature (°C)	B
0− 2	2.6
2− 6	2.5
6− 9	2.4
9−14	2.3
14−17	2.2
17−22	2.1
22−27	2.0
27−32	1.9
32−37	1.8
37−44	1.7
44−51	1.6
51−56	1.5
56−64	1.4
64−72	1.3
72−82	1.2

(C)

Calcium Hardness (ppm CaCO₃)	C
10− 11	0.6
12− 13	0.7
14− 17	0.8
18− 22	0.9
23− 27	1.0
28− 34	1.1
35− 43	1.2
44− 55	1.3
56− 69	1.4
70− 87	1.5
83− 110	1.6
111− 138	1.7
139− 174	1.8
175− 220	1.9
230− 270	2.0
280− 340	2.1
350− 430	2.2
440− 550	2.3
560− 690	2.4
700− 870	2.5
880−1,000	2.6

(D)

Methyl Orange Alkalinity (ppm CaCO₃)	D
10− 11	1.0
12− 13	1.1
14− 17	1.2
18− 22	1.3
23− 27	1.4
28− 35	1.5
36− 44	1.6
45− 55	1.7
56− 69	1.8
70− 88	1.9
89− 110	2.0
111− 139	2.1
140− 176	2.2
177− 220	2.3
230− 270	2.4
280− 350	2.5
360− 440	2.6
450− 550	2.7
560− 690	2.8
700− 880	2.9
890−1,000	3.0

[a]$pH = (9.3 + A + B) - (C + D)$.

Source: From Ref. 3.

Control of bacteria is conducted using 1–2% hydrogen peroxide, but if metals or other substances adhere to the membrane surface, the oxidation force of the hydrogen peroxide is enhanced, adversely influencing the polymer RO membrane, which is less resistant to oxidants. If such a membrane is used, careful monitoring is necessary for metal.

Table 7 summarizes the main control items for the downstream RO module. Prevention of oxidation deterioration of the RO membrane varies depending upon its oxidant resistance. Especially when NaClO is injected to restrict the propagation of bacteria, the

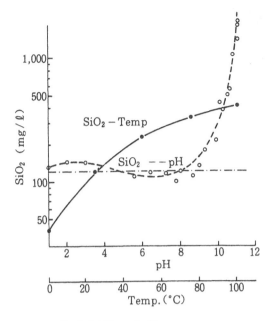

Figure 15   Solubility of silica.

oxidation force changes according to the pH and the purity of the raw water. To prevent deterioration due to oxidation, operational control must be conducted strictly according to the instructions of the manufacturer.

V.   Developments in the Near Future

The performance targets for the RO membrane used in the ultrapure water plant are summarized as follows:

1.   High removal ratio
2.   Higher permeating flux ratio (lower pressurization)
3.   Improvement in chemical resistance
4.   Reduction in substances eluted from the RO module material

These targets should be achieved through development and improvement of existing technologies, which need further refinements that can be made by considering the use of the RO module.

    The targets that follow are the main developments being tackled by various manufacturers.

Upstream RO module: improvement in
    Ion removal
    Organic matter removal
    Chlorine resistance
    Contamination resistance
Downstream RO module and RO module for the subsystem: improvements in
    iron removal in the low concentration field

Table 7   Checklist for the Operation Management of RO Plants

Type of Equipment	Checkup Item	Checkup Frequency		
		Daily	Periodical	
Pretreatment	Raw water analysis (overall)		O	• Check for any seasonal variation in raw water quality.
	Filter pressure differential	O		• Higher pressure differential leads to higher SDI.
	Treated water SDI	O		• In the case of spiral-wound RO membrane, the SDI of pretreated feed water should not exceed 5.
	Treated water residual chlorine	O		• The concentration of residual chlorine should usually fall within a range of 0.2 ~0.5mg/ℓ depending upon the nature of the RO membrane.
	Safety filter pressure differential	O		• Safety filter elements should be replaced before the filtrate flow rate begins to drop.
RO Unit	Feed water analysis (conductivity and pH)	O		• Check for any change in the feed water quality.
	Permeate water analysis (conductivity)	O		• Check for any drop in the salt rejection rate.
	Feed water pressure	O		
	Concentrate pressure	O		
	Permeate pressure	O		• Judge if there is any membrane fouling or plugging.
	Feed water flow rate	O		
	Concentrate flow rate	O		
	Feed pump amperage	O		• Check if the feed pump is in good working order.
	Feed water analysis (overall)		O	
	Concentrate analysis (overall)		O	• Pass overall judgement if the salt rejection performance of the membrane is in good order.
	Permeate analysis (overall)		O	
	Permeate analysis for each membrane vessel (conductivity)		O	• Identify any membrane vessel which produces poor permeate water.
	Supply of motor oil		O	• Supply motor oil at regular intervals.
	Inspection of working condition of pump		O	• Check for any abnormal noise, vibration, oil quantity, ampere value.
	Chemical cleaning of membrane		O	• Remove membrane foulants.
	Replacement of membrane elements		O	• Replace membrane elements when the RO unit cannot exhibit its intended performance due to membrane degradation.

Removal of organic matter of low molecular weight

Oxidant resistance (hydrogen peroxide resistance)

Heat resistance (RO module for which hot water sterilization is possible)

Cleanliness (RO module free from production of eluate and fine particles, RO module structure in which there is no stagnant flow)

Reliability (greater integrity of the RO module and membrane)

Most synthetic polymer composite membranes in extensive use are made mainly of aromatic group or fatty acid polyamide materials, and it can be said that membrane-manufacturing technology is maturing. However, so far, the development of the RO membrane and module has been conducted by stressing desalination and water permeation, but progress is now expected in cleanliness, reliability, and resistance.

REFERENCES

1.   Y. Kamiyama, Reverse osmose composite membrane, Nitto Gihou (Nitto Technical Report), *23*(1), July (1985).
2.   T. Ohmi and T. Nitta (supervisors), "System of feeding ultrapure water/chemicals of high purity," Research Group of Semiconductor Circuit Boards, 1986.
3.   T. Okamoto, K. Goto, and T. Morozumi, Industrial waters and waste water treatment, Gekkan Kogyou Shimbun Co., Ltd., 1972.

2 Membranes

B. Ultrafiltration

Hiroaki Ishikawa
Kurita Water Industries, Ltd., Tokyo, Japan

I. INTRODUCTION

Along with the stunning developments in the electronics industry since the latter half of the 1970s, there have been demands for so-called ultrapure water, as close as possible to theoretical pure water, which contains no ion, particle, organic substance, bacterium, colloidal substance, or any other impurity. To meet this requirement, membrane technology was introduced to pure water systems. The technology is composed of microfiltration (MF) membranes, reverse osmosis (RO) membranes, and ultrafiltration (UF) membranes, which were respectively introduced to ultrapure water production systems. Improvements in the technology have progressed following improvements in large-scale integrated (LSI) circuits.

Table 1 shows the development of LSI and the qualitative requirements for ultrapure water. The ultrapure water required lately may contain less than 5 particles 0.1 μm or more in size per milliliter. The quality of the UF membrane module that is used as the final filter is crucial for particle counts in ultrapure water. This section discusses the features, types, construction, and performance evaluation of UF membranes and the reduction in particle counts.

II. FEATURES OF THE UF MEMBRANE

The UF membrane is installed as a final filter in the polishing process of an ultrapure water system. The features and performance requirements of the UF final filter can be summarized as follows.

A. UF Final Filter

With a UF membrane used as the final filter instead of an RO membrane, ultrapure water systems have the following advantages, among others:

Table 1   Ultrapure Water Quality Target Values for Semiconductor Manufacturing in Japan

Integration Scale / Items	64 Kbits	256 Kbits	1 Mbits	4 Mbits
Resistivity (MΩ·cm)	16 ~ 17	17 ~ 18	17.5 ~ 18	> 18
Particle (Counts/mℓ)	0.2 μm 50 ~ 150 (O Microscope)	0.2 μm 50 ~ 150 (SEM)	0.1 μm 10 ~ 20 (SEM)	0.1 μm 5 > (SEM)
Bacteria (Counts/ℓ)	500 ~ 1000	50 ~ 200	10 ~ 50	10 >
TOC (ppb)	50 ~ 200	50 ~ 100	30 ~ 50	10 >
SiO_2 (ppb)	20 ~ 30	10	5	1 >
O_2 (ppb)	100	100	50 ~ 100	50 >
Metals, Ions (ppb)		~ 1	0.1 ~ 0.5	0.1 >

1. Compact system configuration: the UF membrane allows a larger quantity of water permeation per unit area.
2. Less electrical power consumption by the pump: the UF membrane requires a lower working pressure (1–6 kg cm^{-2}).
3. System installation adjacent to the point of use: the UF membrane does not require a high-pressure pump, which causes vibrations.

Compared with the MF membrane, the UF membrane has the following features, among others:

1. Removability of colloidal substances and high-molecular-weight compounds.
2. Less frequent membrane replacement: the UF membrane discharges some of the feedwater as concentrate so that it is free from clogging, which ordinarily occurs with the MF membrane in full flow operation.

When a UF membrane with these features is applied to an ultrapure water system as the final filter, the membrane should have the performances discussed here.

B. Performance Required

The following three factors are the performing requirements for the UF final filter.

1. Less elution of impurities from module elements. Since the UF membrane is installed in the final stage of the ultrapure water system, a decrease in resistivity due to the elution of ionic substances from module constituents should be minimized. It is also required that the elution of nonionic substances or organic substances that are not indicated by resistivity values be limited.
2. Less discharge particles. It is said that particles or large molecules larger than 10–100 Å in diameter do not permeate the UF membrane. When such a membrane is used as a final filter, it is theoretically presumed that particles larger than 0.05 μm are completely removed. In actual operation, however, particles are still found in permeate water. These are the particles that were dissociated from the surface of the

permeate water side of the module. It is therefore crucial to minimize the dissociation of such particles.
3.   Structure without stagnation. The presence of stagnant space in the module not only prevents the rapid rise of water quality in starting up but also interferes with the performance of the sterilizer, which is periodically injected to prevent bacterial growth.

III.   TYPES AND STRUCTURE OF THE UF MEMBRANE

The UF membrane has a symmetrical or asymmetric structure that is composed of a thin skin layer, which removes impurities, and a porous supporting layer. The skin layer has a thickness of 0.1–1 μm. The layer may be regarded as a thin surface film formed by evaporation of solvent in the membrane synthesizing process. Figure 1 shows the cross section of a hollow-fiber UF membrane. Portion A is the skin layer and portion B the bed layer. Because it has a skin layer on both external and internal surfaces, this type of UF membrane is generally called a double-skin structure.

The RO, UF, and MF membranes cannot be applied for practical use by themselves. Each membrane should be included in a configuration for industrial application, and such a configuration is called the membrane module. Membrane modules are classified into four representative groups by configuration:

1.   Plate and frame module
2.   Tubular module
3.   Spiral module
4.   Capillary or hollow-fiber module

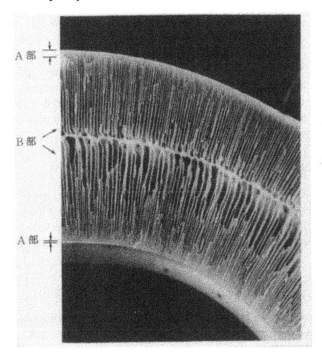

Figure 1   Section of a UF membrane (by SEM).

Because each module has its own advantages and disadvantages, it is necessary to carefully select a module according to its application and fully exploit its advantages. The plate and frame module has a 0.1–3 mm channel gap, and the channels are hollow inside. The spiral module is a rolled plate and frame module with net spacers inserted in the feed channels. The tubular module and the hollow-fiber module, both literally with a cylindrical configuration, are classified into three groups by thickness or diameter:

1. Tubular module: 5.5–25.0 mm diameter
2. Capillary module: 1.0–5.0 mm diameter
3. Hollow-fiber module: 0.1–1.0 mm diameter

In addition to this classification, cylindrical modules are further categorized by internal pressure and external pressure filtration. The cylindrical configuration is strong enough to withstand internal pressures, and the smaller diameter cylinder can bear a higher external pressure. Consequently, small-diameter capillary and hollow-fiber modules are composed only of membrane material that is rigid enough to resist both internal and external pressures. On the other hand, a large-diameter tubular module requires support pipes to protect itself from either internal or external water pressure. Naturally, the smaller diameter module gives the larger unit surface area (specific surface area). A large-diameter tubular module provides a small specific surface area, but the module can process liquid containing suspended solids because it is thick and hollow inside. This is the hollow-fiber type that is widely used in existing ultrapure water systems.

IV.   PERFORMANCE EVALUATION OF THE UF MEMBRANE

The performance of the UF membrane is normally evaluated by the permeation flux of pure water and the rejection rate of impurities, as well as the quality of the product water, indicated by number of particles, total organic carbon (TOC), resistivity, and pyrogen or, in addition to these, chemical resistance, heat resistance, applicable pH range, and pressure resistance. When the UF membrane is used as a final filter to produce ultrapure water, it must fulfill extremely stringent requirements. It is particularly required that the membrane be highly efficient in removing solid particles and bacteria and limited in spontaneous elution from the materials of the module member.

A.   Permeation Flux of Pure Water

Permeation flux is expressed as a permeate water volume per unit time and unit effective pressure at 25°C by unit membrane area. A larger permeation flux indicates a larger permeate water volume and the profitability of a membrane. Since permeate water volume depends on feedwater temperature, it is required that this be compensated for by the viscosity coefficient.

B.   Rejection Test

Standard material is mixed in a solution, and the concentration of the material is measured in feedwater, product water, and concentrate water to calculate the rejection rate of impurities. Rejection rate is expressed by the following formulas, and a rate close to 100% indicates that the membrane has a higher removability.

$$C_p = C_f \times Rec^{-1} \times [1 - (1 - Rec)]^{(1-Rej)} \quad \text{rejection rate 1}$$

$$Rej = \frac{2C_p}{C_b + C_f} \quad \text{rejection rate 2}$$

C_p = concentration of standard reference material in product water
C_f = concentration of standard reference material in feedwater
C_b = concentration of standard reference material in brine
Rec = recovery rate
Rej = rejection rate

Widely used standard reference materials are polyethylene glycol (PEG), globular protein, and insulin, which have molecular weights of several tens of thousands or several hundreds of thousands. However, the standard reference material used for the rejection test differs according to membrane manufacturers and membrane brands, making it impossible to compare rejection rates and cutoff molecular weights between different manufacturers. Equalization of standard reference materials has been strongly desired. For ultrapure water systems, standard particles of 0.1–0.5 μm polystyrene latex (PSL) are occasionally used as an additional means to examine the removal capability of a membrane.

C. Qualitative Analysis of Product Water in Initial Operation

Resistivity, TOC, and number of particles in product water are measured in the initial operation of membrane modules to examine the elution of impurities from the materials of construction. Ultrapure water is fed to modules as raw water, and membrane performance is evaluated from the time required to stabilize the indices of product water and the resulting values.

D. Excitation Test on the Membrane Module

The dissociation quantity of particles adhering to a membrane module on the permeate water side is measured and referred to as an indication of cleanliness of the module on that side. After a module is excited by vibrator, the number of particles in the permeate water is measured by a particle counter. The performance of the module is evaluated by the number of particles during excitation and the fluctuation in count afterward. Since membrane modules are liable to fouling during manufacturing and storage, this test is effective in evaluating their cleanliness on the permeate water side.

V. REDUCTION OF PARTICLES IN PERMEATE WATER

Because the pore size is small, the UF membrane is generally presumed to capture all particles larger than 0.1–0.2 μm. Table 2 shows the number of particles in the permeate water treated by three kinds of membrane modules. Several tens of particles at 0.1 or 0.2 μm per milliliter were found in all water samples. The RO module generated more particles than the UF module, indicating that the pore size of a membrane has nothing to do with the existence of particles in the permeate water.

To study particle rejection, standard particles of polystyrene latex were mixed in the feedwater to different types of membrane modules, and the results are shown in Table 3.

Table 2 Average Particle Counts in an Ultrapure Water System

	Final RO Outlet	Final UF Outlet	Return
0.2 μm (n/ml) (Optical Microscope)	22.7 (35)	9.6 (36)	9.0 (23)
0.1 μm (n/ml) (SEM)	—	15.0 (22)	25.8 (17)

( ) : Number of data

Table 3 Rejection of Particles

Module	Feed n/ml	Permeate n/ml	Rejection (%)
Spiral RO	1×10^8	1100	99.992
Spiral UF	1×10^8	94000	99.906
Hollow fiber UF	1×10^8	< 22	> 99.99998

0.212 μm Polystyrene Latex

Performance evaluation was made by the rejection rate of particles. The hollow-fiber UF module proved its superiority over the spiral UF module. With the number of particles at the UF inlet at actual installation (10^1–10^3 particles ml^{-1}) taken into account, it can be concluded that the particles in feedwater will never leak into permeate water.

Table 4 shows the increments in numbers of particles after vibrating the components of the subsystem of an ultrapure water production system. Compared with piping materials, valves, and other components, the UF membrane module is susceptible to vibration and generates dust.

Table 5 shows the number of the particles adhering to the components of UF membrane modules. As many as 10^{10}–10^{13} particles mm^{-2} were counted on the membrane surfaces, permeate water spacers, and housings of the permeate water side. The

Table 4 Transition of Particle Counts After Impact in Subsystem

	Particle Counts
Cartridge Polisher	1180
Hollow fiber UF A	3970
B	1460
Piping System (PVC)	30 ~ 100
Piping System (PVDF)	13 ~ 60
UF outlet by non-impact	1 ~ 10

Table 5 Particle Contamination of Membrane Modules

Location / Classification	Hollow fiber UF (A)	Hollow fiber UF (B)	Spiral RO
Permeate Side of Membrane	5.3×10^{11}	1.2×10^{13}	8.4×10^{11}
Permeate Side of Spacer Net	1.3×10^{11}	3.4×10^{10}	3.0×10^{11}
Vessel Inside	8.1×10^{10}	1.1×10^{10}	—
Permeate Side Total	7.4×10^{10}	1.2×10^{13}	1.1×10^{12}

Unit ; Particles per module $> 0.05 \, \mu$m

increase in particle counts in permeate water were caused by the dissociation of particles adhering to the wet portions of those components on the permeate water side.

Based on these measurements, UF membrane modules for ultrapure water systems require the following:

1. Fewer particles adhering to the membrane surface on the permeate water side
2. Absence of water pockets on the permeate water side

VI. PARTICLE-FREE UF MEMBRANE MODULE FOR AN ULTRAPURE WATER SYSTEM

The UF membrane module for ultrapure water systems should satisfy the three physical requirements listed here:

1. Absence of pinholes on the membrane surface
2. Fewer particles adhering to both membrane surface and module material on the permeate water side
3. Absence of water pockets on the permeate water side

A. Behavior of Particles in Initial Operation

Figures 2 and 3 show the fluctuation in particle count in permeate water from internal pressure UF and external pressure UF membrane modules, respectively. The internal pressure type (conventional type) still generated a couple of particles per milliliter, with a large fluctuation in quantity at 100 h after the start of service operation; the external type (KU-1010HS) became stable at 1 piece ml^{-1} or less after 100 h. Figures 2 and 3 indicate that the external pressure module is stabilized faster than the internal pressure module.

B. Behavior of Particles by Excitation

With a vibrator fitted to the center of a membrane module, measurement was made of the number of particles in permeate water during excitation, the results of which are shown in Fig. 4. Figure 4A shows measurement by the internal pressure (conventional type) module and Fig. 4B by the external pressure (KU-1010HS) module. The internal pressure type is

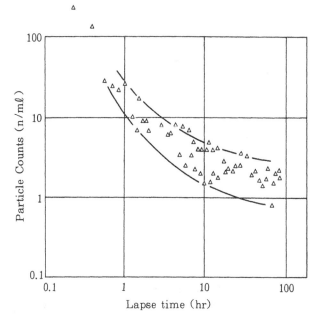

Figure 2   Transition of particle counts (conventional UF).

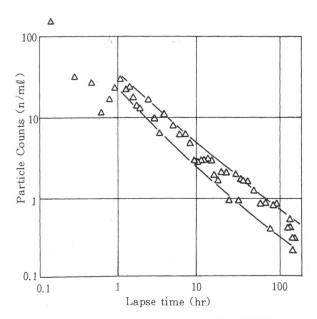

Figure 3   Transition of particle counts (KU-1010HS).

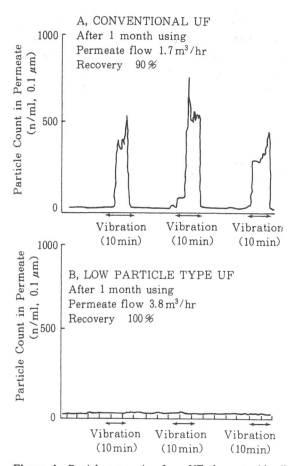

Figure 4   Particle generation from UF element with vibration.

quite liable to external vibration, and the number of particles increased 100 or 1000 times the number obtained in normal service operation. Compared to the internal pressure type, the external pressure type constantly produced quality water with a negligible increase in the number of particles. This is because the external pressure type has less area that contacts the permeate water.

VII.   CONCLUSION

It is predictable that along with the increased density in LSI, the performance requirements of the UF membrane, which is used as the final filter in ultrapure water systems, will become even more stringent. The additional requirements expected of UF membrane are as follows:

1.   Perfect capture of particles
2.   Absence of dust generation
3.   Removability of smaller particles

Besides membrane technology, it is also indispensable for the development of UF membranes to improve the evaluation technology of particles remaining in the ultrapure water. The conventional method of counting the particles captured on a filter by scanning electron microscopy (SEM) has become less satisfactory in terms of analytic accuracy, limits of detection, and the time required for analysis. For further improvement in UF membranes, it is essential to develop a precise, simple, and real-time evaluation technology for particles.

REFERENCE

1. K. Yabe, T. Mizuniwa, and T. Ohmi, Proceeding for the Seventh Symposium of VLSI Ultraclean Technology (Tokyo), (July 1988), pp. 29–50.

2 Membranes

C. Membrane Filtration

Koichi Sawada

Nomura Micro Science Co., Ltd., Kanagawa, Japan

I. INTRODUCTION

The membrane filter (MF) is also termed an organic membrane. The membrane has an absolute pore size and uniform pores of about 0.05–10 μm and is used for eliminating particles in this range. The thickness is about 10–200 μm. In a pure water system, the MF is installed at important points as a final filter (FF) in a terminal or for keeping particle increase in a system under control, especially for eliminating microorganisms that multiply.

In a pure water system, the MF is used as a cartridge filter processed in a pleated form or as a module processed in a hollow-fiber form to increase the filtration area.

Selection of the MF is mainly made on the basis of the following:

1. Pore size
2. Material
3. Filtration area
4. Number of particles in raw water

II. POSITIONING FILTRATION BY MF

Kozima et al. [1] classified the separation of particulate solids in a liquid from the viewpoint of treatment of water into the following:

1. Screening (elimination of bulky suspended matter)
2. Coagulation (elimination of crude particles and colloids)
3. Sedimentation (elimination of particles sedimented by gravity)
4. Flotation separation (elimination of particles floated by gravity)
5. Flotation (elimination by foam separation)
6. Filtration for clarifying (filtration for dilute suspensions)

Filtration by MF is included in the filtration fore clarifying category. Filtration for clarifying obtains clarified filtrate using a filter.

Filtering solids after solid-liquid separation is unnecessary. There are a variety of techniques and classifications, such as rapid sand filtration, slow sand filtration, filtration by a filter medium, such as ceramic, filtration by a filter paper, and filtration by MF. In addition, a filter medium with performance between that of MF and UF exists. Among these MF is excellent in eliminating particles, but it is weak in dealing with a large particle load, and feeding to an MF must be performed after reducing the number of particles to as few as possible by many pretreatment steps. The size of fine particles existing in water and the separation performance [2] by filters are shown in Fig. 1.

III.   FILTRATION MECHANISM

Filtration mechanisms were classified into cake filtration, complete closed filtration, standard closed filtration, and intermediate closed filtration by Hermans and Bredee [3]. In many cases, cake filtration is used for many solids and standard closed filtration is used in dilution.

With regard to filtration using MF, Heertjes [4] devised and confirmed an equation for closed filtration by considering the following three models:

1.   A particle sinks into a pore and completely closes the pore.
2.   A particle sticks on the inner surface of a pore and narrows the pore size, and finally many particles close one pore.
3.   A pore is partly closed by a particle, and a cake is formed there.

IV.   DEPTH FILTER AND MF

A depth filter is composed of irregular nets piled by fiber, such as a filter paper using paper, nonwoven fabric using synthetic fiber, and a wound type wound by fiber thread,

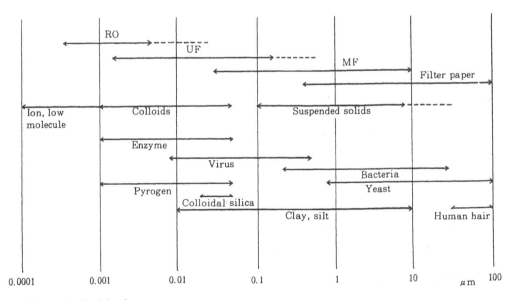

Figure 1   Particle size.

and particles are captured in spaces between the nets and removed from the liquid. Because the spaces are not uniform, however, and have various sizes, some large particles leave the filtrate side through large spaces and some small particles are captured by small spaces and absorbed there. Thus in some cases the captured and adsorbed particles are passed to the filtrate side with variations in pressure and flow during filtration.

As advantages, clogging scarcely arises and many particles are captured because the filter material is thick, compared with that of an MF, and the entire filter material captures particles.

An MF has more uniform pore sizes and a thin membrane and captures particles on its surface. It is possible to display absolute pore size, and all particles larger than that pore size are captured because of the uniform pore size. (Strictly speaking, in special cases, such as fibrous or changing form, the possibility of passage exists.) If the pressure or flow rate changes, particles larger than the pore size are not passed to the filtrate side. Because capture takes place mainly on the surface, the holding amount of particles up to clogging is small and clogging is apt to occur. To use a depth filter as a prefilter and to use an MF as a final filter are reasonable methods utilizing the characteristics of both.

When preparing pure water (used in pretreatment or as a final filter), the amount of solids should be very slight. The selection of filters differs depending on the types and amounts of solids to be eliminated from the water and on the amounts of other solids. For example, particles in the filtrate can be reduced to approximately 0 ml^{-1} by filtering city water (particles over 0.2 μm exist in the range of 10^5–10^6 ml^{-1}) with a 0.2 or 0.22 μm MF, but the filter is required to be exchanged often because of extremely rapid clogging. The removal ratio of particles is only about 50% if city water is filtered with a depth filter of 0.4 μm nominal pore size. By setting the depth filter as a prefilter and installing an MF at a succeeding stage, however, the life of the MF can be doubled and thus economical use is possible.

The appearance of clogging of an MF and a glass fiber filter paper used as a depth filter are shown in Figs. 2, 3, and 4 (Nuclepore general catalog, Nomura Micro Science). The flow rate, because of filtration under constant pressure, is large initially but reduces because of clogging with increasing filtered volume. An integrated filtered volume in 0.22

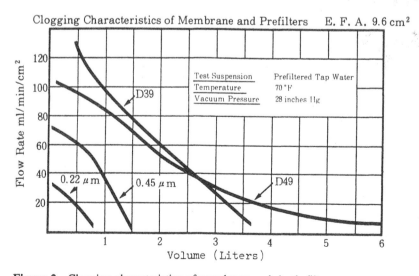

Figure 2 Clogging characteristics of membrane and depth filters.

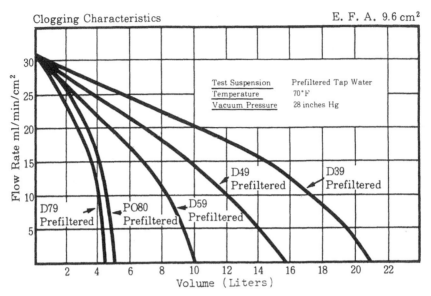

Figure 3   Clogging characteristics of membrane filters following prefiltration.

and 0.45 μm MFs is as small as 11 compared with that using a glass fiber filter paper (Fig. 2). On the other hand, when water filtered by glass fiber filter paper is filtered with a 0.22 μm pore size MF, the volume is increased 20-fold with the D49 (Fig. 3). An integrated filtered volume up to clogging with a 0.45 μm pore size MF is about 25 times (Fig. 3). To increase a filtered volume by MF, filtration with a depth filter with a small nominal pore size is favorable but the selection of appropriate pore size and flow rate must be made by considering the usable life times of the depth filter and the MF.

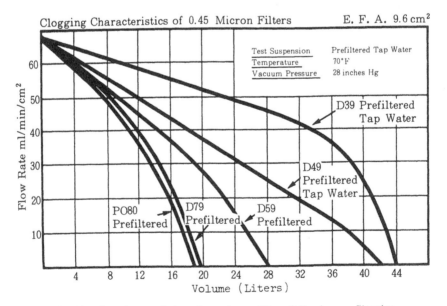

Figure 4   Clogging characteristics of membrane filters following prefiltration.

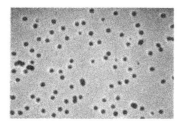

Figure 5 Polycarbonate MF micrograph.

V. MATERIALS AND STRUCTURE

An MF is fabricated by forming fine pores on a polymeric membrane. Such methods as solvent gelatinizing, extracting, sintering, drawing, and particle track etching are used properly according to the different kinds of materials.

Polyvinylchloride, polycarbonate, polyethylene, polysulfone, polyethyleneterephthalate, nylon 6,6, polyvinylidenefluoride (PVDF), polyvinylidenetetrafluoride, polypropylene, acryl polymer, aromatic polymer, polyolefin, and polyvinylalcohol are used as materials for the MF, as well as celluloses, such as cellulose nitrate, cellulose acetate, triacetylcellulose, and regenerated cellulose. The thickness of the membrane is about 10–200 μm.

With regard to the structure of membranes, membranes composed of polycarbonate and polyethyleneterephthalate have cylindrical pores and are a true screen type. Others are, to be exact, depth filters; they have a spongy structure and no pores capable of being measured by microscopic observation. They have extremely uniform pores compared with general depth filters and do not pass particles larger than the indicated pore size.

Photographs of the surface of polycarbonate membrane, cellulose acetate membrane, and cellulosic filter paper are shown in the Fig. 5, 6, and 7. They clearly illustrate that polycarbonate and cellulose acetate membranes are MF filters, but differ in the form of the pores. Examples of the pore size distributions of MFs measured with a 0.2 μm MF (ASTM F316-80) are shown in the Fig. 8 (Nuclepore general catalog, Nomura Micro Science). Figure 8 clearly shows that the uniformity of pore size differs depending on the filter.

The relation between the different filters and removal ratios is shown in Fig. 9 (Nuclepore technical information). This is a general relation in the state in which clogging scarcely arises (with increasing numbers of captured particles, capture by captured particles increases, except capture by a filter; that is, cake filtration begins and the capture

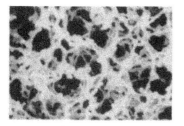

Figure 6 Cellulosic MF micrograph.

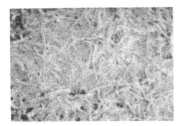

Figure 7   Celluose fiber filter micrograph.

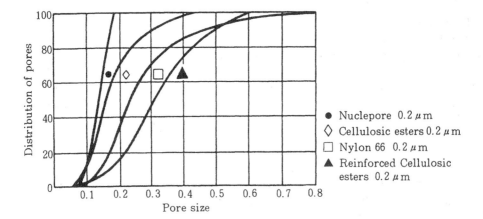

Figure 8   Evaluation of pore size distribution using ASTM F316-80.

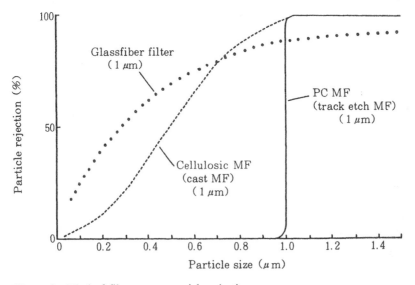

Figure 9   Kind of filter versus particle rejection.

mechanism changes). A polycarbonate membrane with cylindrical pores almost passes particles below the pore size, and cellulosic MFs with spongy pores and relatively thick membranes eliminate 100% of particles larger than an indicated pore size and show fair removal of particles smaller than the indicated pore size. With a depth glass fiber filter paper characterized by a larger membrane thickness, the particle capturing ratio for particles smaller than the indicated pore size is higher than that of cellulosic MFs, and the capture of particles larger than the indicated pore size is incomplete.

As a specific instance, the characteristics of the Zetapor MF are as follows (Zetapor ER EG Grade catalog, Curo Co., Ltd.). A positive charge enables entrapment of particles that are smaller than the pore diameter. When pure water of 13 MΩ·cm containing colloidal silica was subjected to circulating filtration through a nylon 6,6 MF with positive charge and an MF without charge, only the MF with the positive charge could recover against a ratio resistance of 18 MΩ·cm (colloidal silica 0.027 μm in mean diameter could be eliminated). This is shown in Fig. 10. Removal of pyrogen (lipopolysaccharide, LPS) was possible: when 60 ng/ml (appearing to correspond to about 600 EU/ml) of LPS was filtered through a 0.45 μm nylon 6,6 membrane with a positive charge at 13 ml min·cm^{-2}, LPS could not be detected up to 440 ml (less than 0.033 ng ml^{-1}), and a 72% removal effect was seen in 3000 ml filtration.

Although the filter is assumed to be useful if it is used appropriately, when the charge of the MF is saturated by particles and electrolytes, there is no removal by absorption. Thus, caution is necessary because the life of the filter cannot be judged merely by the increase in differential pressure due to clogging.

Particles in water generally have a negative charge. Since the life of a filter cannot be judged only by the increase in filtering, caution is necessary. For the purpose of

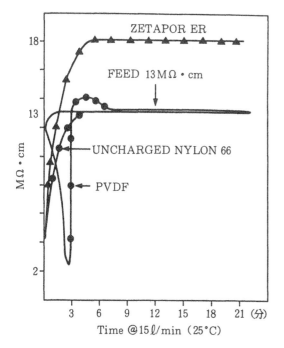

Figure 10   Test for resistivity recovery.

elimination of small particles and pyrogen, it is necessary to follow the instructions of the maker carefully. When eliminating particles smaller than the pore size or pyrogens, users must follow the maker's directions.

VI. FORM OF MF DURING USE

An MF uses a circular disk in filtering a small volume, but requires a larger area in preparing pure water requiring a large filtration flux.

A representative method is a pleated type of cartridge filter that increases the folded filtration area by a rectangular disk in a pleated form. An example is shown in Fig. 11 (Astropore catalog, Fuji Photo Film Co., Ltd.). Various parts other than the filter itself are used. A core for supporting the membrane, placed at the center, a support screen as reinforcing material for the membrane, an outside support net for preventing damage to the membrane and giving strength to the entire assembly, an end cap, and an O ring or a gasket for sealing and installation at a housing are used. With regard to size, the length and diameter of the cylindrical form are 10 in. (in some cases 5, 20, 30, and 40 in.) and about 70 mm, respectively. Liquids flow from the outside of the cylindrical form to the core and flow out from the central hole of the end cap. Other types are a multistage laminated disk type, a spiral type, a tube type, and a hollow-fiber bundle type.

Some cartridges or modules are fabricated in forms that are directly connectable to a line, but a representative cartridge filter is installed in a filtration apparatus (housing). Housings for 1 or 2 as a small size and 72 and 120 cartridge filters as a large size

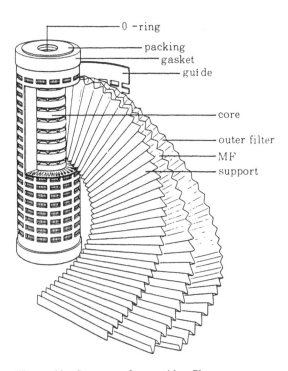

Figure 11 Structure of a cartridge filter.

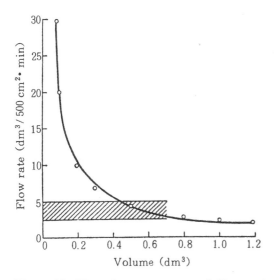

Figure 12 Water flow rate versus total filtrate volume.

(converted to 10 in.) are available. The materials are synthetic resins, such as polypropylene and polycarbonate, and stainless steels (SUS-316 and SUS-304).

The filtration flux per cartridge filter differs according to the effective filtration area of the MF, for example, 0.46 and 2.3 m^2, but is usually determined as about 300 L h^{-1} (5 L minute^{-1}). The relation between flux and integrated filtration volume is shown in Fig. 12 [5]. The relation between flux and the differential pressure (pressure loss) of a cartridge filter is shown in Fig. 13 (Nuclepore catalog).

The number of cartridge filters and the size of the housing are determined on the basis of the treated flow rate of the raw water, the number of particles, and the required life.

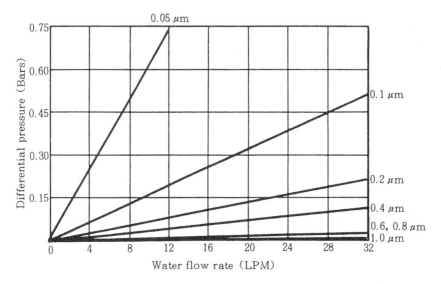

Figure 13 Water flow rate versus differential pressure for 10 in. QR cartridge filters.

VII.   CHEMICAL AND HEAT RESISTANCE

A water-purifying apparatus is periodically sterilized for preventing contamination by microorganisms. Sterilization for an MF is especially required because of the microorganisms captured there. The sterilization methods are limited by the materials of the membrane and the cartridge filter.

A water-purifying apparatus is usually sterilized by hydrogen peroxide or hot water. Formaldehyde, sodium hypochlorite, and steam are used in a water-purifying apparatus used in preparing drugs, but hardly ever used in manufacturing semiconductors. Sterilization by ozone and hydrogen fluoride has also been proposed. The MF membrane itself is resistant to chemicals, but attention should be paid to the parts constituting the cartridge and housing.

VIII.   ELUATE FROM THE CARTRIDGE FILTER

The results of analysis of eluates from four kinds of cartridges are shown in the Table 1 (Nuclepore catalog) [6]. A 10 in. cartridge filter was immersed in 5 L pure water at 30°C for 72 h (stirred lightly for 24 h). Two tests per cartridge filter were performed. Conductivity, absorptiometry (absorbance at 200 nm, and silica), wet oxidation by high-temperature NDIR method (TOC), ion chromatography (negative ion), and atomic absorption spectrometry (metal) were used. Analytic values below the quantitatively lower limit value are indicated by $<$ in Table 1. With regard to PO_4^{3-}, some samples were far below the quantitatively lower limit value as a result of hindrance by other negative ions. Also, blank values of pure water show NO_3^- at 9 and 12 $\mu g\ L^{-1}$ and, these are attributed to cleaning the polyethylene vessel by immersion in nitric acid.

The data suggest that the kinds and amounts of eluates differ extensively depending on the kind of cartridge filter. In the nylon MF cartridge filter of company A, the values for Fe and Zn are extremely large and Al and PO_4^{3-} are also high. In addition, the absorbance at 200 nm is high compared with TOC. In the PVDF MF cartridge filter of company B, the TOC, F^-, SO_4^{2-}, Na, K, Mg, silica, many other negative ions, and alkaline metals (presumed as positive ions) are high, and possibly as a result the conductivity is as high as 3.9 μm. Prevention of F^- elution is difficult because the MF material is a fluorocarbon resin, but naturally other ions must be controlled to a smaller extent. In two kinds of the polycarbonate cartridge filters of company C, substances showing an especially large value are absent, except that alkaline metals and Cl^- were detected.

These tests were performed so that Fe and Cu were barely eluted because pure water was used for immersion and acids and heat were not applied. Only one test did not elucidate whether elution arises only once and after this arises continuously or reduces rapidly. In any event, water is presumably required for flushing after cartridge installation.

IX.   POSITION IN PURE AND ULTRAPURE WATER SYSTEMS

The representative examples of the positions of an MF used in pure and ultrapure water systems are discussed here. Pumps, ultraviolet (UV) sterilizer, and other components were omitted for simplicity.

Table 1 Water-Extractable Matter in 10 in. Cartridge Filters

	μS	Abs at200nm	TOC mgC/mℓ	F⁻ μg/ℓ	Cl⁻ μg/ℓ	NO₃⁻ μg/ℓ	PO₄³⁻ μg/ℓ	SO₄²⁻ μg/ℓ	Na μg/ℓ	K μg/ℓ	Mg μg/ℓ	Al μg/ℓ	Fe μg/ℓ	Cu μg/ℓ	Zn μg/ℓ	SiO₂ μg/ℓ
Blank Pure water No.1	<1.0	0.013	0.10	<1	<1	12	<2	<1	0.06	<0.5	<0.2	<2	<2	<2	<0.5	<10
No.2	<1.0	0.020	0.10	<1	<1	9	<2	<1	<0.05	<0.5	<0.2	<2	<2	<2	<0.5	<10
A. 0.2μmMF (Nylon) No.1	1.1	0.14	0.86	<1	10	11	26	20	2.7	1.5	0.4	13	2.0×10^{2}	<2	1.7×10^{2}	12
No.2	1.2	0.14	0.89	<1	9	2	27	25	4.1	1.5	0.4	14	2.4×10^{2}	<2	1.6×10^{2}	13
B. 0.22μmMF (PVDF) No.1	3.9	0.062	2.3	1.6×10^{2}	1.6×10^{2}	18	<20	69	1.6×10^{2}	72	21	<2	8	<2	1.2	24
No.2	3.9	0.069	2.6	1.5×10^{2}	1.5×10^{2}	20	<20	62	1.4×10^{2}	86	20	<2	9	<2	1.1	24
C. 0.2μmMF (PC) No.1	<1.0	0.034	0.26	<1	9	23	<2	1	3.7	2.7	1.1	<2	<2	<2	1.1	<10
No.2	<1.0	0.040	0.28	<1	12	13	<2	2	6.0	5.4	1.7	<2	<2	<2	1.3	<10
D. 0.1μmMF (PC) No.1	<1.0	0.032	0.40	<1	7	11	<2	6	7.9	4.1	0.6	<2	<2	<2	21	<10
No.2	<1.0	0.034	0.75	<1	2	14	<2	6	5.4	0.8	0.3	<2	<2	3	20	<10

Example 1: raw water → ion exchange → depth filter → MF → use point. This is a system that is unnecessary for highly pure water. Particles in the raw water and fragments of ion-exchange resin are removed with a combination of a depth filter and an MF.

Example 2: raw water → ion exchange → depth filter → tank → ion exchange → MF → use point (→ to tank). This is for reducing more ions and particles than in example 1 by ion exchange twice and circulation, but the load to an MF is still large.

Example 3: raw water → RO → ion exchange → depth filter → tank → ion exchange → MF → use point (→ to tank). This system reduces the large particle load by treating raw water by reverse osmosis (RO). In some cases, an MF is installed behind the depth filter for further reducing particles and stabilizing a system.

Example 4: raw water → ion exchange → depth filter → RO → MF → tank → secondary water-purifying systems. This is a example for intensive treatment of primary purified water and for installing an MF for capturing the small numbers of bacteria produced in the RO unit.

Example 5: primary purified water → tank → ion exchange → MF → UF → use point (→ to tank). This process prevents clogging of the UF unit by removing particles from an ion-exchange resin by installing an MF ahead of the UF.

As shown here, the MF can be positioned (1) behind a position that may produce particles, (2) at the terminus of a makeup water or ultrapure water system, and (3) ahead of a UF membrane. These positions are selected by considering the quality of water and running costs, if a depth filter or a UF membrane could also be used.

X.   VARIATION IN THE NUMBER OF PARTICLES IN PURE WATER SYSTEMS

The appearance of variation in the number of particles in pure water systems differs. Variation in the number of microorganisms in the system shown in Sec. IX is shown in the Fig. 14 [5].

Measurement [7] of the particles produced by filtering pure water with different MFs is shown in Figs. 15, 16, and 17. Distinct differences between the numbers of particles above 0.2 μm are not recognized, but distinct differences exist with regard to the numbers of particles of 0.07 or above 0.1 μm and indicate that the removal ratios of particles differs between these two MFs.

XI.   TEST METHODS

Test methods are summarized here, chiefly the ASTM test methods.

A.   Bubble Point and Specific Pore Size

These are the test methods for properties of MFs of 0.1–15 μm pore size. The maximum pore size of the nonfibrous membrane and the characteristics of a specific pore size are measured.

The bubble point is the pressure at which continuous bubbles begin to rise when a membrane is wet with solvent, such as water, the liquid is added to the secondary side, and pressure is applied gradually by gas from the primary side (requiring a filter support).

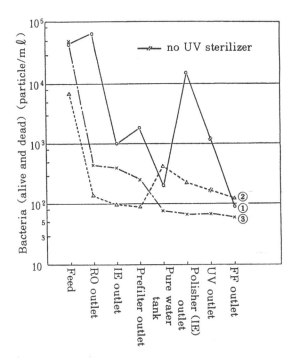

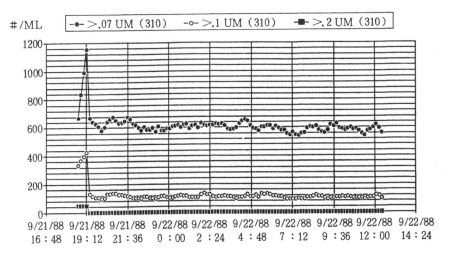

Figure 14   Efficiency of UV sterilizer in pure water system: (1) run to 6 months (no flashing); (2) after flashing (sterilizing by HO); (3) after adding a UV sterilizer.

Figure 15   Particle count after prefilter in systems measured by HORIBA PLCA-310.

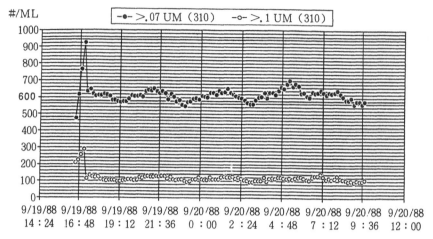

Figure 16   Particle count after final filter bank A measured by HORIBA PLCA-310.

The maximum pore size of a filter can be obtained from the surface tension of the solvent and the bubble point.

$$d = \frac{Cr}{P}$$

where: d = maximum pore size (μm)
r = surface tension (mN m^{-1} or dyn cm^{-1})
P = pressure (Pa or cm Hg)
C = constant (2860 when pressure is Pa, 0.215 when cm Hg)

The pore size distribution is measured as follows. Air is fed by applying pressure to an MF wet with a nonvolatile liquid, and then air passes through when the pressure excels the holding power of fine pores to the liquid. Air passes through a larger pore under lower

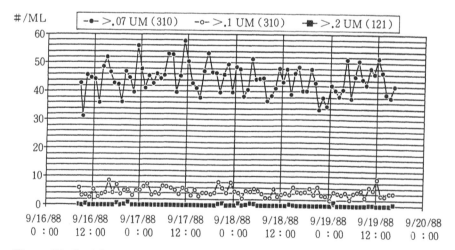

Figure 17   Particle count after final filter bank B measured by HORIBA PLCA-121 and 310.

pressure. The ratio of the contribution of pores in the range of specific pore sizes to the passage of air can be obtained by measuring the flow rate of the air because air passes through a defined pore size under a defined pressure.

The pressure at which a flow rate on wetting reduces to half the flow rate on drying is the mean flow pore pressure, from which the mean flow pore size can be determined. Comparison of pore size distributions between respective MFs can be made by using these values and the figures.

B. Percentage of Porosity

This method is applied to measurement of the percentage of porosity, and the percentage of porosity is determined by defining the difference between a pore volume obtained from the size and thickness of an MF and a volume obtained from the weight and specific gravity of an MF when the specific gravity of the MF material is known.

If the specific gravity is unknown, mercury is introduced into pores under pressure and the percentage of voids is determined from the change in volume.

C. Diffusion

That the flow of a gas passing through a wet filter changes from flow due to diffusion into flow due to diverse causes is the principle used for this test [8]. Liquid or gas from a secondary side that is pushed out by gas is measured. Pressure is applied at 80% of the bubble point.

D. Forward Flow

Gas flow per time is quantitatively measured by applying a definite pressure to a wet filter [8]. The pressure of the gas is equal to a pressure at which conversion from a diffusion air current to a mass release air current arises in a wet filter.

One maker sets less than 8 ml minute^{-1} as a standard under a pressure of 2.8 kgf m^{-2} for a 0.22 μm cartridge filter, and the standard was said to be confirmed by a challenge test using bacteria (technical information, Pall Co.).

E. Bacterial Retention Characteristics With a 0.2 μm MF

This is a test for the ability to retain bacteria with diameter equal to or slightly larger than the of 0.2 μm MF pore size.

A filtration apparatus and a sample MF are sterilized by autoclaving, and a bacterial suspension of *Pseudomonas diminuta* cultured by a defined method (ATCC 19146) is filtered through it. The filtrate is then cultured for 48 h. Whether capture is complete is judged according to the presence of turbidity in the cultured liquid. If turbidity occurs, the bacteria have leaked.

F. Bacterial Retention Characteristics With a 0.4 μm MF

This method is fundamentally similar to that for the 0.2 μm MF. It differs only in using a 0.40–0.45 μm MF pore size for the sample and *Serratia marecescens* (ATCC 14756) as the bacteria.

G.   Log Reduction Value

The log arithmic reduction value (LRV) was proposed by a working section for validation relating to filters for sterilizing liquids for HIMA [9].

The test measures the performance of a filter in removing *P. diminuta* (ATCC 19146). *P. diminuta* cultured under defined conditions is challenged with a filter of 10^7 cm^{-2} effective filtration area, the number of *P. diminuta* in the filtrate is determined, and a calculation is made according to the equation

$$LRV = \log \frac{\text{total number of challenged bacteria}}{\text{number of bacteria in filtrate}}$$

If the effective filtration area is 500 cm^2, 5×10^9 bacteria are challenged. A total bacterial count of 100 in the filtrate corresponds to 7.70 LRV, according to

$$LRV = \log (5 \times 10^9/10^2) = 7.70$$

If the number bacteria in the filtrate under the same conditions is 0, 1 is substituted for the number of bacteria in the filtrate and the value just calculated is displayed; that is,

$$LRV = \log (5 \times 10^9/1) = >9.70$$

H.   Water Flow Rate

This can be applied to a 5.0–0.1 μm MF pore size, and 500 ml water or other solvent is filtered under vacuum with 700 ± 5 mm Hg differential pressure. An effective filtration area is predetermined by filtration of a pigment, for example. The solvent for the test is maintained at 25°C and previously passed twice through an MF. The flow rate (ml minute·cm^{-2}) at 25°C and 700 mm Hg differential pressure is calculated from the filtration time and the effective filtration area.

I.   Quantity of Water-Extractable Matter

An MF is added to boiling water, water-soluble matter is extracted, and the weight of the extracted substances is determined from the difference between the weight of the MF before and after immersion. The same MF not used in extraction is measured at the same time for error correction.

J.   Filterability of Water

A sample of water is filtered under a pressure of 0.7 kg cm^{-2} by a 0.45 μm MF (25 mm diameter). The times required for filtration at 40, 200, and 400 ml are measured, and the FI (filtrability index) is calculated from the following equation. Information related to the amounts of substances (microorganisms, colloidal and gelatinuous substances, and so on that cause clogging can be obtained.

$$FI = \frac{T_3 - 2T_2}{T_1}$$

where: T_3 = time required for 400 ml filtration
T_2 = time required for 200 ml filtration
T_1 = time required for 40 ml filtration

K.   Particle Number

Information related to the form and number of particles can be obtained by a direct microscopic method (see Sec. VII).

XII.   FUTURE EXPECTATIONS FOR MF

As a filter used at the terminal of a ultrapure water system, UF membranes presumed to have smaller pore sizes have often been used as the allowable value (water quality specification) of the required number of particles becomes more stringent. The characteristics of MFs approach those of UFs with reducing pore sizes of the MFs. Characteristics of MF, such as that the MF has a larger filtration flux per area compared with an UF and can be installed at lower cost, must be utilized.

REFERENCES

1. S. Kojima et al. (eds.), Handbook of Water and Waste Water, Maruzen Co., Ltd., (1987), pp. 143–240.
2. Nomura Micro Science Co., Ltd., Technical Information, Purification of Water.
3. P. H. Hermans and H. I. Bredee, J. Soc. Chem. Ind., 56 (1936).
4. P. M. Heertjes, Chem. Eng. Soc., *6*, 190 (1967).
5. H. Ohya (ed.), Technical Handbook of Membrane Application, Saiwai Shobo, p. 741.
6. Nomura Micro Science Co., Ltd., Interoffice Report.
7. M. Tang and D. Tolliver, Ultrapure water particle monitoring for advanced semiconductor manufacturing., Semiconductor Purewater Conference, p. 26-57 (1989).
8. Filtration and Sterilization Process Variation Research Group, GMP Technical Report 3, Variation of Filtration and Sterilization Process, Yakugyo Jihousha Co., Ltd., (1985), pp. 130–134.
9. Health Industry Manufacturers Association Document No. 3, Microbiological Evaluation of Filter for Sterilizing Liquids,

3
Ion-Exchange Resin

Kenji Oda
Nippon Rensui Co., Yokohama, Japan

I. INTRODUCTION

The ion-exchange resin is one of the trailblazing functional polymers. It was first commercialized in Germany in 1938. Since then, various types of ion-exchange resin have been manufactured in many countries. It has been applied to various fields, including water treatment, such as the makeup water to the boiler, the food industry, the pharmaceutical industry, and water reuse [1]. Many books and articles describe the principles of the ion-exchange resin and water treatment technology [2–5]; this chapter touches upon the functions of ion-exchange resin in present ultrapure water production systems and future technical challenges.

II. TYPES AND FUNCTIONS

A. Types

Table 1 summarizes the types of ion-exchange resin used in the ultrapure water production system from the viewpoint of functional group and the material and structure of the matrix. The lines in Table 1 indicate combinations in commercialized products (DIAION, Amberlite, and other catalogs). The ion-exchange resin for industrial use has a spherical shape and a 10–50 mesh diameter. Figure 1 shows micrographs of major ion-exchange resins for industrial use.

Ion-exchange resins can be categorized into cation-exchange and anion-exchange resins. They can also be divided by functional group into strong acid cation-exchange resins containing a sulfonic acid group, weak acid cation-exchange resins containing a carboxylic acid group, strong base anion-exchange resins containing a quarternary ammonium group (type I resin with high basicity and type II resin with slightly low basicity),

Table 1   Kinds of Ion-Exchange Resins for Ultrapure Water Production

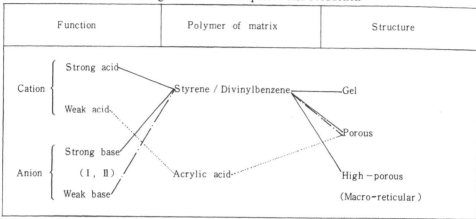

and weak base anion-exchange resins containing a primary and/or secondary and/or tertiary amine. The strong acid cation-exchange resin can split even neutral salts, such as NaCl or CaCl$_2$, as well as bases, such as NaOH, and weak acid salts, such as NaHCO$_3$. A strong base anion-exchange resin can split neutral salts (NaCl) as well as mineral acids (HCl) and weak acids (silicic acid). A weak acid cation-exchange resin can exchange a base (NaOH) or a weak acid salt (NaHCO$_3$). Finally, a weak base anion-exchange resin can exchange mineral acids (HCl) and weak base salts (NH$_4$Cl).

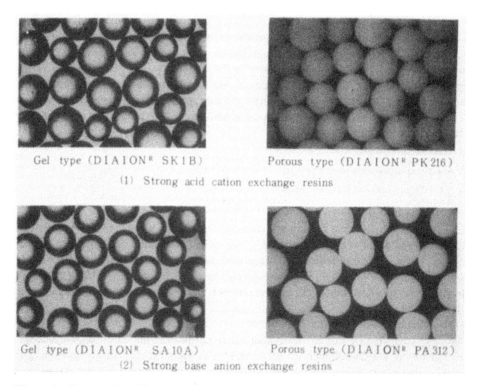

Gel type (DIAION® SK1B)   Porous type (DIAION® PK216)

(1) Strong acid cation exchange resins

Gel type (DIAION® SA10A)   Porous type (DIAION® PA312)

(2) Strong base anion exchange resins

Figure 1   Photographs of ion-exchange resins.

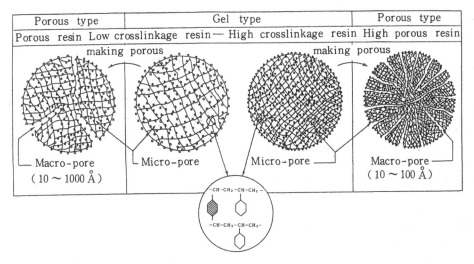

Figure 2   Structure of ion-exchange resin.

An ion-exchange resin is composed of a three-dimensional fine network polymer combined with a functional group. Figure 2 is a schematic diagram of the structure of an ion-exchange resin. In general, the ion-exchange resin is made of a copolymer of styrene and divinylbenzene or a copolymer of acrylic acid and divinylbenzene. The density of the network structure is in proportion to the amount of injected divinylbenzene. Therefore, the percentage of the injected divinylbenzene is the cross-linkage. Usually the standard cross-linkage of strong acid cation-exchange resin is around 8%.

When styrene and divinylbenzene are simply polymerized to produce ion-exchange resin, the completed ion-exchange resin, which is a gel type, is almost transparent and has micropores. On the other hand, when a special polymerization method is applied so that the ion-exchange resin features many pores, the completed ion-exchange resin, which is a porous or high-porous type, is opaque, with macropores. Those two physical types of ion-exchange resin should be recognized as separate.

The exchange capacities of the major strong acid cation-exchange resin and the major strong base anion-exchange resin are 2 and 1 EqL^{-1}, respectively.

In short, the ion-exchange resin can be characterized by a solidified acid or base and a reversible ion-exchange reaction.

B.   Functions

One of the functions of the ion-exchange resin is to split and exchange neutral salt, base, or acid. In other words, it demineralizes the water to be treated. For example, if NaCl solution passes through an H type of strong acid cation-exchange resin, Na is exchanged and removed and the product is HCl solution. If this HCl solution is supplied to an OH type of strong base anion-exchange resin in the following stage, Cl is exchanged and removed from the HCl solution and the product is pure water. This is major function of the ion-exchange resin.

Another function of the ion-exchange resin is based on its surface charge. Electrically the cation-exchange resin has a negative charge and the anion-exchange resin has a positive charge. This surface charge is called the ζ potential. Although this action is not much pursued in the ultrapure water field, in the condensate polishing unit of a power plant the corrosive products derived from metallic materials—*crud*—are removed effectively as a result of this function of the resin. It is thought that because the crud has a negative or positive ζ potential, depending on pH, it is adsorbed and captured by an ion-exchange resin that has the opposite ζ potential [6].

This function can be applied to pure water production. The colloidal materials in raw water usually have a negatively charged ζ potential. Therefore, just like crud, most colloidal materials are considered adsorbed and captured by anion-exchange resin, which has a positive ζ potential. In raw water treatment, the relatively large suspended materials are removed in the pretreatment system, such as a coagulation-filtration tower. The suspended materials sent to the ion-exchange resin tower are therefore very fine, around a few micrometers.

Although the ion-exchange reaction was just described, using the example of an inorganic salt, this reaction can be perfectly applied to organic materials. However, because the organic material in natural water, which is called humic acid, has a relatively high molecular weight, it is sometimes irreversibly adsorbed to the ion-exchange resin, resulting in organic contamination of the ion-exchange resin.

Another function reported for ion-exchange resin is sterilization [7].

III. TYPES OF ION-EXCHANGE UNIT

The ultrapure water production system is usually composed of pretreatment, primary treatment, and subsystem. Figure 3 shows an example of an ultrapure water production system [8]. The demineralization unit employing the ion-exchange resin is placed together with the reverse osmosis unit in the primary treatment system to yield pure water with a resistivity of 10–15 M$\Omega\cdot$cm at 25°C. In the following subsystem, further demineralization is conducted to produce ultrapure water with a resistivity of over 18 M$\Omega\cdot$cm at 25°C.

Table 2 summarizes the types of ion-exchange units and the kinds of ion-exchange resins adopted at present to ultrapure water production. Table 2 also indicates the regeneration method and the quality of the treated water.

A. Primary Treatment System

Major ion-exchange units employed in the primary treatment system are the two-bed, three-tower type, the four-bed, five-tower type, and the mixed-bed type. In the two-bed, three-tower type, the cation-exchange tower is followed by a degasifier (or deaerator) and an anion-exchange tower. In the four-bed, five-tower type, the cation-exchange tower and the anion-exchange tower are added. The mixed-bed type has cation-exchange and anion-exchange resins mixed together to treat raw water in a single tower. A mixed-bed type installed downstream of the two-bed, three-tower type is called the polisher. The two-bed, three-tower type is usually followed by the polisher.

In the single-layer method, one tower is filled with one kind of ion-exchange resin. For example, the cation-exchange tower can be filled with a strong acid cation-exchange resin. On the other hand, in the dual-layer method, one tower is filled with two kinds of ion-exchange resins. For example, the cation-exchange tower contains a strong and a weak acid cation-exchange resin.

As mentioned earlier, the weak acid cation-exchange resin adsorbs weak acid salts and the weak base anion-exchange resin adsorbs mineral acids. Although the components

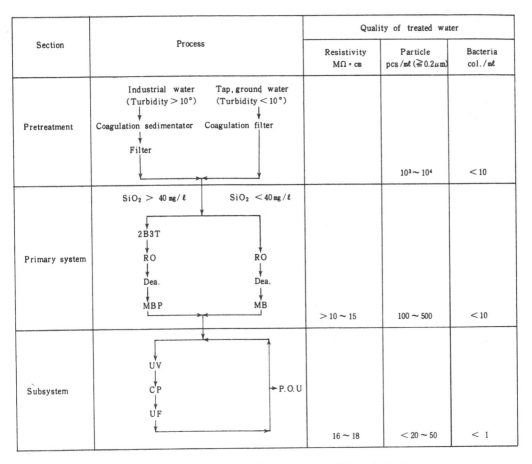

Section	Process	Quality of treated water		
		Resistivity MΩ·cm	Particle pcs/mℓ (≧0.2μm)	Bacteria col./mℓ
Pretreatment	Industrial water (Turbidity > 10°) → Coagulation sedimentator → Filter Tap, ground water (Turbidity < 10°) → Coagulation filter		10³~10⁴	< 10
Primary system	SiO₂ > 40 mg/ℓ : 2B3T → RO → Dea. → MBP SiO₂ < 40 mg/ℓ : RO → Dea. → MB	> 10 ~ 15	100 ~ 500	< 10
Subsystem	UV → CP → UF → P.O.U	16 ~ 18	< 20 ~ 50	< 1

Figure 3 Example of an ultrapure water production system. (From Ref. 8).

Table 2 Ion-Exchange Equipment for Ultrapure Water Production

Section	Type	Ion exchange resin	Regeneration system	Water quality, μS/cm Inlet / Outlet
Primary	2B3T	SK→D→SA WK/SK→D→WA/SA	Counter current regeneration	200/1 ~ 5
	4B5T	SK→D→SA→SK→SA WK/SK→D→WA/SA→SK→SA	Counter current regeneration	200/0.1 ~ 1
	MB	SK/SA	(Inert resin)	200/0.1 ~ 1
	MBP	SK/SA	(Inert resin)	1 ~ 5/0.07 ~ 0.1
Sub	CP	SK/SA	Un−regeneration	1/0.055

2B3T : 2beds 3towers demineralizer	SK : Strong acid cation exchange resin	
4B5T : 4beds 5towers demineralizer	SA : Strong base anion exchange resin	
MB : Mixed −bed demineralizer	WK : Weak acid cation exchange resin	
MBP : Mixed −bed polisher	WA : Weak base anion exchange resin	
CP : Cartridge polisher	D : Degasifier or deaerator	

to be adsorbed by those two resins is restricted to those that can be adsorbed by strong ion-exchange resins, those two weak resins have a large exchange capacity. The adsorbed components are also easily regenerated. Therefore, if raw water contains components that adsorb to weak ion-exchange resins at a high ratio, the regenerant consumption is small, which leads to cost reduction. The dual-layer type is applied to the two-bed, three-tower type and the four-bed, five-tower type.

In ion-exchange treatment, two steps are repeatedly followed: the service step in which raw water is fed to the ion-exchange resin to adsorb salts, and the regeneration step in which the regenerant is fed to the ion-exchange resin to desorb the adsorbed salts.

Ion exchange is conducted in a fixed-bed unit. In the fixed-bed unit, the ion-exchange resin layer in the tower is fixed during the servive and regeneration steps. In other words, the resin is not allowed to fluidize or move during these steps. In this fixed-bed system, downflow is applied to both the service and regeneration steps, and this is called a cocurrent regeneration system. Recently, however, downflow has been applied to the service step and upflow has been applied to the regeneration step, and this is called a countercurrent regeneration system.

Before the details of this countercurrent regeneration system are discussed, we describe the usual, cocurrent system. Neither the cation-exchange resin nor the anion-exchange resin is completely regenerated in this operation. Complete regeneration of a standard strong acid cation-exchange resin, with an exchange capacity of 2.0 EqL^{-1}, requires more than five times as much hydrochloric acid as the exchange capacity can handle. To operate the system in the effective and economical manner, partial regeneration using hydrochloric acid at less than twice the exchange capacity is generally conducted. This means that the adsorbed ions remain in the regenerated ion-exchange resin to some extent.

In the conventional cocurrent regeneration system, the residual ions at the bottom of the resin layer leak into pure water during the service step. As a result, the purity of the treated water does not reach the highest level. In the countercurrent regeneration system in which water flows down during the service step and flows up during the regeneration step, however, most of the adsorbed ions on the bottom resin layer are completely regenerated and no residual ions are found on the bottom. This has made it possible to further improve the purity of the treated water.

Another advantage of the countercurrent regeneration system is that there is no backwashing. In the conventional cocurrent regeneration system, the portion of ion-exchange resin containing adsorbed ions is mixed with the portion not containing adsorbed ions during backwashing, which is conducted before regeneration. In the countercurrent regeneration system, however, the portion with the adsorbed ions remains on the top of the ion-exchange resin layer and the portion without adsorbed ions is located on the bottom. The regenerant can therefore be used effectively, improving the regeneration efficiency. In general, regenerant consumption in the countercurrent regeneration system is less than a half that of the conventional cocurrent regeneration system. The countercurrent regeneration system is applied to both single-layer and dual-layer arrangement in the two-bed, three-tower and four-bed, five-tower units, respectively.

Regeneration in the mixed-bed type is conducted as follows. After backwashing, the cation-exchange resin layer sinks to the bottom and the anion-exchange resin layer rises to the top by making use of the difference in the specific gravity of the ion-exchange resins. Acid is injected from the bottom and alkali from the top, and these regenerants are drained from the collector placed between the two resin layers. This regeneration method works well without causing problems at any point, for example in the quality of the water treated in the mixed-bed ion-exchange unit. To further improve the purity of the treated water, however, the following points should be noted.

In the mixed-bed unit, the regenerant for one resin contacts the other resin at the interface during regeneration. This may lead to cross-contamination, resulting in the leakage of a small amount of ions when water is fed to the unit. To prevent cross-contamination, the condensate polishing unit of a pressurized water reactor (PWR) nuclear power plant, for example, adopts a method of removing resins from around the interface of the two resin layers, where they are more likely to be exposed to the wrong regenerants [9, 10]. Since this method requires a large facility, it is not easy to adapt this method to ultrapure water production. Another possible method of preventing cross-contamination is to use an inert resin whose specific gravity is between that of the cation-exchange and anion-exchange resins and that does not have functional groups. In this method, when the mixed resins undergo backwashing, they are separated into anion-exchange resin, inert resin, and cation-exchange resin layers, from the top down. By draining regenerants from the middle inert resin layer, the cross-contamination phenomenon can be avoided.

The fixed-bed type has been discussed, but for the two-bed, three-tower and mixed-bed types, another method ensures that some portion of the ion-exchange resin to be transferred during the service step is regenerated. In this method, the service and regeneration steps can proceed simultaneously and continuously. Another continuous method of demineralization is also available [11].

The characteristics of ion-exchange units employed in a primary treatment system can be summarized as follows.

1.   Two-Bed, Three-Tower

Since raw water is treated in one cation-exchange resin tower and one anion-exchange resin tower, the quality of the treated water is greatly affected by the salt concentration and the composition of the raw water. The purity of the treated water is intermediate. However, the structure of the unit is relatively simple, and it can be easily expanded in size. The regeneration time required is short.

2.   Four-Bed, Five-Tower

The same process is used as in the two-bed, three-tower type, but the four-bed, five-tower unit removes salt more thoroughly, improving water quality. However, the unit structure and the operation are slightly more complicated than for the two-bed, three-tower type.

3.   Mixed Bed

Since the cation-exchange and anion-exchange resins are mixed together, ions in the raw water are repeatedly removed by both resins. The purity of the treated water is at the highest level among the three types. However, this type requires the mixed resins to be separated in the regeneration step. Because carbonic acid ion, which can be physically removed, is also adsorbed in this type, the resin volume in the tower needs to be relatively large.

The most appropriate ion-exchange unit in the primary treatment system can be determined based on the salt concentration, the composition of the raw water, and the water volume to be treated. The combination of ion exchange and reverse osmosis should be also taken into consideration. The entire primary treatment system should be designed to make the best use of the ion-exchange unit.

B.   Subsystem

The subsystem usually employs an ion-exchange unit called the cartridge polisher (CP), which need not be the regeneration facility. The ion-exchange resin should be replaced periodically.

The resins used for the cartridge polisher are the strong acid cation-exchange resin and the type-I strong base anion-exchange resin. Although they are the same as those used in the primary system, the resins used in the subsystem are regenerated and specially purified in advance, and they are supplied as a mixture. The mixing ratio between the cation-exchange and anion-exchange resins is 1:1 to 1:1.5 in exchange capacity.

The demineralization mechanism of the cartridge polisher is the same as in the mixed-bed type employed in the primary treatment system, but the concentration of impurities adsorbed to the cartridge polisher by the water fed to the subsystem is very low. Therefore the resin can last around 1 year. It should be replaced with purified resin every year because the cartridge polisher is the final ion-exchange unit to maintain high purity of the ultrapure water.

Because the ion-exchange resin for the cartridge polisher is not regenerated in the system, it is called a nonregeneration type.

IV.   FUNCTION IN THE ULTRAPURE WATER PRODUCTION SYSTEM

The functions of the ion-exchange resin and unit in ultrapure water production are discussed here. Some actual examples are presented as well.

A.   Primary Treatment System

The basic function of the ion-exchange resin is demineralization, which can be expressed by the electrical conductivity, the indicator of ion concentration. Table 2 shows the general electrical conductivities at the inlet and outlet of each ion-exchange unit. Because the ion-exchange unit installed in a primary treatment system is basically the same as that installed in a makeup water system for a boiler in terms of the resin and the treatment method, the demineralization performance expressed by the electrical conductivity is well known. The details of demineralization performance are therefore omitted from this discussion. By and large, the electrical conductivity of the treated water is 0.07–0.1 μS cm^{-1} at 25°C.

Figure 4 shows the electrical adsorption force of ion-exchange resin. The behavior of colloidal iron at each unit in the ultrapure water production system is indicated. Colloidal iron decreases by four orders of magnitude in the two-bed, three-tower ion-exchange tower, but it barely decreases in the following units, including the reverse osmosis unit.

Figure 5 shows the behavior of particles and total organic carbon (TOC) in another two-bed, three-tower ion-exchange unit. Since the pretreatment in this two-bed, three-tower unit is filtration only, both the particle numbers and TOC concentration are high at the inlet of the ion-exchange unit. However, the number of particles decreases to less than three orders of magnitude.

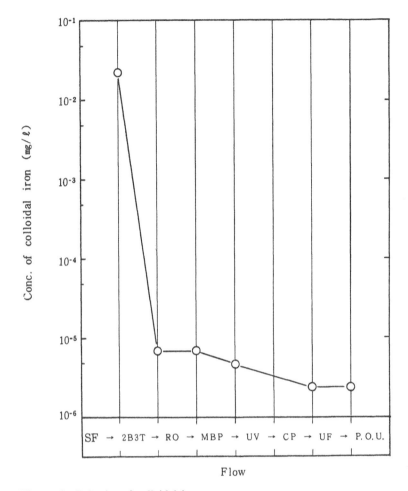

Figure 4 Behavior of colloidal iron.

It is thought that this action is based on the surface charge of the anion-exchange resin, which reduces colloidal materials and particles. This means the resin exhibits functions other than demineralization in the primary treatment system.

B. Subsystem

As mentioned earlier, the ion load on the ion-exchange resin in the subsystem is small. To completely adsorb impurities in the cartridge polisher, which is the final demineralization unit for high-purity ultrapure water, the following are important points: the desorption of impurity ions from the resin and purification of the resin itself.

In high-purity water, such as ultrapure water, which contains impurity ions in extremely low concentrations, the water itself is dissociated into H^+ and OH^-. As a result, these ions can be reacted with the impurity ions adsorbed to the ion-exchange resin to desorb them. This reaction can be expressed by the following chemical equilibrium equation [9]:

Cation resin　　　 : D I A I O N ᴿ　S K 110
Anion resin　　　　: D I A I O N ᴿ　P A 312
Flow rate　　　　　: 25 ℓ/h・ℓ −anion resin
Particle of inlet : $> 10^4$ pcs /mℓ
TOC of inlet　　 : 1.5 mg/ ℓ

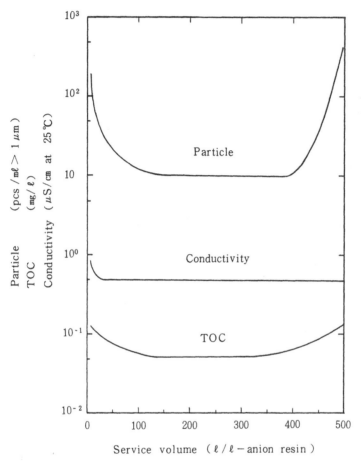

Figure 5　Particle and TOC profiles in the two bed, three-tower demineralizer.

$$
\text{R–M + H}^+ \text{ (or OH}^-) \overset{K}{\rightleftarrows} \text{R–H (or R-OH) + M}^+ \text{ (or M}^-) \tag{1}
$$

where: R　　　　　 = ion-exchange resin
　　　R–M　　　　= concentration of cation (anion) in resin
　　　R–H (R-OH) = concentration of hydrogen ion (hydroxyl ion) in resin
　　　H⁺ (or OH⁻) = concentration of hydrogen ion (hydroxyl ion) in solution
　　　M⁺ (or M⁻) = concentration of cation (anion) in solution
　　　K　　　　　= equilibrium constant

Equation (1) can be rewritten as

$$M^+ \text{ (or } M^-) = K \frac{R\text{--}M}{R\text{--}H \text{ (or } R\text{-OH)}} H^+ \text{ (or } OH^-) \tag{2}$$

It is clear from Eq. (2) that the ion concentration in solution (treated water) is in proportion to the ion concentration in the resin. Therefore, to obtain high-purity water means reducing the ion concentration in the resin.

Figure 6 shows the resistivity as a function of the concentration of various salts. To obtain resistivities higher than 18 MΩ·cm at 25°C, the concentration of each salt should be reduced to less than 0.1 μg L^{-1}. This value of concentration is added to Eq. (2) to calculate the required impurity concentration in the ion-exchange resin. (Na ion and Cl ion are picked up as impurity ions.) For the cation-exchange resin,

$$M^+ = 4.3 \times 10^{-9} \text{ Eq L}^{-1}(0.1 \ \mu g \text{ as Na L}^{-1})$$
$$H^+ = 1.0 \times 10^{-7} \text{ Eq L}^{-1}$$

$$K = \frac{1}{1.5}$$

and thus

$$\frac{R\text{--}Na}{R\text{--}H} = \frac{0.06}{0.94}$$

Therefore the Na ion concentration in the cation-exchange resin should be less than 6%.
For the anion-exchange resin,

$$M^- = 2.8 \times 10^{-9} \text{ Eq L}^{-1}(0.1 \ \mu g \text{ as Cl L}^{-1})$$
$$OH^- = 1.0 \times 10^{-7} \text{ Eq L}^{-1}$$

$$K = \frac{1}{20}$$

and thus

$$\frac{R\text{--}Cl}{R\text{--}OH} = \frac{0.36}{0.64}$$

Therefore the Cl ion concentration in the anion-exchange resin should be less than 36%.

Because the purity of the water treated in the cartridge polisher is determined by the impurity ion concentration in the resin, it is important to maintain the impurity ion concentration at low levels.

One problem with the ion-exchange resin is elution from the resin itself. There are two types of materials eluted from the resin: products formed through chemical decomposition of the functional groups, and other products [12]. A recent study has confirmed that the materials eluted from the cation-exchange resin are benzene sulfonic acid derivative and polystyrene sulfonic acid of various molecular weights (according to the analytic results using liquid chromatography and gel permeation chromatography (GPC)) (see Figs. 7 and 8). The materials eluted from the anion-exchange resin are volatile materials, such as amine, and partially decomposed resin polymer (according to analytic results using gas chromatography and infrared absorption spectroscopy [13].

The materials eluted from the cation-exchange and anion-exchange resins adsorb each other. Figure 9 shows TOC elution from the resin [14]. As shown in Fig. 9, TOC

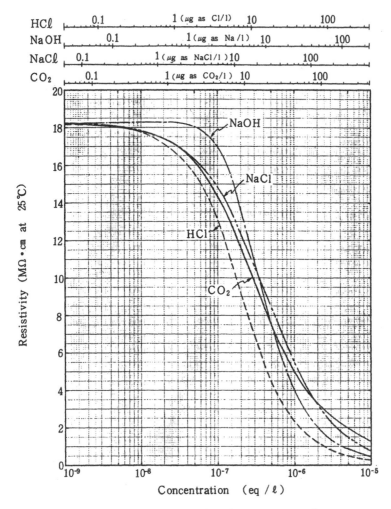

Figure 6   Resistivity versus concentration of ions. (From Ref. 14.)

elution is greater from the anion-exchange than from the cation-exchange resin, but when the two resins are mixed together, the TOC concentration drops considerably. Therefore most of the materials initially eluted from the resins are adsorb to each other.

The causes for this elution have not yet clear. Possible causes include the following [12]:

1. Polystyrene that is not cross-linked during the polymerization process is sulfonated, becoming water soluble, and is eluted.
2. During sulfonation, some of the cross-linked polystyrene chains are cut to form non–cross-linked polystyrene, and this non–cross-linked polystyrene is eluted.
3. As a result of the swelling and shrinkage of the ion-exchange resin during operation, the chains on its polymer are cut.
4. The chains on the polymer are cut as a result of oxidizing materials in the feedwater.

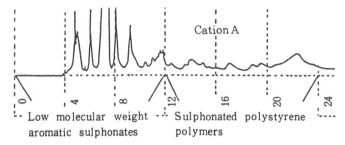

Figure 7 Ion-pair reversed-phase chromatogram of leachables released by a cation-exchange resin. (From Ref. 13.)

Based on these characteristics of the resin (desorption of the impurity ions from the resin, elution from the resin, and others), the performance of the cartridge polisher is as described here.

Table 3 shows the performance of a small-column cartridge polisher [15]. The resistivity, which reflects the demineralization performance of the ion-exchange resin, corresponds to the theoretical value of over 18.2 $M\Omega\cdot cm$ at 25°C. The Na and Cl ion concentrations are lower than 0.01 and 0.05 μg L^{-1} respectively. It is confirmed that the impurity ion concentration in the resin is maintained at sufficiently low levels.

Silica removal is 90%, indicating a sufficient concentration of 0.5 μg L^{-1}. Compared with other ions, however, silica leakage is still high. The reason for this high silica leakage is thought to be as follows. The collodial silica in natural water is difficult to diffuse into the ion-exchange resin and consequently is barely adsorbed. Colloidal silica then passes through the resin and becomes ionic silica as a result of thermal decomposition detected and measured during the analytic process. Another possible reason is that the silica remaining in the ion-exchange resin together with Na and Cl reacts with pure water and is desorbed into water. Further study is required to pinpoint the reasons for the high silica concentration.

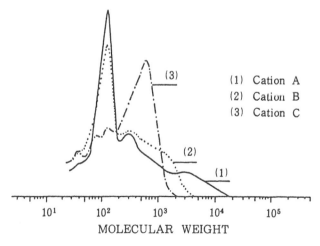

Figure 8 Size-exclusion chromatograms of leachables released by cation-exchange resins. (From Ref. 13.)

Cation exchange resin ： D I A I O Nᴿ SKNUP
Anion exchange resin ： D I A I O Nᴿ SANUP
Immersion time   ： 48 hrs
Immersion temp.   ： 40 ℃

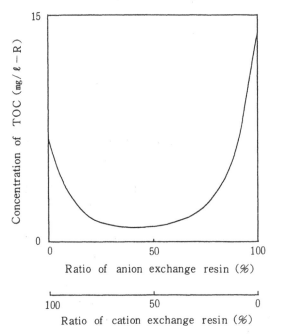

Figure 9 TOC versus mixing ratio of ion-exchange resins by batch test. (From Ref. 14.)

No increase or decrease in particles, TOC, or bacteria, is detected between the inlet and outlet. No impurities other than ions are generated in the cartridge polisher using a purified ion-exchange resin and improved vessel design [15].

Table 3 indicates water quality during the normal operation of a cartridge polisher. Figures 10 and 11 show the behavior of water quality at service start-up. The TOC level at the inlet is different between Fig. 10 and 11. As shown in both figures, the resistivity reaches 18 MΩ·cm at 25°C within a short time. On the other hand, when the TOC level at the inlet is around 20 μg L^{-1}, the level at the outlet approaches that at the inlet within a relatively short time, but when the TOC level at the inlet is around 5 μg L^{-1}, it takes a long time for the level at the outlet to reach the level at the inlet. This indicates that organic materials are eluted from the ion-exchange resin. At the same time, it can be said that the higher the purity level of the ultrapure water, the longer it takes to clean the entire unit at the initial stage. Leakage of impurities is also observed when service is resumed after suspension of operation. To overcome this problem, ion-exchange resins that are free from elution problems need to be developed, or a method of quickly cleaning the entire unit should be devised.

To further reduce the TOC concentration in the subsystem and to prevent the ultrafiltration unit from clogging, an additional ion-exchange resin process has recently been employed [16]. Figure 3 shows the process flow of a conventional subsystem. In the current subsystem, the ultraviolet (UV) unit is equipped with a UV lamp with a low

Table 3 Performance of CP by Small Column

(1) Specification of CP

Column	Material	Stainless steel 316 (Inside ; Electro—polishing)
	Dimension	84.9 I D⌀ × 1,000L (t = 2.1)
	Pressure	max. 5 kgf / cm² G
Ion exchange resin	Brand	DIAION ᴿ SMNUP
	Volume	4 ℓ
Flow rate	100 ℓ/h	

(2) Water quality of CP

Item		Inlet	Outlet
Resistivity	MΩ・cm at 25℃	18.18 ~ 18.20	18.26 ~ 18.28
Na	µg / ℓ	< 0.01	< 0.01
Cℓ	µg / ℓ	< 0.05	< 0.05
Particle	pcs / mℓ > 0.1 µm	0 ~ 3	0 ~ 3
Bacteria	colony / ℓ	0 ~ 1	0
TOC	µg / ℓ	2.0 ~ 3.0	2.5 ~ 3.0
i − Silica	µg / ℓ	5	0.5
Temperature	℃	21.5 ~ 25.1	

Note : Analysis method or equipment

Resistivity	DKK AQ−11
Na	Flameless AAS
Cℓ	Ion chromatography
Particle	HORIBA PLCA−310
Bacteria	Culture method (sampling vol. 1 ℓ)
TOC	ANATEL A−100
i − Silica	Colorimetry

Source: From Ref. 15.

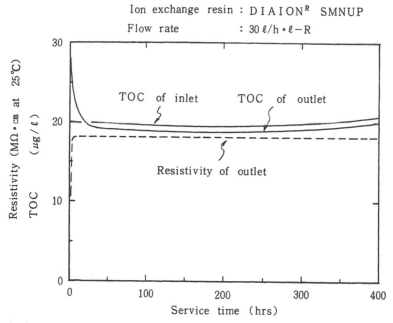

Figure 10   TOC profile in the CP.

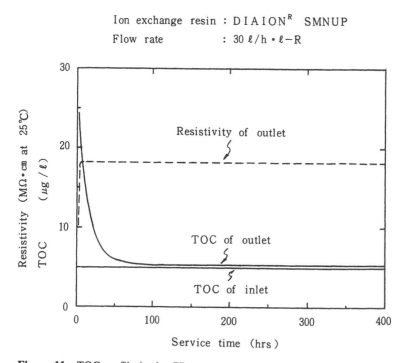

Figure 11   TOC profile in the CP.

wavelength of 185 nm. The anion polisher filled with anion-exchange resin, the cartridge polisher, and the ultrafiltration unit are placed downstream. This system is designed to decompose nondissociative organic materials into dissociative organic materials, such as organic acid and carbonic acid, for adsorbtion onto the strong base anion-exchange resin. The TOC concentration in ultrapure water has been successfully reduced to one-third of that of a conventional subsystem.

C. Wastewater Recovery

Ion-exchange resin is employed to treat hydrofluoric acid-contaminated wastewater, which has a relatively high purity among the various kinds of wastewater drained from the semiconductor production line.

Hydrofluoric acid wastewater usually contains anionic components, such as hydrofluoric acid, sulfuric acid, and nitric acid, and cationic components, such as ammonium. Since it contains more anionic than cationic components, it is acid. Rich hydrofluoric acid wastewater is first treated by coagulation settling and then neutralized and, finally, discharged. On the other hand, dilute hydrofluoric acid wastewater can be demineralized by the following sequence: active carbon, weak base anion-exchange resin, strong acid cation-exchange resin, and strong base anion-exchange resin [17].

Because this wastewater contains organic components, such as surfactants, it should first be treated with active carbon. The anionic components, including fluoride, are then removed by the weak base anion-exchange resin. Next the cationic components are removed by the strong acid cation-exchange resin. Finally, the remaining anionic components are removed by the strong base anion-exchange resin.

Table 4 shows an example of the performance of this system. The treated wastewater can be recovered to the primary treatment system. When the TOC concentration is especially high, oxidation or reverse osmosis may be added.

Wastewater can be treated separately according to the purity level to be recovered. In this way, it is possible to establish a closed system.

D. Trouble Caused by the Ion-Exchange Resin

The trouble caused by the ion-exchange resin is mainly observed in the primary treatment system. There are various causes and effects. Some of the major problems are discussed here.

Table 4 Wastewater Quality After Treatment

Item		Waste water	Treated water
Conductivity	$\mu S/cm$ at 25°C	480	0.2
pH	—	2.9	8.3
F^-	mg/ℓ	53	< 0.1
TOC	mg/ℓ	0.4	0.2

When oxidizing components are contained in the feedwater, the strong acid cation-exchange ion is gradually oxidized and becomes irreversibly swollen, resulting in deterioration of the exchange capacity in the unit. This irreversible swelling can be facilitated when metals work as the catalyst.

The functional groups in the strong base anion-exchange resin become weakly basic, resulting in a decrease in the salt-splitting capacity. However, the total exchange capacity does not drop as much as the salt-splitting capacity. This problem may lead to deterioration of water quality or a decrease in the volume of treated water.

When organic materials are contained in high concentration in the feedwater, the strong base anion-exchange resin is sometimes damaged by organic contamination. This leads to degradation of the quality of treated water and a drop in the treated water volume.

Meanwhile the ion-exchange resin swells or shrinks as a result the change of concentration in the solution. If regenerant at an inappropriate concentration is used, repeatedly, the resin may be destroyed. Oxidation also deteriorates the physical strength of the ion-exchange resin. When this happens, the resin is cracked, causing the ΔP to rise and channeling.

Foreign materials, such as suspended solids (SS) and oil derived from abnormal operation of the pretreatment system, also degrade the performance of the resin.

Degradation of the resin in the primary treatment system may affect the ultrafiltration unit in the subsystem. When impurities and eluted components increase as a result of problems in the primary treatment system and must be removed by the cartridge polisher in the subsystem, residual eluted components can clog the ultrafiltration unit. If the components are eluted from the cation-exchange resin, the clogging in the ultrafiltration unit proceeds gradually. If the components are eluted from the anion-exchange resin, however, the ultrafiltration unit is often clogged within a short time. This is because the components eluted from the cation-exchange resin, which are negatively charged, are repelled from the ultrafiltration membrane surface because the membrane surface is also negatively charged. Components eluted from the anion-exchange resin, which are positively charged, are deposited on the oppositely charged ultrafiltration membrane surface. Another possible cause is the physical difference in components eluted from both resins.

The problems caused by ion-exchange resin should be properly eliminated or controlled to maintain the quality of the ultrapure water.

V. FUTURE CHALLENGES

Because the ion-exchange resin is the most excellent material in terms of the removal of impurity ions, it is essential to ultrapure water production. The ion-exchange unit is also inexpensive. Therefore further improvements are required to supply ultrapure water at higher purity levels.

1. Reduce eluted components. Current resins contain minute amounts of impurities, which can elute from the resin. A method of producing ion-exchange resin that is free from elution or a method of cleaning the eluted components should be developed.
2. Improve durability. The durability as well as the heat resistance should be improved to prevent performance deterioration.
3. Make best use of functions. Not only the demineralization function but also other functions of the ion-exchange resin should be explored.

4. Stabilize the system. Since the ion-exchange resin unit should work together with the membrane filtration units in the primary treatment system in a harmonious manner, it should be designed to maintain stable operation.

REFERENCES

1. Kagaku Kogyo Nippo Co., Chemical Daily, October 24 (1989).
2. H. Kakihana and K. Narita, Saisin Ion Kokan (New ion exchange technology), Hirokawa Shoten Co., 1961.
3. A. Miyahara et al., Jituyo Ion Kokan (Practical ion exchange technology), Kagaku Kogyo Co., 1984.
4. N. Hojo, Chelate Jyusi/Ion Kokan Jyusi (Chelate resin/ion exchange resin), Kodansha Co., 1976.
5. F. Helfferich, Ion Exchange, McGraw-Hill, 1962.
6. H. Kusano, Gensiro-reikyakukei no Mizu-kagaku (Water Chemistry of Nuclear Reactor Systems), Tokyo, Atomic Energy Society of Japan, (1987), p. 191.
7. T. Sakamoto and K. Koike, Kogyo Yosui (Journal of Industrial water), no. 305, 8 (1984).
8. K. Kajikuri, Handotai Process Zairyo Jitumu Binran (Manual of Materials for Semiconductor Manufacturing Process), Science Forum Co., (1983), p. 435.
9. K. Oda, Nenryo oyobi Nensho (Fuels and Combustion), *25*(3), 24 (1988).
10. K. Okukawa, J. Water Re-use Technol., *15*(2), 24 (1989).
11. J. L. White et al., Ultrapure Water, *6*(7), 46 (1989).
12. T. Morita, Proceedings of Ultra Clean Technology Symposium, p. 8–13, September 2nd, 1988 Osaka, Japan.
13. T. Morita, Ultra Clean Technol., *1*(3), 37 (1989).
14. Water Re-use Promotion Center (Japan), Cho-junsui Seizo Gijutsu Kaihatu Hokokusho (Report of Ultra Pure Water Production Technology), 1989.
15. K. Oda, J. Water Re-use Technol., *16*(1), 26 (1990).
16. Tokkyo Kokoku (Examined Japanese Patent), 1-42, 754 (1989).
17. Japanese Patent, no. 1,385,937 (1987).

4
Deaeration Technology

Hitoshi Sato
Hitachi Plant Engineering & Construction Co., Ltd., Chiba, Japan

I. INTRODUCTION

On contact with the air (nitrogen, 78%, oxygen, 21%, and argon, 0.94% by volume) various gases are dissolved in water. The dissolved oxygen seriously affects the corrosion rate, which may be 5–10 times as high as for carbon dioxide in corresponding volume. Dissolved oxygen is regarded as the major cause of corrosion of metal materials. For this reason, dissolved oxygen should be removed from water [1].

Ion-exchange resin is also oxidized by dissolved oxygen, which degrades the resin and shortens its life. In addition, the growth of bacteria in pipes and equipment is accelerated by dissolved oxygen. It was recently pointed out that the performance of semiconductor devices is affected by dissolved oxygen in the rinse water. A method to efficiently remove the dissolved oxygen is urgently required [3–5]. This chapter discusses a deaeration method exclusively for removal of dissolved oxygen, although deaeration generally means the removal of any one of the gases dissolved in water.

A. Solubility of Gas

When water comes in contact with the air, a certain amount of gas is dissolved into the water. The amount of gas dissolved depends on the pressure and the temperature. Henry's law and Dalton's law describe the characteristics of the dissolution of gas. The gas dissolution into solution is in proportion to the partial pressure of the gas phase and in inverse proportion to the temperature of the ambient gas. This does not apply to such gases as carbon dioxide and ammonia, which react with the solution.

The solubility of gas in water decreases as the Cl concentration in water rises and increases as the content of impurities drops. Table 1 shows the solubility of major gases at various temperatures.

Table 1   Solubility of Gases in Water: a Volume (ml) of gas reduced to standard conditions (0°C and 760 mm Hg) dissolved in water (ml) when the total gas-phase pressure (gas partial pressure + vapor pressure of water) is 760 mm Hg at t°C.

Temp. t °C	Air	N_2	O_2
0	0.0286	0.0184	0.0102
5	0.0252	0.0163	0.0089
10	0.0224	0.0145	0.0079
15	0.0201	0.0131	0.0070
20	0.0183	0.0119	0.0064
25	0.0167	0.0110	0.0057
30	0.0154	0.0103	0.0051
40	0.0132	0.0087	0.0045
50	0.0114	0.0075	0.0039
60	0.0098	0.0065	0.0033
80	0.0060	0.0040	0.0021
100	0.0000	0.0000	0.0000

Source: From Ref. 2.

B.   Outline of Dissolved Oxygen Removal

There are three major methods of removing dissolved gas in solution.

The removal of gas from solution by aeration is a very simple method based on Henry's law. According to this law, the amount of gas dissolved in water is in proportion to the partial pressure of the gas in the air or in the gas phase. Therefore, if the oxygen concentration in the air supplied to the water is low enough, it is possible to remove the dissolved oxygen from the water. This method is applied to remove gas that is not contained in the air or is contained in the air at very low concentration, such as carbon dioxide. For dissolved oxygen, however, this method removes the oxygen in the air that is dissolved in water instead of removing carbon dioxide. There are also various impurities in the air, and organic materials and inorganic ions as well as other gases can be dissolved in the water. Thus this method does not contribute to the elimination of the contamination in ultrapure water, which is required for the water to be maintained ultraclean. Lately a new aeration method using high-purity nitrogen gas, which is used in the semiconductor production line, has been put to practical use.

There are two methods of mechanical deaeration: the deaerating heater and the vacuum deaerator. The deaerating heater applies the principle that the solubility of a gas nears zero at the boiling point of water. There are two types of deaerating heater that differ in the contact between the water and the steam: the tray type and the spray type (atomizing type). It is reported that the deaerating heater can remove dissolved oxygen to the level of 0.03 ml L^{-1} (42.8 ppb) [1]. Because the concentration of dissolved oxygen is extremely low, however, a small operation failure or a small mechanical defect can raise the concentration of dissolved oxygen to a level 10 or 20 times the guaranteed value. There is also the possibility that the water is contaminated by factors outside the system.

Vacuum desorption of oxygen from water has long been used for mechanical deaeration. This method removes dissolved gases from a solution into the gas phase under low partial pressure and vacuum. Because this desorption of oxygen is carried out in a closed system that has no contact with the outside, the possibility that the solution will become contaminated is low. Although this method has been widely used as an efficient way of removing dissolved oxygen from ultrapure water, there are problems. To raise the degree of vacuum, the height of the deaerator should exceed 10 m, and the vacuum unit is very noisy. Also, in terms of performance, the lowest level of dissolved oxygen concentration gained with this method is 40–50 ppb, which will not satisfy future demands for ultrapure water. Therefore possible combinations of this method and others should be considered.

A new vacuum deaerator method using a gas separation membrane has been developed [6]. In this method, the dissolved oxygen is removed by passing the water through a hydrophobic high-molecular-weight membrane through which the water molecules cannot pass but the gases can. A vacuum pump is employed on the permeated side. In the past, this method was applied to the degassing of eluant for liquid chromatography and to degassing of boiler water [7–9].

The chemical deaeration method chemically reduces the dissolved oxygen with deoxidizers, such as sodium sulfite (Na_2SO_3) and hydrazine (N_2H_4). The chemical deaeration method is generally applied to the deaeration of boiler water. It is sometimes used together with the deaeration heater or vacuum deaeration. Because this method introduces impurities, such as conductive materials and organic materials, into the solution, it has not been applied to the deaeration of ultrapure water.

A new chemical deaeration method has been developed [13,14] that uses ion-exchange resin with palladium as the catalyst. The dissolved oxygen and deoxidizer react with each other on the catalyst to change the oxygen to water (H_2O) for removal. This new chemical method has a quite different mechanism from conventional chemical methods. It is able to efficiently achieve a low concentration level of dissolved oxygen, and it features a very high performance that can meet future quality standards for ultrapure water.

This is merely an outline of conventional and new deaeration methods. The remainder of this chapter discusses the following units: the vacuum deaerator, nitrogen exposure deaerator, membrane deaerator, and catalyst resin deaerator.

II. VACUUM DEAERATOR

Vacuum desorption has been widely used to remove dissolved oxygen in ultrapure water. Figure 1 shows the configuration of the vacuum deaerator. In general, the steel plate bath is packed with Raschig rings or perforated plates are inserted in the bath. The liquid is atomized from the upper part of the bath through the distributor. In this way, the area in contact with the liquid increases and the deaeration efficiency is improved. The pressure in the bath is reduced by the vacuum unit to a degree of vacuum lower than the vapor pressure at a certain temperature of the liquid. Generally a vacuum pump and an ejector are used as the vacuum unit. Although the vacuum pump has good performance in terms of the ultimate pressure, it consumes a great deal of electrical power, especially the large pump, and the costs are high. Therefore the large type is generally applied to the multistage arrangement of a steam ejector.

The theory of gas absorption is usually applied in the design of the vacuum deaerator. Basically the performance of the deaerator depends on the degree of vacuum and the height of the packed layer. The economical height of the packed layer is calculated by

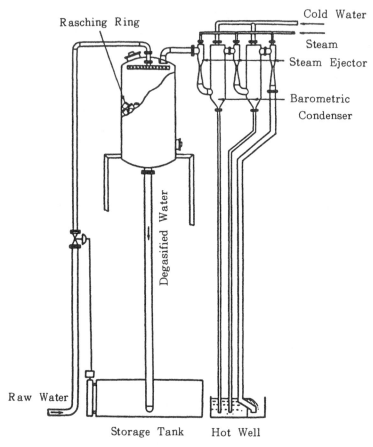

Figure 1   Configuration of a vacuum deaerator. (From Ref. 1.)

using the overall height of the transfer unit (HTU) [10,11], which is frequently calculated based on experiment.

$$\text{NTU}_{OL} = 2.3 \log \frac{X_i - X_e}{X_o - X_e} \qquad\qquad Z = \text{NTU}_{OL}\, \text{HTU}_{OL.}$$

where: HTU_{OL} = overall height of the transfer unit using a liquid-phase driving force (m)

NTU_{OL} = number of transfer units based on overall driving force in the liquid-phase concentration unit (m)

Z = height of the packed layer (m)

X_o = concentration of dissolved oxygen at the tower outlet (ppm)

X_i = concentration of dissolved oxygen at the tower inlet (ppm)

X_e = concentration of dissolved oxygen in equilibrium with the gas phase in the tower (ppm)

In an experiment with a vacuum deaerator with a three-layer steam ejector, an ultimate pressure of 10–40 torr and a concentration of dissolved oxygen in the treated water of

Table 2 Specifications for Highly Purified N_2 Gas

Gas	Content
N_2	99.999 %
CO_2	< 0.1 ppm
O_2	< 0.5 ppm
H·C	< 0.5 ppm
H_2O	< 0.001 mg/ ℓ

40–50 ppb were obtained when the initial concentration of dissolved oxygen was 8 ppm. Although it is possible to further improve performance by improving the vacuum level and raising the temperature of the water, this concentration is considered a limit from the viewpoint of economics.

It is difficult for the vacuum deaerator to meet the stringent demands for water quality by itself. However, this is an efficient method for pretreatment before nitrogen gas bubbling, and a catalytic resin system after this pretreatment helps to reduce the chemicals to be used in the system. An appropriate combination is required to reduce costs as much as possible.

III. NITROGEN GAS BUBBLING METHOD

The nitrogen gas bubbling method is an aeration method in which high-purity nitrogen gas with an extremely low concentration of oxygen is used. Henry's law can be expressed as

$$C_{O_2} = \frac{P_{O_2}}{H}$$

where: C_{O_2} = equilibrium concentration of DO
$\quad\quad P_{O_2}$ = partial pressure of oxygen gas
$\quad\quad H$ = Henry's constant

High-purity nitrogen gas with a low partial pressure of oxygen reduces the equilibrium concentration or the saturation concentration of oxygen to remove the dissolved oxygen. Table 2 shows the composition of the high-purity nitrogen gas, and Fig. 2 shows the relationship between the purity of the nitrogen gas and the concentration of dissolved oxygen is in equilibrium with the nitrogen gas.

To keep ambient air away from air containing a high concentration of oxygen, a closed bubbling tower is employed in the nitrogen gas bubbling system. The difference between the partial pressure of oxygen in the liquid and that in the gas is used as the driving force to diffuse oxygen from the liquid into the nitrogen gas, which is removed. The diffusion of oxygen can be expressed as

$$\frac{dN}{dT} = K_{La}(C - C^*)$$

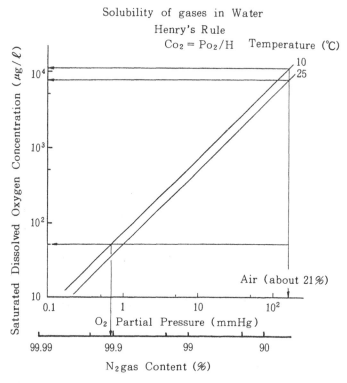

Figure 2   Relationship between the saturated DO (dissolved oxygen) concentration and the partial pressure of O_2 in highly purified N_2 gas.

where: K_{La} = mass-transfer coefficient (h^{-1})
      C^* = equilibrium concentration corresponding to the partial pressure of oxygen gas in nitrogen (mgL^{-1})
      C = concentration of dissolved oxygen in liquid (mgL^{-1})

Since K_{La}, an important parameter in designing the bubbling tower, is affected by the conditions and the configuration of the nitrogen gas supply system, it is determined in experiments by applying the equation

$$K_{La} = \frac{C_i - C_o}{V/Q_L(C_o - C^*)}$$

where: V = effective volume of bubbling tower (L)
      Q_L = amount of water to be fed (Lh^{-1})
      C_o = concentration of dissolved oxygen at the bubbling tower outlet (mgL^{-1})
      C_i = concentration of dissolved oxygen at the bubbling tower inlet (mgL^{-1})

The mass-transfer coefficient K_{La} increases as the diameter of the nitrogen gas bubble decreases. Therefore it is desirable to supply nitrogen gas with a smaller bubble diameter. In practice, the more nitrogen gas contained in the bubbling tower, the more dissolved oxygen is removed. However, because it is important to effectively remove the dissolved oxygen to the required level with a smaller amount of nitrogen, it is necessary to conduct simulations to evaluate the efficiency in advance.

Figure 3 shows the fluctuation of K_{La} as a function of apparent gas flow velocity with a liquid depth of 2 m, a tower diameter of 100 mm, and a ceramics diffuser with a pore size of 100 μm. The water was kept at room temperature, and the concentration of dissolved oxygen at the inlet was 7–8 mgL^{-1}, about the same as the saturation concentration. Based on this result K_{La} can be expressed as 11.9 $U_g^{0.67}$.

For the bubbling tower, it is clear that a multiphase system is efficient in terms of deaeration performance. When bubbling towers with the same capacities are combined, the concentration of dissolved oxygen in the towers can be expressed as

$$C_o = \frac{C_i}{(1 + K_{La}V/Q_L)^n}$$

where n = number of towers.

Figure 4 indicates the relationship between the simulated values based on this equation and the measured values. The theoretical values almost match the actual values. It has been proved that the bubbling tower can be designed without difficulty if the given K_{La} is appropriate in terms of the configuration of the equipment.

For the efficient use of nitrogen, a nitrogen gas circulating system can be employed. As shown in Fig. 5, the bubbling towers are divided into two sections, and the nitrogen gas exhausted from the second section is recycled to the inlet of the first section. The exhausted nitrogen gas can be reused for sealing purposes.

Figure 6 shows the simulated values based on the preceding equation and the measured values in the two-section system. To reduce the concentration of dissolved oxygen to 10 ppb, no more than 0.65 is required for the gas-liquid ratio of nitrogen gas to

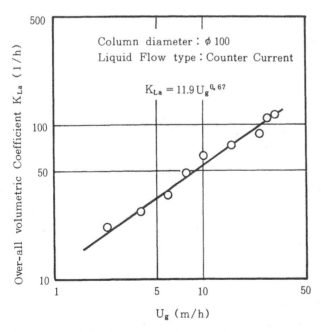

Figure 3   Relationship between the K_{La} value and U_g.

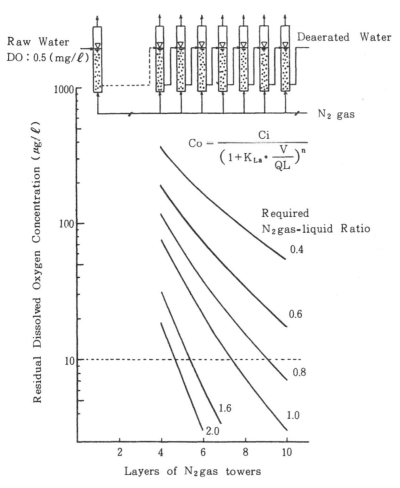

Figure 4   Relationship between the layers of N_2 gas towers and the residual DO concentration at different gas-liquid ratios.

the water to be processed. With a gas-liquid ratio of 0.8, the concentration of dissolved oxygen can be reduced to the level of 3–6 ppb.

IV.   MEMBRANE DEAERATION

Membrane deaeration is a newly developed method that applies to a gas separation membrane. This is a hydrophobic high-molecular-weight membrane made of Teflon or silicon gum. The dissolved gas can be removed by making the gaseous phase vacuum. When the liquid is water, some of the carbon dioxide as well as oxygen and nitrogen can be removed. This is why this method has been applied to the eluant deaerator for liquid chromatography.

Table 3 indicates the specifications for the gas separation membrane available on the market, and Fig. 7 shows the basic principle of removing the dissolved oxygen. In general, the gas separation membrane is a hollow-fiber membrane. The dissolved oxygen

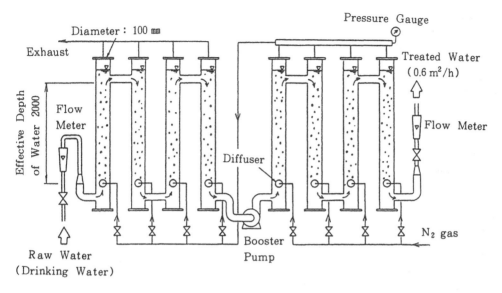

Figure 5 Advanced N₂ gas bubbling system.

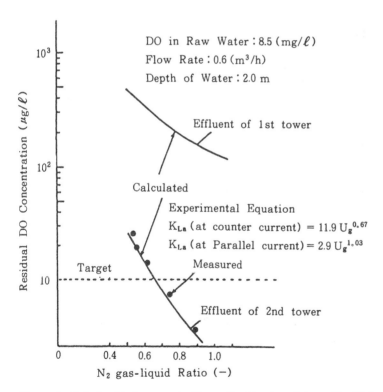

Figure 6 Relationship between the N₂ gas-liquid ratio and the residual DO concentration using an advanced N₂ gas bubbling system.

Table 3 Specifications for the Membrane Deaeration Module

	Membrane A	Membrane B	Membrane C	Membrane D
type of membrane	Hollow Fiber	Spiral	Hollow Fiber	Hollow Fiber
material	complex (poly-sulphon)	—	PTFE	Silicon
diameter in/out	0.3 / 0.4 mm	—	— / 1.0 mm	—
size of Module	—	$\phi\,100 \times 1{,}000$ L	$\phi\,370 \times 1{,}000$ L	$\phi\,265 \times 1{,}400$ L
packing percentage	50 %		—	—
performance (DO) *	0.1 ppm (at -730 mmHg)	0.5 ppm —	0.2 ppm (at -730 mmHg)	0.2 ppm (at -730 mmHg)
feed capacity	108 ℓ/h		1 m³/h	1 m³/h
operating pressure	—	—	3 kg/cm²(Max)	1.6 kg/cm²(Max)
pressure loss	—	—	0.6 kg/cm²(Max)	0.8 kg/cm²(Max)

* : The Value means minimum DO concentration under the condition showing vacuum level

Source: Nitto Denko Co., Ltd., Torey Co., Ltd., ERMA Co., Ltd., and Japan Gore-Tex Co., Ltd. technical reports.

in the pure supply water is transferred to the membrane surface to be diffused into the membrane in the form of gas molecules. The oxygen gas passing through the membrane is exhausted by the vacuum pump, which is employed in the gas phase. The performance of this method depends on the gas permeability of the membrane and the vacuum level. As shown in Fig. 8, the pH effect is small.

Figures 9 and 10 indicate the results of a study in which the deaeration characteristics of two kinds of membrane modules were investigated. Figure 11 shows the outline of the experimental apparatus. It is clear that the lower the partial pressure of oxygen, in the gaseous layer in contact with the pure water, the lower the concentration of dissolved oxygen in the pure water. In other words, the higher the vacuum level is raised, the lower the ultimate concentration of oxygen. To decrease the partial pressure of oxygen, it is possible to supply high-purity nitrogen gas to the exhaust side. The ultimate concentration

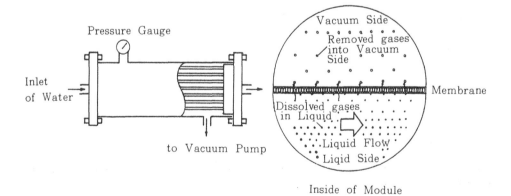

Figure 7 Membrane deaerator module.

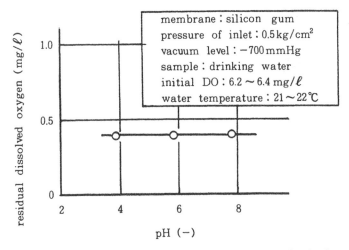

Figure 8   Relationship between the pH and the residual dissolved oxygen concentration using the
membrane deaerator.

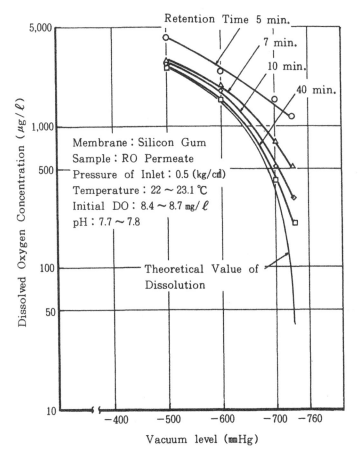

Figure 9   Relationship between the DO concentration and the vacuum level at various retention
times.

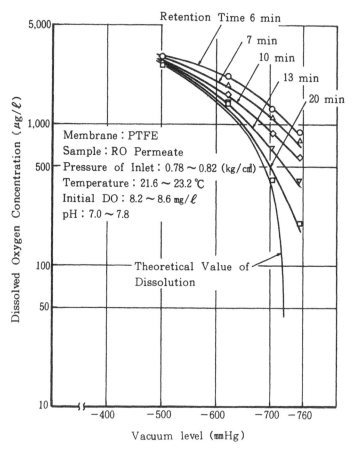

Figure 10   Relationship between the DO concentration and the vacuum level at various retention times.

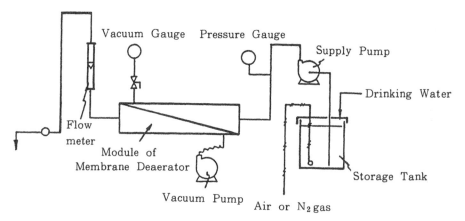

Figure 11   Test equipment diagram of the membrane deaerator.

of dissolved oxygen is affected by the time required for the pure water to pass through the module containing the gas separation membranes. Therefore, the optimum conditions of vacuum level and amount of water to be processed should be worked out in terms of economics.

There are two types of vacuum pump: the diaphragm type and the water-sealing type. The water-sealing type is superior in terms of the level of vacuum; the small water-sealing type can obtain a vacuum level of 10 torr. When processing pure water at 1 m^3 h^{-1}, the ultimate pressure under practical conditions is 30–50 torr, and the concentration of dissolved oxygen is 200–500 ppb when the initial concentration is 8 ppm, very close to the saturation level. It is possible to reduce the concentration level to 100 ppb by providing deaerated pure water by membrane deaeration, although the cost increases. This membrane deaeration method has not yet been widely applied, so the cost of the module is still high. In a small ultrapure water production unit, however, this method can be applied with good cost performance.

The ultimate concentration of the dissolved oxygen is 100 ppb by membrane deaeration, although present requirements are 10 ppb. To achieve this low level, the membrane deaeration method should be combined with other methods. In the following membrane deaeration is combined with nitrogen gas bubbling. Figure 12 shows the ultrapure water manufacturing process, and Fig. 13 indicates how much dissolved oxygen is removed. It was found that the dissolved oxygen can be efficiently removed with a small amount of nitrogen gas by applying membrane deaeration as the pretreatment to reduce the concentration of dissolved oxygen to 500 ppb or less followed by nitrogen gas bubbling.

In membrane deaeration, some carbon dioxide can be removed if the ultrapure water can be controlled at a certain pH level. As shown in Fig. 14, dissolved carbon dioxide changes its form according to pH. At pH 5 or less, most of the carbon dioxide is dissolved in the form of molecular carbonic acid. In this case, the carbon dioxide can be removed by controlling the pH. With a vacuum level of 10 torr and pH 5.0, about 90% of the carbon dioxide can be removed.

AC : charcoal tower
RO : Low-pressure RO equipment
MD : Membrane deaerator
MBP : Mixed bed column
ND : Nitrogen deaerator

ST : Ultrapure water storage tank
UV : Ultraviolet ray oxidation equipment
CP : Cartridge polisher
UF : UF equipment

Figure 12 Configuration of ultrapure water manufacturing equipment including a two-stage deaeration system.

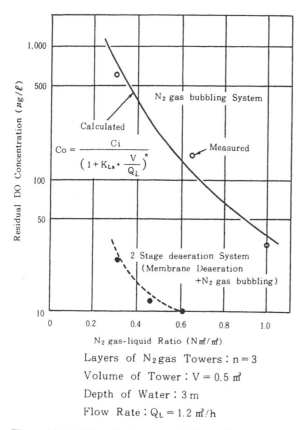

Layers of N_2 gas Towers : $n = 3$

Volume of Tower : $V = 0.5$ m³

Depth of Water : 3 m

Flow Rate : $Q_L = 1.2$ m³/h

Figure 13 Relationship between the N_2 gas-liquid ratio and the residual DO concentration using a two-stage deaeration system.

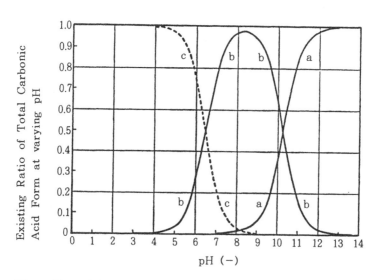

Figure 14 Relationship between the pH of water and the CO_3^{2-}, HCO_3^-, and H_2CO_3 (curve a, CO_3^{2-}; b, HCO_3^-; c, H_2CO_3). (From Ref. 12.)

Table 4 Characteristics of the Catalytic Resin

Item	Note
material	polystyrene
appearance	grey
shape of particles	true sphere
particle diameter range	$0.5 \sim 1.25$ mm
effective particle diameter	0.63 mm
uniformity coefficient	1.6 (maximum)
apparent density	$620 \sim 680$ g/ℓ
true specific weight	1.02
moisture content	$50 \sim 55$ wt %
operating temperature	120°C (maximum)
operating pH range	$5 \sim 14$
permissive pressure loss of packed bed	under 1.5 kg/cm²

Source: From Refs. 14 and 15.

V. DEAERATION WITH CATALYTIC RESIN

Bayer AG in Germany recently developed a new method of removing dissolved oxygen by applying palladium catalyst mounted on the surface of an anion resin [13–15]. The development of palladium catalytic resin has made this method available because this catalytic resin can facilitate the reaction between oxygen and hydrogen to yield water in water at room temperature. Table 4 shows the characteristics of this catalytic resin. It is excellent both mechanically and chemically. It has also been reported that little TOC is leaked from this resin: a TOC elution of less than 1 ppb at a flow rate *SV* of 50 h^{-1} or more [15]. Figure 15 indicates the chemical reaction taking place on the surface of this catalytic resin. For ultrapure water, hydrogen gas or hydrazine is used as the reducing agent to be reacted with the dissolved oxygen. The reaction can be expressed as [14,15]

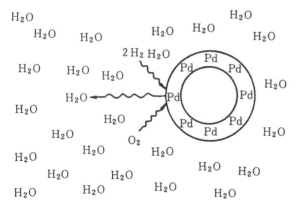

Figure 15 Deoxidation mechanism on the catalytic resin surface. (From Refs. 14 and 15.)

$$O_2 + 2H_2 \xrightarrow{\text{catalytic resin}} 2H_2O$$

$$O_2 + N_2H_4 \xrightarrow{\text{catalytic resin}} 2H_2O + N_2$$

The reaction proceeds in a theoretical manner. For the amount of reducing agent, 1 ppm hydrogen or 8 ppm hydrazine is theoretically required to remove 8 ppm dissolved oxygen. It is reported, however, that 10–15% more reducing agent than the theoretical amount is usually used to secure the reaction in practice [15]. In terms of performance as a reducing agent, hydrogen gas and hydrazine are equivalent. For hydrogen gas, however, because it needs to be perfectly dissolved in ultrapure water, a special dissolution unit should be prepared. Hydrogen gas is also explosive and combustive. Therefore hydrazine is often used in the experiments reported in the literature [14,15]. Sulfuric acid and hydrogen

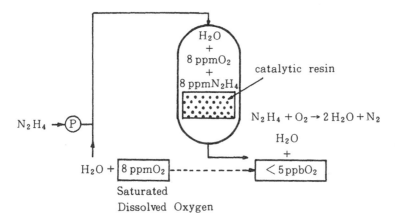

Figure 16 N_2H_4 reduction system using catalytic resin. (From Ref. 15.)

Table 5   Deoxidation Performance of a Catalytic Resin System

flow rate (m^3/h)	SV (1/h)	N_2H_4 feed (ppm as N_2H_4)	DO of treated ppb as O_2
0.3	15	35	<5
0.5	25	20	<5
1	50	10	<5
2	100	5	60
3	150	8	<5
3	150	4	$2,000 \sim 3,000$

condition
    catalytic resin   bed volume : 20 ℓ
                                height of bed : 400 mm
    sample   resistivity : 18.0 MΩ • cm
                                dissolved oxygen : 7~8 ppm
    measuring method   KEMETs
Source: From Ref. 15.

sulfide, which are often used as reducing agents, should not be used because they damage the catalytic action of the palladium.

Figure 16 shows the configuration of the unit, and Table 5 shows its performance. The configuration is very simple, composed of a reaction tower packed with catalytic resin and reducing agent injector. As shown in Table 5, when the reducing agent is in sufficient supply, it is possible to remove dissolved oxygen to the level of 5 ppb or less at a flow rate SV of 150 h^{-1} or less.

This method does not employ mechanical deaeration but chemical deaeration, which means the reaction proceeds almost in a theoretical manner. In this sense, the stable reduction of dissolved oxygen can be expected to securely maintain the low concentration of dissolved oxygen. Moreover, the unit can be compact because it can withstand the high flow rate. It is also reported that this method can be applied to sterilization with hot water because the resin has sufficient heat resistance [14,15].

REFERENCES

1. Water and Waste-water Handbook, Maruzen Co., Ltd., 1971.
2. Nippon Kagaku-kai, Chemical Handbook, Basic Volume II, Maruzen Co., Ltd., 1966.
3. T. Ohmi, M. Morita, E. Hasegawa, M. Kawakami, and K. Suma, 175th Electrochemical Society Spring Meeting Proceedings, 1989, p. 227.
4. M. Morita, T. Ohmi, E. Hasegawa, M. Kawakami, and K. Suma, Appl. Phys., *55*, 562 (1989).
5. M. Morita, T. Ohmi, E. Hasegawa, M. Kawakami, and K. Suma, 1989 Symposium on VLSI Technology, 1989, p. 71.
6. N. Hasimoto et al., Semiconductor Purewater Conference Proceedings, Vol. 1, 1990, pp. 2–26.
7. Y. Yasojima, Hokkai Kanyu, *12*, 11–14 (1988).
8. Y. Yasojima, K. Shirato, et al., Research Report of Government Industrial Development Laboratory, (Hokkaido, 1988).
9. K. Sirato et al., Kenchiku-Setubito Haikankoji, *4*, 112–116 (1989).
10. A. W. Kingsbury and E. L. Phillips, J. Eng. Power, *10*, 331–338 (1961).
11. E. L. Knoedler and C. F. Bonilla, Chem. Eng. Prog., *3*, 125–132 (1954).
12. H. Oya, Membrane Application Technology Handbook, Saiwai Syobo Co., Ltd., 1976, p. 27.
13. K. Yabe et al., Kagaku Kogyo, *6*, 530–535 (1985).
14. T. Kawauti, Chem. Eng., 21–26 (1990).
15. M. Hurukawa, Semiconductor World, 2 (1990).

5
Sterilization Technique

Masao Saito
Shinko Pantec Co., Ltd., Kobe, Japan

I. INTRODUCTION

Hot water, steam, and chemicals, such as formalin, are normally used to sterilize ultrapure water for the pharmaceutical and medical industry. Even the ultrapure water system for semiconductor production has conventionally used sterilization in the distribution system, including the reverse osmosis (RO) unit, the ultrafiltration (UF) unit, and the ultrapure water subsystem at periodic intervals to decrease the microbe concentration level within the system. Recently, there have been increased demands for reduced numbers of microbes in the ultrapure water along with the increased density of the semiconductor circuit and continuous operation of the semiconductor plant for 24 h, 365 days a year.

Primary obstacles to such achievements are sterilization within the ultrapure water production plant and reduction in equipment downtime accompanied by subsequent water quality recovery.

Practical sterilization methods at present include continuous stream sterilization using ultraviolet, thermal sterilization using hot water, chemical sterilization using hydrogen peroxide, formalin, and other compounds, and sterilization using ozone. It is important to select materials for piping and equipment that are suitable to each application to create the optimum system.

II. GROWTH OF MICROBES IN ULTRAPURE WATER

A. Microbe Counting

The microbe count normally referred to as the guaranteed water quality level for ultrapure water is the general bacteria count, not the number of microbes in the sample.

Bacteria count is expressed by colony-forming units (CFU). The ASTM F60-68 membrane filter culture method developed in the United States is the method generally adopted to test bacteria under poor culture condition, such as in ultrapure water. The culture conditions specified in the ASTM standard include M-TGE medium, a temperature of 35°C, and a culture time of 24 h.

It is readily possible to measure the number of bacteria using a commercially available disposable measuring kit, provided that precautions to prevent contamination from outside are observed. On the other hand, it is reported that the culture conditions need to be reexamined because the number of bacteria to be counted in ultrapure water changes greatly depending on M-TGE culture medium concentration, culture temperature, and culture time. For example, the results of measurement of bacteria number using the improved method by Sasaki and Saito [1] and the ATSM method are compared in Table 1.

In a sample in which no bacteria are detected with the ASTM method, bacteria may be detected under different culture conditions. The minimum size of bacteria is approximately 0.2 μm, and this is counted as a "particle" contained in the water. Bacteria that are not detected by the ASTM method may deposit on the wafer surface and cause pattern defects.

B.   Microbe Growth Rate

The relationship between bacterial propagation and temperature within stationary ultrapure water is shown in Fig. 1. The number of bacteria was measured in accordance with the ASTM method. According to Fig. 1, the initial number of bacteria (6 counts per 100 ml) is observed to be suddenly increased 20–60 h thereafter at any incubation temperature. The propagation speed responds to the water temperature within the temperature range of

Table 1   Dependence of Bacteria Number on Different Conditions of Incubation

(C. F. U. / mℓ)

	ASTM method	Improvement method
Agar concentration	M − TGE	10 times dilution of M − TGE
Culture temperature (°C)	35	25
Culture time (H)	24	120
RO inlet	0.04	15.8
	N. D	15.5
RO permeate	N. D	0.01
	N. D	0.29
UV effluent	N. D	0.06
	N. D	0.06
Point of use	N. D	0.01
	N. D	0.02

Source: From Ref. 1.

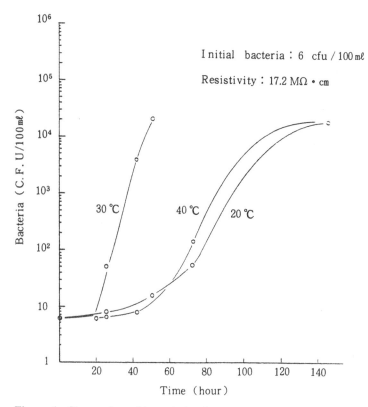

Figure 1   Propagation of bacteria in ultrapure water.

20–30°C. At 40°C, however, almost the same tendency as that at 20°C is observed. Many bacteria living in pure water enjoy the optimum temperature of 25–30°C [2].

At a temperature of over 35°C, a greater part of such bacterial growth is inhibited. The temperature of the ultrapure water normally supplied and circulated is approximately 25°C, and such a temperature is thought to be the optimum propagation temperature for bacteria.

III.   ULTRAVIOLET STERILIZATION

A.   Sterilization Effect of Ultraviolet

The ultraviolet sterilizer is incorporated in the ultrapure water supply system (used to supply the ultrapure water continuously to each use point) to decrease the bacterial level within the system. The ultraviolet sterilizer is installed at the RO unit inlet, which is impervious to the continuous feeding of disinfectant to prevent bacterial growth on the RO membrane, or the sterilizer is submerged in the ultrapure water tank to sterilize the bacteria within the tank. Figure 2 shows an example of installation of an ultraviolet sterilizer in an ultrapure water production system. The ultraviolet sterilizer is widely used for ultrapure water sterilization because it has the following advantages over other methods:

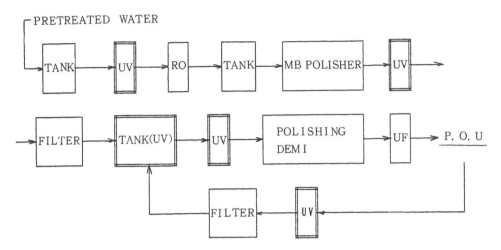

Figure 2 UV sterilizer location in UPW production line.

1. The sterilization effect is limited only to the irradiation period, and it is not necessary to add impurities.
2. It is possible to sterilize feedwater continuously.
3. The sterilizer has little selectivity against the type of bacteria.
4. The ultraviolet sterilizer requires only an electrical power supply. It can be operated easily and provides excellent safety and economic efficiency.
5. The sterilizer requires only periodic lamp replacement and can be maintained easily.

B. Mechanism of Sterilization by Ultraviolet

The sterilization mechanism using ultraviolet can be regarded as a photochemical reactions. Figure 3 shows the wavelength characteristics of relative sterilization effects.

The sterilization energy of ultraviolet provides a specific peak at a wavelength of approximately 260 nm. On the other hand, the ultraviolet ray absorption curve for nucleic acid and protein is shown in Fig. 4, in which the ultraviolet absorption maximum is near 260 nm, which quite closely resembles the wavelength characteristics of the sterilization effect shown in Fig. 3. In other words, there is good absorption of ultraviolet at 260 nm by the nucleic acid. A dimer is formed within the nucleic acid by the absorbed energy, which is thought to be the main cause of living cell death.

C. Sterilization Activity

The sterilization activity of ultraviolet is represented by the amount of ultraviolet irradiated against the amount of water treated (μW·s cm^{-2}). This is obtained by multiplying the irradiation intensity of the ultraviolet by the irradiation time: Amount of UV μW·s cm^{-2} = irradiation intensity (μW cm^{-2}) × irradiation time (s). The unit indicating the amount of ultraviolet is shown in Table 2. The amount of ultraviolet required to kill various types of bacteria is listed in Table 3, although this may vary depending on the experimenter, bacterial activity, environmental temperatures, and other factors.

Compared to the UV irradiation of *Escherichia coli*, the Gram-negative bacteria, such as *Salmonella enteritides*, *Salmonella paratyphi*, and blue pus bacillus, require

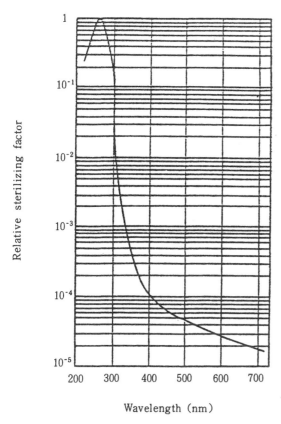

Figure 3 Sterilizing effect at different wavelengths. (From Ref. 3.)

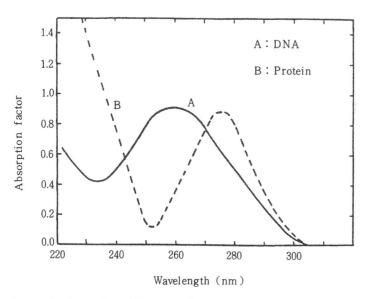

Figure 4 Absorption of UV waves by DNA and protein. (From Ref. 3.)

Table 2 Units of Ultraviolet Irradiation

Irradiation intensity :	$1\,\mu W/cm^2 = 0.01\,mW/cm^2$
	$= 0.01\,W/m^2$
Irradiation amount :	$1\,\mu W \cdot min/cm^2 = 60\,\mu W\ sec/cm^2$
	$= 600\,erg/cm^2$
	$= 0.6\ J/m^2$

Table 3 UV Irradiation Amount for Sterilizing Microorganisms

(99.9 % destruction)

	$\mu W \cdot sec/cm^2$
(Gram − negative Organism)	
Escherichia coli	6,600
Shigella dysenterie	3,400
Pseudomonas aeruginosa	10,500
(Gram − positive Organism)	
Bacillus subtilis	11,000
Bacillus subtilis spores	22,000
Streptococcus hemolyticus	5,500
Staphylococcus aureus	6,600
Staphylococcus albus	5,700
(Yeast)	
Saccharomyces sp.	17,600
Saccharomyces cerevisiae	13,200
(Mold Spores)	
Penicillium roqueforti	26,400
Penicillium expansum	22,000
Aspergillus niger	330,000
Aspergillus flavus	99,000
Rhizopus nigricans	220,000
(Virus)	
Influenza	6,600
Polio Type I	6,000
Coxsachie A 2	4,800
Adenovirus Type III	4,500

Source: From Ref. 4.

almost the same amount, but the Gram-positive bacteria, such as *Bacillus anthracis*, *Bacillus subtilis*, and *Bacillus megatherium*, require 2–5 times more and the mold spores requires 4–50 times more UV.

D. Types of Ultraviolet Ray Sterilization

1. Stream Sterilizer

There are two types of stream sterilizers: (1) in the external irradiation type the water is passed through the inside of a pipe and the ultraviolet is irradiated from the outside; and (2) in the internal irradiation type the sterilization lamp is mounted inside the pipe and the ultraviolet is irradiated from the inside.

In a large-capacity ultrapure water production system where the water temperature changes little, the internal irradiation sterilizer is often used. An example of the construction of an internal irradiation sterilizer is shown in Fig. 5. In this case, the number of sterilization lamps required for the flow rate are installed in the protective pipe (outer pipe) within the pipe through which water flows.

2. Immersion Sterilizer

Since the bacteria are suspected to grow in the ultrapure water storage tank, the tank is designed to reduce retention time. However, the bacteria grow easily because the flow velocity at the tank wall is set to an extremely low level compared with that inside system equipment and piping. To sterilize and prevent bacterial growth, a sterilizer (Fig. 6) in which the sterilization lamp is immersed in the tank is sometimes used. A fiber-reinforced plastic (FRP) or polyethylene tank is not advisable for this purpose; a stainless steel or Teflon-coated tank is used.

When arranging the submerged sterilization lamp, it is necessary to determine the number of sterilization lamps needed in terms of tank size and retention time, since the

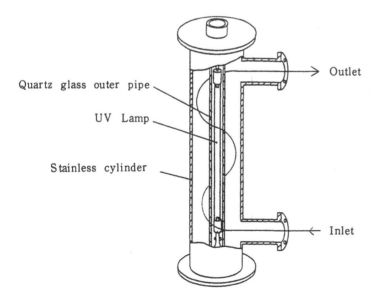

Quartz glass outer pipe

UV Lamp

Stainless cylinder

Outlet

Inlet

Figure 5 Stream UV sterilizer with inner lamp. (From Fact file 11-011, Photoscience Japan Corporation.)

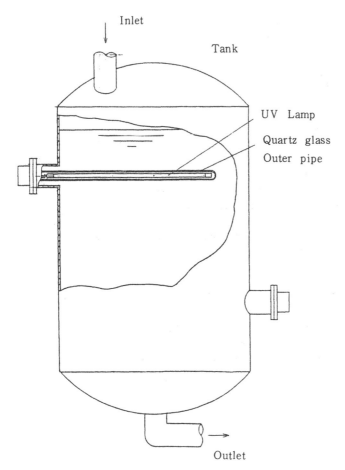

Inlet

Tank

UV Lamp

Quartz glass

Outer pipe

Outlet

Figure 6   Submerged sterilizer. (From Fact file 11–011, Photoscience Japan Corporation.)

effective range of irradiation per lamp is limited. When the water level within the tank can be changed, it is important to maintain a minimum water level, and flickering of the lamp by the level of water should be avoided. This is because flickering of the lamp can shorten the life of the lamp, and time is required for the desired irradiation intensity to be reached after the lamp has been turned on.

E.   Points to Note on the Use of the Ultraviolet Sterilizer

The ultraviolet sterilizer is readily installed and is an extremely effective means of sterilization within the system and for preventing bacterial growth. However, a few points should be noted when selecting and installing the sterilizer.

First, the standard water quantity to be treated and the rated output indicated by the ultraviolet sterilizer are in many cases intended for *Bacillus subtillis* (spore). It is necessary to confirm water quantity and irradiation intensity before selection.

The ordinary sterilization lamp provides the maximum relative sterilization output at a pipe wall temperature of approximately 40°C. In the general ultrapure water system in which the water treatment temperature is 20–25°C, however, the pipe wall temperature of the lamp is more than 50°C. For this reason, modifications are made on the gas to be filled

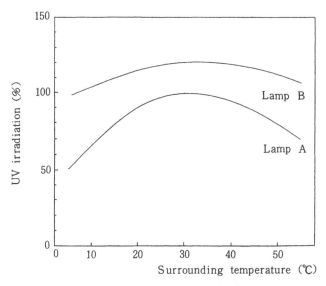

Figure 7 UV irradiation by UV lamp at different temperatures. (From Fact file 11–011, Photoscience Japan Corporation.)

in the pipe, the shape of pipe, and so on. A lamp that ensures maximum output at 50°C is generally used.

If the water temperature is different, the output is decreased. Recently, however, a sterilization lamp was developed that is hardly affected by the water temperature within a general working temperature range. Such a sterilization lamp is effective when it is installed where there is no water temperature control (see Fig. 7).

The pipe through which the water flows is electropolished on the inner surface to improve the ultraviolet reflection ratio, to prevent the deposition of microbes and particles, and further to prevent metal ion leaching from the stainless steel surface. Such a pipe is effective when it is installed in the ultrapure water subsystem.

Microbes destroyed by ultraviolet irradiation leave fragments, and these are counted as particles in the water. When the sterilizer is installed after the RO system, it is necessary to arrange an approximately 0.2–0.45 μm filter after the ultraviolet sterilizer.

Finally, the ultraviolet output of the sterilization lamp deteriorates gradually in response to the time it is lit. This is because electrode components, for example, deposit on the glass wall surface. An output deterioration curve is shown in Fig. 8.

When the sterilization lamp remains turned on continuously, the output decreases to approximately 65–70% of initial output in 7000–8000 h. To maintain the desired sterilization capacity, it is therefore necessary to replace the lamp periodically.

Care must be taken when the sterilization lamp is turned on and off frequently. Sterilization lamps with output drop monitor mechanisms and off alarms are commercially available, and the use of such sterilization lamps assures safe operation.

IV. INTERMITTENT STERILIZATION

Even in an ultrapure water system designed to minimize supply water retention, in which continuous sterilization is performed using ultraviolet light, it is impossible to avoid the

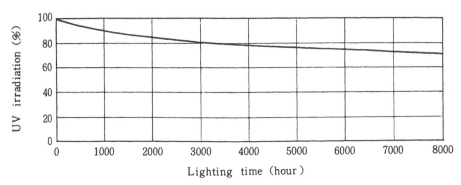

Figure 8   Deterioration curve of UV lamp irradiation. (From Sankyo Electric Co, Ltd. catalog.)

contamination by microbes in the system. For this reason, the equipment is shut down temporarily to perform periodic sterilization using chemicals or hot water.

Such a measure is taken in part to prevent microbe growth by continuous feeding of disinfectant. However, sterilization and rinsing using chemicals must be performed periodically because the bacteria become resistant to the disinfectant.

A.   Sterilization and Cleaning of the RO System

Since the cellulose acetate RO membrane generally applied to the primary pure water system is resistant to free chlorine of approximately less than 0.7 ppm, it is possible to prevent microbe growth on the membrane by continuous feeding of sodium hydrochlorite.

The aromatic polyamide low-pressure synthetic membranes are difficult to sterilize continuously because oxidants, such as hydrogen peroxide and sodium hydrochlorite, cannot be applied to such membranes. Recently, however, it has been reported that combined chlorine disinfectants, such as chloramine-T, can be fed quantitatively and intermittently. It was confirmed that stable performance can be assured for the RO system if the operation conditions are observed carefully.

Formalin is generally used to sterilize or clean the ordinary RO membrane. Reducing agents, such as hydrazine, or caustic soda is used to recover membrane performance after the differential pressure has risen as a result of bacterial slime, for example, generated in the low-pressure synthetic membrane. In any event, it is necessary to use each disinfectant selectively depending on the conditions of contamination.

It is advisable to prepare the postcleaning wastewater treatment method before using formalin. Since the formalin is detected as COD_{Mn} even at low concentrations, it must be treated with due care.

The relationship between 35% formalin (containing 5–15% ethanol) and COD_{Mn} is given by the formula

$$COD_{Mn}(mg\ L^{-1}) = \frac{80.4}{100} \times HCHO(mg\ L^{-1})$$

B.   Sterilization by Hydrogen Peroxide

1.   Sterilization Effect of Hydrogen Peroxide

Hydrogen peroxide is often used for sterilization of the distribution system, including the ultrapure water subsystem. This is because hydrogen peroxide contains no metal ion, compared with other sterilizers, and it decomposes into water and oxygen gas. It is thus less possible to contaminate the inside of the system.

The sterilization mechanism using hydrogen peroxide has many unknowns. However, it is reasoned as follows [5]:

1. Hydrogen peroxide is decomposed quickly by the enzyme contained in bacteria. The oxygen gas generated in this case decomposes and liquefies the waxy membrane of the bacteria.

$$H_2O_2 \rightarrow H_2O + \frac{1}{2}O_2$$

2. The protein in bacteria is changed in quality or decomposed as a result of oxidation by hydrogen peroxide, causing the enzyme activity to decrease.
3. When the bacteria are anaerobic, bacterial growth is restricted by the oxygen generated by the decomposition of hydrogen peroxide. The relationship between the concentration of the hydrogen peroxide and the extinction time of the microorganism is shown in Fig. 9.

Generally, immersion and sterilization are performed for more than 3 h at a hydrogen peroxide concentration of approximately 1–3%. If the immersion time is longer than that shown in Fig. 9, the desired effect can be achieved at a lower concentration.

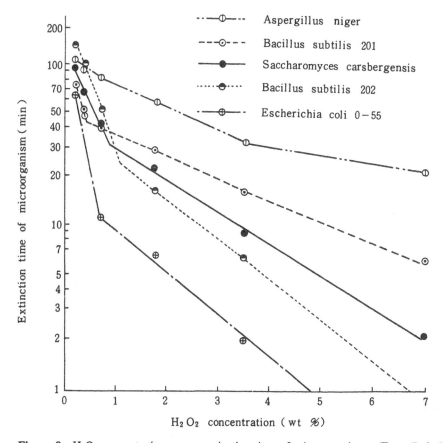

Figure 9 H_2O_2 concentration versus extinction time of microorganisms. (From Ref. 5.)

Table 4 H_2O_2 Standard by EL Grade

Hydrogen peroxide（30%） H_2O_2（EL-UM）   FW: 34.01

Item	Unit	Standard	Item	Unit	Standard
Color	Hazen	< 10			
Content（H_2O_2）	%	31.0 ± 1.0			
Total Solid	ppm	< 10			
Acid（as H_2SO_4）	ppm	< 10			
Cl	ppm	< 0.3	PO4	ppm	< 0.3
NO₃	ppm	< 2	NH4	ppm	< 1
SO₄	ppm	< 0.3	N	ppm	< 0.5
B	ppb	< 10	K	ppb	< 10
Ag	ppb	< 1	Li	ppb	< 1
Al	ppb	< 10	Mg	ppb	< 1
As	ppb	< 5	Mn	ppb	< 1
Au	ppb	< 5	Na	ppb	< 20
Ba	ppb	< 5	Ni	ppb	< 1
Ca	ppb	< 10	Pb	ppb	< 1
Cd	ppb	< 1	Si	ppb	< 100
Co	ppb	< 1	Sn	ppb	< 10
Cr	ppb	< 2	Sr	ppb	< 1
Cu	ppb	< 1	Zn	ppb	< 10
Fe	ppb	< 10			
Ga	ppb	< 10	Particle	part./mℓ	< 50
Ge	ppb	< 10	（$> 0.5 \mu$m）		

Source: From Ref. 6.

2. *Notes on Sterilization*

Care must be taken according to the following in performing the actual work of sterilization:

1. Since the hydrogen peroxide is fed directly to the use point, it is desirable to use a semiconductor (EL) grade of which particle is cut. The applicable standards are shown in Table 4.
2. It is necessary to determine the concentration of hydrogen peroxide considering applicable equipment shutdown time, immersion time, and blowdown time, among other factors.
3. The piping should be designed so that sterilizer is not allowed to deposit in the pipe line, even to the mounting position and the type of valves used.

3. *Treatment for Cleaning Wastewater*

The hydrogen peroxide is washed away by the ultrapure water after completing immersion and sterilization. It is very important to minimize the amount of wastewater discharged

Table 5   Dependence of H_2O_2 Decomposition on Different pH Values After 24 h

Test No.	Added NaOH mg/ℓ for pH adjustment	pH		H_2O_2 mg/ℓ		Decomposition of H_2O_2 %
		t = 0	t = 24	t = 0	t = 24	
1	0	5.6	5.1	3390	3390	0
2	20	8.9	7.8	3390	3390	1.7
3	125	10.1	9.4	3380	3040	10.1
4	1100	11.3	10.4	3320	1230	63.0
5	3000	11.9	11.7	3200	1170	63.5

t : Elapsed time (H)
Source: From Ref. 7.

and to apply the optimum treatment method to minimize the load to the wastewater treatment plant. Hydrogen peroxide is extremely stable, decomposing little within the neutral to acid range. Table 5 shows the hydrogen peroxide concentration measured by preparing test samples after adding NaOH to adjust the pH value to 5.6–11.9 and then leaving it under room temperature (25°C) for 24 h. At a pH value over 10, approximately 60% natural decomposition is observed after 24 h. However, the hydrogen peroxide is not decomposed at all at neutral to acid pH. The hydrogen peroxide decomposition method includes addition of reducing agents, catalysts, and catalyst resins. Treatment using activated carbon is reported to be the most effective [7].

C. Hot-Water Sterilization

Sterilization using hydrogen peroxide requires a long time for chemical feeding, circulation, immersion, and wastewater discharge and has several features that are hard to automate. With this in mind, a thermal sterilization method using high-temperature water has been put into practical use.

In this method, the ultrapure water is heated to a temperature of 80–90°C and is circulated for approximately 1 h. It is then cooled by a heat exchanger and returned to the normal circulating water temperature in approximately 1 h, completing the sterilization. Sterilization using hydrogen peroxide requires the equipment to be shut down for over half a day. With hot-water sterilization, however, it becomes possible to shorten the sterilization time required by automation.

In addition, hot-water sterilization has the advantage that treatment of the wastewater is not required, as in chemical sterilization.

When thermal sterilization is adopted to the sterilization of the distribution system, including the ultrapure water subsystem, heat-resistant material is used for the piping, final filter, and UF module. Heat-resistant material has also been recently developed for low-pressure RO modules used to remove total organic carbon (TOC) in the subsystem, and its application has increased along with thermal sterilization.

V. CONTINUOUS OZONE STERILIZATION

Similar to hydrogen peroxide and hot water sterilization, in ozone sterilization the ozone is fed intermittently to the system. In a recently adopted method, a small quantity of ozone is fed into the outlet of the ultrapure water unit continuously, to maintain the inside of the circulation supply piping system, including the point of use, in an almost bacteria-free condition.

A. Sterilization Effect of Ozone

The physical properties of ozone are shown in Table 6. Ozone is said to provide the second greatest oxidation force next to fluorine. The oxidation-reduction potential of ozone is –2.08 V, whereas that of fluorine is –2.87 V.

The reason that ozone has such a high chemical activity is because it has an unstable electron structure and it is necessary to deprive other compounds of electrons so that it is stabilized. Ozone is said to be decomposed in a process when it reacts with other compounds [9].

A recent report used various indices to describe the effect of ozone on various bacteria and mold spores that can grow in the ultrapure water, an example of which is shown in Table 7 [9,10]. The sterilization force is evaluated using the lethality coefficient A by the formula

$$A = \frac{\ln 100}{Ct_{99}}$$

where A = lethality coefficient
$\quad\quad\ C$ = residual ozone concentration (mg L^{-1})
$\quad\quad\ t_{99}$ = time required to kill 99% of microbes (minutes)

According to this result, the blue pus bacillus, *Serratia marcescens*, and other microorganisms are killed at lower concentrations than *B. subtilis* and fungi, as in irradiation by ultraviolet.

Table 6 Physical Properties of Ozone

Molecular weight	48.0
Boiling point °C (°F)	– 111.9 (– 169.4)
Melting point °C (°F)	– 251.0 (– 420)
Gas density, 0°C, Grams /liter	2.144
Critical temperature °C(°F)	– 12.1 (10.2)
Critical pressure, atm.	54.6
Critical volume, cm/mole	111

Source: From Ref. 8.

Table 7 Lethality of Ozone to Microorganisms at pH 7 and 10–15°C

Organism	A	Ci_{99} : 10
Escherichia coli	4600	0.0010
Streptococcus fecalis	3000	0.0015
Mycobacterium tuberculosis	1000	0.0500
Poliovirus	460	0.0100
Bacillus magatherius (spores)	150	0.1000
Endamoeba histolytica	50	0.0300

Ci_{99} : 10 : Concentration in mg/ℓ for 99% destruction in 10 min

Source: From Ref. 9.

B. Continuous Sterilization

Continuous sterilization is a process developed by a U.S. semiconductor wafer manufacturer for its production line. A typical ultrapure water system flow incorporating continuous sterilization is shown in Fig. 10. When the material used for the ultrapure water tank is not resistant to ozone, the residual ozone can be decomposed by ultraviolet lamp just before the circulation water is returned to the tank. This is because the ozone has a wavelength absorption zone called Hartley bands at approximately 260 nm, which matches the wavelength characteristics of the ultraviolet lamp.

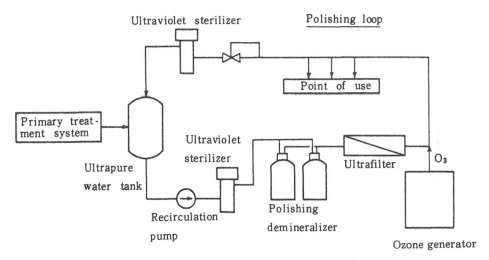

Figure 10 Ultrapure water treatment system with continuous ozonation.

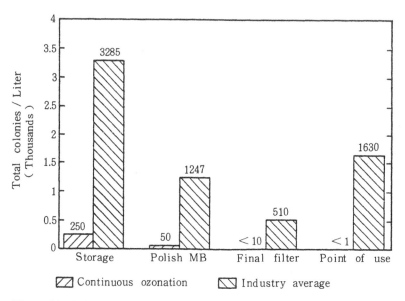

Figure 11 Industry average and continuous ozonation of bacteria. (From Arrowhead Industrial Water, Inc. technical bulletin.)

According to the actual results of the U.S. semiconductor production line, the ozone feed level is reported as follows:

Sterilization effect is not observed, 0–10 ppb.
Microbe growth is restricted, 10–20 ppb.
Sterilization effect is observed, >20 ppb.
Optimum conditions, 40–50 ppb.

Examples of microbes and particles measured in this system are shown in Figs. 11 and 12. More than 80% of the particles in the ultrapure water are said to be organic colloids from

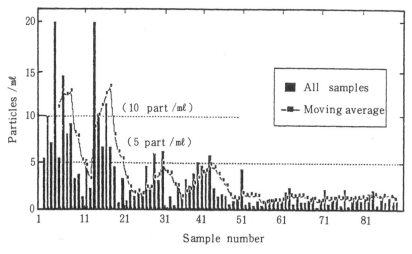

Figure 12 Particle result utilizing continuous ozonation system. (From Arrowhead Industrial Water, Inc. technical bulletin.)

the microbes. The number of particles is greatly reduced by continuous sterilization, considering that the microbes themselves and dead bacteria are also counted as particles.

When using ultrapure water containing ozone, it is necessary to confirm the effect on semiconductor production. If the silicone wafer is rinsed after etching with ultrapure water containing ozone, a monomolecular oxygen membrane forms rapidly on the activated silicone wafer surface. Since this membrane is hydrophilic, it is thought to prevent contaminants in the water from adhering to the wafer and to purify the wafer surface [11].

In a semiconductor production process in which there are problems with ozone in the water, it is necessary to install a small-capacity ultraviolet sterilizer at the ultrapure water inlet to decompose the ozone so that ozone-free water can be supplied to the process. When ozone is fed to a system, the dissolved oxygen contained in the water is increased. In a process in which increased dissolved oxygen is not advisable, the ozone is fed to the water returning from the point of use. It is possible to incorporate a deaeration system using special resins, for example, in the subsystem.

REFERENCES

1. T. Sasaki and M. Saito, Industrial Water, No. 345, 7 (1987).
2. T. Sasaki and M. Saito, Industrial Water, No. 345, 5 (1987).
3. Uragami: Piping & Equipment, P.2 JAN. (1987).
4. Japan Photo-Science, Ultra-violet treatment to develop.
5. Mitsubishi Gas Kagaku, Hydrogen Peroxide Technical Document No. 102.
6. Kantoh kagaku, Chemicals Catalog for Electronic Industries, No. 5.
7. Y. Matsui and S. Kyushin, Shinko Pantec Engineering Report No. *33*(2), 26–29 (1989).
8. N. Khoudary, Pharmaceutical Eng., March–April 30 (1985).
9. C. Nebel and W. W. Nezgod, Ozone, the Process Water Sterilant.
10. R. Sazuka, New Ozone Utilization Technique, (1986), pp. 167–190.
11. G. A. Pittner and P. E. G. Bertler, Ultrapure Water, May/June, 16–22 (1988).

6
TOC Removal Technology

Tetsuo Mizuniwa
Kurita Water Industries, Ltd., Kanagawa, Japan

I. INTRODUCTION

The quality of ultrapure water for large-scale integrated (LSI) circuit production has improved rapidly along with increases in LSI density. Table 1 shows the development of ultrapure water quality [1]. Of the impurities contained in prevailing ultrapure water, the concentration of inorganic substances has fallen below 1 ppb for individual ions, and the total concentration of all ions subject to measurement often does not reach even 1 ppb. On the other hand, the concentration of organic substances is still more than several tens ppb, much higher than that of inorganic substances. For the qualitative improvement of ultrapure water, the most effective means is to reduce organic concentration.

Organic concentration in ultrapure water is indicated by measurement of total organic carbon (TOC). The relation between the impurities in ultrapure water and the defect density of devices manufactured with the ultrapure water containing these impurities has not yet been clarified. The only available reference is the relation of TOC concentration and the relative defect density of devices [2]. According to this reference, with the TOC concentration decreasing below 800–100 ppb, the defect density of devices on wafers manufactured with ultrapure water at that TOC concentration will also decrease, indicating a close correlation. The TOC concentration in prevailing ultrapure water is much lower than this range, but there must be such a relation.

Raw water for the production of ultrapure water is normally underground water or surface water, which contains a variety of impurities. It is impossible to produce ultrapure water by processing raw water containing various impurities with only single-stage water treatment equipment. In other words, an ultrapure water system should be composed of a combination of several water treatment technologies.

Table 1 Ultrapure Water Quality for Semiconductor Manufacturing [1]

Integration Scale		64 Kbits	256 Kbits	1 Mbits	4 Mbits
Resistivity (megohm-cm)		15 ~ 16	17 ~ 18	17.5 ~ 18.0	18.0 ~
Particulates (Counts/mℓ)		0.2 μm	0.2 μm	0.1 μm	0.1 μm
		50 ~ 150	30 ~ 50	10 ~ 20	5 ~ 10
Bacteria (Counts/100 mℓ)		50 ~ 100	5 ~ 20	1 ~ 5	1
TOC	(ppb)	50 ~ 200	50 ~ 100	30 ~ 50	20 ~ 30
SiO_2	(ppb)	20 ~ 30	10	5	5
O_2	(ppb)	100	100	50 ~ 100	50

Table 2 is a list of the elemental water treatment technologies required for the production of ultrapure water, their objects, and the necessary equipment or systems [3]. The equipment in the list that are effective in removing organic substances are the reverse osmosis (RO) unit, the ion exchanger, and the ultraviolet (UV) oxidizer.

As shown in Fig. 1, an ultrapure water system is composed of the following:

1. The primary deionized (DI) water system removes most of the impurities contained in raw water and produces pure water of several MΩ·cm in resistivity, an indication of the total ion concentration of impurities.
2. The subsystem thoroughly removes minute quantities of the impurities still remaining in the primary pure water and produces ultrapure water, the resistivity of which is closest to that of theoretical pure water, with a minimum nonionic impurity concentration.
3. The piping system transports the polished ultrapure water to the points of use.

In some instances, a wastewater recovery system is additionally installed to reclaim wastewater of relatively low contamination for raw water.

The primary DI water system, consisting of a pretreatment system to coagulate and filter raw water, the RO unit, the deaerator, and the ion exchanger, removes suspended solids, mostly ions and organic substances, and dissolved gases from the raw water. The subsystem, consisting of a UV sterilizer (oxidizer), an ion-exchange cartridge filter, and an ultrafiltration (UF) unit, minimizes organic substances and ion concentrations and also removes almost all particles. The piping system is constructed of a material low in elution properties to allow no water stagnation and supply polished ultrapure water to the point of use without recontamination. When it is necessary to recover wastewater for recycling, an ion exchanger, UV oxidizer, and activated carbon filter are additionally installed to remove acid, alkali, and organic substances.

For reducing the TOC concentration in ultrapure water, conventional countermeasures are as follows:

1. Effective removal of TOC in raw water
2. Selection of materials of low discharge of organic materials
3. Persistent decomposition and removal of organic substances in ultrapure water

Table 2 Elemental Technology for Ultrapure Water System [3]

Elemental Technology	Object	Equipment, System
Solid-Liquid Separation	Removal of Turbidity Pretreatment for desalination	Coagulation, Coagulation-flotation
Membrane	Removal of dissolved solid, Organics, particles, bacteria and colloid	Reverse Osmosis, Ultrafiltration Membrane Filter
Adsorption, Ion Exchange	Removal of dissolved solid, trace ions, organics, oxidants	Ion Exchanger, Activated Carbon filter, Adsorption resin
Oxidation-Reduction	Decomposition of organics	UV-oxidation
Gas Separation	Removal of dissolved gas	Vacuum Deaerator Catalyst Resin
Material, Piping	Transportation of water, Prevention of contamination	Piping system, Storage, Valves, Pumps
Monitoring, Controlling	Monitoring of water Quality, Automatic operation	On-line monitor, Sequence-controller
Analysis	Trace analysis of impurities (ions, particles, organics) (ultrapure water, wafer surface)	AA, GC, IC, SEM, XMA, FT-IR, ICP-AES, ICP-MS
Bacteria Control	Sterilization, Prevention of bacteria increase	Sterilization with UV, heat and chemicals

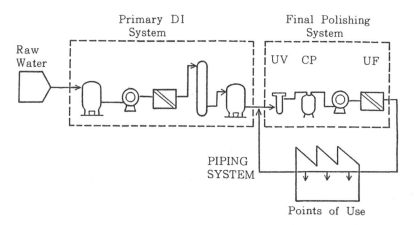

Figure 1 Typical ultapure water system.

Those countermeasures have been executed in actual installations by employing an RO membrane of a high organic removability in the primary DI water system, oxidizing and decomposing organic substances by UV oxidizer in the subsystem, selecting and applying less eluting ion-exchange resins and piping materials, and decomposing all organic substances thoroughly when wastewater is recovered for recycling.

This chapter discusses TOC removal by RO membrane in a primary DI water system, UV oxidation by low-pressure UV lamp, subsequent ion exchange in the final process of an ultrapure water supply system, and UV oxidation by high-pressure UV lamp for wastewater reclamation.

II.   TOC REMOVAL BY RO MEMBRANE

Most TOC in raw water is removed together with inorganic substances by coagulation-filtration, RO, ion exchange, and vacuum deaeration, which are components of a primary DI water system in an ultrapure water production system. Among these components, the RO unit equipped with an RO membrane is the most effective means of TOC removal.

The RO membrane, which was originally developed and used for the desalination of seawater, was introduced to industrial use as a preliminary desalination equipment for an ion-exchange system and has now become an elementary technology indispensable for the production of ultrapure water. Its distinct feature is simultaneous removal of most of the impurities in feedwater. RO is a separation technology using a semipermeable membrane, which is not permeable to the solutes (such as ion and organic substance) in a solution but transmits only solvent (water).

The performance of an RO membrane is expressed by two parameters: rejection rate, which is calculated based on the average impurity concentration in the feedwater (average impurity concentrations of feedwater and concentrate water) and the impurity concentration in the permeate water, calculated by the following formula, and permeation flux, which indicates a permeate water volume through a unit area of membrane under a certain pressure.

$$R = \frac{2C_p}{C_f + C_b} 100$$

R = rejection rate (%)
C_p = impurity concentration in permeate water
C_f = impurity concentration in feedwater
C_b = impurity concentration in concentrate brine

The rejection rate affects mainly the quality of treated water; the permeation flux effects the size of the RO unit.

Although both the spiral and hollow-fiber types are popular RO membrane modules, for stable operation over a long period, the spiral type is superior because it is easy to clean and causes less clogging [4]. The spiral membrane module has the structure shown in Fig. 2; two sheets of flat membrane are glued like an envelope with a permeate spacer inside and their open ends connected to a permeate tube. They are rolled up like photographic film together with a feed spacer around the permeate tube. Since feedwater flows parallel to the membrane surface and the feed spacer causes turbulent flow, this structure prevents condensation of solutes and suspended solids on the membrane surface.

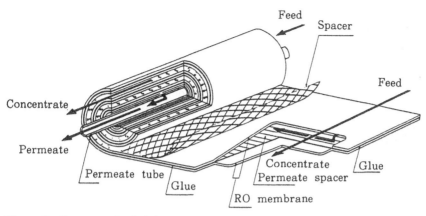

Figure 2 Structure of spiral-wound RO module.

However, manufacturers do not always provide the same construction, which creates differences in clogging tendencies and the cleaning efficiency of the washing function.

Cellulose acetate (CA) membrane was widely used for the RO module in the early days, but polyamide (PA) membrane, with a higher rejection rate and a larger permeation flux, has recently been replacing it. Studies have been made of the relation of various membrane materials and rejection rates of organic substances, because different membranes have different rejection rates. For example, Table 3 shows the organic rejection rates measured in two kinds of polyamide membrane and a cellulose acetate membrane [5]. Both membrane materials demonstrated high rejection rates for high-molecular-weight organic substances but low rates for those with low molecular weight, indicating

Table 3 Rejection of Organic Materials by RO Membrane [5]

	M_w	UTC – 70	BW – 30	CA *
Methanol	32.04	13.9	11.1	5
Ethanol	46.07	54.1	52.7	9
Iso-Propanol	60.10	96.2	90.8	36
Acetone	58.08	69.5	66.9	
Formaldehyde	30.03	32.3	28.9	
Urea	60.06	63.0	58.7	26
Acetic Acid	60.05	55.0	54.4	6
Citric Acid	210.14	99	99	
Ethylenediamine	60.10	95.0	91.0	
Sucrose	342.30	99.79	99.71	>99

Feed Conc. 1000 ppm, Press. 15 kg/cm², Temp. 25 ℃

* M. Kurihara, T. Uemura, Y. Nakagawa and T. Tonomura : *Desalination*, **54**, 75~88 (1988).

that the RO membrane cannot remove low-molecular-weight organic substances, such as alcohol. However, the rejection rate differs depending on the material, and the polyamide membrane has a relatively higher rejection rate of organic substances than the cellulose acetate membrane.

Surface water and subterranean water, which are used as raw water, ordinarily contain organic substances in several ppm. The major component of the TOC in these raw waters is assumed to be so-called humus, including humic acid and fluvo acid. Although yet to be clarified, it is reported that humus has high-molecular-weight functional groups that can be ionized. This is why the RO membrane can remove TOC at a high rejection rate.

Table 4 shows an exemplary qualitative analysis of feedwater and permeate water when municipal water was fed to a polyamide RO membrane as the raw water. The membrane removed more than 99% of the bivalent ions and more than 90% of the monovalent ions and reduced the conductivity from 190 μS cm^{-1} in the feedwater to 12 μS cm^{-1} in the permeate water. Concurrently, the TOC at 600 ppb in the feedwater decreased to 43 ppb in the permeate water, indicating excellent performance of the RO membrane. Recent developments in RO technology have resulted in a two-step RO unit for practical use, a combination of two RO units in series, which can remove more than 98% of organic substances from raw water and produce pure water of 1 or 2 MΩ·cm in resistivity.

In the actual installation of RO units, three to six pieces of spiral membrane module are charged in series in a pressure vessel made of fiberglass-reinforced plastic (FRP). The pressure vessels charged with modules are mounted in parallel on the unit, and a specified quantity of raw water is fed to each vessel. With membrane modules connected in series,

Table 4 Impurity Rejection by Polyamide RO
(Raw water, Atsugi City Water) [5]

Item	RO Feed	RO Permeate
Conductivity (μS/cm)	190	12.0
Cl$_2$ (ppm)	0.5	0
Na$^+$ (ppm as CaCO$_3$)	34.5	4.87
K$^+$ (")	2	0.11
Mg^{2+}(")	11	0.10
Ca^{2+} (")	35	0.32
T. Cation (")	82.5	5.40
Cl$^-$ (")	13	0.47
NO$_3^-$ (")	3	0.24
SO$_4^{2-}$ (")	67	0.62
Carbonate ( )	56	3
SiO$_2$ (")	21	1.5
T. Anion (")	160	5.83
TOC (ppb as C)	600	43
Particles (counts/ml) ($>$0.2 μm)	50,000	380

the flow rate of the feedwater decreases by permeate quantities along with the progress of treatment toward subsequent modules, and eventually the water volume flowing on the membrane surface of the last module is the smallest.

When particles and solvents are removed by RO membrane, those impurities are condensed on the membrane surface of the feedwater side. The structure of the membrane is designed so that feedwater flows parallel to the membrane surface and flushes away the impurities condensed there. However, if the feedwater contains a large quantity of suspended solids or substances that are easily precipitated by condensation, or if the flow velocity is not sufficient to wash out all condensed impurities, some impurities begin to deposit and clog the membrane surface and feed channels. To secure the long stable operation of the RO unit installed in a primary DI water system, it is crucial to practice proper operation of the coagulation filter in the pretreatment system for quality control of the feedwater to the RO unit and to maintain a specified minimum flow rate on the RO membrane surface.

III.   TOC REMOVAL BY LOW-PRESSURE UV LAMP

Medium-purity product water from a primary DI water system has a TOC concentration as low as several tens ppb. However, the concentration must be further decreased to meet the more stringent quality requirements of recent years. In other words, it is required that all organic substances be removed by the subsystem while reducing the concentration of ions and particles to the minimum. Feedwater to the subsystem is the primary pure water that has passed through the RO membrane, so that the organic substances remaining there are those of low molecular weight, although no study has ever clarified their composition.

Table 5 shows the composition of volatile organic substances contained in raw water as well as in treated water from an ultrapure water system, which was analyzed by distillation and headspace gas chromatography [5]. Table 5 indicates that ultrapure water contains methanol and ethanol, whose ratios against TOC occasionally reach several tens percent. This means that these organic substances are probably generated in the ultrapure water system, specifically from ion-exchange resins [6]. Table 6 shows the result of another analysis of an ultrapure water system incorporated with a wastewater reclamation system. Table 6 indicates that if the wastewater contains a large quantity of organic solvents used in the LSI manufacturing process, the wastewater recovery system cannot remove all the solvents, resulting in contamination of the ultrapure water system.

Even if an RO membrane is used as the final filter in the subsystem, because of its low rejection rate it cannot remove such low-molecular-weight organic substances as listed in Tables 5 and 6. As a countermeasure against this problem, a device that decomposes organic substances by UV irradiation was introduced. Since photochemical reaction by UV irradiation enables clean treatment without increasing the quantity of impurities, it is suitable for the treatment of ultrapure water.

There are two methods available for removing organic substances by photochemical reaction. One is to dissociate all interatomic bonds of organic substances and decompose these organic substances into CO_2 and H_2O by light energy, and the other is to dissociate some of the interatomic bonds to form functional groups, whose ionicity is utilized to adsorb the organic substances to ion-exchange resins. When organic substances remain at a low concentration like those in ultrapure water, it is difficult to remove them completely because the reaction rate of UV irradiation with the organic substances is decreased. On

Table 5 Volatile Organics in Raw Water and Ultrapure Water [6]

Sample	Methanol (ppb)	Ethanol (ppb)	Acetone (ppb)	2 - Propanol (ppb)	Diethyl Ether (ppb)	Methyl Ethyl Ketone (ppb)	Calculative TOC (ppb)	TOC (ppb)	Ratio (%)
Industrial Water	< 10	< 5	< 2	< 2	< 5	< 2	—	600	—
MB Product	15	< 5	< 2	< 2	< 5	< 2	5.6	30	19
RO Product	15	< 5	< 2	< 2	< 5	< 2	5.6	20	28
UF Product	20	< 5	< 2	< 2	< 5	< 2	7.5	20	38

Calculative TOC : Concentration of the components in terms of TOC
Analysis by TOC analyzer (whole TOC)
Ratio of Calculative TOC / whole TOC

Table 6   Example of Volatile Organic Concentration in Water-Reclaiming Ultrapure Water System [6]

Sample	Methanol (ppb)	Ethanol (ppb)	Acetone (ppb)	2-Propanol (ppb)	Diethyl Ether (ppb)	Methyl Ethyl Ketone (ppb)	Calculative TOC (ppb)	TOC (ppb)	Ratio (%)
Recovered Water	4100	62	450	4600	<5.0	2.5	4600	5800	79
AC Product	4200	66	450	4900	<5.0	2.5	4800	5500	87
IE Product	4400	66	310	5000	<5.0	2.0	4900	5200	94
UV Oxidation Product	44	<5.0	750	4.5	<5.0	<2.0	480	1100	44
RO Product	56	<5.0	720	5.5	<5.0	<2.0	470	520	90
DI Storage Tank	76	<5.0	120	3.0	<5.0	<2.0	93	190	49
Polisher Product	110	<5.0	110	3.0	<5.0	<2.0	110	170	65
MF Product	120	<5.0	110	2.0	<5.0	<2.0	110	160	69

Calculative TOC : Concentration of the components in terms of TOC
Analysis by TOC analyzer (whole TOC)
Ratio of Calculative TOC / whole TOC

the other hand, the second method has only to cut some of the organic bonds to remove substances in low concentration.

To dissociate organic bonds effectively, it is better to irradiate with short-wavelength UV. As shown in Fig. 3, a low-pressure mercury lamp has the strongest UV irradiation, at 254 nm wavelength, which is highly effective in sterilization. This is why the low-pressure mercury lamp is widely used as a sterilizer. In addition to the 254 nm wavelength, the mercury lamp irradiates using several other UV wavelengths, including the 185 nm wavelength, and visible rays. A UV irradiation device is available that effectively utilizes the 185 nm wavelength UV. With this device installed, it will be possible to irradiate ultrapure water with short-wave-length or high-energy UV.

Table 7 compares UV energies at 185 and 254 nm wavelengths and the energies required to dissociate the interatomic bonds of organic substances. Based on this comparison, the possibilities for dissociating organic bonds by irradiating UV at such wavelengths are also shown in Table 7 (Chiyoda Kohan Co., Ltd. technical information). Table 7 indicates that UV irradiation at 185 nm can dissociate most carbon-carbon bonds, except for double bonds and triple bonds. The UV irradiation has enough energy to dissociate even water; hydroxide radical ($\cdot$OH) is formed from water (H_2O). Since the hydroxide radical is highly active and has enough potential to oxidize organic substances, when ultrapure water containing organic substances is irradiated by UV at 185 nm, the organic bonds are dissociated and the hydroxyl radical oxidizes organic substances, both jointly promoting the decomposition of organic substances. However, regardless of the mechanism of decomposition, following the decrease in concentration of the organic substances to be decomposed, the efficiency of the decomposition reaction is definitely decreased as well. It is therefore difficult to decompose and remove all organic substances thoroughly by UV irradiation alone.

Once organic substances are oxidized after partial dissociation of their interatomic bonds, they form carboxyl groups (—COOH) and are transformed into organic acid, which is ionized. Taking advantage of this reaction, it is possible to remove organic

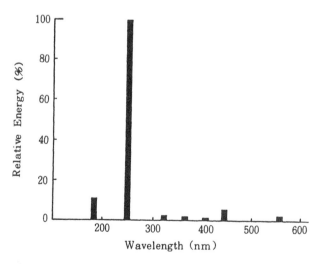

Figure 3   Spectrum of a low-pressure mercury lamp (Nippon Photo Science Co, type AZ-3). Technical data.

Table 7 Dissociation Possibility of Chemical Bonds with 184.9 nm UV

Bond	Dissociation Energy (kcal/ mol)	Maximum Wavelength for Dissociation (nm)	Possibility of Dissociation with 184.9 mm UV (154 kcal)
C−C	82.6	346.1	yes
C=C	145.8	196.1	yes
C≡C	199.6	143.2	no
C−Cl	81.0	353.0	yes
C−F	116.0	246.5	yes
C−H	98.7	289.7	yes
C−N	72.8	392.7	yes
C=N	147.0	194.5	yes
C≡N	212.6	134.5	no
C−O	85.5	334.4	yes
C=O (aldehydes)	176.0	162.4	no
C=O (ketones)	179.0	159.7	no
C−S	65.0	439.9	yes
C−S	166.0	172.2	no
H−H	104.2	274.4	yes
N−N	52.0	549.8	yes
N=N	60.0	476.5	yes
N≡N	226.0	126.6	no
N−N (NH)	85.0	330.4	yes
N−H (NH$_3$)	102.2	280.3	yes
N−O	48.0	595.6	yes
N−O	162.0	176.5	no
O−O (O$_2$)	119.1	240.1	yes
−O−O−	47.0	608.3	yes
O−H (water)	117.5	243.3	yes
S−H	83.0	344.5	yes
S−N	115.2	248.6	yes
S−O	119.0	240.3	yes

From Technological Data of Chiyoda Kohan Co. Ltd.

substances effectively by oxidizing them with UV irradiation to induce ionization and adsorbing the organic acid generated to ion-exchange resins. In other words, UV irradiation in this instance may be just strong enough to ionize most organic substances. Although the ion exchange following the UV irradiation requires only anionic resins, a mixed-bed polisher may also be used for this purpose.

Figure 4 shows the reduction in TOC in ultrapure water by the second method [7]. Ultrapure water of approximately 20 ppb TOC concentration was treated by a UV irradiator equipped with a 0.1 kW low-pressure mercury lamp specially developed for organic decomposition, and then the treated water was fed to a mixed-bed ion exchanger at SV_{50} while its TOC concentration was monitored. Figure 4 indicates the difference in TOC concentration before and after the mercury lamp was turned on. The TOC in the raw water of this system or the TOC in the treated water before the mercury lamp was turned

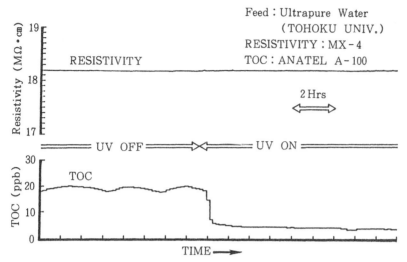

Figure 4   TOC removal using a low-pressure mercury lamp (UV + cartridge polisher). (From Ref. 1.)

on fluctuated around 20 ppb. When the lamp was turned on, this decreased to 5–6 ppb immediately after residence in housing and piping and became stable at this level. It is obvious that both improvement and stabilization in water quality were achieved simultaneously.

In applying this method to the removal of organic substances in ultrapure water, UV exposure amounting to 3–4 kW m^{-3} of treated water is required, and the resins in the subsequent ion exchanger should elute less.

IV.   TOC REMOVAL BY HIGH-PRESSURE MERCURY LAMP

Ultrapure water used in LSI manufacturing is discharged as rinse waste. If rinse waste is separated into waste liquid of high chemical concentration and wastewater of low contamination, including the final rinse water, it is possible to recover less contaminated wastewater for recycling. A wastewater recovery system serves this purpose.

Treatment of recovered rinse waste for the production of ultrapure water is quite different from that of municipal water and other raw water. One of the major differences is TOC removal. Rinse waste contains a larger quantity of low-molecular-weight organic substances, such as alcohol, than raw water, necessitating a special system for their removal. As discussed, since the RO membrane cannot remove organic substances of low molecular weight, photochemical reaction by UV irradiation is applied to their treatment. However, since recovered wastewater occasionally contains TOC in relatively high concentration, amounting to several tens ppm, a high-pressure mercury lamp is used instead of a low-pressure lamp.

A high-pressure mercury lamp with the UV and visible spectrum as shown in Fig. 5 emanates light in a wide range of wavelengths from visible to UV. However, as indicated in Table 7, the light energy of these wavelengths is not strong enough to dissociate the interatomic bonds of organic substances, requiring oxidant as an additional aid. Although hydrogen peroxide, chlorine, and ozone are available as effective oxidants, because the system treats the recovery water that will be used as raw water for the production of ultrapure water, the oxidant to be used must be a substance that neither increases the ion

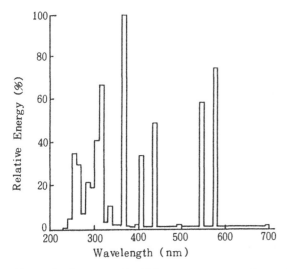

Figure 5 Spectrum of a high-pressure mercury lamp. (Nippon Photo Science Co. type AV-2, technical data.)

concentration nor corrodes the system components. Hydrogen peroxide is the optimum oxidant in this sense. UV irradiation is generally conducted under conditions that allow the coexistence of hydrogen peroxide and organic substances.

The UV oxidation system is composed of a reaction chamber (water tank) in which high-pressure mercury lamps are arranged at regular intervals. After the quantity of hydrogen peroxide required for organic decomposition is injected into the chamber, UV irradiation is applied for a specified time. Aeration stirs up water in the chamber. Figure 6 shows an exemplary composition of the system.

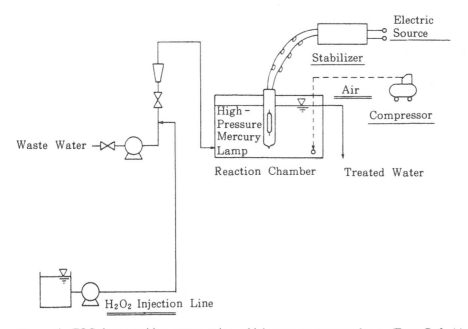

Figure 6 TOC decomposition system using a high-pressure mercury lamp. (From Ref. 4.)

If the organic substance to be decomposed is methanol, the reaction formula is

$$3H_2O_2 \rightarrow 3O + 2H_2O$$
$$\underline{CH_3OH + 3O \rightarrow CO_2 + 2H_2O}$$

$$CH_3OH + 2H_2O_2 \rightarrow CO_2 + 5H_2O$$

The theoretical consumption of hydrogen peroxide is calculated as 3.2 ppm for 1 ppm methanol, but in actual treatment a slightly larger quantity of hydrogen peroxide is necessary.

It has been confirmed that TOC cannot be reduced by either injection of hydrogen peroxide into water containing TOC or UV irradiation of the water without the existence of hydrogen peroxide; that is, both UV irradiation and injection of hydrogen peroxide are required for TOC reduction [7]. TOC removal requires the synergistic effect of UV irradiation and hydrogen peroxide by the following mechanism. UV irradiation encourages the decomposition of hydrogen peroxide, and the highly active hydroxide radicals formed in the decomposition process oxidize organic substances. With UV irradiating the water containing TOC after injection of hydrogen peroxide, the hydrogen ion concentration (pH) decreases first and begins to increase again along with the oxidation, corresponding to the formation of organic acid by the oxidation of organic substances and the reduction of organic acid by further oxidation.

When the water containing TOC is treated by this method, the running costs of the system are mainly for electrical consumption by the mercury lamps. The higher organic concentration requires a longer UV exposure, increasing electrical power rates. From the viewpoint of running costs, the organic concentration applicable to UV oxidation treatment is governed by the kinds of organic substances and their rejection rates. For instance, in the treatment of raw water containing methanol and 2-propanol, TOC decomposition amount per unit electrical consumption changes depending on the TOC concentration of the treated water, as shown in Fig. 7. Figure 7 suggests that the TOC removability of a high-pressure mercury lamp decreases along with the reduction in the TOC concentration of the treated water and that the relation between the TOC concentration of the treated water and the TOC removability by the mercury lamp differs according to the type of organic substance to be decomposed. When methanol is the organic substance to be decomposed and the expected TOC concentration of the treated water is approximately 0.1 ppm, approximately 0.5 ppm is the acceptable upper limit of TOC concentration in the raw water [4].

In designing a system that removes TOC by the method discussed here, it is necessary to clarify beforehand by experiment the relation between the TOC concentration of the treated water and the TOC decomposition amount against the raw water to be fed to the system, by which the quality of the treated water and the operating cost of the system can be estimated precisely.

V.   CONCLUSION

A higher purity of ultrapure water will be required in the future. Since the technology of removing impurities has made rapid progress, more effective technology is expected to be developed. However, developments in removal technology alone are not enough for improvement in the quality of practical ultrapure water. It is also essential to develop

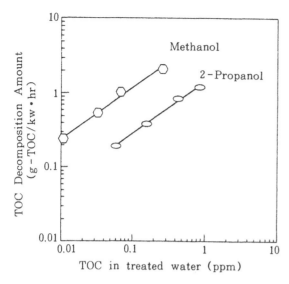

Figure 7 Relationship between TOC decomposition amount and TOC concentration. (From Ref. 4.)

materials and technology that prevent recontamination of ultrapure water by elution from the components. Combinations of all these technologies will result in ultrapure water of high purity.

REFERENCES

1. T. Mizuniwa, K. Yabe, T. Ohmi, et al., Proceedings of Seventh Symposium on ULSI Ultraclean Technology, Tokyo (1988), pp. 27–50.
2. P. A. McConnelee, S. J. Porier, and R. Hanselka, Proceedings Fifth Annual Semiconductor Pure Water Conference, San Francisco, (1986), p. 219.
3. K. Yabe, Water Technol. (Journal of Water Re-use Technology), *14* (2), 21–25 (1988).
4. K. Yabe (Suzuki, Sato, and Hashimoto, eds.), Practical Clean Technology for LSI Factories, Science Forum, Inc., pp. 353–369.
5. T. Uemura and M. Kurihara (compiled by T. Nakagawa): Latest Separation Membrane, CMC, Inc., p. 101.
6. Y. Hirota and Y. Motomura, Proceedings Sixth Annual Semiconductor Pure Water Conference, Santa Clara, CA, (1987), pp. 159–177.
7. S. Naruto and K. Ono, Comprehensive Technical Information of Clean Environment in Semiconductor Factories, Science Forum, Inc., p. 279.

7
Static Electricity Removal Method

Takeo Makabe
Nomura Micro Science Co., Ltd., Kanagawa, Japan

I. INTRODUCTION

In dry winter weather, we encounter shock from static electricity when we remove clothing or touch the handle of a car door. Such electrostatic trouble from electrostatic discharge is also a problem in the manufacture of semiconductors. That is, the breakdown of large-scale integrated (LSI) circuits is caused by trouble that affects the oxide or polycrystalline films of the devices and deteriorates their characteristics.

Consequently, in the clean room of a semiconductor plant, various countermeasures against static electricity are considered for clothes, shoes, and workers' desks and jigs, including interior materials. For example, in the clean room of the Telecommunication Laboratory of Tohoku University, as the result of various countermeasures against static electrical charge, the charge potential is said to be restrained below 30 V [1].

At the beginning of IC manufacturing in the first half of the 1970s, even in the United States, concern about the problems induced by static electricity was barely aroused and only a few specialists recognized the problem. It is said that even in the second half of the 1970s, the degree of the recognition was about 15%. Finally, in the mid-1980s, the degree of recognition reached 80% [2]. Presumably it was because of the difficulty of understanding the phenomenon of examining the problem quantitatively that trouble related to static electricity was not solved. With the rapid escalation of LSI, when the epoch of the megabit began and a pattern dimension in 4M dynamic random access memory (DRAM) reached submicrometer levels, measures to counteract problems due to static electricity have become increasingly important.

The electrical charge phenomena of insulating liquids have long been known. In the petrochemical industries, these phenomena cause accidents, such as explosions and fires. Because the pure water and ultrapure water used in large quantities in wafer cleaning during LSI manufacturing are also insulating liquids, problems induced by static electricity as a result of charge due to contact with solids have arisen.

II. STATIC ELECTRICITY

Static electricity is generally thought to generate more on insulators, such as plastics, than on conductors, such as metals, but on the contrary, static electricity is more readily generated on materials of higher conductivities. In a conductor, because the static electricity generated is not stored in definite portions but soon spreads over the entire surface, it appears not to be generated. On the other hand, in an insulator, because the amount of static electricity generated is small and the charge cannot move but is fixed on the contact surface and stored, it is incorrectly believed that static electricity is readily generated on insulating materials.

Nakamura's intelligible description [3] of the mechanism by which static electricity is generated is followed here.

Strictly speaking all matter has irregularities on its surface. When two electrical insulating materials are rubbed together, heat is generated. The temperature can reach as high as over a thousand degrees. An electrical zone that is ordinarily distributed on the surface of the material, which emits lines of electrical force to the inside or to the surface, begins to adsorb energy and emit lines of electrical force to the surface. The number of surface electrons also begins to increase.

As long as the two materials in contact are not drawn apart, the lines of electrical force that are generated and the increased electrons are held in common between the two materials. They are not discharged and do not adsorb dust. If the materials are separated, the density of the lines of force increases and the electrons are in an excited state in proportion to the distance of separation. When a limit is reached, a discharge phenomenon begins in which electrons are emitted to the unexcited ions in the air. At that moment, a wave of electricity and magnetism appears as a pulse for one several billionth of a second.

The voltage of undischarged, remaining electrons further increases and the electrons become stable by neutralization with the ions in the air. Despite this stable state, however, it is capable of being discharged when approached by another conductor. Such a state is charged or static electricity. It is not clear whether the electricity generated becomes charged. In an insulator, transfer of the electricity generated is slow, the charged state can be held for long periods of time, and thus dynamic problems, such as adsorption of dust, occur. For example, even metallic conductors when held in an insulated state and rubbed continuously store large amounts of electricity and, if this is discharged, lead to the ignition of flammable gas or the malfunction of electronic instruments.

III. ELECTROSTATIC DISCHARGE FAILURE IN SEMICONDUCTOR DEVICES

The resistance of LSI to electrostatic discharge failures can be classified into the following four models by actual breakdown phenomena [4].

1. Human body model. When a human provided with static electricity comes into contact with an LSI device, discharge breakdown occurs in the LSI device. When an electrical charge stored in a condenser of 100 pF with a capacity equivalent to that of the human body is applied to an LSI surface, a resistance of 1.5 kΩ, equivalent to that of a human body, is generated. The discharge constant CR described here is from MILSID 883, IEC 47(CO)955. In an EIAJ standard often used in Japan, when examined at relatively low voltage, C is 200 pF and R is 0 Ω.

2. Charged package model. When an LSI slides in the lane or magazine of an automated handler, a charge is stored between the package and the chip, due to rubbing of the package. Thus the LSI terminal is contacted with metal and broken down.
3. Charged device model. A charge is stored between a chip and the ground via the LSI terminal as a result of handling, for example. The LSI terminal is broken down on contact with a high-capacity metal.
4. Electrical field-induced model. When an LSI is placed in a high electrical field, the conductive part of the LSI acts as an antenna, breaking the oxide film. A method of examining this phenomenon has not yet been defined.

The electrostatic discharge failure of an LSI usually occurs by human body contact. The amount of charge in the human body is important, but the problem is that the human body is a conductor and also a transfer charged body. Because of the lack of perception of the problem, it arises at unexpected times and places, and thus a part that is discharged cannot be specified. A method of preventing human body discharge is described in the following example from a semiconductor plant.

Fiber flocs adhered to a lead after goods were returned from a user. A method of preventing static electricity was developed as a final test process, considering especially discharge from the human body. The conventional method was reexamined and a conclusion [5] was made based on the results of examination of charge amounts due to compulsory charge and operations using a conductive floor, a conductive mat, conductive shoes, and the ground as parameters.

Table 1 [6] shows the (somewhat outdated) aged range of static electricity voltage at which various semiconductor devices may be broken down. Examples of devices that can be broken down by static electricity are shown in Figs. 1 and 2 [2].

The breakdown as a result of static electricity of an LSI chip can occur in three different ways [7].

A. Dielectric Breakdown

In this phenomenon an oxide or nitride film on a chip is broken by an electrical field. The metal short circuits and a diffusion layer arises by breakdown of a capacitor insulating film or gate oxide film in a MOS-FET. Some reports have been published on the

Table 1 Static Susceptibility of Semiconductor Devices

DEVICE TYPE	RANGE OF SUSCEPTIBILITY (VOLTS)
MOS/FET	100 – 200
J-FET	140 – 10,000
CMOS	250 – 2,000
SCHOTTKY DIODE	300 – 2,500
SCHOTTKY TTL	1,000 – 2,500
BIPOLAR TRANSISTORS	380 – 7,000
ECL HYBRID (PC BOARD LEVEL)	500
SCR	680 – 1,000

Source: From Ref. 6.

Figure 1   Magnified view of damaged MOS capacitor on an operational amplifier (X 4300). (From Ref. 2. Photo by Sumitomo 3M Ltd.)

insulating resistance of SiO_2. One example is shown in Fig. 3 [7]. Figure 3 indicates that the insulating resistance of an oxide film increases rapidly below 40 nm film thickness. With reduction of the oxide film and as the electrical field intensity increases proportionally, the breakdown voltage itself decreases with thinning of the oxide film. With improvements in the degree of integration, devices have become increasingly finer and the electrical field intensity applied to a gate oxide film of a DRAM has increased according to this trend. This is shown in Fig. 4 [4].

B.   Breakdown of the Wiring Pattern

Metal melts above its melting point by heat (measured in Joules). In a metal, the breakdown voltage is a function of current density. If a pattern bends or narrows rapidly, there is a point of increasing current density and heat is generated.

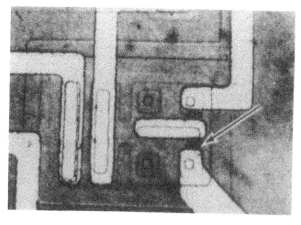

Figure 2   Base emitter short on a bipolar gate. (From Ref. 6. Photo by Sumitomo 3M Ltd.)

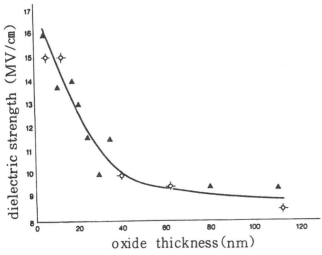

Figure 3 Dielectric strength of thermally grown SiO_2 over silicon. (From Ref. 3.)

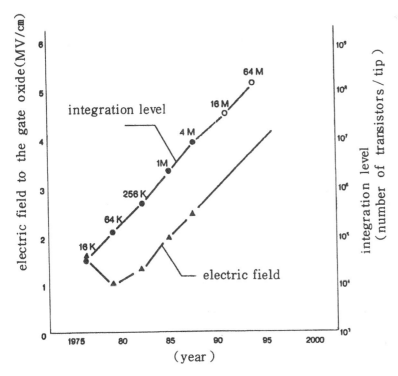

Figure 4 Acceleration in the direction where electrostatic destruction is liable to occur. (From Ref. 4.)

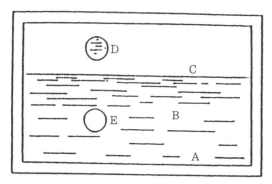

A solid and solid-liquid interface
B bulk
C interface to air
D liquid particle
E bubble

Figure 5 Water and interfaces. (From Ref. 9.)

C. Breakdown of the PN Junction

Sometimes Si melts or the contacted metal passes through a diffusion layer to a substrate. To prevent such static electrical breakdown, a protective circuit that diverts a transient current to a suitable site on a chip can be installed, but in the megabit epoch with minimization of design rules to 1.2 μm, or even to 0.8 μm, a protection circuit alone is not a sufficient solution, and measures related to elemental structures or processes may be required [8].

IV. LIQUID CHARGE

The description of liquid charge by Asano of Yamagata University [9] is discussed here.

The excess electrical charge of a liquid has either a positive or a negative sign. Liquid itself has no shape. The "shape" of a liquid is contributed by the vessel in which it is contained. A gas fills the vessel above the liquid. This means that there is an interface between the solid (the vessel) and the liquid and an interface between the gas and the liquid at the liquid surface. Figure 5 illustrates this concept. An excess electrical charge in the liquid (Fig. 5,B) and the electrical charge at the interface with the gas (C) are different. The interface between the gas and the liquid where liquid particles are dispersed in the gas (D), and the interface between the gas and the liquid where bubbles exist in the liquid (E) also play important roles. The interface between the solid and the liquid (A) is important in streaming charge. For actual problems, liquid charge—bulk particle—must be considered separately.

A. Relaxation of Electrical Charge

In considering an electrical charge in liquids, the relaxation time of the charge is important. The electrical charge density in a liquid changes with the passage of time, as described by the following equations.

$$\rho = \rho_0 e^{(-t/\tau)} \tag{1}$$

$$\tau = \frac{\epsilon}{\kappa} \tag{2}$$

$$\epsilon = \epsilon_r \epsilon_0 \tag{3}$$

where: ρ_0 = initial electrical charge density (cm^{-3})
ϵ = dielectric constant (Fm^{-1})
τ = relaxation time constant (s)
κ = conductivity (Sm^{-1})
ϵ_r = specific dielectric constant of a substance
ϵ_0 = dielectric constant in vacuum, 8.854×10^{-12} (Fm^{-1})

For ordinary insulating liquids, the specific dielectric constants are about 2–5, but water is peculiar, with a dielectric constant of about 80. Thus, dielectric constants change only by one order of magnitude according to the substance, but electrical conductivities can change by 20 orders of magnitude. Consequently, the relaxation time constant τ may safely be said to be determined by conductivity alone.

As Eq. (1) shows, the electrical charge in bulk liquid decreases over time and finally appears on the interface. In a liquid with a very low conductivity, a relaxation time constant may be in the range of minutes to hours. The electrical charge usually encountered in a bulk liquid is in the process of relaxation of the liquid charge.

B. Bulk Charge

In bulk liquid, positive and negative ions arise by electrical dissociation of the molecule—water—and electrical dissociation of contaminants. At the interface with a solid, electrical charges of one sign are selectively adsorbed on the surface of the solid by the difference in electrochemical potential. As shown in Fig. 6, electrical charges of the other sign are diffused and distributed in the liquid. This is an *interfacial electrical double layer* [10] because it is similar to the electrical charge model describing two electrode plates first developed by Helmholtz. In the Helmholtz model, a kind of capacitor is formed, and the capacitor space δ is represented by

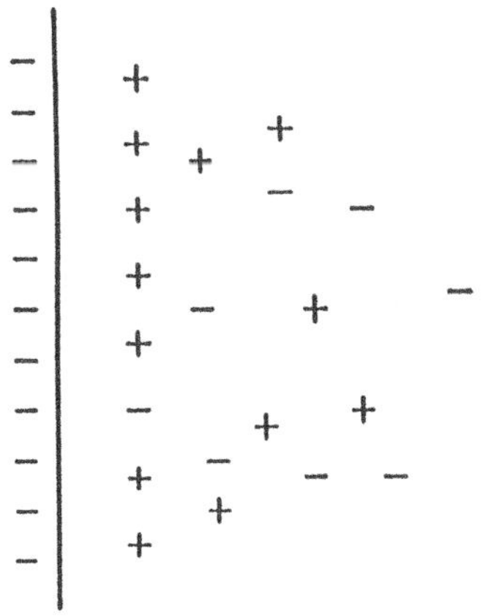

Figure 6 Electrical double layer. (From Ref. 9.)

$$\delta = (D\tau)^{\frac{1}{2}} \tag{4}$$

where D is a diffusion constant. The values of D barely change by substance and thus largely depend on a relaxation time constant.

In a pipe, liquid flowing from the inside of an interfacial electrical double layer at the wall of the pipe has a net electrical charge, and such a liquid is said to be charged because the liquid carries away only the electrical charge from the inside of the pipe. This charge mechanism is the *streaming charge* (11).

C. Particle Charge

The interface between a gas and a liquid is held in place by surface tension, and atomization is possible by applying external force above the surface. Figure 7 shows that at the tip of a nozzle a water column is split to form a drop of water only when the force of gravity overcomes the surface tension.

In atomization, the split is performed by applying an external force or pressure to a liquid. The charge of the liquid split is also called the split charge. The electrical charge in bulk liquid is attenuated according the Eq. (1), and here the relaxation time constant is estimated. The conductivity of ultrapure water, if assumed as a theoretical limit value, is 5.48×10^{-6} Sm^{-1}, so

$$\tau = \frac{80 \times 8.85 \times 10^{-12}}{5.48 \times 10^{-6}} = 1.29 \times 10^{-4} \text{ s}$$

is a very small value, which cannot be compared with the seconds or minutes calculated for hydrocarbons. Even in ultrapure water, the bulk electrical charge is also rapidly relaxed at the interface. If charged liquid particles are dispersed in the air, however, the electrical charge of the particles in the bulk liquid moves onto the surfaces of the particle and must subsequently move through air for further relaxation. The conductivity of a gas, such as air, is extremely low, however, so in fact the electrical charge can be presumed to be held at the interface between the gas and the liquid and the charged particle mass creates an electrically charged space.

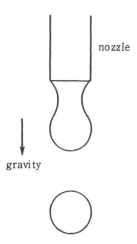

Figure 7 Split of water column from nozzle. (From Ref. 9.)

D. Spray Charge of Water

When a stream of water is split by applying external mechanical force, such as gravity, air, or centrifugal force, to a bulk liquid, fine particles of split liquid droplets are known to be charged. Because the charge mechanism is extremely complicated, however, fundamental research, such as atomization of a definite frequency applied by an artificial, compulsive vibration and charge applied by an outer electrical field, has been performed.

V. ELECTRICAL CHARGE OF ULTRAPURE WATER

With increasing integration of a semiconductor, the quality required for ultrapure water has become increasingly stringent and the specific resistivity is required to be above 18 MΩ· cm, near the present theoretical value. In the dicing, scrubber cleaning, or spinner cleaning processes of a semiconductor manufacturing plant, as ultrapure water with such high resistivity is jetted onto the surfaces of wafers by a high-pressure jet, the ultrapure water is charged. As a result, problems due to static electricity have arisen.

Semiconductor manufacturers are thought to consider countermeasures on the basis of data obtained by various in-house experiments, but few data have been published. On the basis of data measured by Hitachi Yonezawa Denshi, Ltd., shown in the Figs. 8, 9, and 10, the charge mechanism estimated by Asano of Yamagata University [9] is cited here.

Figure 8 shows potentials measured near the tip of a nozzle during cleaning by blowing pure water onto the silicon wafers. Figure 9 shows the potential on the wafers

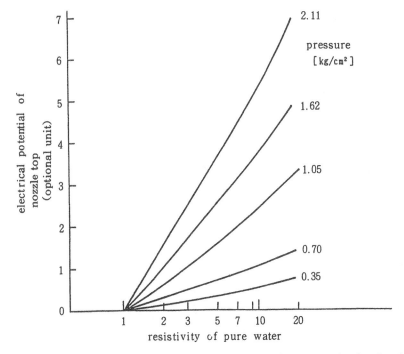

Figure 8 Resistivity and the rate of dependence on the pressure for the electrical potential of a nozzle top. (From Ref. 9.)

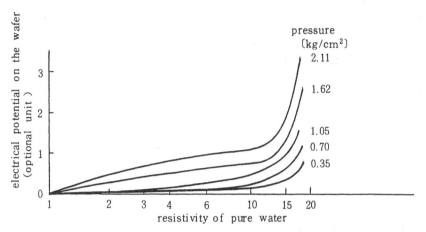

Figure 9   Electrical potential on the wafer. (From Ref. 9.)

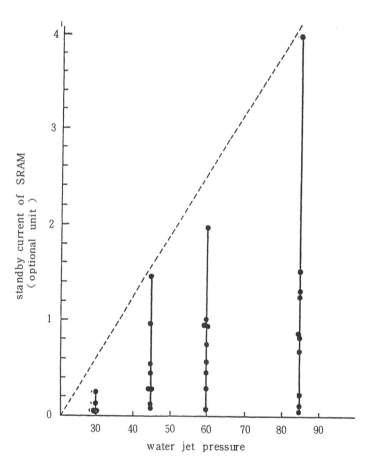

Figure 10   Influence of charge of SRAM standby current. (From Ref. 9.)

immediately after cleaning. Figure 10 shows how the product characteristics are influenced by the charge from jet cleaning. Figures 8 and 9 indicate that the pure water charge increases with the rise in resistivity of pure water and with the increase in the flow rate at the nozzle.

A. Charge Mechanism

Because there are few published data, the concept of charge is difficult to grasp, but an analogy may be drawn to the data relating to insulating liquids, as follows.

The data in Fig. 9 use pressures applied to pure water as parameters. The actual flow rate is unclear, but the mechanism appears to depend on the flow rate because the flow rate increases with the increase in pressure.

Figure 11 is an example measured in the laboratory of Asano et al., where the relation between flow rates and current values was described for jetting of pure water from a 350 μm nozzle and the current was proportional to about two powers of the flow rate. Here, the conductivity was 10^{-5}–10^{-4} Sm^{-1}, with considerable dispersion.

During the jetting of pure water from a nozzle, an electrical double layer is formed in the nozzle and a streaming charge is presumed to be created. In this way, liquid jetted from a nozzle acquires an electrical charge and the charge amount depends greatly on the conductivity of the liquid.

Figure 9 shows the charge mode in ultrapure water. It is especially interesting that charge amounts rapidly increase with conductivities approaching the theoretical limit value. For example, at a resistivity of 18 MΩ·cm, the relaxation time increases 1.2 times

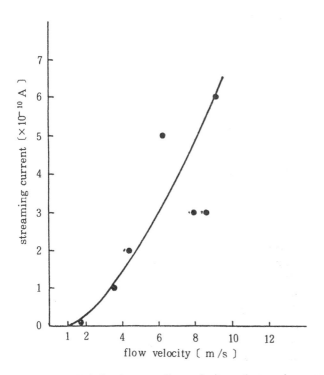

Figure 11 Relation between flow velocity and streaming current. (From Ref. 9.)

that in Eq. (2) compared with the relaxation time at a resistivity of 15 MΩ·cm, and the electrical charge increases 2–3 times that shown in Fig. 9. As discussed in Ref. 9, the collection of fundamental data in this field is required.

B.   Charge-Protecting Method

The preceding data indicate that a large electrical charge is generated and stored on a wafer during cleaning and influences the breakdown and changes in its characteristics. To protect ultrapure water from electrical charge, it is necessary to

1.   Reduce the resistivity to decrease the amounts of the charge.
2.   Rapidly remove the stored charge.

Method 1, in which ionized materials must be added to ultrapure water, which is obtained by removing ions from raw water to reduce the resistivity, is a contradictory method. Consequently, the selection of materials to be injected must be thoroughly examined. In the past, a rod or mesh of metallic magnesium was added to the ultrapure water line, but this method cannot be used because it changes the properties of the wafer body. Organic materials cannot be used because substances remaining on the wafer are subject to carbonization at high temperatures. At present this means that carbon dioxide, which is soluble in water, available in high purity and at a low cost, and additionally can be controlled by injection amount, can be dissolved in ultrapure water to reduce its resistivity.

C.   Dissolution of Carbon Dioxide

Carbon dioxide is dissolved in ultrapure water and dissociated in the reaction with water:

$$H_2O + CO_2 \rightleftarrows H_2CO_3 \tag{5}$$
$$H_2CO_3 \rightleftarrows H^+ + HCO_3^- \tag{6}$$
$$HCO_3^- \rightleftarrows H^+ + CO_3^{2-} \tag{7}$$

The relations between the carbon dioxide concentration dissolved in ultrapure water and the resistivity and between the resistivity and the pH of the CO_2 gas dissolved in ultrapure water are shown in Figs. 12 and 13, respectively.

The degrees to which resistivity can be controlled by the dissolution of carbon dioxide in ultrapure water are determined on the basis of tests by semiconductor manufacturers according to the kinds of devices and the cleaning processes, among other variables. Around 1 MΩ·cm is used at present.

D.   Ionizer

In Method 2, an ionizer removes the charge from materials by improving the conductivity of the air by generating positive and negative ions in the air. This method is often used as a measure of static electricity in manufacturing electronic parts or circuits.

It is also true that when a wafer is processed in a carrier that has good insulating properties during spin-drying after cleaning, the electrical charge on the wafer is held for a long time. An ionizer is useful in processing wafers in such isolated charge states. During the spin-dry process, contamination due to fine particles accompanied by static electricity has been a problem. IPA vapor drying has been examined as a countermeasure, and the experimental results are reported in Ref. 12.

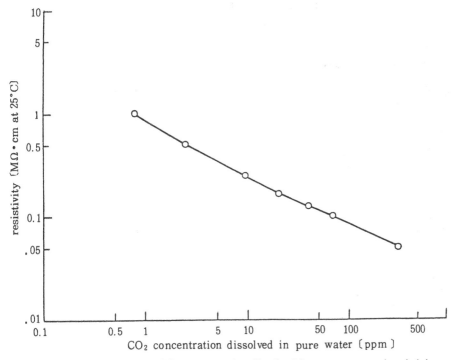

Figure 12   Relation between CO_2 concentration dissolved in pure water and resistivity.

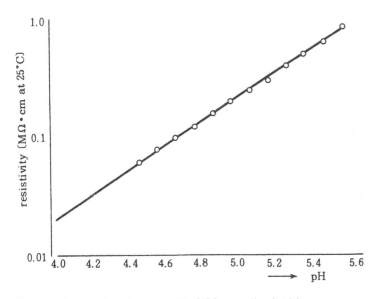

Figure 13   Relations between pH of CO_2 gas dissolved in pure water and resistivity.

VI.   CHARGE-PROTECTING EQUIPMENT FOR ULTRAPURE WATER

Important requirements for equipment to reduce resistivity and prevent charge buildup by dissolving carbon dioxide are as follows (Ulvac Service Co., Ltd. technical information):

1. Avoid including dust in carbon dioxide injected into ultrapure water.
2. Establish a microinjecting system for carbon dioxide because carbon dioxide has a high solubility in ultrapure water.
3. A compact structure enables carbon dioxide injection near the point of use so that the carbon dioxide does not become a nutritive for the proliferation of microorganisms in the ultrapure water.

Equipment actually in use incorporates the following features:

1. In an indirect injection method, carbon dioxide previously dissolved in water at a low resistivity is introduced into the ultrapure water line through a membrane, using a pump.
2. After removing very fine particles, highly purified carbon dioxide is injected directly into the ultrapure water line through a control valve.
3. Carbon dioxide is dissolved in an ultrapure water line through a special hydrophobic membrane.

Modes 2 and 3, which are now used most extensively, are explained here in detail.

A.   Mode for Direct Injection of Carbon Dioxide

The system is shown in Fig. 14. The CO_2 gas is completely gasified through a regulator provided with a heater from a carbon dioxide bomb, and fine particles are removed by a gas line filter microadjusted by a two-stage manual adjustment valve. The injection of CO_2 gas is performed by solenoid valve in response to signals from a resistivity meter. To dissolve the injected carbon dioxide rapidly, a further filter for mixing is installed.

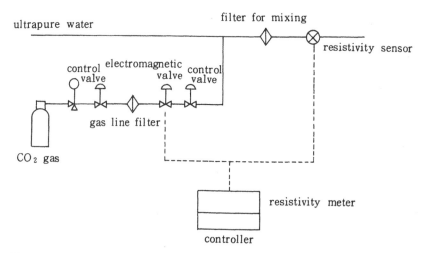

Figure 14   Direct injection of CO_2 gas.

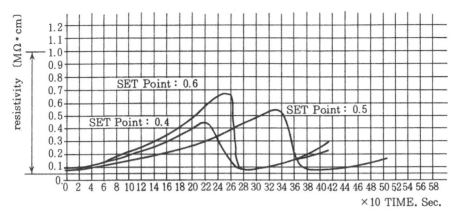

Figure 15 Example of resistivity control.

A method to control the resistivity value is based on feedback of the difference between a preset resistivity value and a value measured by a resistivity meter. The resistivity and the opening and closing times of a solenoid valve are set, and when a measured resistivity value is higher than the set value, the solenoid valve is switched according to a set time to control the amount of carbon dioxide gas with pulse injection. An example of this control method is shown in Fig. 15. The numeric values of the set point—0.4, 0.5, and 0.6—shown in the figure represent set resistivity values. The control resistivity width in this method is as wide as 0.05–1 $M\Omega \cdot cm$ and thus the control performance is inferior. In the dicing process described in the next section, rapid damage to the blades at pH values less than 4.0 has been a particular problem, but some still use the method because of its low cost.

B. Improved Mode for Direct Injection of Carbon Dioxide

The equipment flow is shown in Fig. 16. A method of removing fine particles and a method to increase the dissolution rate after injection are similar to the methods described in Sec. VI. A, but a method of controlling resistivity and a precision valve for injecting carbon dioxide to the ultrapure water line demand ingenuity.

The equipment control system comprises the following three loops.

1. Carbon dioxide injecting loop. The amount of carbon dioxide to be injected into the ultrapure water line can be adjusted by adjusting the amount of CO_2 flow by flowmeter and opening a precision control valve regulated by a controller in response to signals from a microcomputer.
2. Calculating loop for amounts of ultrapure water and carbon dioxide. The amount of ultrapure water flow and the differential pressures of carbon dioxide and ultrapure water in the line are detected by sensors and the amount of carbon dioxide required for a set resistivity value is calculated by a microcomputer. The value is then transmitted to a carbon dioxide controller.
3. Resistivity meter feedback. The difference between the set and the measured resistivity valves is calculated for correction by microcomputer, and the value is transmitted to a carbon dioxide controller.

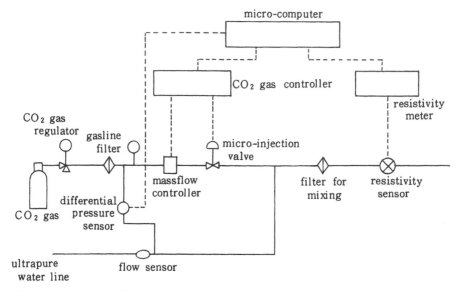

Figure 16 Improved direct injection of CO$_2$ gas.

Feedforward control is carried out based on the ultrapure water flow amount and the differential pressures between the carbon dioxide and the ultrapure water line. Feedback control is based on the differences in resistivity. As shown in Fig. 17, when there is no variation in amount of flow, a control performance as high as ±0.02 MΩ·cm compared to the set resistivity of 1 MΩ·cm can be obtained. A precision injection valve uses a pulse motor so that a rapid, precise response is possible to vary the flow of the ultrapure water. Furthermore, when a diaphragm valve is used, which has a structure that discourages contamination by fine particles, the response to variations in the amount of the ultrapure water is improved by making the amount contained in the valve secondary by integrating the valve body with an injection nozzle. Consequently, as shown in Fig. 17, if the amount

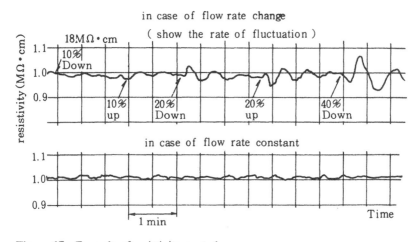

Figure 17 Example of resistivity control.

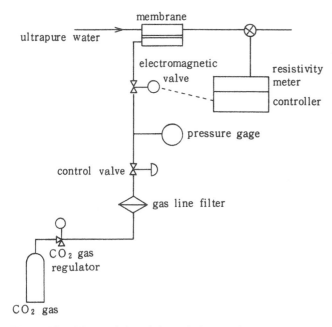

Figure 18 CO_2 gas injected through the membrane.

of flow is changed by 40%, the variation is absorbed within only a few minutes and in this amount of time the deflection width of the resistivity value is only ± 0.08 MΩ·cm from the set value of 1 MΩ·cm.

C.   Method for Dissolving Carbon Dioxide Gas Through a Special Membrane

The equipment flow is shown in Fig. 18. Carbon dioxide gas is fed to a special hydrophobic membrane with a 0.007 μm pore size through a solenoid valve. Fine particles are removed from the carbon dioxide gas by a gas line filter and the amount of fine particles is controlled by a manual adjustment valve. Control is performed by microprocessor by considering the difference between the set and measured resistivity values, the characteristics of the membrane, and other factors, and the high level of control as shown in Fig. 19 has been achieved.

VII.   APPLICATION IN WAFER PROCESSING

As described in Sec. V, in wafer treatment, dicing and scrubbing by high-pressure water jet are briefly described here.

A.   Dicing Process

During wafer treatment, dicing is a finishing process. A patterned wafer processed during pretreatment is cut with an extremely thin grinding wheel (edge thickness of about 20–40 μm) and the LSI circuit is transformed into chips. If during this period, the wafer is

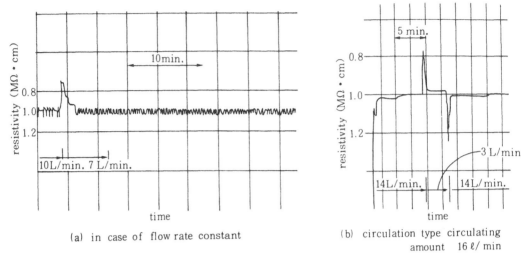

(a) in case of flow rate constant   (b) circulation type circulating
                                             amount   16 ℓ/ min

Figure 19   Example of resistivity control.

damaged by static electricity, the pretreatment operations are rendered fruitless and the loss is enormous.

The resistivity values for the ultrapure water used for removing grinding debris and other contaminants from the dicing machine differ somewhat according to the semiconductor manufacture and the application of the devices. The control value can be determined in the range 0.5–1.0 MΩ·cm. As described earlier, these range values are seldom published by manufacturers. Here, data [4,13] published in June 1989 by Hitachi Seisakusho, Inc. are introduced.

The Hitachi data indicate that if cleaning conditions in dicing are properly selected, variations in the threshold voltage (V) of a metal oxide semiconductor (MOS) transistor and damage to the gate oxide film can be prevented. Furthermore, a mechanism by which the electrical charge contributes to the breakdown of an oxide film was explained on the basis of recovery of the variation in the voltage by ultraviolet (UV).

In this experiment, four parameters of dicing—the form of the nozzle, the jet washing pressure, the resistivity of the pure water, and the cutting direction of the wafer—were adopted, and when the amount of voltage before and after jet washing was above 0.1 V with changes in these conditions, the case was defined as inferior and the generation ratio was determined. The results are shown in Table 2. The optimum conditions are as follows:

1. The resistivity of pure water should be below 1 MΩ·cm. The relation between the electrical potential and the resistivity of pure water is shown in Fig. 20.
    In mass production plants at present, the resistivity is controlled at 0.7 MΩ·cm.
2. A radial jet nozzle is used. The relation between washing water pressure and the rate of generation of inferior voltage θ is shown in Fig. 21. The jet nozzles are shown in Fig. 22.

In addition to the cleaning pressure of the distributing water pipe, the pressures on the wafer produced by the different jet methods were examined. Under 40 kg cm^{-2} of cleaning pressure, using a direct jet, the pressure on the wafer is equal to the cleaning

Table 2 Dependence of Voltage Shift Generation Rates of MOS Transistor on Conditions of Jet Washing

No.	jet type	jet washing pressure (kg/cm²)	resistivity of pure water (MΩ)	wafer cut mode	Vth shift generation rate (%)
		Note 1)		Note 2)	Note 3)
1	direct – jet	90	1.25	return	60
2	↑	50	↑	↑	50
3	radial – jet	↑	0.8	↑	10
4	↑	↑	0.7	↑	10
5	↑	↑	0.5	↑	10
6	↑	40	0.7	↑	0
7	↑	50	↑	down	0

Note 1) show **Fig. 22**
Note 2) show **Fig. 23**
Note 3) Regarding 10 MOS transistors on the 1 wafer per 1 condition, the generation rate of the Vth fluctuation before and after washing exceeds 0.1 V

Source: From Ref. 13.

pressure. If a radial nozzle is used, the pressure on the wafer is extremely low, as little as 0.05 kg cm^{-2}. In addition, the water pressure on the wafer surface is related to the charge potential. At 0.05 kg cm^{-2} of pressure on a surface, the charge potential is about 0 V.

When the cutting direction of the wafer is set in one direction, the charge potential is lower. The cut modes for the wafer are shown in Fig. 23.

A one-direction mode is said to be superior to a reciprocating mode for improving production efficiency. It is presumed that the transfer time for a cutter to begin to cut the next line in a one-direction mode is longer than that in a two-direction mode, and at that time the static electricity generated is discharged.

The resistivity of less than 1 MΩ·cm in this report may be the first value published by device makers.

To prolong the life of the grinding stone used in the dicing process, the resistance must be controlled in specific ranges. Makers seem to recommend 0.5–1.0 MΩ·cm. An extremely thin grinding wheel uses a Ni adhesive to fix diamond powder and is subjected to an electrodeposition process. Ni is melted from the adhesive with decreasing pH, and the life of a blade is shortened. A manufacturer's test results show that below pH 4.5 (about 0.08 MΩ·cm from Figs. 12 and 13), the influence becomes clear. These data are from an immersion test. Because in an actual operation high-pressure water is jetted onto a blade and ground dust is also present, this influence is anticipated to occur above pH 4.5 and so the manufacturer may recommend 0.5–1.0 MΩ·cm.

In actual operation, plural dicing machines are generally used and the amounts of pure water differ between grinding and when dicing machines are on standby. To improve

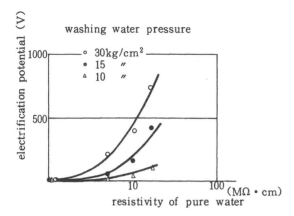

(a)   relations between resistivity of pure
water and electrification potental

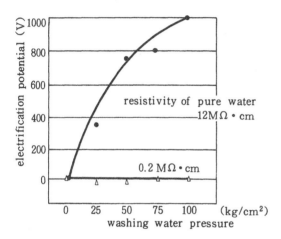

(b)   relations between washing water
pressure and electrification   potential

Figure 20   Optimum conditions. (From Ref. 4.)

the response to this variation in flow amount, to extend the life of a blade, and to reduce
running costs are important goals for charge-protecting equipment.

B.   Scrubber Cleaning Process

A method with a high-pressure water jet is also used in a scrubber used for removing burrs
or dust stuck to a wafer surface after etching. Before this charge-protecting equipment was
used, a method for circulating ultrapure water used in jetting in a system to reduce its
resistivity, for example, may have been designed. According to this method, contamination by particles or microorganisms was a problem. In this cleaning machine, the range of

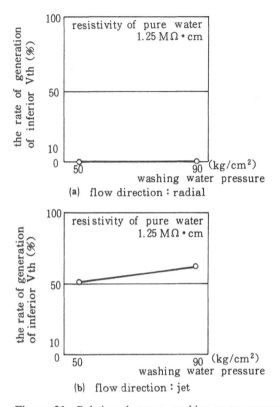

(a)   flow direction : radial

(b)   flow direction : jet

Figure 21   Relations between washing water pressure and the rate of generation of inferior voltage. (From Ref. 4.)

flow during use is as great as around 1 h^{-1} to m^3 h^{-1}. Thus equipment that enables precise control of microflow amount and excellent response to variation in flow is required.

Data for a machine said to be used in other cleaning processes are expected to be published in detail by device manufacturers.

VIII.   CONCLUSION

Charge-protecting equipment to prevent the electrical charge of ultrapure water, to protect a semiconductor device from breakdown due to electrical charge, and to improve the

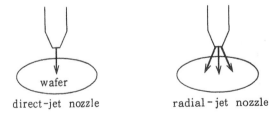

direct-jet nozzle   radial-jet nozzle

Figure 22   Nozzle types. (From Ref. 13.)

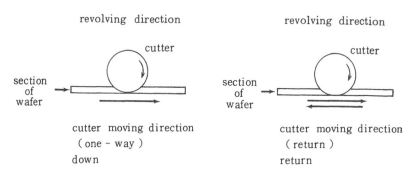

Figure 23   Wafer cutting method. (From Ref. 13.)

yield must be developed in the future for semiconductors integrated to 4–16 Mbit, as follows:

1.   A control system should be established that corresponds to wide variations in flow amount.
2.   The material and structure of the CO_2 gas line should be reexamined.
3.   The smoothness of the inner face of the CO_2 gas line should be improved.

With further improvement in the degree of integration, fine particle diameters of 0.05–0.02 μm are required. Thus cleaning for fine particles stuck to a wafer has become increasingly difficult.

Not only measures to counter the breakdown due to electrical charge and to prevent sticking of fine particles but also the presence of a fine particle-desorbing effect seem to have begun to be examined by developers of this charge-protecting equipment. For this purpose, it will be necessary to develop a technique for confirming the effect of cleaning on the wafer surface by controlling particle size. The technique will also be expected to be used as a cleaning countermeasure for fine particles.

REFERENCES

1.   F. Ohmi, Clean room from which dust is removed by fundamentally reviewing from pipings to dust-proof wears, Nikkei Micro Device, September, 92–95. (1986).
2.   M. Chuck, Prevent breakdown of LSI and adsorption of particles by controlling static electricity, Nikkei Micro Device, October, 89–93. (1983).
3.   Nakamura: Increasing demand for preventive technique of static electricity, Nikkan Kogyo Shimbun, (8), December 11 (1987).
4.   Mizuno, Hidaka, Ikuzaki, Quantitating operational standard and preventing static electricity in assembly process, Nikkei Micro Device, October, 97–103. (1989).
5.   Akabori, CATS exist in Mitsubishi Electric Co., Diamond Co., June, 108–116. (1989).
6.   F. Donald, ESD Considerations for Electronic Manufacturing, American Society of Manufacturing Engineers Westec Conference, March 1983.
7.   D. Hughes, Increasing importance of the countermeasures against electrostatic breakdown, Nikkei Micro Device, November, 131–138. (1986).
8.   The Countermeasures against 0.8–1.2μm static electricity shifted its focus to elementary structure process, Nikkei Micro Device, October, 103–111. (1988).
9.   Asano, Atomization, contact and streaming charge of pure water and ultrapure water and their control, Electrical Society, No. 4, Vol. D108 (1988).

10. Electrostatic Society, Handbook of Static Electricity, Chap. 5, Ohm Co., 1981.
11. Asano, Magazine of Electrostatic Society, 1, 58 (1977).
12. Ohmi, Nitta: Process technique II for high efficiency, Ultra LSI Ultraclean Technology Symposium No. 5, Realize Inc., July 1987.
13. Mizuno, Hidaka, Ikuzaki, Preventive method for obstacles of electrostatic breakdown in LSI assembly process, Announced Theses on Nineteenth Reliability and Maintenance Symposium.

8
Afterword

Yoshiharu Ohta
Nomura Micro Science Co., Ltd., Kanagawa, Japan

From the viewpoints of technology and economics, as described in the Introduction, the elementary techniques used in manufacturing ultrapure water have increasingly extended the range of application. Because analytic techniques for contaminants in very small amounts cannot correspond to new developments, elementary techniques are plagued by difficulties. Elementary techniques used in the final stage of an ultrapure water system especially confirm this. Analytic techniques for contaminants in micro amounts are required.

In addition, future elementary techniques will require cleaning methods. Low-eluting and low-dust methods will be needed. Furthermore, physical means, such as ultrasonic wave propagation and magnetic and electrical field generation, not used yet as elementary techniques, may play important roles. If such physical means are possible, they may be applied, for example, to the cleaning of microscopic deep trenches in ULSI circuitry by modifying the physical properties of the ultrapure water.

As long as the degree of ULSI progresses, the maturity of elementary techniques and development of new elementary techniques must continue.

IV
Equipment and Piping Materials

1
Introduction

Takeshi Shinoda
Hitachi Plant Engineering & Construction Co., Ltd., Tokyo, Japan

The recent progress and rapid increase in the integration of semiconductor devices is so remarkable that manufacturers are getting ready for the mass production of the 4 MDRAM silicon chip, the next generation. Moreover, research and development in the 16 MDRAM chip has advanced so much that it is said that the age of the *megabyte* chip has arrived. Accordingly, in the manufacture of semiconductor devices, ever higher standards of purity and cleanliness for all utilities and for the process environment have been attempted, and the requirements for ultrapure water have become more stringent than ever before.

To satisfy current and future requirements for ultrapure water, improvements in the unit manufacturing operation itself are not the solution: breakthroughs must be made in technical improvements and innovations for all the hardware used in the manufacture of ultrapure water. The quality requirements for ultrapure water are now of the order of ppt from the ppb of only a few years ago, but hardware innovations should not stop with the mere study of the physical construction of equipment. Improvements must be comprehensive enough to include quality studies of the materials used in hardware fabrication.

In this part, the latest technologies in the major equipment and pipe materials used in the ultrapure water manufacturing facility are described, taking into account the trends of the day.

2
Piping

Koichi Yabe
Kurita Water Industries, Ltd., Tokyo, Japan

I. INTRODUCTION

It is required that the purity of water be maintained in the piping system between the ultrapure water production unit and the use points. Ultrapure water is produced in the utility area and is sent to a clean room through the main pipe, which is a few hundred meters long. In the clean room, the main pipe divides into branch pipes. Ultrapure water is ultimately supplied to tens of use points. Because the piping system is so long and complicated, the purity of the water is lowered, mainly as the result of elution and desorption of impurities from the piping materials, the propagation of bacteria in the piping system, and the penetration of air and gases into the piping system. To suppress these effects, low-eluting materials should be selected for the piping, and extra attention should be paid to the structure and the installation of the piping system. During operation, ultrapure water should circulate continuously in the system, to eliminate the entrapment of water.

PVC (polyvinyl chloride) is frequently used for the pipes that carry ultrapure water. PVC is excellent in terms of mechanical strength, chemical stability, and cost performance. The original PVC contained not only vinyl chloride but also tribasic lead sulfate, bibasic lead sulfate, lead stearate, calcium stearate, barium stearate, pigments, including carbon and titanium white, and wax. The ratio of those materials, which affects elution into ultrapure water, depends on the manufacturer. The piping system for ultrapure water uses PVC that has been improved: it elutes less than the original material so that the quality of the ultrapure water does not deteriorate.

In addition to PVC, other materials, such as PP (polypropylene), PVDF (polyvinylidene fluoride), PFA (perfluoroalkoxyvinyl ether), and PEEK (polyether ether ketone), are used for piping ultrapure water. These other materials can be used at high temperatures because of their high heat resistance. The piping for ultrapure water must be sterilized using hot water at 80–90°C, the materials used are heat-resistant PVC, PVDF, and PEEK.

Table 1 lists the characteristics required for the piping materials. For chemical stability, neither the primary material nor the additives should elute into the ultrapure water. The materials should also be impervious to the oxidizing agents, including hydrogen peroxide, ozone, and hot water, that are used to sterilize the piping system.

In terms of mechanical strength, the materials should withstand water pressures of 5–10 kg cm^{-2}, and they should not be subject to creep even when they are used for long periods of time. For heat resistance, these materials must maintain chemical and physical stability at high temperatures and they should have limited thermal expansion. When a large piping system is required, the piping should be installed to absorb the elongation due to thermal expansion.

Nominal pipe and fitting diameters should range from 13 to 150 mm. Quality control during the manufacture of these products should be sufficient to keep the inner surface away from contamination. Table 2 lists the general characteristics of piping materials used in the ultrapure water system.

II. PIPING MATERIALS

A. PVC

Vinyl chloride resin is the most generally used plastic. During production, ethylene and chlorine are mixed to synthesize vinyl chloride monomer, $CH_2=CHCl$, which is then polymerized to make vinyl chloride polymer $(CH_2=CHCl)_n$. Stabilizers, pigments, and wax are added to the polymer for molding at high temperature. The stabilizers are used to increase the heat stability during and after molding. The major stabilizer for general PVC is lead, such as tribasic lead sulfate and lead stearate. These lead stabilizers are added in

Table 1 Requirements for Piping Materials for Ultrapure Water

	Requirement
1. Chemical stability	∘ Neither main component nor additives should be released into ultrapure water ∘ No deterioration with hydrogen peroxide, ozone.
2. Physical stability	∘ Withstanding the water pressure in the range of 5 to 10kg/cm.
3. Heat resistant properties	∘ Having heat resistant up to 100°C. ∘ Less thermal expansion.
4. Product	∘ Inside surface smoothness, particle contamination should be controlled. ∘ A variety of size in pipes, fittings and valves should be available.
5. Cost, Fitting	∘ Material cost is reasonable. ∘ Pipe fitting is a easy work.

Table 2 General Characteristics of Piping Materials

	Unit	Polyvinyl Chloride (PVC)	Polypropylene (PP)	Polyvinylidene fluoride (PVDF)	Perfluoroalkoxy vinylether (PFA)	Poly−ether−ether−ketone (PEEK)
Molecule structure		$\left[\begin{array}{c}H\ H\\ -C-C-\\ H\ Cl\end{array}\right]_n$	$\left[\begin{array}{c}H\ CH_3\ H\\ -C-C-C-\\ H\ \ \ \ \ H\ H\end{array}\right]_n$	$\left[\begin{array}{c}H\ F\\ -C-C-\\ H\ F\end{array}\right]_n$	$\left[\begin{array}{c}F\ F\ F\ F\\ -C-C-C-C-\\ F\ F\ C\ F\\ \ \ \ \ \ \ \ Rt\end{array}\right]_n$	structure
Additives		Stabilizer, pigment	Oxidation inhibitor, stabilizer, pigment	None	None	None
Color		Blue	Blue	Milky white	Milky white	Light brown
Specific gravity		1.43	0.91	1.77	2.17	1.30
Strength against stretching	kg/cm²	500~550	250	500~600	320	930
Elasticity	kg/cm²	2.7×10^4	1.5×10^4	1.4×10^4	—	4.0×10^4
Elongation	%	50~150	400~600	200~300	280~300	150
Expansion const.	1/°C	$6 \sim 8 \times 10^{-3}$	11×10^{-3}	12×10^{-3}	12×10^{-3}	5×10^{-3}
Heat conductivity	kcal/m·h°C	0.13	0.15~0.2	0.11	0.22	0.22
Applicable temp. limit	°C	60	100	140	260	152

powder form. They affect the flatness of the inner pipe surface during molding, and lead ions elute into ultrapure water. Lately PVC has been improved to overcome these problems, and PVC piping systems for ultrapure water have been introduced. For example, Eslon Clean pipe uses tin instead of lead stabilizers. The liquid tin stabilizers can uniformly penetrate vinyl chloride powder with a particle size of 100–150 μm. Unlike the powdered lead stabilizers, tin stabilizers do not gather and adhere to the vinyl chloride particles during molding. When the mixed vinyl chloride and tin stabilizers powder is tempered by screw while the container is heated, the mixture gels to become a fluid unit of 1–10 μm at around 190°C. This resin in gel form is molded by extrusion into a pipe with a flat inner surface.

B. PVDF

Compared with various fluorocarbon resins, PVDF can be easily molded and processed at high temperatures. PVDF can be molded and processed at a resin temperature of 200–260°C, which is almost the same as that for polypropylene. PVDF is a linear polymer, $[-CH_2-CF_2]$. Emulsion and suspension polymerization are used.

The molecular weight distribution affects the characteristics of PVDF. The molecular weight distribution is mainly determined by temperature and the reaction method used for polymerization. The melting point of PVDF crystal is 180–190°C. When it is heated to 350°C or higher, it decomposes, generating HF gas. The wide range between the molting point and the temperature of decomposition in which PVDF can be processed gives PVDF some advantages in terms of the ease of processing. PVDF is molded without additives, such as plasticizers and stabilizers.

C. PEEK

PEEK is a thermoplastic aromatic crystalline resin. It features a glass transition temperature of 143°C and a melting point of 334°C. Among the crystalline resins, PEEK has good heat resistance. The mechanical strength of PEEK starts declining at 150°C. The coefficient of linear expansion of PEEK is about a half that of the fluorocarbon resins. At room temperature, PEEK features high tensile strength, high elasticity, good elongation, and high impact intensity. In this sense, PEEK is a plastic with characteristics very similar to those of metals.

Although PEEK dissolves in concentrated sulfuric acid, it has high chemical resistance against hot water, alkali, and organic solvents. PEEK has good heat resistance, but because the temperature at which thermal decomposition begins is also high, it can be processed easily. Injection, extrusion, compression, and rotational molding, all used for thermoplastic resin in general, can be applied to PEEK.

III. CHARACTERISTICS NEEDED FOR ULTRAPURE WATER USED

A. Inner Surface

Scanning electron microscopy (SEM) is generally used to observe particulate contamination, surface smoothness, and changes in the surface condition when piping is used. Table 3 lists the results of SEM examination of PVC, PP, PVDF, and PEEK used for ultrapure water. Figures 1 through 4 are SEM photographs. Figure 5 shows a PVC surface in use for

Table 3 Immersion Test Results

Material / Maker	Inside Surface	Micropores		
		Before Immersion	After Immersion	
			20°C× 1day	80°C× 1day
Clean PVC	◎	+	+	+ +
Poly −propylene	×	+	+	+
Polyether −ether − ketone	◎	−	−	−
PVDF （Maker B）	○	+	+	+
PVDF （Maker S）	○	−	−	−
PVDF （Maker A）	○	±	±	±

Inside surface smoothness   Micropores
    ◎ Superior   + Many
    ○ Good   − Less
    × No good

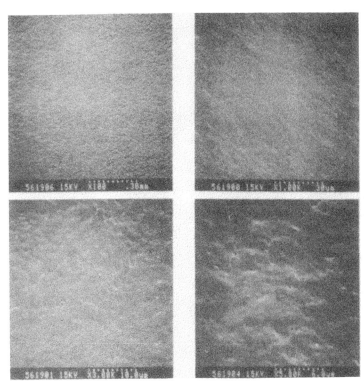

Figure 1 PVC pipe surface (SEM, ×100–5000).

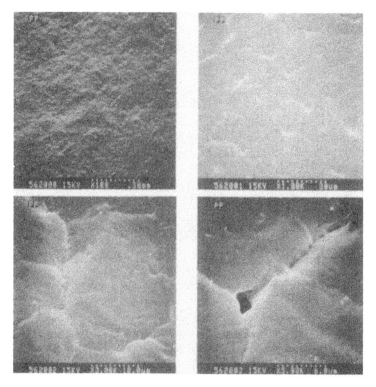

Figure 2 PP pipe surface (SEM, ×100–5000).

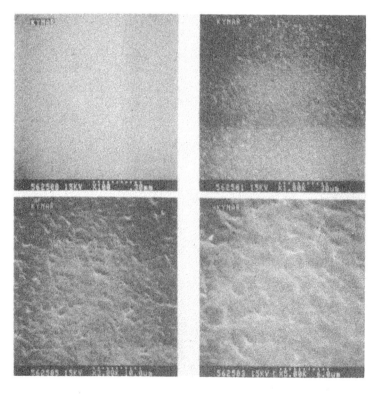

Figure 3 PVDF pipe surface (SEM, ×100–5000).

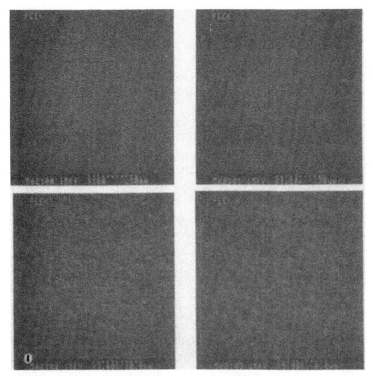

Figure 4 PEEK pipe surface (SEM, ×100–5000).

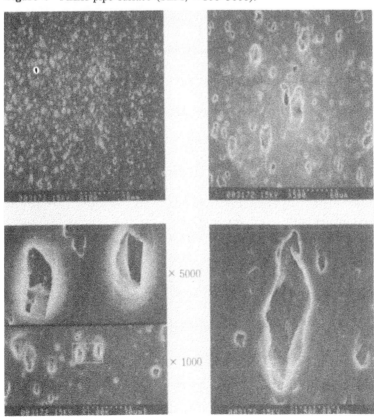

Figure 5 Conventional PVC pipe surface after long use (SEM, ×100–1500).

a long time. Many pores ranging in size from a few to a few tens of micrometers are observed on the surface. Some pores appeared when the PVC pipe was molded; others were generated when materials eluted from the pipe. When there are many pores, chemical displacement is insufficient and the cleaning of contaminants cannot be properly conducted.

Figures 1 through 4 are photographs of PVC, PP, PVDF, and PEEK used in an ultrapure water piping system. PVC and PEEK have a smooth surface that is free of pores. The smoothness varies for PVDF: some PVDF pipes have a smooth surface free of pores, but others have many 0.2–1 μm pores. PP pipes usually have many 0.3–1 μm pores. The surface conditions did not change when these heat-resistant piping materials were immersed in hot water at 80°C.

B. Imprisoning Test for Short Pipes

The short pipe imprisoning test is a method of identifying the ions and organic materials eluted from piping materials and of measuring the amount of elution. Ultrapure water is imprisoned in a short pipe for a certain period of time, and then the eluted ions and organic materials are analyzed. Table 4 lists the procedures for this test.

Inorganic ions and Cl ions eluted from PVC. This elution is inevitable because these ions are the main ingredients. When water was heated to 80°C, Cl elution increased and SO_4, Ca, and Na were detected (Figs. 6 through 14).

Na ion, Ca ion, and SO_4 ion were detected in a PP system, but the amount of elution

Table 4 Imprisoning Test Procedure

Piping

    Size   : 25 mm ϕ × 1000 mm L

    Test piping: three sockets, glued

    cap on bottom and detachable cap on top

Condition

    Imprisoned liquid : Ultrapure water 500 mℓ

    Temperature   : 25 °C, 80 °C

    Period   : 1, 7 and 25 days

Analysis

    Ion chromatography

    (Na, K, F, Cl, SO_4, PO_4, NO_3)

    Flameless AA

    (Fe, Cu, Zn, Ni, Sn, Ca, Mg)

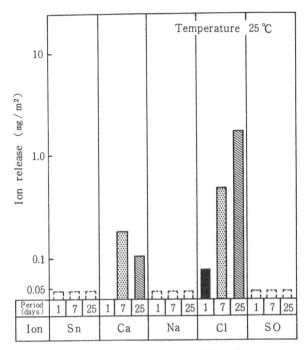

Figure 6 Imprisoning test in clean PVC piping.

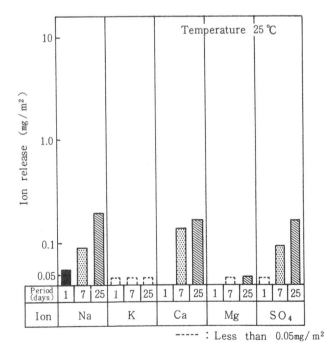

Figure 7 Imprisoning test in polypropylene piping.

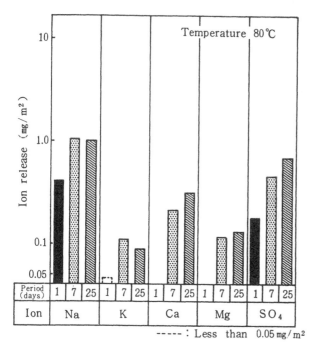

Figure 8   Imprisoning test in polypropylene piping.

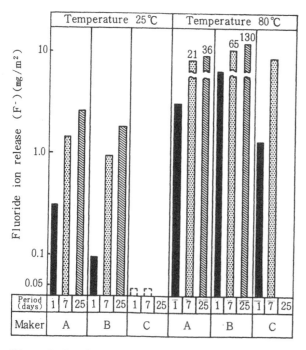

Figure 9   Imprisoning test in PVDF piping manufactured by three different companies.

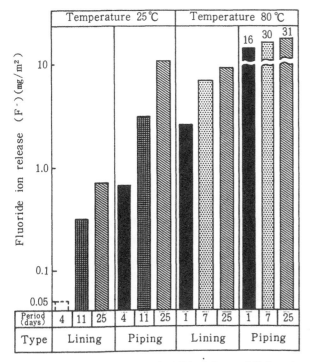

Figure 10 Imprisoning test in PFA piping and lining pipe.

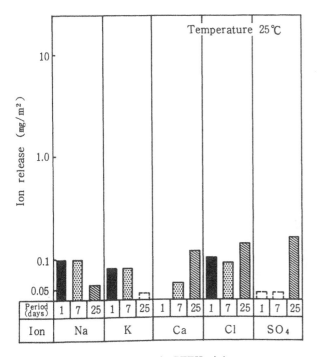

Figure 11 Imprisoning test in PEEK piping.

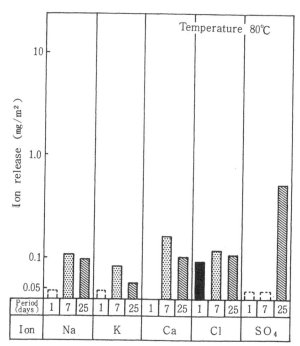

Figure 12   Imprisoning test in PEEK piping.

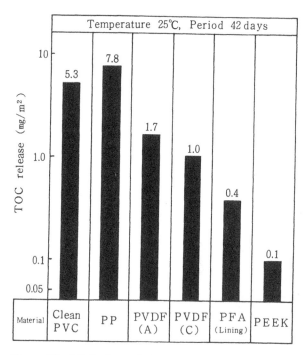

Figure 13   TOC release during imprisoning test.

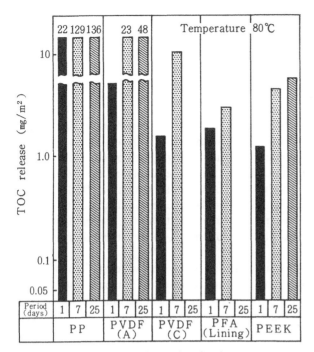

Figure 14 TOC release during imprisoning test.

remained low in every case. When water was heated to 80°C, the elution of Na and SO$_4$ increased somewhat. These ions are thought to elute from the additives. Because the amount of elution did not increase even when PP was immersed for a longer period of time, the elution is believed to be complete early during immersion.

With PVDF, no inorganic ions except F ion was detected. This is because PVDF does not contain additives. The amount of F ion elution varies greatly by manufacturer. The amount of F ion elution at room temperature is the same as Cl ion elution from PVC. When water is heated to 80°C, F ion elution significantly increased a few tenfold on average and around 100 times at maximum. The amount of F ion elution varies with the manufacturer because of differences in the polymerization method. Since F ion always elutes in large quantities from PVDF at high temperature, extra attention should be paid to the effect of F ion on the quality of water when PVDF is used in a system through which high-temperature ultrapure water is distributed.

PFA shows the same characteristics as PVDF: only F ion is eluted. Elution significantly increased at a high temperature of 80°C. For ultrapure water use, PFA has the same chemical stability as PVDF.

PEEK does not contain inorganic additives, and no specific ingredients with significant elution were detected. Small amounts of Na ion, K ion, Ca ion, Cl ion, and SO$_4$ ion were detected, but the amount of elution did not increase even when the immersion time was prolonged. This means that this elution is from contamination on the surface and takes place only at the initial stage. Elution did not increase when the water was heated to 80°C, indicating chemical stability.

PEEK and PFA have the lowest values for TOC elution. When water was heated to 80°C, TOC elution rose considerably with all materials. Since PP particularly elutes a large amount of TOC at 80°C, it cannot be used at high temperatures.

C. Circulation Test Through Long Pipes

The elution of inorganic ions from piping materials gradually decreases day by day. When it is assumed that the elution rate of inorganic ions is stable, it is possible to calculate the concentration of eluted ions in the ultrapure water piping system based on the elution rate of the short pipe imprisoning test. Table 5 is an example of the calculation. When ultrapure water is supplied at a flow rate of 1.0 m s^{-1} through a 600 PVC pipe with a nominal diameter of 25 Å, the water stays in the pipe for 10 minutes. The elution rate of Cl ion measured in the short pipe imprisoning test was 0.1 mg m^2 day^{-1}. The increase in Cl ion was calculated as 0.11 ppb. Conversely, the resistivity dropped 0.11 MΩ cm^{-11}.

Figure 15 shows the resistivity drop as a function of time that water stays in the pipe. The resistivity was actually measured at the inlet and outlet of the ultrapure water piping system. There is a correlation between the drop in resistivity and the time the water stays in the pipe. In an actual branch piping system, various sizes of pipes are combined,

Table 5 Calculation of Resistivity Reduction in Piping

Condition	
Diameter	25 A
Length	600 m
Flow velocity	1.0 m/sec.
Temperature	25 °C
Piping material	PVC (Clean PVC)
Cl⁻ release	0.1 mg/m² •day
Result	
Transit time	10 min.
1) $\triangle R = 0.1\,M\Omega \cdot cm$	from **Fig. 15**
2) Cl⁻ concentration	

$$= 0.1 \times \frac{10}{24 \times 60} \times \frac{47.1}{0.294}$$

$$= 0.11\,mg/m^3$$

Cl 0.11 ppb corresponds resistivity reduction of 0.13 MΩ • cm

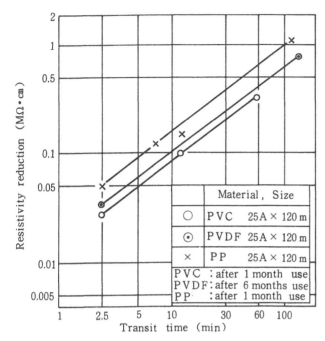

Figure 15 Resistivity reduction with transit time.

complicating the system. Therefore the time that water stays in the system to be delivered to the use points can vary greatly. However, 10 minutes should be long enough to meet any needs. If the piping material, water temperature, and the time the water stays in the system are given, it is possible to estimate the quality of ultrapure water at the use point based on the ion elution rate obtained in the short pipe imprisoning test.

3
Valves and Fittings

Mitsugu Abe
Nomura Micro Science Co., Ltd., Kanagawa, Japan

I. INTRODUCTION

The materials for the valves and fittings used in an ultrapure water system and their structure must be low eluting. Ions, total organic carbon (TOC), particles, and microorganisms should be released as little as possible. Construction and operational performance should be of high quality. Reliable, uncontaminated sampling valves are required for evaluating ultrapure water quality with high precision.

II. FITTINGS

In an ultrapure water system the treatment unit instrumentation and cleaning apparatus are connected to the use points by various pipelines and fittings. The proper selection and use of pipelines and fittings for the points of use enable the economical and effective flow of ultrapure water while maintaining water quality. In addition, efficiencies of construction and maintenance are improved.

A. Types

The kinds and structures of the major fittings used in an ultrapure water system are shown in Fig. 1 (Eslon Valves and Plant Piping Materials, Sekisui Chemical Co., Ltd.). Many other types are available, including connection by pipeline and other materials. Generally, the materials used for the fittings are equivalent to those used for the pipelines.

B. Requirements

Today's pipeline materials for ultrapure water use are required to meet diverse needs, such as steam sterilization and the use of hot pure water, except for upkeep of water quality. These requirements may be summarized.

STRUCTURE	(image)	(image)	(image)	(image)
NAME	REDUCING SOCKET	TEE	ELBOW	SOCKET
PURPOSE	REDUCING	BRANCH	CHANGE of DIRECTION	CONNECTION
METHOD of CONNECT	ADHESIVE SCREW BUTTWELD SOCKETWELD	ADHESIVE SCREW BUTTWELD SOCKETWELD	ADHESIVE SCREW BUTTWELD SOCKETWELD	ADHESIVE SCREW SOCKET WELD

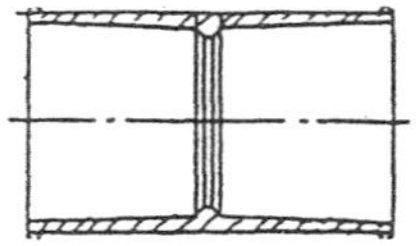

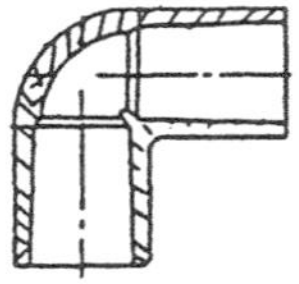

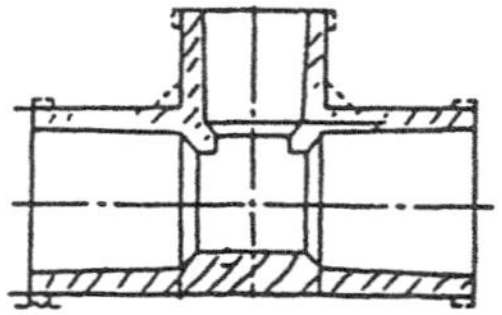

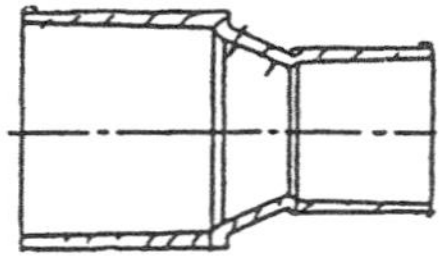

STRUCTURE	(image)	(image)	(image)	(image)
NAME	VICTRIC ®	SWAGE LOK ®	FLANGE	UNION
PURPOSE	CONNECTION (HIGH PRESSURE)	CONNECTION	CONNECTION	CONNECTION
METHOD of CONNECT	GROOVING ON PIPE	DIRECTLY	ADHESIVE SCREW BUTTWELD SOCKETWELD	ADHESIVE SCREW SOCKET WELD

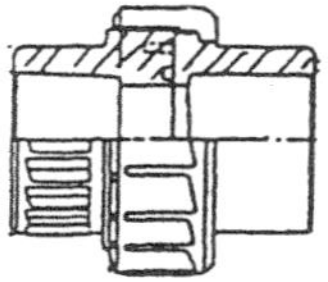

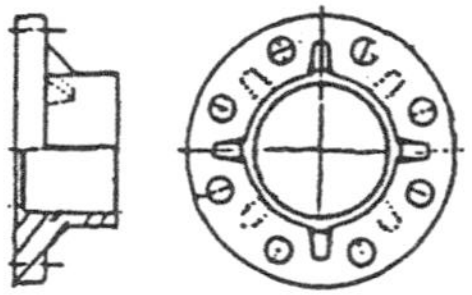

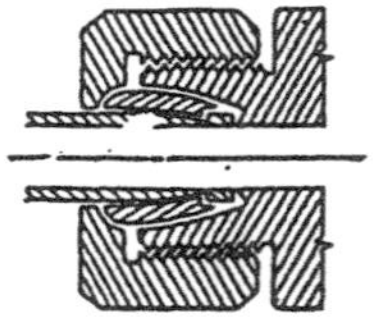

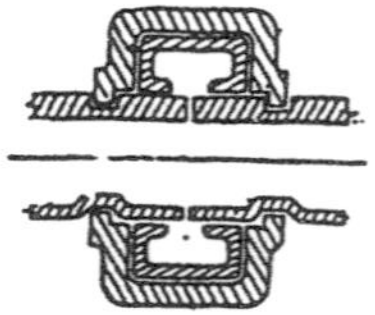

Figure 1 Types and structures of fittings.

1. Very low levels of contaminants, such as various ions and TOC, should be released.
2. Heat resistance is required at temperatures up to 80–100°C.
3. The materials must be resistant to the oxidizing agents used for sterilization, such as H_2O_2, O_3, and ultraviolet (UV).
4. The materials should not peel from the inner walls, and particles should adhere to the walls as little as possible.
5. The inner surface should be smooth. There should be no standing water at connection points.
6. Excellent constructional performance and connecting method for little contamination.
7. The materials are available at low cost and in many sizes.

Corresponding to these requirements, new fluororesins and low-contaminating connecting methods have been adopted.

C. Examination

Here, joining methods for fittings and gaskets are reviewed.

1. Joining Method and Release

In the past, pipelines and fittings were mainly connected by adhesion joining. Recently, weld joining has become the method of choice. In welding, the segments to be connected are heated to above the melting point of the resin, and adhesives are unnecessary.

To compare adhesives and welding, pipelines and joints were connected using different methods and then dipped into ultrapure water and the amounts of TOC released were determined. Clean polyvinyl chloride (PVC) and polypropylene (PP) were used. The results, shown in Table 1, indicate that the quantities of TOC released from both adhesive joints are high [1].

Adhesion is better than welding in constructional performance, however, because a special instrument is unnecessary and the time needed for construction is short. Con-

Table 1 TOC Release During Dipping Test (From Ref. 1.)

① CLEAN PVC

WELD	280 ppb
ADHESIVE	7400 ppb
BLANK	260 ppb

AFTER 20 DAYS

② PP

WELD	210 ppb
ADHESIVE	8000 ppb
BLANK	200 ppb

AFTER 6 DAYS

(TEMPERATURE 25°C)

sequently, in many cases, weld connections are used only in a line that specifically requires high purity.

2.   Structure of Joints

As shown in Fig. 1, a small amount of standing water may be trapped in the flange and union fittings. Recently, the R_{max} (difference of convex and concave) at the inner surface of a pipeline has tended to be below 1 μm. Thus, the volume of standing water is relatively large if a flange and union are as joints. In circular pipelines leading from an ultrapure water purifying apparatus, the use of such joints as the flange and union must be minimized.

3.   Release from Gasket

The amounts of Na^+, Cl^-, S-SiO_2, and TOC released from six kinds of materials often used for gaskets were examined. Cleaned sample pieces were dipped in ultrapure water and heated to 90°C for 1 h, and the difference between the water quality of control and test water samples were measured. The experiments were repeated three times. The structures of the materials and the results are shown in Table 2 and Fig. 2. Before the test, polytetrafluoroethylene (PTFE) was thought to be the lowest eluting. In fact, ions were released that were not present in the material, such as Na^+ and Cl^-. These ions are presumably from contaminants acquired during the preparation of raw materials and molding. Because the amounts of TOC and S-SiO_2 eluted were low, the material is considered excellent for gaskets, if contamination during manufacturing can be avoided.

III.   VALVES

Valves are used for adjusting fluid pressure, changing the amount of flow, and redirecting flow. Uncontaminated valves are required for ultrapure water use.

Table 2   Gasket Types and Structures

NAME	STRUCTURE	PERMISSIBLE TEMP [°C]
TEFLON (PTFE) ®	$-(CF_2)-n$	260
SILICONE (Si)	$-(\overset{R}{\underset{R}{Si}}O)_n$	280
EPT RUBBER (EPDM)	$-(CH_2CH_2)_n-(CH_2\overset{}{\underset{CH_3}{CH}})_m-(\underset{CHCH_3}{\square})_p$	150
VITON ® (FPM)	$-(CF_2CH_2\ \underset{CF_3}{CFCF_2})_n$	200
BUTYL RUBBER (IIR)	$-(CH_2\overset{CH_3}{\underset{CH_3}{C}})_n-(CH_2\overset{}{\underset{CH_3}{C}}=CHCH_2)_m-$	150
NITRILE RUBBER (NBR)	$-(CH_2CH=CHCH_2)_n-(CH_2\underset{CN}{CH})_m$	130

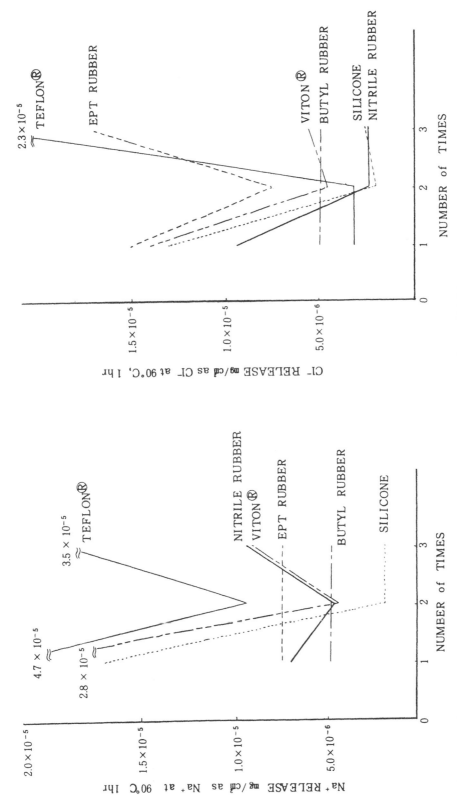

Figure 2 Na⁺ release during gasket dipping test (left). Cl⁻ release during gasket dipping test (right).

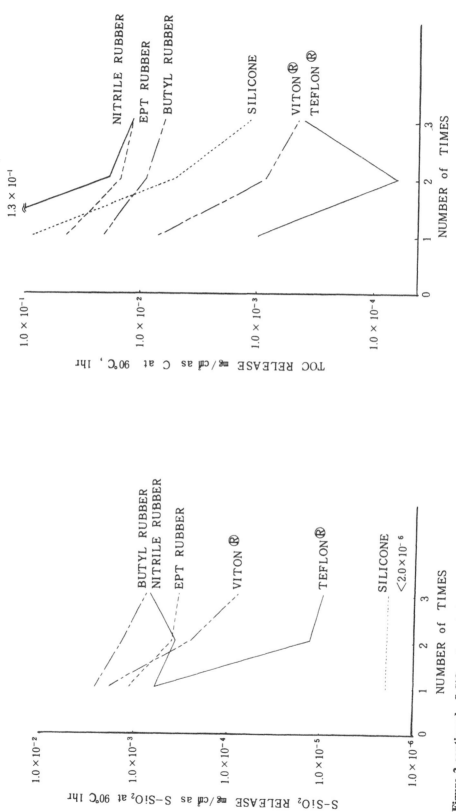

Figure 2 continued S-SiO₂ release during gasket dipping test (left). TOC release during gasket dipping test (right).

A. Types

The kinds and structures of valves are shown in Fig. 3 (Eslon Valves and Plant Piping Materials, Sekisui Chemical Co., Ltd.). The predominant materials are the same as those used for pipelines. For ultrapure water use, fluororesins are used mostly for sheets and gaskets.

B. Requirements

Materials specifications are the same as those required for joints. In addition,

1. Slide valves are used as little as possible to suppress particle contamination.
2. Valves should be manufactured without pockets for standing water.
3. The operation and sealing properties must be excellent.

C. Examination

1. Structure and Potential for Generating Particles and Viable Bacteria

As shown in Fig. 4, various valves were installed in a test unit and the relation between the structures of the valves and the water quality at the outlets was examined. The results (Table 3) indicated excellent values were obtained with diaphragm valves. A ball valve

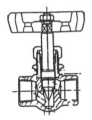

BALL VALVE

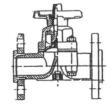

BUTTERFLY VALVE

CHECK VALVE

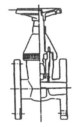

STOP VALVE

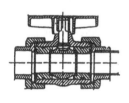

NEEDLE VALVE

DIAPHRAGM VALVE

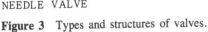

Figure 3 Types and structures of valves.

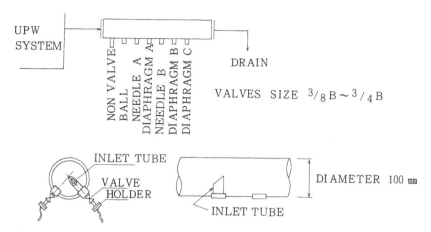

Figure 4   Testing unit for valves.

and a diaphragm valve were then installed in series and allowed to stand in the water for 1 week. The numbers of viable bacteria at the outlet were then measured, as shown in Fig. 5. The number of viable bacteria in the outlet water of the diaphragm valve immediately reached the same level as that using a control (blank) value. This is presumably attributed to the small numbers of pockets and slides in the structure of a diaphragm valve.

The number of viable bacteria in the outlet water of an ultrapure water purifying apparatus was measured with a ball valve used formerly for sampling. As shown in Fig. 6, 270 minutes was required to reach to control values.

Table 3   Number of Particles and Viable Bacteria Released from Valves

TYPE of VALVES \ NUMBER of	$\geq 0.2 \mu$m PARTICLE /mℓ			VIABLE BACTERIA CFU/ 100mL
	MICRO-ORGANISM	OTHER	TOTAL	
BALL	1	6	7	3
NEEDLE A	0	2	2	2
NEEDLE B	1	2	3	2
DIAPHRAGM A	0	0	0	0
DIAPHRAGM B	0	1	1	1
DIAPHRAGM C	0	0	0	0
INLET	0	1	1	1

(TESTING CONDITION)

∘ BLOWING TIME : 10 Min.

∘ NUMBER = OUTLET−INLET

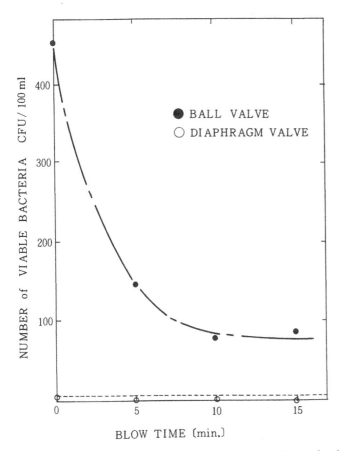

Figure 5   Number of viable bacteria and blowdown time using ball and diaphragm valves.

2.   *Operational Performance and Sealing Properties*

From the viewpoint of a CV value (the relation between working quantity of a handle and an open area), a needle valve is suitable for adjustment of flow amounts.

The ball valve has excellent sealing properties, but complete enclosure is difficult using a needle valve.

3.   *Diaphragm Valve for Ultrapure Water*

A diaphragm valve is thought to be suitable for an ultrapure water system, but diaphragm valves with a small diameter, below ¾ in. are not yet available, and valves suitable for sampling are not manufactured. Recently, diaphragm valves made of PTFE or stainless steel have been used for sampling. These are shown in Fig. 7.

In the ultrapure water purifying apparatus at the Denki Tsushin Laboratory of Tohoku University, a three-port diaphragm designed for preventing standing water in a valve was used. The structure is shown in Fig. 8 [2].

IV.   CONCLUSION

In the past, conventional valves and fittings were modified by ultrapure water use, but recently a variety of new products have been developed specifically for ultrapure water.

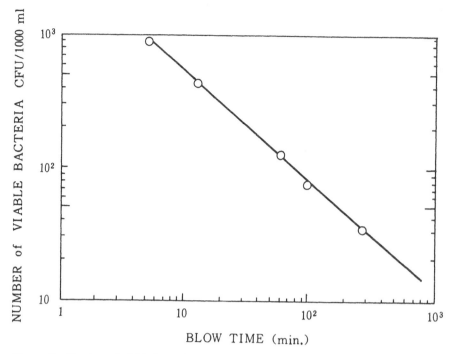

Figure 6   Number of viable bacteria and blowdown time using a ball valve.

Figure 7   Sampling valves for ultrapure water use: left, SUS 316 L; right, PTFE.

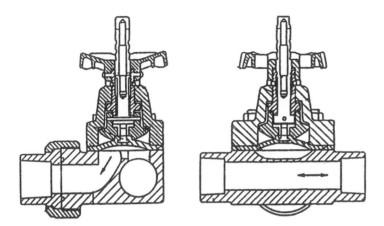

Figure 8 Experimental three-port diaphragm valve.

With further improvements in the quality of ultrapure water in the future, improved performance of valves and joints can be expected.

REFERENCES

1. Y. Yamada, ed., Ultrapure Water and Chemical Supplier, Ultra Clean Society, Realize, Inc., 1986, pp. 2–29.
2. Y. Ohta, J. Soc. Heating, Air-Conditioning, Sanitary Engineers, *62*(1), 35–39 (1988).

4
Pumps

Takashi Imaoka
Tohoku University, Sendai, Japan

I. INTRODUCTION

Now that the impurity concentrations in ultrapure water are required to be suppressed to extremely low levels, it is increasingly important, with regard to the ultrapure water production system, not only to remove impurities from each component unit but also to suppress the particulate elution from the materials themselves.

The ultrapure water production system and distribution system to the points of use are composed of various component units: tanks, pumps, valves, pipes, ion-exchange resins, reverse osmosis membranes, ultrafiltration membranes, and so forth. These component units are made from various kinds of materials. What is immediately necessary is that the cleanliness of the component units must be upgraded. With this goal in mind, every effort is being made in the attempt to reduce particulate elution from the materials.

Among the various component units, the pump has a particularly complicated structure at the liquid-contacting surface. Moreover, the sliding and rotating parts inevitably contact the liquid fed through it. These characteristics of the pump lead to particulate elution as a result of abrasion and scaling or to the prolonged desorption of impurities that adhere to the pump during manufacturing.

One of the major applications of ultrapure water is to clean the wafer surface. Lately ultrapure water has also been applied to the precision cleaning of the peripheral equipment of the semiconductor production process [1]. In precision cleaning, ultrapure water is sometimes pressurized to a few tens of kilograms per square centimeter, which easily leads to the particulate elution from the pump. In the worst case, the pump is not followed by any other unit to remove the impurities that may have been generated but is directly connected to the use points. Therefore, a "clean pump" that suppresses the generation of impurities, such as particles and metallic ions, to the minimum level is essential for the supply pump in the subsystem in the ultrapure water production system and for pumps placed downstream of the subsystem.

This chapter introduces a newly developed pump that suppresses particulate elution better than conventional pumps. Pumps for distributing organic solvents, such as isopropyl alcohol (IPA), are also described. These pumps also require greater cleanliness standards necessary for the pumps used for ultrapure water.

II. LOW-PARTICULATE PUMP AND WATER QUALITY AT THE OUTLET

A. Clean Canned Motor Pump

The canned motor pump is a centrifugal (swirl flow) pump. A thin corrosion-resistant liner is inserted between the motor stator and the rotor, through which a small amount of the delivery water. In other words, this is a water-cooled pump in which water cools and smooths the bearing and motors [2].

The motors in the canned motor pump are protected from the outer air, preventing contamination. This is why the canned motor pump is often used as the motor for the subsystem installed inside the clean room. In this motor, however, the water that has passed through the sliding part of the bearing combines with the jetted water at the later stage. If particles are generated in the sliding part, they are mixed into the delivery water. To prevent this problem, the following improvements have been made:

1. The water supplied to the sliding bearing is guided back to the motor side from the inside of the casing to be drained outside. This is called a reverse-flow system. Particles generated in the sliding part are drained out together with the water.
2. The motor is cooled not by the delivery water but by water supplied separately. In this way, the amount of reverse-flow water can be reduced.
3. The casing and impellers are treated by electropolishing to suppress particulates and particle adhesion.

Figure 1 is an outline of the conventional canned motor pump [2], and Fig. 2 is an outline of the improved canned motor pump.

Figures 3 and 4 show the number of the particles at the inlet and outlet of the improved canned motor pump. The number of particles at the pump outlet was the same as that at the inlet, demonstrating that the improvements just listed prevent particle penetration to the delivery water.

B. Regenerative Turbine Pump

In the general regenerative turbine pump, the motor is cooled by air and the casing and bearing are sealed mechanically. The supplied water does not enter the motor in the regenerative turbine pump, which is different from the canned motor pump, but particulates from the mechanically sealed parts can leak into the casing and mix with the delivery water [3]. To overcome this problem, the following improvements were made to develop a clean regenerative turbine pump with suppressed particulates.

1. The generated particles are drained outside by draining water from the mechanically sealed part, which is the sliding part.
2. In an attempt to suppress particulates, the liquid-contacting parts in the pump are made of SUS 316 L treated by electrochemical buffing.

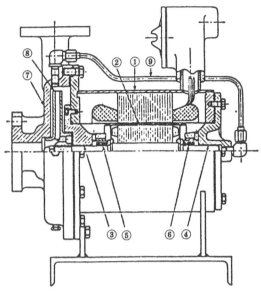

1 : fixed coil   6 : bearing
2 : revolution coil   7 : casing
3 : bearing housing   8 : impeller
4 : bearing housing   9 : cooling water circulation pipe
5 : bearing

Figure 1 Sectional view of a conventional canned motor pump manufactured by Chemi Pump Co., Ltd.

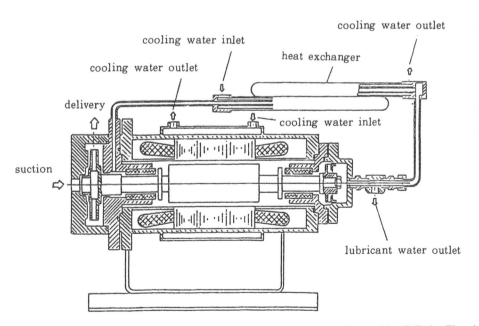

Figure 2 Sectional view of an improved canned motor pump manufactured by Teikoku Electric Mfg. Co., Ltd.

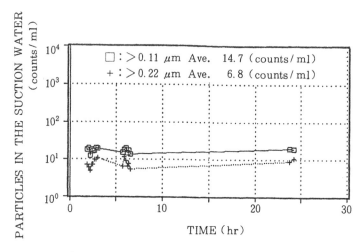

Figure 3 Time dependence of particles in the suction water of the improved canned motor pump. Pump capacity, 1.5 m³ h⁻¹; outlet of lubricant water, 9 L h⁻¹.

3. The area of the liquid-contacting parts is reduced as much as possible.
4. The outside type of mechanical seal is employed so that the rotating circle is positioned with the complicated structure outside the liquid-contacting parts.
5. Some of the water between the casing and the plate and between the cover and the plate is drained to eliminate water entrapment in the casing.

Figure 5 illustrates the air-cooling regenerative turbine pump, which attained a delivery pressure of 30 kg cm⁻². Table 1 compares the number of particles in the delivery water between the conventional pump and the improved pump. Figure 6 compares the resistivity of the delivery water between the conventional pump and the improved pump.

The regenerative turbine pump is used in the precision cleaning unit with high-temperature and high-pressure ultrapure water, which cleans the components of the

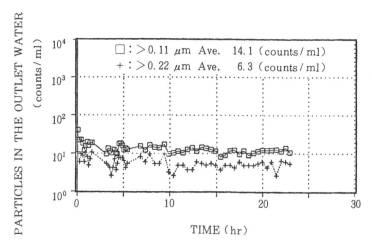

Figure 4 Time dependence of particles in the outlet water of the improved canned motor pump. Pump capacity, 1.5 m³ h⁻¹; outlet of lubricant water, 9 L h⁻¹.

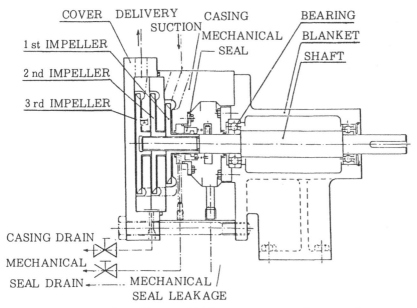

Figure 5 Sectional view of improved high-pressure regenerating pump for ultrapure water manufactured by Nikuni Machinery Co.

process equipment for semiconductor manufacturing. Therefore the high pressure of 30 kg cm^{-2} is used with this pump; the conventional pump employs a delivery water pressure of 6 kg cm^{-2}. The number of particles in the delivery water in the improved pump is 1/10 that of the conventional pump, and the resistivity at the outlet of the improved pump is almost the same as that at the inlet (18.24 MΩ·cm).

C. Clean Magnet Pump

Magnet pumps are often employed in chemical purification and transportation, although they are not used very much for ultrapure water. Compared with the bellows pump and the diaphragm pump, the magnet pump has a less pulsating flow, resulting in a less adverse effect on the filters that follow it. Since the supplied water in the magnet pump is circulated to bearing parts for cooling and smoothing, however, particles generated in the sliding parts are inevitably mixed into the delivery water. To overcome this problem,

Table 1 Particles Generated by Improved High-Pressure Regenerating Pump or Conventional Pump

equipment size range	HIGH PRESSURE PUMP 30 kgf / cm^2	CONVENTIONAL PUMP 6 kgf / cm^2
$>$0. 1 μm	147	1620
$>$0. 2 μm	51	868
$>$0. 5 μm	2	15

VALUES : AVERAGE OF TWO MEASUREMENTS

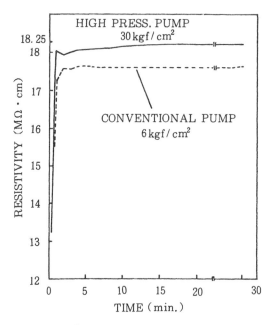

Figure 6 Time dependence of outlet resistivity using an improved high-pressure regenerating pump for ultrapure water.

some of the water is drained from the bearing parts to prevent the generated particles from being mixed into the delivery water [4].

Figure 7 illustrates the improved magnet pump, and Fig. 8 compares the number of particles in the delivery water between the conventional pump and the improved pump when ultrapure water is supplied to the pumps [4]. It is clear that the number of particles in the improved pump is reduced to $\frac{1}{10}$ that of the conventional pump.

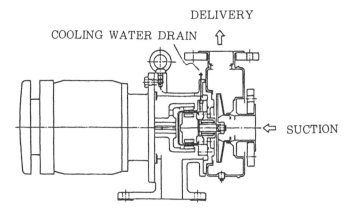

Figure 7 Sectional view of improved magnet pump for ultrapure chemicals manufactured by Ebara, Ltd.

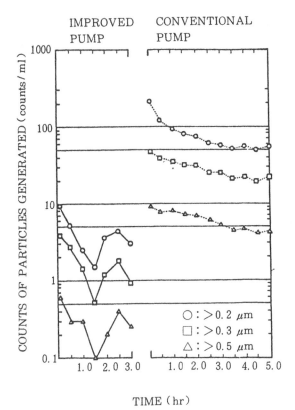

TIME (hr)

Figure 8 Particles generated by improved magnet pump and conventional pump.

III. SUMMARY

The important points in suppressing particulates from clear pumps can be summarized as follows.

1. The surface of the liquid-contacting parts must be smooth.
2. The processed components to be assembled in the clean environment must be clean.
3. As much as is possible, water should be prevented from contacting the sliding parts. The water that contacts the sliding parts should be drained.
4. Low-eluting materials should be selected for the liquid-contacting parts.
5. Particulates from the rotating parts should be prevented from entering the air by using an exhaust duct or by employing the water-cooling pump.

By adopting these improvements, particulates have been markedly reduced and the drop in resistivity has been suppressed as a result of less metal elution. However, problems remain: the amount of water delivered from the system is very great, and particles are still generated for awhile after turning the pumps on or off. Further improvements are expected in design, production technology, and cleaning technology.

To suppress impurities in the water to extremely low levels, particulates from the materials of the water-processing units should be eliminated. Because present ultrapure water production technology has advanced so much that the quality of the water produced is near the detection limit of the measurement instruments, particulates from the materials have a relatively great effect on overall water quality. Therefore it is clear that technologies that do not generate impurities will become more and more important in the future.

REFERENCES

1. T. Ohmi and T. Nitta, Advanced surface cleaning technology for high technology industry, In: Proceedings of 3rd Workshop on ULSI Ultra Clean Technology, Tokyo, January 1990, pp. 1–4.
2. S. Kajiwara, Pump and Its Application, Maruzen, 1989.
3. M. Kawakami, Y. Yagi, K. Sato, and T. Ohmi, Spray cleaning technology using high temperature and high pressure ultrapure water, In: Proceedings of 3rd Workshop on ULSI Ultra Clean Technology, Tokyo, January 1990, pp. 5–22.
4. Taga Plastic Engineering Laboratory, Ebara Corporation, Tokuyama Soda Co., Ltd., and Mitsui & Co., Ltd., Technical Report of TG Magnet Pump, 1988.

5
Tanks

Koichi Wada
Shinko Pantec Co., Ltd., Kobe, Japan

I. MATERIALS

A. Organic Materials

A variety of organic materials are used for tanks such as epoxy, polypropylene (PP), polyethylene (PE), polyvinyl chloride (PVC), and fluororesin. These materials are used independently or as composites (for instance, fiber-reinforced plastic, FRP) or as linings or coatings, for example, depending upon the volume, working pressure, temperature, and other conditions in the tank. Fluororesins are the most suitable organic tank materials for ultrapure water production in the semiconductor production process from the viewpoint of corrosion resistance, heat resistance performance, contamination resistance, and so on.

The fluororesins commercially available at present are listed in Table 1. Fluororesins belong to the thermoplastic resin family, which includes polytetrafluoroethylene (PTFE) and other fluororesins. Resins other than PTFE are called hot-melting fluororesins. Table 2 lists the characteristics of each resin.

B. Metals

The metals used for tanks normally include carbon steel, stainless steel, high-Ni alloy steel, and special metals. These materials may sometimes be used in combination with carbon steel as clad steel for economic reasons.

Carbon steel and stainless steel are mainly used for the ultrapure water system.
Carbon steel is used for linings and coatings.
PTFE is used in most cases for linings and coatings.
Stainless steel is normally used after buffing its surface.

Table 1 Fluoropolymer Product

Fluoropolymer	Trade name	Manufacturer
Polytetrafluoroethylene (PTFE)	POLYFLON TFE ALGOFLON FLUON HOSTAFLON TEFLON TFE TEFLON TFE FLUON	DAIKIN Industry Montefluos ICI Hoechst Du Pont Mitsui Fluoro Chemical Asahi Fluoro Polymer
Tetrafluoro ethylene-perfluoroalkylvinyl ether- copolymer (PFA)	NEOFLON PFA TEFLON PFA HOSTAFLON TFA	DAIKIN Industry Du Pont Hoechst
Tetrafluoro ethylene-hexafluoropropylene (FEP)	NEOFLON FEP TEFLON FEP	DAIKIN Industry Du Pont
Tetrafluoro ethylene-ethylene (ETFE)	NEOFLON ETFE AFLON COP TEFZEL HOSTAFLON ET	DAIKIN Industry Asahi Glass Du Pont Hoechst
Polyvinylidenefluoride (PVDF)	DYFLOR FORAFLON KFPOLYMER KYNAR	Dynamite Novel Products Chimiques Ugine Kuhlman KUREHA Chemical Pennwalt Chemicals
Polychloro trifluoro ethylene (PCTFE)	DYFLON CTFE KEL −F ACLON CTFE	DAIKIN Industry 3 M Allied Fibers & Plastic
Chlorotrifluoro ethylene ethylene (ECTFE)	HALAR	Allied Fibers & Plastic
Polyvinyl fluoride(PVF)	TEDLAR	Du Pont

Table 2 The Properties of Fluorocarbon Polymer

Properties	Unit	PTFE	PFA	FEP	ETFE
Specific gravity		2.13 – 2.22	2.12 – 2.17	2.12 – 2.17	1.70 – 1.76
Coefficient of water absorption	%	< 0.00	0.03	< 0.01	< 0.1
Coefficient of thermal expansion	1 / ℃	10×10^{-5}	12×10^{-5}	$8.3 - 10.5 \times 10^{-5}$	$5 \sim 9 \times 10^{-5}$
Melting point	℃	327	302～310	270	260
Melt viscosity	Poise	$10^{11} \sim 10^{13}$ (380℃)	$10^{4} \sim 10^{5}$ (380℃)	$4 \times 10^{4} \sim 10^{5}$ (380℃)	$10^{4} \sim 10^{5}$ (300～380℃)
Continuous maximum usable temperature	℃	260	260	200	150
Tensile strength	kgf／cm²	140～350	280～315	190～220	410～470
Elongation	%	200～400	280～300	250～330	420～470
Elastic modulus of bending strength	kgf／cm²	5000～6000	6600～7000	5500～6500	9000～14000
Hardness	Shore	D 50～65	D 60	D 55	D 75
Coefficient of static friction		0.02	0.05	0.05	0.06
Dielectric breakdown strength	V/mil	480	500	500～600	400
Volume resistivity	Ω•cm	$> 10^{18}$	$> 10^{16}$	$> 10^{16}$	$> 10^{16}$

Table 3   Dissolution Test Results for Various Materials Used in the Production of Ultrapure Water

	GOLD-EP[1]	PFA[1]	PVDF	PEEK
TOC	0.025	4.8	17	3.5
Na	0.0007			0.15
K	0.003			0.07
Ca	0.012			0.113
Cl	0.0013			0.11

(1) Condition   80 ℃ × 5days
(2) Dissolution data of PFA, PVDF, PEEK are obtained from a paper
     presented at proc. 9th International Symp. contamination control,
     Los Angeles, Sept, 1988.
     Titled "Selection of Plastic Piping Material for Ultrapure water"
     by Koichi Yabe etc.

It is necessary to select the buffing procedure according to each application, because buffing is done in several stages. The buffing grade and corresponding surface roughness for various materials are shown in Table 3. It was recently reported that electrolytic polishing or gold-EP treatment (electrolytic polishing plus high-temperature oxidation) is often done to ensure the desired surface cleaning performance and corrosion resistance.

II.   PRECAUTIONS FOR TANK FABRICATION

A.   Lining Tank

1.   Execution Method

Both adhesive sheet lining and loose lining are available. The adhesive sheet lining method is done by bonding a sheet, to which glass cloth has been bonded on one surface in advance, to the inner surface of the tank using adhesive and then welding the joint (Fig. 1). The joint is beveled to a V shape and welded by melting the round welding rod by hot jet for high-temperature use.

After testing with a pinhole tester, the belt-shaped welding material (normal width 14

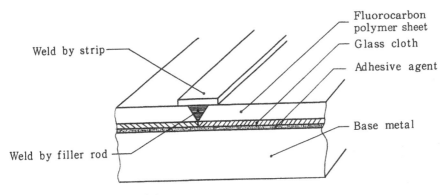

Figure 1   Adhesive sheet lining.

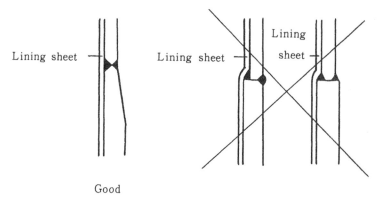

Good

Figure 2   Butt-welding procedure for plates of different thicknesses.

mm) is melted for finishing so that the upper part of the weld just described is covered. The sheet normally used is 2–3 mm thick, and the material used is PTFE, poly-fluoroethylene (PFE), or fluorinated ethylene-propylene (FEP).

In the loose lining method, the sheet is not bonded to the tank. This method is not applicable to tanks whose jackets may be heated because the heat transfer efficiency deteriorates.

2.   Points to Note on Tank Fabrication

a.   Welded Joint.   It is recommended that fillet welding be used as little as possible, and that butt welding be used instead.

Finish the weld bead on the lining execution surface smoothly with a grinder.

In welding plates of different thicknesses, it is important that the lining surface be designed smoothly, as shown in Fig. 2.

b.   Nozzle.   The nozzle must be installed so that it does not protrude into the inner surface of the tank, as shown in Fig. 3. The nozzle fitting must be wedged or chamfered in a large radius (Fig. 4). The nozzle flange must be fabricated in the shape shown in Fig. 5a–c.

c.   Interior.   Complicated parts, such as projections, must not be attached to the inner surface of the tank.

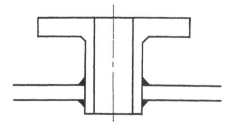

Figure 3   Bad example of nozzle fitting to tank wall.

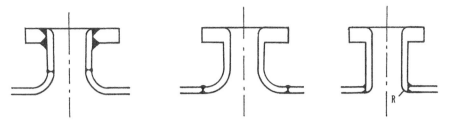

Figure 4 Typical example of nozzle fitting to tank wall.

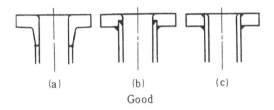

(a)   (b)   (c)

Good

Figure 5 Welding procedure for nozzle flange.

B. Coating Tank

1. Execution Method

In some cases the lining is applied to the tank during manufacturing, but the coating must be applied to the finished product. The coating is applied by electrolytic powder or spraying onto the pretreated surface. The process is shown in Fig. 6.

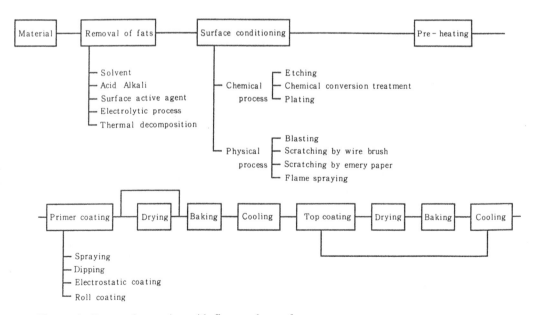

Figure 6 Process for coating with fluorocarbon polymer.

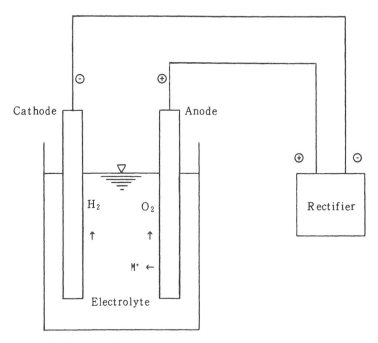

Figure 7   Electropolishing.

2.   *Precautions for Fabrication*

a.   Vessel.   Projections should be minimized and the corners should be chamfered in the same manner as for the lining.

b.   Pretreatment.   Precoating is essential. Thermal decomposition at high temperatures is often used for surface degreasing. Blasting is also used for cleaning and to adjust surface roughness.

C.   Stainless Steel

Stainless steel tanks are used for the entire primary side of the ultrapure water system. It is important that the inner surface be smooth and that there is little elution of metal ion. For this reason, stainless steel tanks are conventionally subjected to high-grade mechanical polishing. However, the conventional method does not satisfy recent requirements for the low amount of metal ion.

Electrolytic polishing and gold-EP are current methods of choice that reduce the amount of elution. In the gold-EP method in particular, the elution of total organic

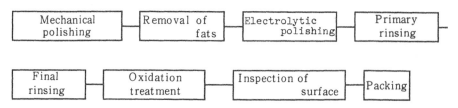

Figure 8   Gold-EP process.

carbons (TOC) in high-temperature ultrapure water is decreased by a factor of 2–3 compared with such plastic materials as perfluoroalkoxyl (PFA), polyvinylidene fluoride (PVDF), and polyether ether ketone (PEEK), as shown in Table 3. A schematic of electrolytic polishing is shown in Fig. 7.

It is desirable that projections be minimized on the inner surface of the tank.

The inner surface of the tank must be finished by mechanical buffing (No. 400 buffing or over).

Different electrolytic polishing methods may be used depending on the capacity of the tank. For a small-capacity tank (2 m^3 or less), electrolytic polishing is executed while the tank is filled with electrolyte under finished product conditions in which all welded parts are attached. For a tank with a capacity of 2 m^3 or more, electrolytic polishing is performed for the end and the shell independently.

In the gold-EP method the oxidative film on the surface subjected to electrolytic polishing is improved in quality; the process is shown in Fig. 8.

6
Heat Exchangers

Kazuhiko Takino
Hitachi Plant Engineering & Construction Co., Ltd., Tokyo, Japan

I. CLASSIFICATION

Heat exchangers can be classified as shown in Table 1. The plate heat exchanger is used in ultrapure water manufacturing, and it has the following features:

1. The overall coefficient of heat transfer is of the order of 4000 kcal $(m^2 \cdot h)^{-1}$, which is approximately five times more than that of the multitube heat exchanger. This means that the plate can be made compact in size.
2. The volume of resident liquid is small and the unit is easy to disassemble, both of which facilitate cleaning.
3. An increase or decrease in surface area is possible, and hence a change in handling capacity is easy.

Table 1 Classification of Heat Exchangers

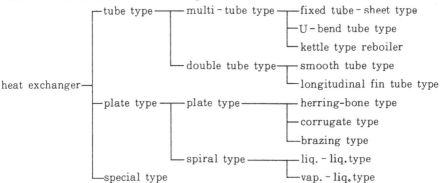

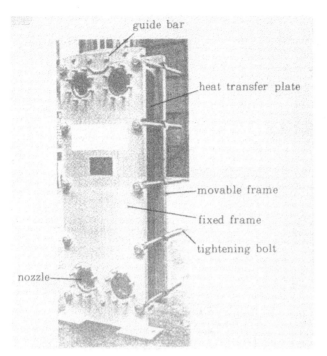

Figure 1   Plate heat exchanger.

II.   PLATE HEAT EXCHANGER

A.   Construction

As illustrated in Fig. 1, the plate heat exchanger comprises a number of heat conduction plates hung on the guide bars side by side between the two end frames. The heat conduction plate is made of stainless or titanium sheet metal (0.6–1.0 mm in thickness), which is press formed and provided with various protrusions and fluid passages. Provided between the plates are gaskets, and the plate assembly is firmly secured between the fixed and movable frames by tightening bolts.

   The spacer connections and fluid passages in the plates permit the hot liquid to flow in one direction through every other plate and the cold liquid to flow in another direction in a similar manner. The flow pattern of the hot and cold liquid in the plate assembly is as shown in Fig. 2.

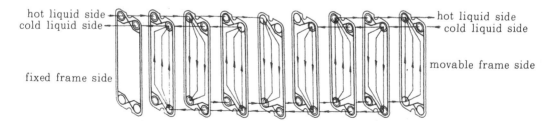

Figure 2   Example of fluid passages.

port hole   gasket

corrugate pattern   herring – bone pattern

Figure 3 Kinds of plates.

The shapes of heat conduction plate are shown in Fig. 3. The preferred material for the plate is SUS 316 L, and electrolytic polishing is becoming popular for the surface finish.

B. Basic Design

Normally, the basic design data for the plate heat exchanger are calculated by computer; however, approximate data can be obtained by hand calculation.

An example is shown here to find the number of 0.4 m² heat conduction plates when the temperature of ultrapure water flowing at 20 m³ h⁻¹ is reduced from 28 to 25°C by passing through cold liquid at 7°C in the plate assembly. The pressure loss of both the hot and cold liquid is assumed to be 0.5 kg cm⁻² each, and the margin for the heat transfer area is 10%. The procedure is as follows.

1. Calorific Value Transferred

$$q = 20 \text{ m}^3 \text{ h}^{-1} \times 1000 \text{ kg m}^{-3} \times 1 \text{ kcal kg}^{-1} \times 28\text{--}25°\text{C} = 60,000 \text{ kcal h}^{-1}$$

2. Outlet Temperature and Quantity of Cold Liquid

Because the acceptable outlet temperature differential of the cold liquid is generally 5°C, the outlet temperature is set at 12 and the quantity of the cold liquid Q is obtained from the equation

$$Q = \frac{60,000 \text{ kcal h}^{-1}}{(12\text{--}7)°\text{C} \times 1000 \text{ kg m}^{-3} \times 1 \text{ kcal kg}^{-1}} = 12 \text{ m}^3 \text{ h}^{-1}$$

3. *Logarithmic Average Temperature Differential* Δt

Ultrapure water	28 → 25°C
Cold liquid	12 ← 7°C
	16 – 18°C

$$\Delta t = \frac{16-18}{\ln 16/18} = 16.98°C$$

4. *Plate Arrangement*

There should be two liquid lines, one hot and one cold, arranged in parallel.

5. *Approximate* **K** *Value*

The flow rate for each line works out as, for ultrapure water,

$$20 \text{ m}^3 \text{ h}^{-1}/2 \text{ lines} = 10 \text{ m}^3 \text{ h}^{-1}$$

For cold liquid,

$$12 \text{ m}^3 \text{ h}^{-1}/2 \text{ lines} = 6 \text{ m}^3 \text{ h}^{-1}$$

The average flow rate is 8 m³ h⁻¹, when the plate heat exchanger allows a larger overall heat transfer coefficient (K value). The empirical K value in this instance is 3400 kcal $(\text{m}^2 \cdot \text{h} \cdot °C)^{-1}$.

6. *Heat Transfer Area*

$$A = \frac{q}{K\Delta t} = \frac{60,000 \text{ kcal h}^{-1}}{3400 \text{ kcal } (\text{m}^2 \cdot \text{h} \cdot °C)^{-1} \times 16.98°C} = 1.04 \text{ m}^2$$

To provide a 10% allowance,

$$\text{Heat transfer area} = 1.04 \text{ m}^2 \times 1.1 = 1.15 \text{ m}^2$$

$$\text{Number of plate conductors} = \frac{1.15 \text{ m}^2}{0.4 \text{ m}^2} = 3$$

7. *Plate Arrangement*

In practice, the last plate at each end does not participate in heat transfer; hence, the total number of the plates becomes five.

$$\text{Spaces between the plate: } 5 - 1 = 4$$

(ultrapure water side, 2; cold liquid side, 2). Because two lines have been provided for both liquids, the number of passages becomes 2/2 = 1, or one passage; hence,

Ultrapure water: two parallel lines × one passage
Cold liquid: Two parallel lines × one passage

8. *Pressure Loss*

The empirical formula for the plate heat exchanger is

$$\Delta P = AQ_R^2 nd\mu^{1/4}$$

where: A = coefficient = 0.008
  Q_R = flow rate per line, $m^3\ h^{-1}$
  n = number of passages
  d = specific gravity
  μ = viscosity, cP
  ΔP = pressure loss, $kg\ cm^{-2}$

$$\Delta P_1 = 0.008 \times \left(\frac{20}{2}\right)^2 \times 1 \times 1 \times 1 = 0.80 > 0.5\ kg\ cm^{-2} \qquad \text{not good}$$

$$\Delta P_2 = 0.008 \times \left(\frac{12}{2}\right)^2 \times 1 \times 1 \times 1 = 0.29 < 0.5\ kg\ cm^{-2} \qquad \text{OK}$$

In view of these results, as the pressure loss of the ultrapure water ΔP_1 exceeds the allowable limit, reexamination of the plate arrangement becomes necessary.

9. Reexamination of Plate Arrangement

The number of lines in parallel for the ultrapure water is increased to three.

Ultrapure water: three lines in parallel × one passage
Cold liquid: Two lines in parallel × one passage

10. Recalculation of Approximate K Value

Ultrapure water:

$$20\ m^3\ h^{-1}/3\ lines = 6.67\ m^3\ h^{-1}$$

Cold liquid:

$$12\ m^3\ h^{-1}/2\ lines = 6 \quad m^3\ h^{-1}$$

The average flow rate is 6.34 $m^3\ h^{-1}$. In step 5, the K value of 3400 kcal $(m^2 \cdot h \cdot °C)^{-1}$ is applied when the average flow rate is 8 $m^3\ h^{-1}$; however, the K value varies in proportion to the average flow rate raised to the 0.75 power; hence,

$$K' = \frac{3400}{(8/6.34)^{0.75}} = 2856\ kcal\ (m^2 \cdot h \cdot °C)^{-1}$$

11. Heat Transfer Area

$$A' = \frac{60{,}000\ kcal\ h^{-1}}{2856\ kcal\ (m^2 \cdot h \cdot °C)^{-1} \times 16.98°C} = 1.24\ m^2$$

By providing an allowance of 10%, the heat transfer area becomes 1.36 m^2.

The required number of six heat conduction plates is obtained by $1.36/0.4 = 3.4 \fallingdotseq 4$; 2 added is for the end plates, which do not contribute to heat transfer but serve to form the liquid passages.

12. Recalculation of Pressure Loss

Ultrapure water:

$$\Delta P'_1 = 0.008 \times \left(\frac{20}{3}\right)^2 \times 1 \times 1 \times 1 = 0.36 < 0.5\ kg\ cm^{-2} \qquad \text{OK}$$

Cold water:

$$\Delta P'_2 = 0.008 \times \left(\frac{12}{2}\right)^2 \times 1 \times 1 \times 1 = 0.29 < 0.5 \text{ kg cm}^{-2} \qquad \text{OK}$$

13.   Total Number of Plates

Extra plates are necessary to provide for the passage of liquid:

Ultrapure water: three lines in parallel $\times$ one passage = three plates
Cold water: two lines in parallel $\times$ one passage = two plates

The total number of plates calculated works out to 6 (3 + 2 + 1), and this number is consistent with that obtained by recalculating the approximate K value and heat transfer area in steps 10 and 11, respectively.

7
Afterword

Takeshi Shinoda
Hitachi Plant Engineering & Construction Co., Ltd., Tokyo, Japan

Descriptions of the construction, materials, and design philosophy for the major equipment and pipes for an ultrapure water manufacturing facility that answers the needs of the so-called megabyte age have been collected in this part.

Remarkably advanced technical studies are being done in each field by users as well as by plant and equipment manufacturers. Technical improvements must be substantiated by technical data, which will have to be collected steadily over a prolonged period of time. Much is expected of those who pursue the research and development of various topics discussed here. By the same token, similar improvements are being carried out on equipment and materials not covered here, and it is earnestly hoped that everyone concerned does his or her best for the betterment of ultrapure water processing technology.

V
Instrumentation

1
Introduction

Tetsuo Mizuniwa
Kurita Water Industries, Ltd., Kanagawa, Japan

As the density of the large-scale integrated (LSI) chip increases, less contamination of the external surface of the wafers is permitted. At the same time, it is necessary to maintain the uniformity of surfaces of a wafer and also uniformity among wafers. To meet these requirements, the environmental purification of film-forming processes and etching processes is indispensible, and perfect wet cleaning of wafer surfaces is also important. This is the ultimate purpose of a stable supply of ultrapure water of the lowest impurity concentration.

In addition to qualitative stability, quantitative stability is required to supply ultrapure water to systems at any time. For the stable supply of ultrapure water, the pressure, flow rate, and the quality of the water produced by the ultrapure water system must be monitored, so that deviations from specification can be detected and countermeasures taken immediately. Because the quantity of impurities in current ultrapure water is extremely low, even a small change instantly causes a fluctuation in water quality. Therefore, instruments to measure water quality must be highly sensitive and promptly responsive.

The quality of ultrapure water is evaluated by resistivity, particle counts, total organic carbon (TOC), bacteria counts, and ion concentrations. Although there have been strong demands for real-time measurement of all these parameters, because of technical restrictions originating from the measuring principles and systems, currently practical real-time measurement is limited.

Many evaluation methods and instruments for ultrapure water are based on novel principles. In some cases, several principles are applied to measure a single parameter. For example, among the instrumentation that measures TOC, a qualitative indication of ultrapure water, several types are available: some oxidate organic substances by ultraviolet light radiation, and others oxidate by chemical reaction at high temperature. Several types of particle counters are on the market that differ in the way they detect scattered light or process data, although all the types share the principle of light scattering. No conspicuous differences in indication may arise between the different types of counters, but since every type has its own features, it is necessary to understand the principles, characteristics, and applicable limits before use.

In ultrapure water production systems, the instrumentation for measuring physical quantities, such as pressure, flow rate, and the water level in the tank, are based on relatively simple principles. However, the construction of such instruments should not allow contamination of ultrapure water: these instruments are in direct contact with ultrapure water distributing lines. As a result, conventional measuring instruments often cannot be used for ultrapure water systems without modifications. This part deals with the principles and features of the instrumentation that monitors the water quality according to stringent requirements or regulates system operation without contaminating the ultrapure water.

2
Instruments for Operational Control

Kazuhiko Takino
Hitachi Plant Engineering & Construction Co., Ltd., Tokyo, Japan

I. FLOWMETERS

The flowmeter is classified as shown in Table 1, depending on the principle of detection. Meters are selected for use in ultrapure water manufacturing based on the following points:

1. The material that comes in contact with the liquid is made of elements that produce the least amount of impurities.
2. Sliding mechanisms are kept to a minimum so that the generation of fine-grained particles is reduced.
3. The liquid is retained in the meter for a minimum amount of time, and it is easy to clean.
4. The meter functions properly even when used in a chemical rinsing or boiling water sterilization operation.
5. The pressure loss is minimal.

Table 1 Flowmeters

kind	principle
floater flowmeter	Bernoulli's theorem
orifice flowmeter	Bernoulli's theorem
paddle flowmeter	No. of revolutions of paddle
ultrasonic flowmeter	Propagation velocity of ultrasonic
doppler flowmeter	Doppler's effect

A flowmeter that satisfies these requirements is the area flowmeter, illustrated in Fig. 1, which is widely used. The area flowmeter measures the quantity of liquid through the position of a float, which is buoyed by a constant pressure differential between up- and downstream created by changing the area of liquid passage through a tapered tube. The volumetric quantity of flowing liquid Q that passes through the tapered tube, which is positioned plumb, and the ring passage around the float may be expressed by the formula

$$Q = CA \sqrt{\frac{2gV_f\,(\rho_f - \rho)}{\rho A_f}}$$

where: Q = volumetric quantity of flowing liquid
C = flow coefficient
A = area of ring passage
V_f = volume of float
A_f = area of float
ρ_f = density of float
ρ = density of liquid
g = gravitational acceleration

For the direct-reading type of flowmeter the tapered tube is made of either Pyrex or silica glass. If there is a possibility that the glass tube may be damaged, a magnetic detector area flowmeter is employed.

The accuracy of the area flowmeter—$\pm 1\%$ over the entire measuring range—is good enough for use in operational control.

Other types of flowmeters, for example the compact paddle flowmeter shown in Fig. 2, are used in conjunction with the measuring instrument or the sampling device. The paddle flowmeter proper is a set of four rotating paddles with a built-in magnet and a pickup coil. According to the flow of liquid in the tube, the paddles rotate, causing the magnetic field to pass across the coil, and ac signals are generated in the coil. By transmitting these ac signals to various meters, the flow rate can be determined. The measuring accuracy of this type of flowmeter ranges from 0.5 to 2% depending on the construction of the paddle.

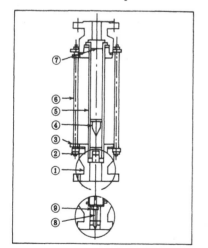

No.	parts name	material
①	main body	SUS 304 , SUS 316
②	packing	NBR, teflon
③	gland nut	SUS 304 , SUS 316
④	float	SUS 304 , SUS 316 , PVC, teflon
⑤	tapered tube	glass
⑥	supporter	SS 41
⑦	float holder	SUS 304 , SUS 316
⑧	guide rod	SUS 304 , SUS 316
⑨	float stopper	SUS 304 , SUS 316

Figure 1   Floater flowmeter.

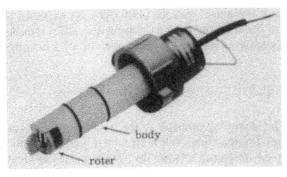

Figure 2 Paddle flow sensor.

The ultrasonic flowmeter shown in Fig. 3 is provided with a paired ultrasonic wave transmitter and receiver that cross a tune. It detects flow rate based on the principle that the propagation velocity of an ultrasonic wave varies with the propagation velocity of the liquid. The detection of variations in ultrasonic propagation by time differential is known as the propagation time differential method, and detection by frequency differential is known as the frequency differential method.

The equation for the propagation time differential method (see diagram in Fig. 3) can be expressed as

$$Q = VD = \frac{\Delta t}{t_0{}^2 \sin 2\theta}$$

where: Q = volumetric quantity of liquid flow
V = velocity of liquid flow
D = inside diameter of tube
Δt = time differential in ultrasonic pulse transmission from detectors P_2 to P_1 and P_1 to P_2
t_0 = average propagation time from detector P_2 to P_1
θ = intersection angle of lines representing direction of liquid flow and ultrasonic propagation passage

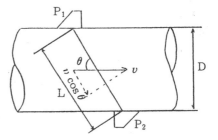

Figure 3 Ultrasonic flow sensor.

The ultrasonic flowmeter has many advantages over the other types: for example, it has no moving parts and no retention of liquid as a result of its construction, and it is made of the same material as the tube. The measuring accuracy is of the order of ±1% over the entire measuring range.

II.   PRESSURE GAGES

The Bourdon gage is widely used. The measuring principle of the Bourdon gage makes use of the displacement of a circular arc Bourdon tube. When the tube is pressurized, the free end changes position according to the difference between the internal and the external pressure. This displacement is converted into the movement of a pointer by means of an amplification mechanism, and the pressure is indicated on a scale plate.

In the conventional Bourdon pressure gage, however, cleaning the inside of the tube is a problem. Moreover, liquid retention is quite large, providing a median in which bacteria can proliferate. Hence, a diaphragm pressure gage—a gage provided with a diaphragm that serves as a partition between the flowing liquid and the tube, which is filled with a sealed liquid—is widely used in ultrapure water manufacturing. A diaphragm pressure gage is shown in Fig. 4. The diaphragm is made of SUS 316 or SUS 316 coated with fluorinated ethylene-propylene–Teflon.

A gage with an accuracy range of 0.5–3% over the entire measuring range is available, but a gage that has an accuracy of 1.5% over the entire range is adequate.

Lately, various types of pressure sensor (a representative sensor is shown in Fig. 5) are being examined for use with the gage. In the strain gage combined with a pressure sensor, the pressure applied to a metal strain gage through the pressure sensor is picked up by means of a bridge circuit as electrical resistance, which is in turn converted into amplified electrical output signals to indicate pressure. This type of strain gage enables stable pressure reading because of its simple construction: it is fixed directly to a flat

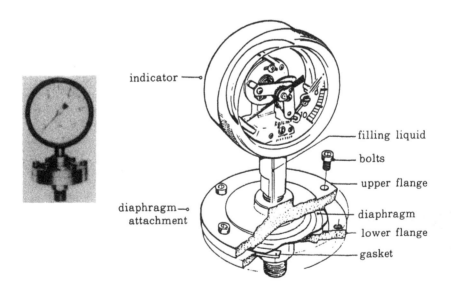

Figure 4   Diaphragm seal pressure gage.

Figure 5 Pressure sensor.

diaphragm and has no moving parts. The measuring accuracy of the metal strain gage ranges from ±0.5 to ±1.0% over the entire measuring range. A semiconductor pressure sensor is also available.

III. THERMOMETERS

Of the thermometers shown in Table 2, the thermocouple and the platinum resistance wire thermometers are widely used in ultrapure water manufacturing. The thermocouple thermometer is provided with two different metal wires joined together to measure temperature by making use of the thermoelectromotive force induced by the temperature differential between the standard contact and the temperature-measuring contact. In Fig. 6, the relationship between the thermoelectromotive force and the temperature of the thermocouple is shown.

Table 2 Thermometers

principle	kind		used temp. (°C)	responsi-bility	control use
expansion	etched-stem thermometer	mercury-in glass	$-50\sim+550$	ordinary	unsuitable
		liquid-in glass	$-100\sim+200$		
	bimetal thermometer		$-50\sim+500$	slow	
pressure	mercury filled thermometer		$-30\sim+600$	ordinary	
	vapor filled thermometer		$-20\sim+350$		
resistance	platinum resistance bulb		$-260\sim+630$	ordinary	
	thermistor thermometer		$-50\sim+350$	fast	Suitable
thermo-couple	thermo-couple thermometer	R (PR)	$0\sim+1554$	fast	
		K (CA)	$-180\sim+1000$		
		E (CRC)	$-180\sim+700$		
		J (IC)	$-180\sim+600$		
		T (CC)	$-180\sim+300$		

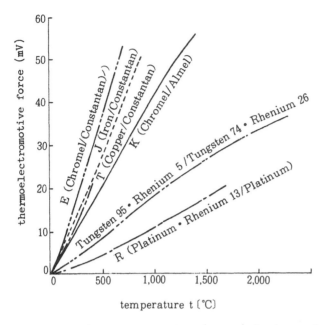

Figure 6   Relationship between thermoelectromotive force and temperature of the thermocouple.

Essentially, the thermocouple thermometer detects temperature differentials; therefore, to measure the actual temperature at a measuring point, it is necessary to make a temperature correction for the standard contact. The bridge circuit corrects the standard contact temperature automatically by measuring the standard contact temperature by resistance wire thermometer. A thermocouple thermometer with a protection tube is shown in Fig. 7.

The platinum resistance wire thermometer measures temperature by taking advantage of the phenomenon that the electrical resistance of platinum varies according to temperature. Figure 8 shows the temperature versus resistance characteristics of platinum.

The construction of the platinum resistance elements is shown in Fig. 9. Because the platinum resistance wire changes resistances as a result of mechanical strain, it is wound on an insulator so that there is no tension or bend. Figure 9a shows a resistance element inserted into a protection tube, which is wound into the serrated grooves of a strip of mica plate and sandwiched by mica insulators. Figure 9b shows a resistance element sealed in a glass tube, which is firmly wound into the grooves of a glass rod with the same thermal

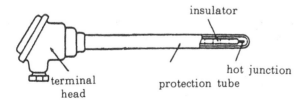

Figure 7   Thermocouple thermometer.

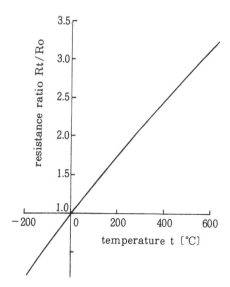

Figure 8   Relationship between resistance ratio and the temperature of platinum.

expansion coefficient as platinum. This instrument is small and quick to respond and strongly resists vibration.

IV.   LEVEL METERS

The level meters used in ultrapure water manufacturing are fitted with a stainless steel electrode level switch or a float level switch (Figs. 10 and 11). Care must be exercised when selecting the electrode level switch for use in an ultrapure water system, because signals may not be conducted using a conventional electrode switch: the extremely low conductivity of the ultrapure water creates a high resistance across the electrodes.

In the float level switch, as the float, which contains a magnet, moves up and down around a shaft with the movement of the water level, the magnet causes a reed switch,

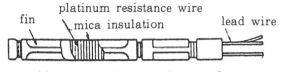

(a) Mica platinum resistance element

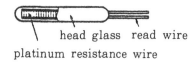

(b) Glass bobbin resistance element

Figure 9   Platinum resistance elements.

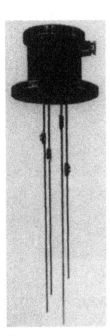

Figure 10   Electrode level switch.

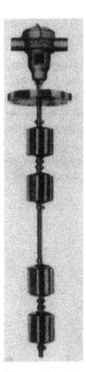

Figure 11   Magnet float level switch.

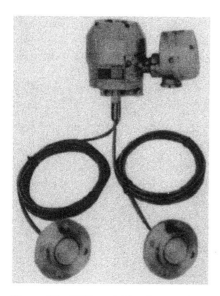

Figure 12 Differential level transmitter.

which is embedded in a polyvinyl chloride shaft at a predetermined position, to close or open.

For the continual monitoring of holdup water volume in tanks, such as ultrapure water reservoirs, a diaphragm differential manometer, as shown in Fig. 12, is used. In the diaphragm differential manometer, the pressure differential between the bottom diaphragm transmitter, which is under constant static pressure, and the upper diaphragm transmitter, which is open to the atmosphere, is utilized. Normally, the diaphragm is made of SUS 304 or SUS 316.

3 On-Line Monitors

A. High-Sensitivity Resistivity Meter

Makoto Saito
DKK Corporation, Tokyo, Japan

I. INTRODUCTION

The electrical resistivity reflects the total volume of electrolytes contained in water, and the resistivity, the reciprocal of electrical conductivity, reflects the purity of the water. It is reported that the resistivity of pure water without electrolyte ions is 18.1–18.3 mΩ·cm at 25°C. With the rapid progress in the integration of semiconductors, the cleaning process must use pure water with a resistivity of 18 MΩ·cm or higher. Consequently, the conventional general-use resistivity meter does not detect water quality with sufficient accuracy. The high-sensitivity resistivity meter shown in Fig. 1 was developed to overcome this problem.

The resistivity of pure water around the 18 MΩ·cm level changes by about 0.1 MΩ·cm as the temperature changes by 0.1°C. It also changes by about 0.1 MΩ·cm as the electrolyte concentration changes by 0.1 μg L^{-1}. The newly developed high-sensitivity resistivity meter features a resolution of 0.01 MΩ·cm. It has been confirmed that the high-sensitivity resistivity meter exhibits an indication stability of less than 0.05 MΩ·cm against stable pure water with a resistivity of around 18 MΩ·cm at 10–30°C.

II. SPECIFICATIONS FOR HIGH-PURITY WATER

High-purity water is now consumed in large quantity in rapidly developing fields, such as semiconductor production lines and nuclear power stations. However, there have been no official standards on quality, and users set up their own requirement specifications.

In Japan, many studies are underway to introduce the Japan Industrial Standard (JIS) to experimental procedures for high-purity water. In the United States, four types are stipulated for electronic-grade water in terms of quality by the American Society for Testing and Materials (ASTM) in ASTM-19 Proposal P-172. Type E-1, the highest,

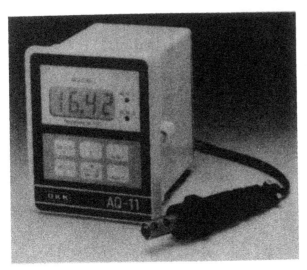

Figure 1 High-sensitivity resistivity meter model AQ-11 and sensor.

requires pure water to exhibit a resistivity of 18 (90% of the time), with 17 a minimum. Table 1 lists the ASTM specifications for water quality.

III. RESISTIVITY OF PURE WATER

Because some water molecules are dissociated into hydrogen ion and hydroxide ion in water, the resistivity of the water never reaches the infinitely high level. Table 2 lists the test data that have been reported with regard to the electrical conductivity of pure water, excluding impurities. The values vary slightly among researchers, and this is thought to be because of differences in dissociation constants and limit ion equivalent conductivities, for example.

Since the resistivity is affected by the temperature, a resistivity meter conducts analog and digital calculations when it receives the resistivity signals, automatically converting the reading value to the value at 25°C. This is called automatic temperature compensation.

Table 1 Electronic-Grade Water Specified by the ASTM

	Type E–I	Type E–II	Type E–III	Type E–IV
Resistivity,minimum,MΩ · cm	18 (90% of time) with 17 minimum	15 (90% of time) with 12 minimum	2	5
SiO_2(total),maximum, μ g/L	5	50	100	1000
Particle count (particles larger than 1 μ m),maximum per millilitre	2	5	100	500
Viable Bacteria, maximum per millilitre	<1	10	50	100
Total organic carbon, maximum, μ g/L	50	200	1000	5000
Copper[A], maximum, μ g/L	<1	5	50	500
Chloride[A], maximum, μ g/L	2	10	100	1000
Potassium[A], maximum, μ g/L	<2	10	200	500
Sodium[A], maximum, μ g/L	<1	10	200	1000
Total Solids[A], maximum, μ g/L	10	50	500	2000
Zinc[A], maximum, μ g/L	5	<20	200	500

[A]Concentrations to be measured periodically at the option of the water system to aid in the diagnosis of system problems
Source: From Ref. 1.

Table 2 Conductivity of Pure Water ($\mu S\ cm^{-1}$)

Reporter / Temperature (°C)	Otten [1]	Rommel [1]	Light, Sawyer [2]	Iverson [3]	Kohlrausch [4]
0	0.01163	0.0114	0.0119		0.0149
5	0.01667	0.0165	0.0168	0.0170	
10	0.0231	0.0228	0.0233	(0.0233)	0.0268
15				0.0315	
20	0.0418	0.0414	0.0421	(0.0422)	
25	0.05483	0.0546	0.0550	0.0548	
30	0.0714	0.0708	0.0709	(0.0708)	
35				0.0893	
40	0.1133	0.114	0.112	(0.0111)	
45				0.137	
50	0.1733	0.174	0.169	(0.166)	0.178
55				0.201	
75	0.409	0.406	0.393		
100	0.788	0.771	0.737		

* 1 K. Rommel, 1985, "Verfahren zur Leitfahigkeitsmessung hochreiner Wässer" VGS KRAFT WERKSTECHNIK 65 Hert
* 2 T. S. Light and P. BSawyer, 1981 "Resistivity of Very Pure Water and Its Maximum Value"
* 3 See Ref. 2
* 4 See Ref. 2
Source: From Ref. 2.

If the reading data used for the calculation differ, as shown in Table 2, the values processed by automatic temperature compensation (converted to the value at 25°C) also differ. In the course of designing the high-sensitivity resistivity meter, many experiments were repeatedly conducted to obtain the proper design data.

IV. AUTOMATIC TEMPERATURE COMPENSATION

In the high-sensitivity resistivity meter, the temperature signals from the temperature element and the resistivity signals from the electrodes are calculated to obtain the resistivity value converted to 25°C, which is displayed after digital calculation in the one-chip central processing unit (CPU). The calculations are described here. To simplify the description, the values are expressed not by resistivity but by electrical conductivity. The ratio between the conductivity at $t°C$ (K_t) and that at 25°C (K_{25}) is called the temperature function (K_t/K_{25}) in the following description. Figure 2 shows the temperature function of major electrolytes and pure water.

The temperature function of an electrolyte $f(t)$ and the temperature function of pure water $g(t)$ can be expressed as

$$f(t) = \frac{K_{it}}{K_{i25}} \qquad g(t) = \frac{K_{wt}}{K_{w25}} \tag{1}$$

where: K_{it} = conductivity of electrolyte
$\qquad K_{wt}$ = conductivity of pure water

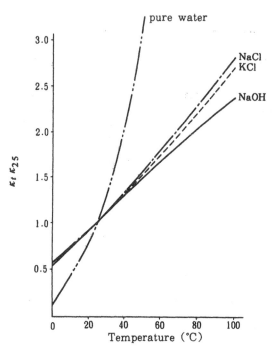

Figure 2 Temperature gradient of conductivity of diluted electrolytes and pure water.

According to Fig. 3, the conductivity of high-purity water at t°C can be expressed as the sum of K_{it} and K_{wt}:

$$K_t = K_{it} + K_{wt} \qquad K_{25} = K_{i25} + K_{w25} \qquad (2)$$

The conductivity of high-purity water at 25°C K_{25} can then be expressed as

$$K_{25} = \frac{K_t + K_{w25}g(t)}{f(t)} + K_{w25} \qquad (3)$$

Therefore, if K_{w25}, $f(t)$, and $g(t)$ are obtained in advance and t and K_t are measured, the resistivity converted to 25°C can be obtained by calculating the reciprocal of Eq. (3).

The high-sensitivity meter uses salt (NaCl) as the electrolyte $f(t)$. However, as shown in Fig. 2, as the temperature differs from 25°C, the difference from other electrolytes increases. This means that when several electrolytes are used, errors may arise in the values indicated by the meter. In operating production lines, the resistivity fluctuation due to the dissolved impurity salts generated in the piping system is greatly affected by construction, materials, temperature, flow rate, water velocity, and other factors. In this case, it is not easy to identify the kinds of impurity salts. Even if the salts that are dissolved under certain conditions can be analyzed and identified, the concentration of impurities that is obtained is not always reproduced when the conditions are changed.

It is therefore desirable to conduct measurements at a temperature as close to 25°C as possible. This is why the range of automatic temperature compensation is set from 10 to 30°C in this high-sensitivity resistivity meter.

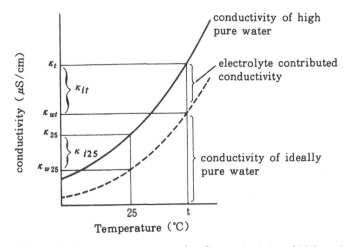

Figure 3   Temperature compensation for conductivity of high-purity water.

V.   DETECTOR

Figure 4 shows the configuration of the detector developed for the high-sensitivity resistivity meter, and Table 3 compares the conventional and newly developed detectors. To upgrade the sensitivity and the performance, the temperature responsiveness and the conversion rate of sample water are improved in the newly developed detector.

A thermistor that records sensitive changes in resistivity against temperature fluctua-

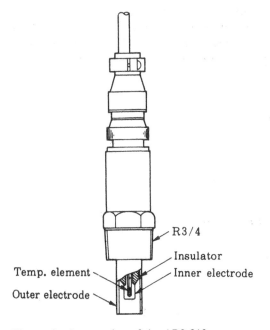

Figure 4   Construction of the AR2-212.

Table 3 Comparison of Sensors

Item \ Model No.	AR 1 - 211	AR 2 - 212
Cell constant	0.01cm^{-1}	0.1cm^{-1}
Electrode Material	Ti	Ti
Insulator	PPO	PPO
Temperature range	0 ~ 100°C	10 ~ 30°C
Pressure range	0 ~ 5kgf/cm^2	0 ~ 5kgf/cm^2
Temperature element	Platinum RTD Class, 0.2	Thermistor Class, 0.1

tion is used for the temperature measurement element. The thermistor is inserted at the end of the inner electrode, setting the dimension of the inner electrode at 1/10 and the cell constant between 0.01 and 0.1 cm^{-1}.

Because the resistivity changes by about 0.1 MΩ·cm as the temperature fluctuates by 0.1°C, high-accuracy temperature measurement is essential to conduct automatic temperature compensation in an accurate manner. Therefore a class 0.1 thermistor was selected. The actual calculation of automatic temperature compensation is performed with a resolution of less than 1/100°C. As each high-sensitivity detector is combined with a particular cell constant, a combined calibration can be conducted by inputting each cell constant into each detector.

VI.   HIGH-SENSITIVITY RESISTIVITY METER

Table 4 compares the conventional resistivity meter and the high-sensitivity resistivity meter. Although both meters have the same measurement range of resistivity (0–20 MΩ·cm), the resolution is set at 0.01 MΩ·cm over the entire range in the high-sensitivity resistivity meter. For transmission output, the arbitrary measurement range of over

Table 4 Comparison of Specifications

Item \ Model No.	AQ-10	AQ-11
Resistivity range	0.00 ~ 9.99 ~ 10.0 ~ 20.0 MΩ · cm	0.00 ~ 20.00 MΩ · cm
Temperature range	0.0 ~ 99.9°C	10.0 ~ 30.0°C
No. of sensors connectable	2-sensor, A & B	2-sensor, A & B
Power supply	AC 100 ± 10V	AC 100 ± 10V
Alarm output	2-output for A	2-output for A
Transmission output	DC 4~20 mA for A	DC 4~20 mA for A Output range is variable

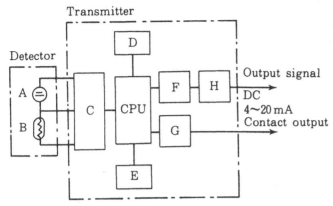

A : Electrode for resistivity measurement
B : Temperature measuring element
C : Input converter
D : Display
E : Switch
F : Isolation circuit
G : Relay
H : Output converter

Figure 5 Block diagram of the AQ-11 resistivity meter.

5 MΩ·cm can be converted to the full scale of 4–20 mA direct current. Therefore, when reading the resistivity with the recorder, it is possible to attain a resolution of 0.01 MΩ·cm.

Although the basic structure of the measurement range in the high-sensitivity resistivity meter is the same as that in the conventional meter, the CPU and the memory chip have been upgraded in an attempt to improve the resolution. Figure 5 is a block diagram of the measurement circuit.

VII. EXPERIMENTAL RESULTS AND MEASUREMENT DATA

Figure 6 shows the experimental apparatus and the procedures, and Fig. 7 shows the experimental results. The purity of water flowing in the experimental apparatus was improved by operating the apparatus constantly for 1 week before the experiment. As a result of this preliminary operation, the resistivity was stabilized at 18.24 MΩ·cm at 25.00°C. Over the course of the experiment, the temperature of the ultrapure water was changed by the heat exchanger and the readings at two high-sensitivity resistivity meters A and B were recorded. The change in the readings was 0.02 MΩ·cm within the temperature range of 10–30°C.

The evaluation test was carried out next by installing the high-sensitivity resistivity meter in a large-scale ultrapure water production plant in actual operation. Figures 8 and 9 show the results. Figure 8 shows the test results when the resistivity at the inlet and the outlet of each unit operating in the plant was measured in the field [3]. Figure 9 shows the results of a test conducted in another plant. In this case, the high-sensitivity resistivity meter was compared with another resistivity meter that had been installed in the plant. The temperature of the sample water was changed to observe the change in the readings of

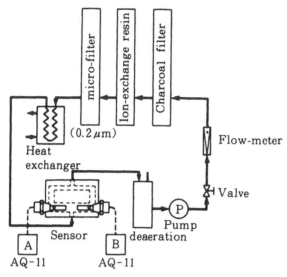

Figure 6 Flow of resistivity meter testing.

the two meters. This test proves that the high-sensitivity resistivity meter has a sufficiently secure temperature compensation to be put to practical use.

VIII.   CONCLUSION

When a trace amount of electrolyte is dissolved in pure water, the resistivity of the pure water is reduced. For example, when the salt concentration in pure water is increased by 0.1 μg L^{-1}, the resistivity drops by 0.07 MΩ·cm. Therefore, the high-sensitivity resistivity meter is expected to be applied not only to the monitoring and inspection of the ultrapure water production system but also to the detection of trace ions and the evaluation of various component units of the ultrapure water production system. Future requirements are to standardize the specifications for ultrapure water and the instrumentation.

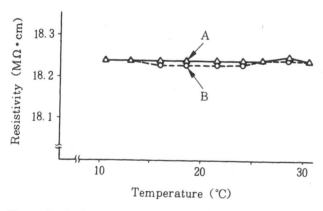

Figure 7 Performance of model AQ-11.

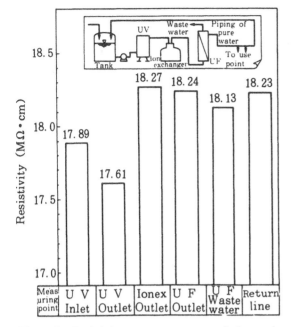

Figure 8   Resistivity measurement example in an ultrapure water process.

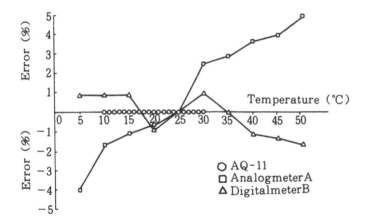

Figure 9   Comparison of resistivity meters.

ACKNOWLEDGMENTS

Author extends gratitude to Mr. Koichi Yabe and Mr. Hiroo Ishikawa of Kurita Water
Industries, Ltd. for their valuable advice given in the course of the development of the
high-sensitivity resistivity meter.

REFERENCES

1. ASTM Section 11, Volume 11.01, Electronic Grade Water, D—19 Proposa, p. 172.
2. T. Kikuchi, Kougyo Yousui, *356* (1988).
3. K. Yabe et al., Proceedings of 5th Symposium on ULSI Ultra Clean Technology (Tokyo), (1987), pp. 1–23.

3   On-Line Monitors

B.   TOC Monitor

Masami Miura
Hakuto Co., Ltd., Tokyo, Japan

I.   INTRODUCTION

The quality of water supplied to the ultra–large-scale integrated (ULSI) chip production line has lately been improved to very high levels because it directly affects the ULSI product yield. The requirements for ultrapure water differ with the intended use: in a nuclear power plant, where corrosion is a critical problem, such electrolytes as chlorine ions are closely watched; in the pharmaceutical industry, bacteria must be carefully removed from the ultrapure water. In the ULSI production line, the highest level of purity is required for the ultrapure water. The resistivity meter has been widely used for a long time as the method to monitor the water quality. Recently monitoring by a total organic carbon (TOC) monitor has gained attention for the accurate measurement of dissolved organic compounds.

It has been reported that the latest ULSI production system is capable of a stable supply of ultrapure water of the quality required for the 16 Mbit DRAM—a TOC level of 10 ppb or less—by adopting both the two-stage reverse osmosis unit and the ultraviolet (UV) oxidation unit with the 184 nm UV lamp. In measuring the concentration of organic materials in ultrapure water, since it is extremely difficult to obtain quantitative data for each organic material, the TOC concentration is detected instead. There are several methods of oxidizing organic materials and detecting carbon dioxide. The oxidation methods include dry oxidation (the combustion method) and wet oxidation (UV oxidation, high-temperature pressurized wet oxidation, and heating oxidation). Detection methods include detecting the change in the electrical conductivity of water caused by the oxidation of organic materials and detecting carbon dioxide formed through the oxidation of organic materials by nondispersive infrared (NDIR) gas analysis.

This part introduces the TOC monitor developed by Anatel Corp. (Boulder, CO). This TOC monitor detects the change in electrical conductivity induced when carbon dioxide formed by UV oxidation is dissolved in ultrapure water. The TOC level is obtained based on the conductivity value at the point at which the oxidation of organic materials is complete. Patents have been granted for the measurement principle, the configuration of the equipment, the structure of the sensor chamber, and the mathematic procedures for the measurements. This TOC monitor can be easily operated and maintained without special expertise. In addition, it is small and light. For convenience of operation, it does not use carrier gas or chemical liquid for oxidation, and frequent calibration is not necessary. As a result of these excellent characteristics and the patents that have been granted, this TOC monitor has been adopted worldwide and features high reliability.

TOC has generally been expressed as total organic carbon, but lately it has been expressed as total oxidizable carbon because it is not clear whether all the organic materials are sufficiently decomposed to carbon dioxide.

II. PRINCIPLE

A. Types and Structures

There are two models of the TOC monitor developed by Anatel: A-100 and A-100P. Model A-100 is installed at several points between the primary demineralization unit and the use points to conduct constant on-line monitoring. The A-100 unit is composed of a sensor chamber, a controller, and a power supply, each of which is packed in a small waterproof box. Model A-100P is a portable TOC monitor with the A-100 placed in one box and equipped with a digital printer for data logging. Model A-100P is used not only in the ultrapure water production system but also at each use point in the clean room. Just by plugging in the power supply and connecting the lines for water supply and water drain, the TOC concentration is measured on-line on the spot. Model A-100P is also widely used for calibrating the A-100 TOC monitors installed at each measurement point (Fig. 1).

The sensor chamber is composed of the UV irradiation unit, the conductivity and temperature measurement unit, and the valve. The controller is composed of the control circuit for the entire unit, the data-processing circuit, various displays, the operation panel, and the input/output interface. The measurement value is digitally displayed with a four-digit ppb (parts per billion) value. The ratio of the measurement value to the alarm level can be displayed as a percentage. In addition to the trend analysis function, whether the current reading is rising or falling during an arbitrary time span is displayed. If the reading is higher than the alarm level, the alarm lamp is turned on. The monitor also features a self-diagnosis function: in the case of malfunctioning or other problems, the specific condition is displayed by numerical code (Fig. 2).

B. Configuration of the Sensor Chamber

The sensor chamber, which is submerged in ultrapure water containing organic materials, is composed exclusively of chemically inert materials: high-purity fused silica (Supersil, Ameresil Co.) and titanium [1]. The seal materials required to trap ultrapure water are made from such durable materials as Viton (registered trademark of duPont), which are placed so that they are exposed to ultraviolet, preventing the degradation due to UV irradiation [2]. The UV lamp used here is a low-pressure mercury lamp. The ultraviolet ray of 184 nm wavelength, which is reported to be especially effective in decomposing and oxidizing organic materials in ultrapure water, accounts for only 3–6% of the irradiation; the remainder is chiefly 254 nm ultraviolet [3]. Therefore a type of glass

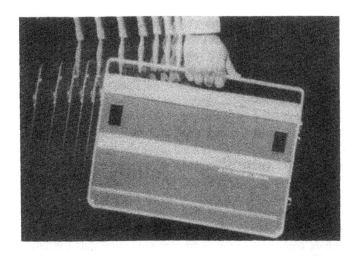

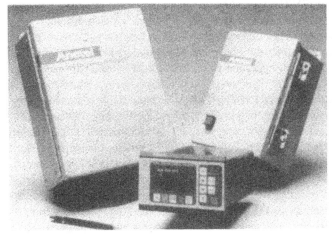

Figure 1   Model A-100P (Portable type; top). Model A-100 (on-line type; bottom).

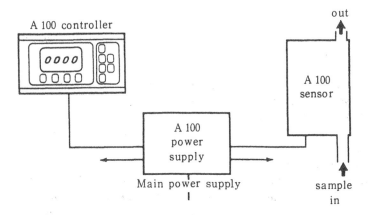

Figure 2   The A-100 standard configuration.

through which the ultraviolet rays of both wavelengths can permeate should be selected for the optical window of the UV irradiation unit. Supersil sufficiently satisfies this condition [4]. Because the 184 nm ultraviolet ray is easily absorbed by oxygen, the UV irradiation unit is filled with nitrogen gas to maintain irradiation efficiency.

The 184 nm ultraviolet ray has two other features: it forms OH free radical with very strong oxidizing action by decomposing water molecules, and it dissociates the chemical bonds of the organic materials dissolved in ultrapure water [5]. As a result, the organic materials in ultrapure water ultimately decompose into water and carbon dioxide [6]. Therefore, even with small amounts of dissolved oxygen in ultrapure water, the oxidation reaction of organic carbon with the 184 nm and 254 nm ultraviolet ray is complete in the low concentration range of organic carbon, 100 ppb or less.

The surface of the titanium electrode used to measure the conductivity is constantly exposed to ultraviolet, which keeps the surface free of organic contamination. Organic contamination on the electrode surface harms the conductivity measurement [4].

An inner and outer electrode pair is used; the outer electrode surrounds the inner electrode at a uniform distance [7]. The fixed thermometer mounted on the inner electrode performs temperature compensation for the electrical conductivity measurement.

When the organic carbon concentration to be measured in ultrapure water is very low, 100 ppb or less, elution from the component materials of the titanium electrodes and other parts, as well as the oxidation of organic materials, is caused by ultraviolet irradiation. This results in a conductivity rise over time and the conductivity is not stabilized [8]. In other words, titanium is constantly eluted and the electrical conductivity seems to increase endlessly [8]. This type of error in the conductivity derived from this equipment has a linear tendency, but the fluctuation in conductivity due to carbon dioxide formed through the oxidation of low-concentration organic carbon exhibits a nonlinear tendency. Those two different types of conductivity fluctuation should be clearly identified [8]. When the organic carbon concentration in ultrapure water is high, however, the error in the conductivity derived from the equipment can be disregarded [8]. The water inlet is located at the bottom of the sensor chamber and the water outlet is placed at the top to facilitate the drainage of bubbles from the ultrapure water. This is because bubbles trapped in the sensor chamber lead to errors in measurement. The valve that operates when ultrapure water is trapped in the sensor chamber is placed at the water outlet (downstream) to suppress contamination, derived from the valve, in the sensor chamber (Fig. 3) [2].

C. Measurement Cycle

Figure 4 shows the measurement cycle. Curve A is the change in the measured values of conductivity. The two lower parts of Fig. 4 illustrate the valve operation sequence and the on-off cycle of the UV lamp.

In step 1, the valve is open while the UV lamp is turned on. During this step, ultrapure water is constantly purged through the sensor chamber and all other water routes. At the same time, the materials remaining in the chamber are oxidized and decomposed by ultraviolet. In short, even trace contaminants remaining in the chamber or the sensor are cleaned and removed by oxidation, which guarantees reliable conductivity measurement with high reproducibility.

In step 2, the valve is open and the UV lamp is turned off. In this step, all the residual carbon dioxide produced in step 1 is drained from the sensor chamber. The valve is closed 10 s after turning off the UV lamp to trap ultrapure water in the chamber. In

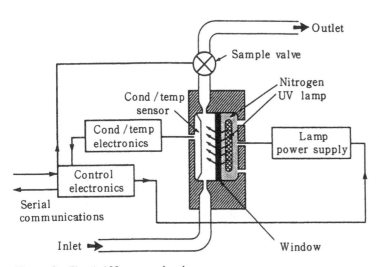

Figure 3 The A-100 sensor chamber.

this state, the background conductivity of ultrapure water, the conductivity before UV oxidation, is measured to perform temperature compensation.

In step 3, the valve is closed and the UV lamp is turned on. In this step, organic materials are oxidized and decomposed to form carbon dioxide. During this reaction, the conductivity gradually rises, nearing the asymptotic value. ΔC, the final conductivity minus the background conductivity, is converted to the concentration of organic carbon contained in ultrapure water. This conversion is conducted based on the well-known correlation between the carbon dioxide concentration in ultrapure water and the conductivity, using, for example, the correlation described in *A New Approach to the Measurement of Organic Carbon* (Poirier et al., American Laboratory DEC, 1978) [8].

In step 4, the valve is turned on to drain the ultrapure water that was treated by oxidation. Step 4 is followed by step 1 (purge mode).

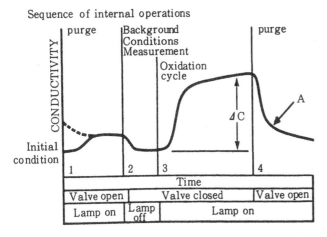

Figure 4 Measurement cycle.

D.   Principle

The TOC monitor calculates the concentration of organic carbon in ultrapure water based on the change in the conductivity of ultrapure water caused when carbon dioxide, which is formed during the decomposition and oxidation of organic materials in ultrapure water by 184 and 254 nm UV irradiation, is dissolved in ultrapure water as carbonic acid ion. The curve in Fig. 5 is an example of the oxidation reaction. The initial value of conductivity of the ultrapure water in this case is 0.055 μS cm^{-1}. The curve represents an idealized change in conductivity: as 50 ppb organic carbon is oxidized by UV, the conductivity nears a certain asymptotic value (0.5 μS cm^{-1}) [4]. In a typical case, the conductivity nears this asymptotic value after 5–10 minutes of exposure of ultrapure water to the UV ray. The line in Fig. 5 shows the change in the background value caused by elution from the material of this equipment, which is peculiar to this equipment [9]. Therefore, the actual change in conductivity is shown by the dotted line, the sum of the values on the curve and those on the line.

   More specifically, the change in the conductivity of ultrapure water is monitored as a function of time. At the point when the results of either the first-order differential or the second-order differential of conductivity against time equals zero and the conductivity fluctuation following that point is almost linear, the oxidation process is considered complete (see Fig. 6) [10].

   If the time for the oxidation process is fixed, accurate readings are sometimes not obtained. For example, when organic materials in ultrapure water are difficult to oxidize or the UV lamp becomes weaker, the readings may be lower than the actual values [11]. On the other hand, when the concentration of organic materials is very low and oxidation is completed earlier than expected, a serious problem may be caused by errors interference derived from the equipment. To avoid these problems, the TOC monitor monitors the oxidation process and accurately and promptly determines the time that the oxidation is complete using with a mathematic method [11], as follows. The time T when the result of the second-order differential of the conductivity reaches zero is multiplied by the result of the first-order differential of the conductivity at T, and then this result (dc/dt T) and the initial conductivity value measured before UV oxidation are deducted from the conductivity value measured at T to obtain ΔC, which is used to obtain the TOC level [10]. As mentioned earlier, since the correlation between the conductivity of ultrapure water and

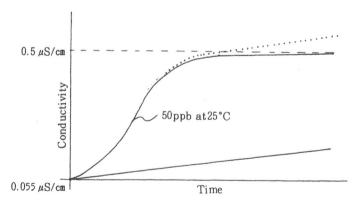

Figure 5   Conductivity change during oxidation of 50 ppb TOC.

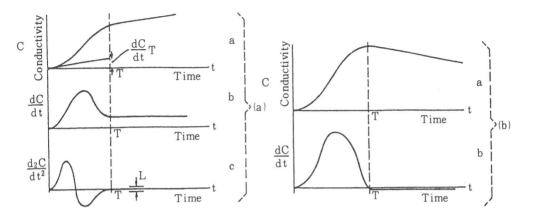

Figure 6   Mathematic technique for determining completion of the oxidation reaction   (Japanese Patent Publication No. 1988-46375).

carbon dioxide is well-known, ΔC can be converted to the TOC level. The value of dc/dt T corresponds to the background conductivity caused by the elution of organic materials or titanium from the equipment.

The background noise and the errors derived from the equipment are not reproducible [12]. Therefore, if the TOC level is measured accurately when the organic material concentration is low, the background conductivity dc/dt T must be precisely measured in each measurement cycle [12].

In this way, the linear change in conductivity caused by the equipment can be determined from the nonlinear change caused by oxidation by observing the second-order differential of the conductivity of ultrapure water against time [13]. As mentioned earlier, when the result of this second-order differential reaches zero within a certain measurement accuracy, the oxidation reaction is considered complete [13]. The first-order differential of the conductivity against time is also monitored. The value of the first-order differential at the point when the result of the second-order differential of the conductivity is zero corresponds to the slope of the background conductivity curve, which can be used as the total amount of background noise [13]. Moreover, when the background noise is deducted from the change in the measured conductivity value, the outcome of this calculation corresponds to the conductivity caused by the oxidation of organic materials in ultrapure water [13].

The major problem with conventional TOC monitors based on other principles is that complicated calibrations must be conducted frequently to deal with errors derived from the equipment and background values [12]. With the newly developed Anatel TOC monitor, however, absolute calibration can be performed merely by accurately calibrating the built-in temperature compensation conductivity meter [14]. Since the Anatel TOC monitor can automatically detect and compensate for the background values, it is not necessary to conduct the calibration as frequently [14].

Based on this basic principle, when ultrapure water containing organic materials is exposed to ultraviolet, the conductivity change due to the oxidation reaction shows the following three patterns over time, depending on the concentration and the type of organic materials (see Fig. 7a) [15]:

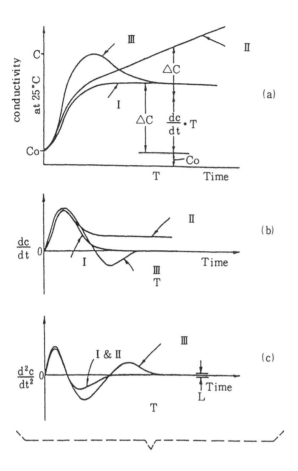

Figure 7   Three kinds of oxidation reaction and the mathematic technique for determining completion of the reaction (Japanese Patent Publication No. 1988-46375).

Type I: after some time the conductivity reaches a certain fixed value or begins to decrease constantly [16].

Type II: the conductivity changes in a nonlinear manner over time, and then it begins to change in a linear manner, gradually increasing [16].

Type III: the conductivity hits its peak and then decreases to attain a stationary value [16].

In these three types, the point at which the oxidation reaction is complete and the TOC concentration are determined using the following methods. In type I, it is obvious that the reaction is complete when the conductivity hits its peak once and then begins to decline [10]. Therefore, the conductivity value at the peak corresponds to the TOC concentration [10]. The oxidation reaction is complete at the point when the first-order time differential value of the conductivity curve reaches zero or below zero (Fig. 7b). The TOC concentration can be calculated based on the conductivity value at this point. In this case, the change in conductivity derived from the equipment is small enough to be ignored [10]. Therefore, type I is the typical case observed when the TOC concentration is high.

In type II, the first-order time differential of the conductivity (dc/dt) stabilizes at a certain value other than zero when the oxidation of organic carbon is complete (Fig. 7b) [10]. Therefore, the conductivity curve is a straight line when its second-order differential result (d^2c/dt^2) is found within the range of L, a small range near zero with the sampling error taken into consideration. This is when oxidation is complete (Fig. 7c) [10]. The conductivity change derived from the carbon dioxide can be calculated merely by deducting the error caused by the equipment ($dc/dt\ T$, where dc/dt is the slope of the curve of the conductivity change caused by the equipment and T is the time when the oxidation is considered complete) from the total change in conductivity. The TOC concentration is calculated based on a conversion table [10]. Type II can be observed when the contribution of organic carbon is smaller than that from the conductivity change derived from the equipment: in general, the TOC concentration is 50 ppb or less.

In type III, the peak appearing at reaction start-up is considered to be caused by the generation of intermediate product as the result of the oxidation of impurities, giving a higher conductivity than the final product of carbon dioxide [16]. For example, acetone and butanol in ultrapure water present similar behaviors [16].In type I and type II, the second-order differential curve for conductivity does not pass the zero point twice. Therefore, the case in which the curve passes the zero point twice is regarded as type III (Fig. 7c) [17]. If the first-order differential of the conductivity is negative when the second-order differential curve of the conductivity is zero for the second time, oxidation reaction type III is underway [18]. The point when the oxidation is complete can be determined by pinpointing when the second-order differential curve of the conductivity passes the zero point twice and enters the range of L from the positive side [17]. The TOC concentration can be obtained with the method mentioned earlier. When oxidation reaction type III is observed continuously in actual measurement, the following mathematic method can be applied to save measurement time. It has been proved that the conductivity change need not be monitored throughout the oxidation process to obtain accurate conductivity values when oxidation is complete. If the initial part (peak value) of the conductivity curve against time is within a certain range of the initial part of the previous sample, it can be assumed that the final part of the curve is also the same as that of the previous sample. In this way, the conductivity at the time the oxidation is complete in the previous sample is read after measuring the peak value [19].

The digital printer prints out the type of reaction (types I, II, or III), together with the conductivity change in each analytic cycle. It is possible to predict the types of organic materials dissolved in the ultrapure water to some extent based on these data. For example, when type I is followed by type III, it is obvious that different organic materials are eluting into the ultrapure water. The TOC monitor developed by Anatel, is the only monitor available on the market that features this analytic function. These are very important data that can be obtained by on-line measurement and applied to water quality and process control.

III.   CALIBRATION

The Anatel TOC monitor employs a measurement principle in which the TOC concentration is converted from the volume of the conductivity change (ΔC) in ultrapure water caused by UV irradiation. The calibration of the monitor attempts to gain a high accuracy of the conductivity sensor. The equipment that must be calibrated is first connected to a high-pressure ultrapure water loop to remove all contaminants from the pipes for ultrapure water, the sensor chamber, and the inside of valves. This is followed by calibration of the conductivity sensor carried out in accordance with the programmed procedures.

The electrical output characteristics of the conductivity sensor and its electrical circuit are adjusted to work together to obtain the output linearity. Therefore, precise adjustment of the conductivity sensor should be necessary only to obtain a resistivity of 2.0 M$\Omega\cdot$cm. More specifically, the conductivity sensor can be adjusted by adjusting the high-accuracy potentiometer placed in the circuit, whose function is to guarantee the slightest change in the cell constant. After adjustment, the monitors should be operated for over 10 h without a break to check stability. The TOC monitor is connected to the ultrapure water loop, and the calibration is then verified across a range of conductivities, with a final long-term test at high resistivities (greater than 17 M$\Omega\cdot$cm), a typical range for instrument operation. The TOC monitor, whose maximum deviation within the entire measurement range is limited to 0.001 μS cm^{-1} or less, clears the first step of this test. The correlation data are then taken with the temperature of ultrapure water varied from 15 to 40°C. In this way, whether the effect of dissolved ionized materials is accurately reflected in the conductivity along with the temperature change is quantitatively confirmed.

The next step of the calibration is measurement of the TOC concentration. A certain measured amount of TOC is injected into the ultrapure water loop. It has been proved through our experience that methyl alcohol dissolves in ultrapure water very well and that the solution is uniform and stable. For this reason, methyl alcohol for high-velocity liquid chromatography is used as the calibration material.

Correlation of low levels of organic contamination is as much statistical as anything else. A low concentration can be created by circulating trace amounts of materials in the ultrapure water loop at high speed to create a uniform suspended state. It has been proved that very little methyl alcohol used in this test is absorbed on the membranes or the inner surface of the piping system of the ultrapure water loop. As a result, a stable low concentration can be maintained. To get the methyl alcohol concentration in ultrapure water to decrease gradually so that the low concentration level is eventually reached, the UV oxidation unit for decomposing organic materials is installed in the middle of the ultrapure water loop. The extremely gradual decline in the methyl alcohol concentration is essential to obtain optimal uniformity in the liquid phase in the ultrapure water loop. This is why the ultrapure water circulating rate in the system, including the UV oxidation unit, is fixed at the high level.

The system for the test is capable of changing the TOC concentration in ultrapure water from 1200 to 0.4 ppb. To measure this concentration fluctuation, the standard unit, which is stringently managed and calibrated, and the equipment that requires calibration are connected to the test system. The resistivity of ultrapure water in the system is maintained at over 16 M$\Omega\cdot$cm. The level of dissolved oxygen in ultrapure water is also detected so that the unit under calibration can completely oxidize high concentrations of organic carbon. This method, in which the measurement data to be analyzed are taken from many points while constantly changing the TOC concentration using the UV oxidation unit to decompose organic materials, is much more effective than the method in which various organic materials in measured amounts are injected into ultrapure water one by one and the calibration is conducted each time. Furthermore, because our method can change the TOC concentration constantly in 5% increments, extremely accurate calibration is possible.

In the final calibration stage, no electrical or physical adjustment is conducted. The TOC concentration in the ultrapure water loop is measured to compare the measurement values with those of the standard unit. In short, the purpose of the final calibration is merely to confirm the analytic accuracy. This confirmation test is conducted at least twice. When the TOC concentration measured by the monitor being calibrated does not fix itself within 5% of the TOC level at the measurement point of the standard unit (in one measurement more than 100 points are picked up, and the same measurement is repeated more than twice), the monitor should be calibrated again.

The data from the calibration process are recorded with the production serial number of the equipment and kept by Anatel. If necessary, some of the stored data can be provided to users.

IV. POSTSCRIPT

About 3000 Anatel TOC monitors have been installed across the world for measuring and monitoring each process in the ultrapure water production system. Recently an increasing number of the TOC monitors have been mounted at the use points in the semiconductor production line.

In the wet process, which is one of the fabrication steps in very large scale integration (VLSI), it is reported that about 2 m^3 ultrapure water is consumed to process one sheet of 6 in. wafer. Since Si wafers go through the wet process repeatedly, to maintain the high quality of ultrapure water in the wet stations and the final rinsing bath it is essential to achieve high yield and quality of the devices. The degraded quality of the ultrapure water is directly reflected in the device. For example, the uniformity of the film thickness of the gate oxide in the CMOS (complementary metal oxide semiconductor) process deteriorates as a result of organic materials that adhere to the wafer surface [20]. The wafer surface is contaminated by the adsorbed organic materials if it is rinsed with ultrapure water immediately after HF etching [21]. Bacterial increase due to organic materials, which results in the contamination of wafers, has an indirect effect on the devices. To prevent these problems, it is critical to maintain the purity of the ultrapure water used in the wet process, which accounts for more than 20% of the entire semiconductor production process. Therefore, TOC measurement, as well as measurement of the resistivity, particles, bacteria, silica, and dissolved oxygen, becomes increasingly important to improve device reliability and yield.

In the 1990s, the volume production of 16 Mbit DRAM will be launched, and more stringent requirements will be presented in terms of the TOC concentrations in the wet process. The detection limit of the highest level TOC monitor available on the market now is 1 ppb. Recently, however, Anatel has developed a dramatically sensitive TOC monitor whose detection limit is 0.05 ppb to satisfy market demands. With the on-line TOC monitor, it now takes 3–10 minutes to complete the measurement, which is far from a real "on-line" monitor, and thus further improvement and development are required to reduce the measurement time.

To introduce an on-line TOC monitor capable of controlling the wet station as efficiently as the resistivity monitor, which as often been used as the on-line monitor, a new idea, free of conventional measurement principles, should be worked out to develop a TOC monitor that has a quick response and a low price and is small in size. At the same time, various problems to be overcome still exist in the TOC monitor for process control, as mentioned earlier. Therefore, for the time being, the most effective way to use the TOC monitor is to measure the TOC concentration in the wet station rinsing bath.

The quality of ultrapure water in the rinsing bath has been controlled only by resistivity. Because of progress in TOC monitor technology, however, it has been revealed that the purity of ultrapure water in the bath affects the yield and quality of devices. If the TOC concentration in the rinsing bath can be grasped accurately, it will be confirmed that the TOC concentration in the rinsing bath is not lowered to the level of ultrapure water purity. This will prove the necessity to improve the rinsing bath, which has not been discussed at all, based on the new idea.

The on-line TOC monitor has lately been adopted in various industrial fields. In the microdevice and pharmaceutical manufacturing industries, it is expected that the TOC monitor will be used more and more for quality control and yield improvement. In nuclear energy facilities, the TOC monitor is expected to play an increasingly important role in safety control.

REFERENCES

1. Japanese Patent Publication No. 1988-46375 (B2), p. 113.
2. Japanese Patent Publication No. 1988-46375 (B2), p. 115.
3. Japanese Patent Publication No. 1988-46375 (B2), p. 115, 116.
4. Japanese Patent Publication No. 1988-46375 (B2), p. 116.
5. T. Sugawara, Photochemical engineering for ultrapure water production, Kagaku Kougaku, *51*, 435 (1987).
6. T. Sugawara, Photochemical engineering for ultrapure water production, Kagaku Kougaku, *51*, 435 (1987).
7. Japanese Patent Publication No. 1988-46375 (B2), p. 106.
8. Japanese Patent Publication No. 1988-46375 (B2), p. 114.
9. Japanese Patent Publication No. 1988-46375 (B2), p. 117.
10. Japanese Patent Publication No. 1988-46375 (B2), p. 118.
11. Japanese Patent Publication No. 1988-46375 (B2), p. 109.
12. Japanese Patent Publication No. 1988-46375 (B2), p. 110.
13. Japanese Patent Publication No. 1988-46375 (B2), p. 112.
14. Japanese Patent Publication No. 1988-46375 (B2), pp. 110, 111.
15. Japanese Patent Publication No. 1988-46375 (B2), p. 130.
16. Japanese Patent Publication No. 1988-46375 (B2), p. 120.
17. Japanese Patent Publication No. 1988-46375 (B2), p. 121.
18. Japanese Patent Publication No. 1988-46375 (B2), p. 122.
19. Japanese Patent Publication No. 1988-46375 (B2), pp. 112, 113, 121.
20. P. A. McConnelee, Water quality improvements and VLSI defect density, Semiconductor International, August 1986.
21. R. C. Henderson, J. Electrochem. Soc., *119*, 772 (1972).

i. Automatic TOC Analyzer: Wet Oxidation at High Temperature and Pressure

Yoshio Senoo
Tokico Ltd., Kawasaki, Japan

I. INTRODUCTION

As large-scale integration (LSI) increases, the purity of the water employed in the production line also increases. Accordingly, the sensitivity of the instrumentation that controls the water quality must be upgraded. In the ultrapure water employed in the production line for the 16 Mbit DRAM, the total organic carbon (TOC) must be suppressed to 10 ppb or less [1].

To measure TOC with high sensitivity, organic materials in the ultrapure water should be completely oxidized and decomposed to CO_2 and the accuracy of the pumps and detectors should be improved.

In the automatic TOC analyzer using wet oxidation at high temperature and high pressure [2], oxidizing agent is added to the sample water and organic materials can be decomposed to CO_2 under strong oxidation conditions at 200°C and 20 kgf cm^{-2}, which makes it possible to obtain TOC measurement values close to the actual values. Also, since the analyzing system has been introduced to constantly measure the sample water, the TOC values can be automatically measured on-line and off-line.

II. PRINCIPLE

A. Oxidation of Organic Materials

Organic materials in ultrapure water ($C_xH_yN_z$) are oxidized by sodium persulfate according to Eqs. (1) and (2) to form CO_2:

$$Na_2S_2O_8 + H_2O \rightarrow Na_2SO_4 + H_2SO_4 + (O) \tag{1}$$

$$C_xH_yN_z + n(O) \rightarrow xCO_2 + y/2H_2O \; zNO_3 \tag{2}$$

Reactions (1) and (2) are facilitated at high temperature. The temperature dependence of Eq. (2), indicating the decomposition of organic materials, is observed as shown in Fig. 1. This temperature dependence depends on the types of organic materials contained in the ultrapure water. When potassium hydrogen phthalate is added to ultrapure water that has been treated by ion exchange after passing through the reverse osmosis membrane, the concentration of CO_2 formed gradually increases up to 120°C and it afterward is stable, as shown in Fig. 1. This means that the organic materials are completely decomposed.

The reactor has a structure capable of constantly supplying sample water with the reaction solution to the heated pipe so that the automatic analysis can be conducted constantly.

B. Configuration of the Measurement Unit

Figures 2 and 3 are the external view and schematic diagram of the TOC analyzer [3]. The upper part of the analyzer is the control unit for displaying and recording the measurement values and sending instructions to component units, and the lower part is composed of

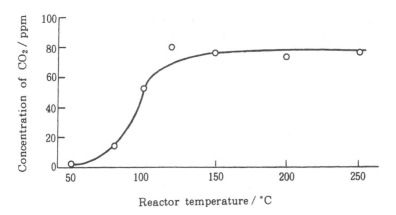

Figure 1   Effect of reactor temperature on the concentration of CO_2 in the extracting gas. Sample water, 13 ml minute^{-1}. Extracting gas, 20 ml minute^{-1}.

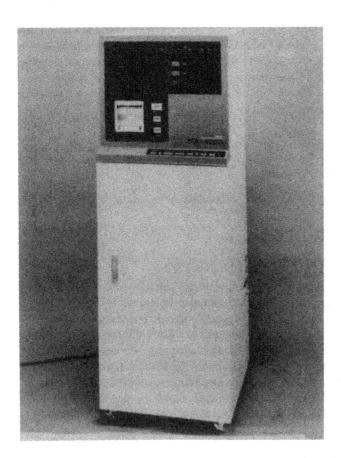

Figure 2   External view of the TOC analyzer.

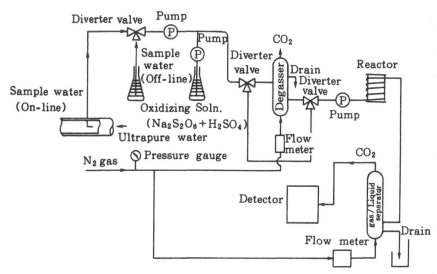

Figure 3 The TOC analyzer.

such core units as reactor and pump. For taking samples, the analyzer can be switched to either on-line operation to take sample water directly from the ultrapure water line or off-line operation to use the sample taken in advance by operating the switching valve.

The inorganic carbon dissolved in the sample water is removed when sulfuric acid and sodium persulfate are added to the sample water and it is exposed to nitrogen gas in the degasser. The pressure of the sample water is then increased to 20 kgf cm^{-2} using the pressure pump to send it to the reactor. The temperature in the reactor is set at 200°C. Organic materials in the sample water react with sodium persulfate in this reactor and are oxidized to CO_2.

CO_2 in the sample water is separated using nitrogen gas in the extractor, and the extracted CO_2 is determined by infrared gas analyzer. The CO_2 concentration is converted to TOC values.

C. Automation of Measurement

To introduce full automation to these analysis procedures, the basic flowchart is as shown in Fig. 4. The monitoring function to detect abnormal measurement values and the automatic calibration function are employed to make it possible to operate the unit constantly.

III. SPECIFICATIONS FOR THE ANALYZER

Table 1 lists the specifications for the analyzer. The major features of the analyzer are as follows:

1. Detection of organic materials that are difficult to decompose. By introducing wet oxidation at high temperature and high pressure, it is possible to conduct TOC analysis for organic materials that are difficult to decompose, such as humic acid.
2. High-sensitivity detection with a 1 ppb detection limit. Wet oxidation at high

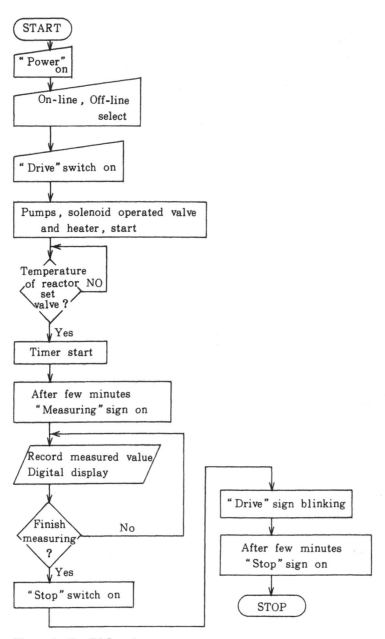

Figure 4   The TOC analyzer.

temperature and high pressure and continuous detection have made high-sensitivity analysis possible.

3.   Continuous automatic analysis. Because the water is sampled continuously and all the TOC analysis procedures are automated, the changes in TOC amount can be measured constantly. The measurement values are displayed in a digital form and recorded in an analog form.

Table 1 Standard Specifications for a TOC Analyzer

Principle	Continuous chemical decomposition in liquid phase at 200° and 20 kgf/cm²
Object of analysis	TOC
Measuring range	0 ～ 50 ppb, 0 ～ 1000 ppb
Measuring mode	Automatic and continuous
Minimum limit of determination	1 ppb
Sampling mode	On-line, Off-line
Sample water	13 mℓ/min.
Oxidizing solution	0.2 mℓ/min. ($Na_2 S_2 O_8 + H_2 SO_4$)
Decarbonating gas	500 mℓ/min. ($>$ 99.99 % N_2 gas)
Extracting gas	20 mℓ/min. ($>$ 99.99 % N_2 gas)
Detector	NDIR
Readout	Both digital display and chart recorder
Linearity	± 2 % F.S.
Repeatability	± 2 % F.S.
Response time	14 min. (90 % response)
Size	(W) 550 × (D) 575 × (H) 1571 mm
Power	AC 100 V , Max 1000 VA

4. On-line and off-line operation. By operating the switching valve, the analyzer can be switched to either on-line or off-line operation.
5. Automatic calibration and reference area. Automatic calibration enables the detector to conduct stable measurement for long periods of time. By setting the reference area in which the TOC values should be found, changes in the water quality can be monitored.

IV. EXAMPLES OF ANALYZING ORGANIC MATERIALS

In these experiments, special grades of chemicals on the market or chemicals of similar grade were used. Ultrapure water (calibration water) was purified in the reverse osmosis unit, the ion exchanger, and by active carbon adsorption and then further purified in a quartz glass still that employs gas-phase oxidation at 900°C. The sample water was produced by dissolving a certain set amount of organic materials in the ultrapure water.

A. Linearity

Figures 5 and 6 indicate the relationship between the reference solution of potassium hydrogen phthalate and the TOC concentration in the measuring ranges 0–50 and 0–100

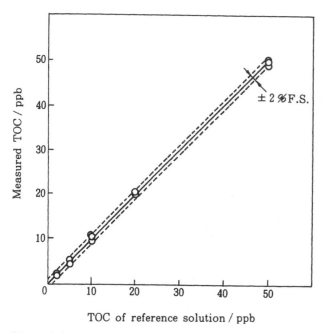

Figure 5   Linearity of the working curve, measuring range 0–50 ppb.

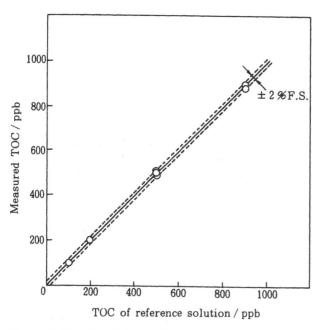

Figure 6   Linearity of the working curve, measuring range 0–1000 ppb.

ppb, respectively [4]. The dotted lines in Figs. 5 and 6 show a range of ±2% FS. In both cases, the results satisfy the ±2% FS requirement.

B. Repeatability

Repeatability [4] was evaluated in reference solutions of potassium hydrogen phthalate with TOC concentrations of 40 and 800 ppb. Table 2 lists the results of three measurements and the difference between these measurement values and average values. In every case, the results satisfy the ±2% FS requirement.

C. Response Time

The response time [4] was measured by using a reference solution of potassium hydrogen phthalate prepared in the same way as before. The measuring ranges were set at 0–50 and 0–1000 ppb. The response time was defined as follows. First it was confirmed whether the measurement values for the sample water were stable enough, and then the sample water was switched to the reference solution of potassium hydrogen phthalate. The time required to obtain a measurement value of 90% was measured as the response time. Figures 7 and 8 show the results in the measuring ranges 0–50 and 0–1000 ppb, respectively. The response time was 14 minutes in each case. It was found that the response time is not greatly affected by the TOC concentration in the sample water.

D. Recovery

After calibrating the analyzer with the reference solution of potassium hydrogen phthalate, the water samples were prepared with certain TOC concentrations by adding various organic materials to determine recovery [4]. Recovery was calculated by Eq. (3):

$$A = \frac{X - b}{C} \, 100 \qquad (3)$$

where: A = recovery, %
X = TOC concentration of sample water, ppb
b = TOC concentration of calibration water, ppb
C = TOC concentration of sample water, ppb

Table 2 Repeatability

		1 st	2 nd	3 rd	Average
Reference solution I	Measured value (ppb)	35	35	34	34.7
	Percent F.S.	+ 0.6	+ 0.6	− 1.4	——
Reference solution II	Measured value (ppb)	807	808	804	806
	Percent F.S.	+ 0.1	+ 0.2	− 0.2	——

Reference solution I : measuring range 0 ∼ 50 ppb
Reference solution II : measuring range 0 ∼ 1000 ppb

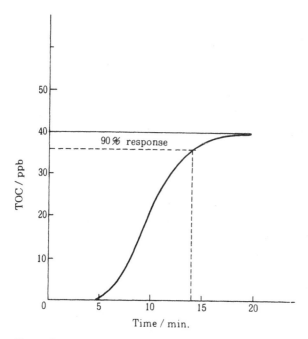

Figure 7 A 90% response time, measuring range 0–50 ppb.

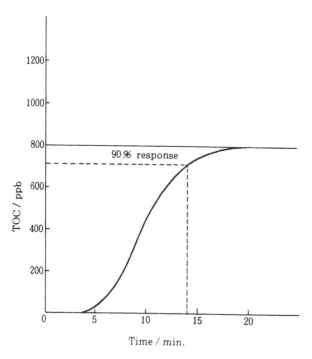

Figure 8 A 90% response time, measuring range 0–1000 ppb.

Table 3 lists the organic materials in ultrapure water used in the semiconductor production line, which have been confirmed by Nishimura and Koyama [5]. Table 4 lists the results of the calculation to determine recovery. A recovery of 100% was obtained in every case: hydroquinone stemmed from humic acid in city water, toluene and anthracene formed through the decomposition of ion-exchange resins employed in the ultrapure water production unit, and methanol, 2-propanol, and acetone are believed to penetrate from the recycling system and to remain even in the purification process.

E. Detection of Organic Materials That Are Difficult to Decompose

To obtain measurement values as close to the actual values as possible in TOC analysis, the oxidation reaction of organic materials should be complete. Recovery was examined for a cyclic nitrogen compound, an azo compound, and a long-chain alkyl compound, all of which are considered difficult to decompose by the conventional wet oxidation method [6,7]. Table 5 lists the results. All compounds showed a high detection ratio of 90% or more in the high-concentration solutions of 100 and 2000 ppb. The strong oxidation action of wet oxidation at high temperature and high pressure was proved in this test.

F. Detection of Volatile Organic Materials

Volatile organic materials with low boiling points can be affected by nitrogen gas exposure in the deaeration process that removes inorganic materials. Some volatile organic materials, such as alcohol, which have sufficient solubility in water, are not

Table 3 Organics in Ultrapure Water

Organics	Structural formula	Source
Methanol	$CH_3\,OH$	Recovery process of ultrapure water at semiconductor manufacturing
2-Propanol	$\begin{matrix} CH_3 \\ CH_3 \end{matrix} > CHOH$	
Acetone	$CH_3 - \overset{\overset{\textstyle O}{\|\|}}{C} - CH_3$	
Hydroquinone	HO —⟨ ⟩— OH	City water
P-Benzoquinone	O =⟨ ⟩= O	
m-Cresol	$CH_3 -$⟨ ⟩- OH	
Toluene	⟨ ⟩- CH_3	Decomposition of ion exchange resin
Anthracene	(structure)	

Table 4 Recovery of Organics

Organics	Recovery (%)	Organics	Recovery (%)
4 - (2 - Pyridylazo) - resorcinol	101	L - Glutamic acid	100
1 - (2 - Thiazolylazo) - 2 - naphthol	93	Acetic acid	98
Hydroquinone	102	Tartaric acid	97
P - Benzoquinone	100	Citric acid	101
Potassium hydrogenphthalate	100	Lactic acid	95
m - Cresol	100	L - Ascorbic acid	100
Sodium sulfobenzoate	101	2 - Propanol	100
Sodium α - naphthalenesulfonate	103	Methanol	102
Anthracene	103	Acetone	100
Carbamide	97	Grape sugar	102
Aminoacetic acid	99	Polyethylene glycol - 1500	103
EDTA	97	Humic acid	103

Table 5 Recovery of Organics Resistant to Decomposition

Organics	Molecular weight	Recovery (%)	
		conc. of samples 100 ppb	conc. of samples 2000 ppb
2, 4, 6 - Tris (2 - pyridyl) - 1, 3, 5 - triazine	312.33	93.6	91.7
1, 10 - Phenanthroline	198.22	95.5	95.2
Methyl orange	327.34	90.9	91.3
Sodium dodecylbenzensulfonate	348.48	100	99.4

greatly affected. On the other hand, others, such as ketone and ether, which are less soluble in water, are exhausted outside the system together with nitrogen gas, which makes the measured values smaller than the actual values.

In this case, it is effective to detect TOC by deducting the concentration of inorganic carbon from the total carbon concentration without employing deaeration. In the ultrapure water production unit in the recycling system [5], the TOC concentration has been detected for diethyl ether and methyl ethyl ketone using this method [8].

G. Detection Limits

The detection limit of this analyzer was determined by the following method [9]. First, reference solutions with TOC concentrations of 2, 5, and 10 ppb were prepared by dissolving potassium hydrogen phthalate into ultrapure water. The analysis was then conducted repeatedly with these reference solutions, and the fluctuation coefficient σX was calculated from the means of each measurement value on the basis of Eq. (4):

$$\sigma X = \frac{\sqrt{\Sigma(X_n - \bar{X})^2 n}}{\bar{X}} \, 100 \qquad (4)$$

where: σX = fluctuation coefficient, %
     X_n = measurement value of TOC in reference solution, ppb
     X = mean of measurement values, ppb
     n = number of measurements

Figure 9 indicates the relationship between the mean of the measurement values and the fluctuation coefficient. With the detection limit defined as the point at which the fluctua-

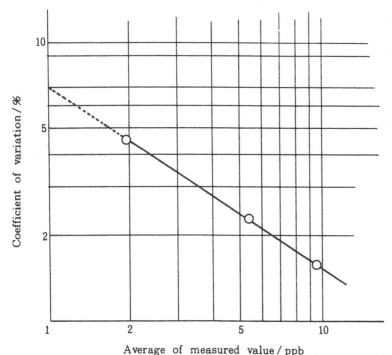

Figure 9   Minimum limit of determination.

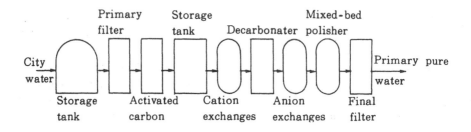

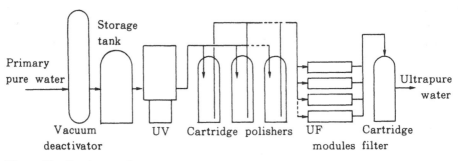

Figure 10   Continuous ultrapure water system.

tion coefficient exceeds 10%, the result of extrapolation is 1 ppb or less. It is therefore considered that the detection limit of this analyzer satisfies 1 ppb.

H.   Measurement at the Use Point

Figure 10 shows the ultrapure water production system of a semiconductor plant located in Tama, Tokyo. This system, composed of a primary pure water system and an ultrapure water system, can produce highly purified ultrapure water. The sample water for the analysis is taken from the use point in the clean room in this plant, and on-line measurement [10] is carried out to protect the sample water from contamination.

Figure 11 shows the result of a 24 h continuous measurement of TOC. The measure-

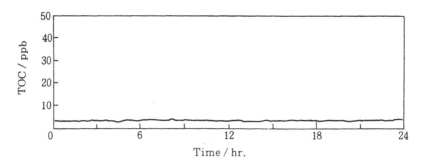

Figure 11   Continuous measurement of TOC at the point of use.

ment values remain stable at the low concentration level of 10 ppb or less, which indicates that this analyzer can be applied to higher level ultrapure water production systems.

V. CONCLUSION

This section has introduced an analyzer for automatically measuring TOC in water with high sensitivity. The wet oxidation method at high temperature and high pressure that is employed in this analyzer features strong oxidizing power to effectively convert almost all the organic materials in ultrapure water for semiconductor production to CO_2. It has also been found that this method exhibits excellent recovery. In addition, the analyzer has achieved a detection limit of 1 ppb, by establishing technologies to maintain cleanliness in the system, to control each component unit, and to automate all analysis procedures.

The water quality evaluation technology as well as the ultrapure water production technology are two of the most important fields in the semiconductor industry. It would be wonderful if this analyzer contributes to improvements in evaluation technology in some way.

VI. SUMMARY

Because the purity of the water employed in semiconductor production is rapidly improving, it is increasingly necessary to develop an analyzer to accurately measure the total organic carbon contained in the water. The method introduced in this section chemically decomposes the organic contaminants in water under strong oxidation conditions at 200°C and 20 kgf cm^{-2} and detects the carbon dioxide formed by infrared gas analyzer. Moreover, such functions as the taking of on-line water samples and automatic and continuous measurement are provided so that the changes in the quality of the water can be monitored constantly. A detection limit of 1 ppb has been achieved.

REFERENCES

1. T. Sakamoto, Mol, *26*, 73 (1988).
2. C. Maegoya and Y. Okazima, Nippon Kagaku Kaishi, 911 (1988).
3. C. Maegoya and M. Nitta, Tokico Review, *31*, 8 (1987).
4. Japanese Industrial Standards, Continuous Total Organic Carbon Analyzen, JIS-K-0805 (1988).
5. M. Nishimura and K. Koyama, Kogyo Yosui, *9*, 22 (1987).
6. T. Sakamoto, Kogyo Yosui, *5*, 6 (1988).
7. M. Nitta, T. Iwata, and T. Koinuma, Proceedings of 24th Annual Publication of Research, Tokyo, (1989), p. 79.
8. T Manabe, Y. Sanui, and T. Iwata, Tokico Review, *33*, 32 (1990).
9. M. Nitta, T. Iwata, and T. Kutsuma, Proceedings of 23rd Annual Publication of Research, Tokyo, 1988, p. 48.
10. T. Iwata, M. Nitta, A. Saiki, and Y. Okazima, Proceedings of the Society for Scientific Research, SDM87-91, Nagoya, (1988), p. 49.

ii.   Ultraviolet-Promoted Chemical Oxidation TOC Analyzer

Yoshiki Shibata
Toray Engineering Co., Ltd., Shiga, Japan

I.   INTRODUCTION

It was in the late 1960s and early 1970s that laboratory and on-line total organic carbon (TOC) meters for comprehensive measurement of the organic material contamination level in water were launched in the marketplace. Most adopted a combustion oxidation method established by Van Hall et al. [1]. The combustion oxidation method was widely introduced to the TOC test of industrial water and industrial wastewater. It was adopted as ASTM D2597 in the United States in 1969, and in Japan it was adopted as JIS K 0101 (testing methods for industrial water) in 1979 and JIS K 0102 (testing methods for industrial wastewater) in 1981.

In the ultrapure water field, which requires measurement of trace TOC, the ultraviolet (UV)-promoted chemical oxidation TOC analyzer, which is capable of oxidizing large sample volumes, began to be used around 1982 [2]. This method was adopted as JIS K 0805 [3] and JIS K 0551 [4], both of which were established in 1988. The 16th edition of *Standard Methods* (1985) [5] also picked up this method, describing it as suitable for TOC measurement in ultrapure water for the semiconductor industry (for rinsing), the thermoelectric power plant (for supply to the boiler), and the pharmaceutical industry (for manufacturing).

II.   PRINCIPLES OF MEASUREMENT

The UV-promoted chemical oxidation TOC analyzer, which is the major TOC meter in the ultrapure water field, is composed of three process steps: IC (inorganic carbon) is removed from the samples, the IC-free samples are oxidized, and the carbon dioxide produced by oxidizing the samples is measured (to measure the TOC level). Figure 1 summarizes the measurement principle [6].

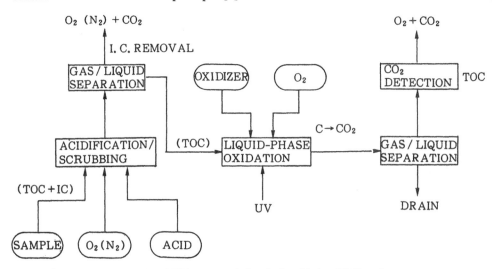

Figure 1   Analytic principle of UV-promoted chemical oxidation TOC analyzer.

First, IC is removed from samples by acidification and scrubbing. As shown in Fig. 1, acid is added to the sample to lower the pH to 2 or less, and then inert gas to exclude carbon dioxide is fed to the sample. In this way, the IC in the sample can be removed by the following reactions:

$$CO_3^{-2} + H^+ \rightleftharpoons HCO_3^- \tag{1}$$

$$HCO_3^- + H^+ \rightleftharpoons H_2O + CO_2 \tag{2}$$

After the IC is removed, organic carbon remains in the sample. When the sample contains volatile organic materials with low solubility, the volatile organic materials are sometimes partially dispersed in the gaseous side, which results in lower measurement values than the actual. However, this does not constitute a serious problem with ultrapure water containing trace amounts of organic materials.

Next the sample, free of IC, is oxidized. As shown in Fig. 1, a powerful UV-promoted chemical oxidation using both oxidizing agent and ultraviolet is used. This method has been widely used in processing general wastewater [7,8]. Basically this method aims at obtaining the synergistic effect of the following two process steps:

1. Oxidation with an oxidizing agent (radical) excited by light. Organic materials are oxidized by the powerful oxidizing action of hydroxy radical (HO·), which is produced when the oxidizing agent absorbs ultraviolet.
2. Decomposition of excited molecules (organic materials) that absorb light. The decomposition reaction of organic materials is triggered by the optical quantum energy, which exceeds the binding energy of organic materials. The decomposition mechanism triggered by light is very complicated and unclear. It has been learned that the lower the wavelength of ultraviolet ray, the more effectively the optical oxidation reaction proceeds. This is because the optical quantum energy is in inverse proportion to the wavelength.

In wastewater processing, to process a large volume of water at lower cost, H_2O_2 or O_3 is used as the oxidizing agent and a high-pressure mercury lamp is used as the UV light source. However, with the TOC meter, which is required to effectively oxidize a small volume of sample within a short time, the decomposition rate of organic materials and the excitation efficiency of oxidizing agent should be enhanced. Therefore sodium persulfate, which has a strong oxidation action, is used as the oxidizing agent and a low-pressure mercury lamp, which has a high irradiation ratio of low-wavelength ultraviolet, is adopted as the UV light source.

The excitation reaction of oxidizing agent in the TOC meter can be expressed as

$$S_2O_8^{2-} + 2H_2O \xrightarrow{\text{UV, heat}} 2SO_4^{2-} + 2H^+ + 2HO\cdot \tag{3}$$

The oxidation reaction of organic materials triggered by hydroxy radical (HO·), which is produced in reaction (3), can be expressed as

$$C_nH_m + (4n + m)HO\cdot \xrightarrow{\text{UV, heat}} nCO_2 + (2n + m)H_2O \tag{4}$$

In this reaction, organic carbon in the sample is converted to carbon dioxide. Since this reaction takes place only in the liquid phase, without changing to the gas phase, high-sensitivity measurement can be conducted as a result of the small limitation imposed by the sample volume.

In the final process the concentration of carbon dioxide (TOC) produced during oxidation of the sample is measured. As shown in Fig. 1, the carbon dioxide concentration in the gas phase, which is separated from the gas-liquid mixture phase, is measured. In the oxidation reaction shown in Fig. 1, if the volume of sample, oxidizing agent, and carrier gas are fixed, the concentration of carbon dioxide corresponds to the TOC concentration. A nondispersive infrared (NDIR) gas analyzer that features high sensitivity, selectivity, and stability is used to measure the carbon dioxide concentration in the carrier gas. This measurement method is called the ultraviolet oxidation-infrared analysis method, in JIS K 0805 [3] and persulfate-ultraviolet oxidation in *Standard Methods* (1985) [5].

III. STRUCTURE AND SPECIFICATIONS

Figure 2 shows the structures of a TOC meter that employs this method. The sample drawn by pump *A* and the phosphoric acid solution drawn by pump *B* are introduced to the IC removal column together with oxygen gas (or nitrogen gas), which is used for bubbling. The gas-liquid mixture flow introduced to the IC removal column is mixed well while passing through the spiral Teflon tube to effectively convert IC in the sample into carbon dioxide. The gas-liquid mixture flow sent out from the IC removal column is next fed to the gas-liquid separation unit to separate the liquid phase from the gas phase. The gas phase containing carbon dioxide is exhausted from the system. This is how IC is removed and organic carbon remains in the sample.

Next the sample from which the IC has been removed, the sodium persulfate solution, which is used as the oxidizing agent, and the carrier gas (oxygen) are fed to the reactor after they are fed at each fixed rate and mixed. In the reactor, ultraviolet is irradiated to continuously oxidize the sample and convert organic carbon to carbon dioxide. The gas-liquid mixture flow sent from the reactor is fed to the gas-liquid separation unit. The separated liquid phase is drained out. The separated gas phase is sent to the NDIR analyzer to measure the carbon dioxide concentration in the oxygen gas.

Since the sample, the oxidizing agent, and the oxygen gas are continuously sent to this system, when the sample (TOC, mg C L^{-1}) fed at a fixed flow rate (F_s, ml minute^{-1})

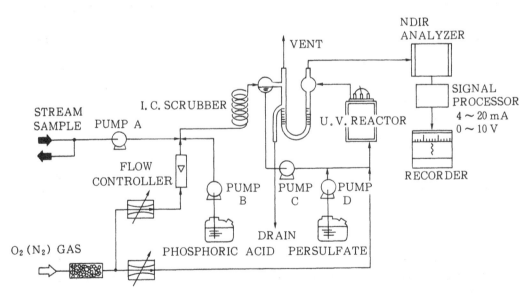

Figure 2 The UV-promoted chemical oxidation TOC analyzer.

is completely oxidized, the carbon dioxide concentration in the oxygen gas flow (F_g, ml minute^{-1}), indicated by the NDIR reading (IR, ppm CO_2), can be expressed as

$$IR = 1.867 \left(\frac{F_s}{F_g}\right) TOC \times 10^3 \tag{5}$$

As shown in Eq. (5), if the flow rates of sample and oxygen gas are fixed, the NDIR reading and the TOC concentration correspond to each other. Therefore, the TOC concentration can be determined from the NDIR readings if the zero-span calibration is done with a standard solution whose concentration is known. Potassium hydrogen phthalate (KHP) solution is used as the standard solution.

In the system shown in Fig. 2, sodium persulfate is used as the oxidizing agent. Since it has a 6–10 times higher solubility than potassium persulfate, the oxidizing action is very powerful. Moreover it can last for a long time under good conditions.

The reason oxygen gas is used as the carrier gas is to take advantage of the ozone produced by UV irradiation as the oxidizing agent. Ozone formation when the ultraviolet is irradiated to the gas-liquid mixture flow, composed of the sample, the oxidizing agent, and the oxygen gas, can be expressed by Eq. (6), and the oxidation reaction of organic materials caused by ozone is expressed by Eq. (7).

$$3O_2 \xrightarrow{UV} 2O_3 \tag{6}$$

$$4C_nH_m + 4(n + m)O_3 \rightarrow 4nCO_2 + 2mH_2O + (2n + 5m)O_2 \tag{7}$$

Figure 3 shows the reactor structure. The pen UV lamp is inserted in the quartz tube. The gas-liquid flow entering from the bottom is exposed to the strong UV irradiation, directly contacting the UV lamp. The outlet of the mixture flow is located at the top. In this system, the reactors are arranged in series to improve the oxidizing efficiency to the level of the combustion oxidation method. Since the structure is simple, with limited

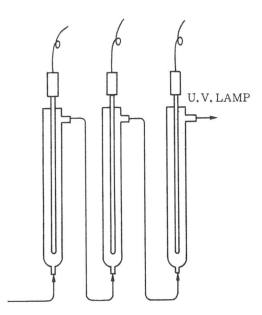

U.V. LAMP

Figure 3 The UV reactor.

entrapment space, the possibility of contamination is kept remarkably low. Therefore the system noise is small enough to be ignored. Table 1 lists the specifications for the UV-promoted chemical oxidation TOC analyzer.

IV.   CALIBRATION AND MEASUREMENT DATA

Because the readings of this system should correspond to the TOC concentration, accurate zero calibration and span calibration should be conducted in advance.

The reading C of the TOC analyzer includes not only the TOC concentration in the sample X but also the TOC concentration due to impurities in the reagents β, the increase in TOC concentration due to contamination in the system r, and the increase in TOC concentration due to system operation, such as the sampling δ. In general, the reading corresponds to $X + \beta + r + \delta$. Therefore, the volume $\beta + r + \delta$ should be compensated by zero calibration so that the signal reflects only the TOC concentration of the sample X.

Because water with a TOC concentration of zero is not available, the common blank compensation method cannot be used. In this system, a special calibration method is worked out on the basis of the fact that the oxidizing efficiency is fixed within the measurement range and that the signal is directly related to the carbon dioxide concentration. The calibration method is as follows. The TOC concentration of water used to prepare the standard solution is determined by the standard addition method, and then the calibration is conducted based on the TOC concentration of the zero-calibration solution and the span calibration solution, both of which are compensated for by the previously determined value. Calibration is conducted by a built-in microcomputer.

Because the TOC concentration measured may be affected by the calibration method in some cases, the significance of the measurement should be fully discussed. (See JIS K 0551 [4] for details.)

Following the zero calibration and the span calibration, the TOC concentration is measured from the readings while supplying the on-line (or off-line) samples instead of the standard solution.

Some of the measurement data obtained with the UV-promoted chemical oxidation TOC analyzer are presented here.

Table 1   Typical Specifications

MODEL	TOC - 710	TOC - 720	TOC - 730
OPERATION MODE	CONTINUOUS	CONTINUOUS	CONTINUOUS
MEASURING RANGE mgC/ ℓ	$0 - 0.1 \sim 0 - 1$	$0 - 0.5 \sim 0 - 5$	$0 - 1 \sim 0 - 10$
READOUT	4 DIGITS, DIGITAL	4 DIGITS, DIGITAL	5 DIGITS, DIGITAL
REPEATABILITY %FS	± 2	± 2	± 3
LINEARITY %FS	± 2	± 2	± 3
STABILITY %FS/D	± 2	± 2	± 3
RESPONSE TIME (90%)	10 min (MAX)	10 min (MAX)	10 min (MAX)

A.   Measurement Data for Zero-Calibration Solution and TOC Standard Solution

Table 2 lists data for the zero-calibration solution and the TOC standard solutions at several different concentrations measured following calibration using the standard addition method. For the TOC concentration in each solution, the calculated values are obtained by adding the measurement value of the zero-calibration solution to the KHP concentration injected in each solution, which are compared with the measurement data. In Table 2, the calculated values match the measurement data well.

B.   Detection Efficiency Data for Pure Materials

Only a few data have been reported for the detection efficiency for samples containing pure materials at 500 μg CL^{-1} or less [3,9]. This may be attributed to the following:

The TOC concentration in the zero-calibration solution used to prepare the standard solution cannot be ignored.
Contamination in the standard solution, which occurs during the preparation, cannot be ignored.
The standard solution is contaminated during storage and measurement.

To obtain accurate detection efficiency data for samples containing trace amounts of pure materials, the authors conducted the following experiment. The standard solution, which had been prepared at high concentration, was continuously mixed with the zero-calibration solution in the special standard solution preparation unit using a closed loop. This mixture was fed to the TOC meter for measurement. Table 3 lists detection efficiency data for pure materials at 50–200 and 500–2000 μg CL^{-1}.

Table 3 was plotted by adding the measured amount of standard solution to the zero-calibration solution step by step (50, 100, 150, and 200 μg CL^{-1} and 500, 1000, 1500, and 2000 μg CL^{-1}). The detection efficiency for each pure material was obtained by comparing these results with the sensitivity coefficient of potassium hydrogen phthal-

Table 2   Determinations of TOC in Various KHP Standard Solutions

Unit (μgC/ l)

Sample		Calculated TOC*	Found TOC
Sample	Added TOC		
Reagent Water	0	35	35
KHP Solution A	100	135	138
B	200	235	240
C	400	435	444
D	600	635	637
E	800	835	835

*Calculated TOC = Added TOC + Found TOC of Reagent Water

Table 3 Recovery of Organic Materials

Recovery (%)

Material added	Added TOC (μgC/1)	
	50 ~ 200	500 ~ 2000
2 - Propanol	102.9	100.1
L-Sodium Gultamate	102.5	102.0
1, 10 - Phenanthroline	98.9	100.7
Tartaric Acid	104.0	101.9
Sodium Dodecylbenzensulfonate	96.0	97.1

ate. The pure materials shown in Table 3 are specified for the detection efficiency test by JIS K 0805 [3] and the Japan Pharmacopoeia [10].

C. TOC Concentration Measurement Data for Reagent Water Available on the Market

Table 4 lists the TOC concentration measurement data for reagent water available on the market [4]. The ultrapure water production system currently available for laboratories can easily produce water with a TOC concentration of around 50 μg CL^{-1}. Since purified

Table 4 Determinations of TOC in Various Reagent Waters on the Market

Unit (μgC/1)

Reagent Water	Found TOC
Distilled Water for General Use, marketed by A Co.	100
Distilled Water for General Use, marketed by B Co.	360
Distilled Water for HPLC, marketed by C Co.	80
Distilled Water for HPLC, marketed by D Co.	50
Distilled Water for HPLC, marketed by E Co.	< 50
Distilled Water for Fluorescence Analysis, marketed by F Co.	160
Distilled Water for Amino Acid Sequenation Analysis, marketed by G Co.	< 50
Deionized/Distilled Water for General Use, marketed by H Co.	110
Ultrapure Water purified by Ultrapure Water Production Unit for Laboratory Use, made by I Co.	70
Ultrapure Water purified by Ultrapure Water Production Unit for Laboratory Use, made by J Co.	50

water is easily contaminated as a result of the air in laboratories and materials eluted from the liquid-contacting parts, the water should be used immediately after it is purified.

V. SUMMARY

To accurately measure the TOC concentration in ultrapure water, maximum attention should be paid to protect the samples from contamination at every stage: sampling, pretreatment, sample injection, and so forth. It is desirable that every step the measurement proceed automatically in a closed environment. The automatic measurement system should be operated continuously. Once the measurement operation is suspended, it takes a long time to regain stable and accurate readings after the re–start-up because of CO_2 or organic material contamination in the system. General points with regard to the measurement of the TOC concentration in ultrapure water are described in detail in JIS K 0551 [4].

REFERENCES

1. C. E. Van Hall, J. Safranko, V. A. Stenger, Anal. Chem., *35*(3), 315 (1963).
2. H. Ohya, general editor, Production Methods of Pure Water and Ultra Pure Water, Saiwaisyobou (Japan), 1985, p. 140.
3. Japanese Industrial Standard, JIS K 0805, Continuous Total Organic Carbon Analyzer (1988).
4. Japanese Industrial Standard, JIS K 0551, Testing Methods for Total Organic Carbon in Highly Purified Water (1988).
5. APHA, AWWA, WPCF, Standard Methods for the Examination of Water and Wastewater, 16th ed., 1985, p. 507.
6. Nippon Kuuki Seijou Kyoukai (Japan), The 6th Textbook of Cleanroom Technology Seminar Series, 1985, p. 67.
7. M. Kazama, S. Ohno, and Y. Kenmoku, PPM (Japan), *18*(2), 43 (1987).
8. T. Sugawara, J. Chem. Eng. (Japan), *51*, 435 (1987).
9. T. Sakamoto, Kagakugijutushi Mol (Japan), *26*(1), 73 (1988).
10. Nippon Kouteisyo Kyoukai (Japan), Explanatory Supplement Note to Japan Pharmacopoeia, 11th. revision. Hirokawasyoten (Japan), 1988, p. D-10.

3 On-Line Monitors

C. Particle Counter

Hirotake Shigemi and Toshio Kumagai
Kurita Water Industries, Ltd., Kanagawa, Japan

I. INTRODUCTION

The ultrapure water system has played a vital role in the rinsing process of very large scale integrated (VLSI) circuit production. In the era of mass production of megabit LSI, there have been demands for ultrapure water that satisfies more stringent qualitative requirements. This section deals with measurement of the particles contained in ultrapure water. Generally, it is said that the size of a particle to be removed from ultrapure water is ¹⁄₁₀ the LSI pattern size. For a 4 megabit LSI, for example, whose pattern size is approximately 0.8 μm, the size of particles to be removed is 0.08 μm. In this part, a measuring instrument that can monitor the behavior of the particles is introduced. We discuss the principle, development, application technology, and features, as well as evaluation by demonstration test, of the high-sensitivity particle counter.

II. MEASURING PRINCIPLE

The particle counter introduced here employs what is called a 90° sideways light-scattering system with an argon ion laser (λ = 488 nm) as the light source.* Figure 1 illustrates its basic structure. A laser beam irradiates the sample water, which flows out of the nozzle in the measuring cell. If a particle crosses the beam, it emanates pulsating scattered light, which is detected by a photomultiplier as electrical pulse signals. The particle counter indicates the number of particles per milliliter after discriminating pulse signals on three levels by particle size.

---

*The high-sensitivity particle counter was developed jointly by Kurita Water Industries, Ltd. and Horiba Manufacturing Co., Ltd. The model numbers are K-LAMIC-100 (Kurita) and PLCA-310 (Horiba).

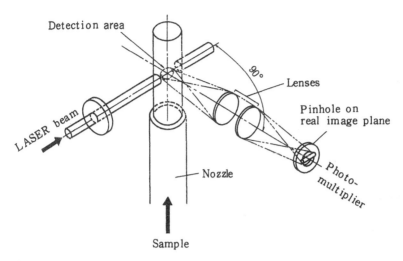

Figure 1   Basic structure of the laser particle counter.

III.   MEASURING SYSTEM AND FEATURES

Conventional particle counters that are based on light-scattering principles are equipped with a helium neon laser so that their lower limit of detection is 0.2 μm. As a result of our efforts to maximize the discrimination of the light-scattering method, the newly developed high-sensitivity particle counter is designed to be an on-line monitor of particles in ultrapure water and can detect a particle 0.07 μm in size. Here we introduce the features of the measuring system and discuss the manner in which the sensitivity was increased [1].

A.   Adoption of the Argon Ion Laser as Illuminant

According to Rayleigh's theory, with a particle of approximately 0.3 μm or less, the strength of scattered light that is emitted when a light hits the particle is in proportion to the sixth power of the particle size, in inverse proportion to the fourth power of the wavelength of the light, and in proportion to the strength of the light.

    To improve the detection sensitivity, we first paid attention to the illuminant and studied the applicability of a laser with a short wavelength and a large output. The particle counter needed was not one for scientific experiment but a practical tool for monitoring the quality of the ultrapure water used for LSI production. Taking into consideration the practicalities of the particle counter, such as cost, service life, physical size, and the time required for measurement, we selected an argon ion laser with a 10 mW output and a 488 nm wavelength as an acceptable light source. For illuminant output, we also studied the possibility of using a laser with a small output, several milliwatts, and increasing its energy density by reducing the beam diameter. However, such an illuminant reduced the measurable volume of sample water. As a result, it took a long time to count number of particles per milliliter accurately, failing to meet the requirement of real-time measurement. It was eventually determined to contract a laser beam of 10 mW output to approximately 20 μm diameter.

B. Improvement in the S/N Ratio by Devising Cell Construction

The requirements for the measuring cell are (1) to reduce the reflection of stray light and (2) to avoid the effect of particle dissociation from the internal surface of the cell or the accumulation of particles (contamination of sample water in the cell). The increase in sensitivity means improving the ratio of signal to noise (S/N ratio). Optical noise, the major component of the noise that causes problems in measuring the strength of scattered light, depends largely on the stability of the optical axis and the size of the dark photomultiplier current on the scattered light detection side. It is therefore necessary to be careful in selecting and installing cell components. Concerning the reflection of stray light, the question is how to minimize its level on both internal and external surfaces of the glass windows provided at the inlet and the outlet of the laser beam.

Figure 2 shows the conceptual structural design of a measuring cell that meets all the requirements mentioned. The cell construction features the following:

1. A wide space in the cell and a long separation distance between the scattered light detection field (central section) and the cell windows are provided.
2. A black shell is arranged in the cell to minimize the light reflected to the central section.
3. The sample water is channeled from the nozzle to the central section in a vertical upward counterflow and is enveloped in sheath water.

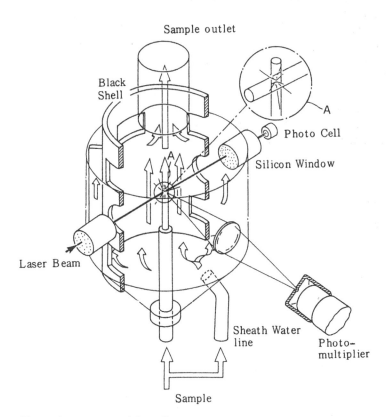

Figure 2 Structure of the cell.

This structure reduces the level of stray light in the central section and prevents particles from adhering to the internal surface of the cell or bubbles from penetrating the cell.

C.   Increase in Sensitivity by Regulating the Flow Velocity of the Sample Water

There seems to be no relation between the flow velocity of the sample water that passes through the measuring field (the volume determined by the diameter of the laser beam and the slit breadth on the observation side) in the cell and the pulse height of the scattered light. However, if a sample of ultrapure water in which standard particles of identified size are intentionally added is monitored by oscilloscope, the flow velocity is found to be inversely proportional to the peak value of pulse signals. For instance, it has been confirmed by experiment that with standard particles of polystyrene latex (PSL), the flow velocity should be reduced by $\frac{1}{32}$ to adjust the signal level of 0.109 μm particles to 0.212 μm.

   The inversely proportional relation of the peak height value of the scattered light and the flow velocity of the ultrapure water can be explained as follows. Along with the increase in the passage time of particles through the measuring field, the energy of the scattered light emanating during that time also increases. When the energy is converted into a quantity of electricity, which is then transformed into pulse signals of a certain time constant by the integration circuit on the input side of the preamplifier, the larger quantity of electricity is converted into the signals with a higher peak value. The particle counter we developed has two sensitivity selection modes, a standard and a high-sensitivity mode. The flow rate in the high-sensitivity mode is reduced to one-fourth of that in the standard mode, which is discussed later.

   Figure 3 is a flow diagram of the particle counter. The aperture of manual needle valves V_1 and V_2 has been set so that the flow rate of the nozzle line to the measuring cell decreases to one-fourth of that in the standard sensitivity mode when the solenoid-operated valve opens automatically in the high-sensitivity mode and bypasses the sample water.

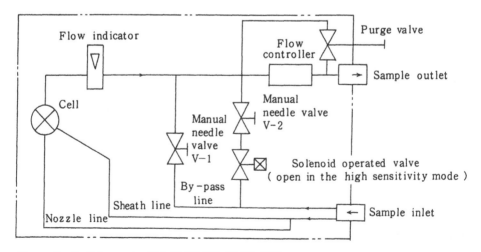

Figure 3   Particle counter K-LAMIC-100 (PLCA-310).

Table 1 Technical Data

Principle : Laser (TE Moo Ar⁺ laser) scattering method Application : Fine particles in ultrapure water Particle diameter ranges :	3) Year, month, day, hour, minute 4) Count in progress (LAP) 5) Setting (LED) 6) Alarm 7) Pulse monitor

	SENS	
Channel	STANDARD	HIGH
No. 1	0.1 μm	0.07 μm
No. 2	0.2 μm	0.1 μm
No. 3	0.5 μm	0.2 μm

Sample conditions :
Pressure : 0.1 - 0.5 MPa (1.0 - 5.0 kg /cm² G)
Flow rate : Approx. 0.6 ℓ/min.
Temperature : 5 - 40 °C, 41 - 104 °F
Measuring time :
Standard sensitivity : 10 min/ml
High sensitivity : 40 min/ml
Displays :
1) Cumulative count by Particle diameter.
   NORMAL (newest value) : 4-digit digital display, max. reading 9999 (particles /ml)
   SMOOTH (mean value) : 5-digit digital display, max. reading 9999.9 (particles /ml)
   Simultaneous counts of channels : No. 1, No. 2 and No. 3.
2) Differential count by particle diameter range.
   Subtraction of channel No. 2 from No. 1 and No. 3 from No. 2 to give number particles in the range.

Output : Simultaneous analog output from channels No. 1, No. 2 and No. 3 ;
SMOOTH output : 1 mV per 0.1 count
NORMAL output : 1 mV per 1 count
Max. 1023 mV
Printout : Cumulative count by particle diameter, differential count by particle diameter range, settings (SENS, alarm levels, INTMT), alarms (COUNT OVER, LASER ALARM, SCALE OVER), date.
Alarm output : Non-voltage," a " contact, 30 VAC/DC, 0.5 A
Power : AC line *, 50/60 Hz, approx. 2.0 kVA
   *Specify 100, 120, 220 or 240 VAC when ordering.
Connections : PT 1/4 internal thread for both sample inlet and outlet.
Dimensions :
   500 (W) × 600 (d) × 320 (h) mm
   19.7 (W) × 23.6 (d) × 12.6 (h) in
Weight : 49 kg, 108 lb
Computer interface : RS - 232C Interface is optionally available.

D. Prevention of Bubble Generation

The production of ultrapure water is normally under a pressure of 2–5 kg cm⁻². On the other hand, the ultrapure water storage tank is normally sealed with pressurized nitrogen gas to prevent atmospheric air intrusion so that the nitrogen gas dissolved in ultrapure water generates fine bubbles under atmospheric pressure. The particle counter of the light-scattering system detects the bubbles as particles and counts them.

For this reason, a constant-flow valve is provided on the downstream side of the measuring cell to keep the pressure of sample water in the cell above atmospheric pressure. The constant-flow valve is equipped with a bypass purge valve through which the largest possible quantity of sample water is flushed into the lines, including the sampling line, at the time of system start-up, thereby washing out the particles adhering to the internal surface of the system.

E. Product Specifications and Features

Table 1 lists the specifications and major features of the particle counter, and Fig. 4 shows its appearance. As a result of the increase in sensitivity just discussed, the minimum

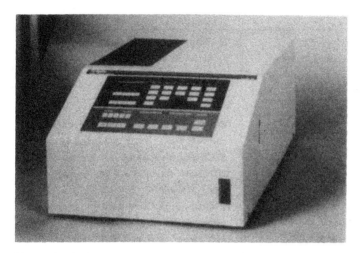

Figure 4   Particle counter K-LAMIC-100 (PLCA-310).

measurable particle size has decreased to 0.07 μm. The measuring time for a 1 ml water sample is 10 minutes by standard sensitivity mode and 40 minutes by high-sensitivity mode, but both modes are designed to update data every 10 minutes. Only 0.25 ml sample water is measured in 10 minutes by the high-sensitivity mode, so that the particle counter indicates and records a total of four counts as the number of particles per milliliter. In other words, the hard copy of data printed out every 10 minutes shows the latest total of four individual counts (0.25 ml per count) as the number of particles per milliliter.

Figure 5 is a sample printout. Counted particles are classified into each channel, 0.1, 0.2, and 0.5 μm size with the standard sensitivity mode or 0.07, 0.1, and 0.2 μm size with the high-sensitivity mode. The 10 printout data show the number of particles larger each class and also the number of each classification as the unit of particles per milliliter. In additional, it indicates the running mean of each number of particles for 10 minutes to one decimal place on the next line.

As discussed earlier, the particle counter uses an argon ion laser as the illuminant to increase the sensitivity. However, the argon ion laser tube is more expensive and has a shorter life than the conventional helium neon laser tube. To compensate for these deficiencies, the particle counter is designed to operate on intermittent measurement mode in addition to continuous measurement mode. There is no need to measure the number of particles every 10 minutes as long as the ultrapure water system is in stable operation at a constant feed rate of raw water with relatively limited fluctuations in load. Under such conditions, if the duty third of the intermittent measurement mode is selected at standard sensitivity, the laser operation time decreases to approximately one-third (although 5 minutes is additionally required to warm up the particle counter after the power switch is turned on), and the counter prints out the data every 30 minutes, by which the service life of laser tube is prolonged by about three times.

IV.   APPLICATION TECHNOLOGY

A.   Calibration

In calibrating the particle counter, PSL standard particles were used to adjust the thresholds of size classification. According to the flow diagram shown in Fig. 6, ultrapure

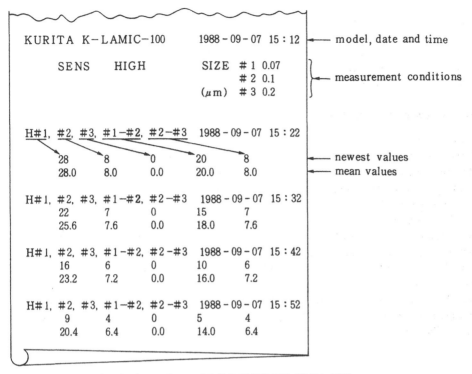

Figure 5   Example of printout by model K-LAMIC-100 (PLCA-310).

water is fed to the particle counter, and then a specified quantity of the diluted standard particle solution whose concentration is verified beforehand by scanning electron microscopy (SEM) is injected continuously into the water by syringe pump. The volume injected by the syringe pump is set so that the particle concentration calculated from the dilution rate is 2000 particles ml^{-1}. This is the concentration that allows easy monitoring of the

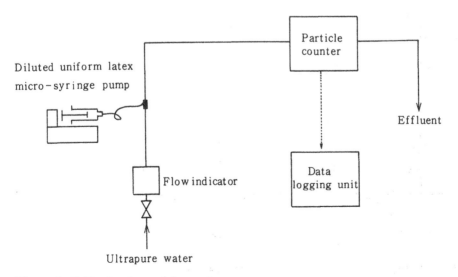

Figure 6   Calibrating the particle counter.

signal wave of standard particles by oscilloscope. Figure 7 shows the signal waves of 0.073, 0.091, 0.212, and 0.497 μm standard particles.

In actual calibration work, we adjusted the thresholds by channel to match the count to the particle concentration calculated from the dilution rate while checking the signal wave of the standard particles.

B.   Demonstration Test

Using the particle counter calibrated in the manner just described, the number of 0.21 and 0.091 μm PSL standard particles in dispersion and actual ultrapure water was counted. The results are shown in Fig. 8 in comparison with values obtained by SEM. Figure 8 confirms that the performance of the particle counter is almost identical to that of SEM, with relative error in a range of $\pm 25\%$.

This counting test with PSL standard particles verified that more than 10 consecutive time series data were distributed normally. This indicates proper distribution of time series data, because the particle counter estimates the particle concentration of the entire water sample by counting the number of particles passing through an extremely small probability space (the measurable volume determined by the sectional area of the laser beam and the slit breadth on the scattered light detection side). The number of particles in actual ultrapure water counted by SEM is corrected by subtracting the number of blank particles originally adhering to the membrane filter that captures the particles.

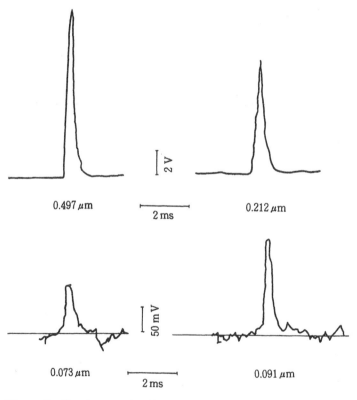

Figure 7   Signal wave of uniform polystyrene latex particles.

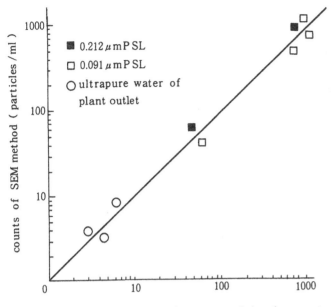

mean values of particles/ml measured by the counter

Figure 8   Comparison of particle counts between the automatic counter K-LAMIC-100 (PLCA-310) and the SEM method.

C.   Practical Application to Plant Water

A large number of particles in discharged into product water soon after start-up of the ultrapure water system. Conventionally, the cleaning operation was performed for a long time, and the end of cleaning was determined, with resistivity and total organic carbon (TOC) additionally monitored by a 0.2 μm particle counter. Figure 9 shows the number of particles contained in rinse waste after start-up of the ultrapure water system, which was monitored by the high-sensitivity particle counter described here. The particle counter, making real-time monitoring of the behavior of particles possible, promises a higher efficiency of the rinsing process than the conventional method, which occasionally allows excessive rinsing.

The product water of an LSI manufacturer's ultrapure water subsystem was monitored by the particle counter described here, and the results are shown in Fig. 10. The subsystem has two feed pumps: one is on standby when the other is on duty. Figure 10 clearly shows that at the automatic changeover of the two feed pumps at midnight, the number of particles increases transiently mainly as a result of vibration of the piping system.

The high-sensitivity particle counter enables real-time monitoring of fluctuations in the number of particles, assuring a higher grade of operational control than ever before.

V.   CONCLUSION

Evaluation technology of higher sensitivity and accuracy is required for the improvement of ultrapure water technology to cope with the higher density of VLSI. Although it is

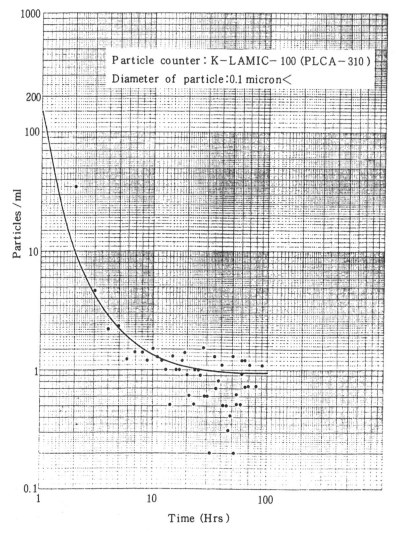

Figure 9   Particles trend in ultrapure water at startup.

technically possible to count the number of finer particles by SEM, filtration to capture
these particles consumes an enormous amount of time. In addition, SEM has many
problems in practical application, such as contamination of the membrane filter or
contamination during filtration and the increase in the number of fields required along
with the increase in the SEM magnification. Therefore, the smaller the size and quantity,
the more a reliable on-line evaluation technology is needed. A light-scattering method and
a laser breakdown method are already available as particle-counting methods with high
sensitivity, but they also have many problems that are yet to be solved, such as reduction
in bubbles, contamination, and background noise level, minimization of measurable
volume, and verification. We keenly feel the urgent requirement of establishing a
high-sensitivity evaluation technology.

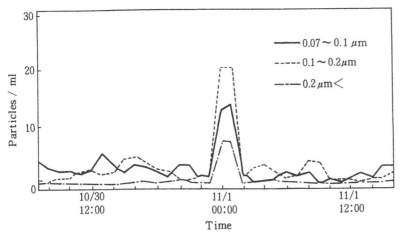

Figure 10 Fluctuations in particle count at the time of exchanging the duty and standby feed pumps.

REFERENCE

1. H. Shigemi, T. Kumagai, S. Akiyama, R. Suzuki, High-sensitivity particle counter for ultrapure water, Monthly Semiconductor World, July 1988.

i. Nano-Sized Particle Analyzer (Nanolyzer)

Takashi Sasaki
Shinko Pantec Co., Ltd., Kobe, Japan

I. PRINCIPLE OF MEASUREMENT

A. Structure and Specifications

Figure 1 shows the principle of measurement. Ultrapure water flows continuously through a Teflon measuring cell with a hollow cylindrical section, and an elliptical laser beam is focused on an intermediate point between the center of the cylindrical part and the wall surface of the cylinder. A particle detection area, which is limited by a light-accepting optical system, is established at the focal point of the irradiated laser beam. Particles in the ultrapure water pass through this dectection area toward the direction of the optical axis of the light-accepting optical system. From the detection area, scattered light is reflected from particles passing through the detection area together with the scattered light coming from water molecules in ultrapure water, for example. This scattered light is detected by a photomultiplier, counted as the number of photon pulses, and preserved as time series data. The time series data are analyzed by special software to calculate the particle size from the scattering intensity of particles and the number of particles that pass, and the

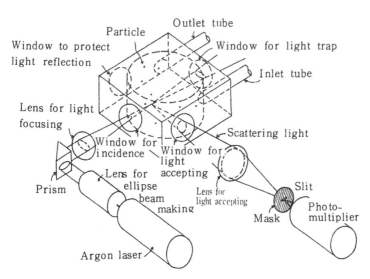

Figure 1 Principle of measurement.

results are displayed as the particle distribution. The specification and configuration of the nanolyzer are shown in Table 1 and Fig. 2, respectively [1,2].

B.   Setting the Particle Detection Area

The relations between the elliptical beam and the particle detection area are as shown in Figs. 3 and 4. A mask is provided on the image formation side of the light-accepting lens to limit the visual field. Particles pass along the optical axis of the light-accepting optical system through the particle detection area (shaded portion), which is limited by the mask on the light-accepting side. The width of the visual field of the limited particle detection area is determined so that the strength of the elliptical beam exceeds 80% of the central strength, and the direction of the elliptical beam is selected so that a large cross section vertical to the particle passage direction is provided. The effect of the setting is illustrated in Fig. 4. As particles enter from direction A into the detection area at positions a_1, a_2, and a_3, the strength of the laser beam applied to particles changes to I_1 (corresponding to position a_1) and I_2 (corresponding to positions a_2 and a_3), but the maximum strength of the beam when passing through the elliptical beam can be limited to the shaded range of I_3, that is, a range in which the beam strength is over 80%.

C.   Automatic Setting for the Mask

It is important that the center of the width of the visual field limited by the mask coincide with the optical axis of the elliptical beam with an accuracy of several micrometers to increase the particle diameter resolving power. The nanolyzer is designed to automatically position the mask regularly to take account of temperature changes and fluctuations from vibration. The principle of automatic control is explained by Fig. 4. When the center of the width of the visual field coincides with the optical axis of the elliptical beam, the background light, which consists of water molecules, increases to the maximum amount, and when these do not coincide, the amount of background light decreases. The

Table 1 Specifications for the Nanolyzer

Principle of measurement	: Argon laser 90° scattering
Method of measurement	: Continuous inline measurement
Measurable particle diameter	: More than 0.07 μm
Measurable particle concentration	: Less than 6×10^4 particle / cm^3
Flow volume	: Less than 3000 ml / min
Indication time	: 5 ~ 10 min interval
Indication output	
particle diameter	: 4 ranges digital indication
	(1) 0.07 ~ 0.1 μm
	(2) 0.1 ~ 0.15 μm
	(3) 0.15 ~ 0.2 μm
	(4) more than 0.2 μm
Indication of condition	: Over-flow of count number
	Abnormality of laser power
	Life of laser
	Impossibility of measurement
Print output	: Self-contained parallel interface
Exterior output	: RS 422 interface
Water pressure	: Less than 5 kg / cm^2
Measurement water temperature	: 0 ~ 45 ℃
Treatment water temperature	: 0 ~ 90 ℃
Power supply	: AC 100 V $\pm$ 10 V (50 / 60 Hz)
Electric power consumption	: Max 1.5 kwh
Materials Tube	: PFA 6 ϕ × 8 ϕ tube
Measurement cell	: PTFE, quarts
Dimension of equipment (HWD)	
Detector	: 220 mm × 685 mm × 515 mm
Power supply	: 160 mm × 460 mm × 175 mm
Analyzer	: 135 mm × 420 mm × 380 mm

Figure 2   The nanolyzer, an ultrafine particle-measuring instrument.

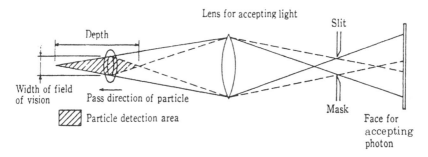

Figure 3   Establishment of the particle detection area.

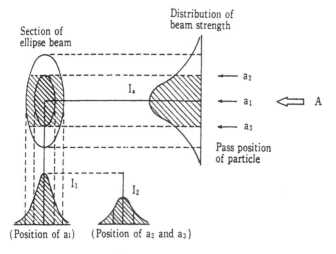

Figure 4   Passage position of the particle and beam strength.

center of the visual field width may therefore coincide with the optical axis of the elliptical beam by scanning the mask along the vertical direction of the optical axis of the elliptical beam within the mask face, setting the mask to the position of maximum acceptable background light.

D. Conditions for Particle Recognition

When measuring particles less than 0.1 μm, an extremely important matter related to the performance of the apparatus is how particles are discriminated from noise. The nanolyzer is set with the following conditions of particle recognition. Based on the time series data obtained from particles that pass vertically through the particle detection area shown in Fig. 5 toward the optical axis of incidence of the elliptical beam, the nanolyzer recognizes as particles only those that have a scattering strength that exceeds a certain level of particle recognition and those whose particle passage width satisfies values in a certain range. The nanolyzer then calculates the particle diameter from the peak value of the scattering strength at this level.

The count value level, expressed by $\bar{B}$, indicates the level corresponding to the mean count value of the background light by water molecules, and $\bar{B} + \sqrt{B}$, exceeding this level by a swing $\sqrt{B}$, is set as the threshold level for particle recognition. The time during which the particles pass through the particle detection area is determined by the standard passing time $t = L/V$, based on the velocity of the particle V, when the beam diameter along the passage direction in the detection area is set to the passage length L. When time exceeding the threshold value $\bar{B} + \sqrt{B}$ is set to W in the time series data, the count value is judged to be attributable to the particle pass when W is equal to t. The diameter of an individual particle can be determined by converting into particle diameter the peak value of the scattered light $\tilde{P}$, exceeding the threshold value $\bar{B} + \sqrt{B}$, of the time series data. Under these conditions of particle recognition, particles are clearly discriminated from noise and nano-sized particles less than 0.1 μm can be measured with accuracy. For calibration of the number and diameter of a particle, we used polystyrene latex particles from 0.067 to 0.177 μm. For calibration of ultrapure water, we used the result of a measurement performed concurrently, in which ultrapure water was filtered through a filter with 0.05 μm apertures and the particles trapped on the filter were counted by SEM (scanning electron microscopy).

E. Light Scattered from Particles and Photon Count Value

The amount of 90° side-scattered light W_p of the spherical particle based on Mie's theory of scattering is expressed by the equation

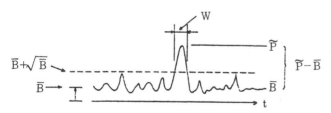

Figure 5 Conditions of particle recognition. $\bar{B}$, mean count value of background light; $\bar{B} + \sqrt{B}$, particle recognition level; $\tilde{P}$, peak value of particle; $P - \bar{B}$, scattering light count of particle.

$$W_p = I_B I_S \pi R^2 \quad \text{mV} \tag{1}$$

where: I_B = center strength at laser beam focal point
$\quad\quad I_S$ = relative strength at 100 mm, the light-accepting distance (function depend-
$\quad\quad\quad\quad$ ing on particle diameter, deflection rate, and wavelength of light)
$\quad\quad R$ = effective radius at the light-accepting position

When the laser beam output is W_o and the elliptical beam diameter at the focal point is $2W_x$, $2W_y$, the following equation is established:

$$I_B = W_o \frac{2}{W_x W_y} \quad \text{mW mm}^{-2} \tag{2}$$

Equation (1) is rearranged by using Eq. (2):

$$W_p = \frac{2W_o I_S R^2}{W_x W_y} \quad \text{mW} \tag{3}$$

When a photon multiplier receives scattered light amount W_p, the output Q is expressed as

$$Q = W_p \sigma \quad \text{A} \tag{4}$$

where σ is cathoded radiation sensitivity (A mW^{-1}).
$\quad$ The photon count P_e is obtained when the count time is T and the elementary electrical charge is e:

$$P_e = \frac{QT}{e}$$

$$= \frac{W_p \sigma T}{e}$$

$$= \frac{2W_o I_S R^2 \sigma T}{e W_x W_y} \quad \text{pulses} \tag{5}$$

II.   EXAMPLE OF MEASUREMENTS

A.   Measurement of Standard Particles

We installed the nanolyzer in an ultrapure water plant to take measurements of particle strength in ultrapure water alone and that mixed with standard polystyrene latex particles of 0.067, 0.091, 0.144, and 0.177 μm, the results of which are shown in Fig. 6. With ultrapure water alone, because of the extremely low particle density, a peal like that seen when the ultrapure water was mixed with polystyrene latex particles was not found, although there were two types of scattered light strength distributions: (1) particles were not found at all, and (2) extremely minute particles were measured. When particles are not measured, as shown in Fig. 7, the distribution of the scattered light is constituted by the light scattered only by water molecules and very minute particles. This is referred to as background light, or blank.

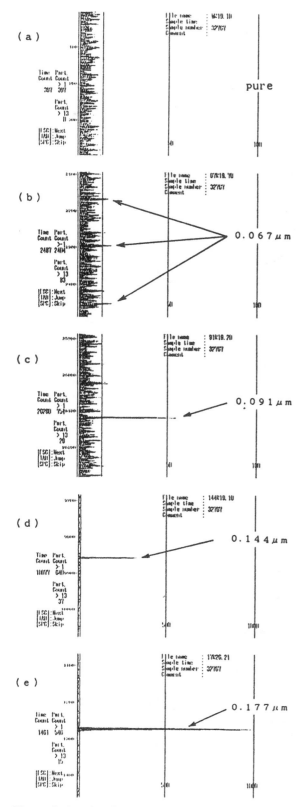

Figure 6 Results of measurement of polystyrene latex particles.

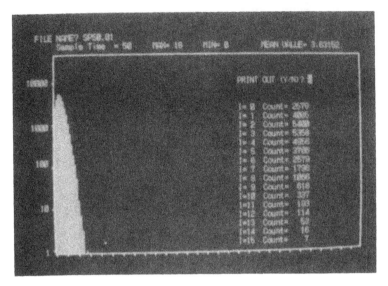

Figure 7   Distribution of scattering light strength for ultrapure water.

In the measurement of ultrapure water filled with latex particles, the scattered light from the particles is added to the distribution of the background light alone, as shown in Fig. 8. The distribution of scattered light is stored and analyzed as time series data, and the number and diameter of particles are calculated based on the particle recognition level and the scattered light strength. The particle diameters are displayed in four ranges.

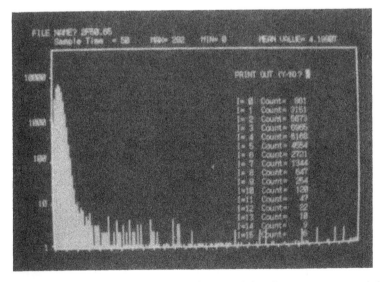

Figure 8   Distribution of scattering light strength for ultrapure water containing polystyrene latex particles.

B. Measurement of Ultrapure Water

1. *Example of Installation*
The nanolyzer can be installed at a use point of an ultrapure water manufacturing plant as an on-line nano-sized particle counter and can also be installed in the secondary pure water piping for continuous monitoring of the water quality, as shown in Fig. 9. For this reason, consideration is given to installation in the secondary pure water piping. The detecting section is separated from the analyzing section and installed near the piping so that the analyzing section can be installed at a position of easy access. It is possible to realize the computer-aided control of an ultrapure water manufacturing plant by transmitting the analyzed data to the host computer.

2. *Measurement of Ultrapure Water*
Figure 10 compares the measurement results using the nanolyzer and SEM. The measurement by a SEM was carried out concurrently by filtering ultrapure water using a 0.05 μm Nuclepore filter. The particles trapped on the filter were examined by SEM. As apparent from Fig. 10, both results agree well in four particle diameter ranges, which verifies the good performance of the nanolyzer.

C. Measurement of Chemical Solutions

The scattered light strength decreases and the particle image blurs when measuring particulates in chemical solutions because the refractive index changes depending on the

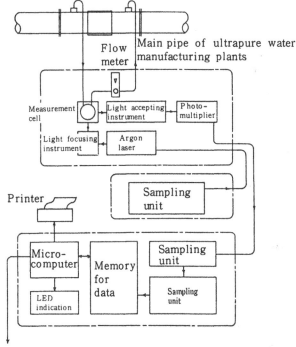

Figure 9 Setting up the equipment.

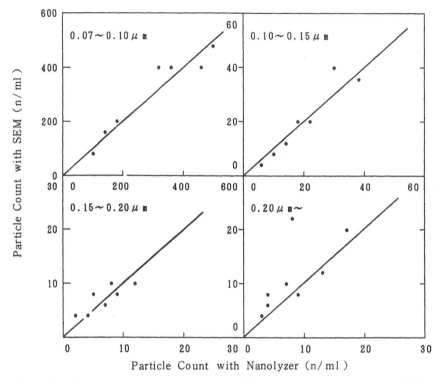

Figure 10   Particle number measured by two different methods: nanolyzer and SEM.

concentration of chemical solutions, which causes the positions of the focal point of the laser beam and the visual field on the light-accepting system to diverge [3].

Considering these problems, the nanolyzer is designed so that if the chemical solution is within the concentrations listed in Table 2, the measurement of particulates in acid and alkali solutions is possible, maintaining particle diameter resolving power.

D.   Direct Detection Method

Generally, the nano-sized particulate measuring apparatus is evaluated in comparison with direct detection using SEM [4]. With this method, a fixed quantity of ultrapure water is filtered through a filter with fixed aperture diameters. Particles in the ultrapure water

Table 2   Measurable Maximum Concentration of Chemicals for the Nanolyzer

Chemicals	Maximum concentration
HCl	2 %
HNO$_3$	4 %
H$_2$SO$_4$	4 %
NH$_4$OH	8.5 %

trapped on the filter are counted in a microscope. Conventionally, an optical microscope was used for counting particles. In the age of megabit memories, measurement of particles less than 0.1 μm has become necessary but the optical microscope cannot measure such minute particles. This in turn called for measurement by SEM, whereas such measurement normally involves extremely difficult problems.

Major factors closely related to accuracy in measurement are blank arrangement, sample quantity, number of visual fields, and magnification, but the standard method is based on the following factors:

1.   Blank B

Generally, a box containing 100 filters made by Nuclepure Co., Ltd. is used for random sampling. A filter selected at random is provided with the same operation and processes as the filter used to filter ultrapure water, and the particles trapped on the filter are counted by particle diameter range in a microscope.

2.   Sample Quantity S

The quantity of the ultrapure water sample filtered by a 0.05 μm Nuclepore filter is over 10,000 ml.

3.   Number of Visual Fields F

At a magnification of 20,000, there are 13 positions from the center of a filter with a 19 mm effective diameter in the horizontal direction of the X axis at a 1.5 mm pitch and 19 positions in the vertical direction of the Y axis at a 1.0 mm pitch, a total of 195 positions. An additional 5 positions are sampled at random to obtain a total of 200 visual fields.

4.   Coating

Pt-Pd ion coating is applied in a clean bench of less than 10 ft^{-3} as 0.1 μm particle including removal of the filter.

5.   Particle Size Range

Particles are counted in a scanning electron microscope in the following four size ranges: 0.05–0.1 μm, 0.1–0.15 μm. 0.15–0.2 μm, and 0.2 μm and over. Based on the standards, the particle count in ultrapure water is calculated by the particle size range as follows:

$$\text{Particle number (number ml}^{-1}) = \frac{X - B}{S} \frac{A}{Fa}$$

where: X = number of excess sample particles observed by microscope
  A = effective filter area, cm^2
  a = area per visual field under microscope, cm^2
  B = blank
  S = sample quantity, ml
  F = number of visual fields

III.   FEATURES

The nanolyzer has the following features:

1.   Particles greater than 0.07 μm are displayed by diameter range. Particle number ml^{-1} is displayed in digits in four particle diameter ranges.

2. On-line continuous monitoring is possible with a simple piping system, and daily maintenance is hardly required.
3. Heat disinfection of the sampling line at 80°C is possible using heat-resistant materials.
4. A completely automatic compensating mechanism is incorporated in the nanolyzer. Long and steady monitoring is provided by the automatic compensating mechanism for setting the amount of light and the optical axis.
5. The nanolyzer is connected to the host computer for control. Various data management and system control features are available using a host computer.

REFERENCES

1. T. Sasaki, Shinko-Pfaudler Technical Report, *31*(1), 8–13 (1987).
2. T. Sasaki, Proceedings of 5th Symposium on ULSI Ultra Clean Technology Symposium, Advanced Process & Technology II, Ultra Clean Society, (1987), pp. 27–40.
3. H. Shigemi et al., Semiconductor World, *8*(3), 134–139 (1989).
4. K. Koyama, Water Purification and Liquid Wastes Treatment, Committee of Japan Water Purification and Liquid Wastes Treatment, (1986), pp. 17–23.

ii.   Particle Counter for Fine Particles in Ultrapure Water

Toshiki Manabe

Japan Organo Co., Ltd., Saitama, Japan

I.   INTRODUCTION

The semiconductor industry is in the stage of mass production of the 1 MDRAM, and 4, 16, and even 64 MDRAM are under development. Prototypes of most such advanced devices have been already produced, but many problems are reported, which are caused by water, chemicals, and gas. Among them, in particular, the contamination caused by fine particles is the most significant factor for a considerable reduction of yield in manufacturing (rate of satisfactory products). For the upcoming "half-micrometer" age, more strict quality control will be required for water, chemicals, and gas. This section discusses a measurement procedure for fine particles in ultrapure water based on a particle counter using a beam of air-cooled argon ion laser (wavelength 488 nm). In addition, the results of the measurement are reported here.

II.   MEASUREMENT PRINCIPLE

With most widely used conventional fine particle counters, the measurement capability is limited: that is, the detectable minimum particle size (diameter) is in the range 0.3–0.5 μm, with the particle concentration not less than 10^1–10^2 counts ml^{-1}. The target of our development of a new instrument is to achieve a significant measurement level of 10^{-1}–10^0 count ml^{-1} for particle sizes of about 0.07 μm or larger.

Using measurements by laser radiation under the same conditions, the scattered light intensity is reduced to 0.01–0.001 for 0.07 μm compared with 0.3–0.5 μm particles. The measurable concentration must be about 0.01 of that of conventional instruments. To achieve this target, we originally developed a novel radiation scattering system with an air-cooled argon ion laser beam for scanning. This system is referred to here as the SLPC (scanning laser beam particle counter).

Conventional counters of fine particles suspended in liquid medium using the light-scattering method are classified into two systems according to the relationship between the radiation beam and the sample flow. In one system, the sample flow is directed into the beam; in the other, the beam is directed into the sample flow. The light intensity profile generally has a Gaussian or normal distribution, as shown in Fig. 1. The former system is meant to stabilize the intensity of the light illuminating the particles by exposing the narrow sample flow only to the portion of the laser beam around the light axis, where the light intensity distribution is relatively flat. Although this system has a relatively high accuracy of size measurement, it is not capable of detecting extremely fine sizes because the beam cannot be narrowed freely compared with the width of sample flow. The latter system introduces the laser beam into the sample flow, setting a sensing zone in the beam to receive the scattered light. This system, too, has drawbacks. The narrower beam reduces the detection zone, resulting in a lower receiving efficiency of scattered light and poorer accuracy of size determination. In addition, compared with the sample flow sectional area, the available sectional area of the detection zone should be too small to evaluate low concentration samples satisfactorily.

Considering these problems, the following essential objectives of improvement have been set up:

1. Narrow the beam to the desired degree, with the same laser output, so that the minimum detectable size is reduced as much as possible.
2. Extend the detection area as much as possible, considering extremely low concentrations.
3. While satisfying requirements 1 and 2, maintain a high accuracy of fitness determination.

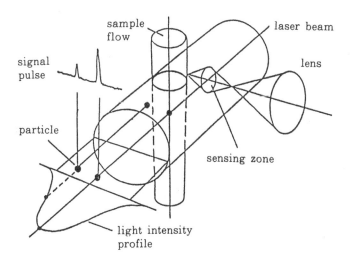

Figure 1   Conventional optical system for a laser particle counter.

Figure 2 illustrates the principle of the newly developed SLPC for satisfying these conditions.

The SLPC is characterized by narrowed beam scanning at high speed in the sample flow. It is provided with an air-cooled argon ion laser beam head (wavelength 488 nm). The beam projected from the head is made, by the action of an audiooptical deflecter, to scan particles linearly. The scanning speed is high enough compared with the velocity of the passing particles. The deflector generates, by ultrasonic wave, a compressional wave in a transparent optical crystal. The laser beam is deflected by means of the diffraction due to the compressional wave. Since the deflection angle is altered by changing the wavelength of the compressional wave, no movable part is required for high-speed scanning.

Fine particles in the sample flow crossing the beam are irradiated a number of times. Light is scattered every time it hits a particle. The scattered light is directed through a condenser lens into a photoelectric amplifying tube, to be converted into electrical pulses. The effective area of the light-receiving system fully covers the beam scanning range. Figure 3 is a signal wave representing light scattered when a single particle passes through the beam scanning range. As shown, the intensity of the scattered light produced during exposure of a particle to the beam traces the light intensity profile of the beam. Therefore, by taking the highest pulse as the scattered light data, the SLPC necessarily obtains data irradiated by the beam near the maximum intensity point. The data for the number of particles are displayed in real time during the measurement. This method enables one to set up, in the sample flow, an irradiation range of the same width as the beam scanning range, where the light intensity is uniformly distributed and is equal to the maximum intensity of a narrowed beam. Hence it satisfies the three requirements mentioned earlier.

Figure 4 shows signal waves from determination by the SLPC of polystyrene standard latex (PSL) particles in pure water. PSL is generally used for calibration of particle counters. One pulse corresponds to one particle. PSL are 0.073, 0.087, and 0.208 μm in

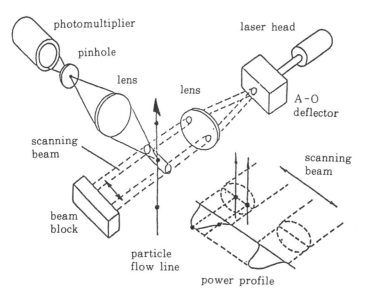

Figure 2   The new optical system with a scanning laser particle counter (SLPC).

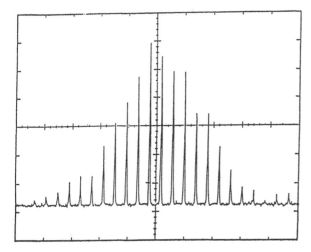

Figure 3 Signal of single particle from the SLPC preamplifier.

diameter. Figure 5 is a characteristic curve used for converting the pulse height into the particle size. For establishing this curve, we created a simulation model based on Mie's theory, which provides the theoretical foundation for the radiation scattering counters in conformity with the optical properties of the SLPC, and collated the values from the model analysis with the experimental results using PSL. In the size range from about 0.2 to 0.4 μm the curve in Fig. 5 is a two valued function, so it is impossible to convert

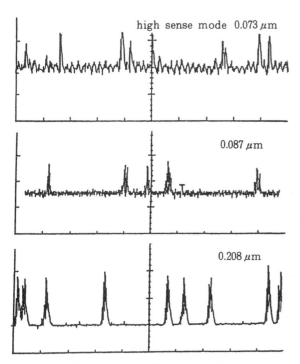

Figure 4 Signal pulses of PSL particles in pure water.

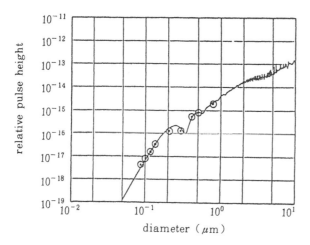

Figure 5 Comparison between observed (points) pulse heights of PSL particles in water and heights calculated (lines) with the argon ion laser.

uniquely the pulse height into the particle size. An optical solution was adopted because of this inconvenience: that is, the laser deflection axis is rotated to obtain monotonously increasing behavior. Hence, when the measurement range corresponding to the particle sizes is selected, conversion data based on another characteristic curve are used.

III.   FEATURES OF THE SLPC

The appearance is shown in Fig. 6 and the specifications in Table 1. The features include:

1.   Measuring range to 0.07 μm
2.   Stable and rapid (3 minutes ml^{-1}) determination of even very low concentrations, that is, several particles ml^{-1}

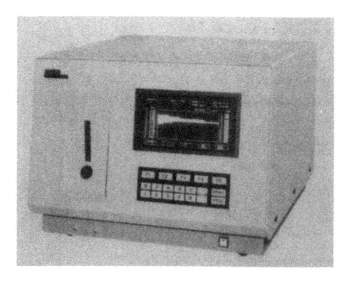

Figure 6   Appearance of the prototype SLPC.

Table 1 Specifications for the SLPC

measurable particle size	$0.07 \sim 10.0\,\mu$m
measuring range	(H) $0.07 \sim \quad \mu$m
	(1) $0.08 \sim 0.17\,\mu$m
	(2) $0.15 \sim 0.70\,\mu$m
	(3) $0.60 \sim 10.0\,\mu$m
	(4) (1) + (2) + (3)
measurable particle concentration	$0 \sim 10^5$ number / ml
measuring time (by 1 ml)	3 min
sample flow rate	10 ml / min
liquid contact materials	PTFE, quartz glass, valflon
display	E. L. display (W 200 × H 104 mm)
out-put	printer, alarm, (RS 232 C : o. p.)
dimensions	W 551 × D 634 × H 392 mm
weight	50 kg
power requirement	AC 100 V 50 / 60 Hz 20 A

3. High accuracy of evaluation of size distribution in a wide range (0.07–10 μm); for low concentrations (in pure water) up to 10^5 particles ml^{-1}

4. An optional sapphire cell that enables assessment of various chemicals comprising fluoric acid

IV. PSL EVALUATION TEST AND MEASUREMENT OF FINE PARTICLES IN ULTRAPURE WATER

A. Calibration Using PSL

There are two ways to use the SLPC. One is to transport sample water directly into the SLPC via a sampling pipe from the main piping. The other is to place the sample in a container (e.g., Teflon or SUS) and feed the SLPC with the sample by means of a specific pressure container or suction sampler.

The former is preferable for the control of extremely low concentrations of particles in ultrapure water. If the latter is selected, care should be exercised to avoid bubbling of sample water. For this purpose, it is recommended when feeding to apply a pressure of 0.5 5.0 kgf cm^{-2} using contaminant-free nitrogen.

In our experiment, liquid containing 0.087, 0.100, and 0.15 μm polystyrene standard latex was diluted with ultrapure water. Each solution was placed into the SLPC, which was pressurized by contaminant-free nitrogen. The size distribution of different solutions determined by the test is shown in Figs. 7, 8, and 9. The dilution operation was implemented in a ultraclean room to avoid contamination.

The peak values of distributions from the test agree well with the PSL sizes. This proves the instrument has a highly accurate capability for determining particle sizes.

B. Measurement of Fine Particles in Ultrapure Water

Figure 10 depicts schematically a typical ultrapure water system for megabit-level semiconductor manufacturing. This system, installed in our research laboratory, is composed

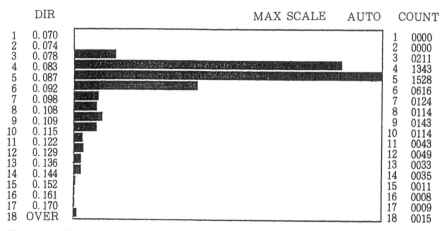

Figure 7   Size distribution output of 0.087 μm PSL particles in pure water.

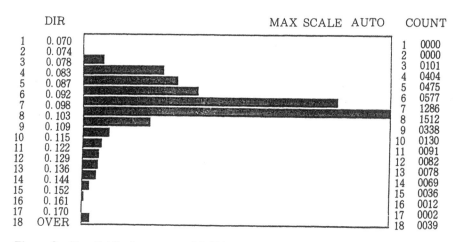

Figure 8   Size distribution output of 0.100 μm PSL particles in pure water.

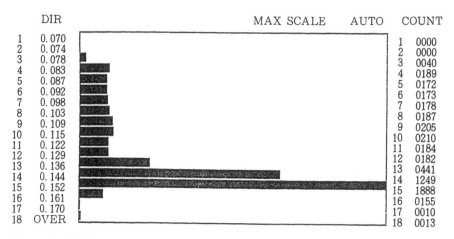

Figure 9   Size distribution output of 0.150 μm PSL particles in pure water.

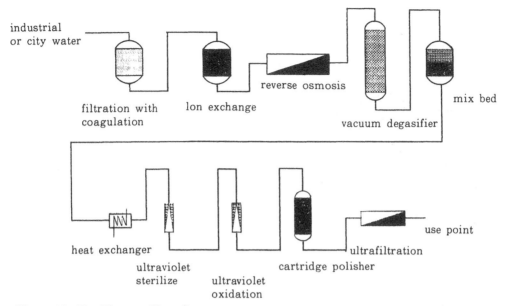

Figure 10   The Ultrapure Water System

of a primary and a secondary line. The primary line is composed of pretreatment, ion exchanger, reverse osmosis unit, and vacuum degasifier, and the secondary line contains the UV sterilizer, ultraviolet (UV) oxidation unit, ion-exchange resin, and ultrafiltration (UF) unit.

Ion-exchanged water and UF product water from the secondary line were assessed for fine particles using a conventional type of instrument (PC-1000, Fuji Electric Co., Ltd.; measurable size 0.11 μm or larger) and the SLPC. The particle counts and size distribution at each sampling point are indicated in Table 2 and Fig. 11, respectively. The duration of the measurement was 2 h. A total of 10 values were taken as effective results after the indication on the counter stabilized. The figures in Table 2 are the averages of such values. The variation of 10 measurements is in the range ±0.1 particle ml^{-1}. The result with the conventional instrument was 3 particles ml^{-1} of 0.11 μm or larger, whereas with the SLPC this was 13 counts ml^{-1} of particles 0.07 μm or larger. From this it is known that among suspended material in ultrapure water there are many very fine particles under 0.11 μm.

Table 2   Particle Counts at the Sample Point

(counts / ml)

Sample point	PARTICLE COUNTER		
	SLPC		PC - 1000
	> 0.07 μm	> 0.11 μm	> 0.11 μm
Ion exchange water	19	4	4
UF product water	13	3	3

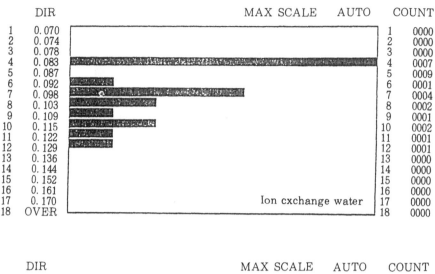

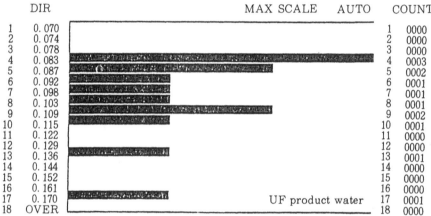

Figure 11   Size distribution output of ion-exchange water and UF product water.

V.   CONCLUSION

The new instrument, SLPC, has been proved to be an effective means for assessing ultrapure water containing an extremely small amount of fine particles. The SLPC can determine the size distribution of particles 0.07 μm or larger. Therefore, the author believes that this instrument, performing rapid and continuous on-line measurement of fine particles in ultrapure water used for megabit semiconductor manufacturing, will be a powerful strategy in quality control.

These are the three principal merits of the instrument:

1.   Measurable size of 0.07 μm or larger: a size distribution of particles larger than 0.087 μm can be obtained.
2.   Measurement of particles larger than 0.07 μm is implemented by the scanning beam in a short span of time, that is, 3 minutes ml^{-1}.
3.   The instrument is provided with a quartz cell for evaluation of ultrapure water. By replacing this type of cell by a sapphire cell, it can be used for chemical solutions. In this case, particles 0.1 μm or larger are detected.

3 On-Line Monitors

D. High-Purity Monitor

Seiichi Inagaki
Nomura Micro Science Co., Ltd.,
Kanagawa, Japan

I. INTRODUCTION

As the fine processing technology for inorganic carbon (IC) advances, the ultrapure water used in the semiconductor industry must be of increasingly higher quality. To meet this demand, ultrapure water production technology has introduced the reverse osmosis unit, the ultrafiltration unit, ion-exchange resin, new materials for the piping system, and advanced installation technology. It is how the ultrapure water production unit is operated and controlled after its installation at the semiconductor manufacturing plant that determines whether the water quality is maintained. In this sense, maintenance of the ultrapure water production unit and the system that monitors the water quality are critical. Based on this understanding, semiconductor manufacturers introduced such continuous monitoring systems as the fine particle counter, the silica monitor, and the total organic carbon (TOC) monitor. In addition, ultrapure water is periodically sampled for analysis.

However, there are still many problems to be overcome. For the in-line monitor, measurable parameters for checking ultrapure water quality are limited. It has also been pointed out that handling of the monitor is not simple: calibration should be conducted regularly. The accuracy is also not sufficient at low concentration levels. Because manual analysis takes a long time, monitoring requires a very long time.

Accordingly, the engineers in charge of controlling ultrapure water in semiconductor plants have demanded equipment that enables them to monitor ultrapure water comprehensively. Specifically, the demands are as follows:

1. Various parameters of ultrapure water quality should be monitored with one piece of equipment.
2. The calibration should be easy and definitive.
3. The reproducibility should be high.

4. The equipment should not be significantly affected by the temperature of water, the pressure of the pure water line, or bubbles.
5. The measurement time should be short (quick response).

The high-purity monitor (HPM) was developed as an ultrapure water in-line monitor to satisfy these demands. This part describes the principles of and specifications for this monitor and introduces actual measurement data to exhibit its accuracy. We also mention the fields to which this monitor can be applied.

II. PRINCIPLES AND SPECIFICATIONS

A. Principles

The principles of measurement in the HPM are as follows. The ultrapure water to be measured is atomized into clean air, the moisture is evaporated at 100°C, the impurities in the ultrapure water are introduced to the measurement unit in a aerosol form, the number of particles is counted with the nuclear condensation counter, and the number of particles is converted to the impurity concentration to be displayed. Figure 1 illustrates the procedure.

The ultrapure water sampled on-line from the ultrapure water line is sent to the atomizer through a stainless steel capillary tube that controls the flow rate. The atomizer constantly atomizes the polydispersive droplet, which is always controlled to maintain certain conditions of the condensed clean air and the capillary tube, into high-temperature clean air. Here it is assumed that two kinds of ultrapure water with different impurity concentrations are atomized. Ultrapure water with the higher impurity concentration is A, and that with the lower impurity concentration is B. Figure 2 shows the particle size distribution in the atomized droplet. A–1 and B–1 exhibit the same distribution regardless of the impurity concentration in the ultrapure water. When the droplets pass through the evaporation tube, however, the particle distribution changes, as shown by A–2 and B–2, because the moisture has evaporated.

Because only one impurity particle can be formed from one droplet, however, the number of particles remains the same. The particles are then cooled and sent to the diffusion battery, where the smaller the particle size, the more particles are removed by further activated Brownian motion. A–3 and B–3 show the distribution after particles pass through this diffusion battery. As the particle size distribution decreases, the number of particles is decreased. The impurity concentration of the ultrapure water can therefore be determined by counting the number of particles that arrive at the detection area.

The principle is described by the following equations [1]. The change in the number of particles at the diffusion battery can be expressed as

$$\frac{n}{N_a} = e^{(-\beta t)} \tag{1}$$

where: N_a = number of droplet particles at the atomizer
   n = number of impurity particles downstream of the diffusion battery
   t = time required for a droplet to pass through the tube
   β = precipitate constant

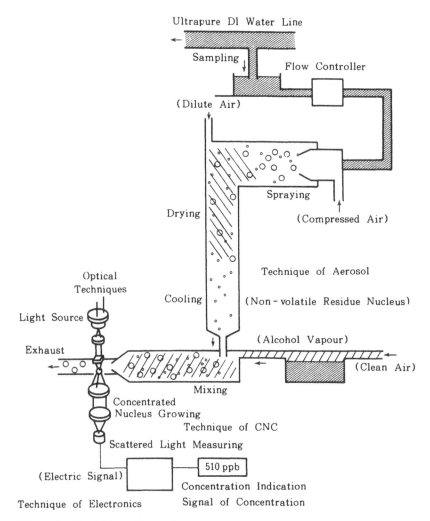

Figure 1   The high-purity monitor.

The precipitate constant β can be expressed as

$$\beta \propto X_a^{-1.22} \tag{2}$$

where X_a = diameter of impurity particle. Here the relation between the diameter of the impurity particle and the diameter of the atomized droplet is assumed as

$$\gamma = \frac{\rho_a}{\rho_w}$$

where: ρ_a = density of impurity
ρ_w = density of atomized droplet (density of ultrapure water)
The diameter of impurity X_a can then be expressed as

$$X_a = \frac{C}{\gamma} \frac{1}{3} X_w \tag{3}$$

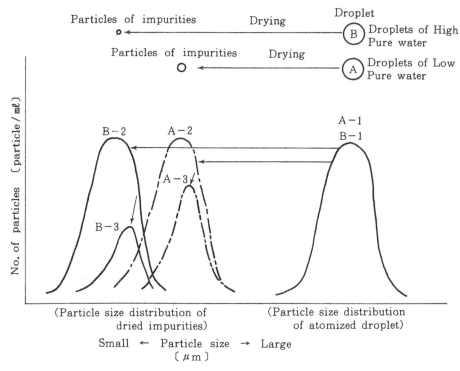

Figure 2 Variation in the particle size distribution.

where: X_a = diameter of impurity particle
       X_w = diameter of atomized droplet
       C = impurity concentration (volume density)

Based on Eqs. (1) through (3), the impurity concentration C can be expressed as a function of the number of impurity particles at the outlet of the diffusion battery. The HPM applies this principle to measure the impurity concentration accurately at low levels within a short amount of time.

B. Specifications for the HPM

Table 1 lists the major specifications, and Fig. 3 is an external view of the HPM. The sample water can be supplied from the ultrapure water pipe in-line to conduct continuous measurement. The measuring range is 0–1000 ppb. The data are displayed on a cathode ray tube (CRT) every minute and recorded on the built-in floppy disk. A KCl solution is employed for calibration conducted at the time of shipment.

III. DETECTION OF IMPURITIES IN ULTRAPURE WATER

Various kinds of impurities were dissolved in ultrapure water, and they were measured with the HPM In the following experiments.

Table 1 Specifications for the HPM

Model	HPM – 1000
Measuring Object	non-volatile residue in ultrapure water
Measuring Method	Atomizing and Drying
Detection Range	0 – 1000 ppb
Sensitivity	1 ppb
Calibration	KCl standard Solution
Condition of Sample Water	FLOW Rate : Less than 500 mℓ/min
	Pressure : Higher than 0.5 kg/cm
	Temperature : 15 – 35 ℃
Response Time	Within one minute
Warm Up Time	Within 30 minutes
Sampling Time	One minute
Electrical Power	AC 100 V ± 10 %, 50/60 Hz 500 W
Condition of Circumstance	15 – 35 ℃, RH ≤ 85 %

A. Measurement of KCl

KCl solution is used for calibration of the HPM. Therefore, the correlation between the chemical analysis value and the value indicated by using KCl is very good. In Fig. 4, the vertical axis shows the HPM data and the horizontal axis shows the KCl concentration (which is measured by frameless atomic absorption spectrometry). In the range of 5–1000 ppb, the HPM data are a good match to the values measured by atomic absorption spectrometry.

Figure 5 indicates the results of an experiment to observe the HPM response against the change in the KCl concentration. The measurement time of the HPM was 1 minute. When the KCl concentration was changed to 20, 50, and 100 ppb, the HPM data followed the changes in about 2 minutes. In addition, the measurement values remained stable when the measurement was repeated at the same concentrations.

B. Measurement of SiO_2

The correlation was examined for SiO_2, which often causes trouble in ultrapure water. Figure 6 compares the HPM values and SiO_2 values analyzed using molybdate blue. It was confirmed that they match each other well at low concentrations of 10 ppb or less.

C. Measurement of Other Materials

$FeCl_3$, CH_3COOH, and KHP were measured using the same method. Figure 7 shows the measurement results for $FeCl_3$, which exhibits high accuracy. Figure 8 shows the measurement results for potassium hydrogen phthalate, which is used for calibration of

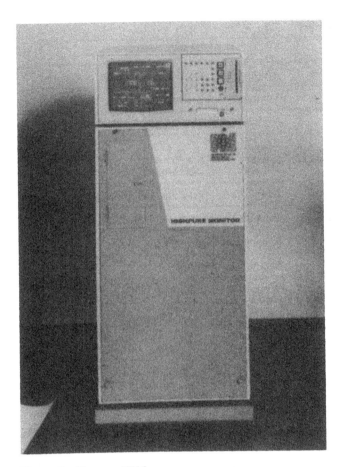

Figure 3 The new HPM.

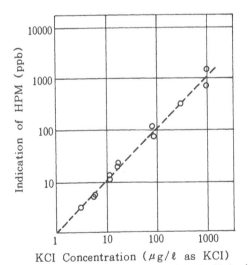

KCI Concentration (μg/ℓ as KCI)
(Chemical Analysis Value + Blank Value)

Figure 4 Correlation between the chemical analysis value and HPM data for KCl.

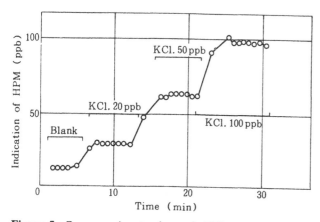

Figure 5 Response time to change in KCl concentration.

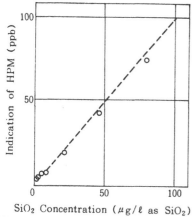

SiO₂ Concentration (μg/ℓ as SiO₂)
(Chemical Analysis Value + Blank Value)

Figure 6 Correlation between chemical analysis value and HPM data for SiO₂.

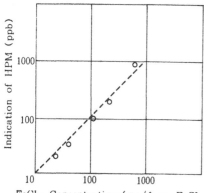

FeCl₃ Concentration (μg/ℓ as FeCl₃)
(Chemical Analysis Value + Blank Value)

Figure 7 Correlation between chemical analysis value and HPM data for FeCl₃.

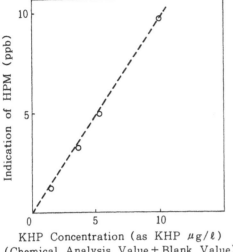

KHP Concentration (as KHP μg/ℓ)
(Chemical Analysis Value + Blank Value)

Figure 8 Correlation between chemical analysis value and HPM data for potassium hydrogen phthalate (KHP).

the TOC analyzer. They are also accurate at the low concentration level of 0–10 ppb. Figure 9 indicates the results of measuring acetic acid. The HPM did not react against acetic acid at all. This is because acetic acid is evaporated together with water in the evaporation tube of the HPM as a result of its low boiling point of 118.1°C.

IV.   MEASUREMENTS IN THE ULTRAPURE WATER PLANT

The measurement results for ultrapure water are described here. The following data were measured by the HPM in an operating ultrapure water plant.

A.   Data at the Outlet of the Mixed-Bed Tower

Figure 10 shows the results measured at the primary pure water unit that supplies city water processed by reverse osmosis to the mixed-bed tower. This mixed-bed tower is recovered with HCl and NaOH after water passes through for 55 h. The purity of the processed water is controlled at 10 MΩ·cm or greater. For the first 40 h of operation, the

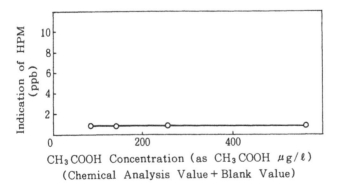

CH$_3$COOH Concentration (as CH$_3$COOH μg/ℓ)
(Chemical Analysis Value + Blank Value)

Figure 9 Correlation between chemical analysis value and HPM data for acetic acid.

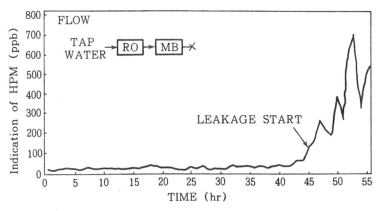

Figure 10   Change in mixed-bed deionizer (MB) effluent over time.

data from the HPM remained stable at 40–50 ppb, but a surge was observed at 42 h. Since there was no change in the resistivity of the processed water, it is thought that silica leakage took place. Although the ion-exchange capacity in the mixed-bed tower is designed with extra attention to safety, silica leakage probably took place as a result of degradation of the resin.

Figure 11 shows a similar phenomenon. In this unit, the leakage is more conspicuous because the well water contained a large amount of silica.

B.   Colloidal Silica from a Two-Bed, Three-Tower Ultrapure Water Production Unit

Figure 12 indicates the data from a primary pure water unit in which water is processed in a cation tower, decarbonator, anion tower (two-bed, three-tower unit), and mixed-bed polisher (MBP). Water sampling for the MBP is conducted for 160 h. In the meantime, the two-bed, three-tower prestage is recovered four times. The two-bed, three-tower unit has two lines that are used alternately. In this primary pure water unit, the conductivity meter is equipped at the outlet of the two-bed, three-tower unit, and the conductivity meter, the silica meter, and the HPM are positioned at the outlet of the MBP. All data

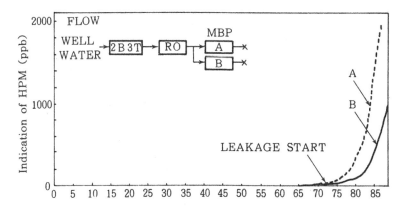

Figure 11   Change in mixed-bed polisher (MBP) effluent over time.

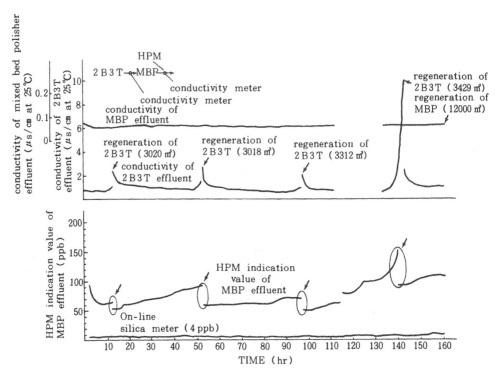

Figure 12   Colloidal silica leakage caused by a two bed, three-tower (2B3T) D1 system.

were taken during one MBP cycle. The conductivity at the outlet of the MBP was stable at 0.06 μS cm^{-1} (16.6 M$\Omega \cdot$ cm). Silica was also stable at 4–5 ppb. However, the HPM indications at the outlet of the MBP were always higher toward the end of water sampling from the two-bed, three-tower unit. Although the conductivity at the outlet of the two-bed, three-tower unit is usually around 1 μS cm^{-1} (1 M$\Omega \cdot$ cm), the quality of water is degraded to 2 μS cm^{-1} (0.5 M$\Omega \cdot$ cm) immediately after recovery. Nevertheless, the HPM values at the MBP outlet decreased. It is thought that colloidal silica leakage took place toward the end of water sampling from the two-bed, three-tower unit. It is not certain whether the colloidal silica was contained in the two-bed three-tower ultrapure water from the beginning or silica in ion form was polymerized to become colloidal. In any case, however, the colloidal silica was not removed in the MBP at all. The silica meter does not react because it measures silica only in ion form. When the water-sampling time from the two-bed, three-tower unit was extended during the fourth trial to obtain a water quality of 10 μS (0.1 MΩ), the colloidal silica leakage became more conspicuous.

C.   Leakage from the Cartridge Polisher of an Ultrapure Water Production Unit

Figure 13 shows the HPM data at the outlet of the cartridge polisher in the ultrapure water production unit. In general, the polisher is replaced every 6 months, but in this plant it was not replaced for a year. Before the polisher exchange, the HPM data were very high, at 900 ppb, but the resistivity was good, at 17 M$\Omega \cdot$ cm. After replacement, the HPM data dropped to the few ppb level and the resistivity rose to 18 M$\Omega \cdot$ cm. It was found that the water quality was degraded 100 times in terms of the HPM data when the resistivity at the outlet of the polisher changed by 1 M$\Omega \cdot$ cm.

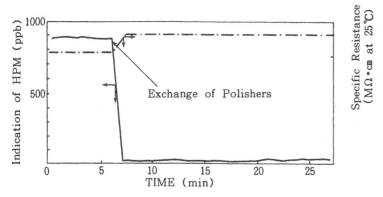

Figure 13 Impurity leakage caused by overload operation of polishers and no leakage after exchange of polishers.

D. Measurement in the Wafer-Cleaning Bath

Figure 14 indicates the results of measurement of ultrapure water sampled directly from the final rinsing bath. The HPM data fluctuated when a wafer was inserted in the bath. Figure 15 shows the measurement data when the filter set in the rinsing bath was replaced. After replacement of the filter, the HPM data surged from the usual value of a few ppb to 350 ppb. This is because of the elution of impurities from the new filter. The elution of impurities was reduced to 50 ppb in 1 h, but it took 13 h to return to the usual level.

E. Degree of Degradation of the RO Module Indicated by the HPM

Recent ultrapure water production systems employ the reverse osmosis unit downstream in an attempt to suppress TOC elution from the system as much as possible. This reverse osmosis membrane cannot be controlled with the demineralization ratio meter that controls the reverse osmosis membrane installed upstream of the ultrapure water production unit because it is exposed to ultrapure water. At present, there is no effective monitor for this reverse osmosis membrane. Very troublesome handling is therefore required to control it: the reverse osmosis membrane is regularly removed from the unit and immersed

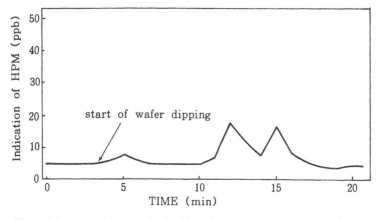

Figure 14 Change in impurity (residue after evaporation) concentration in wafer-rinsing bath with time.

in pure water with NaCl at 1500 ppm, and then the demineralization ratio is measured to control the membrane. Table 2 lists the results of an experiment conducted to find out whether the HPM can be used to monitor the reverse osmosis membrane. Run 1 in Table 2 applied the reverse osmosis membrane with a demineralization ratio of 96.5% for the 1500 ppm NaCl solution. In run 1, city water is filtered with a 0.2 μm filter and then processed in the ion-exchange tower to produce ultrapure water with a resistivity of 17–18 MΩ · cm. This ultrapure water was supplied to the reverse osmosis membrane. The values measured by HPM, TOC meter, and silica meter were compared upstream and downstream of the membrane. In run 2, the same reverse osmosis membrane was immersed in the NaCl solution to degrade the performance of the membrane, and then the same experiment as run 1 was carried out. As a result, the reverse osmosis membrane with a demineralization ratio of 96.5% showed a HPM removal ratio of 97.7%, and the reverse osmosis membrane with a demineralization ratio of 75.2% indicated a HPM removal ratio of 83.2%. The demineralization ratio and the HPM removal ratio were similar, which means that it is possible to employ the HPM as the monitor for this reverse osmosis membrane. We plan to carry out the same test by reducing the slope of the reverse osmosis membrane degradation.

V.   MEASUREMENT IN THE BOILER INSTALLED IN A THERMAL POWER PLANT

For pure water used in the boiler of a thermal power plant, particles are not controlled as stringently as for the ultrapure water used in the semiconductor industry, but trace ions are controlled at the same level. Since the trace ions in steam adhere to the steam turbines, which leads to stress corrosion, they must be suppressed as much as possible. Figure 16 shows the flow in a once-through boiler. The steam turns to water while being cooled in the condenser. Water then passes through the condensate demineralizer. Then, ammonia and hydrazine are added to prevent corrosion. Water passes through the heat exchanger and the deaerator to be sent to the boiler. Water is again turned to steam in the boiler to operate the turbines. It then returns to the condenser. Pure water in this system was measured by the HPM at the inlet and the outlet of the condensate demineralizer. A resistivity meter cannot be applied to this system because it is affected by ammonia, hydrazine, and carbon dioxide. Because the HPM does not measure volatile materials, it

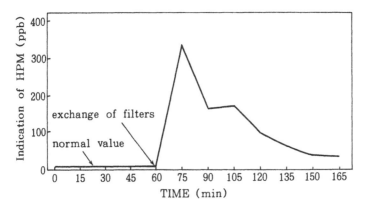

Figure 15   Degradation in pure water quality caused by filter exchange in the wafer-rinsing bath.

Table 2   Correlation Between NaCl Rejection and HPM Values in Pure Water Using Reverse Osmosis

RUN NO.	ITEM	NaCl REJECTION *1 OF TESTED RO MODULE (%)	HPM VALUE (ppb)	ON-LINE SILICA METER (ppb as SiO$_2$)	TOC *2 METER (ppb as C)
1	Ⓐ RO FEED WATER	——	87	9.1	49
	Ⓑ RO PERMEATE	——	2	3.7	20
	REJECTION %	96.5	97.7	59.3	59.2
2	Ⓐ RO FEED WATER	——	155	7.6	50
	Ⓑ RO PERMEATE	——	26	5.5	41
	REJECTION %	75.2	83.2	27.6	18

FLOW OF TEST UNIT

TAP WATER → MF (0.2 μm) → MIXED BED DEIONIZER →✕ Ⓐ → TESTED RO →✕ Ⓑ →

NOTE   * 1   NaCl rejection was tested under conditions 1500 ppm as NaCl, 25 ℃, 25kg/cm², 15% recovery.

     * 2   TOC METER : O.I.Corp. Model 700.

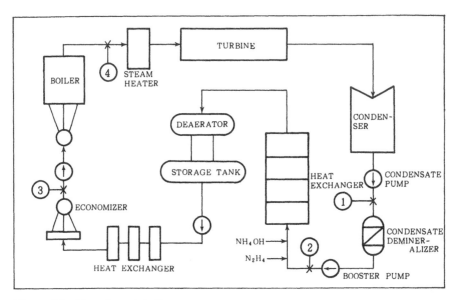

Figure 16   Once-through boiler.

is effective in this system. Figure 17 indicates the results of two tests: one used ultrapure water with ammonia added, and the other used ultrapure water with carbon dioxide added. The HPM did not detect the additive in either case. Figure 18 shows the load of electrical power and the data taken at the inlet and the outlet of the condensate demineralizer. The HPM data were 16–18 ppb at the inlet and 2–5 ppb at the outlet. There is a correlation between the electrical power load and the HPM indications at the outlet of the condensate demineralizer. This is because a trace amount of Na^+ leakage takes place when the velocity in the condensate demineralizer increases as a result of the rise in electrical power. This was also confirmed by the staff of the thermal power plant.

Some thermal power plants enhance the control of resin by suppressing Na^+ leakage to 5 ppb or less. The HPM can be also applied to this operation.

The data recorded when steam to be supplied to the turbine is cooled and directly measured by the HPM were well matched to the data from the manual analysis of Na^+, SO_4^-, SiO_2^-, and Cl^-. The details are not described in this part.

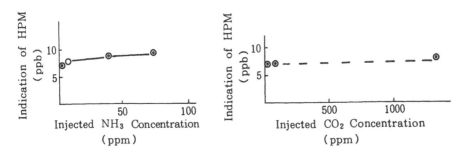

Figure 17   Correlation between HPM value and injected chemicals in a once-through boiler.

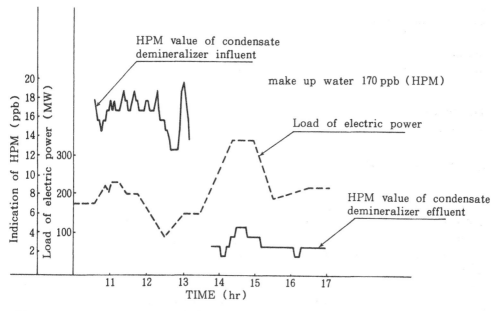

Figure 18 Correlation between HPM value of condensate demineralizer influent, effluent, and electrical power load.

VI. FUTURE CHALLENGE

The first HPM was developed 3 years ago. As innovations of the ultrapure water production system proceed rapidly, some plants have a continuous display of 1 ppb. At present a new HPM is being developed that can deal with measurement of the 0.1 ppb level. Since it has been completed at the research level, further study is now being carried out to commercialize it.

REFERENCE

1. T. Niida, Y. Kousaka, Y. Tanaka, and S. Oda, Abstracts 10th Annual Tech. Meeting on Air Cleaning and Contamination Control, Tokyo, p. 235.

i.   Total Solids Monitor

Sankichi Takahashi
Hachinohe Institute of Technology, Aomori, Japan

Toshihiko Kaneko
Hitachi Machinery & Engineering, Ltd., Kanagawa, Japan

I.   INTRODUCTION

The trace impurities in ultrapure water have a serious effect on the quality and yield of semiconductors. Along with the further integration of semiconductor devices, higher and higher purity levels are required for the ultrapure water used in the semiconductor manufacturing line.

The impurities in ultrapure water can be categorized as follows: by form, trace particulate materials, and dissolved materials and by constituent organic and inorganic materials. At present, particulate materials are evaluated and controlled by the number of particles and bacteria and dissolved inorganic materials are evaluated and controlled by the specific resistivity, the total organic carbon (TOC) concentration, and the silica concentration.

Since the impurity concentration in ultrapure water is extremely small (μg L^{-1} level) and very close to the detection limit for each impurity, it is doubtful whether the five items just mentioned are appropriate to evaluate water quality. In the United States, the volume of residual impurities on evaporation was added to the list. However, its measurement in extremely low concentrations is too difficult to be put to practical use.

To maintain the very high quality that will be required for water in the future, it is necessary to conduct on-line measurement of water quality so that the data can be immediately fed back to the operational control of the ultrapure water manufacturing system. Present measurement methods for the five major impurities are barely able to cope with this requirement.

Hitachi, Ltd. and Hitachi Mechanical Engineering Co. are now studying methods and equipment for measuring trace residual impurities on evaporation in ultrapure water in an attempt to improve reliability in evaluating and controlling the quality of ultrapure water. The idea is to atomize and evaporate ultrapure water simultaneously with clean and heated air or nitrogen gas, flashing the solids contained in ultrapure water and the dissolved materials suspended in clean gaseous flow and measuring their volume by determining their number and size. The equipment is referred to as the total solids (TS) monitor because it detects the concentration of all the impurities in ultrapure water as total solids.

Here we discuss the necessity to develop a TS monitor and its technical significance. Part VII of this book describes the industrial applications of this measurement method.

II.   NEW CONCEPTS FOR EVALUATING THE QUALITY
OF ULTRAPURE WATER

The left side of Fig. 1 shows how ultrapure water is used in the large-scale integrated (LSI) water rinsing process. In the clean room, wafers are rinsed with ultrapure water and

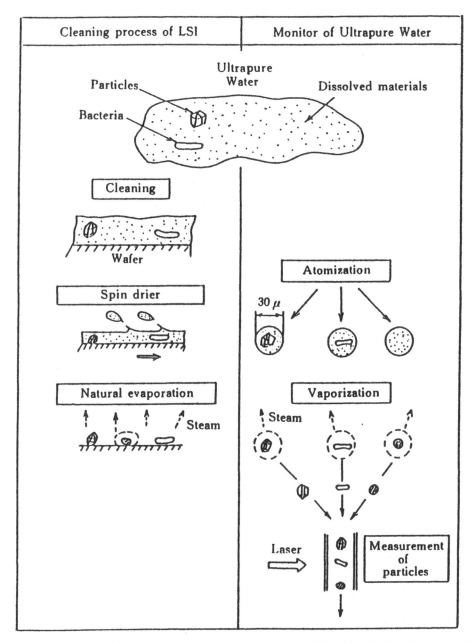

Figure 1 Action of impurities in the LSI cleaning process and idea for estimating ultrapure water quality.

then the ultrapure water remaining on the wafers is removed in a spin drier, for example. The trace volume of ultrapure water remaining is dried. If the impurities in the ultrapure water remain on the wafer surface after drying, they greatly affect device yield. Therefore, to successfully further integrate LSI devices, it is essential to reduce the impurities remaining on the wafer surface after evaporating ultrapure water. This is why the evaluation of ultrapure water quality with the residue on evaporation is considered appropriate.

The present method of measuring the residue on evaporation, described in JIS K 0101 and ASTM D1888-78, requires a long time for evaporation processing and a large amount of water to obtain a measurable volume of residue on evaporation. Consequently contamination cannot be avoided, which means the present method is difficult to apply to measurements at extremely low concentrations of μg L^{-1}.

III.   SPECIFICATIONS FOR EQUIPMENT UNDER DEVELOPMENT

The measurement technology under development is described in Sec. IV. Here the specifications and the features of the Hitachi TS monitor are presented.

Figure 2 is a photograph of the Hitachi TS monitor. This monitor is designed to satisfy the following basic requirements:

1.   Small, light and portable
2.   Easy to mount at any point where measurements are required
3.   No chemicals necessary for measurement

The specification are as follows:

Name: ultrapure water monitor
Type: HITS-100
Substance to be measured: impurities in ultrapure water (including organic materials)
Measurement principle: vaporization of fine water droplets by a nozzle driven by clean
    heated gas
Measurement range: 0–999 μg L^{-1}
Range of indicated concentration: full scale, 99.999 μg L^{-1}, two-stage
Volume of water required for measurement: 100 ml minute^{-1}
Pressure of water required for measurement: 5 kg cm^{-2} or less
Temperature of water required for measurement: 70°C or lower

Figure 2   The TS monitor HITS-100, Hitachi, Ltd.

Response: less than 1 minute
Preliminary heating time: less than 30 minutes
Operating conditions: room temperature, humidity of 85% RH (relative humidity) or less
Power supply: ac 100 V, 50/60 Hz, about 300 W
Weight: portable
Dimension: $380W \times 380H \times 580D$

When the sample is taken from the ultrapure water piping system or when the sample is connected to the monitor, extra attention should be paid to prevent contamination. The Hitachi TS monitor features a high-accuracy monitoring system to meet requirements for the latest monitor.

IV.   MEASUREMENT OF TRACE RESIDUE ON EVAPORATION

The technical research carried out in the course of developing the Hitachi TS monitor are described here.

A.   Measurement Method

On the dried wafer surface shown on the left side in Fig. 1, deposited particles, including sodium, microorganisms, and organic materials, remain. To understand the quantitative relationships of the residue mixture is very important in establishing the measurement method.

Table 1 lists the standard water quality of ultrapure water for 256 kbit LSI devices and the measurement range for each measurement method. The TS value to be calculated by converting each measurement value to the weight concentration of the residue on evaporation exhibits the following characteristics:

1.   Organic materials constitute the majority of the residue.
2. SiO_2 and dissolved sodium (related to resistivity) are the second largest in volume.
3.   The volume of microparticles and microorganisms is very limited. It is therefore essential to select a measurement method that is suitable for measuring these impurities.

Table 1   Water Quality Specified for Each Evaluation Parameter for LSI Pure Water and TS Values Calculated from Each Parameter

Specified value for each items			TS-value	Gauging methods	
Item	(unit)	Spec. Value	(ppb)	Present methods	New
Spec. Resist.	(MΩ·cm)	≤=17~18	≤=2	Spec. resist. meter	TDS-meter / TS-monitor
Particle	(No./cm³)	≤=20~50 (>=0.1μm)	≤=6*10⁻⁵	Particle counter	
SiO_2	(μg/l)	≤=0.01	≤=10	Silica monitor	
TOC	(μg/l)	≤=0.05 ~ 0.2	≤=200	TOC analyzer	
Bacteria	(No./cm³)	≤=0.02 ~ 0.1	≤=6*10⁻⁵	Count. aft. fertil.	

Conditions in calculation of TS :
  Specific weight of particle : 2.16g/cm³ (d>=0.1μm)
  Specific weight of Bacteria : 1.0g/cm³ (d>=1μm)
  Electrolytic matter in LSI pure water is just NaCl.
• TOC is the greatest component of TS. Other impurities are negligible.

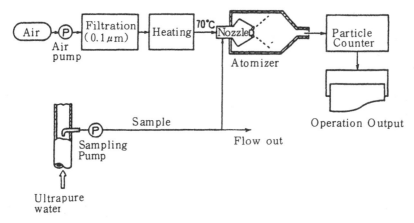

Figure 3   Basic concept of the TS monitor and the on-line monitoring system.

The authors set the following basic policies. First, to prevent methanol, whose evaporation temperature is the lowest (about 64°C) among organic materials, from being volatilized, the temperature for the evaporation processing of water is set at 60–70°C.

Second, although there are several ways to control the particle sizes formed after evaporation, a method that does not use chemicals and optimizes the water jet nozzle structure is employed. This is partly because the majority of impurities are organic and partly because the actual process during which impurities are left on the wafer surface in the LSI production line, as shown on the right side of Fig. 1, must be re-created as closely as possible.

B.   Measurement System and Structure of the Apparatus

Figure 3 shows the basic concept of the on-line measurement system of the TS monitor. Although pumps to supply air and nitrogen gas are shown in Fig. 3, it is possible to use air and nitrogen supplied to the clean room, which makes the apparatus compact enough to be installed in the clean room.

The major technical challenges in the course of developing this trace impurity monitor for ultrapure water can be summarized by processing step, as follows:

1.   The atomization of ultrapure water greatly affects the accuracy of particle count measurement in the next step and the economy of the measurement apparatus. The water atomizer is therefore required to produce spherical particles with high uniformity. This is why the water atomization nozzle for gaseous flow and the water jet were selected.
2.   In the water droplet transportation process, it is important to prevent the atomized water droplets from adhering to the inner surface of the pipelines. For this reason, the pressurized gas is heated to a predetermined temperature to evaporate the water droplets immediately after they are atomized. In this way, the water droplets are transported in the gas phase.
3.   During the evaporation process, the water droplets must be thoroughly evaporated and the optimum velocity for preventing trace residual impurities from adhering to the inner surface of the appartus must be used. The jet water volume and the temperature and flow rate of heated pressurized air are optimized.

Based on these considerations, the new monitoring system with a water atomization nozzle was developed by employing atomization and evaporation with heated gas.

V. SELECTION OF NOZZLE TYPE AND IMPROVEMENT IN ATOMIZATION PERFORMANCE

A. Selection of Nozzle Type

Various water atomization nozzles are available. The following three nozzles were studied in the course of nozzle selection: a water atomization nozzle without an air collision nozzle driven by clean heated gas; a water atomization nozzle with an air collision nozzle driven by clean heated gas; and an ultrasonic atomizer. The following points were noted:

1. The ultrasonic atomizer features a high atomizing performance but it is difficult to control the flow rate.
2. The atomization nozzle without the air collision nozzle and the atomization nozzle with the air collision nozzle are almost equal in terms of atomizing performance, but the particle size distribution of the double type is smaller than that of the atomization nozzle without the air collision nozzle, which is more suitable to the complete evaporation of water droplets.

B. Characteristics of Atomization with the Air Collision Nozzle and Application to TS Measurement

As water droplets evaporate in the air flow and the particle size decreases, the impurities in the water droplet are condensed and deposited when the point of solubility is passed. These impurities represent the residues on evaporation and they float in the air.

The size, number, and volume of the water droplets and the particles that form were studied. It was assumed that one spherical particle is formed when one droplet is evaporated. The symbols used in the following equations are defined as follows:

a = constant in Eq. (3)
b = constant in Eq. (3)
c = concentration of impurities in ultrapure water (UPW), $\mu g\ L^{-1}$; (c = TS)
d = diameter of dried particle, μm
D = diameter of water droplet, μm
N = number of dried particles, particles cm^{-3}
v = volume ratio of particles per water
γ = ρ/ρ_w
ρ = specific weight of impurities, $g\ cm^{-3}$
ρ_w = specific weight of ultrapure water

The diameter of dried particle d is expressed using the diameter of the water droplet D and the concentration of impurities in UPW c as

$$d = D \sqrt[3]{\frac{c}{\gamma}} \times 10^{-3} \qquad (1)$$

In practice, the particle size of the water droplet formed by the nozzle is not uniform and has the following general distribution:

$$\frac{\Delta N}{\Delta D} = aD^2e^{-bD} \tag{2}$$

Based on these two equations, the particle size distribution of trace particle droplets formed from water droplets is expressed as

$$\frac{\Delta N}{\Delta D} = 1000\sqrt[3]{\frac{\gamma}{c}}\,\frac{\Delta N}{\Delta D} = 1000\sqrt[3]{\frac{\gamma}{c}}\,aD^2e^{-bD} \tag{3}$$

Based on Eqs. (1) and (2), the total volume of trace particle droplets is expressed as

$$V = \int dv = 0.5236\,\frac{c}{\gamma}\,a \times 10^{-18} \int D^5 e^{-bD}\,dD \tag{4}$$

Coefficients a and b are determined by the performance of the nozzle that is used to form the trace particles. Therefore, the relationship between the concentration of impurities in UPW c and the volume of trace particle droplets formed can be calculated based on Eq. (4).

If the measurement limit of particle size is d_{min}, the water droplet size at the start of integral D_{min} in the course of calculating the total volume of trace particles can be obtained from Eq. (1).

Therefore, the total volume of trace particles with diameter d_{min} or greater that are formed when ultrapure water with impurity concentration c is atomized and evaporated can be expressed by the equation

$$V = \int dv = 0.5236\,\frac{C}{\gamma}\,a \times 10^{-9} \int D^5 e^{-bD}\,dD$$

By obtaining the total volume of trace particles, the concentration of residue on evaporation in ultrapure water c can be calculated using Eq. (5):

$$V = 0.5236\,\frac{C}{\gamma}\,a \times 10^{-9} \int_{Dmin}^{Dmax} D^5 e^{-bD}\,dD \tag{5}$$

C. Atomization Characteristics

1. *Experimental Apparatus and Method*

Figure 4 is a schematic diagram of the experimental apparatus, composed of the air distribution unit, the water distribution unit, the atomizing unit, the evaporation unit, and the gas flow particles counter. Air is supplied to the experimental apparatus after passing through a filter to remove particles of 0.1 μm or greater (99.5%). The atomizing and evaporation units are made of transparent acrylate so that the interior can be observed. They are wrapped with heat-insulating material so that they remain warm. The particle counter is a laser particle counter (detection limit $\geq$ 0.1 μm; air volume = 300 ml minute^{-1}). The air distribution unit and the water distribution system are carefully kept clean.

To evaluate the performance of the nozzle in producing trace particles, the size, number, and particle size distribution of water droplets are studied by microscope after the water droplets that form are captured by the silicon oil film that forms on the glass surface.

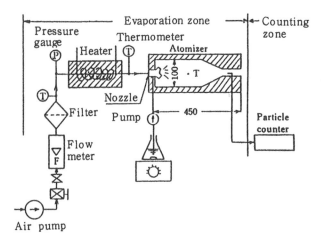

Figure 4   The experimental apparatus.

2.   Formation of Water Droplets

Figure 5 shows the particle size distribution of the nozzle; uniform water droplets are formed with a particle size peak at 10 μm. In summary, based on Eq. (2), the experimental results approximate the line shown in Fig. 6. The coefficients in Eq. (2), which exhibits the formation trace particle performance, are

$$a = 0.5 \qquad b = 0.21$$

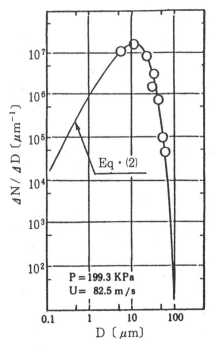

Figure 5   Diameter distribution of water droplets.

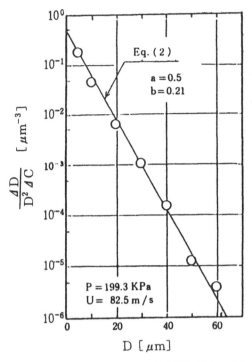

Figure 6   Relations between $\Delta N/(D^2 \, \Delta D)$ and D.

The solid line in Fig. 6 shows the values calculated from Eq. (3).

3.   *Formation of Trace Particles by Atomization and Evaporation*

The evaporation of water droplets is governed by the droplet size, the temperature of the heated air, and the gas-liquid ratio. In the newly developed system, the water droplet size is determined by the nozzle structure and the gas-liquid ratio is determined by the relative humidity after evaporation. The temperature of heated air is set as 60–70°C in an attempt to prevent water droplets from evaporating in the nozzle and to suppress the evaporation of organic constituents.

Figure 7 shows the measurement results of trace particles formed by atomizing and evaporating a sample solution containing NaCl in known concentration dissolved in ultrapure water. The solid line in Fig. 7 represents the values calculated from Eq. (3) with coefficients a and b. The calculated values are a good match with the experimental results.

The trace particle size values measured by the laser particle counter were calibrated using the relationship between the diameter of a spherical polyethylene particle and the scattering intensity. The shape of particles produced by atomization and evaporation could therefore cause particle size measurement error. To avoid such an error, it is important to confirm the shape of trace particles that are formed. The scanning electron microscope (SEM) image of these trace particles captured on the filter surface indicates that their shape is almost spherical. It is thought that the possibility of measurement error is small.

4.   *Impurity Concentration and Volume of Particles Formed*

Figure 8 shows the relationship between the impurity concentration and the total volume of particles of 0.1 μm or larger in the sample solution containing NaCl in a known

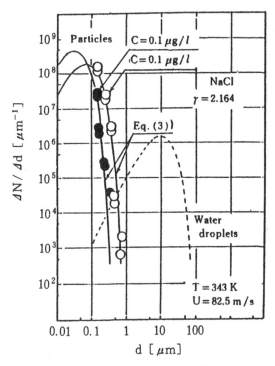

Figure 7 Diameter distribution of NaCl particles after evaporation of water droplets.

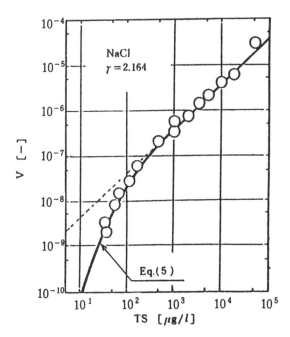

Figure 8 Relations between concentration of impurities (NaCl) and volume of particles from evaporated water droplets.

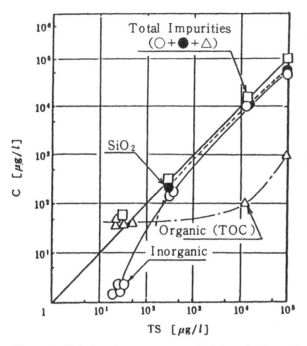

Figure 9 Relations between new method (atomization and evaporation) and present method.

concentration dissolved in ultrapure water. The solid line represents the calculated values obtained from Eq. (5), which are a good match with the experimental values. The dotted line shows the total volume of particles formed calculated from Eq. (4). The gap between the dotted line and the measurement values is in the area below the measurement limit of the current particle counter. Controlling the size of water droplets by optimizing the nozzle structure, with the measurement limit taken into consideration, is the best countermeasure from the standpoints of both measurement accuracy and economy.

The basic concept and technology of this measurement method has been proved to work theoretically and experimentally. The concentration of trace impurities in ultrapure water can be measured on-line by measuring the size and number of particles formed through atomization and evaporation.

VI. CORRELATION WITH CURRENT INDICATORS OF WATER QUALITY

Figure 9 shows the correlation between the TS value and current indicators of water quality (see Table 1). Particles and bacteria are negligible in terms of weight concentration. As a result, the sum of the weight concentration of inorganic materials, organic materials (TOC), and SiO_2, that is, the total impurity concentration, matches with TS value. Since the concentrations of inorganic materials and TOC (in the low concentration areas) are below the measurement limits, the measurement values below the measurement limit exhibit a tendency different from those within the measurement limit. As mentioned earlier, the majority of impurities in ultrapure water are inorganic materials in high concentration (TS > 100 μg L^{-1}), but TOC, which is hard to remove, is present in low

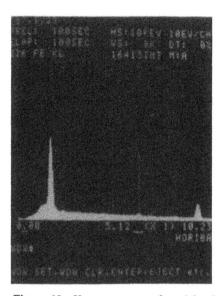

Figure 10   X-ray spectrum of particles from evaporated ultrapure water droplets.

concentration (TS < 100 μg L^{-1}). As predicted by converting the weight concentration from ultrapure water quality specifications, the TS value is almost equal to the organic material (TOC) concentration within the range of TS < 100 μg L^{-1}. Figure 10 shows the elemental analysis of formed particles: it is obvious that the majority of residues on evaporation are organic materials.

Because the TS value is stable even at low concentrations, it is considered suitable for the comprehensive evaluation of ultrapure water quality.

The newly developed Hitachi TS monitors are mounted on the operating ultrapure water production system and the LSI production line to conduct on-line measurement of fluctuations in water quality and elution from component materials. The TS monitors have proved to work effectively in this trial. The operating data are presented in Part VII of this book.

VII.   SUMMARY

An outline of the Hitachi TS monitor and the results of technical studies have been described in this section. The important points are summarized as follows:

1.   The impurity concentration in ultrapure water can be measured on-line by simultaneously atomizing and evaporating ultrapure water with heated clean air and nitrogen gas and measuring the size and number of the residual particles that form on evaporation.
2.   The correlation between the TS values measured by this method and current specifications for water quality (resistivity, TOC, silica, and so on) was studied. This method proved to be adequate as a comprehensive water quality indicator.
3.   On-line monitoring with this method was conducted to confirm that it is capable of evaluating ultrapure water quality.

3 On-Line Monitors

E. Silica Analyzer for Ultrapure Water

Makoto Satoda
DKK Corporation, Tokyo, Japan

I. MEASUREMENT PRINCIPLE

A. Principle

The molybdate blue absorptiometry principle is the most widely used to measure low concentrations of silica. It is used in most of automatic measurement instruments. Even when trace amounts of silica are measured by condensation through solvent extraction [1], the molybdenum blue absorptiometry principle is still used in the final stage.

The molybdenum blue absorptiometry principle can be summarized as follows [2]. Silicate ion is reacted with ammonium molybdate in acid solution to produce yellow dodecamolybdosilicic acid. The production of dodecamolybdosilicic acid is then stopped by adding tartaric acid or oxalic acid. Then, a reducing agent is added to reduce the dodecamolybdosilicic acid to molybdenum blue to measure its absorbance.

B. Determination

The determination method is as follows. A 25 ml sample and 3 ml mixture of 5% ammonium molybdate and 2 N sulfuric acid are placed in the reaction vessel, which is preheated to 40°C. After stirring for 240 s, dodecamolybdosilicic acid is produced. Next, 3 ml of 10% tartaric acid is added and the reaction is stopped after 30 s of stirring. Then, 3 ml of 2.5% ascorbic acid is added for reduction to molybdenum blue after 100 s of stirring. The absorbance is measured and compared with a working curve obtained in advance to calculate the concentration.

Tartaric acid or oxalic acid is used to decompose dodecamolybdophosphoric acid in measuring boiler water. Although phosphate ions are though not to exist in ultrapure water, tartaric acids is injected to act as a coloring stabilizer.

In developing the automatic analyzer, it was desirable to simplify the configuration as much as possible while satisfying the specifications. This is why the solvent extraction method was not used.

C.   Calibration

Since water of higher quality than ultrapure water is not available at present, the blank calibration uses a method different from zero silica content water measurement. Generally the blank is set by measuring the gap between the absorbance of the normal measurement and that using a double volume of molybdate reagent [3].

In this case, however, tartaric acid is added first and then the coloring reagent (mixture of ammonium molybdate and sulfuric acid) is used. This method obtains water that does not contain silica while suppressing the reaction between silica in the sample and molybdate. Since it is necessary to lower the silica concentration in the coloring reagent as much as possible, ultrapure water is used for reagent adjustment.

In span calibration, the standard solution is measured using the normal method.

II.   REQUIRED SPECIFICATIONS

The specifications for the silica analyzer as a monitor of the silica concentration in ultrapure water are described here.

A.   Measurement Range

Although the silica concentration in ultrapure water is usually less than 5 μg L^{-1}, it rises to 100 μg L^{-1} if the ion-exchange resin deteriorates. Therefore the measurement range should be set at 0–10 μg L^{-1} for monitoring under normal operating conditions and at 0–200 μg L^{-1} in resin deterioration.

B.   Accuracy

Since the silica analyzer should perform constant monitoring, it is expected to feature the accuracy of industrial measurement instruments, or higher. In general, industrial instrumentation has a reproducibility of ±5% FS (full scale) or less. Therefore the silica analyzer should satisfy this level.

The accuracy of the silica meter depends on the traceability of the standard solution since the measurement is conducted based on the working curve. The characteristic that determines the analyzer accuracy is therefore linearity. The deviation of the standard solution measurement from the standard value should be suppressed to ±5% FS or less if possible.

C.   Utility

It is desirable that the required sample volume be small and that special utilities are not required. The silica meter is designed to be operated simply, using only an ac power supply, instrument air (or nitrogen gas), and the sample.

D.   Maintenance

Reliability and easy maintenance are sought. In an attempt to increase reliability, the liquid-contacting valves are pneumatic. A light-emitting diode (LED) is employed in the colorimeter to reduce maintenance and to improve reliability.

Table 1 Main Specifications for the Silica Analyzer

MEASURING RANGE	0 – 10, 0 – 500 ug / L DOUBLE RANGE (AUTOMATIC SWITCHABLE)
PRECISION	LESS THAN ±2%FS
LINEARITY	LESS THAN ±2%FS
UTILITY	POWER SUPPLY ; AC 100 V, 50/60 Hz, 500 VA INSTRUMENT AIR ; 4 – 7 kgf / cm^2 DRAIN ; OPEN PIT
PERIOD OF MEASUREMENT	10 – 999 min (PRESETTABLE)

III. OUTLINE OF THE APPARATUS

The analyzer was designed based on the required specifications. The specifications are listed in Table 1, Figure 1 is the flow sheet for the meter, and Fig. 2 lists measurement procedures.

A. Colorimeter

The colorimeter is one of the most critical elements for accuracy. It is composed of a flow cell with an optical pathlength of 100 mm, an LED light source with a wavelength of 860

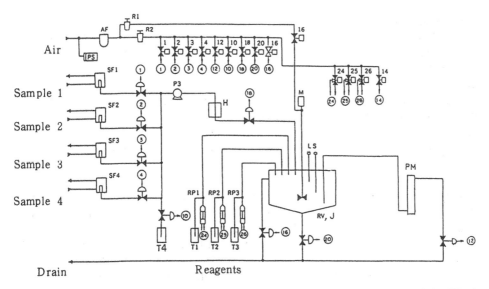

Figure 1 The silica analyzer: AF, air filter; R1, R2, regulators; PS, pressure switch; SF1–4, sample filters; P3, diaphragm pump; M, mixer; H, thermostated bath; RP1–3, reagent pumps; T1–4, tanks; RV, J, reaction vessel with jacket; PM, photometer.

```
WASHING
    ↓
ADDITION OF REAGENT 1
    ↓
VOLUMETERING OF SAMPLE SOLUTION
    ↓
REACTION 1
    ↓
ADDITION OF REAGENT 2
    ↓
REACTION 2
    ↓
ADDITION OF REAGENT 3
    ↓
100 % T ADJUSTMENT OF PHOTOMETER
    ↓
MEASUREMENT
    ↓
WASHING
    ↓
WAIT
```

Figure 2 The measurement procedure.

nm, a photodiode photoreceptor, and a temperature adjuster. The temperature adjuster is mounted to keep the peripheral temperature of the LED unchanged so that the fluctuation in light intensity is suppressed.

An LED with a wavelength of 860 nm is used, although the peak wavelength of molybdenum blue is around 815 nm. This is because the LED has many advantages, even though it has reduced sensitivity. The advantages of the LED are as follows: (1) because the LED lifetime is almost infinite, it need not be replaced; (2) the LED does not require a monochromator; (3) because the LED does not run hot, the fluctuation in cell length due to thermal expansion does not occur; and (4) because the electrical current consumption of the LED is small, it requires a small power supply.

Among the disadvantages of the LED, its sensitivity is low and interfering substances at 860 nm are not studied. In measuring ultrapure water, however, interfering substances are negligible and this problem has not occurred. A sufficiently high level of sensitivity is obtained under conventional conditions.

If an LED with a wavelength close to the peak is developed in the future, it will be possible to detect a much smaller volume of silica merely by replacing the LED.

Because the colorimeter is required to be maintenance free, it is designed to compensate for contamination on the cell during every measurement cycle. The transmittance of cleaning water (sample water) introduced to the coloring cell immediately before measurement is itself measured and regarded as 100% T.

B. Reaction Vessel

The reaction vessel is made of polymethyl methacrylate (PMMA) resin and is used not only to contain the coloring reaction but also as the sample measurement instrument and the unit for transporting the sample water to the colorimeter.

The sample is measured by a level sensor using an electrode made of SUS 316. Because the conductivity of the sample itself is too low to be detected, the reagent is injected in advance. Voltage is applied to the level electrode only when the measurement is conducted. The measurement accuracy deteriorates as the water surface to be measured is agitated. To keep the water surface calm, the sample injection nozzle is placed below the water surface.

The water is fed to the colorimeter under pressure, which also prevents the generation of bubbles. This is why the reaction vessel and the colorimeter employ a closed system.

C. Thermostated Bath

To reduce the coloring reaction time, a jacket is mounted on the reaction vessel and hot water at 40°C is circulated between the reaction vessel and the separate thermostated bath. The thermostated bath is a closed structure to prevent evaporation of the hot water.

D. Reagent Pump

The reagent is injected by syringe pump. The piston material is polytetrafluoroethylene (PTFE), and the cylinder material is PMMA or hard glass. The pump cylinder for the mixture of ammonium molybdate and sulfuric acid is made of PMMA. The pump cylinder for the other two reagents is made of hard glass. The piston employs an air-driven method using an air cylinder, and it moves back and forth within a certain distance. The volume of pumped reagent is fixed at 3 ml per pump stroke for all three pumps.

A check valve is mounted at the top of the cylinder. It is made of acid-resistant isobutylene-isoprene rubber.

E. Controller

The controller has both sequencer and calculation functions. It is equipped with an 8 bit microcomputer. The central processing unit (CPU) is a Z80, and the clock frequency is 4 MHz. The minimum time unit to create a program is 1 s. The analog-digital conversion element for converting the signals from the colorimeter to digital signals is 12 bit. The measurement results are displayed on the liquid crystal display while being transmitted as output signals.

F. Output Signal

To transmit the measurement results, a 4–20 mA analog transmission output and a printer output are provided. An alarm contact output is also provided to warn of abnormal concentrations and equipment failures. The RS-232 C interface is provided to transmit the measurement results and warnings to a higher level computer.

IV. MEASUREMENT DATA

Equipment
   Silica measurement unit, SLC-1605 type (Denki Kagaku Keiki)
   Ultrapure water unit, miLLI-Q (Millipore)
Reagents
   Source solution for silicon standard, 1000 mg Si L^{-1} (Kanto Kagaku)
   Silicon standard solution, diluted 100 times from the source solution with ultrapure
      water (prepared when necessary)
Coloring solution: mixture of the following two solutions in equal volume
   Ammonium molybdate: 10 wt/vol% aqueous solution
   Sulfuric acid: 4 N aqueous solution
Mask solution: tartaric acid, 10% wt/vol aqueous solution
Reducing solution: L-ascorbic acid, 2.5% wt/vol aqueous solution

Ultrapure water is used for dissolving all reagents.

A. Linearity Test

Figure 3 shows the results of a test in which the silicon standard solution is added to
ultrapure water. The sample is adjusted in the following way.

   Ultrapure water (1 L) is transferred from the ultrapure water unit to the polyethylene
measurement flask. The silicon standard solution, 10 mg Si L^{-1}, is added to ultrapure
water in a known volume using a 4780 micropipette (Eppendorf). Ultrapure water is
poured into the flask to the standard level. The measurement is conducted immediately
after the flask is shaken to mix the solutions.

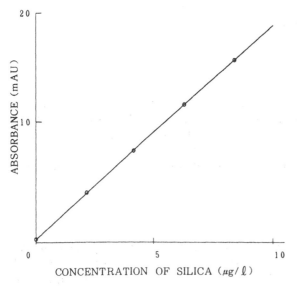

Figure 3 Linearity test of the silica analyzer: wavelength = 860 nm; light path = 100 mm.

Table 2   Repeatability Test Results

SAMPLE	ULTRAPURE WATER	DEIONIZED WATER	STANDARD SOLUTION
MEASURED VALUE	0.41 ug /L 0.47 0.47 0.41 0.47 0.47 0.34 0.41 0.41 0.47	26.71 ug /L 26.80 26.64 26.48 26.57 26.48 26.73 26.36 26.73 26.72	8.83 ug /L 8.76 8.90 8.76 8.90 8.76 8.90 8.97 8.97 8.83
n	10	10	10
MEAN	0.433	26.62	8.858
S. D	0.044	0.143	0.082
CV (%)	10.16	0.54	0.93

B.   Reproducibility

Table 2 lists the measurement results for ultrapure water, ion-exchange water, and standard solution.

The samples are prepared in the following way. Ultrapure water is transferred to the polyethylene contained from the same ultrapure water unit as before. The standard solution is adjusted in the same way. Although the standard value is 8.56 μg L^{-1}, it becomes 8.99 μg L^{-1} when the silica concentration in the ultrapure water is considered. The ion-exchange water is transferred from the G-10 cartridge pure water unit (Organo) to the polyethylene container. Each sample is measured immediately after being taken or adjusted.

REFERENCES

1. P. Pakalns and W. W. Flynn, Anal. Chim. Acta, *38*, 403 (1976).
2. Muki Ouyou Hishoku Bunseki Henshuu Iinkai ed., Muki Ouyou Hishoku Bunseki, No. 5, p. 40, Kyouritu Shuppan (1976).
3. JIS:B8224–1986, p. 110.

3 On-Line Monitors

F. Dissolved Oxygen Meter

Shin'ichi Akazawa
DKK Corporation, Tokyo, Japan

I. INTRODUCTION

Most dissolved oxygen (DO) meters currently used in the ultrapure water production line employ a membrane electrode developed on the basis of the Clarke electrode. The DO meter has been used for more than 20 years. In this time, it has been modified to meet the requirements of various applications, including the gas-bubbling tower for processing wastewater and the fermentation plant.

The DO meter for pure water has also been greatly improved in durability and sensitivity. The current DO meter is capable of measuring an extremely low concentration level of dissolved oxygen. Further study is required to improve the sensor function of the DO meter.

This part presents the present status of the DO meter in terms of the low concentration levels of dissolved oxygen in pure water.

II. MEMBRANE DO ELECTRODE

A. Operational Principle

Figure 1 shows the principal configuration of the membrane electrode. The membrane, which is made of polyethylene or Teflon, for example, for their high permeability to oxygen, is placed at the end of the support pipe. The inside of the support pipe is filled with electrolyte, and the cathode and anode are attached to the membrane.

When the appropriate voltage is applied between the cathode and the anode to operate the electrode, the dissolved oxygen is first reduced and consumed at the thin electrolyte layer between the membrane and the cathode and then the electrical current is stopped. Consequently the oxygen is supplied only through the membrane at a velocity in proportion to the partial pressure of oxygen (PO_2) contained in the sample.

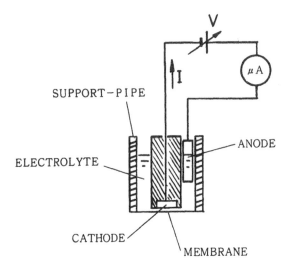

Figure 1   Construction of the electrode.

At this point, a small amount of dissolved oxygen in the top of the electrolyte moves to the cathode side because of the large diffusion resistance at the site of adhesion.

The oxygen-reducing electrical current in this stationary state can be expressed as the voltage-current curve (polarogram) shown in Fig. 2. If the voltage applied to the cathode and anode is set within the range of −0.5 to −0.9 V, which corresponds to the plateau of the curves in Fig. 2, the DO concentration can be obtained from the volume of the electrical current. Gold or platinum is used for the cathode, silver is used for the anode, and potassium chloride solution is used for the electrolyte, as usual.

If a lead anode and alkali electrolyte are used together, no voltage must be applied from outside because of the electromotive force that is generated. In other words, the curves shown in Fig. 2 are shifted to the left to locate the flat portion at 0 V.

The first is called the polarograph type and the second is called the galvanic cell type. There is no major difference between the two types in terms of the principles of operation.

When the reducing current runs, the following reactions take place at the cathode and the anode, respectively [1]. Cathode reaction:

$$O_2 + 2H_2O + 4e^- \rightarrow 4OH^- \tag{1}$$

Anode reaction, polarograph type:

$$4Cl^- + 3Ag \rightarrow 4AgCl + 4e^- \tag{2}$$

Galvanic cell type:

$$2Pb \rightarrow 2Pb^{2+} + 4e^- \tag{3}$$

$$2Pb^{2+} + 4OH^- \rightarrow 2Pb(OH)^2 \tag{4}$$

In every reaction, electrolyte replacement and grinding of the cathode, together with membrane replacement, are required because of the decrease in the anion concentration in the electrolyte and the deposition of reaction products.

Various kinds of membrane DO electrodes are currently manufactured based on these principles. Because the configuration is simple, each manufacturing adopts various

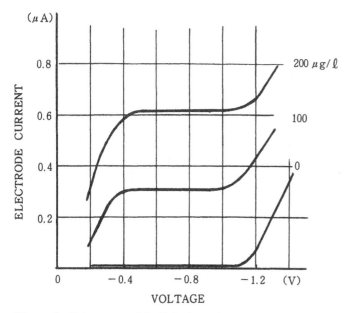

Figure 2   Polarogram of the DO electrode: sample, O_2/N_2 gas; temperature, 25°C.

modifications in materials and the structure of the detailed parts and in the way the membrane is supported.

B.   Output of the Electrode

When the operation of the membrane electrode is in the stationary state, the output electrical current can be expressed as [2]

$$i = NFAP_m \frac{1}{L} PO_2 \tag{5}$$

where: N = number of electrons required to reduce 1 mol oxygen (4)
  F = Faraday constant (96,500 C (g equivalent)$^{-1}$)
  A = area of cathode, cm^2
  P_m = oxygen transmission coefficient of membrane, ml at STP, cm ($cm^2 \cdot s \cdot cm$ Hg)$^{-1}$
  L = thickness of membrame, cm
  PO_2 = partial pressure of oxygen contained in the sample, cm Hg

Since all the variables in Eq. (5) except PO_2 are fixed at the fixed temperature in the stationary state of the electrode, the output electrical current is in proportion to PO_2. The volume of the output current is in proportion to the area of the cathode and the oxygen transmission coefficient of the membrane and is in inverse proportion to the membrane thickness. Therefore, if the area of the cathode is larger and the membrane is made as thin as possible, a high-sensitivity electrode can be developed relatively easily.

The output electrical current is compensated for in practice by using a temperature element, such as a thermistor. This is because the oxygen transmission coefficient of the membrane features a temperature characteristic around 3%/°C and because the DO saturation level changes along with the temperature fluctuation even when PO_2 remains the same.

III.   DO METER FOR PURE WATER

The DO meter for pure water enjoys the following favorable operating conditions (the only unfavorable condition is that it is expected to measure very low concentration levels):

1.   The temperature remains almost the same.
2.   Since the sample is not contaminated very much and its velocity is low, it is difficult for contaminants to adhere to the membrane and no stress is applied to the membrane.
3.   Because of the low DO concentration, the electrode is barely consumed or degraded.

Compared with the operating conditions for DO meters used for general wastewater and in the fermentation plant, these conditions are extremely favorable. The load on the electrode is very light.

A.   DO Electrode

Figure 3 shows the configuration of the DO meter electrode that Denki Kagaku Keiki uses for pure water. This electrode was developed to measure low DO concentrations in pure water and boiler supply water. The major features are as follows.
    The cathode diameter is increased from the conventional size of 2 mm to 8 mm to obtain a high output of about 2.5 μA per 1 mg L^{-1} (16 of DO times as much as the

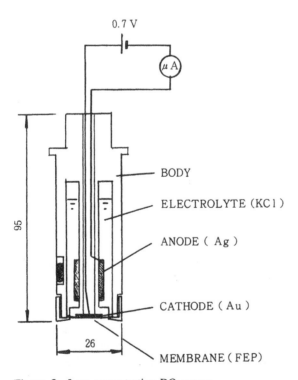

Figure 3   Low-concentration DO sensor.

conventional, according to the comparison done by Denki Kagaku Keiki). A high sensitivity is also obtained: detection limit of about 1 μg L^{-1} and a minimum measurement range of 0–20 μg L^{-1}.

Based on tests to determine the best combination of anode and electrolyte, silver and potassium were selected. This combination exhibits safety in handling as well as durability. At the same time, by replacing the electrolyte and the membrane, the initial characteristics can be easily re-created.

Because heat-resistant materials are employed for the support pipe of the electrode, the cathode support, and the adhesives, the DO meter can withstand the high temperature of 60°C at maximum. In addition, because of the low residual electrical current, a stable zero-point measurement is possible despite the high sensitivity.

The accuracy of the temperature compensation is improved by mounting thermistors for both the oxygen transmission coefficient of the membrane and the saturation DO level of the sample, both of which fluctuate along with temperature.

In principle, the electrode is the polarograph type. The advantages of the polarograph electrode are as follows:

It does not operate unless the voltage is applied from outside.
The anode is silver, which is resistant to corrosion, and thus the electrode is not consumed
    or degraded at all when operation is suspended or it is stored.

B.   Characteristics of the Electrode

The electrode shown in Fig. 3 has the following characteristics.

1.   Linearity

Figure 4 shows the working curve confirmed with O$_2$/N$_2$ standard gas. When water at 25°C is saturated with gas containing 21% oxygen, the DO level reaches about 8 mg L^{-1}. Therefore if gas containing 0.05% oxygen is used, the DO level reaches 20 μg L^{-1}, which is confirmed relatively easily.

The DO meter exhibits very good characteristics in the wide range from the microgram to milligram per liter level.

2.   Response

Figure 5 shows the response confirmed by using O$_2$/N$_2$ standard gas and N$_2$ gas. The results shown in Fig. 5 were obtained at 25°C and in the same measurement range as before. The 90% response takes about 2 minutes. If the sensor is exposed to high concentrations of oxygen, however, the 90% response to the zero point takes 20–40 minutes.

The response characteristic is affected by the oxygen transmission coefficient of the membrane and by the temperature. If the membrane is made of the same material, the thinner the membrane is, the quicker the response. In addition, the higher the temperature is, the quicker the response.

How firmly the cathode and the membrane are attached to each other affects the response. If they are loosely attached, the electrolyte layer becomes thicker to slow the oxygen transfer, resulting in a slow response. The dissolved oxygen in the electrolyte at the top also moves to the cathode surface, which increases the residual electrical current and raises the zero point.

Degradation of the response is an effective indicator of when the membrane needs replacement.

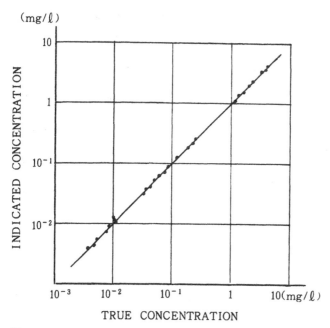

(mg/ℓ)

TRUE CONCENTRATION

Figure 4 Linearity of the DO sensor.

3. Effect of Sample Flow Rate

The electrode constantly consumes oxygen during measurement. Therefore, unless the sample on the membrane surface runs at a certain flow rate, oxygen shortage takes place at some portion, leading to a decrease in the electrode output. Figure 6 shows the relationship between the sample flow rate and the electrode output.

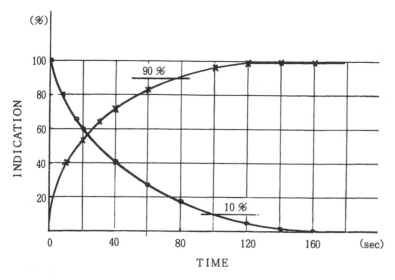

Figure 5 Responses of the DO sensor: temperature, 25°C; range, 0–100 μg L^{-1}.

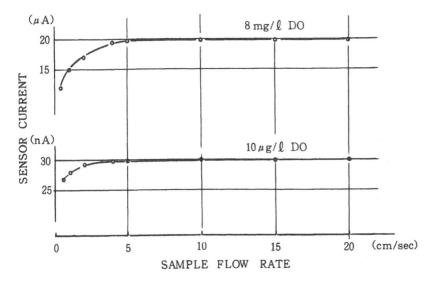

Figure 6   Influence of sample flow rate. The sample is pure water.

When water with a viscosity of less than 1 cP is used, a flow rate of 5–10 cm s^{-1} or more is sufficient to conduct the measurement properly, regardless of the DO concentration. As the viscosity increases, however, a flow rate of 20–40 cm s^{-1} or more is required.

When the volume of the sample is limited or the viscosity is high, the flow rate is usually fixed by stirring the sample.

4.   *Effect of Solute Concentration [1]*

As mentioned earlier, the output of the membrane electrode is in proportion to the partial pressure of oxygen (PO_2). To obtain the DO level from this output, the DO level should fluctuate along with the PO_2 fluctuation and the oxygen saturation in the sample should be detected. If the DO level changes without fluctuation in the PO_2, the output of the sensor does not change in proportion to the DO level.

For example, if salt is added to pure water saturated with air, the DO level gradually drops but the PO_2 level remains unchanged at 160 mm Hg. In this case, the electrode output is measured unchanged. On the other hand, measurement based on the Winkler method detects the absolute value of the DO level in the sample. The measurement values of the two methods can differ when the concentration of solute in the sample reaches a certain level. This is an important point to be noted to use the membrane electrode properly.

C.   **Configuration of the DO Meter**

Figure 7 shows the configuration of the DO meter. Basically it is composed of a DO electrode and a converter. The structure of the measurement cell and the head tank can be modified to cope with the characteristics of the electrode.

The sample is fed to the meter from the process line at 0.5–1 L minute^{-1}. After the pressure is adjusted in the head tank, about half of the fed sample is introduced to the measurement cell. It is drained after the measurement is completed.

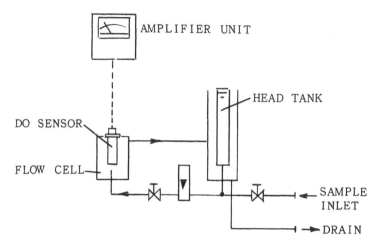

Figure 7   The DO meter system.

D.   Factors Affecting Measurement and Countermeasures

Major factors that affect the measurement values and the characteristics of the electrode during actual DO measurement are the pressure, temperature, flow rate, and composition of the samples. The countermeasures adopted in the DO meter are described here.

1.   Sample Pressure

Since the membrane electrode senses the partial pressure of oxygen, the air pressure, which affects the partial pressure of oxygen and the pressure of the sample at the time of measurement, has an effect on the measurement results. The oxygen saturation level also changes with the pressure. To protect the electrode from the process pressure of the sample, the head tank is placed at the inlet of the measurement cell. Furthermore, the outlet is open to the air to maintain the pressure difference between the time of calibration and the time of measurement.

2.   Sample Temperature

The temperature fluctuation of the sample affects the oxygen transmission coefficient of the membrane and the oxygen saturation level of the sample. As the temperature of the sample rises, the saturation level decreases but the transmission coefficient increases. Consequently the electrode output rises, increasing the error. In practice, the sample is not saturated by O_2 gas and the DO level does not change even if the temperature fluctuates. Therefore the error is detected so that the DO meter readings fluctuate despite that the DO value remains unchanged. To compensate for these two phenomena caused by the temperature change, the DO meter is equipped with exclusive thermistors to conduct high-accuracy temperature compensation.

3.   Sample Flow Rate

To raise the sample flow rate on the membrane surface with limited sample consumption, the sample is jetted directly onto the electrode with a nozzle placed in the measurement cell. A flow rate of 0.5 L minute^{-1} is therefore sufficient. This countermeasure is also effective in suppressing contaminant deposition on the electrode surface.

4. Sample Composition

When the sample is pure water or a solution with a low solute concentration, the effect of sample composition is negligible. However, the types and concentration of the solute greatly affect the oxygen saturation level.

The saturation level of alcohol is five to six times as much as that of water, and the saturation level of oils is about three times as much as that of water. However, because such a change in saturation level is not accompanied by a change in the partial pressure of oxygen (PO_2), the current DO meter cannot detect it.

As for the relationship between the salt concentration in pure water and the saturation level, conversion to the DO level is possible because detailed data have been reported.

IV. ANALYTICAL METHOD FOR DISSOLVED OXYGEN

Table 1 lists analytic methods for dissolved oxygen. Compared to the DO meter, which indicates the DO level converted from the measured PO_2 level, an analytic method is a way to measure the DO level directly. In this sense, an analytic method is significant.

On the other hand, there are some disadvantages of DO analysis: it indicates the DO

Table 1   Methods of DO Analyzing

Analysis method	Determination range	Accuracy	Description in standard	
			JIS K 0101	JIS B 8224
Winkler Sodium azide method modified	Not less than 0.1 mg/ℓ		◯	—
Miller method modified	Not less than 1 mg/ℓ		◯	—
Winkler Starch indicator method	Not less than 100 µg/ℓ (5 µg/ℓ)	± 30 µg/ℓ	◯	◯
Winkler Amperometric titration	Not more than 10 µg/ℓ	± 2 µg/ℓ	—	◯
Winkler Potentiometric titration	Not more than 10 µg/ℓ	± 2 µg/ℓ	—	◯
Indigo carmine method	Not more than 60 µg/ℓ	± 2 µg/ℓ	—	◯

JIS K 0101 − 1986 Testing Methods for Water

JIS B 8224 − 1986 Testing Methods for Boiler Feed Water and Boiler Water

level only at a certain moment, although the DO level constantly fluctuates, and DO analysis takes a long time to produce results.

In the course of DO analysis, it is most important to keep the sample away from the air when the reagent is injected. The reliability of the analytic result greatly depends on this point. However, it is extremely difficult to keep the sample away from the air when analyzing the low concentrations of dissolved oxygen.

The minimum deviation of the analytic methods shown in Table 1 is ± 1 to ± 5 μg L^{-1}. There is no perfect way to measure absolute values or to calibrate the DO meter.

At present, it is necessary to select the better method on each occasion based on a full understanding of the features of the two methods: measurement using the DO meter and analysis of the DO level.

V. CONCLUSION

To cope with future demands to measure the absolute value of the DO level, the analytic technology and the DO meter should be further improved. The membrane electrode method is an effective way to measure the DO level. However, a new way should be developed by examining the characteristics of the membrane electrode method. For the time being, the membrane electrode and the DO meter should be studied in detail. It would be very much appreciated if the users of the electrode and the DO meter could point out the areas that need to be improved.

REFERENCES

1. Shuichi Aiba, Optimum Measurement and Control for Ferment Process, Science Forum Co., 1983.
2. DKK Domestic Technical Report.

3 On-Line Monitors

G. ATP Monitor

Toshiki Manabe
Japan Organo Co., Ltd., Saitama, Japan

I. INTRODUCTION

As the degree of integration rises in the large-scale integrated (LSI) chip manufacturing process, increasingly stringent requirements have come to be placed on water quality. In particular, requirements for regulating the bacteria in ultrapure water are becoming more and more stringent every year. Table 1 lists changes in the standard value for bacteria and the degree of integration. The generally accepted standard value for the bacteria level for the 1 Mbit chips is 0.1–0.01 CFU (colony-forming unit) ml^{-1}. The requirements will become increasingly stringent as the degree of integration increases in the future.

Currently, bacteria are measured mainly by the ASTM (American Society for Testing and Materials) standard. According to this method, the sample water is passed through a field monitor kit (marketed) equipped with a membrane filter (with 0.45 μm diam holes) and the captured bacteria are cultured at 35°C for 24 h. The bacteria are then counted as colonies. The number of bacteria present in ultrapure water is extremely low, however, and the bacterial growth rate is very slow. Indeed, complete colonies cannot form within the conventional culture time period. Each manufacturer therefore alters the culture temperature and time period to cope with this problem. As a result, there are greater or lesser variations in the measured value depending on the filter and the culture temperature and time period.

Table 1 Bacteria Quality Requirements

Degree of integration	64K	256K	1M	4M	16M
Bacteria (cfu / 100 ml)	<25	<10	<1 ~10	<1	<0.1~0.5

In the future, technology enabling measurement within short and medium periods will be demanded because, within these periods, control of bacteria is said to exert a critical effect on the product yield.

The use of the Microsure-100, which is introduced here, has achieved the technology to count, within a short period, bacteria caught by the 0.2 μm filter. This part describes an example of the measuring principle and the measurement itself. The Microsure-100 is referred to here as the MS-100 and the detection method as the ATP method. The appearance of this system is shown in Fig. 1.

II.   PRINCIPLE OF MEASUREMENT

Generally, bacteria use ATP (adenosine triphosphate) as a kinetic energy source. Figure 2 shows the structural formula of ATP. ATP is a nucleotide, which comprises three successive molecules of phosphate coupled to the fifth hydroxyl ribose of adenosine. ATP therefore contains two high-energy phosphate couplings in one molecule.

The chemical formula is $C_{10}H_{16}N_5O_{13}P_3$. The ATP content of different bacteria is shown in Table 2. This means that measurement of the ATP content in the sample can lead to confirmation of the presence of bacteria if any are present in the sample water.

At present, the technology for determining the ATP content from the amount of light emitted utilizing bioluminescence phenomena (for example, firefly bioluminescence) has been established. It is possible to estimate the number of bacteria from this ATP content. This system calculates the number of bacteria on the basis of the average ATP content per cell of aqueous bacteria.

Normally, the bioluminescence reaction of a living organism involves a reaction between luciferin, ATP, and oxygen in the presence of luciferase and Mg^{2+}. Figure 3 shows the bioluminescence mechanism of the firefly. The reaction between firefly

Figure 1   The Microsure-100.

$C_{10}H_{16}N_5O_{13}P_3$ MW = 507.8

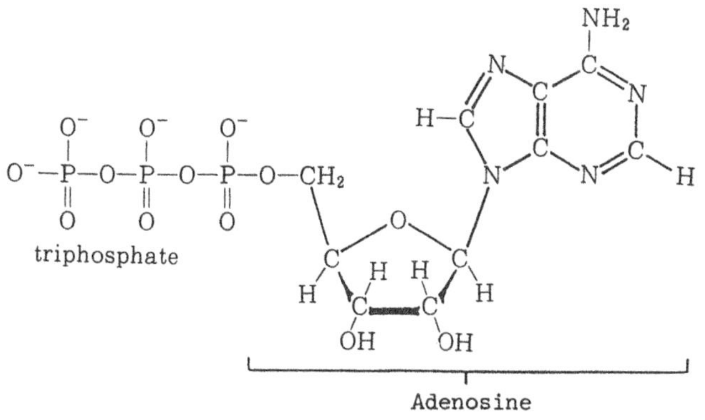

Figure 2 Adenosine 5-triphosphate.

luciferin, luciferase, and ATP has been utilized and applied to the MS-100. The measuring principle of the MS-100 is briefly illustrated in Fig. 4.

III. PRACTICAL MEASUREMENT EXAMPLES

A 0.2 μm Nuclepore membrane filter was installed at each point of an ultrapure water production system, and the sample water was passed through the filter for filtration.

Bacteria caught by the filter were counted with the MS-100. Similar counting was also made according to the current ASTM method. The samples used included industrial water, city water, pure (deionized) water, reverse osmosis (RO) product water, and ultrafiltration (UF) product water. For the blank, a filter through which no sample water was passed and a filter through which water sterilized at 120°C for 15 minutes in an autoclave had been passed were used. The measurement results for the sample passage amount at each point are listed in Table 3.

The count value of <34 shown among the counting results (Table 3) from the MS-100 indicates that the count is below the lower limit because the bioluminescence of the bacteria in the sample was less than 1.2 times the bioluminescence using reagent only.

Table 2 Bacterial ATP Content

Bacteria	ATP (* 10^{-10} μg ATP / cell)
Escherichia coli	1.2
Streptococcus faecalis	1.1
Staphylococcus aureus	1.8
Staphylococcus epidermidis	8.2
Proteus mirabilis	4.3
Klebsiella	2.8
Pseudomonas	0.5
Proteus rettgeri	1.4
β - streptococcus	1.5

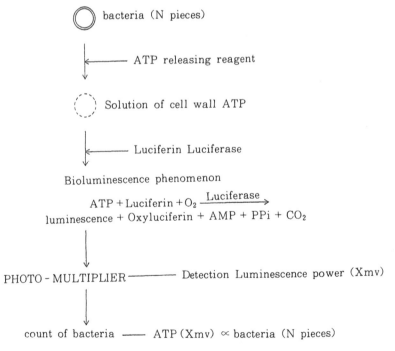

Figure 3   Bioluminescence of a firefly.

bacteria (N pieces)

— ATP releasing reagent

Solution of cell wall ATP

— Luciferin Luciferase

Bioluminescence phenomenon

$$ATP + Luciferin + O_2 \xrightarrow{\text{Luciferase}}$$
$$luminescence + Oxyluciferin + AMP + PPi + CO_2$$

PHOTO - MULTIPLIER ——— Detection Luminescence power (Xmv)

count of bacteria —— ATP (Xmv) ∝ bacteria (N pieces)

Figure 4   Principle of the MS-100.

Table 3 Bacteria Counts at Sample Point

Sample	Q (ml)	MS - 100 (Bacteria cells)	ASTM (cfu)
Blank 0.2 μm Filter	0	<34	0
Blank 0.2 μm Filter	0	<34	0
Blank Sterilized water	100	41	0
Blank Sterilized water	500	<34	3
Indust water	1	175	101
Indust water	5	758	396
City water	100	<34	0
City water	500	<34	2
Pure water	100	119	208
Pure water	100	140	203
Pure water	500	480	790
Pure water	500	480	770
Pure water	1000	893	1700
Pure water	1000	889	1620
RO product water	1000	<34	17
RO product water	1000	<34	24
RO product water	2000	77	32
RO product water	2000	113	40
UF product water	20000	60	143
UF product water	20000	105	497
UF product water	30000	200	4120
UF product water	40000	275	5781
UF product water	60000	339	9984
UF product water	70000	420	10200
UF product water	80000	505	13216

Also, in view of the detection sensitivity of this system, the amount of water was changed at each point so that 100 or more bacteria can be caught on the filter. For ultrapure water, however, the sample water passage rate was set at 20–80 L because the bacteria count was extremely small.

Figures 5 and 6 are graphs that compare measurement results with pure water and UF product water between the MS-100 and the ASTM method. By conversion using a certain factor, the ASTM value may be estimated from the value measured by the MS-100. It is known from comparison of measured values that the number of colonies obtained according to the ASTM method is relatively larger than the number of cells obtained with the MS-100. This result may occur if the ATP content of individual cells in pure and ultrapure water is smaller than that of aqueous bacteria in general because the value measured by the MS-100 is converted using the ATP content of aqueous bacteria in general.

IV. CONCLUSIONS

The following advantages are expected using this system. First, the conventional culture method suffers from variations in the measured values depending on the culture time and temperature. Also, only bacteria cultured and selected under certain conditions could be

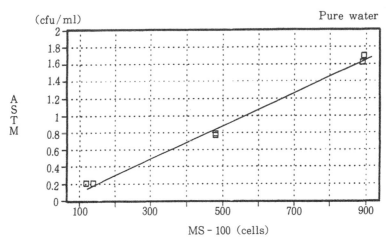

Figure 5   Relationship between bacteria counts in pure water with the ASTM method and the MS-100.

detected. However, the ATP method allows correct counting through counting of all bacteria by means of ATP content.

Second, indication of the bacteria count by means of the ATP content allows rapid judgment of an increase or decrease in bacteria from the ATP content through periodic measurement at each point.

In view of these advantages, the MS-100, which enables rapid counting of bacteria in ultrapure water, will prove helpful in controlling the water quality. The method is also intended to employ the new means of determining the bacteria count through ATP conversion by utilizing the advantage of detecting the ATP content of all bacteria in the sample water.

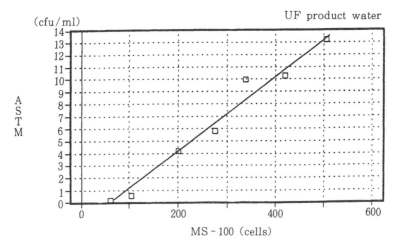

Figure 6   Relationship between bacteria counts in UF product water with the ASTM method and the MS-100.

3 On-Line Monitors

H. Ozone Monitor

Akira Yamada

Shinko Pantec Co., Ltd., Kobe, Japan

I. INTRODUCTION

Ozone injection into ultrapure water has received attention not only for sterilization but also for wafer cleaning performance. A study by the Arrowhead/Monsanto Co. proved that ultrapure water injected with a slight amount of ozone can minimize the organics, particulates, and so on, that adhere to the wafer surface when rinsing the wafer [1].

Ozone sterilization is characterized by continuous sterilization without interrupting the large-scale integrated (LSI) chip production line, a stable sterilization effect, and no requirement for heat-resistant materials, for example, compared with other batch-type sterilization methods (H_2O_2, hot water, and others).

Ultrapure water into which a small quantity of ozone is injected is effective as a cleaning liquid used in part of the semiconductor production process; on the other hand, the ozone concentration microanalysis technique and continuous monitoring instruments are not sufficient.

Since the applications of ozone are expected to increase in the ultrapure water field in the future, this part summarizes the validity, principles, and performance of the ozone monitor.

II. WAFER CLEANING WITH OZONATED ULTRAPURE WATER

The point to be noted in the study made by Arrowhead/Monsanto is that ozone is not used merely as a substitute for disinfectants, such as hydrogen peroxide or hot water, but as an injection agent that makes it possible to improve the yield when producing integrated circuit and semiconductor material. In fabricating the integrated circuit on the wafer, the wafer surface is cleaned effectively by conditioning with pure water that contains ozone at more than 10 ppb (optimal range of 20–90 ppb). This procedure cleans the bare silicone surface and various interfaces of the circuit construction. Such conditioning serves to minimize contamination by reducing the adhesion of organics and further minimizes particulates, including bacterial particles.

Figure 1 is a block diagram of a water treatment process using ozone for the fabrication of integrated circuits on wafers. Ozone is injected into the ultrapure water. Purified oxygen gas is converted to $O_2 + O_3$ gas with a pressure of approximately 0.7 kgf cm^{-2} by the ozone generator. The gas is further compressed to 6 kgf cm^{-2} by a compressor, fed to a gas injector, and mixed with ultrapure water. A bleed line is also provided to remove the air bubbles generated within the piping downstream of the gas injector. The ultrapure water injected with ozone is fed directly to the use point. An ozone monitor is arranged to detect the concentration of ozone. The ozonated ultrapure water is recirculated, and the residual ozone is decomposed by ultraviolet (UV) lamp as it is returned to the ultrapure water tank.

Figure 2 contains sectional views of wafers at various stages in the integrated circuit fabrication process. Figure 2A shows a wafer immediately after fabrication. The oxidized film has a thickness of 20–30 Å. The wafer is then oxygenated in a furnace, providing an oxidized film with a thickness of 100–400 Å (Fig. 2B). The silicone nitride overlayer is provided in a furnace (Fig. 2C), and a photoresist layer is placed over the silicone nitride layer (Fig. 2D). The photoresist is illuminated and then etched (see Fig. 2E), leaving portions of the silicone nitride uncovered.

The wafer shown in Fig. 2E is then conditioned by rinsing with ozonated ultrapure water containing between 20 and 90 ppb ozone. After this rinsing, the portions of the silicone nitride that are no longer covered by photoresist are etched, resulting in the wafer illustrated in Fig. 2F, in which the photoresist has been removed and portions of the oxide are exposed.

The wafer is again conditioned by rinsing it with ozonated ultrapure water. Thereafter, the wafer is placed in a furnace to provide a thick film of oxide, approximately 2000 Å thick, which grows where the nitride has been removed (Fig. 2). The wafer is then

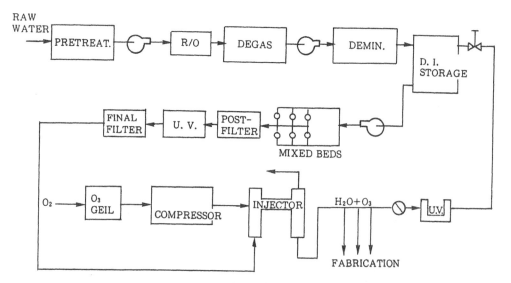

Figure 1   A water treatment process used to provide ozonated water.

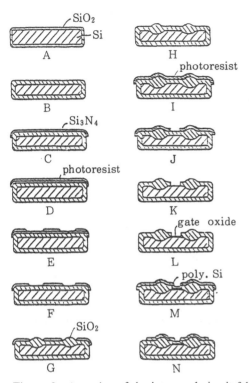

A

B

C

D

E

F

G

H

I

J

K

L

M

N

Figure 2 A portion of the integrated circuit fabrication process.

conditioned by rinsing it with ozonated water and the silicone nitride is removed (Fig. 2H). The wafer is again rinsed with ozonated water, and then a photoresist is added (Fig. 2I), illuminated, and then etched (Fig. 2J). The wafer is again conditioned by rinsing with ozonated water, and the photoresist is removed to grow a gate oxide 150–300 Å thick (Fig. 2L).

The gate oxide is typically the most critical oxidation in the entire fabrication of the integrated circuit. It is absolutely essential that the bare silicone be properly conditioned for growing the gate. The ozonated water rinse treats and conditions the bare silicone and provides effective rinsing for the growth of gate oxide film. After the polysilicone is deposited (Fig. 2M), another photoresist layer is rinsed with ozonated water, and then the uncovered polysilicone layer is etched and rinsed with ozonated water, resulting in the wafer illustrated in Fig. 2N.

Although a part of the typical integrated circuit production process has been described, the process includes other rinsing steps using ultrapure water. The rinsing process performed immediately after etching of material that is not covered by photoresist is particularly important to properly control succeeding formation process of the gate oxidation film by using 20–90 ppb ozonated ultrapure water.

III. OZONE MONITOR

The ozone monitor used in the semiconductor production process needs to provide the following functions and characteristics:

1. Excellect stability in operation and construction. Since there is no standard ozone solution (known concentration of the solution), it is hard to calibrate the monitor using test solutions before use.
2. Wide measurement range and high repeatability.
3. Short measurement interval (or continuous) and quick response.
4. Hardly affected by fluctuations in flow rate and pressure.
5. Compact construction.
6. Easy maintenance.

The monitors generally used at present for monitoring ozonated ultrapure water include a membrane electrode method and an ultraviolet light absorption method. In addition, other types of monitors apply iodometric titration or color identification for analytic measurement. The following details the principles of operation and construction of each method.

A.   Membrane Electrode Method

The detector contacting the ozone water through a thin membrane detects as a current value the ozone passing through the membrane. An example of a polarographic membrane ozone sensor is shown in Fig. 3. The gold cathode, silver anode, and sealed electrolyte are isolated from the sample water by the semipermeable membrane. The ozone that permeates the sensor through the membrane is reduced by the cathode, during which a reduction current proportional to the amount of ozone flows between the cathode and the anode. The electrolyte contains a catalytic substance that enhances reduction. It is possible to detect a slight amount of ozone by adjusting the load voltage even when using ozonated water into which a small amount of oxygen is mixed.

   The reaction within the sensor generates a reduction current in linear proportion to the amount of ozone generated. The point at which the linearity is disrupted within the low-concentration range is determined by the residual current of the sensor (Fig. 4). The sensor is provided with several improvements, such as protective electrodes, so that ozone

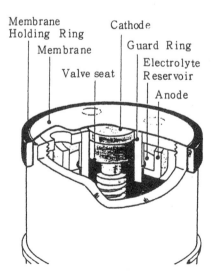

Figure 3   Polarographic membrane electrode ozone sensor.

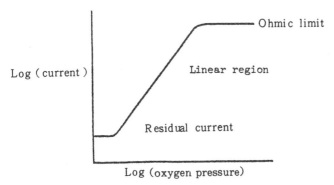

Figure 4   Ozone concentration versus current generated on the polarogram.

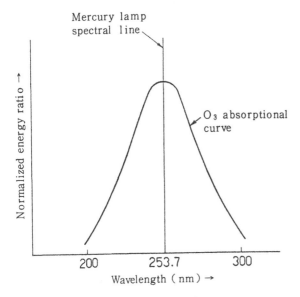

Figure 5   Ozone absorption curve [2].

at low concentrations can be measured by reducing the residual current. The membrane method requires calibration to perform the concentration measurement, because the signal current is generated in proportion to the partial pressure of the subject of measurement. However, ozone monitors with unique calibration functions have also been developed.

B.   Ultraviolet Light Absorption Method

As shown in Fig. 5, ozone intensively absorbs light of approximately 245 nm. On the other hand, the following relationship exists between the light absorption amount and its concentration:

$$C + k \log \frac{I_o}{I_x}$$

where: C = ozone concentration
   k = conversion coefficient
   I_o = amount of light permeating the ozone-free ultrapure water
   I_x = amount of light permeating the ozonated ultrapure water

Consequently, the ozone concentration can be determined by alternately passing the ozonated water and ultrapure water through light of 245 nm applied to the sample measurement point (cell) and by measuring the amount of light that permeates. Even when change occurs in the sensitivity of the light sensor (photoelectric tube), the value does not change because the change in sensitivity affects both I_o and I_x equally. The same method can be applied even when the sample measurement point is stained. Accordingly, frequent calibration is not needed.

An example of actual equipment flow is shown in Fig. 6. When the solenoid valve is changed, the reference water (ozone-free ultrapure water) and ozonated ultrapure water flow past the sample measurement point alternately and I_o and I_x can be measured.

C. Iodometric Titration Coulometry

This method is based on the iodometric titration, which has been often used conventionally as an ozone analysis method and has been put to practical use using coulometric titration.

First, the ozone-free ultrapure water is mixed with electrolyte containing sodium thiosulfate and potassium iodide for measuring the full-scale concentration. When the ozonated ultrapure water is mixed with a similar electrolyte, the ozone oxidizes the potassium iodide, generating iodine. The iodine immediately reacts with the known amount of sodium thiosulfate contained in the electrolyte. As a result, sodium thiosulfate corresponding to the ozone concentration is consumed:

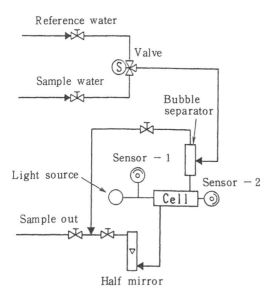

Figure 6 UV absorption ozone monitor.

$$O_3 + 2KI + H_2O \rightarrow I_2 + 2KOH + O_2$$
$$I_2 + 2Na_2S_2O_3 \rightarrow 2NaI + Na_2S_4O_6$$

The remaining sodium thiosulfate is coulometrically titrated by the iodine generated by iodine ion electrolytically oxidized in the electrolyte:

$$2I^- - 2e^- \rightarrow I_2 \quad \text{electrolytic oxidation}$$
$$I_2 + 2Na_2S_2O_3 \rightarrow 2NaI + Na_2S_4O_6$$

As shown, it is possible to measure the ozone concentration by reverse titration of the sodium thiosulfate after the reaction. The measurement point is illustrated in Fig. 7.

D. Color Identification (DPD)

The color identification method uses DPD (N, N-diethyl-p-phenylenediamine) indicator and sodium iodide. The iodide is oxidized by the ozone contained in the ultrapure water, and the oxidized iodide reacts with the DPD, yielding a dark purple color. Since the concentration of coloring substance is in proportion to the ozone concentration, the absorption coefficient is measured at the maximum absorption concentration using a spectrophotometer, and the ozone concentration is determined from the amount of absorption.

Figure 8 is a flow diagram of the equipment, and Table 1 compares commercially available ozone monitors.

IV. CONCLUSION

When used as a continuous monitor for ozonated ultrapure water used in cleaning wafers, commercially available ozone monitors are hardly sufficient from the viewpoints of lower

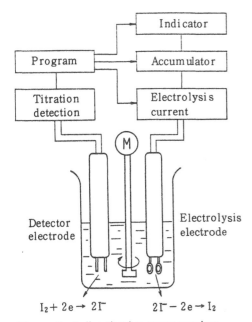

Figure 7 Iodine titration ozone monitor.

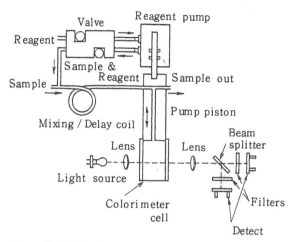

Figure 8   DPD colorimetry ozone monitor.

Table 1   Specifications for the Ozone Monitor

Manufacturer	Orbisphere Laboratories	Griffin Technics INC.	Ebara jitsugyo CO., LTD.	Hiranuma sangyo CO., LTD.	Hach Company
Model	27501	D 03 − 01	EL − 2001	AZC − 15	32000
Method	polarographic probe	polarographic probe	ultraviolet absorption	iodine titration	DPD coloring
Range	0 − 2 0 − 200 ppm	0 − 1 0 − 20 ppm	0 − 5 0 − 20 ppm	0 − 15 ppm	0 − 0.5 0 − 1 ppm
Repeatability	± 1 %	± 50 ppb	± 1 %	± 10 ppb	± 1 %
Linearity	± 5 ppb	± 1 %	± 1 %	± 50 ppb	± 5 %
Response time	30 sec. (90 %)	60 sec. (95 %)	20 − 60 sec.	180 sec.	120 sec. (90 %)

Remarks :   Above values are collected from each manufacturer's brochure ╱ catalog.
          Repeatabilities and linearities depend upon operating conditions.

measurable limit and accuracy. Also, comparison with the analysis value is required to verify the ozone monitor results. There is currently no authentic low-concentration ozone measuring method. The methods widely used at present to measure the ozone include iodometric titration [3], absorptiometry [3], and the indigo method. Above all, the indigo method has been introduced as having a lower measurable limit of 10 ppb and an accuracy of ±2 ppb [4].

REFERENCES

1.   Official patent report (A)(Hei-1)–132126 (1989).
2.   R. Satsuka, New ozone utilization technique, Sanshyu Shoboh, 175 (1986).
3.   Japan Water Works Association, Potable water test method, (1985), pp. 327–332.
4.   H. Barder and J. Hoigne, Determination Aspects of Ozone Treatment of Water and Wastewater, Lewis Publishers, (1986), pp. 153–156.

4
Afterword

Tetsuo Mizuniwa
Kurita Water Industries, Ltd., Kanagawa, Japan

Analysis and evaluation technology is an essential means of producing ultrapure water of a quality higher than currently available. Measuring instruments that monitor water quality in real time contribute to the production of such water to a great extent.

This part has discussed a water quality measuring instrument that may have the highest current sensitivity and an apparatus that may be the most effective in upgrading the purity of ultrapure water. Such instrumentation has sufficient sensitivity to monitor the quality not only of present ultrapure water but of future ultrapure water as well. However, developments and improvements in the instrumentation to increase their sensitivity have been actively carried out so that those introduced here will not be the only methods used in the future. Rather, it is expected that instrumentation with a higher performance than that of today will be developed.

There have been demands for high-sensitivity and real-time measuring instruments for the monitoring of water quality. However, some instruments available at present consume a great deal of time in operation from the introduction of the sample water to the indication of the measured value. Apparently more time is required if the concentration of impurity decreases. This is contradictory to the desire for quick detection and measurement. It is therefore hoped that a highly sensitive and instantly responsive water quality measuring instrument will be developed that is not restricted by existing principles but is based on new principles.

It is definitely impossible to maintain stable operation of an ultrapure water system without the proper functioning of the instrumentation discussed here. In other words, stable function and proper indication of the instrumentation are prerequisites for the stable operation of ultrapure water systems. There is no uniform maintenance standard available for newly developed instrumentation, because there are too many types. However, when the measuring principles and operation mechanisms are clarified, the items to be controlled will be identified. Observance of manufacturer's operating and maintenance instructions is the minimum requirement. In addition, the user's positive attitude toward the instrumentation helps to obtain correct information.

VI
Fabrication and Construction Technologies

1
Foreword

Akihiko Hogetsu
Shinko Pantec Co., Ltd., Kobe, Japan

Ultrapure water supply systems are composed of a variety of components, such as vessels, piping, pumps, instruments, and the ultraclean technology to manufacture and to fabricate these components. The ultrapure water supply system should be systematically constructed under controlled superclean conditions with expertise on machinery, electricity, instrumentation, architecture, chemistry, chemical engineering, and so on.

In general, when complicated plants are built, the most efficient organization is set up to achieve the most appropriate design and construction economically in a limited amount of time. The budget, schedule, labor-hours, and quality control of the projects are conducted in accordance with technical criteria for design, manufacture, quality, operation, and maintenance. The construction of ultrapure water supply systems is no exception. However, the requirements specific to ultraclean water supply systems are to avoid contamination with pollutants and the leaching of metal and organic substances from the materials used in the system.

Considering these points, overall attention should be addressed to ensure that these ultraclean technologies specifically applied to materials, parts, and components in the layout and installation of ultrapure water supply systems function correctly.

In this part, specialists in this field describe features of vessels, piping material, valves, pumps, and instruments used in ultrapure water supply systems, rinsing methods for these materials, and matters that demand special attention in the layout and installation of the system.

2 Cleaning Methods

A. Cleaning and Installation of Polyvinyl Chloride Piping

Yukio Hamano
Sekisui Chemical Co., Ltd., Shiga, Japan

I. CLEANING PROCESS AND COMPARISON OF PARTICLE COUNT AND TOC LEVEL BETWEEN CLEANED AND UNCLEANED PIPE

Ultrapure water is distributed to use points through the ultrapure water distribution pipeline after it is purified in the ultrapure water production system. The high purity of ultrapure water is easily deteriorated if the pipeline is contaminated with dust or oil. To maintain the purity of ultrapure water, the ultrapure water distribution line is cleaned with such chemicals as hydrogen peroxide after completion of piping installation. Although dust and particles can be removed in this cleaning step, oils are not cleaned. In addition, if the dust content in the ultrapure water is very high, the cleaning efficiency is degraded.

A. Necessity for Cleaning

In general, dust and oil are deposited on the inner surface of polyvinyl chloride (PVC) pipes and fittings. They are contained in the cooling water used in the molding process. Dust and oil in the air also adhere when the runner is cut, during temporary storage of fittings, and when packing. Another problem is sprue, which is generated when resin is injected into the mold. To eliminate these problems, the cooling water and the process air in the steps from production to packing must be cleaned. Considering the many types of fittings, however, it is very difficult to achieve a high level of cleanliness in practice. This is why degreasing is required to eliminate dust and oil on fittings.

 Various kinds of contaminants are deposited on the inner surface of pipes: dust contained in the air from the production process to the packing process, sludge and swarf generated when pipes are cut, lubricants contained in the raw materials of PVC, and organic materials in pigments. To suppress contaminant adhesion on the inner surface of pipes, several countermeasures should be studied: a sludge-free degreaser to cut the oils, methods to prevent dust in the air from adhering to the inner surface, and the selection of clean raw materials for the PVC pipe.

The Esloclean pipe for ultrapure water was developed based on such studies of methods to suppress contaminant deposition. Table 1 shows the performance of Esloclean pipe, including particles deposited on the inner surface, total organic carbon (TOC) elution, fluctuations in conductivity, and heavy metal elution. No oil was detected in any sample.

Oil deposition was also not observed in either the degreased pipe or in the untreated pipe. There was no difference in the particle count. However, the treated pipe is superior in terms of TOC elution and conductivity fluctuation. TOC elution from the untreated pipe is high. This is thought to be because lubricants are vaporized and adhere to the inner surface of pipes when high temperatures are applied to fuse PVC. The degreasing treatment is required to achieve rapid start-up of resistivity after piping installation is complete.

B. Cleaning by Cutting the Oils (Degreasing)

Figure 1 shows the procedure for the degreasing treatment. The instruments, jigs, and containers used in the treatment should be cleaned with cleaning solution in advance. Operators should wear clean medical-grade rubber gloves and dust-free garments throughout the process, from cleaning to packing, without failure. The process from cleaning through packing should be conducted in the degreasing treatment room (clean room). The details of each step in the degreasing treatment for fittings and pipes are described here.

Table 1 Evaluation Results

Test items	Esloclean Pipe		Manufacture A	Notes
	After Cleaning	Without Cleaning	After Cleaning	
TOC (ppb)	60	170	90	Sealed with DI waters for 24 hours at room temperature
Micro-Particle (numbew / m.l.)	57	53	190	
Electric Conductivity (μs / cm)	0.95	1.2	1.5	Sealed with DI water for 15 days at room temperature
Ca (ppb)	48	81	190	
Mg	ND	ND	ND	
Zn	ND	ND	ND	
Sn	ND	ND	ND	
Pb	ND	ND	ND	
Surface Smoothness (μm)	0.3	0.3	2.0	JIS B0601

Notes : 1. Test specimin of pipe used was 25 mm ¢ × 1 meter.

2. Comparison among Cleaned Esloclean pipe, Esloclean pipe without cleaning, and Cleaned pipe produced by Manufacturer A.

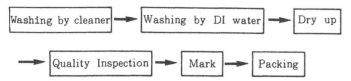

Figure 1   Cleaning method.

1.   Fittings

a.   Cleaning with Cleaning Solution.   The ultrasonic cleaning bath is filled with the cleaning solution, a phosphorus-free neutral detergent diluted to 5%. Fittings are immersed in this solution and cleaned by ultrasound. The liquid temperature should be kept at 10–30°C.

b.   Rinsing with Ultrapure Water.   After the cleaning with solution is completed, the fittings are withdrawn from the ultrasonic cleaning both. They are transferred to another ultrasonic cleaning bath for ultrasonic cleaning with ultrapure water. The ultrapure water should be constantly allowed to overflow the ultrasonic cleaning vessel during rinsing.

c.   Drying.   Fittings are dried in a clean manner in the treatment room.

d.   Quality Inspection

1.   Purpose of quality inspection: Confirm whether oils remain at the liquid-contacting surface of fittings.
2.   Criteria for quality inspection: No oils should be detected when the cleaned surface is wiped with clean cotton.
3.   Method of quality inspection: Five samples are taken from one lot. More than two parts on the liquid-contacting surface of a fitting are wiped with cotton soaked in ethyl alcohol to check whether oils are detected.

e.   Labeling.   A label indicating "degreased" is attached to the outer surface of fittings, which does not contact liquid.

f.   Packing.   A determined number of fittings are placed in a polyethylene bag, and the bag is sealed with vinyl tape. A predetermined number of sealed bags are packed in a carton.

The degreasing treatment for pipes is basically the same as that for fittings. Because pipes are longer than fittings, they require a somewhat different procedure, briefly described here.

2.   Pipes

a.   Cleaning with Cleaning Solution.   The prepared cleaning solution is jetted onto the inner surface of pipes at high pressure.

·b.   Rinsing with Ultrapure Water.   The preliminary rinsing of pipes is with city water. Ultrapure water is then jetted onto the inner surface to rinse off the cleaning solution.

c.   Drying.   The drying step is the same as that used to dry fittings.

d.   Quality Inspection.   The quality inspection procedure is the same as that for fittings.

e.   Packing.   End caps are placed on the two ends to seal the pipes before they are packed.

   f.   Labeling.   The label indicating "degreased" is placed on the outside of pack-
ages.

II.   ASSEMBLING AND PACKING OF CLEANED PVC VALVES

A.   Introduction

Just like fittings, valves need to be cleaned. There are two ways to clean valves: after
assembling and before assembling. In general, as for PVC valves, the components are
first cleaned and then they are assembled. This is because the degreasing treatment and
dust removal cannot be conducted perfectly after assembly: the sliding part in the
liquid-contacting surface must be thoroughly cleaned.

B.   Assembly Process for Cleaned Valves

There are five steps from cleaning to packing of PVC valves: (1) component cleaning, (2)
assembly, (3) inspection, (4) labeling, and (5) packing.

1.   Component Cleaning

The cleaning of components is conducted in the same way as the cleaning of fittings.

2.   Assembly

The assembly process should be conducted in a clean environment free of dust. Jigs used
in the assembly process should be degreased. Operators must wear clean rubber gloves
and dust-free garments. No lubricants, such as silicon and grease, should be used. The
assembled valves should be kept in a clean environment until they are packed. They
should not be handled with bare hands.

3.   Inspection

Valve inspection should include the external appearance, the dimensions, the pressure
resistance of the valve housing, and leakage from the valve seat. The pressure resistance
of the valve housing and leakage from the valve seat are the important items to be
checked.
    For inspecting the pressure resistance of the valve housing and leakage from the valve
seat, the JIS (Japanese Industrial Standard) water pressure test is used and the gas pressure
test is used as a substitute test. The water pressure test is not suitable for cleaned valves
because bacteria may breed in water remaining inside the valve after the test. The gas
pressure test is therefore employed for cleaned valves.
    The gas pressure test should be conducted in a clean, dust-free environment. The gas
and the instruments for the test should be clean as well.
    The valve housing pressure resistance test should be conducted as follows. The valve
is half-open or fully open to apply pressure to the entire liquid-contacting surface of the
valve. A pressure of 6 kgf cm^{-2} is applied to the inner surface of the valve for a certain
amount of time, and then the external leakage is checked.
    The valve seat leakage test is as follows. The valve is closed, a pressure of 6 kgf cm^{-2}
is applied from one side for a certain amount of time, and then the sealing ability of the
valve seat is checked.
    The gas pressure test usually uses filtered clean nitrogen gas.

4. Labeling

Valves that pass inspection have the production sequence numbers and the "degreased" label on the outer surface.

5. Packing

After labeling, each valve is placed in a clean polyethylene bag. To suppress dust deposition on the valves, the bag is sealed with heat. Together with the inspection reports, the heat-sealed bags are packed in a carton labeled "degreased."

2 Cleaning Methods

B. Polyvinylidene Fluoride Pipes, Fittings, and Valves

Hiroto Fujii and Shosuke Ohba
Kubota George Fischer Ltd., Osaka, Japan

Junsuke Kyomen
Kubota Corporation, Osaka, Japan

I. CLEANING METHOD

In most cases, polyvinylidene fluoride (PVDF) piping materials for ultrapure water are cleaned by the degreasing method. This is required to remove the contaminants and oils that adhere to the materials during the production process. The cleaning should be conducted in the clean room, which is isolated from the production plant. The most common cleaning method is described here.

A. Preparation

The valves to be cleaned should be divided by type and size. The assembled valves should be disassembled before cleaned.

B. Scat (Rapid) Cleaning

Scat 2DX-PF (phosphorus free) #1, produced by Daiichi Kogyo Pharmaceutical, is diluted with tap water to 5%. Valve components are immersed in this solution in the ultrasonic cleaning unit and cleaned for 10–15 minutes.

C. Rinsing with Tap Water

Valve components treated by Scat cleaning are rinsed with tap water. Water should run constantly during rinsing. Tap water should be circulated and allowed to overflow constantly. Rinsing takes 20–30 minutes of bubble cleaning.

D. Rinsing with Pure Water

The resistivity of the pure water should be more than 5 MΩ·cm. The resistivity should be measured immediately after the pure water inlet. Valve components rinsed with tap water are placed in the bubble cleaning unit filled with pure water for rinsing. Pure water should be replaced when the resistivity decreases to less than 0.1 MΩ·cm.

E.   Drying

The rinsed valve components are dried with clean hot air on a clean bench.

F.   Inspection of Cleaning Efficiency

1.   Inspection A

No contamination should be detected when the surface of the valve and the inner surface of the ultrapure water line are wiped with clean white cloth. The light level in the inspection room should be over 300 lx. Contamination is checked by macroscopic inspection.

2.   Inspection B

When the white cloth used in inspection A is dipped into hot water at 60–80°C, no oil film should be detected on the surface of the water.

G.   Inspection

After assembling the valves, the sealing ability is inspected. Oil-free nitrogen gas filtered through a 0.2 μm filter is supplied at a predetermined pressure for a predetermined amount of time to check for leakage.

H.   Packing

Each valve product is packed in a polyethylene bag (Fig. 1). The sealed bags are placed in a carton.

I.   Labeling

The cleaned products should be labeled "degreased." The same label should be placed on the carton.

J.   Notes

Garments free of particulates, such as dust-free garments, should be worn. (Fig. 2). Operators should wear cleaned rubber gloves (Fig. 3). The external and internal surfaces of pipes should be free of scratches and cracks.

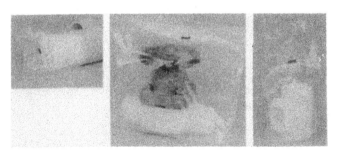

Figure 1   Form of delivery.

Figure 2   Cleaning process.

K.   Summary

The most common cleaning method is described here. There are various other methods, including those specified by piping material manufacturers and users. Even if the cleaning of each product is conducted perfectly, contamination during installation is inevitable. Blowing with air after installation is therefore necessary, and a longer blow time is desirable.

II.   PVDF, HP SERIES

The HP (high-purity) series SYGEF is manufactured under stringent conditions of cleanliness from the very beginning: metals and organic carbon are carefully kept away from the

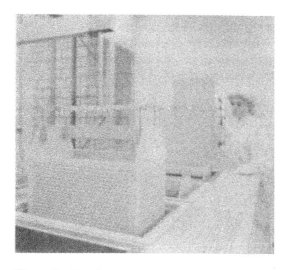

Figure 3   Cleaning process.

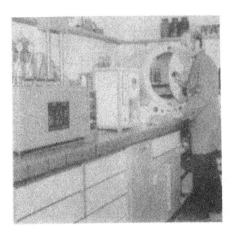

Figure 4   Quality control.

raw material polymerization process, and raw materials of special grade are employed to provide the inner surface of pipes with a smooth finish. In the production of pipes, fittings, and valves, special equipment and facilities are installed so that the high-purity raw materials are not contaminated.

Pipe manufactured by continuous molding is cut using a special method that is free of sludge and swarf. The pipe is packed in a dust-free manner immediately after the cutting process with its two ends capped.

Since fittings and valves are slightly contaminated during injection molding, finishing, and assembly, they are cleaned in a way similar to the standard SYGEF. The cleaning process is conducted in a completely separate line from the standard products, so there is no concern that the HP series is mixed with standard products.

Before the packing process, the fittings and valves of the HP series are subjected to 100% microscopic inspection (Fig. 4). A penlight is also used to confirm that no harmful foreign materials are mixed in the resin. After this inspection, they are packed in double polyethylene bags. Figure 1 shows pipes, fittings, and valve packed for delivery. Of course, the bags should only be opened immediately before installation.

A comprehensive quality control program should be developed for all components of the piping system. In addition to standard inspection, various items are checked closely, observing the surface characteristics of the liquid-contacting area in an effort to maintain the high level of purity.

Table 1 shows the product line of the PVDF standard series and the PVDF HP series.

Table 1   Applications for SYGEF PVDF and PVDF HP

Nominal diameter (mm)			10	15	20	25	32	40	50	65	80	100	125	150	200
Pipe			◎	◎	◎	◎	◎	◎	◎	◎	◎	◎	◎	◎	◎
Fitting	Elbow / Tee / Socket / Reducer		◎	◎	◎	◎	◎	◎	◎	◎	◎	◎	◎	◎	◎
	Union		◎	◎	◎	◎	◎	◎	◎						
	Flange adaptor			◎	◎	◎	◎	◎	◎	◎	◎	◎	◎	◎	◎
Diaphragm Valve	Spigot			◎	◎	◎	◎	◎	◎	◎	◎	◎			
	Flange			◎	◎	◎	◎	◎	◎						
Pneumatically actuated Diaphragm Valve	Spigot			◎	◎	◎	◎	◎	◎	◎	◎	◎			
	Flange			◎	◎	◎	◎	◎	◎						
Ball Valve	Socket		○	○	○	○	○	○	○						
	Flange			○	○	○	○	○	○						
Pneumatically actuated Ball Valve	Socket		○	○	○	○	○	○	○						
	Flange			○	○	○	○	○	○						
Electrically actuated Ball Valve	Socket		○	○	○	○	○	○	○						
	Flange			○	○	○	○	○	○						
Ball Check	Socket		○	○	○	○	○	○	○						
Valve	Flange			○	○	○	○	○	○						
Butterfly Valve	Wafer								○	○	○	○	○	○	○

◎ : PVDF Standard and PVDF–HP

○ : Only PVDF Standard

Other Products : Sheet (Thickness : 1–10 mm) , Block (Thickness : 15–80 mm). Rod (φ 15– φ 200 mm)

2 Cleaning Methods

C.   Polyether Ether Ketone Pipes and Fittings

Katsuhiko Ito
Mitsui Toatsu Chemicals, Inc., Tokyo, Japan

This part describes the cleaning process and other manufacturing processes for polyether ether ketone (PEEK) pipes and fittings.

I. PRODUCTION

The raw material for PEEK resin is a heat-resistant crystalline polymer. The pellet form of PEEK is used to produce pipes and fittings. In pipe production, the raw pellet material is first heated to close to 400°C in the extruder for fusing. Pipes are manufactured by extrusion from dies. Because PEEK resin exhibits excellent thermal stability, no stabilizer to improve thermal stability is added. Figure 1 is an outline of pipe production. From the step in which pellets are supplied to the hopper to cutting, the inner surface of the pipe is kept away from the outer environment. This suppresses contamination of the inner surface. The dust generated during the cutting process is removed by vacuuming. Before packing, the inside of the pipe is blown with nitrogen gas to prevent contamination. Aluminum laminate bags are employed for packing to prevent the penetration of outside air.

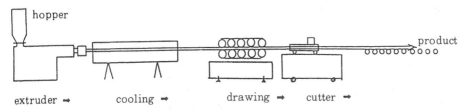

Figure 1 Pipe production.

Fittings are fabricated by injection molding. They are cleaned with organic solvent, dried, and then packed in the aluminum laminate bag.

II. SPECIFICATIONS

Table 1 lists the specifications for PEEK pipes and fittings available on the market. Fittings are designed to have less dead space and to be joined by butt fusion, which does not use adhesive. Because the mechanical strength of PEEK resin is high, products made with PEEK can be designed thinner than those made with other materials. Reducers and bosses are now manufactured from dowels by cutting.

Table 1   Specifications for Pipes and Fittings

Pipe   in : mm

SIZE	D	t	APP ID	Length
13 A	18±0.15	1.5±0.10	15	4,020
16 A	22±0.25	1.9±0.15	18	4,020
20 A	26±0.25	1.9±0.15	22	4,020
25 A	32±0.25	2.4±0.20	27	4,020
40 A	48±0.35	3.0±0.20	42	4,020
50 A	60±0.40	3.0±0.25	54	4,020
65 A	76±0.40	3.0±0.25	69	4,020
80 A	89±0.45	3.5±0.25	82	4,020
100 A	114±0.45	3.5±0.25	107	4,020

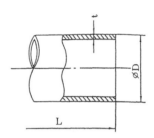

Elbow (90°)   in : mm

SIZE	D	H	t
13 A	18±0.10	60±0.20	1.5±0.10
16 A	22±0.20	65±0.20	1 9±0.10
20 A	26±0.20	60±0.20	1.9±0.10
25 A	32±0.20	58±0.20	2.4±0.10
40 A	48±0.30	84±0.30	3.0±0.10
50 A	60±0.30	96±0.40	3.0±0.10
65 A	76±0.40	120±0.50	3.0±0.20
80 A	89±0.45	110±0.50	3.5±0.25
100 A	114±0.50	125±0.50	3.5±0.30

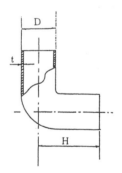

Stub Ends   in : mm

SIZE	D	D_1	D_2	H	H_1	H_2	t
13 A	18±0.20	34±0.20	27±0.20	70±0.10	8±0.10	8±0.10	1.5±0.10
16 A	22±0.20	34±0.20	27±0.20	70±0.10	8±0.10	8±0.10	1.9±0.10
20 A	26±0.20	41±0.20	33±0.20	70±0.10	8±0.10	8±0.10	1.9±0.10
25 A	32±0.20	50±0.20	41±0.20	70±0.10	8±0.10	12±0.10	2.4±0.10
40 A	48±0.30	73±0.25	61±0.25	70±0.10	8±0.10	13±0.10	3.0±0.10
50 A	60±0.30	90±0.25	76±0.25	70±0.10	10±0.10	13±0.10	3.0±0.10
65 A	76±0.40	106±0.30	90.5±0.30	70±0.10	10±0.10	20±0.10	3.5±0.20
80 A	89±0.45	125±0.35	109±0.35	70±0.10	12±0.10	21±0.10	3.5±0.25
100 A	114±0.50	150±0.35	131±0.35	70±0.10	12±0.10	23±0.10	3.5±0.30

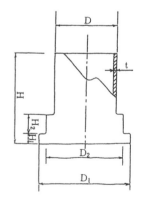

Table 1   Continued

Tees   in ∶ mm

SIZE	D	t	t_1	H	H_1	H_2	H_3
13 A	18±0.10	3.0±0.10	1.5±0.10	120±0.20	60±0.20	40±0.10	60±0.20
16 A	22±0.20	4.0+0.15	1.9±0.10	130±0.20	65±0.20	40±0.10	65±0.20
20 A	26±0.20	4.0+0.15	1.9±0.10	130±0.20	65±0.20	25±0.10	65±0.20
25 A	32±0.20	4.8+0.15	2.4±0.10	116±0.20	58±0.20	25±0.10	58±0.20
40 A	48±0.30	6.0±0.10	3.0±0.10	156±0.30	78±0.30	29±0.10	78±0.30
50 A	60±0.30	6.0±0.20	3.0±0.10	180±0.40	90±0.40	30±0.10	90±0.40
65 A	76±0.40	7.0±0.20	3.5±0.20	220±0.50	110±0.50	35±0.10	110±0.50
80 A	89±0.45	7.0±0.25	3.5±0.25	180±0.50	90±0.50	5±0.10	90±0.50
100 A	114±0.50	7.0±0.30	3.5±0.30	246±0.50	123±0.50	14±0.10	123±0.50

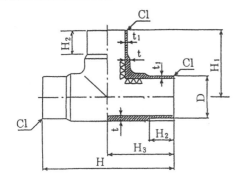

Reducing Tees

SIZE	D	D_1	t	t_1	t_2	H	H_1
40A / 25A	48±0.30	32±0.20	6.0±0.10	2.4±0.10	3.0±0.10	120±0.20	67±0.20
50A / 25A	60±0.30	32±0.20	6.0+0.20	2.4±0.10	3.0±0.10	130±0.20	73±0.20
50A / 40A	60±0.30	48±0.30	6.0+0.20	3.0±0.10	3.0±0.10	130±0.20	90±0.20
65A / 25A	76±0.40	32±0.20	7.0+0.20	2.4±0.10	3.5±0.20	116±0.20	100±0.20
80A / 40A	89±0.45	48±0.30	7.0±0.25	3.0±0.10	3.5±0.25	156±0.30	111±0.30

in ∶ mm

SIZE	H_2	H_3	H_4
40A / 25A	32±0.10	73±0.15	17±0.10
50A / 25A	35±0.10	81±0.15	12±0.10
50A / 40A	30±0.10	90±0.15	30±0.10
65A / 25A	47±0.10	100±0.20	23±0.10
80A / 40A	5±0.10	110±0.20	23±0.10

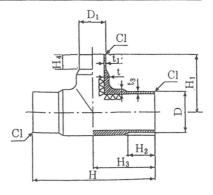

III. INNER SURFACE OF PIPES

From scanning electron microscopy (SEM) of the surface, summarized in Table 2, the surface of PEEK pipes is smooth and has fewer pores. Even after immersion in water at 20 and 80°C, no degradation is detected. Because it is difficult for bacteria and dust to adhere to PEEK pipes before as well as after installation, PEEK pipes are easily cleaned before installation.

IV. CLEANING

Although not much information is available on the cleaning of PEEK pipes, it can be concluded that cleaning with ultrapure water can lower elution from the pipes to acceptable levels.

Table 3 lists the effect of cleaning the pellets, the raw material of the pipe. Total organic carbon (TOC) and elution of metals is decreased by cleaning with ultrapure water. If the pellets are immersed in 5% hydrochloric acid for 1 day before cleaning, elution can be lowered further. Since the elution of TOC is also further decreased in this case, this may be the effect not of the hydrochloric acid but also of the immersion in ultrapure water.

According to the elution data from the PEEK pipes available at present, pipes can be precleaned with ultrapure water. Figures 2 through 5 show the elution level in ultrapure water trapped in pipes. The pipes are precleaned with running ultrapure water for 30 minutes. The elution of metals and anions is low even on the first day at both 25 and 80°C, which proves cleaning is sufficiently. TOC elution from PEEK is low at 25°C but it is higher at 80 than at 20°C, although it is lower than the TOC elution level from other materials. This indicates that cleaning with hot ultrapure water is effective in reducing the initial TOC elution.

There are no data for cleaning with surfactants, but this is not thought to be an effective method. This is because contamination with oil is not expected and surfactants themselves constitute contaminants to be cleaned.

Table 2 Immersion Test Results

Material / Maker	Inside Surface	Micropores		
		Before Immersion	After Immersion 20°C × 1 day	80°C × 1 day
Clean PVC	◎	+	+	++
Poly-propylene	×	+	+	+
Polyether-ether-ketone	◎	−	−	−
PVDF (Maker B)	○	+	+	+
PVDF (Maker S)	○	−	−	−
PVDF (Maker A)	○	±	±	±

Inside surface smoothness   Micropores
    ◎ Superior   + Many
    ○ Good   − Less
    × No good
Source: From Ref. 1.

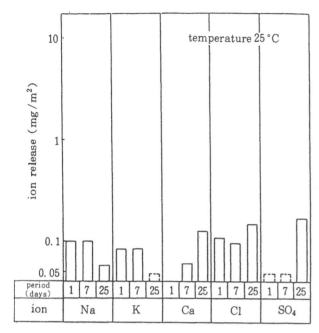

Figure 2   Imprisonment test for PEEK pipe. (From Ref. 2.)

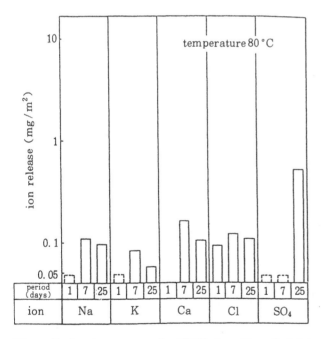

Figure 3   Imprisonment test for PEEK pipe. (From Ref. 2.)

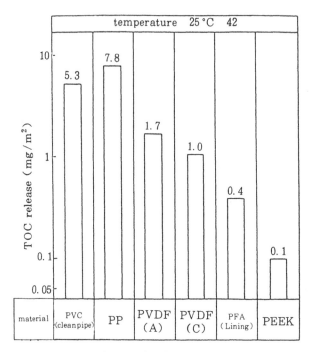

Figure 4   TOC release during imprisonment test. (From Ref. 2.)

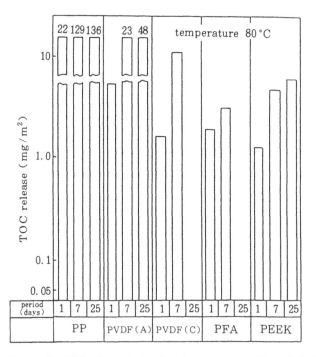

Figure 5   TOC release during imprisonment test. (From Ref. 2.)

Table 3 PEEK Pellet Immersion Test

in (ppb)

washing method Element	without washing	washed with pure water	washed with HCl
TOC	340	230	130
Na	6.3	1.5	2.7
K	3.3	<2.0	<2.0
Fe	<0.5	<0.5	<0.5
Cu	<0.5	<0.5	<0.5
Zn	1.4	0.5	<0.2
Ni	<2.0	<2.0	<2.0
Sn	<2.0	<2.0	<2.0
Ca	780	200	51
Mg	12	4.6	<0.4

Test procedure

1) Washed with pure water

150g of PEEK pellets were put into 1 ℓ pp container and washed 5 times with ultra-pure water, then immersed in 800 ml ultra-pure water for 7 days. Extracted elements in this water were measured.

2) Washed with HCl

150g of PEEK pellets were immersed in 5 % HCl solution for one day, then washed 5 times with ultra-pure water as above, and immersed in 800 ml ultra-pure water for 6 days.

3) TOC Measurment : photo-chem method

Elution from PEEK pipe can be suppressed by cleaning with ultrapure water. The effect is enhanced when hot ultrapure water is employed.

V. FUTURE CHALLENGES

PEEK resin demonstrates a high resistance against water, including hot water, and elution into ultrapure water from PEEK resin is lower than that from other materials. However, PEEK resin has not yet been used to its best advantage. The following are challenges for the future:
1. Discover the cause of material elution.
2. Prevent the contamination from these substances.
3. Develop the optimum cleaning method.
Because the present elution levels are extremely low, advanced technology will be required to overcome these obstacles. However, it is expected that there is a possibility of lowering the elution infinitely close to zero.

REFERENCES

1. K. Yabe et al., Proc. 9th. Symp. Contamination Control, Los Angeles, September 1988, p. 69.
2. K. Yabe et al., Proceedings for 7th Symposium on ULSI Ultra Clean Technology, Tokyo 1988.

2 Cleaning Methods

D. Synthetic Resin Valves

Tomoyuki Ueda
Asahi Yukizai Kogyo Co., Ltd., Miyazaki, Japan

I. INTRODUCTION

The design of semiconductor devices is further refined year by year. Integration has exceeded 1 Mbit and the submicrometer processing level has been reached. As the level of integration increases, reliability and performance are required. It is generally said that particles measuring 0.1 of the design rule should be eliminated. In most manufacturing steps, if not all, close attention must be paid to particles and other contaminants. Synthetic resin valves mounted on the ultrapure water piping system should be free of oil and contamination.

II. MATERIALS FOR SYNTHETIC RESIN VALVES

Various materials are used for the synthetic resin valves mounted on the ultrapure water production system. Polyvinyl chloride (PVC) is the most popular material for pipelines at room temperature. PVC can sometimes decrease the resistivity of the ultrapure water, and it is not the best material. PVC can be easily processed, however, and it exhibits excellent mechanical strength and is inexpensive.

Polyvinylidene fluoride (PVDF) is the most widely employed synthetic resin material in the wide temperature range from low to high (120°C). Compared with PVC, metallic ion elution from PVDF is minute, which is important for the piping material used for ultrapure water.

The highest quality material used in the ultrapure water production systems is polyether ether ketone (PEEK) resin. It exhibits excellent heat resistance (withstanding temperatures of up to 150°C), and the surface of the molded product is very smooth. In addition, PEEK resin demonstrates much lower total organic carbon (TOC) elution than other engineering plastics. In general, metallic ion elution from PEEK resin is less than

0.05 that from PVDF resin, and TOC elution is less than one-third. For other components, C-PVC and polypropylene (PP) are employed.

III.   TYPES OF SYNTHETIC RESIN VALVES

Although there are many types of synthetic resin valves, it is important to select valves with the least dead space where the liquid can be trapped. The ball valve is widely used. This is a straight and full flow type with a small resistance against fluid, allowing a very high flow rate. It is also a blocking type and is easily cleaned because it can be easily mounted and removed from the valve body.

Another type is the diaphragm valve. This does not have entrapment space, so no bacteria are generated. This is the optimum valve for an ultrapure water piping system.

Other valves include the butterfly valve and the gate valve.

IV.   CLEANING FACILITIES

A.   Clean Room

The clean room is defined as a space where the number of airborne particles is controlled at a predetermined level and the environmental conditions, such as temperature, humidity, and pressure, are controlled if necessary. The cleaning of synthetic resin valves is conducted in a clean room meeting these requirements.

According to the FS 209B standard, the clean room is controlled at Class 1000 or less in terms of airborne particles > 0.5 μm. The clean room entrance for personnel is completely separated from the entrance for goods. People should enter through the air shower room. Goods should be transported through the pass box.

B.   Clean Room Personnel

All the people entering the clean room should have the proper attitude to achieve and maintain the high cleanliness level of the room. The minimum number of personnel approved to work in the clean room should be allowed to enter the clean room. The following personnel should not enter the clean room:

1.   Those who are not approved to work in the clean room, that is, personnel who are not granted approval in advance from the chief manager of the clean room.
2.   Those who do not wear clean room clothing using the proper procedures.
3.   Those who exercised strenuously during a break and perspired.
4.   Those who smoked or ate within the previous 30 minutes.
5.   Those who have a cold or a cough.
6.   Those who wear cosmetics, including nail polish.

C.   Clean Room Clothing

Clean room clothing should allow people to move around easily and should prevent hair and skin from falling off. The overall is desirable (dust-free cap, dust-free gloves, dust-free overall, dust-free shoes, and dust-free mask). Garments should also be cleaned by a laundry that specializes in the cleaning of clean room garments.

D.   Pure Water for Cleaning in the Clean Room

Pure water should be kept overflowing during cleaning. A 0.1 μm filter should be used to maintain the resistivity of the pure water at 5 MΩ·cm or higher.

E.   Cleaning Solution

Scat 20X-PF (manufactured by Daiichi Kogyo Pharmaceutical) is widely used as the cleaning solution for synthetic resin valves mounted on ultrapure water piping systems in Japan. The major constituents of Scat 20X-PF are a phosphorus-free builder, a special anionic surfactant, and a non-ionic surfactant. It is very effective in cleaning various contaminants that adhere to plastic, glass, and metal.

For normal cleaning, a 2–5% solution is used. A 10–20% solution is used for additional cleaning. The cleaning time is 2–6 h in general. Ultrasonic cleaning takes as little as 5–15 minutes.

V.   CLEANING PROCESS

Figure 1 shows the cleaning procedure: (1) clean jigs; (2) prepare valve components to be cleaned; (3) conduct precleaning; (4) conduct cleaning with cleaning solution; (5) conduct prerinsing with pure water; (6) conduct rinsing with pure water; (7) dehydrate the cleaned components; (8) dry the cleaned components; (9) assemble valves; (10) conduct quality inspection; (11) label valves; (12) seal valves; and (13) pack valves. All the procedures in Fig. 1 should be conducted in the clean room. Steps 1–5 and 13 require Class 5000 or less; steps 6–12 require Class 1000 or less.

VI.   DETAILS OF CLEANING

A.   Jigs

All the jigs for assembly, tests, and inspections and the containers used in the clean room should be cleaned before they are carried into the clean room. Jigs and containers should be made of rust-free materials. In the cleaning process, jigs are first immersed into a solution of Scat 20X-PF diluted with ultrapure water to 3% for over 2 h. They should then be washed with a physicochemical brush. The cleaned jigs should be immersed in pure water for more than 2 h and then brushed again. In the meantime, pure water should be kept overflowing from the bath.

B.   Preparation of Valve Components

The components of synthetic resin valves for the ultrapure water piping system should be prepared. For the diaphragm valve, for example, all components—body, bonnet, compressor, sleeve, spindle, handle, diaphragm, bolts, nuts, washers, gasket, and gage cover—should be prepared.

C.   Precleaning of Valve Components

The valve components are rinsed with industrial water (with a resistivity of 1600 Ω·cm or higher) to remove large foreign materials and contaminants. The rinsed components are placed on a stainless steel net.

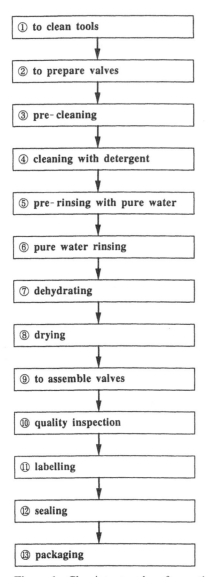

Figure 1 Cleaning procedure for synthetic resin valves.

D.   Cleaning of Valve Components with Cleaning Solution

All the prerinsed components are immersed in the 3% Scat 20X-PF solution for more than 3 h. They are then brushed. The liquid-contacting surface should be carefully cleaned. In particular, the parts of body that cannot be brushed should be cleaned with gauze for inorganic carbon (IC). The gage cover should be cleaned carefully as well, although it is not a liquid-contacting area. The cleaned components are placed on the stainless steel net.

The 3% Scat solution should be checked by pH measurement. When the solution is dirty, it should be replaced. When employing the ultrasonic cleaning method, the immersion time is about 10 minutes. Too long a washing may lead to surface roughness of the synthetic resin valve.

E. Prerinsing of Valve Components with Pure Water

To obtain a resistivity of 5 MΩ·cm as soon as possible after pure water rinsing, the cleaned valve components are rinsed in the prerinsing bath. Prerinsing takes about 10 minutes. Pure water with a resistivity of 5 MΩ·cm or higher should be used. The prerinsed components are to placed on the stainless steel net.

F. Pure Water Rinsing of Valve Components

The prerinsed components are immersed in pure water for more than 2 h. They are then brushed carefully. In particular, the liquid-contacting surface of the body and diaphragm should be rinsed well.

G. Dehydration of Valve Components

The components placed on the stainless steel net are dehydrated while being blown with nitrogen gas. When they are naturally dehydrated, water spots remain on the surface, and these are subject to the dust adhesion as a result of the long exposure to the ambient air. High-purity nitrogen gas blown through a 0.01 μm filter should be used.

H. Drying of Valve Components

After the nitrogen gas blowing, the components are loaded on the hot-air drier. The drier is made of stainless steel. The hot air is heated to 55 ± 2°C after passing through the 0.01 μm filter. The hot air is circulated. The humidity at the exhaust should be kept at 5% or lower. Drying takes 1 h.

I. Assembly of Valves

The dried components are arranged in the order of assembly. Valves are assembled according to the assembly procedures. O rings and gaskets to be mounted on the sliding part and sealing part should not be greased. The rubber for the ball valve should be wet with pure water with a resistivity of 5 MΩ·cm or higher before the union component with the O ring is inserted into the body to prevent the rubber from being damaged.

J. Quality Inspection

Visual inspection, operation inspection, and pressure resistance inspection are conducted. The light level in the inspection room should be kept at 1000 lx or higher.

1. Visual Inspection

A 100% visual inspection, is conducted. The surface and liquid-contacting area are wiped with gauze for IC. No contamination should be detected during the macroscopic inspection. Rubber products should be wiped gently before being checked.

For the sampling inspection, three products are taken from a single lot (by type, by size, and by total volume to be inspected in a day) and wiped with gauze for IC. No oil film should be detected when they are immersed in 80°C water. This is also a macroscopic inspection.

2. Operation Inspection

A 100% operation inspection is carried out. The open-and-close operation of the handle must meet specifications. Since the valves are processed with by complete degreasing, the open-and-close operation force needed is 20–30% larger than that of most valves on the market.

3. Pressure Resistance Inspection

A 100% pressure resistance inspection is carried out. High-purity compressed nitrogen gas through a 0.01 μm filter is used in this inspection. The inspection is divided into the valve housing pressure resistance inspection and the valve seat pressure resistance inspection.

In the valve housing pressure resistance inspection, the valve is opened and nitrogen gas at 6 kgf cm^{-2} is supplied from one side of the valve for 1 minute. The indication of the pressure gage should not be lowered at all. In the valve seat pressure resistance inspection, the valve is closed and nitrogen gas at 6 kgf cm^{-2} is supplied for 1 minute. During the test, the indication of the pressure gage should not be lowered at all.

K.   Labeling

The valves that pass the quality inspection should be labeled "degreased" on the inspected part. The tag inspection clearing is also placed on the valve handle. Synthetic resin valves labeled "degreased" must be used for ultrapure water piping systems.

L.   Sealing

After labeling, each valve is sealed in a polyethylene bag. To keep valves free of contamination before they are installed, a polyethylene bag that can be sealed must always be used.

M.   Packing

Each sealed product is placed in a carton. The carton is sealed and tied with a PP band. It should bear a stamp, "degreased product."

2 Cleaning Methods

E. Electropolished Pipe

Shigeharu Nakamura
Kobe Steel Ltd., Shimonoseki, Japan

I. INTRODUCTION

In the semiconductor production process, various high-purity gases, including inert gases, highly reactive gases, and toxic gases, are used. The distribution piping system for high-purity gas must be contamination free: no particles or deposited materials are allowed. To meet this requirement, electropolished SUS 316 L, with extra smoothness on the inner surface and high cleanliness, is used. To realize contamination-free electropolished pipe, in addition to the manufacturing technology, from melting to electropolishing to yield the extremely smooth surface, the technologies for cleaning and packing following electropolishing are very important. This part describes these technologies.

II. CLEANING AND PACKING PROCEDURES

Figure 1 lists the procedures for the cleaning and packing process. After electropolishing, water rinsing, high-pressure jet cleaning, and hot pure water rinsing are carried out to completely eliminate residual electropolishing solution. After this step, hot ultrapure water rinsing is conducted in a stainless steel automatic unit installed in a Class 1000 clean room equipped with a demineralizing filter in an effort to remove alkali metals from the air. Hot ultrapure water at 80°C flows through the pipe and is released under pressure to the outside. The pipes are then dried by blowing nitrogen gas through them. The optimum volume of ultrapure water to be supplied and the optimum time for nitrogen gas blowing should be determined so that the operation can be automated. Nitrogen gas is then purged from the pipes, and both ends of the pipes are capped with caps rinsed with ultrapure water.

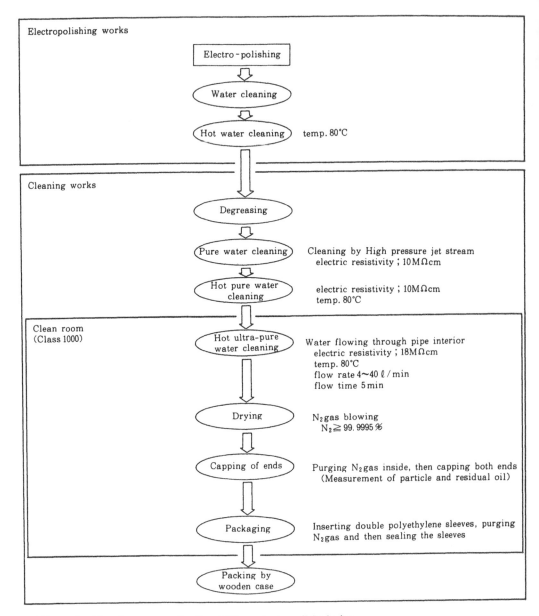

Figure 1   Cleaning process for Excel Clean electropolished pipe.

Pipes are packed in double polyethylene sleeves and stored until immediately before welding to prevent contamination. Nitrogen gas is imprisoned in each sleeve, and the bags are then sealed with heat. The outer sleeve is removed immediately before the pipes are carried into the clean installation spot. The inner sleeve should be removed immediately before welding.

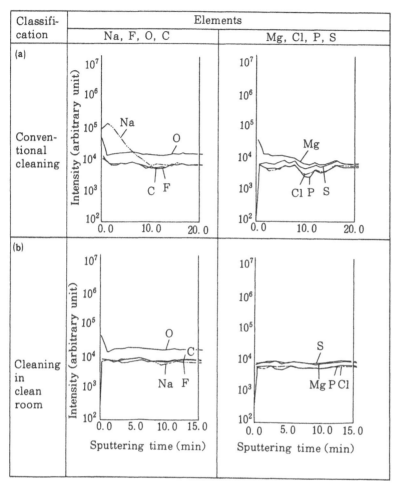

Figure 2   In-depth profile of elements deposited on the inner surface of EP pipe (IMA).

III.   ANALYSIS OF MATERIALS ADHERING TO THE INNER SURFACE

Figure 2 shows the result of IMA analysis of the inner surface of the electropolished pipe. Conventional cleaning (cleaning in air, Fig. 2a) removes residue from electropolishing, such as P and S, but cannot remove alkali metals, such as Na. Hot ultrapure water rinsing in the clean room (Fig. 2b) removes alkali metals completely. Moreover, the particulation is extremely low—one 0.1 μm particle per CF.

2 Cleaning Methods

F. Gold-Electropolished Pipe

Katsumi Yamazoe
Shinko Pantec Co., Ltd., Kobe, Japan

I. INTRODUCTION

Gold-electropolishing (gold EP) is a surface treatment technique in which stainless steel is subjected to heat for the formation of a rigid and passivated coating on the surface after electropolishing. This method not only serves to assure a smooth stainless steel surface similar to electropolished pipe but also minimizes the elution of metal ions to the ultrapure water. More precise cleaning than ordinary electropolishing is necessary. The outline of this cleaning method is described later.

II. MATERIAL

The raw pipe for gold EP treatment uses BA pipe or buffed SUS 316 and SUS 316 L pipe taking its corrosion resistance into consideration.

Because electropolishing on equivalent materials may differ because of differences in the fabrication methods, it is necessary to confirm the fabrication method. It is desirable to remove the oil or grease content and buffing chips, for example, that adhere to the inner surface.

A special abrasive may be used to produce a mirrored finish on the outer surface; however, it should be noted that such a method may adversely affect the gold EP treatment.

III. ELECTROPOLISHING

Electropolishing is a chemical cleaning method. The electrolyte used for gold EP employs a strongly acid electrolyte, and the electrochemically finished surface is smooth, even, and glossy. Metal particles and oil and grease are dissolved in the liquid. Accordingly, the surface appears to be chemically clean with little ion elution after electropolishing.

IV.   PRECLEANING

It is necessary to remove electrolyte that adheres to the pipe surface after electropolishing. Wash off the electrolyte using high-pressure pure water of over 10 $M\Omega\cdot cm$.

Because washing with water alone may generate an insoluble coating in water containing metal ions, the pipe is dipped for a fixed amount of time in weak acid diluted with ultrapure water, and a surface suitable for gold EP is formed.

The cleaning is repeated using ultrapure water of 18 $M\Omega\cdot cm$ until the specific gravity remains unchanged before and after cleaning.

The pipe is then naturally dried in the clean room or dried forcibly using high-purity nitrogen gas (over 99.999%).

V.   HEAT TREATMENT

The pipe is sealed in the container in the clean room, and the air within the container is replaced with fresh air or pure oxygen for heat treatment within the furnace to decompose organic materials absorbed on the surface and to form the passivated film.

VI.   POSTCLEANING

The pipe is removed from the container and dipped in weak acid diluted with ultrapure water. This treatment is intended to further dissolve the uneven parts to ensure overall quality and to check the film for corrosion resistance performance.

When no change is found in corrosion resistance performance, the pipe repeatedly washed with ultrapure water of 18 $M\Omega\cdot cm$ until there is no change in the specific resistance of the water.

After washing with water, the pipe is dried naturally within the clean room or forcibly using high-purity nitrogen gas (over 99.999%).

VII.   PACKING

After drying, the air within the pipe is replaced with high-purity nitrogen gas, and the pipe is capped at both ends.

It is then packed in two polyethylene bags filled with high-purity nitrogen gas and shipped with both ends sealed by heat.

2   Cleaning Methods

G.   Oxidation-Passivated Stainless Steel Pipe

Yasuyuki Yagi
Hitachi Plant Engineering & Construction Co., Ltd., Chiba, Japan

Motohiro Okazaki
Toray Industries, Inc., Shiga, Japan

I.   INTRODUCTION

The ultrapure water distribution system installed in semiconductor plants mainly employs such organic materials as polyvinyl chloride (PVC) and polyvinylidene fluoride (PVDF). Metals, in particular stainless steel, have seldom been used because of metal elution and particulation. Electropolishing, however, which finished the inner surface of stainless steel pipes to an excellent smoothness level, as well as oxidation passivation technology for metal surface treatment, have recently been greatly promoted for significantly upgrading the cleanliness of stainless steel pipe. Consequently, stainless steel pipes employing these technologies have been introduced in some production lines.

Because stainless steel pipes are excellent in terms of mechanical strength and abrasion and heat resistance, they are expected to be adopted as ultrapure water piping material in the near future when efficient surface treatment technologies are introduced. This part mainly describes the metal elution characteristics from stainless steel pipe passivated using pure oxygen gas.

II.   MANUFACTURING PROCESS

The oxidation passivation treatment forms an amorphous and Cr-condensed oxide on the stainless steel surface. Figure 1 is an example of the manufacturing process for oxidation-passivated stainless steel pipe [1]. SUS 316 L purified by vacuum deoxidation is frequently employed. In oxidation passivation, the film quality is easily damaged by trace amounts of contaminants on the inner surface of the pipe. This means that oxidation passivation requires the metal surface to be kept at a high level of cleanliness by completely eliminating various impurities that adhere during rolling and electropolishing. In the

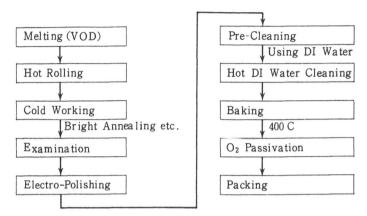

Figure 1 Example manufacturing process for stainless steel pipe.

manufacturing process shown in Fig. 1, hot ultrapure water cleaning and baking to eliminate outgassing are combined to clean and dry the pipe.

It has been reported that 0.2–0.5 μm pinholes exist on the electropolished surface. To reduce the number of pinholes by one order of magnitude, the vacuum double-dissolution method (VIF + VAR) is now being studied [2].

III. METAL ELUTION CHARACTERISTICS OF THE TEST SAMPLE

Table 1 shows the results of a metal elution test conducted on oxidation-passivated stainless steel and electropolished pipe. In this test, the amount of metallic ions eluted from SUS 316 L immersed in imprisoned hot water at 80°C was measured. In the electropolished sample, Fe, Ni, and Mn eluted into water at high concentrations. On the

Table 1 Effects of Immersion Test Using SUS 316 L Electropolished (EP) Pipes and SUS 316 L Electropolished O_2-Passivated Pipes

	IMMERSION CONDITION	PIPE SIZE	DISSOLVED METAL ION CONCENTRATIONS (ppb)								
			Fe	Ni	Cr	Mo	Mn	Na	Si	K	T_1
SUS316L ELECTRIC POLISH	80°C– 33days	7.5∅×353mm	19000	6000	1.1	0.18	820	86	500	12	1.4
SUS316L ELECTRIC POLISH and O_2 PASSIVATION	ROOM– TEMP. –33days	7.5∅×358mm	4.5	2.6	0.5	0.02	11	18	24	9.1	0.10
	80°C– 33days	7.5∅×357mm	16	330	13	0.02	16	33	270	26	0.29
SAMPLE WATER			<0.1	<0.1	<0.02	<1	<0.01	<0.01	<1	–	–

MEASURED BY : ATOMIC ABSORPTION METHOD and ICP– MS METHOD

other hand, metallic ion elution was reduced to 0.01–0.001 in the oxidation-passivated sample.

The amount of metallic ion elution from the oxidation-passivated sample was almost the same between room temperature and high temperature. This means hot water can be supplied through the oxidation-passivated stainless steel pipe.

The excellent characteristics of oxidation-passivated stainless steel pipe for the semiconductor process gas distribution pipe have been published [3]. As mentioned, because oxidation-passivated stainless steel pipe has very limited elution of impurities into ultrapure water even at high temperatures, it can be applied to the ultrapure water distribution system of the semiconductor production line, which requires very high cleanliness levels. Particularly because it can handle hot ultrapure water, it is expected to be widely adopted for the production line because the organic polymer materials mainly used for the ultrapure water distribution system at present cannot deal properly with hot ultrapure water. Further comprehensive study of oxidation-passivated stainless steel pipe is required, including the wafer cleaning characteristics during the wet process.

REFERENCES

1.   K. Fujiwara et al., Application of electropolished stainless steel pipes to semiconductor industry, Kobe-Steel Engineering Reports, *37*(3) (1987).
2.   H. Sato et al., Structure of surface oxide film of electropolished stainless steel and its corrosion resistance, Kobe-Steel Engineering Reports, *39*(1) (1989).
3.   K. Sugiyama et al., High-Performance Technology for ULSI Manufacturing Processes, Ultra Clean Society, (1989), pp. 215–223.

2   Cleaning Methods

H.   Stainless Steel Valves

i.   Common SUS Valves

Mamoru Torii
Fujikin Incorporated, Osaka, Japan

I.   SUBSTANCES TO BE CLEANED FROM THE VALVES

In the process of manufacturing valves, the following materials must be cleaned from the valves:

1. Oils used in machining (oleaginous and water soluble)
2. Oils and alkalis derived from bare hands
3. Acids from electropolishing, chemical polishing, and passivation
4. Burrs and scraps from cutting
5. Dust deposited while handling products in the air
6. Residual solvent from the cleaning process

II.   CLEANING SOLUTION

In general, various organic solvents, pure water, and ultrapure water are mixed to make the cleaning solutions. Organic solvents must be handled carefully to protect the health as well as the environment. In many cases, the cleaning solution is heated and used in the ultrasonic cleaning bath to further enhance the cleaning effect. An ultrasonic frequency of 28–100 kHz is widely used in ultrasonic cleaning. It is reported that as the frequency rises the cavitation effect is degraded, but ultrasound is applied to a wider area of the part being cleaned.

In cleaning with water, hot water at around 80°C demonstrates an excellent cleaning effect. In cleaning with organic solvent, shower cleaning and vapor cleaning following ultrasonic cleaning is effective. In the past, ultrasonic cleaning was used to remove deposited particles, but it is now used to remove residual chemicals. Baking is effective in removing gaseous constituents adsorbed on the surface. Higher temperatures are prefer-

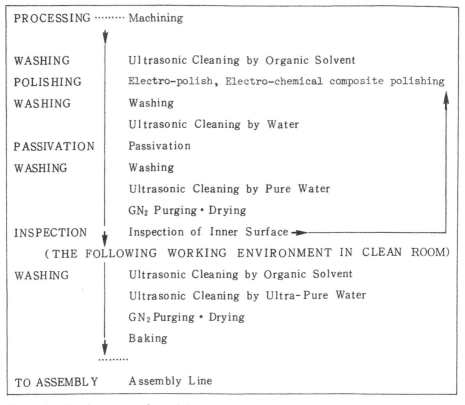

Figure 1   Cleaning process for stainless steel valves.

able in baking, but the optimum temperature should be determined by taking the heat resistance of the valve materials (e.g., fluoroplastics) into consideration.

III.   CLEANING PROCESS

Figure 1 shows a common procedure used to clean SUS valves.

ii.   Metal Diaphragm Valves

Yoh'ichi Kanno
Motoyama Engineering Works, Ltd., Miyagi, Japan

The metal diaphragm valve used for ultrapure water is a modified version of a valve with a metal seat structure developed for gas. It is cleaned using the same procedures as for the metal diaphragm valve for gas.

Table 1 Final Rinse Process

	Process	Solvent	Method	Temperature	Remarks
1	Degreasing	Methylene chloride	Ultra-Sonic	15 ℃	Splash
2	ditto	ditto	Vapor	40 ℃	
3	Water Rinse	Ultra-Pure Water	Ultra-Sonic	60 ℃	Splash
4	ditto	ditto	Dip	Room Temp.	
5	Dehydration	Isopropyl Alcohol	Ultra-Sonic	ditto	Splash
6	ditto	ditto	Dip	ditto	
7	ditto	ditto	Vapor	82 ℃	
8	Dry	Infrared Rays		$\leqq 100$ ℃	

Solvent : All is Semiconductor Use
Ultra-Pure Water : 18 MΩ・cm

The cleaning of components is conducted in the final stage of the manufacturing process to upgrade the cleanliness of the liquid-contacting surface. Harmful contaminants deposited on the surface of components during fabrication or transportation must be removed with appropriate cleaning solutions. Furthermore, the cleaning solution should be completely removed from the component surface.

The method of cleaning components, including the cleaning solution, the cleaning procedure, and the cleaning unit, should be determined based on study of the component materials and the characteristics of contaminants deposited on the component surface.

Semiconductor-grade cleaning solution containing limited amounts of contaminants (metallic ions and particles) must be selected to suppress the contamination caused by the solution. Methylene chloride, ultrapure water, and isophthalic acid (IPA) are usually used as cleaning solutions.

The most advanced cleaning method is described here. The component of the metal diaphragm valve passes through a three-bath cleaning machine employing methylene chloride. The final precision cleaning is conducted in a seven-bath automatic cleaning unit. Clean air filtered with a HEPA filter is introduced into this automatic cleaning machine to maintain the cleanliness level inside the machine. Table 1 lists the procedures for the final precision cleaning.

Because the cleaned components are brought directly into the clean room (0.1 μm, Class 10), recontamination after cleaning is carefully prevented. Operators in the clean room should wear gloves and masks when handling the cleaned components to prevent contamination in the clean room.

2 Cleaning Methods

I. Stainless Steel Pumps

i. Ultraclean Regenerating SUS Pumps

Rokuheiji Satoh
Nikuni Machinery Industry Co., Ltd., Kanagawa, Japan

I. PERFORMANCE

There are several ultraclean pumps: one for high-pressure cleaning with a waterhead pressure of 300 m, one for midpressure cleaning with a pressure of 100 m, a pump for the filter, and a pump for 40 m transportation. The pump can be designed so that the volume varies from a few liters to 300 L minute^{-1}. Figures 1 and 2 show the exterior dimensions and the appearance of the ultraclean pump. Figure 3 shows the major performance characteristics of the ultraclean pump.

Table 1 lists fluctuations in resistivity when ultrapure water with a resistivity of 18.2 MΩ·cm is fed to the ultraclean pump at Tohoku University. Table 2 compares the particulation from the ultraclean regenerating pump and the improved canned motor pump with the reduced contamination achieved by mounting the drain. This measurement was performed at Tohoku University.

The major characteristics of the ultraclean pump are as follows:

1. The resistivity is not reduced. The resistivity of ultrapure water measured at the pump inlet was 18.25 MΩ·cm, close to that of theoretically pure water. At the outlet, it is kept at 18.25 MΩ·cm.
2. Start-up time. It takes as little as 60 minutes to achieve these resistivity conditions. This is a great difference from conventional systems, whose start-up time requires a few weeks.
3. Particulation. The particle count of the ultraclean pump is $1/15$ that of the pump for super-ultrapure water measured at Tohoku University.

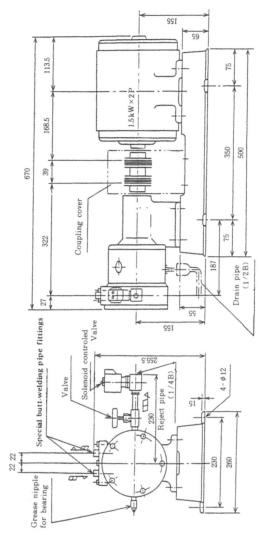

Figure 1   Exterior dimensions of the regenerating pump.

Figure 2   An ultraclean pump.

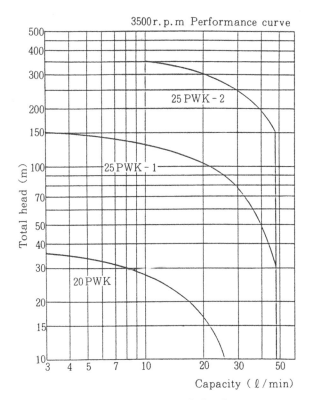

Figure 3   Standard performance of ultraclean pumps.

Table 1   Changes in the Resistivity of Ultrapure Water Before and After the Pump is in Service

		Resistivity at pump outlet (MΩ • cm)	Capacity (ℓ/min)
Pump inlet		18. 20	3. 2
Pre-washing		18. 11	0. 5
Elapsed time after service	1 min	17. 35	3. 2
	5 min	18. 13	3. 2
	10 min	18. 15	3. 2
	15 min	18. 16	3. 2
	30 min	18. 19	3. 2
	60 min	18. 20	3. 2

II.   CONFIGURATION

By using rolled SUS 316 L for the liquid-contacting surface of the pump, particulates from pits have been reduced. The liquid-contacting surface is treated by precision electrochemical buffing and finished with a mirrored surface. An advanced SiC/SiC mechanical seal is used, which features high abrasion resistance. The regenerating pump structure is small, and it is easy to achieve a high lift. In particular, the area of the liquid-contacting surface is reduced. All packings for the pump and packings between the pipe and the liquid-contacting surface use Ni-plated metal rings. The drain is mounted to prevent particulation from the mechanical seal, which is a sliding part.

III.   ADVANTAGES OF THE STAINLESS STEEL REGENERATING PUMP

The regenerating pump is suitable for small volumes and high lifts. Compared with the centrifugal pump and other pumps, the pump can be designed to be very small. Figure 4 compares the performance between the regenerating pump and a centrifugal pump with the same external dimensions. The motor rotation is set at 1750 rpm in both cases. To obtain a pressure of 10 kgf cm^{-2}, the regenerating pump needs only one impeller but the centrifugal pump must either use a multiimpeller system or it must be large.

The area of the liquid-contacting surface in the centrifugal pump is greatly increased, resulting in a greater possibility of contamination. Also, in terms of the impeller structure, the regenerating pump is more suitable for use with ultrapure water than the centrifugal pump because the regenerating pump impeller has its gears exposed completely and it is easy to machine and clean (see Fig. 5).

Table 2   Particle Counts (Number per Milliliter)

Unit (μm)	0. 07∼0. 1	0. 10∼0. 15	0. 15∼0. 20	0. 20 ≦	Total	Reject water
Pump inlet	80	0	0	0	80	
Regenerative pump outlet	81	72	25	7	185	100 mℓ/min
Canned pump outlet	590	543	273	200	1606	1300 mℓ/min

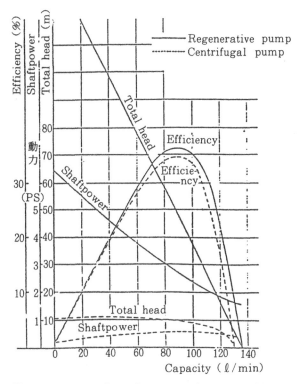

Figure 4   Comparison of a regenerating pump with a centrifugal pump.

On the other hand, the centrifugal pump gears are located inside two disks. This makes the cleaning of contaminants very difficult.

IV.   CLEANLINESS

The pump manufacturing process is as follows:

1. Materials
2. Machining
3. Cleaning 1 (hot water to remove oils)
4. Electrochemical buffing
5. Cleaning 2 (to remove buffing agents)
6. Cleaning 3 (three-bath alcohol cleaning)
7. Assembly
8. Performance test
9. Cleaning 4 (with pure water)
10. Vacuum drying
11. Sealing

Process steps 6–11 are conducted in the clean room to suppress contamination. By carrying out precision cleaning in many steps, particles and oils are completely removed.

The final step is a performance test and final cleaning. The assembled pump is

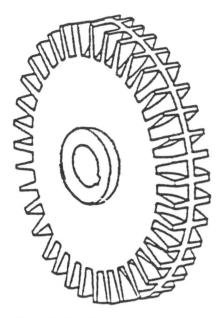

Figure 5 Impeller.

inspected using primary pure water. In this case, pure water plays the role of an excellent detergent to remove inorganic materials. In other words, turbulent flow at sufficient velocity is generated in the pump, which eventually cleans the pump. Following this step, the pump is operated with secondary pure water to clean the inside of the pump. The moisture in the pump is completely dried by vacuum drying to prevent bacterial propagation. Finally, the pump is double sealed.

ii.   Canned Motor Pumps

Kotaro Karita
Teikoku Electric Manufacturing Co., Ltd., Hyogo, Japan

I. INTRODUCTION

It is important that the liquid-contacting surface of the units employed in the ultrapure water production system be clean and that particulation or elution not be allowed. Extra attention is required to the pump, which has a rotating part. This section presents the treatment for the canned motor pump.

II. CANNED MOTOR PUMPS

The canned motor pump is widely used in Japan mainly because it is free of leaks. Lately it has been used increasingly as the pump for the ultrapure water production system.

As shown in Figure 1, the canned motor pump applies various structures according to different applications. For the ultrapure water production system, the basic types (F and FA) and the multistage types (F-M) are often adopted. The standard canned motor pump is equipped with a liquid-contacting surface made of stainless steel. The wearing ring is not placed inside the pump, and the pump is designed so that the impeller rotates with a gap. This structure suppresses the particulation caused by abrasion. In this sense, the canned motor pump is adequate for ultrapure water production.

III. TREATMENT FOR REMOVING OIL

The removal of oil should be thought of not only to remove the oil and contaminants that adhere during component fabrication but also for comprehensive production control and quality control.

Although stainless steel is used in the pump, the cast metal surface is not completely free of contamination: impurities may adhere to the surface and may penetrate the cast metal. Therefore it is necessary to consider comprehensive quality control with regard to the cast metal, including the control of cast sand and the casting method. Furthermore, because visual inspection of the surface is essential, the inner structure of a large pump should take this point into consideration. Some components might be redesigned differently so that cast metal is not used.

Some pumps follow the same process as used for products in general industry. They are assembled first to check the performance. Next they are disassembled to conduct component cleaning, and then they are reassembled for shipping. To enhance the reliability of the product, however, it is better to conduct the final cleaning for the precleaned components, assemble them, perform the pump performance test using pure water followed by the leakage test in the clean room and then ship the pumps.

The following methods are used to remove oil. In practice, these methods are combined.

1. Solvent vapor cleaning is applied to the machined components to remove adhering oil.
2. Acid cleaning is applied to the cast metal components to remove impurities from the surface and to passivate the surface.
3. During ultrasonic cleaning with a cleaning solution, the components are dipped into the cleaning solution in a bath. Foreign materials on the surface or in the components gaps can be removed. This method should be followed by rinsing in running water to eliminate residual solution.
4. Cleaning with detergent and pure water is applied to such components as gaskets and O rings. They are immersed, and the surface is cleaned with a white cloth.

IV. INSPECTION

The inspection to check whether oil is thoroughly eliminated is conducted after the final rinsing step. One of two methods is selected, the hot water method and the white cloth method, depending on the required cleanliness level. Although judgment is diffcult in the two methods, they basically work well if the contamination in each cleaning step and contamination between steps are suppressed completely.

F−TYPE (BASIC TYPE)

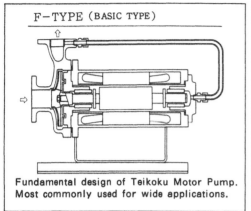

Fundamental design of Teikoku Motor Pump. Most commonly used for wide applications.

FA−TYPE (BASIC TYPE)

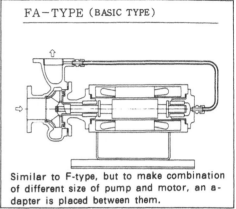

Similar to F-type, but to make combination of different size of pump and motor, an adapter is placed between them.

R−TYPE (REVERSE CIRCULATION TYPE)

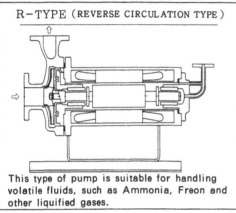

This type of pump is suitable for handling volatile fluids, such as Ammonia, Freon and other liquified gases.

RA−TYPE (REVERSE CIRCULATION TYPE)

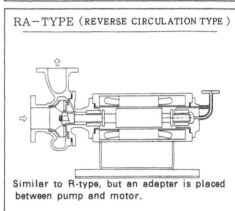

Similar to R-type, but an adapter is placed between pump and motor.

K−S TYPE (FULL−STEAM−JACKET TYPE)

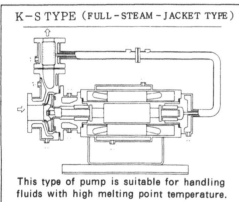

This type of pump is suitable for handling fluids with high melting point temperature.

K−TYPE (FULL−STEAM−JACKET TYPE)

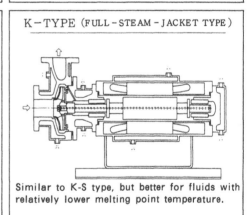

Similar to K-S type, but better for fluids with relatively lower melting point temperature.

Figure 1A   Basic versions of the canned motor pump.

B-TYPE (HIGH-TEMPERATURE
               INSULATION TYPE)

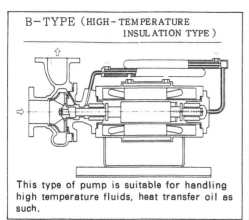

This type of pump is suitable for handling high temperature fluids, heat transfer oil as such.

D-TYPE (SLURRY SEAL TYPE)

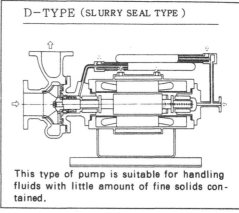

This type of pump is suitable for handling fluids with little amount of fine solids contained.

G-TYPE (SELF-PRIMING TYPE)

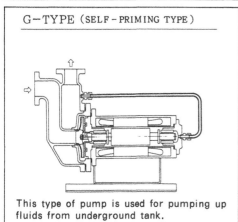

This type of pump is used for pumping up fluids from underground tank.

X-TYPE (GAS-SEALED SLURRY TYPE)

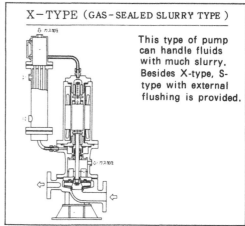

This type of pump can handle fluids with much slurry. Besides X-type, S-type with external flushing is provided.

FA-M TYPE (MULTI-STAGE TYPE)

Higher head, higher efficiency pump. Besides FA-M type, R-M (Reverse Circulation) type and B-M (High Temp-Insulation) type are provided.

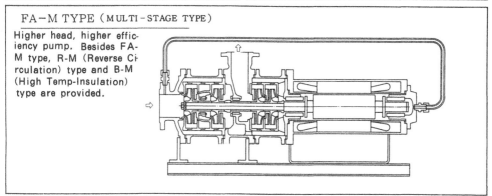

Figure 1B Basic versions of the canned motor pump.

V. FUTURE DEVELOPMENTS

A higher level of cleanliness will be required for units in the ultrapure water production system as well as the ultrapure water itself. Particulation in the canned motor pump, which has not been a problem, will be addressed if the pump is mounted close to the use points.

A clean pump for ultrapure water that does not allow generated particles to move into the process has been developed. It is in actual operation and the evaluation test is complete. This pump is free of cast metal and is designed to have the least amount of liquid holdup space. Also, particles generated in the pump are drained together with a very small amount of water.

2 Cleaning Methods

J. Towers and Tanks

Katsumi Yamazoe
Shinko Pantec Co., Ltd., Kobe, Japan

I. INTRODUCTION

Polyvinyl chloride (PVC) and polypropylene are among the resins often used for materials of towers and tanks in the ultrapure water system. High-grade materials, such as polyvinylidene fluoride (PVDF), perfluoroalkoxyl (PFA), and gold-electropolished (EP) stainless steel (passivation of stainless steel surface), are also used at present. To utilize such materials for ultrapure water, it will be necessary to satisfy the following requirements:

1. Limited elution of metal ions, total organic carbon (TOC), and other contaminants.
2. A smooth inner surface that does not absorb impurities or encourage the proliferation of microbes.
3. Easy machining or connection
4. Provided with necessary mechanical properties
5. Excellent heat resistance
6. Economical

Taking these points into consideration, different materials are compared in Table 1. When selecting these materials, it is necessary to fully examine the grade, heat resistance, and mechanical strength for use in the ultrapure water system.

Particularly when using large-capacity towers and tanks, it is necessary to fully control quality during the production process, because material of uniform quality will be the essential factor.

II. ORGANIC MATERIAL

Organic materials normally include polymers produced by adding stabilizer or pigment, for example, to the monomer for the polymerization reaction. The elution of such

Table 1 Properties of Materials Used in the Ultrapure Water System

	PVC	Clean PVC	PP	PVDF	Ti	Electropolished stainless steel
Mechanical strength	○	○	○	○	◎	◎
Heat resistance	×	×	○	○	◎	◎
Impurity elusion	△	○	○	○	◎	◎
Surface roughness	△	○	○	○	○	◎
Workability	◎	◎	○	○	×	○
Cost	◎	◎	○	×	×	△

additives into water differs greatly according to the method used for production and mixing. The elution of additives from organic materials into water is generally thought to cause insolubles to remain throughout such processes as permeation of water into material, dissolution, and elution from material.

Recently, a solution to such problems made it possible to produce material that could reduce the elution of TOC and impurities to such an extent as not to be practically harmful. To use these organic materials, however, it is necessary to rinse off the impurities before use. This rinsing requires caution, because even equivalent materials have different elution characteristics. The rinsing method is generally determined by the manufacturer. Figure 1 is a graph showing the amount of TOC eluted when the epoxy resin used for fiber-reinforced plastic (FRP) is dipped in pure water. As evident from the graph, the amount of FRP that elutes is not increased even if the dipping time is increased,

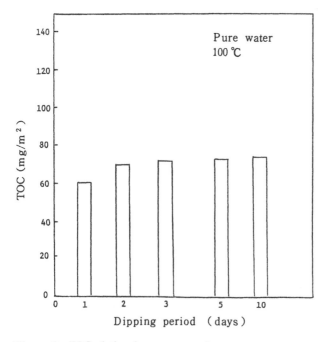

Figure 1 TOC elution from epoxy resin.

and the elution is thought to occur only at the initial stage. Such a tendency is commonly seen with the other organic materials.

If a rinsing method is used in which the material is dipped in high-temperature ultrapure water in advance to fully draw out the eluting component, it is possible to reduce the TOC when the material is in actual use.

III.   STAINLESS STEEL

Although stainless steel provides mechanical strength when fabricating large-scale towers and tanks and heat resistance when performing thermal sterilization and it is possible to execute butt welding, which does not create fine holes where microbes can live, it is not normally used for ultrapure water towers and tanks because there is concern that metal ions will be eluted in the pure water as a result of fine metal corrosion. However, stainless steel has recently been applied to ultrapure water systems because such techniques as electropolishing (EP) and gold EP were developed, which made it possible to stabilize the stainless steel surface.

Electropolishing is a treatment in which the metal surface is polished electrochemically. It is a surface treatment technique that ensures a smoothness of less than 0.01 μm by its unique electrolytic film and further creates a Cr-enriched rigid passivated film.

On the other hand, gold EP is a surface treatment technique intended to minimize ion elution by heat treatment on the electropolished stainless steel to remove the organic material that adheres to the surface and by forming a passivated film.

Figure 2 shows the data for corrosion reduction in high-temperature ultrapure water,

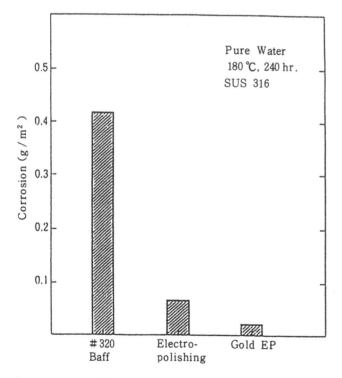

Figure 2   Corrosion resistance by different surface treatments. (From Ref. 1.)

demonstrating that the electropolishing and gold EP methods are excellent techniques for the ultrapure water system.

Ordinary buffing creates many active spots where ion elution may easily occur as a result of fines and metal particles, for example, resulting from the buffing mesh. In addition, the contamination due to abrasives is inevitable in specular glossing.

As a result, it is recommended that stainless steel over grade SUS 316 and SUS 316L be used for ultrapure water production, taking its corrosion resistance into consideration and after performing surface treatment, such as electropolishing or gold EP. The following describes stainless steel subjected to electropolishing and gold EP as a rinsing method for stainless steel towers and tanks.

A.   Electropolishing

When electropolishing, the condition of the surface finish may be different according to differences in the method of fabricating the stainless steel. Therefore, casting of uneven quality should not be used.

The raw material is generally buffed. This removes scars, grease, and mill scale. For simple electropolishing, #400 buffer or greater is generally used.

The buffing is performed beginning with the coarse buffing grade and continuing in sequence without neglecting any mesh grade. Simple, cylindrical constructions steel without concave parts is ideal from the viewpoints of electropolishing and rinsing.

Electropolishing is a rinsing technique intended to smooth the material surface and to remove impurities, to create an extremely clean surface. Rinsing after the electropolishing is complete is essential.

The electrolyte contains a large amount of metal ion that may elute, and it segregates metallic salt sludge. Some strongly acid and base electrolytes are highly viscous and may be difficult to rinse off completely using ordinary rinsing methods. The high-pressure pure water at several dozens of kg cm^{-2} is generally used to rinse the liquid or sludge that adheres to the steel.

It is necessary to rinse concave parts very thoroughly because the rinsing liquid may be easily deposited in such parts.

Various methods may be used to clean large-capacity towers and tanks, depending on the shape. An insoluble thin film may adhere to the stainless steel and is not fully rinsed off with pure water, depending on the electrolyte used.

Stainless steel in complicated shapes is more easily joined with adhesives, and a small amount of sludge may remain. In such a case, the stainless steel is dipped in weak acid diluted with pure water to dissolve the impurities on the surface. It is then rinsed in the clean room or a sealed system using high-pressure ultrapure water (18 MΩ·cm). This process is repeated until the specific resistance of ultrapure water is equal before and after the rinsing.

The towers and tanks are then closed, and the air within the towers and tanks is replaced with high-pressure nitrogen gas and then dried.

Stainless steel subjected to this process is packed and shipped.

B.   Gold Electropolishing

Since heat treatment is executed after electropolishing with a strongly acid electrolyte, strict care must be taken during the rinsing process. Bear in mind that incomplete rinsing after electropolishing may cause impurities to affect the gold EP film.

Various modifications are needed for large-scale, complicated towers. Rinsing is

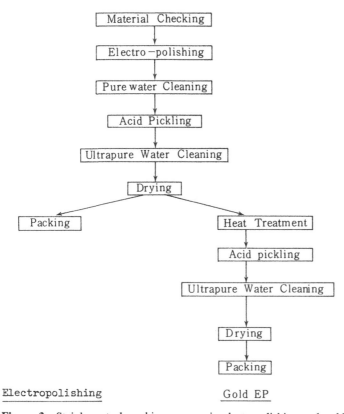

Figure 3   Stainless steel washing process in electropolishing and gold electropolishing.

performed before heat treatment in the same way as after electropolishing. To ensure a uniform quality of the gold EP film, however, it is absolutely necessary to perform pickling after high-pressure ultrapure water rinsing. Heat treatment is performed to remove the organic impurities absorbed on the surface and to further harden the passivated film. It is desirable to perform uniform heating for heat treatment and to practice strict temperature control. The heating must be performed gradually, because the towers and tanks are constructed of thick plates and their shape is complicated.

Heat treatment is performed within a sealed purified air or pure oxygen system. After heat treatment, the material is dipped for a fixed amount of time using a weak acid prepared by diluting with ultrapure water. This process also dissolves the uneven portions of the surface for stabilization after heat treatment and checks whether the gold film is seated firmly.

The stainless steel is rinsed with ultrapure water of 18 MΩ·cm at the final stage until the specific conductivity is virtually the same before and after the treatment. It is then dried with high-purity nitrogen gas for shipment. To prevent contamination by air, it is necessary to perform the final process within the clean room or in a sealed system. The stainless steel electropolishing and gold EP rinsing process is shown in Fig. 3.

REFERENCE

1. K. Yamazoe, Kinzoku, *58*(10), 101 (1988).

2   Cleaning Methods

K.   Gages and Meters

i.   Pressure Gages

Shigenori Hokari
Nagano Keiki Seisakusho Co., Ltd., Ueda, Japan

In general, the Bourdon pressure gage (JIS B7505) is widely used to measure pressure. However, the inside of the Bourdon pressure gage is difficult to clean and microorganisms propagated easily in the large dead zone where water is trapped.

The diaphragm seal pressure gage is more suitable for use in pure water. The process medium is separated from the Bourdon tube with a diaphragm, and the pressure transmission liquid fills the diaphragm seal pressure gage. Figure 1 shows the diaphragm seal

Figure 1   Diaphragm seal pressure gage (ISO 2852 clamp connection).

Figure 2   Diaphragm seal pressure gage (wetted material, FEP coating).

pressure gage (ISO 2852 clamp connection), which is easily mounted or taken off and cleaned. However, the liquid-contacting surface is made of SUS 316 (JIS).

When the pressure gage must be metal free, the diaphragm seal pressure gage with a liquid-contacting surface treated with a fluorinated ethylene-propylene (FEP) coating is used, as shown in Fig. 2.

Figure 3 shows the dry pressure gage, which does not use liquid at all, but employs a bellows. The entire liquid-contacting surface, including the bellows, is made of polytetrafluoroethylene (PTFE). Each component is precision cleaned. In this way, it is further upgraded from the diaphragm seal pressure gages. The gage shown in Fig. 3 can be also treated by sterilization with ethylene oxide gas. Because it is sterilized and shipped

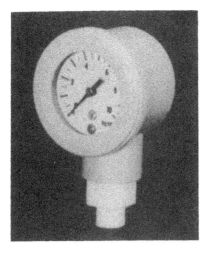

Figure 3   Dry bellows pressure gage (wetted material, PTFE).

sealed in a polyethylene bag, there is no concern that it is contaminated during transportation. Sterilization with ethylene oxide gas is one of the most reliable sterilization methods, along with autoclaving sterilization. It can even kill spores.

ii. Flowmeters

Yoshiaki Hashimoto
Tokyo Keiso Co., Ltd., Kanagawa, Japan

I. INTRODUCTION

Along with advancements in the integration of semiconductor devices, higher water quality is now required in Japan. Flowmeters are used for various applications in the primary and secondary ultrapure water production systems and at points of use. This section presents the flowmeter used for ultrapure water and methods of cleaning them.

II. FLOWMETER FOR ULTRAPURE WATER

The prerequisites for the flowmeters used for ultrapure water are as follows:

1. The materials must cope with the water quality level.
2. The dead zone should be limited and easily cleaned.
3. The flowmeter must withstand hot water sterilization while maintaining function and performance.
4. Sliding parts should be limited and the flowmeter should be free of secondary contamination and particulates.
5. Pressure loss should be limited and to the process load should be suppressed.

The variable area flowmeter is one of the flowmeters that satisfies these requirements. The basic flow rate detector is composed of a couple of taper tubes and a float. The position in which the float is stationary is determined by the flow rate. This is how this flowmeter detects the flow rate value. For the indication of flow rates, two types are available: in the first the calibration is registered directly on the transparent taper tube, which displays the flow rate; in the second a magnet placed in the float conveys the flow rate fluctuation by means of a magnetic coupling. Figures 1 and 2 show the major flowmeters used for ultrapure water.

SUS, polyvinyl chloride (PVC), polyvinylidene fluoride (PVDF), and polytetrafluoroethylene (PTFE) are available materials suitable for ultrapure water. The direct indication flowmeter is made of Pyrex, quartz glass, or perfluoroalkoxyl (PFA).

Another type is the impeller flowmeter. This is designed compact enough to be mounted on the sampling unit. The magnet inserted in the impeller during resin molding works together with the outer coil to obtain a pulse that reflects the flow rate. It is equipped with a contact alarm.

The UL600R ultrasonic flowmeter has no sensor in contact with the fluid. It sends out an ultrasonic signal from outside the pipe and measures the flow rate by detecting the time

Figure 1   Flowmeter for ultrapure water.

gap between the outgoing signal and the returning signal. Although there are limitations to the required straight tube length, this ultrasonic flowmeter is very promising for the Mbit era because it can be installed in the same manner as the pipes.

III.   CLEANING METHOD

Somewhat different cleaning methods are required depending on the material, the shape, and the size. The following are methods of cleaning the major materials. In principle, the cleaning process is conducted in three steps: primary cleaning immediately after molding, secondary cleaning of components, and tertiary cleaning in the clean room immediately before installation.

Figure 2   Flowmeter for ultrapure water.

For SUS,

1. Primary cleaning: remove oils with solvent.
2. Secondary cleaning: remove impurities with acid solution, to stabilize the metal surface, conduct the hot water cleaning (70°C, 70 kgf cm^{-2} G), and dry and seal products.
3. Tertiary cleaning: in the clean room, repeat the ultrasonic and alcohol cleaning steps.

For resin and glass,

1. Primary cleaning: remove oils with alcohol.
2. Secondary cleaning: clean with pure water and neutral detergent.
3. Tertiary cleaning: in the clean room, repeat the ultrasonic and alcohol cleaning steps.

Figure 3 lists one of the major cleaning procedures.

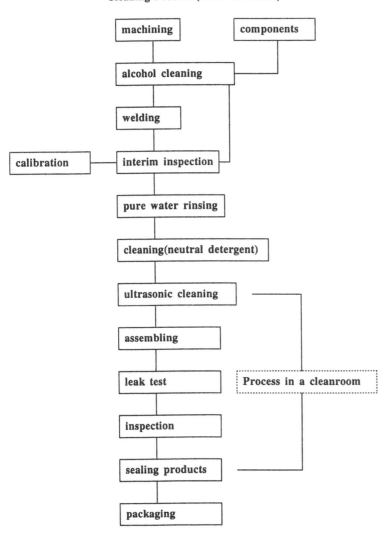

Cleaning Process (Resin materials)

Figure 3   Cleaning process for resins.

3 Fabrication and Construction

A. Polyvinyl Chloride Pipe Installation

Yukio Hamano
Sekisui Chemical Co., Ltd., Shiga, Japan

I. INTRODUCTION

Esloclean pipe for ultrapure water is packed with both ends capped. Polyvinyl chloride (PVC) fittings, 1–10 degreased (oil-cut–treated) fittings per polyethylene bag, are packed in a carton. To obtain the required quality of ultrapure water within a short time, pipes and fittings must be handled with extra attention.

The PVC pipe for ultrapure water can be installed in two ways—by adhesion joining and by fusion, connecting the outer surface of the pipe with the inner surface of the fitting using heat.

II. IMPORTANT POINTS SHARED BY ADHESION AND FUSION JOINING

The important points concerning installation with adhesion and fusion are as follows [1]:

1. Pipes and fittings should be stored indoors. Pipes should be carefully stored so that they are not bent.
2. Operators should wear clean and dust-free clothing.
3. Jigs used for the installation should be free of oil.
4. Operators should wear clean gloves while working.
5. The caps on the pipes should be removed at the very last minute.
6. Only the necessary number of fittings should be removed from the polyethylene bag.
7. Pipes should be cut with a jig that does not generate sludge or swarf. Pipes must be cut at right angles.

8.   Connecting the pipes should be conducted in a clean environment, for example in a clean booth.

III.   ADHESION JOINING

Figure 1 shows the procedure for adhesion joining.

A.   Preparation of Jigs

A pipe cutter that is able to cut pipes at right angles without generating sludge or swarf should be selected. The followings are recommended jigs:

1.   Scissor cutter for cutting pipes with diameter 50 A or less.
2.   Rotating bite cutter for cutting pipes with diameter 40 A–150 A.
3.   Rotary cutter with an edge for plastics only for cutting pipes with diameter 13 A– 150 A.

A chamfering machine is used to chamfer the outer surface of the pipe ends.
Oily marking ink is used to draw the reference line for insertion.
A scale is needed for measuring dimensions.
The Eslon is used for inserting pipes with diameter 150 A or less.
Oil-free gauze is used to clean the pipe surface.
Operators should wear dust-free gloves while handling pipes.
A cleaning agent is used for cleaning the joined area of pipes.
A cloth is used to wipe off the extra adhesive after joining is complete.

B.   Pipe Cutting

The pipe cutter is cleaned. Pipes are cut at right angles to the pipe axis. The roughness of the cut surface is finished with a knife.

C.   Joining

The outer surface of the pipe end is chamfered at level C1–2. The inner surface of the pipe end is chamfered.
The reference line indicating the length of the joined portion of fittings is drawn on the pipes.
The portion of pipes to be joined is cleaned with oil-free gauze.
Adhesive is applied to the portion of pipes and fittings to be joined.
A fitting is inserted at the reference line drawn on the pipe. After insertion, the joined part should be supported for over 1 minute in summer and for over 2 minutes in winter.
Extra adhesive is wiped off.
No extra stress, such as bending, pulling, or twisting, should be applied to the joined part before the adhesive has dried.

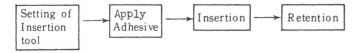

Figure 1   Adhesive joining process.

IV.   FUSION JOINING

Figure 2 lists the procedures for fusion joining.

A.   Preparation of Jigs

The fusion machine should be selected from the following according to the diameter of the pipes.

1.   Type 30 is a manually operated fusion machine for joining pipes with diameter 13A–30A. In principle, two people should operate this machine.
2.   Type 75 is a manually operated fusion machine for joining pipes with diameter 40A–75A. In principle, two people should operate this type.
3.   Type N75 is a hydraulic fusion machine for joining pipes with diameter 13A–75A. In principle, one person should operated this type.

A surface thermometer is used for measuring the temperature of the heater.
A stopwatch measures the fusion time.
A monkey wrench is used to adjust the fusion machine.
A pipe cutter, knife, marking ink, scale, oil-free gauze, gloves, and cleaning agent are used as in the adhesion joining method.

B.   Pipe Cutting

The method is the same as that used in adhesion joining.

C.   Preparation for Joining

Confirm and check the area for the joining operation.
Prepare the power supply. Each fusion machine needs a different power supply:

1.   Types 30, 75, and N75: single-phase, 100 V.
2.   Type 150: three-phase, 200 V.

Install the heater face. The heater face is mounted on the heater. The heater is then turned on to adjust the temperature at 260 ± 10°C.
Install the liner. The required length of liner is mounted on the fusion machine.
Adjust pipes and fittings. Pipes and fittings are loaded on the fusion machine for adjustment.
Finish the cut surface. The roughness on the cut surface is finished.

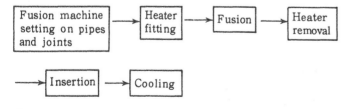

Figure 2   Fusion joining process.

D.   Joining

The portion of the pipes to be joined is cleaned with oil-free gauze. Pipes and fittings are loaded on the fusion machine. The aligning marks of pipes and fittings are confirmed. Pipes and fittings are fused and joined.

1.   The heater is positioned.
2.   Pipes and fittings are placed close to the heater face, and the timer is turned on.
3.   Pipes and fittings are promptly inserted into the heater face to fuse them for the determined time.
4.   Pipes and fittings are removed from the heater face.
5.   The heater is put away.
6.   Pipes and fittings are joined to each other immediately.
7.   The joined part is supported and cooled without stress for more than 2 minutes.
8.   The joined pipes and fittings are unloaded from the fusion machine.

Note: Steps 4–6 should be completed within 5 s.

V.   COMPARISON OF PERFORMANCE BETWEEN ADHESION AND FUSION JOINING

A.   Strength at the Joint

Pipes and fittings are joined to each other using the two methods. Then the joined piece is cut in half along the pipe axis to prepare the test samples. A sample is flattened to the level of half the outer diameter of the pipe with the universal tester. The results is as follows:

1.   Adhesion joining: the sample is cut at the joined part.
2.   Fusion joining: no problem is seen in the joined part.

In the water pressure rupture test, pipes are broken and no problem is seen in any sample. It can be concluded that both joining methods have enough strength to be used in the actual production line. However, the fusion method is superior to the adhesion method in terms of safety.

B.   Resistivity at Start-Up

Esloclean pipe, 13A (length 18.6 m; number of joined parts, 41), is installed and the time to raise the resistivity at the outlet to 18.2 MΩ·cm is measured. (See Ref. 2.) In fusion joining, it takes about 30 minutes; in adhesion joining, it takes about 3 h. There is a great difference between the two methods in terms of the start-up time of water quality. The adhesion method needs a longer time to reach the required resistivity at start-up because organic materials contained in the adhesive are dissolved in the water. The fusion joining method does not use adhesive at all, but heat.

REFERENCES

1.   Sekisui Chemical Co., Ltd., Esloclean Installation Handbook.
2.   N. Mikoshiba and T. Ohmi, Special Issue on Superclean Room Technologies, Annual Report of Laboratory for Microelectronics, Research Institute of Electrical Communication, Tohoku University, 2-6, Ultra pure water Supplying Technology for VLSI, p. 140.

3 Fabrication and Construction

B. Polyvinylidene Fluoride Installation

Hiroto Fujii and Shosuke Ohba
Kubota George Fischer Ltd., Osaka, Japan

Junsuke Kyomen
Kubota Corporation, Osaka, Japan

I. SOCKET FUSION INSTALLATION

Polyvinylidene fluoride (PVDF) pipes and fittings are joined using a fusion method in which the outer surface of pipes and the inner surface of fittings are fused. The fitting for fusion joining must be carefully selected: the thickness of the fitting should be determined based on considerations of wedge stress, the fitting should not have an outer diameter taper, and it must be easily chucked.

Both large (for 10A–100A) and small (for 10A–40A) machines are available for fusion joining. The small machine is used at the site of installation; the large machine is used to produce prefabricated pipes. This part describes installation procedures using the large fusion machine.

A. Preparation of Jigs

Jigs should be kept clean. In particular, oils must be carefully removed.

B. Preparation of Pipes and Fittings

Based on the effective length of pipes and fittings, how long the pipes should be is calculated. The pipes are then cut with a lead pipe cutter. When a scissor vinyl chloride cutter is used, pipes cannot be cut at right angles in some cases. After cutting, the pipe ends are chamfered with the chamfering machine.

C. Preparation of the Fusion Machine

The heating bush, pipe clamp, and fitting clamp of appropriate size are mounted on the heater (see Fig. 1). By adjusting the thermostat, the temperature is set at 250–270°C. Next the pipe and fitting are inserted in the heating bush and the length is adjusted with a stopper. It should be also confirmed whether the axial center matches between the pipe and the fitting.

D.   Cleaning of Articles to be Fused and Heating Bush

The surface to be joined is cleaned with cleaning agent (e.g., acetone) and cleaning paper. The heating bush should be checked after every fusion operation so that no residue adheres to it. The coating should be carefully handled so that it is not damaged.

E.   Joining

By rotating the transfer handle, the pipe and fitting are inserted in the heating bush. When the insertion is complete, the heating time begins to be measured. After heating for the determined time, the pipe and fitting are promptly removed from the heating bush, the heater is carefully lifted, and then the pipe is inserted into the fitting until the stopper hits it. After joining, the joined part is supported and cooled longer than the heating time.

II.   EXPANSION

Because PVDF, just like other plastics, is expanded by heat, the thermal expansion problem should be addressed when a long straight pipeline is planned.

The coefficient of linear expansion a of PVDF is 0.12 mm $(m \cdot {}^\circ C)^{-1}$.

Figure 1   Bench fusion machine.

$$\Delta l \text{ (mm)} = L \text{ (m) } \Delta T \text{ (°C) } a \text{ [mm (m·°C)}^{-1}]$$

$$a_{\text{PVDF}} = 0.12 \text{ mm (m·°C)}^{-1}$$

where: Δl = length of expansion
$\quad\quad L$ = length of pipeline
$\quad\quad \Delta T$ = pipe temperature difference between installation and during operation

For a 10 m straight pipeline, for example, the length of expansion Δl is

$$\Delta l = 10 \times 70 \times 0.12 = 84 \text{ mm}$$

To absorb this expansion length, a bent pipeline like that shown in Fig. 2 (generally called an elbow) should be used.

In this example, an elbow of about 110 cm is required, assuming the pipe size is 50A. If there is not enough space for the elbow, expandable fittings should be employed. The material of the expandable fitting must have the same chemical resistance and the purity as PVDF.

III. SPACE BETWEEN BRACKETS

The space between the brackets for PVDF pipe is determined by the mean temperature of the pipe surface, the density of the fluid, and the outer diameter and thickness of the pipe. Table 1 lists the distances between the brackets. (The maximum acceptable deflection is assumed to be 0.25 cm.)

IV. BEAD AND CREVICE FREE (BCF) FUSION METHOD

Socket fusion and butt fusion are currently used as PVDF joining methods. In both methods, however, the beads remain inside. In this sense, the two methods are not perfect for installing an ultrapure water piping system. This problem was overcome for the first time by the technology developed by George Fischer, Ltd., the BCF fusion method. This method is essential for Mbit devices.

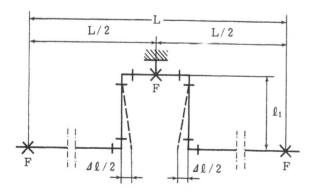

Figure 2 Expansion loop.

Table 1   Distance Between Brackets for SYGEF (PVDF) Pipe

Nominal Diameter	Distance between brackets （cm）						
10	85	80	75	70	65	60	55
15	95	90	80	75	70	65	60
20	100	95	90	85	80	75	70
25	110	100	95	90	85	80	75
32	125	115	110	100	95	90	80
40	140	130	120	115	110	100	95
50	150	140	130	120	115	105	100
60	165	155	140	130	125	115	110
80	180	165	155	145	135	125	120
100	200	185	175	160	155	140	130
Temperature （℃）	20	40	60	80	100	120	140

Spacing based on a fluid specific gravity of 1. 0.

A.   System

(See Fig. 3.)

Pipe: SYGEF (PVDF) HP series pipe
Fitting: a spigot fitting with a positioning shoulder exclusively for the BCF method (the
　　positioning shoulder is used to fix the fitting securely in the clamp)
Valve: diaphragm valve, SYGEF HP series
Fusion machine: a newly developed machine exclusively for the BCF method

B.   Fusion Process

(See Fig. 4). Pipes and fittings are melted by applying heat to the outer surface with a
half-shell heater. At the same time, a bladder inflated with the air supports the portion to
be fused from inside, which prevents the generation of beads. The mechanical strength of
the fused part in withstanding pressure is almost the same as that of the injection-molded
fittings and extrusion-molded pipe.

　　The fusion machine is fully automated. Therefore, perfect fusion can always be
achieved regardless of the abilities of different operators.

C.   Fusion Procedure

The fusion machine is turned on. The temperature lamp and the pressure lamp are turned
on.

　　The clamp and a heater of appropriate size are mounted on the machine. At this point,
the size of articles to be fused is input to the machine.

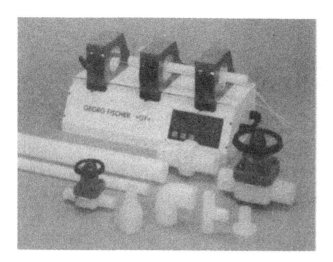

Figure 3 SYGEF HP BCF system.

The pipe and fitting are loaded so that their ends are located on the center of the heater. The gap between the pipe and fitting should be less than 1 mm.

The bladder is mounted directly above the part to be fused.

The clamp and the heater are closed.

The start key is hit. The heater temperature, the bladder pressure, and the time are programmed by size in the machine. Immediately after heating is complete, cooling air begins to blow over the fused part.

When the temperature reaches 70°C, the lamp "complete" light turns on.

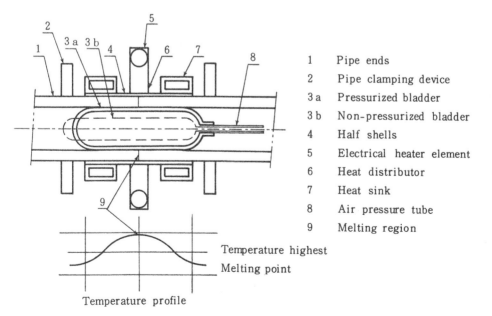

1	Pipe ends
2	Pipe clamping device
3 a	Pressurized bladder
3 b	Non-pressurized bladder
4	Half shells
5	Electrical heater element
6	Heat distributor
7	Heat sink
8	Air pressure tube
9	Melting region

Temperature highest
Melting point

Temperature profile

Figure 4 BCF fusion joining process.

Figure 5 BCF joining cross section.

The fused article is unloaded while opening the clamp and the heater.

Figure 5 shows a sample fused with the BCF method. The BCF method is certain to contribute greatly to the future development of the semiconductor industry by achieving ultrapure water piping systems free of water entrapment.

3 Fabrication and Construction

C. Polyether Ether Ketone

Yoshito Motomura
Kurita Water Industries, Ltd., Tokyo, Japan

I. INTRODUCTION

Polyether ether ketone (PEEK) has impeccable characteristics as a piping material used in semiconductor manufacturing, including light weight, high strength, high degree of accuracy even after injection molding, high heat resistance and chemical resistance, and extremely limited generation of dust and elution of impurities into ultrapure water even after friction. Because of the high melting temperature as a result of its high heat resistance, the butt fusion which is not used for polyvinylidene fluoride (PVDF) and other types of pipe, can be used. PEEK pipe requires special fusing equipment for connection instead of the conventional fusing equipment usually needed.

In this part, PEEK piping techniques are introduced with examples of actual piping executed for the LSI Laboratory of Nippon Telegram & Telephone Co., Ltd. (NTT); Atsugi City, Kanagawa Prefecture, 1987 and the mini superclean room at Tohoku University, Faculty of Engineering (Sendai City, Miyagi Prefecture, 1989).

II. PIPING METHOD

A. Fusing Equipment

Because of the high melting temperature and the application of butt fusion which has not been used for other kinds of pipes, conventional fusing equipment is useless for connecting PEEK pipes. Figure 1 shows the fusing equipment and heater used for PEEK piping work at the site. The fusing equipment is composed of a special heater that melts the connecting end faces of the pipes and a fixture that supports the pipes against their dislocation from alignment during fusing work. The fixture is replaceable depending on the size of pipes.

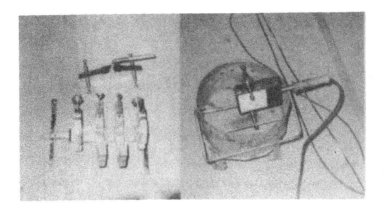

Figure 1   Sealing machine.

B.   Fusion Procedure

The following is the procedure for PEEK piping work [1]:

1.   Cutting of pipes: Cut the pipes using a pipe cutter so that the cut faces are perpendicular. A metal lead pipe cutter is suitable for cutting.
2.   Fixing of pipes: Fix the pipes to the fusing equipment. Replace the fixture according to the pipe diameter.
3.   Heating of cut faces: Check the heater by surface pyrometer or temperature probe to see if it has reached 550°C or above. Set the heater between the end faces of two pipes, and heat them.
4.   Fusion of pipes: Remove the heater, and fuse the end faces quickly upon discoloration (dark brown) and melting. The heating time is approximately 45 s, for instance, with pipes 25 mm in diameter.

Since the wall thickness of PEEK pipe is less than that of other kinds of pipes and it is connected by butt welding, there should not be even a slight dislocation at the contact surface. This is the reason the fixture shown in Fig. 1 is used to support the pipes during fusion. Figures 2 and 3 show the welding at the site. To minimize field work and increase

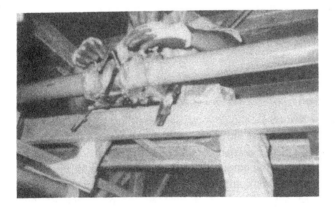

Figure 2   Fusion.

Figure 3 Sealing.

work efficiency at the site, welding is occasionally carried out beforehand outside the site. Figure 4 shows welding in a clean booth away from the site. Figure 5 shows part of a PEEK piping system for the point of use after welding and Fig. 6 shows the internal surface of a welded PEEK piping, demonstrating remarkably neat sealing.

III. STRENGTH OF WELDED POINTS

Despite its small specific gravity and light weight compared with polyvinylidene fluoride and perfluoroalkoxyl piping materials, PEEK is definitely superior in physical strength, including tensile strength and modulus of longitudinal elasticity [1]. It also has a smaller coefficient of linear expansion than other materials so that its expansion by hot water is quite small. To confirm those features, the following tests were conducted on PEEK piping.

Figure 4 Sealing in a clean booth.

Figure 5   The last step in sealing (PEEK piping system).

A.   Hydraulic Destruction Test

Increasing hydraulic pressure was applied to PEEK pipe 600 mm long and 25 mm in diameter with a welded joint in the middle, and the pressure was monitored until destruction of the joint, the results of which are shown in Table 1. The PEEK pipe could withstand a water pressure of approximately 190 kg cm^{-2} at ordinary temperatures, indicating its excellent pressure resistance.

B.   Tensile Strength Test

Increasing load was applied by universal tester to the bottom edge of a PEEK pipe, which had the same specifications as the pipe used in the hydraulic destruction test, and the pipe strength was monitored until destruction, the results of which are shown in Table 2. The PEEK pipe withstood a load of more than 600 kg cm^{-2}, indicating its excellent tensile strength.

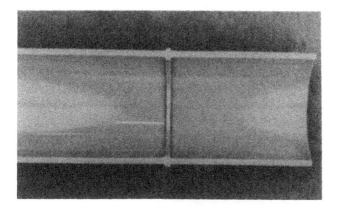

Figure 6   Inner surface and sealing point of a PEEK pipe.

Table 1 Destruction Pressure at the Weld Spot

Pipe	Water Pressure of Destruction (kg/cm²)	Water Temp.
PEEK No. 1	190	17 ~ 18 ℃
PEEK No. 2	187	″
CVP	138	23 ℃
HTCVP	150	″

  Sample
    Pipe size 25 A , L = 600 mm

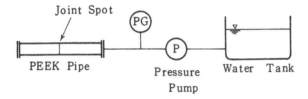

    Joint Spot

  PEEK Pipe   Pressure   Water Tank
                      Pump

Table 2 Breaking Strength of Joined Pipe

Pipe	Breaking Strength (kg/cm²)
PEEK No. 1	646
PEEK No. 2	716
CVP No. 1	429
CVP No. 2	427

Sample
  Pipe size 25 A , L = 600 mm

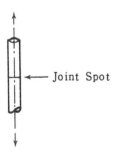

Joint Spot

Table 3   Displacement of PEEK Pipe

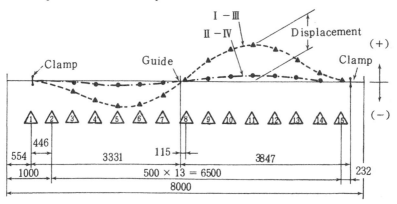

Test \ Point		1	2	3	4	5	6	7
Initial	9 ℃							
I	82 ℃	0	− 10.5	− 41.5	− 77.4	− 105	− 102	− 55
II	9 ℃	− 0.5	− 5	− 10.5	− 11.9	− 15	− 12.5	− 6
III	84 ℃	0	− 11.5	− 42.5	− 77.4	− 105.5	− 99.5	− 56
IV	9 ℃	0	− 6	− 12.5	− 13.9	− 17.5	− 14.5	− 7

Test \ Point		8	9	10	11	12	13	14	15
Initial	9 ℃	0							
I	82 ℃	13.5	83.5	132.5	151.5	131	87.5	38.5	4.5
II	9 ℃	2	12.5	21.5	25	23.5	18.5	10.5	2.5
III	84 ℃	12	82.5	134.5	151	131.5	88	38.5	5.0
IV	9 ℃	2	12.5	22	26	24.5	19	10.5	2.5

（Displacement，mm）

IV. DEFORMATION BY HEATING

To study the thermal deformation of a PEEK piping system, hot water and cool water were alternately fed to a PEEK pipe 8 m long and 25 mm in diameter with a guide in the middle [1–3]. The extent of pipe deformation was monitored at 15 measuring points (indicated by triangles in Table 3) provided between the clamped ends of the pipe, the results of which are shown in Table 3.

Deformation started at the guide as a supporting point in the middle of the pipe, and the maximum displacement reached approximately 150 mm. The test result indicates the necessity of widening the distance between guides if no bend is provided. However, the best solution is to provide a bending section to compensate for the expansion and contraction of the pipe, since the test showed a large deformation in the horizontal direction.

Table 4 shows the results of another deformation test conducted on PEEK piping with a bend in the middle. The pipe is almost free of deformation because of this bend. Several bends are provided in the actual piping system to suppress piping deformation as a result of expansion.

Figure 7 is a flow diagram of an existing distribution system. This distribution system, which is called a reverse return system, was installed under the floor of the clean room in the NTT LSI Laboratory. Several bends are provided along the linear part of the piping. The PVDF piping system has more bends, or the whole system is mounted on a tray without fixing the pipe to it.

Table 4 Displacement of PEEK Pipe

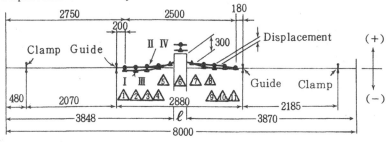

Test	Point	1	2	3	4	5	6	7	8	9	10	11	ℓ
Initial	9 ℃						0						282
I	87 ℃	− 5	− 5.5	− 4	0.5	9.5	15.5	9	2.5	0.	− 0.5	0	258
II	9 ℃	0.5	2	3	5	7	8.5	9.5	8	7.5	4.5	3	282
III	89 ℃	− 5	− 5.5	− 4	0.5	9	16	9	3	− 1	− 2	− 1.5	258
IV	9 ℃	0	2	2.5	4.5	6.5	8	8	7	5	3.5	1	282

(Displacement, mm)

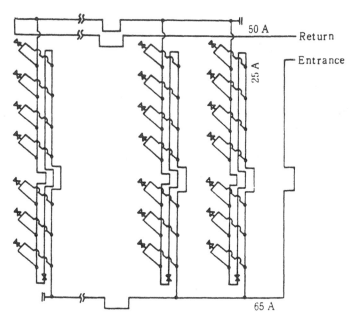

Figure 7   PEEK piping system (reverse return system).

V.   CONCLUSION

The semiconductor manufacturing process has begun to use a rinsing method with hot ultrapure water in addition to conventional rinsing with ultrapure water. Compared with other piping materials, PEEK pipe is a superb material because of its low elution of impurities into ultrapure water, high heat resistance, and long durability. As a piping material for hot water distribution lines and hot water feed lines, for example for a hot water sterilization system, PEEK is expected to become popular in a wide range of fields.

REFERENCES

1. H. Harada, T. Motomura, et al., PEEK Ultrapure Water Supplying System, Proceedings of the Seventh Symposium of ULSI Ultra-Clean Technology, organized by Ultra Clean Society, Tokyo, 1988.
2. T. Motomura, Ultrapure Water Distribution System, Haikan to Souchi, *1* (1989).
3. T Motomura, Designing of Ultrapure Water Piping System and Material Selection, Haikan to Souchi *1* (1987).

3 Fabrication and Construction

D. SUS Piping

Katsumi Yamazoe
Shinko Pantec Co., Ltd., Kobe, Japan

I. INTRODUCTION

A large amount of ultrapure water is used for rinsing chemicals and other contaminants that adhere to the wafer during the large-scale integrated (LSI) water production process, and the quality required for such water has become much more stringent along with increased semiconductor integration. The piping used to supply ultrapure water to each use point will require sophisticated techniques to minimize deterioration in the water quality as well as techniques to produce high-purity ultrapure water.

This part describes stainless steel piping treated with gold electropolishing (EP passivated), which has recently received attention as an ultrapure water piping material.

II. DESIGN MAIN POINTS

A. System Flow Design

The piping flow is designed to allow an even supply of ultrapure water to each necessary use point without accumulation in dead zones. The piping should employ the reverse return system so that the minimum flow velocity is over 0.3 m s^{-1}. Since the design is complicated, the use of computer programming facilitates the work.

B. Considerations for Thermal Sterilization

The thermal expansion coefficients of SUS are smaller by one digit compared with other organic piping, such as polyvinylidene fluoride, and the deformation of SUS piping as a result of thermal sterilization is not as great as in organic piping. However, thermal expansion does in fact occur, and thus the piping route and supports, for example, are designed to relieve stress against such deformation.

Where personnel may come in contact with the pipe, heat insulation is provided to prevent burns.

C. Design for Electropolishing and Gold EP

Electropolishing and gold EP may be accompanied by various restrictions, because these methods are applied to all liquid-contacting parts, including welds. It is therefore necessary to check the following points, described later:

1. Diameter of piping.
2. Length of piping.
3. Shape of piping: Up to one bend is allowable for each L-shaped part. Pipe bent at right angles is recommended.
4. Shape and material of the connections: the connector should be round and simple in shape so that it is easily electropolished. Piping attached using a different metal is not recommended from the viewpoint of treatment. In either case, construction should be simple and free of concave parts. Piping that can be easily electropolished or gold EP treated is suitable for ultrapure water piping.

III. MAIN POINTS FOR EXECUTION OF ULTRAPURE WATER SUS PIPING

Since welded pipe is used in practice for electropolishing and gold EP, the welded surface must also be finished smoothly and cleanly. To ensure a smooth welding surface when electropolishing and gold EP are performed, it is necessary to control the welding.

Since such defects as insufficient penetration, irregular beads, and segregation of carbide Cr may not only adversely affect the electropolishing and gold EP but also result in the elution of impurities into the ultrapure water during operation, it is essential to standardize the welding in detail and to follow such standards faithfully.

A. Welding

Since butt welding using an automatic welder is employed to weld piping that will be subjected to electropolishing and gold EP, the pipe must be of the same material and thickness. These points must be controlled strictly on the basis of material receiving, inspection, and storage standards. It is also important to avoid any defects, such as slag fusion, piping eccentricity, irregular beads, cracks, blowholes, and penetration failures, by using a dedicated cutter or dedicated drill, for example, to cut and bevel the pipe.

If the piping center may deviate when a tack welding or fixing stand is used, such a method should not be used.

It is necessary to use an automatic welder for each size independently, to perform the operation economically.

Note that welding failures or other defects may be generated unless the sealed gas and electrode within the welding head are strictly controlled.

B. Assembling

Pipe that is electropolished and gold EP is assembled using a join, because it cannot be welded in the field.

Even when the pipe is made in a factory atmosphere of approximately 10,000 (normal class), there will not be any problems as long as the material is fully cleaned and the material storage control is satisfactory.

3 Fabrication and Construction

E. Support Structure

Nobuyuki Hirose
Shinko Pantec Co., Ltd., Kobe, Japan

I. METAL SUPPORT

The metal piping support is normally attached to the structure of a building. The following are points to note on the fabrication and installation of the metal support structure.

1. Design the metal support to fully resist the total weight of piping, fluid flowing in the piping, and accessories, such as flanges, valves, and heat-retaining materials.
2. Design the metal support to fully resist vibration and shock. A double nut should be used for each bolt, particularly if there is any possibility of vibration.
3. Construct the metal support so that any piping vibration is not transmitted to the structure.
4. Design the metal support to be fully compatible with the expansion or shrinkage of piping as a result of changes in temperature.
5. Support the piping around the equipment so that the load is not applied to the equipment.
6. Support the valves and flowmeters, for example, where they are arranged.
7. Support all curved, rising, falling, and branching parts of the piping.
8. Maintain a proper support interval against piping deflection.
9. Install the metal support through an insulator when stainless steel piping is support by steel.

II. PIPING SUPPORT INTERVAL

The standard support interval for each pipe size is shown in Table 1. When the support interval is short using small-aperture horizontal piping, such as polyvinyl chloride (PVC), polyether ether ketone (PEEK), or polyvinylidene fluoride (PVDF), the use of continuous support by shaped steel, for example, is more effective and economical.

Table 1 Piping Support Distances

(Unit : m)

Size Material	15A	20A	25A	32A	40A	50A	65A	80A	100A	125A	150A	200A
SUS	2.0	2.0	2.0	2.0	2.0	3.0	3.0	3.0	4.0	4.0	4.0	4.0
PVC, PEEK	1.0	1.0	1.0	1.0	1.5	1.5	2.0	2.0	2.5	2.5	2.5	2.5
PVDF	0.8	0.8	1.0	1.0	1.0	1.3	1.3	1.5	1.5	—	—	—

An example of continuous support is shown in Fig. 1.

III. SUPPORT OF MULTIPLE PIPELINES

When a common metal support is used to support multiple pipelines, the steel support material is determined depending on the operation load. Among the applicable support intervals, the minimum support interval is basically used.

IV. SUPPORT OF EXPANSION AND SHRINKAGE OF PIPING

When supporting piping that may expand or shrink, horizontal piping is supported by an extension hook or support roller, for example, so that the portion that is expanding or shrinking can be relieved. A support example is shown in Fig. 2.

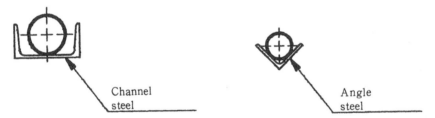

Channel
steel

Angle
steel

Figure 1 Piping supported by shaped steel.

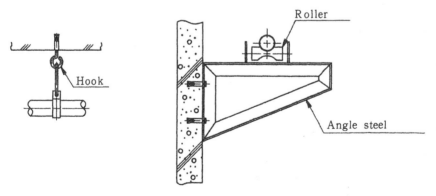

Roller

Hook

Angle steel

Figure 2 Loose support.

For vertical piping, the support point is tightened lightly to allow the piping to expand or shrink freely in the vertical direction.

V.   SUPPORT OF VERTICAL PIPING

Vertical piping may be supported by either the antisway or the fixed method. When only the antisway method is used to support vertical piping, the weight of the vertical pipe is supported by the horizontal pipe. It is therefore necessary to support the point where the horizontal pipe changes to the vertical pipe or the vertical pipe changes to the horizontal pipe.

When the vertical pipe is fixed at its halfway point to support the load, it is fastened with a flat steel band, for example, to support the load using friction [1].

VI.   FIXING AND SUPPORT OF PIPING

When fixing or supporting the piping, it is firmly fixed to the structure at the piping branch point or expansion or shrinkage point, for example, taking the expansion or shrinkage, lateral vibration, deflection, and other factors into consideration. A fixing and supporting example is shown in Fig. 3.

VII.   HANGER BOLT

The allowable support force of a hanger bolt, when this is used for support, is determined by the method of the setting hanger bolt rather than by the allowable tensile force of the hanger bolt itself.

When piping is installed by an insert, hole-in-anchor, or drill anchor, for example, the allowable support force is determined by the strength of concrete or is generally proved to be less than 200 kg at 10 mm, less than 400 kg at 13 mm, and less than 600 kg at 16 mm. When the hanger bolt is welded firmly to reinforcing steel or to a steel frame, the allowable support force is roughly determined by the allowable stress of the hanger bolt [1]. A support example using a hanger bolt is shown in Fig. 4.

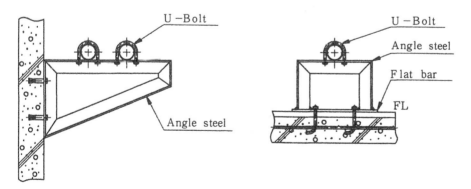

Figure 3   Rigid support.

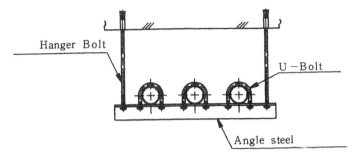

Figure 4   Hanging support.

VIII.   FIXTURES

The fixtures to secure the piping to the piping supports include U bolts, U bands, saddle bands, and pipe clamps.

REFERENCE

1.   Piping Work Installation Technique, Chiiki Kaihatsu Kenkyuusho (Laboratory for Local Development), 1988.

3 Fabrication and Construction

F. Equipment Installation

Nobuyuki Hirose
Shinko Pantec Co., Ltd., Kobe, Japan

I. GENERAL

Since the transportation and installation of the main equipment can greatly influence other work processes, it is necessary to fully examine the shape, weight and time, route, and method of transportation of the equipment in advance, thoroughly coordinate with the building contractor, electrical contractor, and other relevant contractors, set up a detailed plan of execution, and, further, execute the work in accordance with the execution plan.

A. Confirmation of Equipment

Confirmation is made that the workmanship (performance and shape, for example) of the equipment produced within the company is consistent with the design document before the equipment is transported and installed, so that it should not require correction after transportation.

B. Installation Position Requirements

The installation position is the position specified in the execution drawing. Confirmation is made that this installation position allows easy equipment operation, inspection, repair, and so on.

C. Access for Equipment

The passage or opening is examined fully before installation, and measures to prevent damage are taken against the finished portions, such as the surrounding wall and floor.

D. Foundation

The foundation should fully resist the equipment load, and such external forces as earthquakes, and the support surface must be sufficient for installation. The installation surface is finished horizontally, and the equipment anchor bolt holes are drilled correctly.

E. Installation

The equipment is installed firmly and horizontally with equipment characteristics taken into account. When countermeasures against vibration must be taken, depending on the surrounding conditions, vibration-proof materials, such as rubber and springs, are used. In such a case, a stopper, for example, is installed to prevent the lateral deviations due to earthquakes.

   The vibration-proof material must provide excellent noise insulation performance, limited resonance, and high durability, with the natural vibrating frequency, rotating speed, operation load, and other factors of the equipment taken into consideration. An example using vibration-proof material is shown in Fig. 1.

II. BLOWER

The blower level is based on the shaft [1]. If the desired level is not assured when the blower installed temporarily on the foundation is checked with a level, the anchor bolts are tightened evenly while checking for horizontal by inserting liners between the foundation and the common rack. If the anchor bolts are tightened unevenly, the shaft center may deviate, causing the casing bearing stand to be distorted, which may result in vibration and a shortened blower life.

   The points to be noted in driving the blower with a V belt are as follows:

1. Centering between the blower and the motor side pulley is executed by adjusting the irregularity with a ruler or leveling string, for example, applied to the outer surface.
2. The initial tension is adjusted so that the V belt can be twisted approximately 90°, that is, deflected by approximately the thickness of the V belt when the belt is pressed with a finger, or elongated approximately 0.5% when the gage mark is marked at the proper intervals.
3. An ordinary V belt made of rubber is hardly damaged by moisture; however, strict care must be taken with oil and grease, because rubber does not resist such substances. It is important to wipe off dust and dirt.

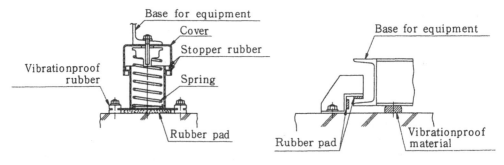

Figure 1   Vibration-proof base.

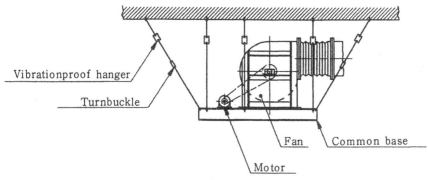

Figure 2   Hung vibration-proof base.

When the blower is coupled directly to the motor, the check is made using a ruler and clearance gage so that the shaft center is horizontal and on a single straight line.

The clearances at the blower, motor, and foundation are adjusted by inserting liners as required.

When a large-capacity blower is installed where vibration or noise may be a problem, vibration-proof rubber or a vibration-proof spring is used as described previously.

When the blower is installed by suspending it from the ceiling, the blower rack is installed with an anchor bolt fixed to the slab-reinforcing steel. Where vibration-proof treatment is necessary, a vibration-proof hanger is used. An example using a vibration-proof hanger is shown in Fig. 2.

III.   PUMPS

The pump is assembled on the common floor base together with the motor and centered at the factory before shipment. However, it is necessary to check for shaft center deviation that may occur during transportation and installation.

In adjusting the shaft center, first the pump and motor are checked for horizontal and then the shaft coupling flange surface for the outer edge and the clearance (see Fig. 3). Normally it is necessary to restrict the outer edge deviation to less than 0.03 mm and the clearance error to less than 0.1 mm.

For a belt pump, the checks are performed in the same manner as for the blower.

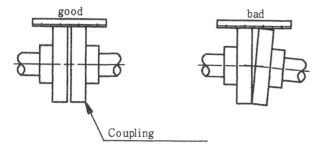

Figure 3   Appropriate setting for coupling.

IV. TANKS

In lifting the tank for transportation, square lumber or a thick plate is attached to the bottom plate of the tank or a cushion is inserted where the wire rope may make contact, so that the tank main unit is not subject to direct force.

The foundation is constructed using a concrete or steel frame. The tank is installed horizontally after checking that the foundation provides sufficient support surface so that the bottom plate is not deformed by weight when the tank is completely filled with water.

REFERENCE

1. Piping Work Installation Technique, Chiiki Kaihatsu Kenkyuusho (Laboratory for Local Development), 1988.

3 Fabrication and Construction

G. Clean Room Piping

Nobuyuki Hirose
Shinko Pantec Co., Ltd., Kobe, Japan

I. SELECTION OF MATERIALS

Materials are selected that satisfy the requirements for pressure proofing, heat proofing, corrosion proofing, sanitation, and so on.

The materials to be used for ultrapure water include clean polyvinyl chloride (PVC), polyvinylidene fluoride (PVDF), polyether ether ketone (PEEK), and stainless steel. When using stainless steel, the materials are subjected to electropolishing or gold EP (oxidation passivation) to reduce the amount of metal ion elution.

II. ARRANGEMENT

The piping is arranged in an orderly fashion not only to satisfy the purpose for which it is designed but also to include the space required for maintenance. It is also necessary to consider that the installed piping should not greatly change the flow of air-conditioning air. Consequently, the piping is installed along the floor, ceiling, and wall surface as much as possible.

III. CUTTING AND CONNECTION

When cutting the piping, the cutting method is chosen to suit the material. A dedicated pipe cutter is always used. A high-speed electrical cutter should be avoided, because it may cause slag to fuse on the inner surface of the pipe.

Heat fusion should be used to connect PVC, PVDF, and PEEK. A bonding agent may be used for PVC, but this depends on the grade of ultrapure water: the use of a bonding agent may be accompanied by total organic carbon (TOC) elution.

Welding is used to connect stainless steel pipe. An automatic welding machine is used to finish the inner welding surface smoothly and cleanly. To finish the inner welding surface to much higher levels, it is necessary to determine the optimum conditions, such as beveling current and voltage, the purity of the sealed gas, and the welding time, in detail.

IV.   DESIGN AND EXECUTION

Minimize the piping dead space. It is desirable to use a three-way valve for the change-over valve, and a spigot valve or a valve with tees, for example, for the main valve at the branch points (Fig. 1).

It is fundamentally important not to remove piping and valves. When they must be removed, make it a rule to minimize the part exposed to the air, to prevent contamination.

The piping diameter is determined so that the flow velocity within the pipe is more than 0.3 m s^{-1} to prevent the bacterial growth within the pipe. Particularly for the subsystem, it should be noted that the flow velocity is decreased toward the terminal.

The reverse return system is used for the use points. The main valve is inserted for each area at the branches leading from the main loop to the sub-loop and to each piece of equipment, so that repairs can be made on each unit. A flow example for the reverse return system is shown in Fig. 2.

Air must not be allowed to deposit in the piping. Air vents are installed at points where it is difficult to discharge air.

Prefabricated piping is used as much as possible to reduce connections in the field.

When connecting piping in the field, it is necessary to determine the piping route to ensure proper work posture.

V.   PRECAUTIONS FOR EXECUTION

Only piping materials and parts that have been washed are used, and they are also washed after completing prefabrication of the piping. After washing, they are sealed and packed and filled with high-purity nitrogen gas for prevention of contamination by air.

The piping materials and parts are stored indoors after sealing and packing to avoid

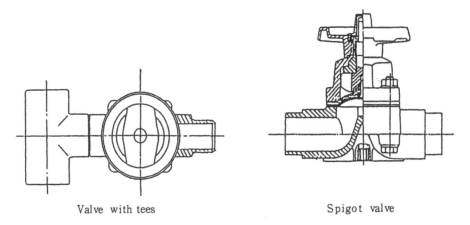

Valve with tees   Spigot valve

Figure 1   Valve with tees and spigot valve.

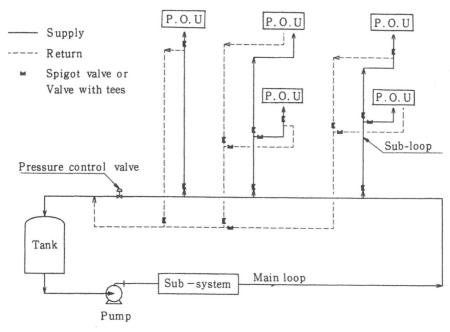

Figure 2   Reverse return piping system.

damage, and in a place where there is little dust, for example, even after transporting them to the field.

Each joint is confirmed for secure connections at each block before assembling the prefabricated piping.

Special care needs to be taken during the assembly of piping so that pipes and joints are not installed with any unusual shock or deflection, for example.

When tightening the flange, special care must be taken not to under- or overtighten the flange.

When a metal gasket, such as a metal C ring, is used for EP pipe or gold EP pipe, a torque wrench is used to tighten it.

The tools used should be as new as possible. The tools are washed as required before use. When working in the clean room, dedicated tools are used, and ordinary tools should not be used in the clean room.

When working in the clean room, designated clean room garments are worn and an adequate air shower is taken when going in or out of the clean room. The hands are kept clean at all times.

When it is necessary to bring the materials, parts, or tools into the clean room, the number of such items should be minimized.

Any work that may generate vibration or dust, such as machining of material, must not be performed in the clean room. When such work must be performed, a vacuum cleaner is used. A separate work station must be used for this kind of activity to prevent dust from scattering.

A vacuum cleaner dedicated to the clean room is used, which incorporates a HEPA filter.

When machining material in the field, a temporary clean booth is arranged to perform the work within the booth. In this case, the atmosphere class is less than approximately 10,000.

Static dissipative paint is applied whereever static electricity may be generated, such as the resin.

The work is performed in accordance with the predetermined process, observing safe work standards.

4
Afterword

Akihiko Hogetsu
Shinko Pantec Co., Ltd., Kobe, Japan

This part was contributed by professionals who have kindly disclosed the latest technical information in their fields on rinsing methods and manufacturing ultrapure water supply systems.

The demand for higher integration in semiconductor chips has become more stringent in recent years. In compliance with this demand, more stringent requirements for cleanliness are being demanded for ultrapure water manufacturing systems. The systems and technologies are daily becoming more sophisticated. Every component of the system, including piping materials and parts, is fabricated and rinsed under clean conditions at the manufacturer's plant and delivered in a sealed package to the customer's plant, where they are assembled in a clean working environment to create a complete system.

However, the embodiment of the design philosophy of clean technology basically depends on those who are involved in all stages related to the ultrapure water manufacturing system. We will be happy if this part serves as a textbook or reference for these individuals in pursuing careers in this field.

VII

Analytical Technology for Ultrapure Water

1
Foreword

Katsumi Koike
Japan Organo Co., Ltd., Tokyo, Japan

Ultrapure water is used in large quantity as wash water in the electronics industry, playing a vital role in the manufacture of semiconductor devices.

The quality of ultrapure water must meet increasingly stringent requirements as the integration of devices increase and demand has always been placed on improvement and development of analysis technologies to assess water quality. Methods generally applied to the quality control of ultrapure water have involved direct measurement of the resistivity, suspended fine particles, total organic carbon (TOC), bacteria, and silica. However, at present, there is no official water quality standard for ultrapure water because the effect on elemental devices, as well as the relationship with the water quality, is not fully defined and each device manufacturer has its own particular standard.

The United States has guidelines and established testing methods for ultrapure water established by the American Society for Testing and Materials (ASTM) and the SEMI. In Japan, Japanese Industrial Standards (JIS) requirements for the testing methods for ultrapure water have been set forth recently in line with the trend toward standardization of analysis and evaluation technologies.

The JIS specifies testing methods for bacteria count, total organic carbon, electrical conductivity, metallic elements, suspended fine particles, silica, and anions.

This part introduces analysis and evaluation technologies for impurities in ultrapure water, but certain technologies are not refined enough to assess the quality currently required for ultrapure water.

The impurities contained in ultrapure water are extremely small in quantity. Practically, we determine the content of such impurities by subtracting a blank test value several times larger than the required value. The exact absolute value is therefore very difficult to obtain, numerical values varying depending on the measuring method. Consequently, it is necessary for water quality evaluation to define the measuring method.

What is required for most analysis and evaluation technologies is to analyze directly and identify impurities that adversely affect devices, thereby contributing to improvement in treatment systems for obtaining ultrapure water. From this viewpoint, it is essential to improve the evaluation technology.

A recent report also stipulates that impurities on the wafer surface must be controlled to 10^8 atm cm^{-2} or less, and simple analysis of the impurities in ultrapure water cannot meet this requirement. As a result, water quality evaluation methods must be related to phenomena occurring on the wafer surface. Approaches that measure the water mark, colloidal materials, and the residue of evaporation have been proposed. These analysis and evaluation methods are introduced here, together with the conventional parameters mentioned earlier.

2
Resistivity

Akira Yamada
Shinko Pantec Co., Ltd., Kobe, Japan

I. INTRODUCTION

"Resistivity" is used as one of the quality indexes for ultrapure water. Resistivity is the electrical resistance possessed by a solution between the electrodes with a cross-sectional area of 1 cm^2, which oppose each other at a distance of 1 cm; ($\Omega \cdot$cm) is used to express resistivity. The inverse of the electrical resistance is the electrical resistivity and is expressed in siemens per centimeter (S cm^{-1}). The lower the concentration of an ion in water, the lower is the electrical conductivity (resistivity is high), except in the case described later.

II. RESISTIVITY AND ELECTRICAL CONDUCTIVITY OF PURE WATER

Pure water is dissociated as expressed by Eq. (1):

$$H_2O \rightarrow H^+ + OH^- \tag{1}$$

Consequently, the electrical conductivity of pure water containing no ions other than hydrogen ion and hydroxide ion is not zero, but indicates the total electrical conductivity value for both ions generated as a result of dissociation. The relationship between the electrical conductivity and resistivity of pure water is shown by the following equation:

$$k = (\Lambda_H[H^+] + \Lambda_{OH}[OH^-])\rho 10^{-3}$$

$$R = \frac{1}{k} \tag{2}$$

715

where: k = electrical conductivity, S cm^{-1}
Λ_H = ultimate molar conductivity of hydrogen ion, S·cm^2 mol^{-1}
Λ_{OH} = ultimate molar conductivity of hydroxide ion, S·cm^2 mol^{-1}
$[H^+]$ = concentration of hydrogen ion, mol 1000 g^{-1}
$[OH^-]$= concentration of hydroxide ion, mol 1000 g^{-1}
ρ = density of water, g cm^{-3}
R = resistivity, Ω·cm

$$[H^+] = [OH^-] \tag{3}$$

$$K_w = [H^+][OH^-] \tag{4}$$

where K_w is the dissociation constant of water Equation (2) can thus be rewritten as Eq. (5):

$$\frac{1}{R} = k = (\Lambda_H + \Lambda_{OH})K_w^{1/2}\rho 10^{-3} \tag{5}$$

The ultimate molar conductivity of hydrogen ion and hydroxide ion, the dissociation constant of water, and the density of water vary with temperature. Table 1 shows an example of the relationships for Λ_H, Λ_{OH}, $pK_w(=-\log K_w)$, ρ, resistivity, and temperature. These values may differ slightly, depending on the publication [2]. The resistivity of pure water is higher with lower water temperature, and therefore resistivity is normally expressed by the value converted to that at the normal temperature of 25°C.

Table 1   Properties of Water Versus Temperature

Temp.	Λ_H	Λ_{OH}	pKw	ρ	R
°C	S · cm^2/mol		—	g/cm^3	MΩ · cm
0	224. 2	127. 8	14. 944	0. 99987	84. 2
10	275. 5	156. 2	14. 535	0. 99973	42. 9
18	315. 6	178. 8	14. 236	0. 99859	26. 6
20	325. 5	184. 4	14. 167	0. 99823	23. 8
25	349. 8	197. 8	13. 997	0. 99707	18. 25
30	373. 7	211. 6	13. 833	0. 99567	14. 1
40	419. 5	237. 2	13. 535	0. 99224	8. 98
50	462. 6	261. 4	13. 262	0. 98807	5. 98
60	502. 5	284. 2	13. 017	0. 98324	4. 17
70	545. 8	324. 2	12. 813	0. 97781	3. 00
75	565. 2	344. 1	12. 712	0. 97489	2. 56
80	582. 8	364. 8	12. 613	0. 97183	2. 20
90	612. 5	405. 9	12. 431	0. 96534	1. 67
100	634	447	12. 265	0. 95838	1. 31
200	824	701	11. 289	0. 865	0. 334
300	894	821	11. 406	0. 712	0. 413

Source: From Ref. 1.

III. RELATIONSHIP BETWEEN TRACE ION AND RESISTIVITY

When a small quantity of another trace ion is contained in pure water as impurity, the electrical conductivity is calculated as a sum of the value given by multiplying each ultimate molar conductivity by the molar concentration of each ion [2]. In other words,

$$\frac{1}{R} = k = \Sigma(\Lambda_1[I])\rho 10^{-3} \tag{6}$$

where: $\Lambda_1 =$ ultimate molar conductivity of each ion, $S \cdot cm^2 \, mol^{-1}$
$[I] =$ concentration of each ion, mol 1000 g^{-1}

A. Resistivity of Pure Water Containing a Small Quantity of Strong Electrolyte

When pure water contains a small quantity of strong electrolyte as impurity, such as sodium chloride, the electrical conductivity can be calculated by the following equation using Eqs. (5) and (6):

$$\frac{1}{R} = k = (\Lambda_H[H^+] + \Lambda_{OH}[OH^-] + \Lambda_{Na}[Na^+] + \Lambda_{Cl}[Cl^-]) \, \rho 10^{-3}$$

$$\tag{7}$$

$$= \{(\Lambda_H + \Lambda_{OH}) \cdot K_w^{1/2} + \Lambda_{Na}[Na^+] + \Lambda_{Cl}[Cl^-]\}\rho 10^{-3}$$

B. Resistivity of Pure Water Containing a Small Quantity of Acid or Alkali

When a small quantity of acid or alkali is contained in pure water as impurity, the resistivity can be calculated by the following equation using Eq. (6), provided that X is substituted for a small quantity of ion other than hydrogen and hydroxide ions:

$$\frac{1}{R} = k = (\Lambda_H[H^+] + \Lambda_{OH}[OH^-] \pm \Lambda_X[X]) \, \rho \, 10^{-3} \tag{8}$$

where: $\Lambda_X =$ ultimate molar conductivity of X, $S \cdot cm^2 \, mol^{-1}$
$[X] =$ ion concentration of X, mol 1000 g^{-1}

Note: $[X]$ may be positive (cation) or negative (anion). The plus (+) symbol indicates a cation and the minus (−) an anion. If the following relationship is taken into account,

$$[H^+] - [OH^-] + [X] = 0 \tag{9}$$

the resistivity of water containing a small quantity of ion X is expressed as a function of $[X]$ by the following equation derived from Eqs. (4), (8), and (9):

$$\frac{1}{R} = k$$

$$= \frac{1}{2} \{(\Lambda_H + \Lambda_{OH})([X]^2 + 4K_w)^{1/2} - (\Lambda_H - \Lambda_{OH} \mp 2\Lambda_x)[X]\} \, \rho 10^{-3} \tag{10}$$

Table 2 lists the ultimate molar conductivities Λ for each ion.

Table 2 Ultimate Molar Conductivities [$(S \cdot cm^2)/mol^{-1}$] at 25°C

Ion	Λ_{25}
$1/2\,Ca^{2+}$	59.8
K^+	73.5
$1/2\,Mg^{2+}$	53.3
Na^+	50.1
Cl^-	76.3
HCO_3^-	44.5
NO_3^-	71.5
$1/2\,SO_4^{2-}$	80.0

Source: From Ref. 3.

The following can be obtained from Eq. (10):

$$\frac{dk}{d[X]} = \frac{1}{2}\{(\Lambda_H + \Lambda_{OH})([X]^2 + 4K_w)^{-1/2}[X] - (\Lambda_H - \Lambda_{OH} \mp 2\Lambda_x)\}\,\rho10^{-3} \quad (11)$$

When the anion [X] is lower than 0, the following relationship is always established:

$$\frac{dk}{d[X]} < 0$$

For a cation, on the other hand, the solution of $dk/d[X] = 0$, or of Eq. (13), exists under the conditions of Eq. (12):

$$\Lambda_H - \Lambda_{OH} - 2\Lambda_x > 0 \quad \therefore \quad \Lambda_x < \frac{\Lambda_H - \Lambda_{OH}}{2} \quad (12)$$

$$[X] = (\Lambda_H - \Lambda_{OH} - 2\Lambda_x)[(\Lambda_H - \Lambda_x)(\Lambda_{OH} + \Lambda_x)]^{-1/2}K_w^{1/2} \quad (13)$$

This is because there is a range in which the resistivity becomes greater when a small amount of cation exists in pure water than when the water is pure. This means that the concentration of ion at the maximum resistivity can be given by Eq. (13). For example, Fig. 1 shows the resistivity when Na^+, Ca^{2+}, Cl^-, and SO_4^{2-} exist independently in pure water. The values shown in Tables 1 and 2 were used for these calculations.

The resistivity is generally used widely as an index of ultrapure water quality. As demands increase for much higher water quality it becomes necessary to meet the following requirements:

1. Improvement in the accuracy of the resistivity meter
2. Correct measurement value for the ultimate molar conductivity and dissociation constant for each ion
3. Improved measuring technique for trace ions
4. Established relationship between trace ion amount and resistivity

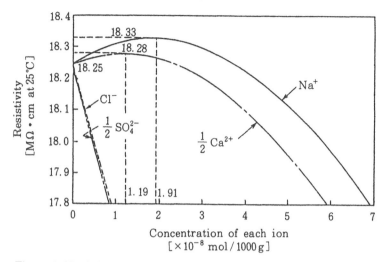

Figure 1 Resistivity of ultrapure water containing trace ions.

REFERENCES

1. S. Truman, Light, Analytical Chemistry, *56*(7), 1138 (1984).
2. T. Ohmi and T. Nitta, Process high-performance technique-II in LSI production, Realize, Inc., 1989, p. 13.
3. Japan Chemical Association, Fundamentals of Chemistry, Chapter II, Maruzen, 1989, p. 460.

3 Particles

A. Measurement of Particles

Tetsuo Mizuniwa
Kurita Water Industries, Ltd., Kanagawa, Japan

I. INTRODUCTION

Since particles in ultrapure water cause pattern defects in the lithography and etching process and may become the origin of impurities on the wafer surface during heat treatment, particles whose size exceeds one-tenth of the design rule must be removed from ultrapure water. Along with reductions in the design rule, the size of particles to be removed has become increasingly microscopic, and technology for the measurement of particles has also been further developed.

There are two methods widely used to count particles in ultrapure water: one is monitoring by on-line particle counter, which employs laser scattering or an acoustic wave, and the other is microscopic counting in which the particles captured on a membrane filter are magnified for visual counting [1–3]. The particle counter was described in a previous chapter; here we discuss direct microscopy.

There are also two methods available for microscopic counting: the optical microscope has mainly been used to count particles larger than 0.2 μm in diameter, and the scanning electron microscope (SEM) is useful in counting the particles smaller than 0.1 μm. Although these methods share many fundamental properties, there are differences. The following discusses the principle of analysis and the points of sampling common to both methods, and also the analysis procedures for each method.

II. PRINCIPLE OF PARTICLE MEASUREMENT

Direct microscopy is a method of determining the number of particles contained in a water sample; sample water is filtered through a membrane filter, which is capable of capturing particles for counting. The filter is then magnified by microscope, and the particles on the filter surface are counted visually. The membrane filter used for sample filtration must have a structure that can trap all particles of the size subject to measurement on its surface. If particles penetrate the filter, it is impossible to count them visually. Because of this restriction, a membrane filter of a specified pore size is used for sample filtration. Figure 1 is a microscopically observed membrane filter with particles on its surface. The Nuclepore filter is the most popular membrane filter.

An ultrapure water system is equipped with a sampling valve to take samples for measurement. One end of a sampling tube is connected to the valve and the other end to a filter holder to which a membrane filter is attached. Taking advantage of piping pressure, the laboratory filter filters sample water and captures particles on the surface of the membrane while keeping external contamination to the minimum. The filtration volume is determined by the estimated number of particles in the water sample, the analytic level, and the number of particles originally adhering to the surface of the membrane filter. Japan Industrial Standard (JIS) K0554, Testing Method for Concentration of Fine Particles in Highly Purified Water, provides that the filtration volume of the sample water should be enough for a membrane filter 25 mm in diameter to capture approximately 50,000 particles for measurement by optical microscope and approximately 500,000 particles for measurement by SEM [4].

It is known that in the membrane filter of 0.2–0.1 μm pore size generally used for the measurement of particles, the number of particles originally adhering to the surface reaches 10^5–10^6 pieces for a disk [5]. It is impossible to distinguish such particles from those taken from the water sample. Since the number of such particles is not a negligible quantity when measuring the number of particles in present-day ultrapure water, it should be taken into account.

For instance, in filtering a sample of ultrapure water, using a membrane filter, that originally contained 10^5 particles, the filtration volume should be so set so that the number of particles newly captured by the filter is approximately 10 times greater (10^6 particles) than the original particle count before filtration. If not, the analytic reliability is drastically reduced. In this instance, the filtration volume should be more than 100 L for a water sample that may contain around 10,000 particles L^{-1} or more than 1000 L for a water

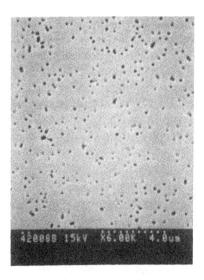

Figure 1 Surface of filter for particle counting.

sample containing about 1000 particles L^{-1}. Such a large filtration volume controls the influence of particles originally adhering to the filter to less than 10%, thereby securing the reliability of the analytic result.

The analytic result may also be affected by contamination of the membrane filter during filtration and particle counting. As a measure to prevent this, JIS K0554 provides that the blank test should be conducted separately by filtering another water sample whose volume is equivalent to 5–10% of the original filtration volume and subtracting the number of particles obtained from that in the original sample water. The particle count on the filter surface differs depending on the membrane filter. Differences as large as one figure are found among filters taken from the same manufacturing lot [5]. Caution is therefore required in subtracting the blank test value from the original number of particles.

Errors in particle counting should also be taken into consideration. Only part of the membrane filter that captures particles is observed by microscope to estimate the total number of particles over the filter. Consequently, the reliability of the count depends on the number of visual fields and the number of particles to be counted [4]. It is necessary to count a minimum of 10 particles to keep the confidence interval of the count in the range ±60% of the median. JIS K0554 provides that the number of visual fields should cover more than 0.01% of the effective filtration area of a membrane filter for SEM and more than 0.1% for the optical microscope.

For measuring the number of particles accurately by direct microscopy, it is important to take into account the relation of the number of particles newly captured by filtration with a membrane filter, the number of particles originally adhering to the filter, and the quantity of contaminants attracted to the filter during particle counting, as well as errors in particle counting.

III. CAPTURE OF PARTICLES

In counting particles by microscope, it is impossible to discriminate the original particles contained in the sample water from foreign particles that contaminate the membrane filter during filtration and counting. To accurately determine the particle concentration in the sample water, the quantity of foreign particles should be minimized as much as possible. Since the membrane filter during filtration is most liable to contamination, it is not an exaggeration to say that the extent of minimizing the contamination in this process governs the accuracy of particle measurement. The following discusses the sampling instruments and procedure to minimize such contamination.

A. Instruments and Reagent

1. Sampling Valve

The sampling valve, the valve that removes a water sample for the measurement of particles, should have a simple structure made of corrosion-resistant material, enabling direct connection of a laboratory filter. It is desirable that the sliding part of the valve not contact the water sample and that the distance between the valve and the piping is as short as possible. Actual configurations are shown in the references [6, 7].

2. Other Instruments

1. A membrane filter with a pore size smaller than the particles to be captured and a structure capable of capturing all the particles on its surface

2. A laboratory filter capable of holding the membrane filter mentioned, with a structure that allows direct connection to the sampling valve for filtration under the piping pressure of the ultrapure water system
3. A portable clean bench capable of creating a Class 5 clean environment, JIS B9920, or equivalent to Class 100 of Federal Standard 209D, in the filtration area.
4. Tweezers that have a construction making it easy to handle the membrane filter
5. A container for the membrane filter, which is cleaned thoroughly for storing a membrane filter that contains the filtered water sample
6. Other equipment, such as an ultrasonic cleaner, syringe, and chemical disinfectant

B.   Filtration Procedure

It is desirable that ultrapure water constantly flow from the sampling valve. If the valve is kept closed except for sampling, before sampling rinse the internal and external surfaces of the nozzle with 3% hydrogen peroxide solution and keep discharging ultrapure water for more than 5 minutes at a flow rate of approximately 5 L minute^{-1} or more. Continue discharging the water for more than 30 minutes at a flow rate of approximately 1 L minute^{-1}.

Adjust the flow rate of the sample water so that it is convenient for sampling, and keep discharging the water for more than 15 minutes.

Connect one end of a sampling tube to the nozzle of the sampling valve and the other end to the filter holder to which a membrane filter is attached, and feed sample water into the filter to purge air.

Filter a specified volume of sample water, which is measured properly. Disconnect the sampling tube from the filter holder after filtration.

Extract the water sample remaining in the filter from the outlet using a syringe at a clean bench. Remove the membrane filter and place it in a container. Count the particles as soon as possible after filtration.

IV.   PARTICLE COUNTING BY OPTICAL MICROSCOPY

The optical microscope is acceptable for the measurement of particles larger than 0.2 μm. Calculate the number of visual fields from the visual area of the microscope to cover more than 0.1% of the effective area of the membrane filter.

A.   Chemicals

1. Water: distilled water or ion-exchange water filter by membrane filter with a pore size smaller than the size of the particles to be measured; used for preparation and use of reagents.
2. Immersion oil.
3. Fuchsin (acid) solution (0.2% wt/vol): dissolve 0.20 g fuchsin (acid) in water to 100 ml, and filter the solution with a 1G4 glass filter.
4. Methylene blue solution (0.02% wt/vol): dissolve 20 mg methylene blue in water to 100 ml.
5. Fuchsin-methylene blue dye solution: mix and stir 8 ml fuchsin-phenol solution, 10 ml alkaline methylene blue solution, and 10 ml water.

B. Instruments

1. Optical microscope with more lens greater than ×1500 by immersion objective for observation: the objective lens is ×100, and the ocular lenses are ×15 or 20
2. Slide and cover slip
3. Petri dish
4. Filter paper
5. Tweezers

C. Preparation of Sample

1. Desiccate the membrane filter that captured the particles.
2. Place filter paper soaked with fuchsin-methylene blue coloring solution on a petri dish, set the membrane filter face up on the filter paper, and leave it for 6 minutes to dye the particles.
3. Place the membrane filter on filter paper soaked with water to remove extra coloring solution.
4. Desiccate the membrane filter.
5. Drip several drops of immersion oil on a slide, place the membrane filter face up on the slide, and place a cover slip on the membrane filter.

D. Particle Counting

Set the preparation on the stage of an optical microscope.

Observe the preparation at ×1500 or ×2000 while sliding it over a certain distance. Observe the entire area of the membrane filter, and record the number and size of particles in each visual field. The number of visual fields is determined so that the total observation area exceeds 0.1% of the effective area of the membrane filter, and it is desirable to continue observation until a minimum of 10 particles are detected [5]. Shift visual fields mechanically: do not select them depending on the presence of particles.

Sum the number of particles counted in each visual field, and calculate the average number per visual field.

Convert the average number of particles per visual field into the particle concentration of the ultrapure water by the formula

$$N = 1000(N_s - N_b) \frac{A/a}{V_s - V_b}$$

where: N = number of particles in 1 L sample water, particles L^{-1}
$\quad\quad\quad N_s$ = number of particles per visual field
$\quad\quad\quad N_b$ = number of particles per visual field in blank test
$\quad\quad\quad A$ = effective filtration area, cm^2
$\quad\quad\quad a$ = area of a visual field, cm^2
$\quad\quad\quad V_s$ = filtration volume, ml
$\quad\quad\quad V_b$ = filtration volume of blank test, ml

V. PARTICLE COUNTING BY SCANNING ELECTRON MICROSCOPE

The scanning electron microscope is suitable for measuring particles larger than 0.1 μm. The number of visual fields is calculated from the visual area of the microscope to cover more than 0.01% of the effective filtration area.

A. Instruments and Equipment

1. A scanning electron microscope, capable of detecting 0.1 μm particles on the filter surface at approximately $\times 10,000$
2. A vacuum evaporator capable of evaporating noble metals, such as gold, or carbon and depositing the metal on the filter surface to provide electrical conductivity so that the sample can be observed; evaporator and spatter coat are available
3. A metal sample holder capable of holding the sample filter flat; should be free of magnetic properties
4. A sample holder: Several methods are available, including holding the filter with double-sided adhesive tape or electrically conductive paste, and a ring weight; when using adhesive tape or conductive paste, select a material that neither contaminates the sample nor discharges gas
5. An ultrasonic cleaner, clean bench, tweezers, and other equipment

B. Procedure

Clean and dry the sample holder, tweezers, and other instruments.

Fix a dried sample membrane filter flat on a SEM sample holder by some appropriate method.

Deposit an electrically conductive material on the sample surface by vacuum evaporator. When the sample is set on the holder after metal deposition, ensure the electrical conductivity between the sample surface and the sample holder.

Set the sample holder in the SEM sample chamber, and observe the surface.

While shifting visual fields over a distance, observe the sample surface at magnifications that can detect the expected particle size (approximately $\times 8000$–10,000 for 0.1 μm particles). Observe as large an area as possible over the sample surface, and record the number and size of particles in each visual field. The number of visual fields should cover more than 0.01% of the effective filtration area. For the improving measurement reliability, it is necessary to examine visual fields until a minimum of 10 particles are counted [5].

Sum the number of particles counted in each visual field, and calculate the average number of particles per visual field.

Conduct the blank test using the same procedure, and calculate the number of particles in ultrapure water by the formula

$$N = 1000(N_s - N_b)\frac{A/a}{V_s - V_b}$$

where: N = number of particles in 1 L sample water, particles L^{-1}
  N_s = number of particles per visual field
  N_b = number of particles per visual field in blank test

A = effective filtration area, cm^2
a = area of a visual field, cm^2
V_s = filtration volume ml
V_b = filtration volume of blank test, ml

VI.   COMPOSITION OF PARTICLES

In addition to particle counting, if the quality of particles is clarified, their origin can be identified and preventive countermeasures can be taken. Particle dyeing for particle counting by optical microscope was originally a method of dyeing fungi [8]. This method therefore selectively observes particles derived from microorganisms. On the other hand, in particle counting with the electron microscope, secondary electrons with information on the surface shape are detected. In other words, electron microscopy can detect particles that form convex and concave shapes on the sample surface, but it cannot identify the quality of these particles. However, by taking advantage of characteristic x-rays generated simultaneously with secondary electrons by the elements on the sample surface, it is possible to determine the composition of the particles. All that is required for this measurement is to equip an ordinary electron microscope with an x-ray analyzer.

Figure 2 shows an analytic result for the composition of approximately 0.4 μm particles by energy-dispersive x-ray analyzer (EDX). Besides the peak of the gold deposited as a conductive film, the characteristic x-ray of aluminum is detected, indicating that the sample contained aluminum. EDX measurement was applied to several particles captured from ultrapure water, the results of which are shown in Table 1. The composition of these particles is indicated semiquantitatively. As discussed earlier, a combination of observation principles like those utilized for particle counting and supplementary analyzing methods makes it possible to qualitatively and semiquantitatively

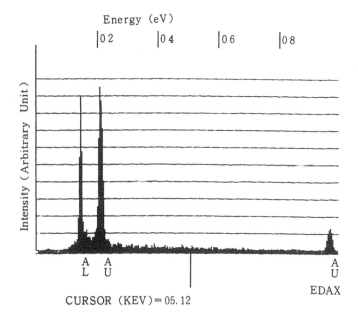

Figure 2   EDX spectrum of 0.4 μm particles. The sample is sputter coated with Au.

Table 1 EDX Analysis Data for Particles Found in an Ultrapure Water System

Sampling Point	Approx. Diameter (μm)	Element
Primary D I	0.6	Si > Cu > Fe
Primary D I	0.7	Si > Al > Mg
Ultrapure Water	0.5	Si > Fe > Cr
Ultrapure Water	0.4	Si > Al > Fe

analyze the particle composition, thereby producing valuable information on the origin of the particles so that their quantity can be reduced.

VII. CONCLUSION

Among the available methods of measuring particles in ultrapure water, this part has discussed a method called direct microscopy. Since speed is required for the qualitative evaluation of ultrapure water, it is not practical to think that particle measurement can be managed solely by direct microscopy. However, future demands for the measurement of extremely fine particles, smaller than 0.1 μm, are expected. In this connection, a real-time measurement particle counter theoretically may not cope with these demands. On the other hand, electron microscopy can detect fine particles of around 0.01 μm. It is therefore necessary to preserve particle observation and counting technology by electron microscopy. In other words, selective use of particle monitoring methods is required: an automatic on-line particle counter is used to constantly monitor the behavior of particles, and direct microscopy is periodically exercised to measure the number of particles and calibrate the particle counter.

The optical microscope is relatively easy to operate in particle counting, and it can selectively detect microorganisms or particles derived from microorganisms, thereby producing useful information for the stabilization of ultrapure water quality.

As described, the two microscopic investigation and particle counting methods differ from each other in principle. As a result, even with the particles of identical size, these methods may give different results, and it is pointless to compare these different counts. It is therefore necessary to select one of these methods based on their principles, advantages, and disadvantages and also estimation of the reliability of their analytic results.

REFERENCES

1. T. Mizuniwa, SEMICON News, 5(9), 75–80 (1985).
2. K. Yabe (ed. by T. Ohmi and T. Nitta), Ultrapure Water High-Purity Chemical Feed System, Realize, Inc., 1986, pp. 167–180.
3. K. Yamashita, Proceedings of the Sixth Study Meeting of Air Purification and Contamination Control, pp. 231–234.
4. Japan Industrial Standard (JIS) K0554, Particle Counting Method in Ultrapure Water.
5. T. Mizuniwa, Text for SEMI Semiconductor Technology Seminar, SEMI Japan, December 1988, pp. 57–61.

6. R. Hango, B. Eldred, and B. Frith, Proceedings of the Third Annual Semiconductor Pure Water Conference, San Jose, January 1984, pp. 127–136.
7. H. Kano, H. Kakizaki, Y. Kubokawa, and S. Matsumoto, Proceedings of the Eighth Annual Semiconductor Pure Water Conference, Santa Clara, January 1989, pp. 3–25.
8. Sakurai, Yosui to Haisui, *11*(9), 11–12 (1969).

3 Particles

B. Automatic Microscopy Employing Image-Processing Technology

Hisanao Kano
Nippon Rensui Co., Yokohama, Japan

I. NECESSITY FOR IMAGE PROCESSING

In Japan, particles in ultrapure water are measured mainly by microscopy. Depending on the particle size, either the optical microscope or the scanning electron microscope is used:

Use the optical microscope for ≥ 0.2 μm.
Use the scanning electron microscope for ≥ 0.05 μm.

Microscopy has several advantages, not only in measuring the size and number of particles but also in observing their shape. On the other hand, it has some disadvantages:

1. A long time is required for measurement.
2. Researchers are forced to bear heavy burdens in terms of time and mental effort.
3. Measurement values vary among researchers.

As shown in Table 1 [1], increasingly stringent water quality is needed along with the increased integration of LSI (large scale integrated) devices. As the minimum particle size decreases, microscopy with higher magnifications will be required. Consequently the microscopic field will decrease. Therefore, to maintain measurement accuracy, microscopic examinations are repeated so that many fields can be observed, which will further worsen these disadvantages.

To address this problem, automation of microscopy employing image-processing technology is now attracting attention. Unfortunately, the history of this field is very short, and the number of published or announced reports is very limited. This part presents an example of applying image processing to optical and scanning electron microscopy, quoting mainly from our recent report [2].

Table 1   Ultrapure Water Quality Target Value for ULSI (Ultra-LSI) Manufacturing

DRAM Integration Scale	1M bit	4M bit	16M bit
Resistivity MΩ•cm at 25°C	>18	>18	>18
Particles (counts/ml) (Min. particle size, μm)	<10 (0.1 − 0.2)	<10 (0.05 − 0.1)	<10 (0.03 − 0.05)
Bacteria (counts/ml)	<10−50	<5−10	<1−5
TOC (μg/l)	<30−50	<20−30	<10−30
Silica (μg/l)	<5−10	<3−5	<3−5
DO (μg/l)	<100	<50−100	<50

1) H. Kano : Haikan Gijutsu (The Piping Engineering) Japan, **29** [14], 59, (1987).

II.   APPLICATION OF IMAGE PROCESSING TO OPTICAL MICROSCOPY

A.   System

Figure 1 is an external view of the optical microscope–particle image analyzer (OM-PIA), and Fig. 2 shows its configuration. The basic specifications of this system are as follows:

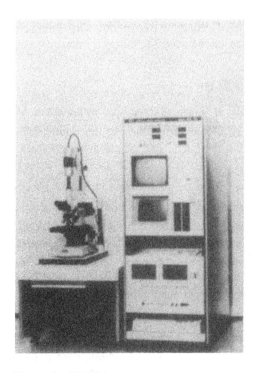

Figure 1   OM-PIA system.

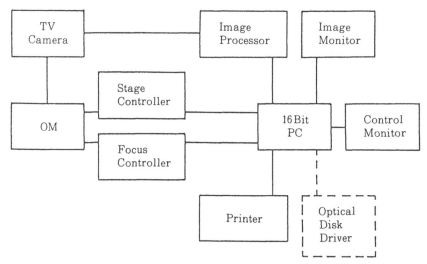

Figure 2 Configuration of the OM-PIA system.

1. Optical microscope: Nikon Metaphoto
2. Particle image analyzer: Mitsubishi Chemical GX-10
3. Magnification of field lens: ×100 (dry)
4. Membrane filter manufactured by Nuclepore: diameter 25 mm, pore size 0.2 μm

As shown in Fig. 3, this particle image analyzer automatically proceeds with the measurement procedures when all the required parameters are input.

B. Detection of Particles

This particle image analyzer detects particles by recognizing the difference in brightness among particles, which are indicated in black or white, and the surface of the membrane filter, which is shown in gray on the binary-coded display. Since OM-PIA uses transmitted light, the particles are shown in black on the display. To detect particles with high accuracy, it is important to make a clear contrast between particles and the surface of the membrane filter: particles should be indicated in a darker color, and the surface of the membrane filter should be shown in a lighter gray.

C. Measurement Operation

Figure 4 shows the procedure used for OM-PIA compared with the manual method. OM-PIA differs from the manual method. To enhance the contrast between light (membrane filter surface) and shade (particles), the following actions are taken. Because the particles are captured on the surface of the membrane filter, extra stain is spread over it. To remove the extra stain from the filter, OM-PIA employs ethanol (50% by volume) to enhance the removal efficiency. The manual method uses water.

Particles are stained with fuchsin and methylene blue into red. A green filter is placed between the light source of the optical microscope and the stage so that the red color of the stained particles is displayed as dark black on the screen.

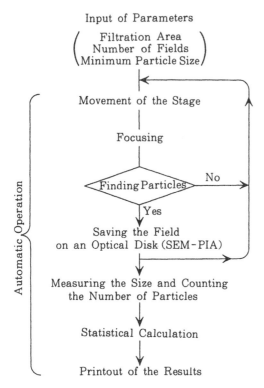

Figure 3 Operation steps of PIA.

So that automatic focus adjustment is easy, the membrane filter containing captured particles is fixed on the filter holder, which holds the filter in flat. (In the manual method, a specimen is prepared on a slide from the particles captured by the filter.) The filter holder is made of synthetic resin. At this stage, the edge of the membrane filter is clamped with a ring to hold the membrane flat.

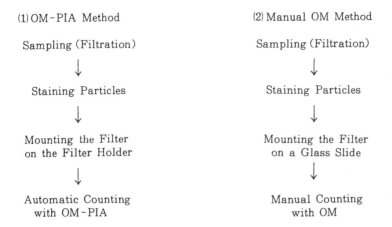

Figure 4 Measuring procedures of the OM-PIA method.

D.   Particle Count

The particle count with OM-PIA is about half that of the manual method, as shown in Fig. 5 [2]. Still, there is quite a good correlation between the two particle counts. The lower particle count with OM-PIA is thought to be because OM-PIA counts only the particles detected in a single focal plane but the manual method adjusts the focus for each particle on the filter during counting.

E.   Measurement Time

When the measurement area ratio is 0.2%, it takes 30–60 minutes to count the particles on one sheet of membrane filter using the manual method. (The time greatly depends on the skill of the researcher. With OM-PIA, however, counting takes as little as 4 minutes.)

III.   APPLICATION OF IMAGE PROCESSING TO SCANNING ELECTRON MICROSCOPY

A.   System

Figure 6 is an external view of the scanning electron microscope–particle image analyzer (SEM-PIA), and Fig. 7 shows its configuration. The basic specifications of this system are as follows:

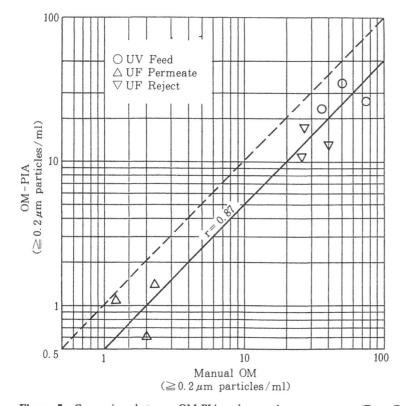

Figure 5   Comparison between OM-PIA and manual measurements. (From Ref. 2.)

Figure 6   SEM-PIA system.

1. SEM: Hitachi S-510
2. Particle image analyzer: Mitsubishi Chemical GX-20
3. Magnification: ×10,000
4. Membrane filter: metal-coated membrane filter with a uniform pore size distribution: diameter 25 mm; pore size 0.08 μm

Just like the GX-10, the GX-20 of this PIA automatically proceeds with the measurement procedure, as shown in Fig. 3, after the required parameters are input.

The system is connected to an optical disk unit in which one optical disk can store up to 2000 observation fields. It is therefore very easy to store or retrieve data. In addition, the optical disk unit can also be connected to the OM-PIA system.

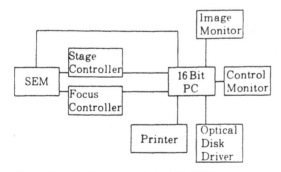

Figure 7   Configuration of the SEM-PIA system.

B.   Detection of Particles

This system applies the same principle for detecting particles as the OM-PIA system.

In the usual method of treating the membrane filter surface with a metal coating, the contrast between particles, particularly bacteria, and the filter surface is not sufficiently clear, resulting in degraded accuracy. To address this problem, a new method has been developed in which sample is filtered using a membrane filter with a gold coating. Particles captured on the gold-coated filter are divided into two groups: one group is displayed in white because of the high density of secondary electrons emitted by the particles, and the other is shown in dark because there is little emission of secondary electrons. Using this phenomenon, the system detects particles by recognizing the difference in brightness among white particles, black particles, and the gray filter surface.

C.   Measurement Operation

Figure 8 compares the SEM-PIA and the manual methods in terms of the measurement procedure. Table 2 [2] shows an example of outputs obtained with SEM-PIA. SEM-PIA differs from the manual method in the following points:
1. In an effort to make the automatic focus adjustment easy, the filter containing captured particles is dried and fixed to the filter holder, which holds the filter flat. The filter holder is made of metal (nonmagnetic), such as stainless steel and brass. The edge of the membrane filter is clamped with a ring to hold it flat.
2. Since the gold-coated membrane filter that is used is electrically conductive the metal coating is not applied after particles are captured. (The filter containing captured particles is observed by microscope as is.)

D.   Particle Count

As shown in Fig. 9 [2], the particle count using SEM-PIA is comparable to that using the manual method.

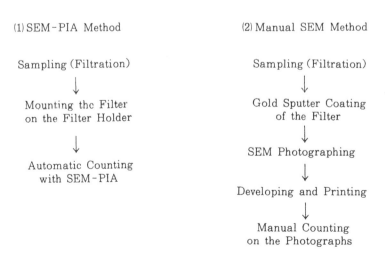

Figure 8   Measuring procedures for the SEM-PIA method.

Table 2   Example of SEM-PIA Output

SAMPLE : UF PER. Lot No. : 1 CONDITION : MAX. LENGTH OF PARTICLE DATE 88. 9. 8.			

SAMPLE : UF PER.   Lot No. : 1   CONDITION : MAX. LENGTH OF PARTICLE   DATE 88. 9. 8.

    FILTER AREA   284 mm 2   OBJECT VIEW   500 VIEWS

    MAGNIFICATION   10000   EFFECTIVE VIEW   493 VIEWS

                                OBSERVED AREA   0. 055 mm 2

                                AREA RATIO   0. 0194 %

                                AVERAGE SIZE   0. 214 micrometers

MAX. LENGTH	NO.	TOTAL NO.	HISTOGRAM
0. 10 − 0. 20	29	149566	61. 70 % I ********************************
0. 20 − 0. 30	12	61890	25. 53 % I *************
0. 30 − 0. 40	5	25787	10. 64 % I ******
0. 40 − 0. 50	0	0	0. 00 % I
0. 50 − 0. 60	0	0	0. 00 % I
0. 60 − 0. 70	0	0	0. 00 % I
0. 70 − 0. 80	0	0	0. 00 % I
0. 80 − 0. 90	0	0	0. 00 % I
0. 90 − 1. 00	1	5157	2. 13 % I *
TOTAL	47	242400	

Source: From Ref. 2.

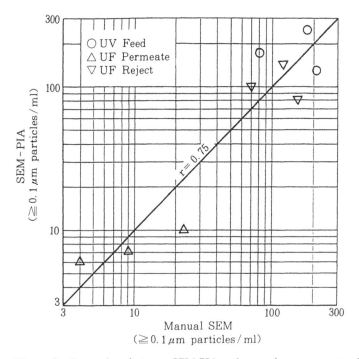

Figure 9   Comparison between SEM-PIA and manual measurements. (From Ref. 2.)

E.   Measurement Time

When the measurement area ratio is 0.02%, it takes 3–5 days overall to complete measurement with the manual method—for microphotography, development of films, printing the photographs, and counting the particles. The SEM-PIA method takes as little as 1 h to complete.

IV.   FUTURE CHALLENGES

The efficiency of the OM-PIA and SEM-PIA methods has been recognized, but they should be further improved.

A.   Membrane Filter

The multipore membrane filters for particle counting available on the market have several problems: linear defects and chained pores of the filter affect particle detection, for example. The problems can be addressed to some extent by modifying the image-processing method, but the problems are still not completely overcome. In addition, the measurement time is sometimes lengthened as a result of modifications in image processing. It is therefore necessary to develop a new membrane filter that is free of defects that affect image processing.

B.   OM-PIA Method

To make the OM-PIA particle count approach the particle count using the manual method, more appropriate focal planes should be studied.

C.   SEM-PIA Method

The filtration rate after gold coating drops to 60% of that before gold coating. Such a drop in the filtration rate leads to longer sampling times. To shorten the sampling time, the volume of filtered sample should be reduced. This requires study to suppress the contamination on the membrane filter surface at the time of filter production and sampling.

REFERENCES

1.   H. Kano, Haikan Gijutsu, *29*(14), 59 (1987).
2.   H. Kano, H. Kakizaki, Y. Kubokawa, and S. Maltsumoto, Semiconductor Pure Water Conference, Santa Clara, 1989.

4

Ultrapure Water and Bacteria

Minami Tsuchizaki
Japan Microbiological Clinic Co., Ltd., Kanagawa, Japan

I. INTRODUCTION

Recent large-scale integrated (LSI) circuit production technology is achieving incredible success in integrating the devices to even higher levels. With such developments, it is essential to upgrade the peripheral materials, including pure water, reagents, solvents, and gases, all of which are consumed in large quantity in the LSI production process.

Water in particular is consumed in huge volumes, and it should be purified to the level of so-called ultrapure water. The impurities to be removed in the course of purifying water are ions, colloidal materials, organic materials (total organic carbon, TOC), dissolved oxygen, microorganisms, and microparticles. As ultrapure water production technology has developed, so have these impurities come to be removed more easily.

Except for microorganisms (bacteria), none of these impurities increase again once they are removed. However, microorganisms can breed even in pure water. Therefore bacteria should be regarded as living foreign substances [1]. Living bacteria increase to 0.4–0.6 × 1.0–1.3 μm, and dead bacteria of equivalent size also affect the performance of LSI devices.

The total bacteria count measurement method [2], which counts both living and dead bacteria, is considered useful. Since this method is described elsewhere in this book, this chapter discusses bacteria that breed in ultrapure water.

II. MEASUREMENT OF BACTERIA IN ULTRAPURE WATER

The general method of measuring bacteria in liquid, including water, is described first. Second, the Japanese Industrial Standard JIS method of measuring bacteria in ultrapure water [3] is presented. In addition, the results of experiments are described.

A. Methods of Counting Bacteria in Pure Water

1. Agar Dilution

Sample water (1 ml) is applied to three to five sterilized petri dishes (around 90 ϕ) aseptically. Agar medium that has been treated with high-pressure sterilization (121°C, 15–20 minutes) and cooled to 40–45 °C is poured into the petri dishes aseptically. After stirring the agar well, the petri dishes are left undisturbed while the agar cools and solidifies. When the agar has solidified, the petri dishes are placed in an incubator to incubate bacteria at an appropriate temperature. If the bacteria count in a sample is expected to be high, the water sample is diluted with sterilized water to 1/10, 1/100, and so forth, to prepare the incubation water sample. After incubation (24–120 h), the number of bacteria that form colonies is counted to obtain the bacteria count in ultrapure water.

In this experimental method, various parameters, including the type of culture medium and the concentration, pH, incubation time, and incubation temperature, can be set to meet the purpose of the experiment. It is therefore important to discuss and investigate different conditions appropriate to the experiment.

2. Membrane Filter

For a sample that has an extremely low bacteria count, such as ultrapure water, the agar dilution method of counting the number of bacteria in a 1 ml water sample cannot be employed. Table 1 shows the result of examining the same sample with two different methods, the 1 ml incubation method (the agar dilution method) and the 100 ml incubation method (the membrane filter method). Table 1 compares the two methods. The 1 ml incubation method yields a zero bacteria count. The 100 ml incubation method, however, yields much larger bacteria counts for the same sample.

Pseudomonas can propagate even in ultrapure water. Therefore the membrane filter method using a 100 ml sample or more is suitable for studying the bacteria in ultrapure water. The membrane filter procedure is briefly described here.

A membrane filter with a diameter of 47 mm and pore size 0.2–0.45 μm is placed on the funnel and sterilized. A water sample of 100–1000 ml is accurately measured with this filter, and then the sample is filtered. Following filtration, the membrane filter is moved

Table 1 Number of Viable Cells Using Different Incubation Methods

Samples	Agar dilution method (cells/ml)	Membrane filter method (cells/ml)
1	0	3
2	0	7
3	0	1.2×10
4	0	4.3×10
5	3	1.7×10^2
6	5	9.2×10^2
7	9	6.8×10^2
8	1.2×10^2	TNTC*

medium : Nutrient agar

Incubation : 30 ± 1 °C. 120 hrs.

*TNTC : TOO NUMEROUS TO COUNT

onto the agar medium solidified in a sterilized petri dish. Bacteria are incubated at a constant temperature. After a certain time, the number of colonies appearing on the membrane filter is counted to obtain the bacteria count in a strictly measured sample volume.

This method shares some procedures with the agar dilution method: it provides bacteria with the necessary nutrients, temperature, and time for propagation, and it evaluates the number of colonies formed.

Various membrane filters are available on the market, and it is necessary to select the appropriate filter in terms of the recovery rate (the bacteria count to be detected).

A bacteria analysis monitor that uses a method similar to the membrane filter method is available. Because this monitor requires advanced technology to prevent contamination during sampling, using this monitor is not a simple procedure.

3.   MPN (Most Probable Number) Method

This method was originally employed to study coliform bacilli in liquid foods and beverages. Many bacteria increase more rapidly in liquid medium than on solid media. Also, bacteria are more easily recognized in liquid because the turbidity of the liquid is changed by the presence of bacteria.

In the MPN method, samples are first diluted by 1/10 several times. The diluted water sample is then taken in five lactose bouillon fermentation tubes. In this way, gas generated and the level of turbidity are detected in the liquid as indicators of coliform bacilli. Based on this result, the number of coliform bacilli is calculated by probability, and the calculation result is expressed by the MPN.

For pure water, three different samples are prepared aseptically: undiluted sample water, 1/10 diluted sample water, and 1/100 diluted sample water. Each 1 ml water sample is separately injected into liquid media. Each liquid medium with a water sample is divided into five test tubes that have been sterilized and cooled. They are kept at a certain temperature for incubating bacteria. After awhile, the number of test tubes whose liquid medium is turbid as a result of bacterial growth is counted for each dilution level. Based on the most probable number table, a probability with a reliability limit of 95% is obtained.

Generally, because the bacteria count in ultrapure water is very low, the number of bacteria sometimes cannot be detected using the dilution method. Therefore another method is recommended for ultrapure water: the water sample is condensed 10 or 100 times with a membrane filter, and the membrane filter is inserted in a test tube with liquid medium to incubate bacteria.

The MPN method should be conducted by skilled operators. It is important as a reference when the type of culture medium and the incubation conditions are studied for the agar dilution and membrane filter methods.

B.   JIS Method of Examining Bacteria in Ultrapure Water

This method is based on almost the same concept as the membrane filter method [3]. This method uses two different time settings: short-term incubation (24 ± 2 h) and long-term incubation (around 120 h).

1.   Short-Term JIS Incubation Method

a.   Outline.   A membrane filter containing bacteria is laid on an absorption pad containing M-TGE liquid medium or standard liquid medium. Bacteria are then incubated at 35 ± 1°C for 24 h. By counting the number of colonies formed on the membrane filter, the number of bacteria in the 1 ml water sample can be calculated.

b.   Medium.   *M-TGE Liquid Medium.*   Tryptone, 10 g, meat extract, 6.0 g, and glucose, 2.0 g, are added to 1 L water. The water is then heated to dissolve these substances. After sterilization, the pH should be 7.0 ± 0.1.

Standard Liquid Medium.   Peptone (use peptone produced by hydrolyzing casein with pancreatin), 5.0 g, yeast extract, 2.5 g, and glucose, 1.0 g, are added to 1 L water. The water is then heated to dissolve these substances. After sterilization, the pH should be 7.0 ± 0.1.

2.   Long-Term JIS Incubation Method

a.   Outline.   A membrane filter containing captured bacteria is laid over M-TGE liquid medium or standard liquid medium. The bacteria are then incubated at 25 ± 1°C for 5 days. By counting the number of colonies that form on the membrane filter, the number of bacteria in a 1 ml water sample is calculated.

b.   Medium.   The medium used in the long-term incubation method is 15 g powdered agar added to the two kinds of media used in the short-term incubation method.

C.   Problems

Generally the incubation method used to measure the bacteria count. The incubation method multiplies bacteria that cannot be detected by macroscopic observation by growing the bacteria under appropriate conditions (including nutrient, temperature, oxygen, pH, and incubation time).

As mentioned, a 1 ml water sample is placed in a petri dish in which the water sample is mixed with medium containing agar and other nutrients. In this way, bacteria in the water sample are spread over the petri dish and increase over time, forming colonies that can be recognized by macroscopic inspection. The bacteria count in the water sample under certain conditions (the type of medium, the temperature, and the time) can be calculated based on the colony count. The membrane filter method can be applied to an ultrapure water sample containing very small numbers of bacteria.

Bacteria can be divided into the following three types in terms of their ability to synthesize essential organic metabolic products:

1. Autotrophic bacteria: bacteria that synthesize nutrients, Sources of carbon: CO_2 and HCO_3^-. Sources of nitrogen: NO_3^-, NO_2^-, and NH_4^+. Sources of energy: inorganic materials and light.
2. Heterotrophic bacteria: bacteria that are able to synthesize enough nutrients by themselves and must obtain some nutrients from outside.
3. Hypotrophic bacteria: bacteria that are unable to synthesize nutrients at all and must obtain all nutrients from outside.

Table 2 shows the incubation results of bacteria in ultrapure water. For this incubation, three types of media were prepared: medium A for heterotrophic bacteria and medium B and medium C for autotrophic bacteria. The composition of each is as follows:

1. Medium A, nutrient agar: 0.5% meat extract, 0.5% peptone, 1.3% agar, and pH 7.1 ± 0.1. This medium is diluted to ½ and ¹⁄₁₀. In each case, however, the agar content is 1.3%.
2. Medium B, Winogradsky medium: 0.1% $(NH_4)2SO_4$, 0.1% K_3PO_4, and 1.3% agar.
3. Medium C, Czapek-Dox medium: 0.1% K_2HPO_4, 0.05% $MgSO_4 \cdot 7Ag$, 0.2% $NaNO_3$, 0.05% KCl, 0.001% $FeSO_4$, and 1.3% agar. No sucrose is added.

Table 2   Number of Viable Cells in Individual Media

Sample	Incubation Time (hr)	Medium A			B		C	
		No dilution	1 / 2 conc.	1 / 10 conc.	No dilution	1 / 5 conc.	No dilution	1 / 5 conc.
1	96	1	1	1	0	0	0	0
2	96	1	2	1	0	0	0	0
3	96	1	1	0	0	0	0	0
4	96	6	3	4	0	0	0	0
5	96	2	5	2	0	0	0	0
6	96	2.6×10	2.3×10	3.0×10	6	4	0	4
7	96	1.0×10	1.3×10	1.2×10	2	2	0	1
8	96	0	1	1	0	0	0	0

Agar plate method : $30 \pm 1\ ^{\circ}\text{C}$

Ultrapure water is considered to contain extremely small amounts of organic nutrients. As shown in Table 2, however, heterotrophic bacteria, which need organic nutrients, are most often detected in ultrapure water. It is important to further explore the optimum conditions so that as many bacteria can be detected in ultrapure water as possible.

III.   BACTERIAL SEPARATION FROM ULTRAPURE WATER

When producing pure water, such bacteria as *Pseudomonas, Flavobacterium, Alcaligenes, Achromobacter, Bacillus, Corynebacterium, Micrococcus, Aeromonas,* and *Microcyclus* are detected, but in ultrapure water with a resistivity of around 18 MΩ·cm, the types and numbers of bacteria decrease markedly. In our research, bacteria including *Pseudomonas, Flavobacterium, Alcaligenes,* and *Achromobacter* are detected in ultrapure water. *Pseudomonas* in particular accounts for 80–90%. One strain of *Pseudomonas* is taxonomically discussed here, including its propagation in ultrapure water and its heat resistance.

A.   Taxonomic Characteristics of *Pseudomonas* Separated from Ultrapure Water

Ultrapure water (100 ml) was removed a semiconductor production line and filtered using a sterilized membrane filter (Nuclepore membrane filter, pore size 0.2 μm, diameter of 47 mm). Bacteria were then incubated on standard agar medium (0.5% meat extract, 0.5% peptone, 1.3% agar, and pH 7.2 $\pm$ 0.1) at 30 $\pm$ 1°C for 4 days.

The colonies that formed on the filter were found to be a single type of bacteria by macroscopic and microscopic observation. One strain of bacteria was picked up and separated using a standard microbiological method.

Since the separated bacteria were found to be pure, they were stored in a slope culture on standard agar medium for subcultivation. The bacteria were then tested for the following microbiologic characteristics:

Gram stain: negative
Motility: positive
Flagellum: polar flagella (one to two pieces)
Colony: complete, convex, and smooth
Pigment: None
Catalase: positive
Oxidase: positive

Indol formation: negative
H$_2$S formation: negative
Urease formation: positive
Starch decomposition: positive
Gelatin decomposition: positive
Nitrate reduction ability: positive

The results prove the bacteria are *Pseudomonas*.
Next the assimilation of various carbon sources was studied.

Glucose: positive
Maltose: positive
Xylose: positive
Mannitol: positive
Sorbitol: negative to negative or positive
Arabinose: negative to negative or positive
Adonitol: positive
Erythreitol: positive
Lactose: positive
Galactose: positive
Fructose: positive
Saccharose: positive or negative to positive

Based on the study of the assimilation of carbon sources, these bacteria have been found to be very similar to *Pseudomonas cepacia*. Although more characteristics should be studied, the bacteria are regarded as *P. cepacia*.

B.   Growth of *P. cepacia* in Ultrapure Water

Various articles have reported that bacteria can propagate even in high-purity water. According to the study by Favero [4], when *Pseudomonas aeruginosa* (a Gram-negative bacillus) was maintained in distilled water at 25°C for 24 h, it increased from 4.3×10^3 to 1.1×10^6 ml^{-1}.

Kruger [5] studied how bacteria in distilled water live and propagate. Carson [6] studied the *P. cepacia* separated from distilled water to determine its morphologic characteristics, taxonomic characteristics, and growth in distilled water.

Researchers have identified bacteria separated from ultrapure water used in the electronics industry as *P. cepacia*, as mentioned earlier. We then studied how *P. cepacia* grows in ultrapure water with a resistivity of 18 MΩ·cm.

Well-cleaned flasks with a capacity of 1 L were filled with ultrapure water. A certain amount of separated *P. cepacia* was injected into the flasks. The temperature was set at

Table 3   Growth of *P. cepacia* in Ultrapure Water

Strain	Temp.	Start*	15 hrs	24 hrs	48 hrs	72 hrs
Pseudomonas	25 °C	2 コ/ml	2.0×10	2.5×10^2	2.8×10^5	4.0×10^5
cepacia	30 °C	3	1.4×10^2	3.5×10^4	5.5×10^5	6.5×10^5
	37 °C	5	5.0×10^2	2.5×10^5	6.0×10^5	7.0×10^5

Medium : Nutrient ager   Incubation : 30 ± 1 °C, 120hrs

*Pieces/ml
Medium: Nutrient agar   Incubation: 30 ± 1 °C, 120 hrs

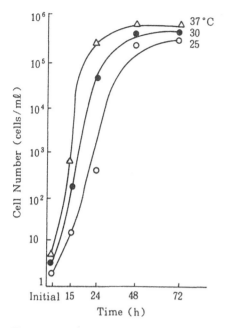

Figure 1 Growth curve of *P. cepacia* in ultrapure water.

25, 30, and 37°C. The flasks were then kept for 15, 24, 48, and 72 h. After each time period, a small amount of water was removed from the flasks to incubate the bacteria in it using the agar dilution method. The colonies that formed were counted after 96 h of incubation. The results are shown in Table 3 and Fig. 1.

The bacteria grew rapidly at 25–37°C. The bacteria count was less than 10 ml^{-1} when the sample was taken. In the 48 h incubation, it increased to the order of 10^5 ml^{-1} , and after this the bacteria count remained flat. Considering that bacterial grow to the order of 10^7–10^8 ml^{-1} under rich nutrient conditions, the poor nutrient conditions in ultrapure water are thought to affect bacterial growth.

C. Heat Resistance of *P. cepacia*

As mentioned, the ultrapure water production line can also be contaminated by bacteria. A contaminated line should be treated with chemical sterilization or hot-water sterilization. Chemical sterilization is not desirable because it may affect the quality of the ultrapure water flowing through the line. Hot-water sterilization is therefore preferred. The temperature level and the length of sterilization should be determined. We studied the heat resistance of *P. cepacia*.

A well-cleaned flask with a capacity of 1 L was filled with ultrapure water. A certain amount of *P. cepacia* was injected and stirred into the ultrapure water. The ultrapure water containing *P. cepacia* was placed in well-cleaned test tubes, 10 ml each. Several thermal treatments were then conducted. After thermal treatment, a 1 ml water sample was aseptically removed from the test tubes for incubation using the agar dilution method at 30 ± 1°C for 96 h. The medium was standard agar. Table 4 lists the results. As shown in Table 4, *P. cepacia* was completely killed by thermal treatment at 60°C for 30 minutes. The ultrapure water production line should therefore be heated at 60–65°C for 30 minutes to completely sterilize bacteria.

Table 4   Heat Resistivity of *P. cepacia* Isolated in Ultrapure Water

Heating Condition		Cell Number Before heating	Cell Number After heating	Bactericidal Ratio
60 °C	15 min	5.6×10^3 (piece/mℓ)	1.2×10^2	98.2 %
60 °C	30 min	8.5×10^3	0	100
70 °C	15 min	7.2×10^3	0	100

IV.   CONCLUSION

Very high levels of ultrapurity are required for water used in the semiconductor industry. In particular, microorganisms should be kept at extremely low levels. Because microorganisms, particularly bacteria, contain metallic ions, other ions, and organic materials as nutrients, microorganisms seriously affect the production yield and the quality reliability of semiconductor devices. Living bacteria in particular are unique since their numbers can increase even in ultrapure water. Even after sterilization, dead bacteria remain as particles. It is therefore very important to properly control bacteria in the pure water production line.

To control bacteria properly, their quantity and quality should be measured accurately at each point. Without this study, excessive facilities and costs may be introduced. Microbiologic inspection plays an important role in the proper production of ultrapure water. Microorganism control technology will become more and more important as semiconductor devices are further integrated.

REFERENCES

1. M. Tsuchizaki, Semiconductor World (1983).
2. M. Tsuchizaki and Y. Saito, Nippon Boukin Boukabi Gakkaishi, *3*(1) (1975).
3. JIS-K-0550 (1988).
4. M. S. Favero, Science, *173*, 27 (1971).
5. D. Kruger, Drug Made in Germany, *20*, 162 (1977).
6. L. A. Carson, Applied Microbiology, March 1973.

5
Total Organic Compounds

Sachio Satoh
Japan Organo Co., Ltd., Tokyo, Japan

Toshiki Manabe
Japan Organo Co., Ltd., Saitama, Japan

I. INTRODUCTION

Even ultrapure water, so-called highly refined water, contains residues of organic compounds. Although referred to as "ultrapure water," the raw material is natural water, and ultrapure water is produced by refining natural water to a high degree using various refining processes.

Accordingly, ultrapure water contains residues, although extremely small in quantity, or organic compounds of relatively low molecular weight in raw water and of low-molecular-weight compounds that elute from the component materials (e.g., piping) of the refining units. The concentration of organic compounds in water is generally very low (and much lower in ultrapure water), and it is extremely difficult to perform fractionation and determination of each chemical species. As an alternative, therefore, determination by functional groups or determination of the total amount of organic compounds is employed.

A method of determining the total amount of organic compounds is to quantify the amount of a specific element (carbon, for example) comprising organic compounds. If this is used as an index, the carbon is referred to as an organic carbon.

This chapter briefly describes organic compound test methods for total organic carbon (TOC), chemical compounds with low boiling points, low-molecular-weight halogenated carbon hydroxide, and organic acids.

II. MEASUREMENT OF TOTAL ORGANIC CARBON

Organic carbon is the carbon in organic compounds in water. A quantitative analysis method was developed by Van Hall et al. [1, 2] from 1963 to 1965. According to this method, impurities in the water are burned (at 900–950°C) to transform the carbon in organic compounds to carbon dioxide. The total carbon (TC) amount is determined by infrared absorption by carbon dioxide. Then, at a temperature (150°C) at which organic compounds are not decomposed, carbons of inorganic compounds (hydrogen carbonate and carbonate) are transformed to carbon dioxide and the inorganic carbon (IC) amount is determined by infrared absorption by carbon dioxide. The IC amount is subtracted from the total carbon amount, giving the amount of total organic carbon.

However, this method of burning impurities in water suffers restrictions in the sample charge amount and in the detection limit (to 200 μg C L^{-1}) because a large amount of sample may cause a temperature drop in the combustion furnace, resulting in degradation of the decomposition rate. Recently, a high-sensitivity version of this method was developed that successfully lowers the detection limit to around 50 μg C L^{-1}.

The initial TOC determination method was a combustion type and thus difficult to apply to samples with extremely low TOC concentrations. As a result, a determination method using wet oxidation (addition of persulfate and ultraviolet, UV, irradiation) was developed in the late 1970s.

Compared with the combustion method, this method may suffer low oxidation rates depending on the chemical structure and molecular weight of the organic compounds. However, it is adequate for samples containing extremely low concentrations of organic compounds, such as ultrapure water used in the semiconductor field.

A. Automatic Combustion Infrared Oxidation TOC Measurement

Acid is continuously added to the sample supplied to the measuring instrument to reduce the pH to below 2, and then air is blown into the sample to remove inorganic carbons. A certain amount of this sample is supplied to a TOC measuring oxidation catalyst at high temperature together with a carrier gas to measure the concentration of the carbon (TOC). The detection range is 50–1000 μg C L^{-1}, which varies depending on the system and measuring conditions.

B. Automatic UV Oxidation Infrared TOC Measurement

Acid is continuously added to the sample supplied to the measuring instrument to reduce the pH to below 2, and then air is blown into the sample to remove inorganic carbons. A certain amount of the sample is fed, together with persulfate and carrier gas, into a TOC-measuring oxidation reaction tube with a built-in ultraviolet source. The carbon in organic compounds is transformed into carbon dioxide, and its concentration is measured with a nondispersive infrared gas analyzer to determine the concentration of total organic carbon.

The detection range is 10–1000 μg C L^{-1}, depending on the system and measuring conditions.

The results of a study of the decomposition rate using this method are shown in Tables 1 and 2 [3, 4]. Decomposition rates are compared by comparing the concentrations of two kinds of organic carbon. One is determined from a solution containing a known carbon concentration (target set at about 2 mg L^{-1} and about 4 mg L^{-1}) for highly pure organic compounds (high-purity reagents) whose chemical structure was determined using an automatic TOC measuring instrument of the wet UV oxidation type used with an oxidant. The other is determined using a combustion-oxidation TOC analyzer.

Table 1 TOC Determination of Various Organic Compounds

Organic Compound	Chemical Formula	TOC by Combustion-Infrared Oxidation (mg C/L)						TOC by Persulfate-UV Oxidation (mg C/L)						Average Rate of Oxidation
		1	2	3	4	5	Averag	1	2	3	4	5	Average	(%)
Methanol	CH_4O	1.8	1.7	–	–	–	1.8	1.70	1.71	–	–	–	1.71	95.0
		3.7	3.7	–	–	–	3.7	3.45	3.38	–	–	–	3.42	92.4
Ethanol	C_2H_6O	1.7	1.7	–	–	–	1.7	1.59	1.65	–	–	–	1.62	95.3
		3.5	3.4	–	–	–	3.5	3.27	3.29	–	–	–	3.28	93.7
2-propanol	C_3H_8O	2.1	1.7	1.9	1.9	1.8	1.9	1.77	1.82	1.87	1.85	1.81	1.82	95.8
		3.7	3.8	3.8	3.9	3.6	3.8	3.65	3.65	3.70	3.72	3.67	3.68	96.8
Citric acid	$C_6H_5O_7Na_3$	2.0	1.9	2.0	–	–	2.0	2.06	2.00	2.01	–	–	2.02	101
		3.9	4.0	4.1	–	–	4.0	3.98	3.95	4.00	–	–	3.98	99.5
Benzoic acid	$C_7H_6O_2$	2.3	2.3	2.2	2.1	2.1	2.2	2.04	2.02	2.02	2.09	1.99	2.03	92.3
		3.9	4.0	4.1	–	–	4.0	4.02	3.98	3.95	–	–	3.98	99.5
Potassium hydrogen phthalate	$C_8H_5O_4K$	2.0	2.0	1.9	–	–	2.0	1.99	2.02	2.00	–	–	2.00	100
		4.0	4.0	–	–	–	4.0	3.98	3.99	–	–	–	3.99	99.8
Acetone	C_3H_6O	1.9	1.8	1.9	–	–	1.9	2.00	1.95	1.93	–	–	1.96	103
		3.9	3.8	–	–	–	3.9	3.87	3.78	–	–	–	3.83	98.2
L-sodium glutamate	C_5H_8ONNa	2.0	2.0	–	–	–	2.0	1.99	1.99	–	–	–	1.99	99.5
		3.9	4.0	–	–	–	4.0	3.96	3.99	–	–	–	3.98	99.5
4-aminobenzene sulfonate	$C_6H_7O_3NS$	2.0	2.0	2.2	1.9	2.0	2.0	1.92	2.00	1.97	2.03	2.00	1.98	99.0
		4.1	4.1	4.2	4.2	4.1	4.1	3.94	3.91	4.02	3.99	4.01	3.98	97.1
D-glucose	$C_6H_{12}O_2$	1.8	1.7	–	–	–	1.8	1.82	1.70	–	–	–	1.76	97.8
		3.8	3.8	–	–	–	3.8	3.79	3.80	–	–	–	3.80	100
8-quinolinol	C_9H_7ON	2.0	2.0	–	–	–	2.0	2.02	1.97	–	–	–	2.00	100
		4.0	4.1	–	–	–	4.1	4.02	4.02	–	–	–	4.02	98.0
4-aminoantipyrine	$C_{11}H_{13}ON_3$	1.9	1.9	1.8	–	–	1.9	2.07	2.00	1.93	–	–	2.00	105
		3.9	4.0	4.2	–	–	4.0	3.96	3.97	3.96	–	–	3.96	99.0
2,4,6-tri(2-pyridyl)-1,3,5-triazine	$C_{18}H_{12}O_6$	1.9	1.8	2.0	1.9	2.0	1.9	1.69	1.61	1.59	1.58	1.59	1.61	84.7
		4.0	4.0	3.9	3.6	3.7	3.8	3.48	3.50	3.30	3.28	3.15	3.34	87.9
1,10-phenanthroline	$C_{12}H_{10}ON_2$	2.0	2.0	1.9	–	–	2.0	1.94	1.90	1.92	–	–	1.92	96.0
		3.9	4.1	4.0	–	–	4.0	3.75	3.89	3.84	–	–	3.83	95.8
Methylene Blue	$C_{16}H_{18}N_3ClS$	1.8	1.8	1.7	–	–	1.8	1.72	1.80	1.77	–	–	1.76	97.8
		3.6	3.4	3.5	–	–	3.5	3.51	3.50	3.54	–	–	3.52	101
Methyl Orange	$C_{14}H_{14}O_3N_3SNa$	2.3	2.4	2.3	–	–	2.3	2.25	2.18	2.20	–	–	2.21	96.0
		4.4	4.3	4.4	–	–	4.4	3.92	4.07	4.00	–	–	4.00	90.9
Orange II	$C_{16}H_{11}O_4N_2SNa$	1.5	1.5	1.5	–	–	1.5	1.48	1.51	1.41	–	–	1.47	98.0
		3.0	2.9	2.9	–	–	2.9	2.93	2.81	2.77	–	–	2.84	97.9

Note : Figures in the upper and lower columns were obtained when the sample chemical concentration was about 2 and about 4 mg C/ℓ, respectively.
Source: From Refs. 3–5.

As is seen from Tables 1 and 2, the decomposition rate is low in chemical compounds with high molecular weights, such as cyclic nitrogen compounds [2,4,6-tri-(2-pyridyl)-1,3,5-triazine and 1,10-phenanthroline], azo compounds (methyl orange and orange II), peptone, and humic acid. This method may not achieve complete oxidation decomposition depending on the organic compound.

However, it appears that the wet UV oxidation method used with an oxidant is applicable to highly refined water in which the concentration of organic compounds is extremely low and the molecular weight is low.

Table 2 TOC Determination of Various Organic Compounds

Organic Compound	Chemical Formula	TOC by Combustion-Infrared Oxidation (mg C/L)						TOC by Persulfate-UV Oxidation (mg C/L)						Average Rate of Oxidation (%)
		1	2	3	4	5	Average	1	2	3	4	5	Average	
Decyl sodium sulfate	$C_{10}H_{21}O_4SNa$	2.1	2.1	–	–	–	2.1	2.03	2.04	–	–	–	2.03	96.7
		4.0	3.9	–	–	–	4.0	3.89	3.84	–	–	–	3.87	96.8
Dodeyl sodium sulfate	$C_{12}H_{25}O_4SNa$	2.0	2.0	–	–	–	2.0	2.01	2.06	–	–	–	2.04	102
		4.0	3.9	–	–	–	4.0	3.97	3.92	–	–	–	3.95	98.8
Tetradecyl sodium sulfate	$C_{14}H_{29}O_4SNa$	2.0	2.1	–	–	–	2.1	2.04	2.00	–	–	–	2.02	96.2
		3.8	4.0	–	–	–	3.9	3.90	3.85	–	–	–	3.88	99.5
Ethane sodium sulfonate	$C_2H_5O_3SNa$	1.9	1.9	–	–	–	1.9	2.00	1.96	–	–	–	1.98	104
		3.6	3.6	–	–	–	3.6	3.79	3.80	–	–	–	3.80	106
1-butane sodium sulfonate	$C_4H_9O_3SNa$	2.0	2.0	–	–	–	2.0	2.04	2.05	–	–	–	2.05	103
		3.7	3.7	–	–	–	3.7	3.97	3.95	–	–	–	3.96	107
1-hexane sodium sulfonate	$C_6H_{13}O_3SNa$	1.7	1.8	–	–	–	1.8	1.98	1.94	–	–	–	1.96	109
		3.4	3.2	–	–	–	3.3	3.74	3.68	–	–	–	3.71	112
1-octane sodium sulfonate	$C_8H_{17}O_3SNa$	2.0	1.8	–	–	–	1.9	2.01	2.02	–	–	–	2.02	106
		3.6	3.9	–	–	–	3.8	3.86	3.84	–	–	–	3.85	101
1-decane sodium sulfonate	$C_{10}H_2O_3SNa$	1.9	1.8	–	–	–	1.9	2.03	2.04	–	–	–	2.04	107
		3.7	3.5	–	–	–	3.6	3.90	3.89	–	–	–	3.90	108
Dodecyl benzene sodium sulfonate	$C_{18}H_{29}O_3SNa$	1.9	1.9	–	–	–	1.9	1.92	2.15	–	–	–	2.04	107
		3.7	3.7	–	–	–	3.7	3.75	3.71	–	–	–	3.73	101
Peptone	–	2.5	2.2	2.3	–	–	2.2	1.99	2.04	2.03	–	–	2.02	91.8
		4.3	4.3	4.2	–	–	4.3	3.99	4.07	4.03	–	–	4.03	93.7
Humic acid	–	2.7	2.6	–	–	–	2.7	0.75	–	–	–	–	0.75	27.8
		4.9	4.9	–	–	–	4.9	1.45	–	–	–	–	1.45	29.6
		4.6	4.3	4.5	–	–	4.5	2.41	2.33	2.20	–	–	2.31	51.3
Fulvic acid	–	3.5	3.0	3.2	3.3	3.2	3.2	1.86	1.68	1.69	1.72	1.60	1.71	53.4
		5.8	5.3	5.5	5.3	5.1	5.4	3.30	2.98	2.99	3.02	3.13	3.08	57.0

Note : Figures in the upper and lower columns were obtained when the sample chemical concentration was about 2 and about 4 mg C/ℓ, respectively.

Source: From Refs. 3–5.

C. UV Oxidation: Specific Resistance TOC Analysis

The sample supplied continuously to the measuring instrument is fed to a TOC-measuring oxidation reaction tube with a built-in ultraviolet source to decompose the carbon in organic compounds into organic acid and carbon dioxide. The change in the specific resistance of the water sample is then determined and converted into TOC concentration.

The detection range is 1–1000 μg C L^{-1}, which varies depending on the system and measuring conditions.

D. Other TOC Testing Methods

Recently developed is an automatic measuring instrument that performs continuous decomposition under conditions nearly equivalent to those (high-temperature pressurization with oxidant) of the combustion-oxidation method and enables determination of the carbon in organic compounds with low boiling points. The TOC testing method using this instrument is described here [5].

The sample is acidified using persulfate and is fed into a reactor with a temperature of 200°C and pressure at about 20 kg cm^{-2} to transform the carbon in organic compounds into carbon dioxide. The sample is then fed, together with carrier gas, into the nondispersive infrared gas analyzer to measure the concentration and thus the total carbon concentration. Sample is also separately acidified and fed into the pressurization reactor at a temperature (about 130°C) at which organic compounds are not decomposed. The carbon dioxide produced is fed, together with carrier gas, into the nondispersive infrared gas analyzer to measure the concentration and thus the concentration of inorganic carbon. The concentration of inorganic carbon is subtracted from that of total carbon to determine the TOC concentration (Fig. 1).

E. Conversion to an Automatic Measuring Instrument

Organic compounds remaining in highly refined water are extremely low in quantity, and the amount of component carbon is naturally very small, too. Accordingly, the concentration of organic carbon falls in the extremely small concentration range of 1–10 μg C L^{-1}. For determination within this range, either the sample is taken from the main pipeline to the sample container to be brought back to a laboratory for testing with an analyzer, or the sample is introduced directly from the main pipeline via a sampler to an automatic measuring instrument. Comparison of these two methods shows that the latter offers highly reproducible test results. Accordingly, various automatic TOC measuring instruments have been marketed for continuous measurements to determine organic carbon in ultrapure water in the semiconductor industry.

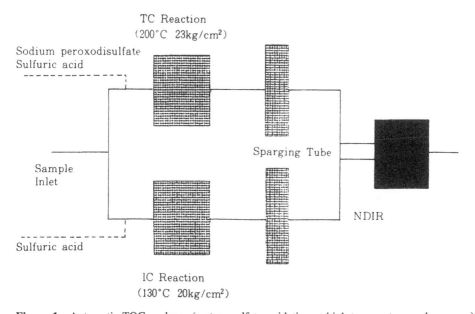

Figure 1 Automatic TOC analyzer (wet persulfate oxidation at high temperature and pressure).

III. ANALYSIS OF OTHER ORGANIC COMPOUNDS

A. Analysis of Organic Compounds with Low Boiling Points

The percentage of organic compounds is generally low in ordinary natural water but is often high in water recovered for use from wash water in the electronics industry. The organic compounds with low boiling points referred to in this section are organic compounds with boiling points below that of water.

These organic compounds with low boiling points have low molecular weights (carbon numbers 1–6) and include alcohols (methanol, ethanol, and 2-propanol), ketone (acetone and 2-butanone), ester (ethyl acetate), ether (diethyl ether), halogenated carbon hydroxide (carbon tetrachloride, 1,1,1-trichloroethane, trichloroethylene, and Freon), and organic acids.

Since the amount of these organic compounds with low boiling points in recovered water is relatively small, gas chromatography is applied to determine the amount of these compounds. In this case, the sample is injected into a gas chromatography unit directly or immediately after distillation and condensation, or a certain amount of gas phase separated according to the head space method [6] is injected. Note [6] that when the sample (mostly the liquid) is stored in a closed container, the space above the sample liquid surface in the container is called the head space. When enough sample to leave a space in a container of a certain volume is removed and the remainder is allowed to sit at a certain temperature for a certain amount of time, the gas and liquid enter an equilibrium state. In the head space method, a certain amount of the gas phase in this gas-liquid equilibrium state is sampled to determine the content of the gas phase.

The detection limit of a method in which the sample is directly injected into the gas chromatograph is 50 μg L^{-1} for methanol, ethanol, acetone, 2-propanol, dimethyl ether, and 2-butanone. The detection limit for head space gas chromatography is nearly equal to that attained by directly injecting the sample into the gas chromatograph.

Table 3 shows the detection range of distilled and condensed samples using gas chromatography. The range may vary depending on the system and measuring conditions.

B. Analysis of Halogenated Carbon Hydroxides with Low Molecular Weights

These are obtained by replacing the carbons in carbon hydroxides with carbon numbers 1 and 2 by halogen. Chlorinated carbon hydroxides include 1,1,1-trichloroethane, trichloroethylene, tetrachloroethylene, and carbon tetrachloride; trihalomethanes include trichlo-

Table 3 Detection Limits for Samples Concentrated by Evaporation

	Detection Limits
Methanol	$0 \sim 500 \, \mu$g CH$_3$OH/ ℓ
Ethanol	$5 \sim 500 \, \mu$g C$_2$H$_5$OH/ ℓ
Acetone	$5 \sim 500 \, \mu$g CH$_3$COCH$_3$/ ℓ
2-propanol	$5 \sim 500 \, \mu$g CH$_3$CH (OH) CH$_3$/ ℓ
Diethyl Ethene	$5 \sim 500 \, \mu$g C$_2$H$_5$OC$_2$/ ℓ
2-butanone	$5 \sim 500 \, \mu$g CH$_3$COC$_2$H$_5$/ ℓ

Table 4 Detection Limits of Organic Solvent Extraction–Gas Chromatography

	Detection Limits
1, 1, 1-trichloroethylene (CH$_3$CCℓ_3)	0.01 ~0.2ng
Trichloroethylene (C$_2$HCℓ_3)	0.04 ~0.75ng
Tetrachloroethylene (C$_2$Cℓ_4)	0.01 ~0.2ng
Carbon Tetrachloride (CCℓ_4)	0.0025 ~0.05ng

romethane (chloroform), bromodichloromethane, dibromochloromethane, and tribromomethane (bromoform). These eight materials are generally termed low-molecular-weight halogenated carbon hydroxides.

These halogenated carbon hydroxides with low molecular weights exist, although in extremely small amounts, in environmental water and plant effluent. Chlorinated carbon hydroxides are widely used, mainly in detergents. Among these materials, trichloroethylene and tetrachloroethylene are subject to discharge regulation covered by the Water Quality Pollution Prevention Act.

Trihalomethanes are generated during chlorine disinfection in the final stage of water purification when natural water is supplied as city water.

Chlorinated carbon hydroxides may coexist, although in extremely small quantity, in water recovered in the semiconductor industry.

To test halogenated carbon hydroxides with low molecular weights, the organic solvent extraction–gas chromatography method or the head space–gas chromotography method is applicable. Generally, the former method is recommended. Table 4 lists the detection range, which may vary depending on the system and measuring conditions.

C. Analysis of Trihalomethanes

The solvent extraction–gas chromatography method is applicable to test for trihalomethanes. In this method, trihalomethanes in the sample are extracted with hexane, and the hexane layer is charged into the gas chromatograph for determination. Table 5 shows the detection range, which may vary depending on the system and measuring conditions.

D. Analysis of Dissolved Organic Halogen (DOX)

The amount of halogen in the total quantity of halogenated carbon hydroxide of low molecular weight absorbed on activated carbon is determined as an equivalent amount of chlorine and expressed in milligrams or micrograms in 1 L of sample.

Table 5 Detection Limits for Trihalomethanes

	Detection Limits
Trichloromethane (CHCℓ_3)	0.025 ~ 0.5ng
Bromodichloromethane (CHBrCℓ_2)	0.006 ~0.12ng
Dibromochloromethane (CHBr$_2$Cℓ)	0.01 ~ 0.2ng
Tribromomethane (CHBr$_3$)	0.05 ~ 1 ng

In this method, halogenated carbon hydroxide of low molecular weight in the sample is absorbed on a small amount of activated carbon, and halide ions (mainly chloride ion) absorbed on the activated carbon are washed and removed with a weak acid (pH 2) solution of potassium nitrate. The total amount of activated carbon is then burned in the combustion furnace and transformed into hydrogen halide and absorbed into absorptive solution. The hydrogen halide thus produced is subjected to coulometric titration to determine the halide ion as chloride ion. In this way, the amount of organic halogen is determined. This method was first defined as TOX in the 1985 Standard Methods (16th ed.) but was renamed DOX in the 1989 Standard Methods (17th ed).

The method is intended to determine, for chlorine, the total amount of halogen in the low-molecular-weight halogenated carbon hydroxide absorbed on activated carbon. Different from the gas chromatography, therefore, this method cannot perform fractional determination of low-molecular-weight halogenated carbon hydroxide.

The detection limit of this method is 8 μg Cl L^{-1}, assuming that 50 ml sample is passed through a column filled with a small amount (for example, 40 mg) of particulate activated carbon. Since even the quality activated carbon used for absorption contains chloride at around 10 μg Cl g^{-1} activated carbon (about half is left even after washing with the pH 2 potassium nitrate solution), the no-load test value becomes around 4 μg Cl L^{-1}. The detection sensitivity is therefore inferior to that of gas chromatography.

The analyzer is required to have a combination of a combustion furnace and coulometric titration unit and an absorption column. Table 6 shows the results of measuring total organic halide at each point in an ultrapure water production line.

E. Organic Acid

This section describes test methods for low-molecular-weight fatty acids (including volatile acids), with carbon numbers 1–6. To determine organic acids, ion-exchange chromatography, high-performance liquid chromatography, and steam distillation-titration are applicable. Among these methods, the one that ensures determination with the highest sensitivity is the ion-exclusion ion-exchange chromatography.

This ion chromatography exclusion (ICE) method was developed by researchers at Dow Chemical, as was ion-exchange chromatography [6, 7]. This screening method is based on the fact that ionic substances pass rapidly through exclusion from anion- or cation-exchange resin and nonionic substances are caught by ion-exchange resin and are thus slow to migrate.

Rich et al. [8] at Dionex introduced this method in 1979 to ion-exchange

Table 6 Concentrations of DOX at Strategic Points in an Ultrapure Water Production Plant

Sample Water	DOX
Industrial Water	24.2 (μgCℓ^-/ℓ)
Filtered Water	197.4 (μgCℓ^-/ℓ)
Ion Exchanged Water	10.8 (μgCℓ^-/ℓ)
RO Permeate Water	8.3 (μgCℓ^-/ℓ)
Ion Exchanged Water (mixed – bed polisher)	4.5 (μgCℓ^-/ℓ)
UF Permeate Water	6.0 (μgCℓ^-/ℓ)

Table 7 Detection Range of Ion Chromatography

	Detection range
Formic Acid (HCOOH)	$0.2 \sim 4\,mg/\ell$
Acetic Acid (CH_3COOH)	$0.2 \sim 4\,mg/\ell$
Succinic Acid $[CH_2COOH]_2$	$0.2 \sim 4\,mg/\ell$
Tartaric Acid $[CH(OH)COOH]_2$	$0.2 \sim 4\,mg/\ell$
Propionic Acid (CH_3CH_2COOH)	$0.25 \sim 5\,mg/\ell$
1-methylpropionic Acid ($(CH_3)_2COOH$)	$0.5 \sim 10\,mg/\ell$
Butyric Acid $CH_3(CH_2)_2COOH$	$0.5 \sim 10\,mg/\ell$
1-methylbutyric Acid ($(CH_3)_2CHCH_2COOH$)	$0.5 \sim 10\,mg/\ell$
Valeric Acid ($CH_3(CH_2)_3COOH$)	$0.5 \sim 10\,mg/\ell$
Phthalic Acid ($C_6H_4(COOH)_2$)	$0.5 \sim 10\,mg/\ell$
Benzoic Acid (C_6H_5COOH)	$2 \sim 40\,mg/\ell$

chromatography to develop a method of fractional determination of weak acids (e.g., organic acids) by application of the theory of Donnan membrane equilibrium between the ion-exchange resin and the migration phase and incorporation of a method of separating strong from weak electrolytes.

The detection range of organic acids in the sample using ion-exchange (exclusion) chromatography is shown in Table 7. The reproducibility is 3–10%, with variations in methods taken into account.

IV. CONCLUSIONS

Described here are some of the analytic methods used for organic compounds. Apart from these methods, gas chromatography–mass spectroscopy (GC-MS) and liquid chromatography–mass spectroscopy (LC-MS) can also be used to analyze individual organic substances.

Subjects for future analysis of organic compounds in ultrapure water are as follows:

1. The water quality demanded of ultrapure water is increasing along with enhancement of the degree of integration, with metallic elements present in the ppt. Study must be focused on organic compounds so that detection of extremely small amounts becomes possible.
2. Faster analytic methods need be established for product control.

REFERENCES

1. C. E. Van Hall, J. Safranko, and V. A. Stenger, Anal. Chem., *35*, 315 (1969).
2. C. E. Van Hall, D. Barth, and V. A. Stenger, Anal. Chem., *37*, 769 (1965).
3. T. Sakamoto and T. Tomisaka, Synopsis of Lectures in the 21st Presentation of Japan Industrial Water Association, 1987, p. 76.
4. T. Sakamoto, Japan Industrial Water, *356*, 6 (1988).
5. T. Sakamoto, Journal of Chemistry, MOL, *308*, 73 (1988).

6. R. M. Wheaton and W. C. Bauman, Ind. Eng. Chem., *45*, 228 (1953).

7. D. A. Simpson and R. M. Wheaton, Chem. Eng. Prog., *50*, 45 (1954).

8. W. Rich, F. Smith, Jr., L. McNeil, and T. Sidebottom, Ion Chromatographic Analysis of Environmental Pollutants, Vol. 2 (J. D. Mulik and E. Sawicki, eds.), Ann Arbor, MI, 1979, p. 14.

6
Trace Metals and Ions

Yukinobu Sato
Kurita Water Industries, Ltd., Atsugi, Japan

I. ION-EXCHANGE CHROMATOGRAPHY

Ion-exchange chromatography has been applied extensively to analyses of the environment as well as to analyses in geochemistry, food chemistry, pharmaceutics, and laboratory experiment. Standardization of ion-exchange chromatography has been a challenge in each field. This chapter deals with the application of ion-exchange chromatography to the analysis of anions in ultrapure water. Titration, absorptiometry, and nephelometry have been conventionally used to determine the quantity of ions, such as chloride ion, sulfuric ion, fluoride ion, nitric ion, and phosphoric ion. However, these analytic methods have imposed burdens on those carrying out the analysis, since analysis of each ion must be done, requiring relatively large amounts of various chemicals, not to mention the complicated analytic operation and precautions required against contamination during operation. In contrast, ion-exchange chromatography consumes fewer chemicals and operation is simple. It is also possible to analyze these ions consecutively in a short amount of time, 15 or 20 minutes. In addition, with a sample condensed in advance by concentrator column, chromatography can easily analyze low concentrations of 1 μg L^{-1} or less.

A. System Composition

Figure 1 shows the composition of a system for the analysis of anionic components [1,2]. The system is mainly composed of eluant tank, feed pump, sample feeder, separator column, suppressor column, and detector. There are two systems available for the suppressor column, which removes electrolyte from the eluant discharged from the separator column: one is a double-column system equipped with a suppressor column, and the other is a single-column system that does not need the suppressor column. For trace analysis, a suppressor column should be installed.

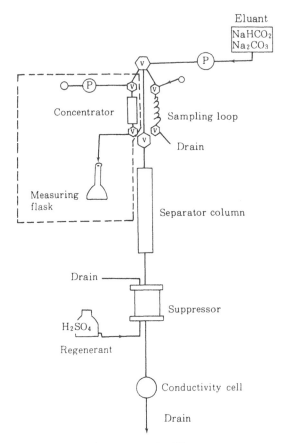

Figure 1   Ion chromatography [1].

1.   Sample Feeder

A syringe is used to take a sample and inject it into the sampling loop (normal capacity of 50 or 100 μl) through an injection valve. Generally, a certain quantity of sample held in the sampling loop is fed to the separator column by operation of the valve. In analyzing a large quantity (up to around 100 ml) of sample, a specified quantity of sample is fed to a concentrator column by metering pump and concentrated there beforehand. Eluant is then fed to the concentrator column by valve switching operation to elute the object ion from the column and send it to the separator column together with the eluant.

2.   Separator Column

The separator column for anion analysis uses a surface-coated pellicular anion-exchange resin whose capacity is as low as approximately 0.01–0.05 mEq g^{-1}. Figure 2 shows the structure of the resin [3]. Fine particles (0.1–0.5 μm size) of styrene-divinylbenzene copolymer coat the surface of a particle (5–15 μm size) made of the same copolymer. The surface of the coated particle is first chloromethylated and then aminated, whem quaternary ammonium base (I) is introduced.

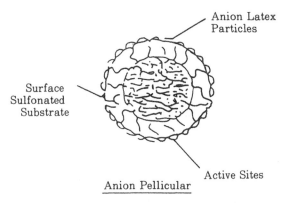

Figure 2 Latexed anion-exchange particle [3].

3. Suppressor Column

The suppressor column aims at removing the background electrolytes of the eluant that has passed through the separator column, converting the electrolytes to water or a substance with a low electrical conductivity and feeding only the target ion to the detector of the conductivity meter, thereby increasing the sensitivity of the detector. Primitive suppressor columns used a column containing cation-exchange resin. However, its frequent regeneration and onerous operation required improvements. As a result, the continuous regeneration ion-exchange membrane (hollow-fiber) and micromembrane were developed, which are widely used today. Figure 3 shows the principle of the hollow-fiber ion suppressor column. In anion analysis, regenerant flows on the external

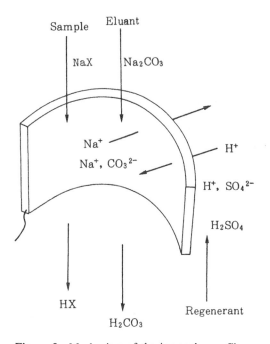

Figure 3 Mechanism of the ion-exchange fiber.

surface of a strongly acid cation-exchange membrane while the eluant flows on the internal surface.

When sodium carbonate is used as eluant, sodium ion (Na^+) is exchanged with hydrogen ion (H^+) so that sodium carbonate is transformed into carbonic acid with a lower electrical conductivity. Sodium hydroxide as the eluant is likewise converted to water (H_2O). The anions subject to analysis are sent to the detector for the measurement of their electrical conductivity, in the form of H^+X strong acid (X: F^-, Cl^-, NO^{2-}, NO^{3-}, SO_4^{2-}, and so on).

B.   Principle

The following is an example of the reactions in anion analysis. The eluant (Na_2CO_3) and sample (objective ion Na^+X^-) passing through the separator column dissociate ions successively by ion-exchange chromatography. The reaction formula is

$$\text{Resin } N^+HCO^{3-} + Na^+X^- \rightleftharpoons \text{resin } N^+X^- + Na^+HCO^{3-}$$

where X is F^-, Cl^-, NO^{2-}, NO^{3-}, SO_4^{2-}, and so on. The suppressor column then removes the counter ions by ionic reaction:

$$\text{Resin } SO_3H^+ + Na^+HCO_{3-} \rightarrow \text{resin } SO_3Na^+ + H_2CO_3$$

$$\text{Resin } SO_3H^+ + Na^+X \rightarrow \text{resin } SO_3Na^+ + H^+X$$

The ions and eluant flow into the detector in the form of H^+X (strong acid) and H_2CO_3, respectively, where the electrical conductivity of H_2CO_3 + H^+X (strong acid) and H_2CO_3, is measured and indicated on the detector. Figure 4 shows an ion chromatogram.

C.   Detection Limit

The detection limit generally differs depending on the apparatus and measuring conditions. Table 1 shows the values obtained with the Dionex Model 2020.

II.   FLAMELESS ATOMIC ABSORPTION SPECTROSCOPY

Atomic absorption spectroscopy and inductive coupling plasma analysis (ICP) are ordinarily used for the analysis of heavy metals. There are two types of atomic absorption spectrocopy: flame atomic absorption spectroscopy utilizes an acetylene flame and air or dinitrogen monoxide for atomization, and flameless (furnace) atomic absorption spectroscopy conducts atomization in a graphite tube electric heating furnace [4,5]. Because the analysis of heavy metals in ultrapure water requires high sensitivity, flameless atomic absorption is often employed. The following sections describe this method.

A.   Principle

If a sample solution containing metallic salt is sprayed into the flames (2000 or 3000°C), it is vaporized and transformed into fine particles and finally reduced to atoms by thermal decomposition. Since a majority of the atoms formed in this process are in the ground state, if light of a specific wavelength is irradiated to a specific atom in the ground state, the light is absorbed by the atom. For atomization, flameless atomic absorption spectroscopy utilizes Joule's heat, which is generated by charging a graphite tube heating furnace with electricity instead of flame.

$$\text{ResinSO}_3\text{H}^+ + \text{Na}^+\text{HCO}_3^- \longrightarrow \text{ResinSO}_3\text{Na}^+ + \text{H}_2\text{CO}_3$$

$$\text{ResinSO}_3\text{H}^+ + \text{Na}^+\text{X} \longrightarrow \text{ResinSO}_3\text{Na}^+ + \text{H}^+\text{X}$$

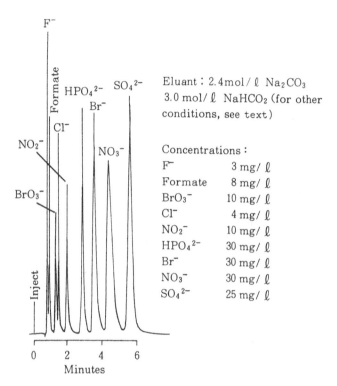

Eluant : 2.4 mol/ℓ Na₂CO₃
3.0 mol/ℓ NaHCO₂ (for other conditions, see text)

Concentrations :

F⁻	3 mg/ℓ
Formate	8 mg/ℓ
BrO₃⁻	10 mg/ℓ
Cl⁻	4 mg/ℓ
NO₂⁻	10 mg/ℓ
HPO₄²⁻	30 mg/ℓ
Br⁻	30 mg/ℓ
NO₃⁻	30 mg/ℓ
SO₄²⁻	25 mg/ℓ

Figure 4   Ion chromatogram.

Table 1   Detection Limits

ion	loop method	concent method
Cl⁻ (μg/ℓ)	50	0.05
NO₂⁻ (μg/ℓ)	50	0.05
PO₄³⁻ (μg/ℓ)	100	0.5
Br⁻ (μg/ℓ)	100	0.1
NO₃⁻ (μg/ℓ)	100	0.1
SO₄²⁻ (μg/ℓ)	100	0.1
amount of the sample (mℓ)	0.05	50

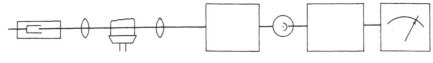

light source   atomization part   optical system detector   readout

Figure 5   Design of the atomic absorption spectrophotometer.

B.   System Composition

Figure 5 shows the basic composition of an atomic absorption spectrophotomer. The system is basically composed of light source, atomization section, spectroscope, and photometer.

1.   Light Source

The light source is a hollow cathode lamp whose structure is shown in Fig. 6. As shown, an anode and a hollow cathode are sealed in a glass tube whose beam window is made of quartz, and the tube is charged with rare gas, such as argon or neon, at low pressure (4 to 10 torr). The cathode is made of the heavy metal subject to analysis, or its alloy. When 200 or 450 V is applied between the electrodes, the argon is ionized and positively charged (Ar^+); it then collides against the cathode under acceleration. The energy of the collision excites the metal of the cathode to atoms. If these atoms in the ground state collide with argon, they are excited. When the excited atoms return to the ground state, they emit light of a wavelength peculiar to the element. This light is used as an illuminant.

2.   Atomizer

Flameless (furnace) atomic absorption spectroscopy generally uses a graphite tube for atomization. As shown in Fig. 7, the sample is injected into a molded graphite tube and heated in three phases. The first phase is drying at around 100°C to remove water from the sample. The second phase is ashing at 400 or 1000°C to remove organic substance. The last phase is atomization of the target element at 1400 or 3000°C. Normally, the operation repeats these three phases for atomization. Besides the electric heating furnace, a metal furnace made of tungsten or tantalum is also available for flameless atomization. Although the metal furnace is characterized by a rapid rise in temperature and no carbide generation, because of restrictions in the combination of target elements and furnace material, it is regarded as inferior to the graphite furnace from the viewpoint of general applicability.

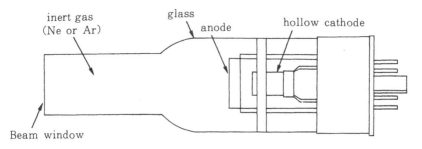

Figure 6   Hollow cathode lamp.

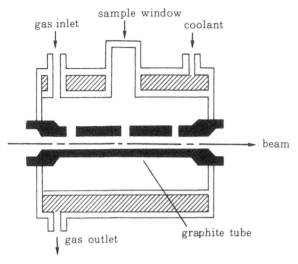

Figure 7 Flameless atomizer.

C. Requirements for Measurement

1. Prevention of Contamination

For the analysis of trace metals, it is necessary to take precautions against contamination by the instruments, tools, and chemicals used in the analysis, and also the environment.

1. Conduct acid rinsing of the sample container beforehand to prevent contamination. Do not use the container for sodium, calcium, magnesium, or zinc, which can be contaminated by glass.
2. Use purified reagents.
3. Use a clean laboratory for pretreatment and quantitative analysis. It is better to use a glove box or a clean bench.

2. Requirements for Analytic Operation

1. Do not increase the drying temperature drastically or the sample solution will splash out, resulting in fluctuation in the analytic values.
2. Monitor the ashing time and temperature, or volatile elements, such as lead, zinc, cadmium, and copper, will vaporize at 500°C or above.
3. Set the optimum atomizing temperature for individual elements. The optimum temperature depends on the composition of the sample (type and concentration of salts or of acid), except for ultrapure water.

D. Detection Limit

Table 2 shows the general lower limits of detection for major elements, although the limits are affected by the analysis system and conditions.

E. Inductively Coupled Plasma–Atomic Emission Spectrometry (ICP-AES) and Inductively Coupled Plasma–Mass Spectrometry (ICP-MS)

ICP-AES is an analytic method using high-frequency inductively coupled plasma, which causes the elements in the sample to emit light whose wavelength and strength are

Table 2 Detection Limits for FLL-AA

element	wavelength (nm)	D.L. (μg/ℓ)	element	wavelength (nm)	D.L. (μg/ℓ)
Na	589.0	1.0	Fe	248.3	5.0
K	766.5	0.8	Co	240.7	10
Ca	422.7	2.0	Ni	232.0	25
Mg	285.2	0.4	Cu	324.8	5.0
Ba	553.6	25	Zn	213.9	0.2
Cr	357.9	2.0	Cd	228.8	0.1
Mn	279.5	1.0	Al	309.3	5.0

measured to identify the type and concentration of the elements. The detection limit is between those of flame atomic absorption and flameless atomic absorption, but for some elements, ICP-AES demonstrates a sensitivity higher than that of flameless atomic absorption spectroscopy. The method is characterized by less chemical interference and a simple measurement principle. ICP-MS is an analytic method using high-frequency inductive coupling plasma as the source of ions, which ionizes the elements in the sample. This method uses a mass spectrometer as a detector to analyze the ionized elements, and isotopes can also be analyzed. Since both analytic methods can manage the simultaneous analysis of multiple elements, they will be applied to the analysis of trace elements in ultrapure water. ICP-MS has a sensitivity higher than that of ICP-AES, so that it will be utilized for the analysis of ultratrace concentrations at the level of nanograms or picograms per liter, although the equipment is expensive.

REFERENCES

1. Y. Muto and K. Oikawa, Ion Chromatography, Kodansha, Inc., 1983.
2. H. Oya, Pure Water and Ultrapure Water Systems, Elemental Technology and Application System, Sachi Publishing, 1985, p. 137.
3. T. Nomura, Summary of lecture for Dionex Ion Chromatography Seminar, Abe Trading, 1984.
4. T. Takahashi and H. Daidoji, Furnace Atomic Absorption Analysis (Probing ultra trace quantity), Japan Spectrum Society, 1984.
5. Shimadzu Atomic Absorption Spectroscopy Seminar, Principle of Atomic Absorption Spectroscopy, Shimadzu Manufacturing Co., Ltd.

7

Quantitative Determination of Silica in Ultrapure Water

Ikuo Shindo

Japan Organo Co., Ltd., Saitama, Japan

I. DETERMINATION OF SILICA

The following methods are used to determine ion silica in water:

1. Molybdenum yellow photometry [1]
2. Molybdenum blue photometry [1]
3. Molybdenum blue extraction photometry [2,3]

Ion silica is a silica that reacts with ammonium molybdate to produce heteropoly compounds, $H_4[SiO_4MO_{12}O_{36}]^{4-}$. Silica is determined (1) by measuring the yellow absorbance of the heteropoly compound produced at a wavelength of around 410 nm, the yellow method; and (2) by measuring the absorbance of the heteropoly compound produced at a wavelength of around 815 nm after reduction of the heteropoly compound into molybdenum blue, the blue method. These methods are set forth in JIS K0101 and ASTM D859 [4]. Molybdenum blue extraction photometry is a method newly established in JIS K0555. The conditions for generating heteropoly compounds are the same in this method, except that about 1.3 mol L^{-1} of sulfuric acid molybdenum blue is generated and extracted to 1-butanol, and its absorbance is measured at a wavelength of 800 nm. This method has a higher sensitivity than the blue method. The operation procedures for each measurement method is outlined in Fig. 1.

The detailed procedure and precautions for each measurement method are omitted here because of limited space, and the pertinent documents [1–4] should be referred to. This chapter describes points to be considered when extremely small amounts of silica are measured in ultrapure water by referring mainly to the molybdenum blue extraction method.

Automatic silica measuring instruments are also manufactured by many companies [5]. Measuring instruments for low concentrations are mostly based on the blue method.

Also developed recently is a measuring instrument with built-in condensation column to measure the condensed silica in the sample according to the yellow method.

The detection limits of the measurement methods are shown in Table 1.

II. CONSIDERATIONS DURING MEASUREMENT OF EXTREMELY SMALL AMOUNTS OF SILICA

Table 1 shows the detection limits for the measurement methods, which can be further lowered by lowering the standard solution concentration during preparation of the detection line and by using ultrapure water with a low silica content for preparation of the standard solution.

If water to be used for preparation of the standard solution and reagent is totally free of silica, the normal no-load test is sufficient. Actually, however, water (e.g., water obtained by refining ion-exchange water in a stainless steel distiller) always contains

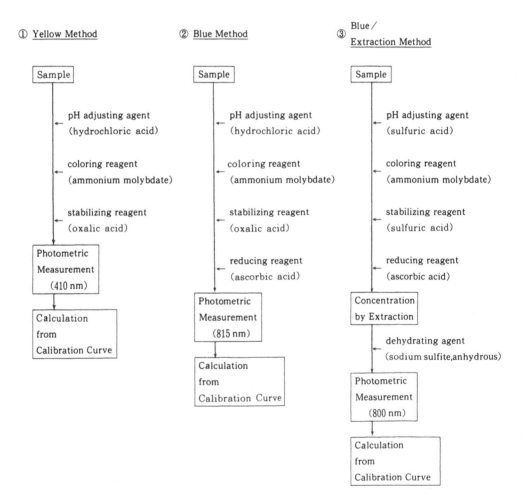

Figure 1 Procedure for silica measurement.

Table 1   Minimum Detection Concentration

Method	Yellow	Blue	Blue/Extraction
Minimum Detection Concentration	2 mgSiO$_2$/1	10 μg SiO$_2$ /1	2.5 μg SiO$_2$/1

around 0.5–1.5 μg SiO$_2$ L^{-1} of silica, and it is necessary to perform a correction by carrying out the special no-load test. This method is described here.

To determine silica in extremely small amounts, 1 μg SiO$_2$ L^{-1}, the special no-load test must be followed by molybdenum blue extraction photometry. Determination can be made according to Fig. 1 (for details, refer to JIS, ASTM, or Ref. 3). Compensation must be made in this case for the reagent while taking account of the following two points:

1.   Effect of silica in the reagent (silica compensation for the reagent)
2.   Effect of absorption of the reagent (coloration compensation for the reagent)

The true silica concentration in the reagent is determined by the following compensation equation with due consideration of the preceding points:

True silica concentration in reagent = measured value – silica compensation value in reagent – coloration compensation of reagent

A.   Silica Compensation Value for the Reagent

The silica concentration when reagents are added in a quantity twice as large as the standard is designated C_w. When the silica concentration with addition of reagents in standard quantity is designated C_s, $C_w - C_s$ is the silica concentration in the reagents.

B.   Coloration Compensation for the Reagent

This is determined from the absorbance obtained by adding reagents to ultrapure water according to the reverse procedure. Normally, reagents are added in the order of coloring agent, stabilizing agent, and reducing agent. If reagents are added in the opposite order—stabilizing agent, coloring agent, and reducing agent—no reaction occurs even when silica exists in the water. The absorbance in this case is considered absorption by the color of the reagent itself and is converted to the silica concentration. The figure thus obtained is used as a coloration compensation value for the reagent. This process is not considered necessary for the molybdenum blue extraction method.

The compensation value must be recalculated when new reagent is prepared. Typical compensation values for molybdenum blue extraction photometry and molybdenum blue absorption photometry are shown in Table 2.

The silica compensation value in reagent is high at 3.0–3.3 μg SiO$_2$ L^{-1} with molybdenum blue absorption photometry because the silica content is very high in the hydrochloric acid used for pH adjustment. The effect of hydrochloric acid amounted to 2.5–2.8 μg SiO$_2$ L^{-1}. There is almost no effect with the blue extraction method because sulfuric acid is used instead of hydrochloric acid. Since sulfuric acid is almost completely free of silica, it is advisable to replace hydrochloric acid by sulfuric acid in the molybdenum blue absorption method. Indeed, the use of sulfuric acid instead of hydrochloric acid could lead to a silica compensation value for the reagent nearly equivalent to that using the blue extraction method. The effect of silica contained in ammonium molybdate amounted to around 0.5 μg SiO_2 L^{-1} with other reagents in both methods.

The coloration compensation value for reagents is 0.3–0.6 μg SiO_2 L^{-1} with molybdenum blue absorption photometry. Similar compensation is necessary for various automatic silica measuring instruments.

III. CONCLUSIONS

Extremely small amounts of ion silica can be measured according to the blue extraction method.

The detection limit of this method is estimated as 0.5 μg SiO_2 L^{-1} in view of various considerations.

The absorbance of 0.001 with this method is equivalent to a silica concentration of 0.009 μg SiO_2 L^{-1}. That is, detection is possible to 0.1 μg SiO_2 L^{-1}. Considering that the sum of compensation values is actually 0.5–0.6 μg SiO_2 L^{-1}, the detection limit is currently around 0.5 μg SiO_2 L^{-1}. Below this level, reliability may not be assured.

When the silica concentration of actual plants is measured, many plants have a capacity well below the detection limit. To lower the detection limit further, the following means may be considered (an example of silica measurement in an actual plant is shown in Fig. 2):

1. Increase the sample amount to raise the condensation magnification through extraction according to the blue extraction method.
2. Extend the cell beam channel during measurement of absorbance to raise the sensitivity of the absorbance.

Table 2 Calibration for Reagents

absorbance ; (–logt)
concentration; μgSiO_2/1

Method		Silica in Reagents	Color in Reagents	Total
② Blue	absorbance	0.010 ~ 0.011	0.001 ~ 0.002	0.011 ~ 0.013
	concentration	3.0 ~ 3.3	0.3 ~ 0.6	3.3 ~ 3.9
③ Blue / Extraction	absorbance	0.005 ~ 0.006	0.000 ~ 0.001	0.005 ~ 0.007
	concentration	0.5	0.0 ~ 0.1	0.5 ~ 0.6

Industrial

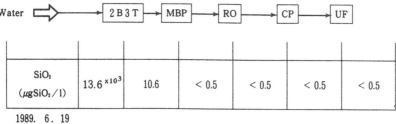

| SiO₂ (μgSiO₂/l) | 13.6 ×10³ | 10.6 | < 0.5 | < 0.5 | < 0.5 | < 0.5 |

1989. 6. 19

2 B 3 T	two bed three tower	RO	reverse osmosis unit
	demineralizing equipment	CP	cartridge polisher
MBP	mixed bed polisher	UF	ultrafiltration equipment

Figure 2 Results of measurement.

3. Reduce reagent addition to alleviate the effect of reagents because measurement is made in the low concentration range.

It is expected that the required water quality in terms of silica will be directed toward the lower concentration (e.g., 0.1 μg SiO$_2$ L^{-1}). Improvement and progress in the determination method, as well as development of a system to cope with such a trend, are essential.

REFERENCES

1. Japan Industrial Standard, JIS K0101.
2. Japan Industrial Standard, JIS K0555.
3. E. Nakamura, R. Sasai, and H. Namiki, Analytical Chemistry, *38*, 729 (1989).
4. ASTM D859.
5. DKK, Hach, Orion, Toray Engineering, Catalogue and Literatures for silica measuring instruments.

8
Colloids

Takaaki Fukumoto, Masaharu Hama, and Motonori Yanagi
Mitsubishi Electronic Corporation, Kumamoto and Hyogo, Japan

I. INTRODUCTION

Trace colloidal materials in ultrapure water are one of the causes of water spots (haze) sometimes observed as residues on the silicon wafer surface in the cleaning and drying steps of ultra–large-scale integrated (ULSI) circuit production. This chapter presents a relatively permeating water check method capable of detecting trace colloidal materials (ppt level) with high sensitivity. This method measures the membrane-clogging ratio by comparing the flux after a certain period of operation with the initial flux when a water sample of some 10s of $L \cdot day^{-1}$ is filtered through an ultrafine membrane. Furthermore, this chapter introduces some examples of the application of this method: to maintain the ultrapure water production system, to predict trouble, and to analyze the cause of trouble.

To eliminate all the characteristic problems of 0.1 μm design-ruled ULSI devices, the total impurity concentration in ultrapure water must be suppressed to below 2 ppt. It was recently reported that the current level of ultrapure water with a total impurity concentration of a few ppb cannot cope with 0.1 μm devices [1]. It was also reported that native oxide is formed when a silicon wafer is cleaned in ultrapure water containing some dissolved oxygen and that the silicon is dissolved in the ultrapure water in the native oxide growth process [2].

In the wet etching process of current ULSI production, the formation of water spots (haze) very sensitively reflects the reaction of the silicon surface and ultrapure water. Photographs of water spots (haze) formed on the silicon wafer are shown in Fig. 1. A water spot seems to be formed when the trace impurities (a few ppb or less) in ultrapure water are condensed and left on the silicon wafer surface. The trace amount of colloid in ultrapure water is considered one of the constituents of water spots. This is why colloidal materials are now often studied as one of the important analytic parameters in ultrapure water production for ULSI. This chapter describes the details of colloidal materials, the

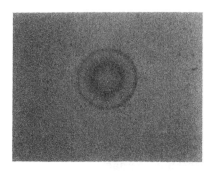

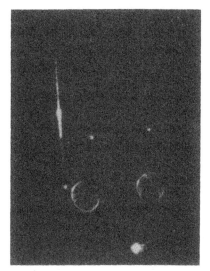

Figure 1 Haze on substrate surface.

colloid evaluation method, and the application of this evaluation method to ultrapure water production.

II. DEFINITION OF A COLLOID

There are mainly three kinds of colloid in ultrapure water: colloidal silica, colloidal organic materials, and colloidal metals. The size of a colloid is within the range 0.001–0.1 μm. Because a colloid is a combination of an ion or molecule and a particle, it changes form along with the change in the state of the ultrapure water, and it is unstable. Figure 2 shows the detectable size of materials by several analyzers. Among the currently available analytic monitors, the silica meter, the total organic carbon (TOC) meter, and a meter that measures the residue on evaporation can detect colloids at the ppb level.

Because of the instability of colloids, operation of the current ultrapure water production system is still suffering from unexpected trouble and ultrapure water purification technology is not capable of reducing the total impurity level below a few ppb. To be more specific,

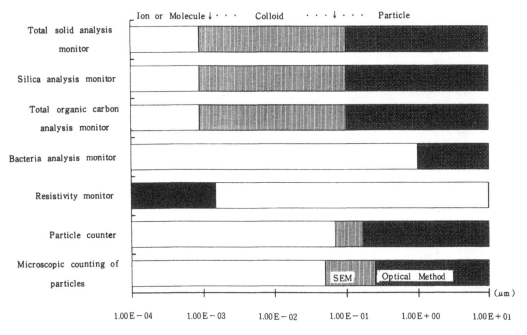

Figure 2   Colloid size range detectable by different monitors.

1. Insufficient technology of membrane separation (reverse osmosis, RO, and ultrafiltration, UF)
2. Clogging of membranes (RO and UF)
3. Release of the substances, which are immediately absorbed on the ion-exchange resin
4. Elution from the system components, including membrane modules (RO, UF, and membrane filtration, MF), ion-exchange resin, valves, distribution piping system, pumps, and flowmeters

A specific problem with the ultrapure water production system is now described: UF membrane clogging, which is thought to be caused by elution, was re-created in an experiment. In this experiment, two types of ion-exchange resins manufactured by the same company were arranged in parallel, and ultrapure water was supplied to the resins. Downstream of the ion-exchange resins were placed UF membranes manufactured by the same company. Initially the same pressure level was set and the same volume of water was supplied to the two lines. The fluctuation in the flux ratio of the UF membranes was then recorded. Figure 3 shows the experimental results. For the type B ion-exchange resin, the flux ratio began to decrease at the point where the feed water volume topped 50 m³. Although TOC, silica, and particles in the product water were measured constantly, no change in the ppb level could be detected at this point.

In this experiment, it was found that the degree of UF membrane clogging is affected merely by the type of ion-exchange resin. The possible cause of this phenomenon is that trace colloidal materials eluted or desorbed from the ion-exchange resin are gradually deposited on the UF membrane.

From a study of these phenomenon, which actually took place in an operating ultrapure water production system, the "permeating water check method" was developed

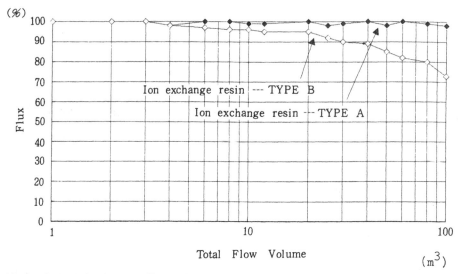

(%)

Flux

Total Flow Volume

(m^3)

Figure 3   Relation between flux and total flow volume in a laboratory UF Unit.

to evaluate trace colloidal materials. This method observes the degree of clogging of the MF membrane with a small pore size (0.1, 0.05, 0.03, and 0.01 μm) and the UF membrane when permeated with water.

III.   EVALUATION OF COLLOIDAL MATERIALS

In principle the permeating water check method is almost the same as the fouling index (FI) method and the silt density index (SDI) method, both of which are commonly used. The procedure is discussed briefly here.

The components used in this method are as follows:

1.   A membrane filter with a pore size of 0.8 μm
2.   A filter holder
3.   A valve to adjust the flow rate
4.   A small UF membrane exclusively used in laboratories
5.   Two tools to measure the volume of permeating water (e.g, measuring cylinder)

As shown in Fig. 4, these components are mounted on the sampling valve of the ultrapure water distribution piping system to be evaluated. The filter holder with a 0.8 μm MF and the valve unit for adjusting the flow rate are mounted on the right side in Fig. 4 for the purpose of compensating for the pressure fluctuation in the distribution pipe. The laboratory UF unit is mounted on the left side in an attempt to capture colloids. After mounting all the components, the volume of water permeating through the membrane filter is adjusted to about $\frac{1}{2}(Q_1 = \frac{1}{2}Q_2)$ of that permeating through the laboratory UF unit with the flow rate adjustment valve. The volume of water permeating through the MF and the laboratory UF unit (i.e., Q_{1a} and Q_{2a}) is then measured simultaneously to obtain the initial values 30 minutes after mounting the components. After 24 h, Q_{1b} and Q_{2b} are measured at the same time. The measurement data are recorded as the permeating water ratio Q_1/Q_2. The ratio is calculated to eliminate the effect of pressure fluctuations.

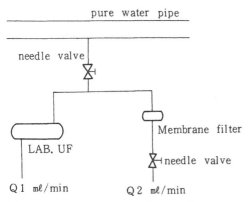

Figure 4   Measuring transmittance of pure water.

The number obtained when Q_{1b}/Q_{2b} is divided by Q_{1a}/Q_{2a} is regarded as the reference value for maintaining the ultrapure water production system. For example, the figure can be set at 80% or more [3]. Figure 5 shows the measurement results and the reference example. It is clear that the the the permeating water ratio b/a decreases as the water quality deteriorates.

This evaluation method is capable of measuring colloids, but it also has some problems. The advantages and the disadvantages of this method are listed here.

Advantages
   Detects trace colloids with high sensitivity that before were undetectable (detection of
      colloids as particles below the 0.07 μm in diameter is difficult).
   Able to predict trouble in the ultrapure water production system.

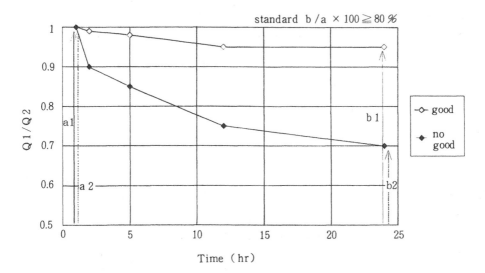

Figure 5   Relation between Q_1/Q_2 and time.

Disadvantages

Difficult to identify the trace colloids. To identify trace colloids, another analysis technology should be employed, which lowers the detection sensitivity (a few ppb). Therefore, in many cases, trace colloids cannot be identified.

Because it takes 24 h to complete the measurement, this method cannot be applied to detect trouble on the spot. (It is possible to conduct semicontinuous measurement by using many pieces of experimental equipment together.)

In addition to this permeating water check method, the silica meter, the TOC meter, and the meter that measures residue on evaporation can detect colloids. For comparison with the permeating water check method, the measurement principle of these meters is described here.

1. Silica meter (SLC-1605 manufactured by DKK). By adding ammonium molybdate to the water sample, silicomolybdic acid is formed. Phosphorus is masked next by adding tartaric acid. The water sample is then reduced to molybdate blue with ascorbic acid. This water sample is sent to the colorimeter to measure the absorbance at 860 nm. Based on a working curve that has been obtained in advance, the ionic silica concentration is calculated automatically. The detection limit is 0.2 ppb.

2. TOC meter (TOC-1000 manufactured by Tokico Co., Ltd.). The reaction solution (potassium persulfate + sulfuric acid) is added to the water sample. The water sample is then subjected to nitrogen gas bubbling to remove inorganic carbon. After this process, organic carbon in the water sample is oxidatively decomposed to CO_2 under high pressure and high humidity.

3. Nonvolatile residue meter (total solid; HPM-1000 manufactured by NMS) [4]. The water sample is atomized into the clean air and heated to about 140°C. In this way, the volatile materials in the water sample are evaporated and the nonvolatile materials form particles. The sample is then measured using a CNC (condensation nucleus counter). The colloid concentration is automatically calculated based on the working curve obtained earlier using a KC1 solution. The detection limit is 1 ppb.

It has been found that the values obtained with the meter that measures residue on evaporation have a very good correlation with those obtained with the permeating water check method. Figure 6 shows this correlation. In the permeating water check method in Fig. 6, the permeating water ratio when 50 L water is permeated is shown.

The remainder of this chapter presents examples of a situation in which this permeating water check method proved its effectiveness in maintaining the ultrapure water production system and in predicting trouble.

IV. APPLICATION OF PERMEATING WATER CHECK METHOD: HANDLING TROUBLE

An unprecedented problem occurred in the ultrapure water production system from October 1984 through February 1985 [5]. The details of the problem are as follows:

1. During this period, the front-stage RO membrane was clogged rapidly. The pressure difference exceeded the standard value in about 1 month.

2. The phenomenon did not change even after replacing the membrane.

3. When the quality of pure water hit the lowest level, the wafer surface was contaminated when it was cleaned with pure water.

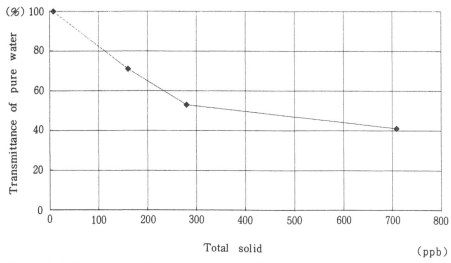

Figure 6   Relation between TS and transmittance of pure water.

Regular monitoring employed the high-sensitivity TOC meter, the permeating water check method, scanning electron microscopy (SEM), and XMA analysis. Figure 7 shows the TOC values, Fig. 8 shows the results of the permeating water check, Fig. 9 shows the SEM image of the membrane surface after the permeated water was checked, and Fig. 10 shows the results of XMA analysis at the SEM observation area. The findings of the analysis are as follows:

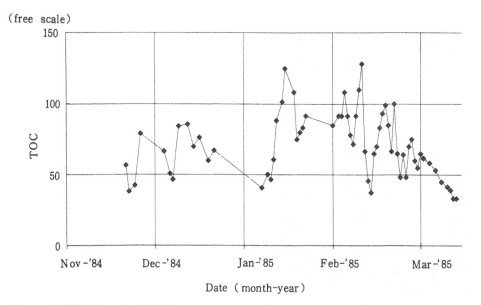

Figure 7   TOC data obtained during pure water production system trouble.

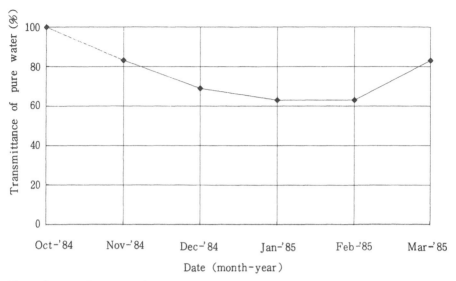

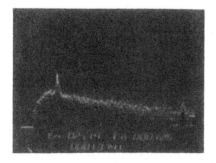

Figure 8   Data for the transmittance of pure water during pure water production system trouble.

Figure 9   Colloidal substance in pure water.

Figure 10   Colloidal substance on a 0.2 μm microfiltration filter (XMA).

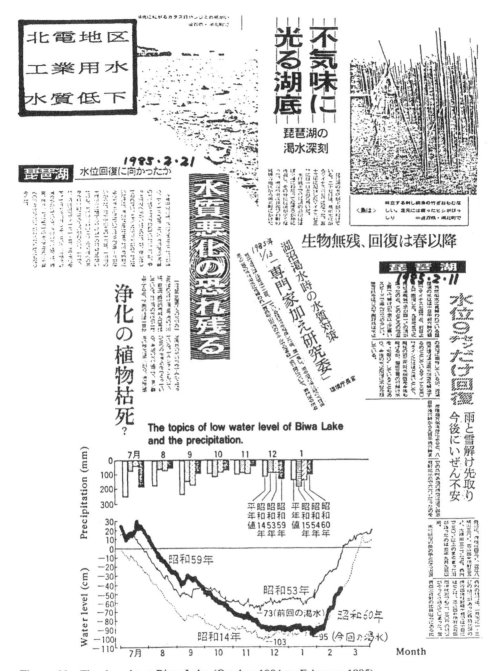

Figure 11 The drought at Biwa Lake (October 1984 to February 1985)

1. The measured TOC values were two to four times as much as the conventional TOC value.
2. The measured permeating water ratio was worse than the conventional value.
3. By SEM observation, substances similar to colloids were found on the membrane surface.

4. XMA analysis did not detect any metallic elements from the colloids.
5. Around that time, the water level in Lake Biwa was falling. The quality of supplied industrial water decreased. Figure 11 shows the articles from newspapers at the time.

Based on these results, it was concluded that organic materials and bacteria in the industrial water supply greatly increased because of the water shortage in Lake Biwa and the limitations on the water supply to the Yodo River, with which the pretreatment system of the ultrapure water production system could not cope. Thus organic materials penetrated the ultrapure water production system in large volume, and colloidal organic materials, which were especially hard to remove, caused membrane clogging and contamination of the wafers.

As countermeasures, the membrane was cleaned with caustic soda, the frequency of membrane flushing was increased, and the membrane was replaced.

The water shortage in Lake Biwa was overcome around February 1985. Membrane clogging ceased at the same time.

V. CONCLUSION

As a method of evaluating colloids, the permeating water check method is described here, together with its applications to the ultrapure water production system. Because this method uses a membrane for ultrapure water purification, not all colloids are captured. However, it is effective in detecting changes in the water quality with high sensitivity in practice. Since trace colloids are considered to affect both the ULSI production process and the ultrapure water production system, it is required to develop qualitative and quantitative evaluation methods for colloids at the ppt level [6].

REFERENCES

1. T. Mizuniwa et al., ULSI Ultra Clean Technology Symposium No. 7, Ultra Clean Society, Realize, Inc., 1988, p. 31.
2. T. Ohmi et al., ULSI Ultra Clean Technology Symposium No. 8, Ultra Clean Society, Realize, Inc., 1989, p. 169.
3. Japanese Patent Application 60-211572.
4. Y. Ota et al., ULSI Ultra Clean Technology Symposium No. 5, Ultra Clean Society, Realize, Inc., 1987, p. 65.
5. M. Yanagi, M. Hama, and T. Fukumoto, Senjo-Sekkei, *37*, 10 (1988).
6. M. Yanagi et al., Chemical Times, *116* (2), Kantoukagaku, 8 (1985).

9
Dissolved Oxygen

Nobuko Hashimoto
Hitachi Plant Engineering & Construction Co., Ltd., Chiba, Japan

I. INTRODUCTION

Dissolved oxygen (DO) is molecular oxygen dissolved in water. In both pure water and ultrapure water, DO is a potent influence on the reproduction of bacteria in the pipe and the formation of native oxide film on the wafer. The volume of DO in water depends on the water temperature and the partial pressure of oxygen in accordance with Henry's law, and the concentration of dissolved salt as well.

II. SAMPLING

The samplers and containers must be washed clean with the water to be sampled before use, and the containers must be run over the top opening and capped quickly so that air bubbles do not enter the water sample.

When the DO concentration is in the range from 1 to 10 μg O_2 l^{-1}, the sample container must be of the type that enables the water to fill it without the water being exposed to the atmosphere. When a portable DO meter is used, the electrodes must be submerged at the depth specified for direct measurement. Records of the place of sampling, temperature, atmospheric pressure, and humidity, among other parameters, must be kept.

III. PRECAUTIONS

Because DO is consumed by reductive materials, by the decomposition caused by organic substances or bacteria, and by the respiration of aquatic organisms, the DO concentration varies greatly according to a number of factors, for example, if the water was exposed to the open and how, temperature variations, and the quantities and properties of substances dissolved in the water. So much so that, when testing ultrapure water with an extremely low DO concentration, the water sample must be taken directly and tested as soon as possible.

IV. FACTORS INFLUENCING DO

The oxygen in the open air dissolves in water in accordance with Henry's law, and among the factors that influence its solubility are temperature, pressure, and the concentration of dissolved salt in the water.

A. Pressure

The change in the volume of DO in water caused by pressure can be expressed by the equation

$$S' = \frac{P}{760} S$$

where: S = solubility at standard atmospheric pressure (760 mm Hg)
S' = solubility when the atmospheric pressure changes to P (mm Hg)

B. TEMPERATURE

The volume of DO in water decreases according to increases in temperature, and the relationship between the volume and temperature is shown in Fig. 1.

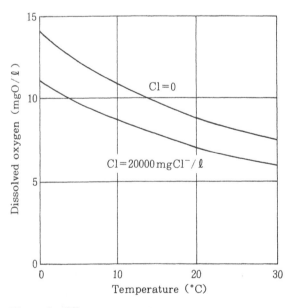

Figure 1   Effect on temperature.

C.   Water Vapor Pressure

Water vapor pressure, because it influences the overall pressure of the atmosphere, can be expressed by the following equation, taking humidity into account:

$$S'' = \frac{P - P''}{760 - P} S$$

where: P = saturated vapor pressure, mm Hg
P'' = vapor pressure of water obtained from the relative humidity

V.   QUANTITATIVE DETERMINATION

Methods for the quantitative determination of DO include Winkler's method (titration), colorimetry, and electrochemistry. For pure water and ultrapure water, whose DO concentration is so low, Winkler's method (titration) requires a great deal of skill in processing, and on occasion the measurement readings can vary greatly among individuals.

A.   Colorimetry

As an application of colorimetry, DO measuring devices in ampoule form are available. Based on the measuring principle derived from the following reaction formula, these devices determine the DO concentration by comparing the color reading of isolated iodine in the transformation of Winkler potassium permanganate against the standard color band or by comparing colors using acid indigo carmine, which enables DO measurement at less than 1 ppb.

$$MnSO_4 \cdot 6H_2O + 2NaOH \rightarrow Mn(OH)_2 + Na_2SO_4$$

$$2Mn(OH)_2 + O_2 \rightarrow 2MnO(OH)_2$$

$$MnO(OH)_2 + 2KI + 2H_2SO_4 \rightarrow MnSO_4 + K_2SO_4 + I_2 + 3H_2O$$

B.   Electrochemistry

Currently, the diaphragm electrode is widely used. The diaphragm electrode, which consists of an anode, a cathode, and an electrolytic solution (internal liquid), is an electrode separated from the sample liquid by a thin diaphragm. The principle of the structure is shown in Fig. 2. The permeable diaphragm, which is influenced by the partial vapor pressure, is made of a thin film of Teflon or polyethylene. When the diaphragm electrode is submerged in the water sample, the DO in the water diffuses and reaches the cathode after penetrating the diaphragm by virtue of the permeability of the diaphragm, generating a reduction current. The strength of the reduction current is in proportion to the volume of reduced oxygen or that of the oxygen reaching the cathode during each unit of time.

In addition, because the number of oxygen molecules reaching the cathode is related to the diffiusion rate of the diaphragm film, the ratio between the volume of oxygen reaching the cathode and that of the DO in the sample water is constant. In other words, electrical current proportional to the volume of DO in the water is caused to flow; hence, the DO volume can be determined by measuring the strength of the current.

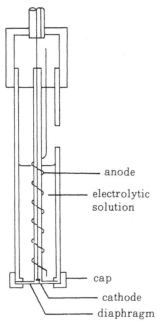

Figure 2 Diaphragm electrode.

The diaphragm electrode is further divided into two categories: polarographic and galvanometric. In the former, the voltage is applied externally, whereas in the latter, the electrolytic bath itself serves as a cell.

1. Polarographic Type

Suppose a polarogram of oxygen, as illustrated in Fig. 3, is obtained using a diaphragm electrode with a cathode made of gold or platinum and an anode made of silver filled with

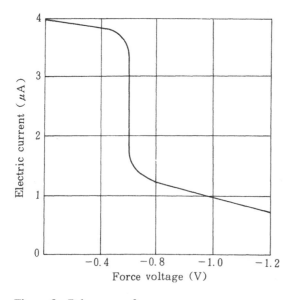

Figure 3 Polarogram of oxygen.

electrolytic solution (such as potassium chloride solution). It is noted that the curve flattens in the applied voltage range from -0.5 through -0.9. When voltage in this range is applied constantly across the two electrodes, the following electrode process takes place, causing electrical current to flow in proportion to the volume to DO:

Cathode process: $O_2 + 2H_2O + 4e \rightarrow 4OH^-$

Anode process: $4Cl + 4Ag \rightarrow 4AgCl + 4e$

The magnitude of voltage applied across the electrodes depends on the type of electrode, but normally it is in the range from -0.5 to -0.8 V.

2. *Galvanometric Type*

In the diaphragm electrode, when precious metals, such as gold, platinum, and silver, are used for the anode, and base metals, such as lead and zinc, for the cathode, the base metals dissolve to supply electrons, which causes an electrical current to flow in proportion to the volume of DO. When lead is used with alkaline electrolytic solution, the electrode process can be expressed by the formula

Cathode process: $2Pb \rightarrow 2Pb^{2+} + 4e$

$$2Pb^{2+} + 4OH^- \rightarrow 2Pb(OH)_2$$

$$2Pb(OH)_2 + 2KOH \rightarrow 2KHPbO_2 + 2H_2O$$

Anode process: $O_2 2H_2O + 4e \rightarrow 4OH^-$

10

Total Solids Monitor: Measurement of Residue on Evaporation

Sankichi Takahashi
Hachinohe Institute of Technology, Aomori, Japan

Toshihiko Kaneko
Hitachi Machinery & Engineering, Ltd., Kanagawa, Japan

I. INTRODUCTION

Impurities in ultrapure water have a serious effect on the performance of large-scale integrated (LSI) devices, as well as the LSI production process. In particular, it is reported that the total organic carbon (TOC) value greatly affects the relative defect density of LSI circuits. Figure 1 shows an example: the TOC value is in almost direct proportion to the relative LSI defect density. The organic material concentration in ultrapure water needs to be carefully monitored and controlled.

As mentioned earlier, the total solids (TS) monitor mainly detects the organic materials in ultrapure water. The TS monitor aims to contribute to LSI manufacture and quality control by conducting proper on-line measurement of organic materials.

II. CONFIGURATION OF THE MONITORING SYSTEM

The monitor is automatically started up and turned off. Data recording is also automated. Therefore no extra attention needs to be paid to the operation of the monitor. However, the connection between the ultrapure water piping system and the TS monitor should be made carefully to prevent contamination. In this sense, a special exclusive pipe should be employed that is elution and contamination free. The Hitachi TS monitor is equipped with a special pipe to maintain the integrity of the comprehensive measurement tool to achieve high accuracy and reliability.

Figure 2 shows the configuration of the monitoring system. From the ultrapure water pipe, ultrapure water at 5–10 cm^3 $minute^{-1}$ is removed constantly while ultrapure water at 0.5 cm^3 $minute^{-1}$ is fed to the monitor. In this way, ultrapure water about to be fed to the use point can be constantly monitored and contamination in the sampling line can be prevented.

From P. A. McConnellee ;
Semiconductor International, 82 (1986)

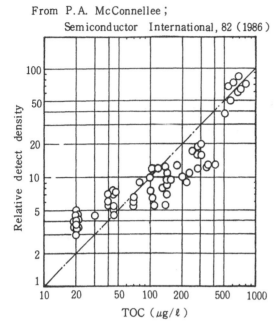

Figure 1 Relation between TOC and relative defect density. The correlation between TOC and the relative defect density is good.

To evaporate atomized water droplets, heated air or nitrogen is used. Therefore the temperature of the heated air or nitrogen greatly affects the evaporation of water droplets. In general, the monitor is equipped with a desiccating agent to remove humidity, and some attention should be paid to maintain this desiccating agent.

III.   MEASUREMENT RESULTS IN ACTUAL OPERATION

A.   Correlation Between TS Value and Specific Resistivity

Figure 3 shows the measurement results for the specific resistivity and TS values. In the ultrapure water manufacturing process, including the use point pipe (about 150 m, made of vinyl chloride), the specific resistivity and the TS value are measured simultaneously at

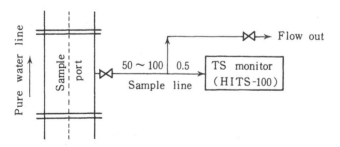

Figure 2 TS monitor setting with special sampling port. Keeping the sampling line from contamination is essential for highly accurate monitoring.

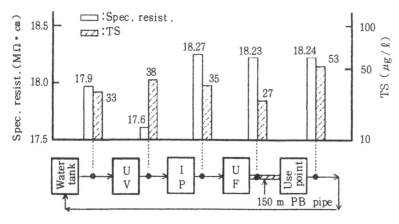

Figure 3   Measurement of specific resistance and TS in LSI pure water manufacturing equipment. No correlation was found between the specific resistivity and TS.

each measurement point (filled circles). Because the TS monitor detects organic materials, the TS values fluctuate greatly downstream of the ion polisher. However, the specific resistivity remains almost unchanged. As shown in Fig. 3, there is no correlation between the specific resistivity and the TS value. Particularly, the TS value is doubled between the inlet and the outlet of the use points as a result of organic material elution from the use point piping material while the specific resistivity remains the same.

On-line measurement of the specific resistivity is easy, but it is not effective in detecting organic materials. In other words, if monitoring of the water quality relies only on the specific resistivity measurement, it is possible to overlook organic materials and the wafers may be cleaned with water containing a high organic material concentration. This certainly affects the performance of LSI devices as well as the LSI production process.

B.   Correlation Between TS and TOC

Figure 4 shows the measurement results in the ultrapure water manufacturing process with the HITS-100. The ultrapure water manufacturing process is also described in Fig. 4. The length of the use point piping system (polyvinylidene fluoride, PVDF) is about 150 m. The TS and TOC values (Anatel TOC meter) are simultaneously measured at each measurement point, shown as a black dot. Based on this experimental result, the correlation between TS and TOC, the change in water quality at each water treatment point, and elution from the use point piping system are studied.

First, TS and TOC are very well correlated. The TOC level measured with the TOC meter decreases at each water treatment step. In particular, the TOC level at the ultrafiltration (UF) outlet is the lowest. At the outlet of the use point piping system, however, the TOC level rises. This is thought to be because of organic material elution from the pipe.

The TS value also drops at every treatment point, reaching the lowest level at the outlet of the UF unit. The TS value fluctuation in Fig. 4 is exactly the same as that in Fig. 3.

The TOC level decreases by about 2 μg L^{-1} at the outlet of the UF unit and the TS level drops by 10 μg L^{-1}. For elution from the use point piping system, the TOC value increases by about 2 μg L^{-1} and the TS value increases by about 20 μg L^{-1}. This is

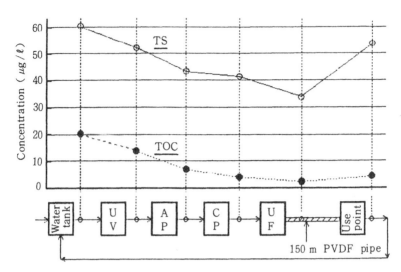

Figure 4   Results of measurement of TS and TOC in LSI pure water manufacturing equipment. TS has a great dependence on TOC, but the TS monitor seems to sense even organic matter not detected by the TOC analyzer.

thought to be because the TS meter can detect organic materials the TOC meter cannot detect. However, it is not clear whether this is the real reason. To detect the actual TS concentration at the μg L $^{-1}$ level is a technological challenge for the future.

Still, it is obvious that TS measurement is quite sensitive to detect organic materials (impurity) and is able to reflect constant fluctuations in the water quality.

IV.   CONCLUSIONS

This chapter has presented measurement data taken with the Hitachi TS Monitor (HITS-100) and important points in terms of TS monitor applications.

1.   The TS monitor can measure and evaluate the comprehensive quality of ultrapure water on-line.
2.   The TS monitor can detect what the TOC monitor cannot detect. Therefore it is suitable for monitoring the quality of ultrapure water for LSI.
3.   The TS monitor can be effectively applied to the research and development of the water treatment process and the materials to be employed by the ultrapure water production system.
4.   The on-line measurement of the quality of ultrapure water with the resistivity meter and the TS meter is very promising.

11

Ultrapure Water Evaluation by the Water Spot Method

Michiya Kawakami
Mitsubishi Gas Chemical Company, Inc., Tokyo, Japan

Makoto Ohwada
Alps Electric Co., Ltd., Fukushima, Japan

Yasuyuki Yagi
Hitachi Plant Engineering & Construction Co., Ltd., Chiba, Japan

Tadahiro Ohmi
Tohoku University, Sendai, Japan

As mentioned earlier, many water quality evaluation technologies directly measure such parameters as resistivity, particles, total organic carbon (TOC), and silica using analytic instruments. The measurement capability of these instruments has been improved to measure impurities at less than the 10 ppb level.

To comprehensively evaluate the quality of ultrapure water, unlike the method of measuring specific trace impurities, the water spot method is now being studied. This method is based on the fact that ultrapure water remains on the wafer surface during the cleaning process: if no residue is found after the water is dried, the quality of the ultrapure water can be considered very high. The evaluation method is relatively simple: a droplet of ultrapure water is placed on the wafer surface and dried to observe any residue.

When an ultrapure water droplet of high purity (shown in Table 1) is placed on a bare Si wafer surface treated with dilute hydrogen fluoride (DHF) to remove the native oxide film, a very clear water spot can be observed after drying (see Fig. 1). On the other hand,

Table 1 Quality of Ultrapure Water (Tohoku University Mini Super Clean Room)

	RESISTIVITY (MΩ・cm)	TOC (μg/ℓ)	SiO$_2$ (μg/ℓ)	Dissolved Oxygen (μg/ℓ)	TOTAL RESIDUE (μg/ℓ)	PARTICLES (COUNTS/mℓ >0.07μm)
SYSTEM 1	18.25	<1	<1	3~5	<1	1~2
SYSTEM 2	18.25	<1	<1	6~10	<1	1~2

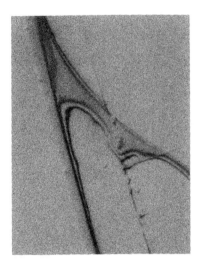

Figure 1   The water marks left on a bare Si wafer surface after drying the ultrapure water.

when the same ultrapure water droplet is placed on a SUS 316 L surface treated by electrochemical buffing, which has limited elution of impurities, no water spot is observed after drying. Taking these two experiments into consideration, it can be said that the water spot observed when the ultrapure water droplet is placed on the bare Si wafer does not accurately reflect the quality of the ultrapure water.

The native oxide growth when a Si wafer is immersed in ultrapure water has been studied. In the course of this study, it was confirmed that native oxide growth in ultrapure water accompanies Si elution from the Si substrate. Figure 2 shows the time dependence

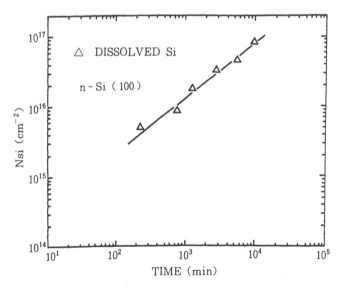

Figure 2   Time dependence of the number of dissolved Si atoms for wafers immersed in ultrapure water. N_{Si} is defined as the number of Si atoms per unit area on the wafer surface.

of Si elution from the n-Si(100) wafer with its native oxide removed in DHF when it is immersed in 100 ml ultrapure water (the concentration of dissolved oxygen is 9 ppm) at room temperature. N_{Si} in Fig. 2 is the number of Si atoms eluted from a unit area on the wafer surface.

Based on the data in Fig. 2, the amount of Si eluted when ultrapure water on a wafer is dried for 5 h is estimated roughly. This calculation is based on the following assumptions:

1. When ultrapure water is dropped on the wafer, it forms a hemispherical shape.
2. The size of the ultrapure water droplet decreases as it dries. The diameter of the area where the ultrapure water contacts the wafer surface becomes half the initial diameter, on average.

$$\tfrac{4}{3}\pi r_0 3 \times \tfrac{1}{2} = 1 \qquad r_0 = 0.78$$

where r_0 = diameter of 1 ml ultrapure water immediately after it is dropped onto the wafer.

According to Fig. 2, the amount of Si elution after drying (N) can be expressed as

$$N = 5 \times 10^{15}\ \pi \times \left(\frac{0.78}{2}\right)^2 \times \frac{28}{6.02 \times 1023} = 1.11 \times 10^{-7}\ g$$

On the other hand, if the amount of impurities in ultrapure water is assumed based on Table 1 as 5 ppb, the residual impurities in ultrapure water N' can be expressed as

$$N' = 5 \times 10^{-9}\ g$$

Therefore it is found that $N \gg N'$.

As is clear from this rough estimate, most of the residues found when ultrapure water is dropped on the wafer surface with its native oxide removed with DHF are Si eluted from the Si substrate.

Although evaluation using the water spot is unique compared with conventional evaluation methods, its reliability is very poor at present.

Table 2 shows the amount of Si eluted from Si wafers treated with different types of oxidation processing. As shown in Table 2, the Si elution is very limited when the wafer surface is treated by thermal oxidation or by H_2SO_4-H_2O_2 cleaning to form native oxide in advance. When the water spot is used to evaluate the quality of the ultrapure water, wafers used in the evaluation should at least be treated by thermal oxidation to eliminate Si elution during the evaluation process.

Table 2 Dissolved Si Weight for Wafers Immersed in Ultrapure Water (DO, 9 ppm) After Various Oxide Treatments

SAMPLES	HF ACID CLEANING	H_2SO_4 - H_2O_2 CLEANING	AIR EXPOSING (4 DAYS)	THERMAL OXIDE (940 Å)
DISSOLVED Si WEIGHT (μg/cm^2)	4. 03	< 0. 05	0. 14	< 0. 05
DIPPING PERIOD (DAYS)	7	7	9	15

As the quality of ultrapure water nears the detection limit of current microanalysis equipment, it is hardly evaluated appropriately. In this sense, new evaluation technologies for ultrapure water using fresh approaches, such as water spot measurement and evaluation of electrical characteristics of metal oxide semiconductor diodes, will become important in the future.

12
Afterword

Katsumi Koike
Japan Organo Co., Ltd., Tokyo, Japan

Introduced in Part VII were technologies for the analysis and evaluation of impurities in ultrapure water. Among the various impurities, suspended fine particles in particular have for some time been controlled to particle sizes about ⅒ of the minimum pattern dimension of the devices. This required level has been higher for other impurities, however. For example, the quantity of impurities is as small as 10 pg L^{-1} even when assuming the number of particles as 10 particles of 0.1 μm ml^{-1} at a specific gravity of 2. It is evident that the required concentration of fine particles is extremely low compared with that of other impurities, which are of the order of micrograms or nanograms per liter. Also, fine particles are readily affected by sampling, and this sensitivity rises with decreasing particle size.

It is therefore extremely important that ultrapure water be evaluated in terms of fine particles and other impurities while being kept away from contamination. The use of measuring instruments as much as possible for on-line measurement will prove advantageous in the prevention of secondary contamination and will also lead to savings in labor.

Although the quality standards for ultrapure water are rising tandem with increasing integration of the devices, the enhancement of accuracy of the instrumentation used to evaluate water purity cannot be achieved within a short period of time. It is therefore urgent that evaluation methods related to phenomena on the wafer surface be incorporated in current technologies and that such methods be standardized. This part covers new evaluation technologies, although areas already covered in the Japanese Industrial Standards (JIS) are not detailed completely and pertinent specifications must be referenced when necessary.

VIII
Wet Processing

1
Foreword

Michiya Kawakami
Mitsubishi Gas Chemical Company, Inc., Tokyo, Japan

As ultralarge-scale integration (ULSI) patterns are refined, the electrical charge of the signal decreases and is more sensitive to noise, such as dark and leakage currents. The effects of minute quantities of contaminants, not previously a problem, must be suppressed. This requires high-performance process technology that satisfies the following three principles [1]:

1. Ultraclean process environment
2. Ultraclean wafer surface
3. Process parameters: perfectly controlled equipment

The process environment (ultrapure water, chemicals, gases, clean rooms, and so on) must be clean to maintain an extremely clean wafer surface. The equipment must perfectly control the factors that govern wafer surface phenomena. Wet processing plays a significant role in realizing these three principles. Impurities on the wafer must be completely, uniformly, and consistently removed.

First, basic ultraclean technologies, such as ultrapure water, chemicals, and cleaning and drying equipment, are discussed. Then, the role of wet process phenomena in governing the factors that control these phenomena is reviewed.

For example, even if a wafer is cleaned with very clean ultrapure water and chemicals, when it is left in the clean room atmosphere, native oxide [2, 3] is formed, impurities in the atmosphere are adsorbed on the wafer, and the cleanliness deteriorates. This problem must be treated comprehensively, by wafer cleaning, drying, and transportation in high-purity nitrogen gas.

In a high-purity gas distribution system, stainless steel electrochemical buffing technology is important to achieve a particle-free and outgas-free system. High-concentration $NaNO_3$ is used in this buffing process, however, and the success of the buffing technology depends on the complete removal of $NaNO_3$ by precision cleaning [4]. This precision cleaning will become more important as semiconductor production attains higher performance levels. Precision cleaning technology for components and chambers used in the gas distribution system, the equipment, and the ultrapure water distribution system is a great challenge for the future, along with technology to improve the cleanliness of the wafer surface.

This part describes the conventional cleaning equipment widely used in wet processing, IPA (isopropyl alcohol) drying, and spin drying. We also describe the physical and chemical details of etch cleaning, including cleaning with NH_4OH-H_2O_2 and buffered HF. Water adsorption on the wafer and native oxide suppression are discussed as future challenges. A new cleaning technology using high-temperature, high-pressure ultrapure water spray cleaning is introduced. Finally, the future of wet processing is outlined.

REFERENCES

1. T. Ohmi, Ouyou Butsuri, *58*, 193 (1989).
2. M. Morita, T. Ohmi, E. Hasegawa, M. Kawakami, and K. Suma, Control Factor of Native Oxide Growth Silicon in Air or in Ultrapure Water, Appl. Phys. Lett, *55*(6), 562 (1989).
3. M. Morita, T. Ohmi, E. Hasegawa, M. Kawakami, and M. Ohwada, Growth of Native Oxide on a Silicon Surface, J. Appl. Phys., *68*(3), 1272 (1990).
4. M. Kawakami, Y. Yagi, K. Sato, and T. Ohmi, Spray Cleaning Technology Using High Temp. and High Pressure Ultrapure Water, Proceedings of 3th Workshop on ULSI Ultra Clean Technology, Tokyo, 1990.

2
Outline

Michiya Kawakami
Mitsubishi Gas Chemical Company, Inc., Tokyo, Japan

Tadahiro Ohmi
Tohoku University, Sendai, Japan

As circuits are further integrated, more importance has come to be attached to technologies for raising the cleanliness of the silicon wafer surface. Dry-cleaning technologies are making solid progress, with advancements in the removal of organic materials using photoexcited O_3, the removal of heavy metals by laser-excitation Cl radicals [1], and the selective removal of native oxides using dilute HF gas [2]. A technology for removing alkali metals, including Na and K, the most critical contaminants in the semiconductor production process in the gas phase, has not yet been developed. Moreover, because dry cleaning does not use physical force, it is difficult to completely remove particles from the wafer. This is why wet processing is still highly important. In addition, because ultrapure water and chemical cleanliness has been significantly improved, wet processing, with its very high selectivity, is even more important.

Current methods in semiconductor production include wafer cleaning, photoresist development, photoresist removal, and wet etching. Wafer cleaning proceeds according to five steps:

1. Acid cleaning
2. Alkali cleaning
3. Dilute HF cleaning
4. Ultrapure water cleaning
5. Dry processing

Several reagents can be used for acid cleaning (step 1), including HCl-H_2O_2 ($HCl|H_2O_2|H_2O = 1:1:6$–$1:2:8$ at 85°C) and H_2SO_4-H_2O_2. This method is used primarily to remove heavy metal and organic contaminants on the wafer surface. The major alkali cleaning method (step 2) uses NH_4OH-H_2O_2 ($NH_4OH|H_2O_2|H_2O = 1:1:5$–$1:5:20$ at 75–85°C), which is remarkably efficient in removing particles absorbed on the wafer surface. Dilute HF cleaning (step 3) removes the native oxide film that form on the wafer. Ultrapure water cleaning (step 4) removes the chemical residues left on the wafer after the

Table 1   Wafer-Cleaning Process

1. $H_2SO_4 : H_2O_2 = 4 : 1$	5 min	3 times
OVER-FLOW RINSING	5 min	
2. ETCHING (HF : $H_2O = 1 : 100$)		
OVER-FLOW RINSING	10 min	
3. $H_2SO_4 : H_2O_2 = 4 : 1$	5 min	
OVER-FLOW RINSING	5 min	
4. ETCHING (HF : $H_2O = 1 : 100$)		
OVER-FLOW RINSING	10 min	
5. RCA CLEANING		
(1) $NH_4OH : H_2O_2 : H_2O = 1 : 4 : 20$	10 min	
(2) HOT WATER DIPPING (90°C)	10 min	
(3) OVER-FLOW RINSING	10 min	
(4) $HCl : H_2O_2 : H_2O = 1 : 1 : 6$	10 min	
(5) HOT WATER DIPPING (90°C)	10 min	
(6) OVER-FLOW RINSING	10 min	
6. ETCHING (HF : $H_2O = 1 : 100$)		
7. OVER-FLOW RINSING	10 min	
8. N_2 BLOWING		

chemical treatment of high-grade ultrapure water. Drying (step 5) includes spin drying, N_2 blow, and drying with organic solvents (e.g., IPA drying), all of which completely dry and remove ultrapure water remaining on the wafer surface.

Each cleaning process has been studied and its procedures specified to suppress contaminants to the lowest possible levels. Table 1 lists the procedure used by an advanced research institution. As shown, contamination of the wafer can be eliminated by repeatedly removing a contaminant using a chemical and then rinsing with ultrapure water. Drying at the end of the process cycle completely removes the ultrapure water. The native oxide [3] formed during chemical processing in the air and in ultrapure water can be removed using dilute HF during cleaning.

Clearly, wet processing should completely and uniformly remove particles and organic and metal contaminants from the wafer surface, but it is also very important that the wafer be protected from stress and damage. Therefore, in Chap. 4, the physics and chemistry of etching are reviewed and the wettability of the wafer by different chemicals is also discussed.

Future wet processing will require the perfect and consistent removal of impurities and maintenance of the wafer surface at a high level of cleanliness. A new cleaning method for three-dimensional devices, for example the trench capacitor, will be needed. The chemical and ultrapure water quality and the reproducibility and uniformity of wet processing need further improvement.

REFERENCES

1. T. Ito, R. Sugino, and S. Watanabe, Proceedings 1st Int. Symp, p. 114 (1989).
2. N. Miki, H. Kikuyama, I. Kawanabe, M. Miyashita, and T. Ohmi, Gas-Phase Selelctive Etching of Native Oxide, IEEE Transactions on Electron D Devices, *37* (1), (1990).
3. M. Morita, T. Ohmi, E. Hasegawa, M. Kawakami, and M. Ohwada, Growth of Native Oxide on a Silicon Surface, JAP, *68*(3), (1990).

3 Wet Etch Cleaning

A. Cleaning Unit

Hiroyuki Horiki
Nisso Engineering Co., Ltd., Tokyo, Japan

Takao Nakazawa
Haruna Incorporated, Tokyo, Japan

I. INTRODUCTION

Although dry processing is now used more widely, it is expected that at least prediffusion cleaning will continue to employ wet processing for the time being. As semiconductor devices are increasingly integrated, particulation caused by operators must be eliminated. The total automation of the clean room has been realized.

In an automated system, the cleaning unit, which is controlled by a large computer, must be equipped with an automated wafer transportation system and automated distribution and drainage systems for chemicals and pure water. The efficiency of cleaning will greatly depend on the development of technology of the cleaning unit, especially in the megabit era.

This chapter describes the chemical cleaning unit used for preprocessing by diffusion (equipment using organic solvents is not discussed here).

II. CONFIGURATION OF THE CLEANING UNIT

Figure 1 shows the external appearance of the cleaning unit, and Fig. 2 shows the configuration [1]. The cleaning unit includes a loader where wafer carriers can be placed, three or four process stations, consisting of chemical processing baths and a water bath, a final rinse bath, a drier, and an unloader. The automated wafer transfer system is installed between the loader and the unloader. Wafer transportation is directed by a control unit. The unit usually also controls the automated distribution and drainage of chemicals and pure water.

A. Loader and Unloader

There are two types: in one an operator loads and unloads the wafers; the other is operated completely by robots using transfer boxes.

Figure 1 A wet station.

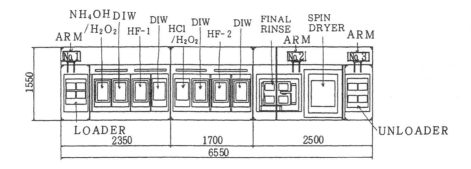

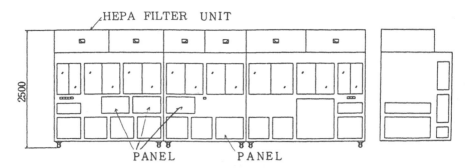

Figure 2 Construction of a wet station.

B.   Process Station

The peripheral technologies, including the chemical processing baths and water bath, are critical to the cleaning unit. Perfluoroalkoxyl (PFA), and high-quality quartz are used for the baths. Because the internal construction of the cleaning bath has a great effect on the cleaning efficiency, various ideas are being studied to ensure immediate and uniform liquid contact with the water surface. To keep the particles from readhering to the wafer after removal, however, a simple configuration is desirable.

The cleaning unit usually combines various cleaning technologies to handle different impurities and particles on the wafer surface. For example, ultrasound is widely used in chemical cleaning to remove particles, and indirect irradiation is preferable to suppress contamination. To reduce attenuation of the ultrasonic waves, a simple configuration is desirable in this case as well (see Fig. 3). The bath is surrounded by sensors to control the liquid surface and to find carriers. Since quite a few problems are caused by these sensors, the light receptor should be kept clean and corrosion of the internal components should be avoided (see Table 1).

Control of the drainage system should be coordinated with the cleaning process: the highly concentrated chemicals used during the initial stage of cleaning and the more dilute chemicals used in later stages should be drained separately. In general, because drained HF is processed differently, the HF drainage is constructed independently of the lines for other acids. When different lines are used (as for HF effluent) the cleaning unit must be equipped with additional valves. Because the aperture of a conventional valve is quite large, it is not possible to install all the necessary valves on the unit. In an attempt to solve this problem, a special switching valve was developed for draining liquids and installed at the bottom of a plenum (see Fig. 4).

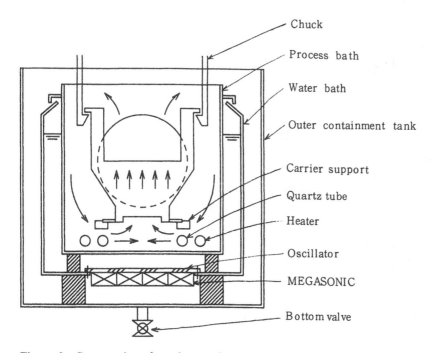

Figure 3   Cross section of an ultrasound process bath (MEGASONIC).

Table 1 Level Sensor Features

	erosion-proof	contamination	precision	life	stability
N₂ gas purge type level switch	excellent	good	excellent	good	good
Electrostatic capacity type level switch	excellent	excellent	fair	good	good
Photoelectric level switch (non−dip−type)	good	excellent	fair	good	good

It is also critical to handle the mist exhausted from the processing bath. For example, the mist is discharged through slits of the cover at accelerated velocity. If necessary, an inert gas is used to seal the mist. Ammonia mist and hydrochloric acid mist react with each other, producing by-products. To suppress this reaction and its consequent particle generation, the two types of mists should be carefully separated (see Fig. 5).

Extra attention should be paid to the dangerous chemicals used at high temperatures in the plenum. In a sulfuric acid line at 150°C, for example, a plenum made of heat-resistant polyvinyl chloride (PVC) should not directly contact the circulating pipe and the surface of a plenum must be cooled with overflow-water from an overflow bath (see Fig. 6).

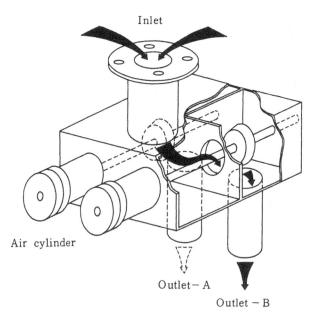

Figure 4 Bottom outflow valve (PVC, HPVC, and PTFE) with outer containment tank.

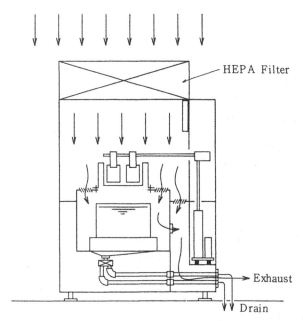

Figure 5 Closed wet station with HEPA filter.

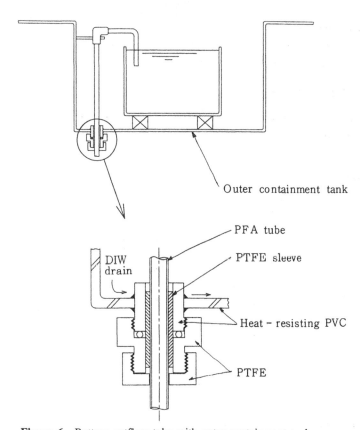

Figure 6 Bottom outflow tube with outer containment tank.

C. Final Rinse Bath and Drier

This is the most important part of the cleaning unit: the wafer can be recontaminated in these components if they are not properly constructed.

The final rinse bath is used not only to rinse wafers with water but also to temporarily store wafers when the drier is occupied. Wafer carriers are transported from cleaning process baths through water. There must be no particulation in this final rinse (see Fig. 7). Isopropyl alcohol (IPA), drying, and centrifugal dehydration with a spinner are widely used, and described in detail later.

D. Cabinet

There are two types of cabinets: an open cabinet used in a clean ambient atmosphere with downflow, and a clean bench cabinet with a HEPA filter installed at the top (see Fig. 8). The construction should ensure a favorable cleaning environment by combining local and general ventilation.

Tohoku University developed an energy-saving clean bench to meet the increasing need for further cleanliness of the ambient air and to cut costs. This clean bench, with a horizontal curtain of air above the chemical containers, is outstanding in cutting total gases exhausted to one-third of the conventional type and in suppressing the exhaustion of chemical vapor [2].

The cabinet is constructed of iron and PVC or polypropylene to suppress particulation and corrosion: all the iron supports are completely covered with PVC or polypropylene. Furthermore, because PVC with good corrosion resistance generates poisonous chlorine gas when burned, some cabinets employ heat-resistant polypropylene instead of PVC.

E. Wafer Transportation System

The wafer transportation system widely used in Japan is a motor-driven transportation system using rack and pinion and ball and screw joints. It is composed of a body that

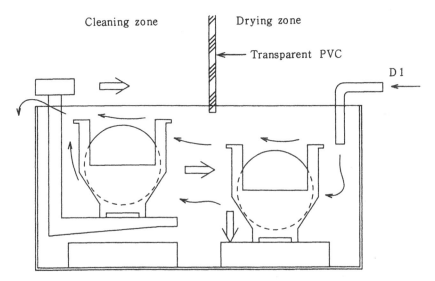

Figure 7   Final rinse bath.

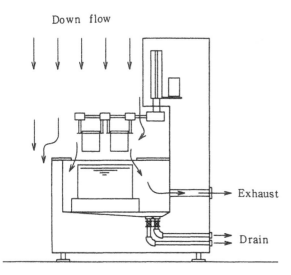

Down flow

Exhaust

Drain

Figure 8   Open wet station.

slides horizontally, an arm mounted on the body, which moves vertically, and a chuck placed at the end of arm for grasping a carrier.

Most recent grips apply the open and close chuck to automatically load and unload carriers at the drier. Because the chuck directly contacts the wafer carrier, extra attention must be paid to particulation and corrosion. Also, the reliability of the chuck is also very critical in the safe transport of wafer carriers. Many chucks are constructed of fluorocarbon resin. The chuck should be made of fluorocarbon resin different from that used for the carrier because when the same resin is used, they may slip. The chuck is placed at the edge of the carrier to grasp and lift it. The edge should be designed with a certain tolerance against chuck-positioning errors (see Fig. 9). The chuck is usually rinsed with water before holding the carrier because it is dipped in chemicals during the cleaning process.

The motor-driven transportation system should stop at each station with sufficient accuracy. Although most conventional systems employ positioning sensors to set the stop positions, some new systems use numerical control by servomotor. The number of problems caused by sensors are decreasing. Overrun sensors are also used, but only for safety reasons.

A general-purpose robot, which slides in the X, Y, and Z directions, has been modified by applying countermeasures against dust and corrosion. It is mounted in front of an open station (see Fig. 10).

The transportation system can be mounted under, above, in front of, or behind the process station. Because with machinery, unlike humans, it is fairly easy to pinpoint the point of particulation, countermeasures should be worked out immediately. In practice, however, space is the limiting factor and so the transportation system should be mounted under the processing baths. Several transportation systems are mounted on one cleaning unit to function efficiently and harmoniously (see Fig. 2).

The metal components of the transportation system must be very accurately machined to maintain stability of operation over a long period. In particular, the sliding parts require precise attention to improve the reliability of the entire system.

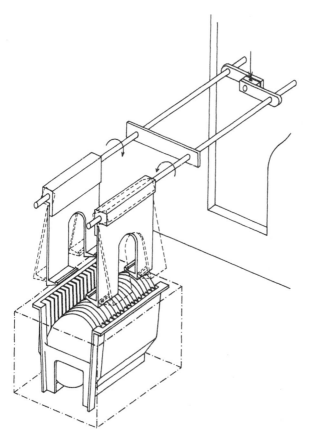

Figure 9 Wafer carrier chuck. The mechanical part that opens and closes the chuck is located away from the outer containment bath so that the process bath is free from its dust.

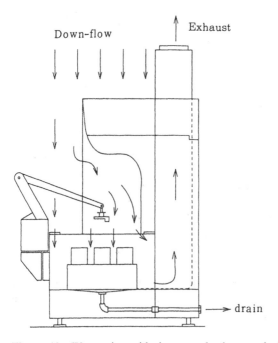

Figure 10 Wet station with the general robot attached to the front.

F. Control Unit

The control units using semiconductors malfunctioned in the past as a result of noise caused by various units in the clean room, but they are now modified so that they are not affected by such noise.

Assisted by a programmable sequencer and a personal computer, recent control units are capable of controlling the entire unit, communicating with the computer for process automation. The computer also handles interlock controls for safety purposes.

The control unit has such wide-ranging functions as the selection of cleaning process recipes, the execution of cleaning, the storage and output of information, including temperature, time, and error, information display, the supply and discharge of chemicals and pure water, and the transportation of wafers. SECS I and II can be used for the external communication system.

III. COMPONENTS AND EQUIPMENT TECHNOLOGY

Because the wafer-cleaning unit is composed of such various components as fittings, pumps, valves, sensors, and transportation system, the design concept, materials, structure, and reliability of those components are very critical. Moreover, the overall design of the unit (the piping technology for delivery and discharge of chemicals and pure water, the ventilation method, and the construction of stations) greatly affects particulation and contamination. It was on this basis that we developed all the components of the cleaning unit, and some of the major components and equipment design are described here.

A. Piping Components

As the purity of chemicals rises, better performance in terms of contamination and particulation are required for piping. At present, polytetrafluoroethylene (PTFE) and perfluoroalkoxyl (PFA) are considered the best fluorocarbon resins.

1. Fittings

In terms of sealing efficiency, the line seal is superior to the surface seal (see Fig. 11).

Fittings should be checked for performance under torque, high tensile strength and high pressure, resistance to heat, convenience and reliability when fittings are repeatedly connected and disconnected, and contamination during installation. The skill of installation, the time required for installation, and the difficulty of acheiving the necessary tube size should also be checked. Fittings are currently evaluated by testing for leakage, heat cycle, tensile strength, and ability to withstand a side load (see Fig. 12). Comprehensive evaluation confirmed that the cup-shaped fitting performs best.

2. Valves

Generally many diaphragm valves, needle valves, and cock valves made of PTFE or PFA are used for chemicals and pure water. Bellows valves have been introduced for stringent pressure conditions. As shown in Table 2, the particulation from needle valves, ball valves, and cock valves is significant, and diaphragm valves cannot withstand high pressure or reverse pressure. All four valves have a short useful life. Bellows valves were developed to overcome those weak points. The most recent bellows valves have further improved the sealing ability and reverse pressure resistance (see Fig. 13).

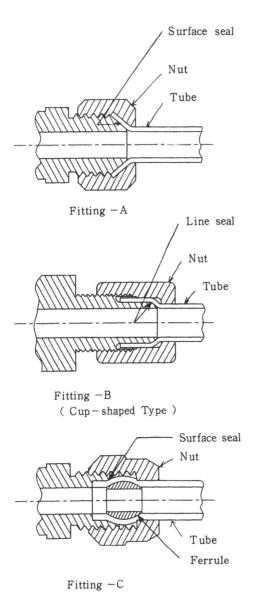

Surface seal

Nut

Tube

Fitting —A

Line seal

Nut

Tube

Fitting —B
(Cup—shaped Type)

Surface seal

Nut

Tube

Ferrule

Fitting —C

Figure 11   Typical fittings (PTFE and PFA).

3. *Pumps*

Generally, diaphragm pumps and bellows pumps made of PTFE or PFA are used. Because diaphragm pumps have problems in terms of bending fatigue to destruction, shortening their useful life, bellows pumps, which have a small amount of distortion per unit, have been increasingly used. The progress of the technology used in machining the bellows has also helped to improve bellows pump performance. The latest bellows pumps can withstand severe pressure conditions during operation without a chemical or water supply and high temperatures. In many horizontal bellows pumps chemicals

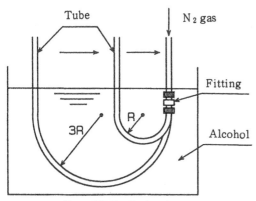

R :   Minimum bend radius

Figure 12   Side-load test.

Table 2   Safety and Reliability of PTFE Valves

	Bellows valve	Diaphragm Valve	Ball valve
Sealing to normal direction	good	good	good
Sealing to reverse direction	good	fair	good
Temp. and press. resistance	good	fair	good
Dust free	good	excellent	fair
Life	excellent	fair	fair
Inside and outside leakage	good	good	fiar

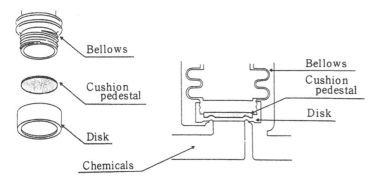

Figure 13   Sealing point of the bellows valve (NSVH).

circulate in the bellows. Countermeasures against contamination have been taken. Because the vertical bellows pump was developed for applications at high temperatures, problems caused by the specific gravity of liquid chemicals and the elimination of bubbles have been overcome (see Fig. 14).

B. Piping Technology

Because high-temperature chemicals are circulated at use points through pipes, various factors must be considered in piping besides using reliable fitting valves and pumps under this circumstance: the expansion and shrinkage of pipe, and piping without reverse inclines. The piping installation system is also very important: maintaining the cleanliness of cleaned components ensures prompt and easy start-up of the cleaning unit. Additionally, the piping system should be equipped with bypass lines, which make it possible to sterilize with high-temperature pure water or hydrogen peroxide at any time.

A special automatic drain switching valve was developed to drain liquids (see Fig. 4). This system enables the cleaning unit to operate in a fully automated manner.

C. High-Temperature Circulating Filtration and Cleaning Units

Many cleaning units are equipped with filtration units containing circulating chemicals for removing particles from the wafer surface, the chemicals, and the processing baths. This circulating filtration unit can be installed on the cleaning unit or can be set up separately. Each has advantages and disadvantages in terms of safety and efficiency.

The Nison series high-temperature chemical circulating filtration unit has drawn recent attention [3]. It includes heat-resistant components (fittings, valves, pumps, and so on), a heat-resistant precision filter made of PFA, and a line heater.

To handle the chemicals at 160°C, higher than the maximum temperature for PTFE and PFA in practice, the safety and reliability of the entire system, as well as the heat resistance of each component, should be confirmed. The introduction of this system has greatly contributed to improvements in device quality control, reducing the particles on the wafer by $1/10$ to $1/20$ (see Fig. 15). It has also facilitated cost savings because the useful life of chemicals has been increased. This system has greatly improved the conventional hot H_3PO_4 process in terms of the selectivity between the nitride film and the oxide film and the rate and uniformity of etching (see Table 3).

IV.   FUTURE CHALLENGES

A cleaning method to remove particles smaller than 0.2 μm should be developed and modified in accordance with changes in the process. The superiority of batch processing or single-wafer processing should be established. A larger wafer will be handled in the future, and this should be accommodated. For conventional units, measures to cut cost and delivery time should be worked out by making the cleaning unit smaller. It is also required to manufacture process stations separately so they can be combined in different ways to meet different production needs. Furthermore, a fully automated cleaning unit is needed to improve methods of cleaning and transporting carriers.

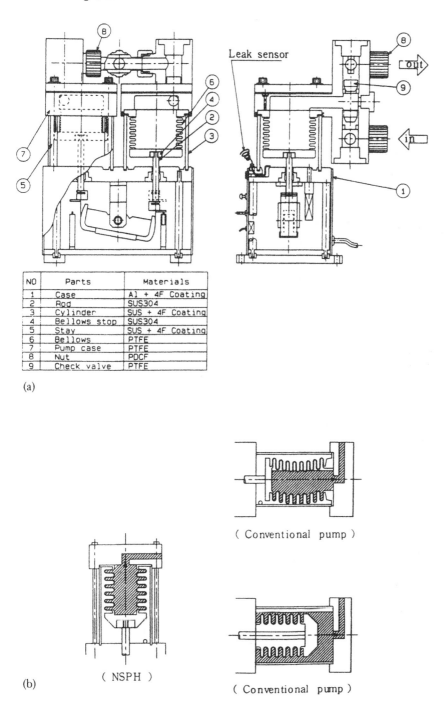

NO	Parts	Materials
1	Case	Al + 4F Coating
2	Rod	SUS304
3	Cylinder	SUS + 4F Coating
4	Bellows stop	SUS304
5	Stay	SUS + 4F Coating
6	Bellows	PTFE
7	Pump case	PTFE
8	Nut	PDCF
9	Check valve	PTFE

(a)

Leak sensor

(Conventional pump)

(NSPH)

(Conventional pump)

(b)

Figure 14   (a) Vertical bellows pump (NSPH). The bellows are placed vertically. If the bellows are placed horizontally, the weight of chemicals may cause the bellows to rub against the cylinder wall, resulting in damage as shown in b. (b) Bellows structure.

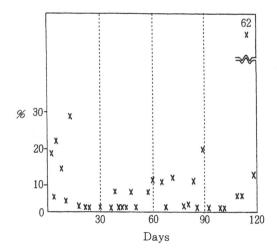

$$
\text{Relative number of particles on the wafer } (\%) = \frac{\text{Number of particles on the wafer after process with NISON}}{\text{Number of particles on the wafer after process without NISON}} \times 100
$$

As shown above, data counted over a 120-day period
in a Japanese production line (resist removal).

Figure 15   Effective data for the hot acid filtration system (Nison series).

Table 3   Typical Data for the Nison 1800

(FOR THE CASE OF Si_3N_4 FILM REMOVAL)		
TEMPERATURE (℃)	150	160
ETCH RATE (Å / min)	37	55
UNIFORMITY (%)	± 1.2	± 0.9
SELECTIVITY (Si_3N_4 / SiO_2)	> 40	> 40
PARTICLE COUNT (pcs) (PARTICLE SIZE 0.3 μm)	25 ~ 50	30 ~ 60

CHEMICAL   : H_3PO_4 (85 %)
FILTER   : 0.2 μm FILTER
WAFER SIZE : 6″μm

REFERENCES

1. K. Fujinaga, K. Yano, H. Harada, and E. Arai, Denshi Tsushin Gakkai Ronbunshu, Vol. J66-C, No. 3, 252 (1983).
2. T. Ohmi and J. Inaba, NiKEI Microdevices, 51, 115 (1989).
3. Y. Nishikata, T. Kaji, and H. Horiki, Proceedings of 9th Symposium on ULSI Ultra Clean Technology, Tokyo, (1989), pp. 287–305.

3   Wet Etch Cleaning

B.   Isopropyl Alcohol Vapor Drying Technology

Hiroyuki Mishima
Tokuyama Soda Co., Ltd., Yamaguchi, Japan

I.   INTRODUCTION

Although dry processing is applied widely, wet processing, which uses ultrapure water and high-purity chemicals, still plays an important role in dust-free and cleanliness technology for semiconductor devices. Rinsing with ultrapure water should be followed by drying. Even if all impurities are removed from ultrapure water, a water spot forms when the wet wafer is dried, as it is after rinsing with ultrapure water. Because this water spot leads to deterioration of the device, a drying technology that protects wafers from drying naturally and leaves no contaminants is needed. This part introduces a vapor drying technology using isopropyl alcohol (IPA).

II.   PRINCIPLE

Figure 1 illustrates the concept of IPA vapor drying. The unit is composed of IPA bath, heater, cooling coil, and container to collect condensed IPA. IPA is heated indirectly to the boiling point. The evaporated IPA is cooled in a tube located in the upper part of the bath and returns to the bath after condensation. This cycle forms a layer of vapor above the liquid IPA. The temperature of this vapor layer is the same as the boiling point of IPA. When the rinsed wafer is inserted in the vapor layer, the IPA condenses and liquefies on the wafer surface. The wafer surface is cleaned by dissolving the water that adheres to it. The condensed IPA solution containing water is immediately drained out of the system. When the wafer is as hot as the vapor, the condensation phenomenon stops. The wafer is then removed from the bath and the drying process is complete.

III.   OUTLINE OF THE VAPOR DRYING UNIT

An IPA vapor drying technology free of particle contamination requires a particle-free drying system [1,2]. The system must be operated carefully. Some of the parameters that

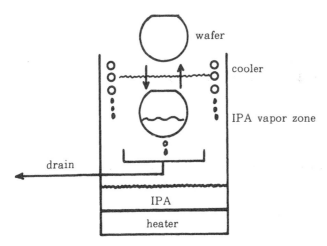

Figure 1   Model of an IPA vapor drying system.

affect drying efficiency have been specified. Safety issues must also be considered: IPA is flammable.

A.   Development of the Clean Unit

SUS 316 L is used for an IPA bath in which IPA vapor and liquid are in contact. The surface is treated by ultraclean processing. The IPA delivery system is automated. IPA is delivered by pressurized high-purity nitrogen gas. The piping is electropolished. IPA is supplied to the IPA bath after membrane filtration. The wafer transportation tunnel is filled with clean air filtered through a ULPA filter. The number of particles and metallic ions in the IPA liquid during vapor drying was found to be smaller by one order of magnitude than in the IPA liquid in the bath. This means that the IPA evaporation process purifies very effectively and the IPA vapor is kept very clean.

B.   IPA Quality Control

IPA quality must be monitored during drying to maintain stable drying performance. For this purpose, a small resistivity meter has been developed that is capable of continuously measuring the resistivity of liquid IPA. This meter is capable of the on-line measurement of ionic impurities in liquid IPA at the ppb level (Tokuyama Soda technical information, 1989).

A recent study found that the resistivity of IPA condensed on the carrier surface sometimes drops sharply when the carrier, made of fluorocarbon resin (mainly perfluoroalkoxyl, PFA), undergoes vapor drying after chemical cleaning and pure water rinsing (Tokuyama Soda technical information, 1989). Further study revealed that this phenomenon is closely related to the carriers and the chemical processing methods used in the past. In other words, this phenomenon is created by the ability of fluorocarbon resin to adsorb chemicals during chemical cleaning (chemical carryover [3]) and by the desorption of adsorbed chemicals during vapor drying. Since the desorbed chemicals contaminate the wafers, it is important to control the cleanliness of carriers to maintain wafer cleanliness.

C. IPA Gas Monitor

Safety control is vital before IPA vapor drying technology can be used widely. In addition to the alarms mounted on each utility line, an IPA gas monitor has been developed that has high reliability over a long period. This gas monitor was mounted on the exhaust line of a vapor drier to expose it continuously to IPA. This test proved the high reliability of this monitor: the sensitivity was not at all degraded after 6 months of exposure (Tokuyama Soda technical information, 1989).

D. Cleanliness of the Wafer Surface

The laser particle counter is generally used to measure the surface cleanliness of mirror-polished wafers. Table 1 lists the results of measuring residual particles on the wafer after vapor drying. Only a few particles larger than 0.5 μm were found on a 5 inch wafer. It was proved that IPA vapor drying is an excellent drying method that is free of particulate contamination and water spots. To achieve a high level of cleanliness of the wafer surface, high-purity IPA, a clean unit, proper operation of the process, and adequate control of liquid IPA quality are necessary. In addition, the cleanliness of all components that contact the wafers, such as carriers, chemicals, and ultrapure water, must be maintained at high levels.

E. Removal of Static Electricity

The IPA vapor drying method electrically neutralizes not only the wafers but also the PFA carriers, which are easily charged (Tokuyama Soda technical information, 1989). IPA itself is an insulator, although the mechanism has not yet been defined. Experimentally, however, this same phenomenon is observed with Ga-As wafers as well as Si wafers. When the charge on a solid surface, such as a wafer or carrier, is completely removed, neither electrostatic discharge nor electrostatic adsorption of particles takes place. The ease of charge removal from PFA, which is an extremely efficient electrical insulator and attains a rapid negative charge, will contribute greatly to the production of submicrometer and quarter-micrometer devices.

Table 1 Particle Count on the Si Wafer Surface After IPA Vapor Drying

Cleaning System	Defect Counts (>0.5 μm) (Wafer^{-1})		
	D = 75mm	100 mm	125 mm
$H_2SO_4 - H_2O_2$	0	0	0
$H_2SO_4 - H_2O_2$ $\downarrow$ HF	0.6	1.0	1.0
$HCl - H_2O_2$	1.0	2.0	3.5

(D : Wafer Diameter)

IV. SUMMARY

IPA vapor drying technology, which features such excellent characteristics as freedom from particulates and charge buildup, is certain to be crucial in the production of submicrometer devices. To achieve high levels of wafer surface cleanliness with this technology, comprehensive cleanliness control must be practiced throughout the wet processing, including the drying and precleaning steps.

REFERENCES

1. H. Mishima, T. Yasui, T. Mizuniwa, M. Abe, and T. Ohmi, Proc. of 9th Int. Symp. on Contamination Control, Los Angeles (1988), pp. 446–456.
2. H. Mishima, T. Yasui, T. Mizuniwa, M. Abe, and T. Ohmi, IEEE Transactions on Semiconductor Manufacturing, Vol. 2, 69–75 (1989).
3. J. Goodman and L. Mudrak, Solid State Technology (in Japanese) December, 16–19 (1988).

3 Wet Etch Cleaning

C. Spin Drying

Seiichiro Aigo
Alpha Science Laboratory Co., Ltd., Chiba, Japan

I. INTRODUCTION

The ultrapure water used in the semiconductor production line was developed for rinsing wafers. Although it is as close to the theoretical limits as possible, the purity greatly depends on the cleaning and drying methods. This part describes the present status of wet processing and the spin-drying method.

II. PRESENT STATUS OF WET PROCESSING

During wet processing, various chemicals are used, including H_2SO_4, HF, HCl, NH_4OH, and H_3PO_4. A cleaning process that uses chemicals should be followed by rinsing with ultrapure water and spin drying. When these are not adequate, residual ions, such as SO_4^{2-} and PO_4^{3-}, and thin films, such as water marks and stains, remain on the wafer surface, which depresses the yield. Such inadequate processes can be attributed to two major causes: insufficient ultrapure water quality and inadequate construction of the units. The remainder of this part deals with the latter issue.

The cleaning unit includes the chemical processing baths, the rinsing bath, and a robot to transfer the wafer carriers. Organic and inorganic impurities are generated both inside and outside each bath, and these impurities adhere to the wafers when the wafer carriers are transported from bath to bath. In other words, impurities readhere to wafers being cleaned in a bath. This is mainly because the rate of flow on the wafer surface is slow and impurities float in middle of the chemical surface that overflows from each bath. In a chemical or QDR bath that uses punched plates, 10–25 particles are counted on average on a 6 inch wafer with a 0.3 μm laser particle counter. The production process requires 5 particles or less, however.

Residual ions, water marks, and haze on the wafer surface are caused by a lack of uniformity during chemical dissolution and slow dissolution during rinsing. These prob-

Figure 1   Particles, including both organic and inorganic materials.

lems, which can be observed after spin drying, often appear in an ordinary bath equipped with a punched plate system. Improvements in each component are required based on detailed study of wet processing.

Figure 1 shows organic and inorganic impurities.

III. DIFFERENT TYPES OF SPIN DRYING

There are two types of spin drying in terms of construction: one with the center of rotation on the wafer (the single-spin drier, (Fig. 2) and the other with wafers placed symmetrically on both sides of the rotational axis (the multispin drier, Fig. 3).

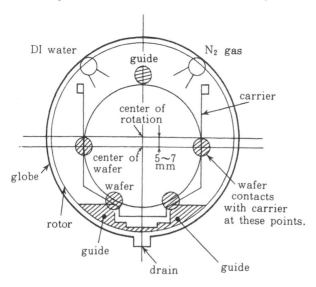

Figure 2   Center of rotation cannot be dried.

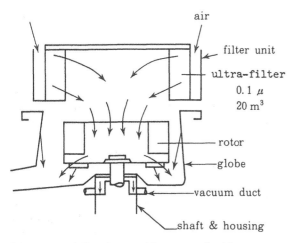

Figure 3 Airflow in a multicassette spin drier.

A. Single-Spin Drying

The single-spin drier is equipped with a cradle in the middle of which a wafer carrier is loaded. The drier is designed so that the center of the wafer is 5–7 mm off the center of rotation. This prevents the wafer from being thrown from the center during the rotation by imposing a centrifugal force in one direction. Unlike a photoresist coating, the rotor start-up takes a long time and the rotational speed is very low, around 3000 rpm, which generates a 20–30 μm undried spot in the center of the rotor. Therefore, when wet processing is repeated with a single-spin drier, an insufficiently dried spot with a diameter of 5–7 mm forms, resulting in 4–8 defective chips depending on the chip size.

Because the single-spin drier cannot maintain dynamic balance, vibration is generated continuously. As a result of this vibration, wafers in the carrier scratch the V grooves of carriers, which leads to the generation of particles and microcracks. The built-in nozzles that supply ultrapure water and nitrogen gas send these particles to the wafer surface. The yield is further decreased by this mechanism (Fig. 4).

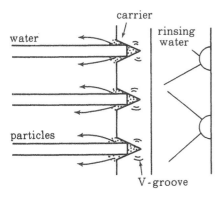

Figure 4 The wafer strikes the V groove of the carrier and generates particles (small pieces of carrier material). These particles flow onto the wafer surface with the rinse water.

Although the single-spin drier is compact, it causes the adhesion of 60–100 particles on insufficiently dried chips.

B.   Multispin Drying

The configuration of the multispin drier is indicated in Fig. 3. All the weaknesses of the single-spin drier have been overcome. Vibration during rotation was eliminated by placing cradles symmetrically in the rotor to maintain dynamic balance. To dry wafers sufficiently, a centrifugal force should be maintained during rotation by setting the center pitch of the cradles at more than 300 mm.

The Verneuille effect was applied to completely dry the U-shaped trenches. For this purpose, the filters were modified to supply a large amount of clean air and N_2 gas. During rotation, wafers vibrate because of the reversed airflow (the airflow hits the bottom of the cradle and reverses direction). The vibrated wafers scratch the V grooves of the carrier, generating particles and microcracks. To stop this reversed airflow, the cradle was modified as shown in Fig. 5. Figure 3 indicates the airflow described by the Verneuille effect. The degree of negative pressure in the middle of the rotor depends on the rotational speed. To stabilize the pressure, air and N_2 should be supplied through a filter. The air passing between the wafers in the carrier reaches the globe surface without generating reversed airflow. Because the globe surface is tilted downward, the air is sent downward and exhausted through a duct. Because the airflow is continuous and uniform at each point. only a very small number of particles adhere. (see Figs. 6, 7, 8). Five particles are counted on average with a 0.3 μm laser particle counter.

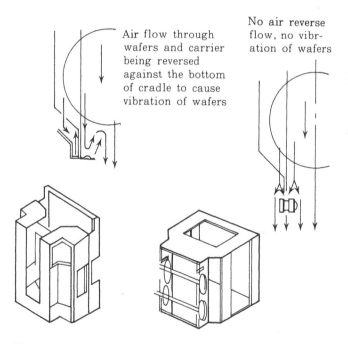

Figure 5   Modification of the cradle to eliminate reverse airflow.

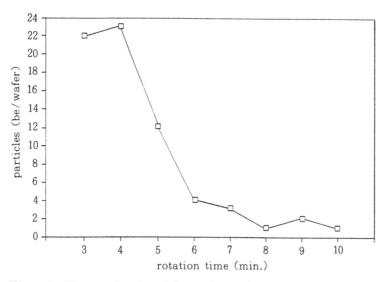

Figure 6   Water mark and particle test for rotational time.

IV.   EVALUATION OF PARTICLES AND WATER MARKS

As mentioned earlier, quartz and PTFE baths equipped with punched plates had problems with flow rate and floating particles. Our company specially developed a PFA bath using a jet flow system, which is patented. Compared with the punched plate baths, for a 6 inch wafer, the jet flow system can increase the flow rate on the wafer surface 12- to 20-fold and improve the particle transportation speed 3- to 6-fold. (See Figs. 6 through 9.)

To further improve the efficiency of ultrapure water cleaning, all the other units as well as the spin drier must be modified on the basis of the overall system concept.

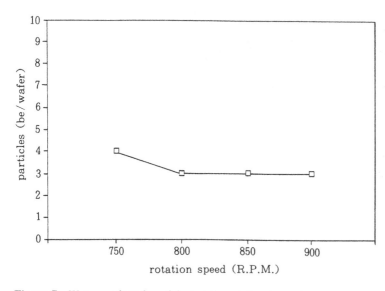

Figure 7   Water mark and particle test for rotational speed.

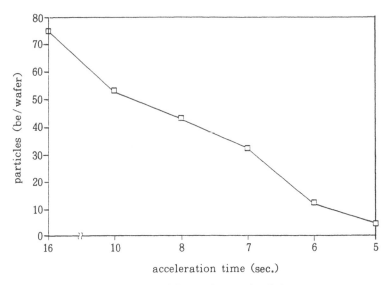

Figure 8   Water mark and particle test for rotational time.

Figure 9   Complete line of the wet station spin drier.

ACKNOWLEDGMENTS

Author thanks Prof. Tadahiro Ohmi of Tohoku University for valuable instruction during development. Credit also goes to Samsung Electronics which kindly provided some of the figures presented here.

3 Wet Etch Cleaning

D. Water Spot Growth Conditions

Michiya Kawakami
Mitsubishi Gas Chemical Co., Inc., Tokyo, Japan

Yasuyuki Yagi
Hitachi Plant Engineering & Construction Co., Ltd., Chiba, Japan

Tadahiro Ohmi
Tohoku University, Sendai, Japan

Motohiro Okazaki
Toray Industries, Inc., Shiga, Japan

I. INTRODUCTION

As the device patterns are increasingly integrated and refined, wet processing is required to achieve higher performance levels. Various studies are underway to improve the chemicals and ultrapure water used in wet processing and to upgrade the cleaning and drying processes [1]. One of the issues being worked on now is the water spot that appears on the wafer during drying and degrades the device. The mechanism of its growth [2] and the relationship with the ultrapure water quality are being studied. Furthermore, making positive use of the water spot, a method of evaluating the ultrapure water quality is now being developed. It is thought that the growth of a water spot depends on such factors as the drying method, the drying time, and the drying environment. Besides, as mentioned earlier, the elution of Si from the wafer is also believed to affect the growth of the water spot.

This part describes the growth of a water spot when ultrapure water is dried under various conditions on thermal oxide, without Si atom elution.

II. DRYING CONDITIONS AND WATER SPOT GROWTH

In this test, a certain amount of ultrapure water is left on the thermal oxide so that there is no Si elution from the wafer.

As shown in Fig. 1, the thermal oxide forms lattice patterns during oxidation (1100°C, 40 minutes, and a film thickness of 940 Å) and photolithography. Because of the hydrophobicity of the bare Si surface and the hydrophilicity of thermal oxide, a certain amount of ultrapure water is left on the thermal oxide lattice.

The drying unit used in the test was capable of controlling the temperature and environment. Temperatures of 20 and 70°C were used in N_2 and air. The water spot that

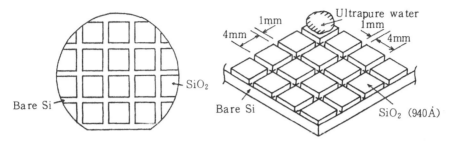

Figure 1 The water mark test.

Table 1 Conditions for Water Spot Generation

ATMOSPHERE	N₂	N₂	AIR
DRYING TEMP. (℃)	20	70	70
DRYING TIME (min)	10	2～3	2～3
EYESIGHT	N.D.	N.D.	N.D.
MICROSCOPY	DETECTION	N.D.	N.D.

N. D. : Not Detection of water-mark

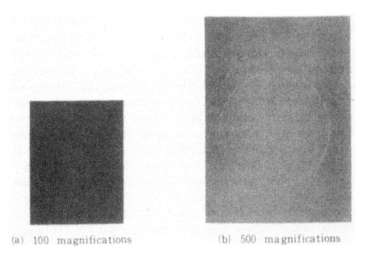

(a) 100 magnifications   (b) 500 magnifications

Figure 2 Microphotography of water mark formation.

formed was studied by macroscopic inspection and optical microscopy. Table 1 summarizes the results. In every case, the water spot was not recognized by visual inspection. By optical microscopy, however, the water spot was observed in the sample dried at 20°C for 10 minutes. Figure 2 shows this water spot observed by optical microscopy. The water spot appears as a dot on one point of the lattice. This means that the temperature and time of drying greatly affect water spot growth. Also, considering the problems of adsorbed impurities and native oxide, which affect the device as well as the water spot, high-purity

N_2 gas should be the best environment, although the test results show no difference in terms of drying between N_2 and air.

III.   SUMMARY

It was found that it is necessary to dry the ultrapure water on the wafer promptly at high temperatures to suppress the growth of water spots. A bare Si wafer with the native oxide etched by dilute HF must be treated more carefully because there are problems with Si elution and surface roughness, as mentioned in Chap. 4.

Because, in addition to Si and SiO_2, various other materials, such as polysilicon and A1, are used for semiconductors, further study is required to determine the characteristics of water spot growth on these different materials.

4   Chemistry of Wet Etch Cleaning

A.   Physical Chemistry of Etch Cleaning

Tetsuo Mizuniwa
Kurita Water Industries, Ltd., Kangawa, Japan

Mitsuo Abe
Tomco Manufacturing Ltd., Tokyo, Japan

Hiroyuki Mishima
Tokuyama Soda Co., Ltd., Yamaguchi, Japan

I.   INTRODUCTION

As devices decrease in size and chips are more sensitive to contamination, it is required that the residual contaminants and damage to the water surface be reduced significantly. To realize the future ultralarge-scale integration (ULSI) of wafers, every unwanted impurity must be eliminated from the prodution process [1].

Although dry processing has been introduced to various wafer-processing steps, wet cleaning with water and chemicals cannot be replaced by dry processing yet. This is because the characteristics of an aqueous solution, ensuring that a mild and uniform reaction or dissolution is performed at low temperatures, are essential to wafer processing. Also, wet cleaning can deal with all the impurities to be eliminated: trace particles, inorganic ions, organic materials, and native oxide [2].

Now that LSI production is being improved to achieve the contamination- and damage-free processing, improvements in wet cleaning are also required to guarantee cleaner wafers with less damage.

The RCA cleaning method [3], which was developed about 20 years ago, is still widely employed. The automation of cleaning equipment is the major current approach to improving cleanliness.

Conventional cleaning methods of removing trace particles from the silicon wafer surface have been studied. This part describes a cleaning method that causes less damage to the wafer.

II.   SAMPLES AND EXPERIMENTAL PROCEDURES

A.   Samples and Chemicals

In these cleaning tests, particles of identified materials were caused to adhere to silicon wafers. These wafers were then treated by different cleaning methods. In this way, it was possible to understand the relationship between particle materials and cleaning and to obtain information on the mechanism of cleaning.

For contaminant particles, polystyrene latex was used as an organic material and silica latex was used as an inorganic material. The polystyrene latex used in this experiment measured 0.43 μm. The silica latex was manufactured by Tokuyama Soda Co., Ltd. and was 0.5 μm on average. The original concentrations of these particles were 7×10^{10} and 7×10^9 ml^{-1}, respectively. They were diluted with ultrapure water to 10^3–10^4 ml^{-1}. A few milliliters of dilute particle solution was dropped on the wafer. After 30 s, the ultrapure water was eliminated by spin drying, depositing latex on the wafer. In this way, wafers on which 500–2000 particles were deposited were prepared as samples for the experiments.

For the wafers, 6–8 Ω·cm CZ-P(100) and 3–5 Ω·cm CZ-N(100) were used. The containers and apparatus used in the experiments were made of quartz or perfluoroalkoxyl (PFA). Table 1 lists the manufacturers and grades of high-purity chemicals [4–6] used in the experiments.

B. Cleaning Procedures

The RCA cleaning method was compared with a cleaning method that uses H_2SO_4–H_2O_2. Table 2 lists the cleaning procedures. All cleaning steps were conducted manually. During the drying process that follows the cleaning process, IPA vapor drying with a quartz vapor drier or spin drying was used.

C. Evaluation of Cleaning

To count trace particles on the wafer, the wafer inspection was carried out by Aeronca model WIS-100. The wafers are transported by a belt in the horizontal direction. A laser scans in a direction perpendicular to the wafer movement, and then the light scattered by adhering particles is detected and the number and size of particles is checked. This unit can inspect one wafer in about 10 s. The measurement results are expressed as the number and location of sections in which particles or damage 2 μm or larger, 2–0.5 μm, and

Table 1   Chemicals for Cleaning Tests

Chemical	Maker	Grade
Sulfuric Acid	Mitsubishi Kasei Corp.	EL
Hydrochloric Acid	Mitsubishi Kasei Corp.	EL
Hydrofluoric Acid	Hashimoto Chemical Industries Co., Ltd.	SA
	Daikin Industries Ltd.	--
Ammonia Solution	Mitsubishi Kasei Corp.	EL
Hydrogen Peroxide	Santoku Chemical Industries Co., Ltd.	Super High Purity
Iso-Propanol	Tokuyama Soda Co., Ltd.	SE

Table 2 Cleaning Procedures

Treatment	Procedure			
$H_2SO_4 - H_2O_2$ Cleaning	$H_2SO_4 : H_2O_2 = 4 : 1$			5 min.
	Overflow Rinsing			10 min.
$NH_4OH - H_2O_2$ Cleaning	$NH_4OH : H_2O_2 : H_2O$			
	$= 0.1 - 1 : 1 : 5$	80°C		10 min.
	Hot Water Rinsing	90°C		10 min.
	Overflow Rinsing			5 min.
HF Treatment	$HFI : H_2O = 1 : 100$			1 min.
	Overflow Rinsing			10 min.
$HCl - H_2O_2$ Cleaning	$HCl : H_2O_2 : H_2O = 1 : 1 : 6$	80°C	10 min.
	Hot Water Rinsing		90°C	10 min.
	Overflow Rinsing			5 min.

0.5 μm or smaller (haze) is detected. The detection sensitivity of this unit was confirmed by using a reference wafer with trace pits on silicon oxide, which generate the same scattering light as the smallest polystyrene latex measuring 0.36 μm.

When wafer cleanliness was evaluated with this equipment, the results were often affected by contamination from the measurement environment. In this experiment, it was confirmed that no increase in particles was detected on the wafer surface even after a wafer was placed horizontally near the inspection equipment for 14 h.

D. Surface Observation

Two methods are applied to observe the wafer surface: one method detects areas lit by the optical microscope when the flatness of the wafer surface is deteriorated, and the other method uses scanning electron microscopy (SEM; Hitachi S-570).

III. EXPERIMENTAL RESULTS AND DISCUSSION

A. Comparison of Polystyrene Latex Removal Efficiency

Wafers contaminated by polystyrene latex were cleaned with the following five methods to determine the particle removal efficiency.
1. $H_2SO_4 - H_2O_2$
2. $NH_4OH - H_2O_2$
3. DHF (dilute HF)
4. $HCl - H_2O_2$
5. NH_4OH with lower concentration and H_2O_2

Figure 1 shows the results.

Before cleaning, the number of haze particles (0.5 μm or less) that adhered to a single wafer was 400–700, and the number of particles 0.5 μm or larger was 100–600 per wafer. Considering the size of latex particles, it was assumed that some particles combined on the wafer surface.

$H_2SO_4 - H_2O_2$ cleaning exhibited a high removal efficiency regardless of particle size.

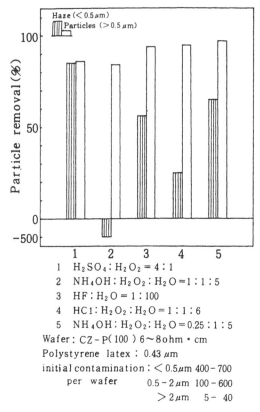

Figure 1   Comparison of particle removal efficiencies of various cleaning methods when poly-styrene latex spheres adhered to silicon wafers.

Cleaning with low-concentration NH_4OH and H_2O_2, as well as DHF cleaning, also exhibited high removal efficiencies. With $NH_4OH-H_2O_2$ cleaning ($NH_4OH/H_2O_2/H_2O = 1:1:5$), however, the haze count increased after cleaning. This is thought to be because the wafer surface was etched non-uniformly by ammonia.

B.   Comparison of Silica Latex Removal Efficiency

Figure 2 compare the efficiency of five cleaning methods (the same methods just described) in cleaning silica latex.

Before cleaning, the number of haze particles of ($0.5~\mu m$ or less) that adhered to a single wafer was 400–800, and the number of particles $0.5~\mu m$ or larger was 300–1000 per wafer. $NH_4OH-H_2O_2$ cleaning exhibited the highest removal efficiency of the five methods, followed by DHF cleaning. It was found that $H_2SO_4-H_2O_2$ and $HCl-H_2O_2$ cleaning were not as effective. For silica latex, $NH_4OH-H_2O_2$ cleaning ($NH_4OH/H_2O_2/H_2O = 1:1:5$) did not increase the haze count.

C.   Cleaning Mechanism

With regard to polystyrene latex removal, $H_2SO_4-H_2O_2$ cleaning is thought to oxidize and decompose polystyrene latex with its strong acidity. $NH_4OH-H_2O_2$ cleaning re-

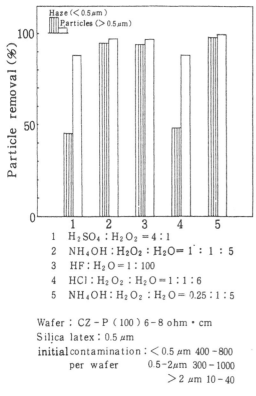

1    $H_2SO_4 : H_2O_2 = 4 : 1$
2    $NH_4OH : H_2O_2 : H_2O = 1 : 1 : 5$
3    $HF : H_2O = 1 : 100$
4    $HCl : H_2O_2 : H_2O = 1 : 1 : 6$
5    $NH_4OH : H_2O_2 : H_2O = 0.25 : 1 : 5$

Wafer : CZ – P (100) 6–8 ohm • cm
Silica latex : 0.5 μm
initial contamination : < 0.5 μm 400 – 800
     per wafer   0.5–2μm 300 – 1000
                  > 2 μm 10 – 40

Figure 2 Comparison of particle removal efficiencies of various cleaning methods when silica latex spheres adhered to silicon wafers.

moves polystyrene latex from the wafer surface while etching the native oxide. DHF exhibits good cleaning efficiency because it also dissolves oxide.

On the other hand, the following mechanisms are thought to apply to silica latex. DHF is effective in removing silica latex because it directly dissolves latex. $NH_4OH-H_2O_2$ cleaning etches oxide while dissolving some silica latex. With $H_2SO_4-H_2O_2$ and $HCl-H_2O_2$ cleaning, however, the oxide is not etched, which is why the removal efficiency is not good.

D. Cause of Haze Increase

Although $NH_4OH-H_2O_2$ cleaning is effective in removing particles, this cleaning method may cause the fatal damage to the wafer surface under certain conditions if it etches the wafer surface. Figure 3 shows the change in the wafer surface when it was treated repeatedly with $NH_4OH-H_2O_2$ and DHF cleaning. It was learned that the number of particles 0.5 μm or larger increases in DHF cleaning and decreases in $NH_4OH-H_2O_2$ cleaning. However the haze count (0.5 μm or smaller) increases with both cleaning methods. In particular, $NH_4OH-H_2O_2$ cleaning greatly raises the haze count: thousands of such particles were reported after a few rinses with $NH_4OH-H_2O_2$. It has been reported that $NH_4OH-H_2O_2$ cleaning sometimes increases the haze count: this was confirmed in this experiment.

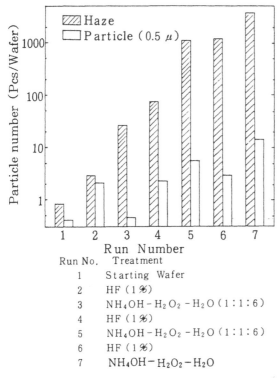

Run No.   Treatment
   1   Starting Wafer
   2   HF (1%)
   3      $NH_4OH - H_2O_2 - H_2O$ (1:1:6)
   4   HF (1%)
   5      $NH_4OH - H_2O_2 - H_2O$ (1:1:6)
   6   HF (1%)
   7      $NH_4OH - H_2O_2 - H_2O$

Figure 3   Particle count change on a silicon wafer by alternating between $NH_4OH - H_2O_2$ cleaning and dilute HF treatment.

Figure 4 shows the light reflected from the wafer surface after $NH_4OH - H_2O_2$ cleaning. No light spot is observed on the mirrored surface. When the haze count increases as a result of $NH_4OH - H_2O_2$ cleaning, however, light spots can be observed on the wafer surface. The light intensity increases as the haze count rises. This means the haze count does not reflect an increase in trace particles that adhere to the wafer surface but is caused by deterioration in the surface. When the surface becomes rougher, light causes the diffused reflection.

Figure 5 shows SEM images of a wafer surface on which the haze count increased significantly. It is clear the surface is roughened and is not etched in uniformly. Because of this etching caused by the $NH_4OH - H_2O_2$, the cleaning is not consistent. This problem is difficult to address.

E.   Cleaning Efficiency of Ammonia Solution at Low Concentration

The preceding experiments made it clear that $NH_4OH - H_2O_2$ cleaning is very effective in removing contaminants but it may damage the wafer surface by nonuniform etching. It was also confirmed that low-concentration ammonia solution exhibits satisfactory cleaning efficiency. Therefore it was necessary to determine to what extent the ammonia concentration can be lowered without deterioration of the cleaning efficiency.

Figures 6 and 7 show the number of residual particles when wafers contaminated by polystyrene latex and silica latex were cleaned with $NH_4OH - H_2O_2$ solutions with

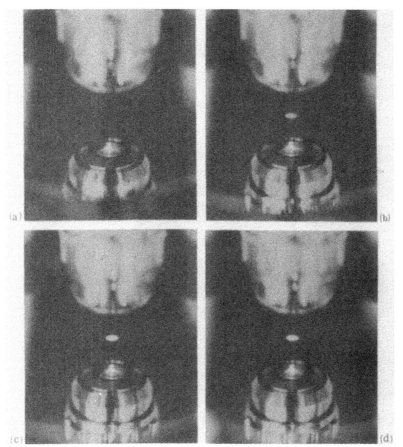

Figure 4   Light reflection from the water surface after $NH_4OH-H_2O_2$ cleaning. (top, left) No treatment (haze count = 1), (top, right) treated once (haze count = 30), (bottom, left) treated twice (haze count = 1000), (bottom, right) treated three times (haze count = 3500).

different ammonia concentrations. For both polystyrene latex and silica latex, cleaning with the $1:1:5$ $NH_4OH/H_2O_2/H_2O$ solution raised the haze count. No great difference was observed with the other concentrations in terms of removal efficiency. This means that satisfactory cleaning can be achieved even if the ammonia concentration is reduced to 1/10 of the present level.

Figure 8 shows the result of the same experiment using wafers on which dust from the clean room air were deposited. For comparison, the results of $H_2SO_4-H_2O_2$ cleaning are also indicated. The removal efficiency for particles 0.5 μm or larger was very good. Even when the ammonia concentration was to 1/10 of the present level, the removal efficiency was maintained.

Figure 9 shows the results of an experiment in which organic contaminants were deposited in large quantities on the wafer. The wafers were coated with a positive photoresist. When the ammonia concentration was lowered to $NH_4OH/H_2O_2 = 0.1:1:5$, the number of residual particles increased. When the concentration was $NH_4OH/H_2O_2/$

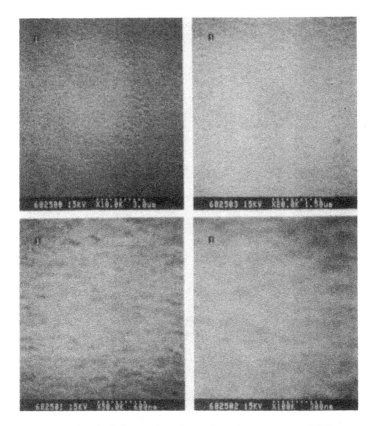

Figure 5   SEM of damaged wafer surface (haze count = 3500).

$H_2O = 0.25:1:5$ or higher, however, the removal efficiency was maintained at the same level as with the $1:1:5$ concentration.

IV.   CONCLUSIONS

It was confirmed that $NH_4OH-H_2O_2$ cleaning is effective in removing particles that adhered to silicon wafers regardless of the materials. It was found, however, that the solution with a concentration of $NH_4OH/H_2O_2/H_2O = 1:1:5$ that has been commonly used, raises the haze count on the wafer surface.

Since the diffused reflection of light was observed on a wafer with increased haze count and SEM images of those wafers showed poor uniformity on the surface, it was learned that the increase in the haze count was not caused by the increase in particles but by the poor surface uniformity as a result of etching.

To prevent the wafer surface from etching by $NH_4OH-H_2O_2$ cleaning, the ammonia concentration should be lowered. The efficiency of removing trace particles can be maintained even if the ammonia concentration is reduced to 1/10 of the present level. When the amount of adhering particles is large, however, the number of residual particles increases. Therefore $NH_4OH/H_2O_2/H_2O = 0.25:1:5$ is considered appropriate. For

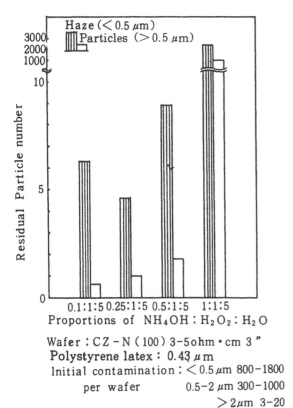

Wafer : CZ – N (100) 3–5ohm • cm 3 ″
Polystyrene latex : 0.43 μm
Initial contamination : < 0.5 μm 800–1800
      per wafer   0.5–2 μm 300–1000
                  > 2μm 3–20

Figure 6 Influence of NH₄OH ratio on particle removal efficiencies when polystyrene latex spheres adhered to wafers.

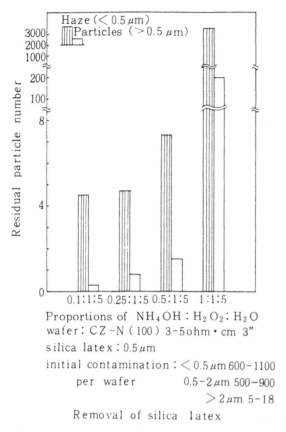

Proportions of NH₄OH : H₂O₂ : H₂O
wafer: CZ -N (100) 3-5ohm・cm 3"
silica latex : 0.5μm
initial contamination : < 0.5μm 600-1100
      per wafer   0.5-2μm 500-900
                       > 2μm 5-18
Removal of silica latex

Figure 7   Influence of NH₄OH ratio on particle removal efficiencies when silica latex spheres adhered to wafers.

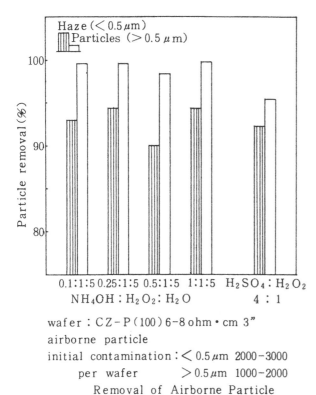

Figure 8 Influence of NH₄OH ratio on particle removal efficiencies when airborne particles adhered to wafers.

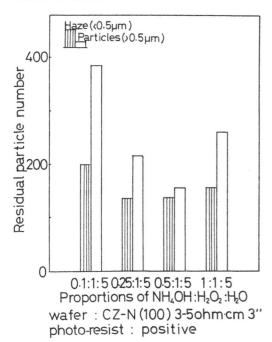

Figure 9 Removal of photoresist.

solutions with ammonia concentrations of $0.5:1:5$ or lower, no significant increase in haze, is observed, which means the damage to the wafer surface is limited.

REFERENCES

1. T. Ohmi, N. Mikoshiba, and K. Tsubouchi, Proceedings of the First International Symposium on Ultra Large Scale Integration Science and Technology, Pennsylvania, May 1987.
2. T. Ohmi and T. Nitta, Preface to Ultrapure Water and Ultrapure Chemicals Supplying System, Realize, Inc., 1986.
3. W. Kern and D. A. Puotinen, RCA Review, 31, 187 (1970).
4. H. Matsuzaki, Proceedings of the 2nd ULSI Ultra Clean Technology Symposium, Tokyo, 1986, p. 299.
5. A. Fukumura and H. Ushida, Proceedings of the 2nd ULSI Ultra Clean Technology Symposium, Tokyo, 1986, p. 214.
6. N. Miki, Proceedings of the 2nd ULSI Ultra Clean Technology Symposium, Tokyo, 1986, p. 278.

4 Chemistry of Wet Etch Cleaning

B. Surface Chemistry of BHF

Nobuhiro Miki and Hirohisa Kikuyama
Hashimoto Chemical Corporation, Osaka, Japan

I. ETCHING CHARACTERISTIC OF BHF

A. Etching Rate and Chemical Composition

BHF is a HF solution containing ammonium fluoride that is used to etch SiO_2. In general, it is composed of 40% ammonium fluoride solution and 50% HF solution. The etching rate is adjusted by the composition ratio of the two solutions. More specifically, the etching rate can be adjusted within a wide range from 1200 to 90 Å minute^{-1} at 25°C by changing the composition ratio between 40% ammonium fluoride and 50% HF from 6:1 to 100:1. To clean the wafer surface, the composition ratio is set to lower the etching rate further. As shown in Table 1, the etching rate of BHF depends on the concentration of HF.

Table 1 Interfacial Tension and BHF Etching Rate

BHF	Composition		Contact Angle (degree)	Surface Tension (dyn/cm)	Etching Rate (Å/min)
	HF (%)	NH$_4$F (%)			
6 : 1	7.1	34.3	69	85.6	1170
9 : 1	5.0	36.0	73	87.8	780
10 : 1	4.5	36.4	70	87.8	730
20 : 1	2.4	38.1	73	90.1	370
30 : 1	1.6	38.7	70	91.4	280
100 : 1	0.5	39.6	75	92.0	90
63	6.0	30.0	73	86.9	1050

Instead of HF, a mixture of ammonium fluoride and HF is used as an etching agent. This is because the photoresist is not chemically resistant to HF: it swells and peels off the wafer. In short, the photoresist film is protected by the ammonium fluoride.

Since wet etching is a chemical reaction in aqueous solution, many factors control this process, such as composition ratio, etching temperature, conditions for stirring, and the number of wafers to be processed. The effect of the composition ratio on the etching rate is particularly complicated, which is described in detail here, together with the chemical reaction of etching.

B.   Chemical Reaction Between BHF and SiO_2

The reaction between SiO_2 and HF or between SiO_2 and BHF can be stoichiometrically expressed by Eqs. (1) and (2):

$$SiO_2 + 6HF \rightleftharpoons H_2SiF_6 + 2H_2O \tag{1}$$

$$SiO_2 + 4HF + 2NH_4F \rightleftharpoons (NH_4)_2SiF_6 + 2H_2O \tag{2}$$

$$SiO_2 + 3H^+ + 3HF_2^- \rightleftharpoons 2H^+ + SiF_6^{2-} + 2H_2O \tag{3}$$

However Eqs. (1) and (2) do not accurately express the actual etching process as an ionic reaction. The ion species that reacts chemically with SiO_2 is HF_2^- [1], and the chemical reaction is expressed in Eq. (3). It is therefore necessary to understand the ion dissociation and the species of ions in HF or BHF solution.

1.   Dissociation in HF

As expressed in Eqs. (4) and (5), F^- is dissociated via the transient complex of $[H_3O^+ \cdot F^-]$ in HF [2]. HF_2^-, the dominant ion in the etching reaction, must be formed by the equilibrium as shown in Eq. (6). With HF, however, stable molecules are formed by strong bonds between H and F, and the hydrogen bond between two molecules is strong, which makes the degree of dissociation in solution very small. Therefore F^- dissociated through processes (4) and (5) is extremely limited. This means the HF_2^- formed in the equilibrium equation (6) is also limited.

$$HF + H_2O \rightleftharpoons [H_3O^+ \cdot F^-] \tag{4}$$

$$[H_3O^+ \cdot F^-] \rightleftharpoons H_3O^+ + F^- \tag{5}$$

$$HF + F^- \rightleftharpoons HF_2^- \tag{6}$$

2.   Dissociation in a Mixture of Ammonium Fluoride and HF

Using a mixture of ammonium fluoride and HF, the reaction between SiO_2 and BHF can be expressed by Eq. (7):

$$SiO_2 + 4NH_4^+ + 4HF_2^- = 4NH_4^+ + SiF_6^{2-} + 2F^- + 2H_2O \tag{7}$$

HF_2^-, which reacts with SiO_2 in this equation, is formed directly in the reaction expressed in Eq. (8), not going through the dissociation equilibrium expressed in Eqs. (4) and (5).

$$NH_4F + HF \rightleftharpoons NH_4^+ + HF_2^- \tag{8}$$

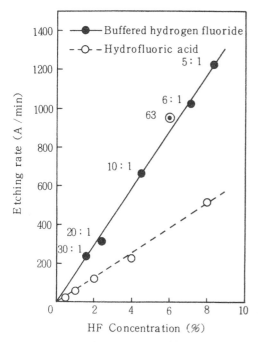

Figure 1 Relation between etching rate and HF concentration.

Since this reaction proceeds much more easily than the reaction in Eqs. (4) and (5), more HF_2^- is formed than by HF alone. Therefore, to obtain the same etching rate, the HF must have a higher HF concentration than the mixture of ammonium fluoride and HF, as shown in Fig. 1. This proves that BHF, the mixture of ammonium fluoride and HF, is more effective in etching.

II. SURFACE ACTIVE BHF

A. Surface Activation of BHF

BHF features a high surface tension and a high contact angle. Table 1 lists quantitative data. The surface tension of water is 72 dyn cm^{-1}, and its contact angle on silicon is 85° at a temperature of 25°C. This means BHF is not peculiar in terms of low wettability on silicon: this characteristic is common to aqueous solutions.

Good wettability on solid surfaces as well as chemical functions are required to deal with submicron and high-aspect-ratio processing. So that solutions uniformly penetrate the fine patterns and precise surface processing is possible, the quality of the solutions must be improved by surfactants [3–6]. The surfactants to be employed in the submicron process must meet the following requirements:
1. Same etching rate as BHF
2. Low contact angle
3. Nonsegregating
4. Nonfoaming
5. Low generation of particulates

Table 2   Relation of Wettability on the Silicon Wafer Surface and Contact Angle of BHF on Si

Table 2   Relation of wettability on Silicon wafer surface and contact angle of BHF / Si

BHF	Contact angle	Wetting area of BHF on silicon wafer surface	Wettability
Conventional	73°		No wetting zone
	52°		A little wetting zone
	37°		A little wetting zone
Improved	30°		The whole wetting surface

6. Low generation of impurities (possibility of purification)
7. Limited particle adhesion to wafer surface
8. No surface residue
9. Excellent surface smoothness
10. High etching selectivity
    The following five basic principles must be satisfied to meet these characteristics:
1. Good solubility in BHF
2. Hydrophilic at the wafer surface
3. No decomposition in BHF
4. No reaction with BHF
5. Sufficient lowering of contact angle at the critical micelle concentration (CMC)

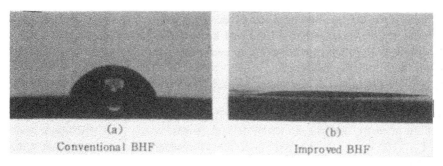

Figure 2   Wettability of BHF on the silicon surface.

These five basic principles are shared by all liquid chemicals.

B.   Wettability of the Silicon Surface

A solid is much more easily wet with a liquid if surfactants are added to the liquid. This is called the wetting action. When a droplet of BHF without surfactant (conventional BHF) and a droplet of BHF with surfactant (improved BHF) are dropped on a silicon surface, the shape of the two droplets appear different, as shown in Fig. 2.

The difference in the droplet shape can be indicated by the contact angle θ. As θ nears $0°$, the surface becomes wetter. The contact angle θ can quantitatively express the wetting action, clearly describing the relationship between the boundary tension and the wetting action.

The following equation is provided by Fig. 3:

$$\gamma_s = \gamma_1 \cos \theta + \gamma_i \tag{9}$$

The following equation can be derived from Eq. (9):

$$\cos \theta = \frac{\gamma_s - \gamma_i}{\gamma_1}$$

Each solid has a fixed value of γ_s, but γ_1 and γ_i decrease when surfactant is added. Therefore $\cos \theta$ increases when surfactant is added. This means the contact angle θ decreases and the wetting action is enhanced. In this way, the contact angle decreases and the wettability of the silicon surface improves if surfactant is added to BHF.

Hundreds of surfactants are available on the market, made from various compounds. Their chemical structures and boundary actions are different. It is necessary to determine which surfactant is suitable to BHF to improve its functions.

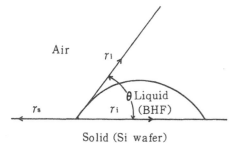

γ_1 : Free energy of liquid surface

γ_s : Free energy of solid surface

γ_i : Free energy of solid−liquid interface

θ : Contact angle

Figure 3   Contact angle at the liquid/solid/air boundary line.

The surface tension and the contact angle on the silicon surface were studied using BHF with fluorocarbon surfactant and BHF with hydrocarbon surfactant. Figures 4 and 5 show the results with cationic and amphoteric surfactants. Similar results were obtained with anionic and nonionic surfactants.

The effect of filtration was also studied. Using BHF with fluorocarbon surfactant, the surface tension drops but the contact angle does not decrease. In most cases, filtration increases the surface tension of hydrocarbon and fluorocarbon surfactants. This is because the solubility of many surfactants in BHF is insufficient, which does not satisfy basic principle 1. As surfactant mixed in BHF to decrease the surface tension is separated in the course of filtration, the filtration increases the surface tension of the BHF.

When fluorocarbon surfactants are added to BHF, the surface tension decreases but the contact angle remains high. This is because the $-CF_3$ or $-CF_2^-$ surfactant base, which is hydrophobic, causes the surface to shed water and oil, which does not satisfy basic principle 2.

Not many surfactants satisfy all five basic principles. More than 350 types of surfactants were examined. The following hydrophilic hydrocarbon compounds were found to meet the requirements:

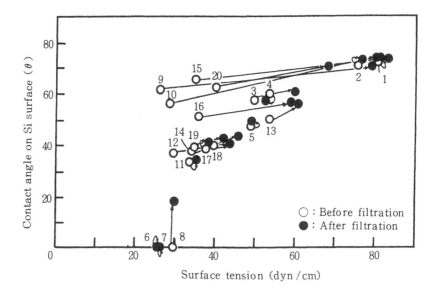

1. Propylamine, 2. Butylamine, 3. Amylamine, 4. Hexylamine, 5. Heptylamine, 6. Octylamine, 7. Nonylamine, 8. Decylamine, 9. Laurylamine, 10. Tetradecylamine, 11. Octadecylamine Acetate, 12. Coconutamine Acetate, 13. Trioctylamine, 14. Stearylamine Acetate, 15. Dodecyl Trimethyl Ammonium Chloride, 16. Lauryl Trimethyl Ammonium Chloride, 17. Alkyl Benzyl Dimethyl Ammonium Chloride, 18. Dimethyl Alkyl Betaine, 19. 2−Alkyl−N−Carboxy Alkyl−N−Hydroxy Ethyl Imidazolinium Betaine, 20. Coconut Alkyl Dimethyl Benzyl Ammonium Chloride

Figure 4 Surface tension and contact angle on a Si surface using hydrocarbon surfactant (cationic and amphoteric) in BHF.

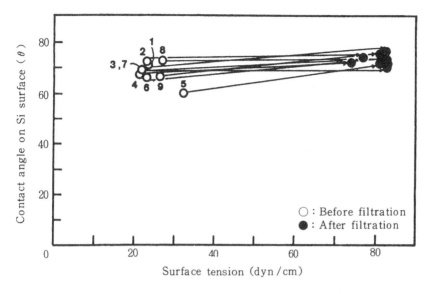

1. Perfluoro Alkyl Trimethyl Ammonium Salt, 2. Perfluoro Alkyl Trimethyl Ammonium Iodide, 3. Perfluoro Alkyl Trimethyl Ammonium Salt, 4. $R_f C_2 H_4 S C_2 H_4 N (CH_3)_3 CH_3 SO_4$, 5. $R_f SO_2 N R C_2 H_4 N^-(CH_3)_3 I^-$, 6 – 9.Perfluoro Alkyl Betaine

Figure 5 Surface tension and contact angle on a Si surface using fluorocarbon surfactant (cationic and amphoteric) in BHF.

Aliphatic amine $(C_7–C_{12})$
Aliphatic carboxylic acid and its salt $(C_5–C_{11})$
Aliphatic alcohol $(C_6–C_{12})$

Mutual solubility, bubbling, and wettability can be controlled more functionally in a mixture of these compounds than with any of these compounds alone. Aliphatic amine greatly improves the wettability but it slightly increases bubbling. When it is mixed with aliphatic carboxylic acid or with aliphatic alcohol, however, the bubbling can be suppressed and all requirements are satisfied.

When a surfactant meeting the five basic principles is added to BHF to improve the contact angle on the silicon surface to 30–40°, sufficient wettability on the silicon wafer can be imparted to the BHF.

To evaluate wettability, although it is common to measure the contact angle, it is easy to observe the condition of the silicon wafer surface after immersing it in BHF and pulling it out vertically. Table 2 shows the relationship between the measured contact angle and the wettability evaluated using this method.

C.   Limited Number of Particles on the Silicon Surface

Figure 6 shows the number of particles in BHF counted by laser particle counter with different surfactant concentrations and treated by circulating filtration under fixed conditions. Generally, the number of particles increased even after filtration as the surfactant concentration increased (a) in Fig. 6). For aliphatic amine, however, it was found that the number of particles was extremely limited at certain concentrations. The causes of this phenomenon have not yet been revealed. It is conjectured, however, that it is caused by a multiple effect of the following: the decrease in filtration resistance as a result of the better wettability of the filter induced by the improved wettability of the BHF and the trace particle aggregation effect caused by surfactant micelles both in the solution and on the surface of filter membrane.

In Fig. 6, the number of particles increases when the surfactant concentration is greater than 80 ppm. If 80 ppm is the critical micelle concentration (CMC) of the added surfactant, it can be deduced that the surfactant forms micelles at concentrations higher than 80 ppm and that more micelles form as the surfactant concentration increases. Therefore the surfactant concentration should not exceed the CMC to limit the number of particles. In other words, the surfactant to be added should give the solid surface sufficient wettability before its concentration increases above the CMC (basic principle 5).

Figure 6 shows the relationship between the surfactant concentrations and the contact angle on a silicon surface. At the CMC of this surfactant (80 ppm), the contact angle drops to 30°, providing the silicon wafer surface with sufficient wettability. It was also learned that excessive surfactant interferes with uniform etching or even decreases the

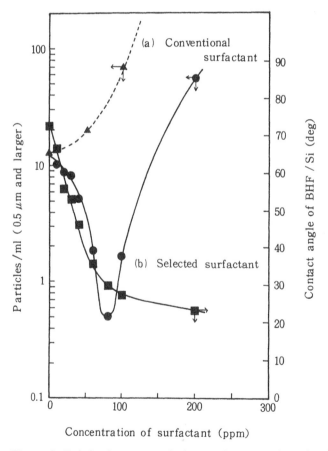

Concentration of surfactant (ppm)

Figure 6   Relation between particulates and concentration of surfactant.

etching rate. The particle level of BHF developed based on these findings is very low. The following experiment was conducted to count the number of particles that adhere to a silicon wafer treated with BHF.

The number of particles adhering to a wafer surface that was treated with three different BHF solutions (no surfactant, CMC, and surfactant above the CMC) was counted using the wafer inspection system WIS-150 (Aeromca).

The experimental procedure was as follows:
1. One batch: 3 inch silicon wafers, 12 sheets
2. $H_2SO_4-H_2O_2$ cleaning (5 minutes)
3. Ultrapure water overflow rinsing (10 minutes)
4. $H_2SO_4-H_2O_2$ cleaning (5 minutes)
5. Ultrapure water overflow rinsing (10 minutes)
6. BHF treatment
7. Ultrapure water overflow rinsing (10 minutes)
8. IPA (isopropyl alcohol) vapor drying
9. Measurement by WIS-150

Figure 7 shows the relationship between the number of adhering particles and the contact angle. The number of particles ($\leqq 0.5$ μm and $\geqq 0.5$ μm) using BHF with surfactant at the CMC is about half that using BHF without surfactant. This is because the wettability of the silicon surface is improved by the surfactant. However, because the excessive surfactant adversely affects the haze count (particles $\leqq 0.5$ μm), it is extremely important to determine the appropriate concentration of surfactant.

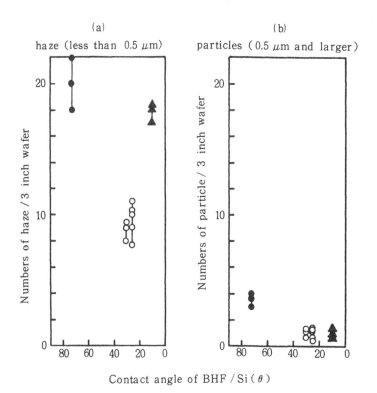

Figure 7   Adhesion of particulates to a silicon wafer. No surfactant, ($\bullet$); CMC ($\circ$); surfactant above the CMC ($\blacktriangle$).

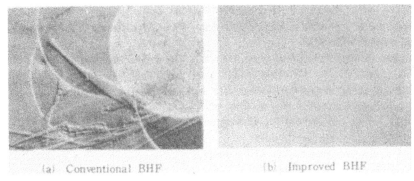

(a) Conventional BHF   (b) Improved BHF

Figure 8   Silicon surface after immersion in BHF for 92 h.

D.   Smoothness of the Silicon Surface

As semiconductors are further integrated, the smoothness of the etched surface becomes more and more critical. The smoothness of the silicon surface after forming a shallow junction is particularly important. To evaluate the roughness of a silicon surface etched with BHF, the following experiment was conducted. Silicon wafers were immersed in two different BHF solutions, BHF without surfactant and BHF with surfactant, for 92 h. Optical microscopy was used to observe the wafer surface (Fig. 8).

Using BHF without surfactant (Fig. 8a), the surface is not uniform. On the other hand, the silicon surface treated with BHF with surfactant (Fig. 8b) is etched uniformly. Further studies are required to determine the cause of this difference.

ACKNOWLEDGMENTS

The authors thank Prof. Tadahiro Ohmi of Tohoku University for valuable advice and instruction during the development of the improved BHF presented in this part. We appreciate the efforts of Mr. Ichiro Kawanabe and Mr. Masayuki Miyashita, researchers at Tohoku University. We also note that Mr. Kiyonori Saka and Mr. Jun Takano of Hashimoto Kasei helped us to prepare this manuscript.

REFERENCES

1.   J. S. Judge, J. Electrochem. Soc., *118,* 1772 (1971).
2.   P. A. Giguere and S. Turrell, J. Am. Chem. Soc., *102,* 5473 (1980).
3.   N. Miki, H. Kikuyama, I. Kawanabe, J. Takano, and M. Miyashita, Process Capability Technology II for LSI Processing, (1989), p. 109.
4.   H. Kikuyama, N. Miki, K. Saka, J. Takano, I. Kawanabe, M. Miyashita, and T. Ohmi, IEEE Trans. Semiconductor Manufacturing, *3*(3), 99, (1990).
5.   H. Kikuyama and N. Miki, 9th International Symposium on Contamination Control, Los Angeles, (1988), p. 378.
6.   H. Kikuyama, N. Miki, J. Takano, and T. Ohmi, Microcontamination, *7*(4), p 25–28, 50–51 (1989).

5
Suppression of Native Oxide

Michiya Kawakami
Mitsubishi Gas Chemical Co., Inc., Tokyo, Japan

Makoto Ohwada
Alps Electric Co., Ltd., Fukushima, Japan

Yasuyuki Yagi
Hitachi Plant Engineering & Construction Co., Ltd., Chiba, Japan

Tadahiro Ohmi
Tohoku University, Sendai, Japan

I. INTRODUCTION

Current semiconductor devices use a multilayer structure of thin films. In this structure, the interface between a thin film and substrate or between two films plays a critical role in operation of the device. Therefore, the formation of high-quality films and an ideal interface are important challenges to further advances. To overcome these obstacles, contamination, stress, and damages to the Si wafer [1] must be controlled perfectly. It is necessary to suppress contaminants on the wafer surface as much as possible, including particles, organic contaminants, metallic contaminants, native oxide, and adsorbed impurities. Ultrapure water and chemicals have been studied actively, and consequently, particles and organic and metallic contaminants have been successfully suppressed to a level lower than the detection limit of the analyzing equipment. In the future, it will certainly be important to reduce these three contaminants further. At the same time, however, it is essential to find a way to reduce the remaining two contaminants. The native oxide formed on wafer surface interferes with the growth of the epitaxial Si thin film at low temperatures and can affect the precise control of film thickness and film quality of extremely thin gate oxides [2]. It may also cause an increase in the contact resistance by microscopic contact holes.

This chapter describes the suppression of native oxide, essential for realizing submicrometer ultralarge-scale integration (ULSI) in the future.

II. TYPES OF NATIVE OXIDE AND THEIR MEASUREMENT

Thermal oxidation at high temperature has been studied in detail to determine the reaction mechanism and the oxidation rate. In contrast, the native oxide has not yet been studied. If the relationship between the oxidation rate and temperature during thermal oxidation is extrapolated to room temperature, no oxide is formed at all. It is important to therefore understand the mechanism of native oxide formation scientifically, to control the growth rate. Native oxide has been defined as oxide formed at room temperature. However, to further upgrade such processes as gate oxidation, oxide formed by chance during the temperature increase in processing chambers must be considered native oxide as well.

Native oxide can be categorized as follows:

1. At room temperature in air, ultrapure water, and chemicals
2. During the temperature increase in processing chambers

The growth of native oxide at room temperature is related to the wet process. In this chapter, native oxide growing in ultrapure water, in air, in chemicals, and during cleaning and drying is described.

For measurement of the native oxide, photoelectron spectroscopy (XPS and ESCA) and ellipsometry are effective to accurately measure the thickness of extremely thin oxides [3]. The measurement procedure is as follows:

1. Measure the film thickness of thermal oxide (70–140 Å) by ellipsometry without fixing the refractive index.
2. Measure the Si_{2p} spectrum of the same thermal oxide by XPS (ESCA).
3. Calibrate the peak area ratio of the XPS spectrum (I_{sio_x}/I_{si}) using the film thickness measured by ellipsometry.
4. Measure the XPS spectrum of native oxide, and determine the film thickness based on the peak area ratio and the calibration data.

III. NATIVE OXIDE GROWING IN ULTRAPURE WATER

A. Amount of Dissolved Oxygen in Ultrapure Water

The dissolved oxygen concentration in the ultrapure water line installed in the superclean room at Tohoku University, which employs a vacuum deaerator to reduce dissolved oxygen, is 0.04 ppm at present. When this ultrapure water is distributed to a cleaning bath exposed to the air, it immediately absorbs oxygen in the air and the dissolved oxygen concentration rises to 0.6 ppm. This means the ultrapure water rinse is usually conducted with a dissolved oxygen concentration of 0.6 ppm. The ultrapure water in equilibrium with the air contains 9 ppm dissolved oxygen at room temperature.

B. Growth of Native Oxide in Ultrapure Water

Figure 1 shows the time dependence of the thickness of the native oxide that grows when wafers are immersed in three types of ultrapure water with different dissolved oxygen concentrations at room temperature. The wafers used in this test were $n-Si(100)$ substrates (10^{15} cm^{-3}). For pretreatment, the wafers were treated with $H_2SO_4-H_2O_2$ and RCA cleaning, immersed in dilute HF (DHF) for 1 minute, and then rinsed with ultrapure water for 10 minutes. (The pretreatment in the following tests was the same.) Figure 1 indicates that the growth of native oxide in ultrapure water is related to the dissolved oxygen and it is possible to suppress native oxide formation by reducing the amount of dissolved oxygen.

The effect of impurity concentration on native oxide growth was also studied. Figure 2 shows the result of a test in which n^+-Si(100) wafers (10^{20} cm^{-3}) were immersed in ultrapure water containing 9 ppm dissolved oxygen. In this test, n^+ wafers were prepared by depositing PSG on samples and treatment at 1000°C for 30 minutes. As shown in Fig.

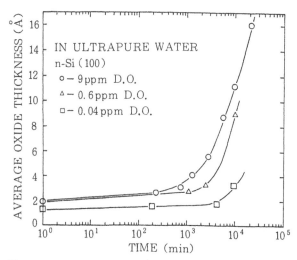

Figure 1 Oxide thickness as a function of immersion time of wafers in ultrapure water at room temperature for different dissolved oxygen concentrations.

2, the growth of native oxide on n^+ wafers stops when the thickness reaches 10 Å but the native oxide grows faster on n^+ wafers than on n wafers in the initial stage. Therefore, extra attention must be paid to native oxide growth when n^+ wafers are cleaned.

Table 1 lists the results of an experiment in which n-Si(100) wafers were immersed in hot ultrapure water (50–90°C) for 8 h. The thickness of the native oxide was about 2.5 Å at room temperature but about 10 Å at 70–90°C. The native oxide grew more quickly when the temperature of the ultrapure water was raised.

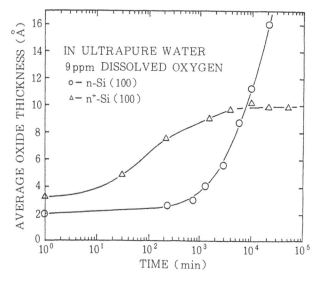

Figure 2 Oxide thickness on n-Si and n^+-Si as a function of immersion time in ultrapure water with 9 ppm dissolved oxygen.

Table 1   Growth of Native Oxide in Ultrapure Water (8 h)

WATER TEMPERTURE (°C)	OXIDE THICKNESS (Å)
50.0	4.5
70.0	11.3
90.0	10.1
ROOM TEMPERATURE	2.5

C.   Si Elution and Increased Surface Roughness in Ultrapure Water

Figure 3 shows the time dependence of the number of Si atoms eluted into ultrapure water or contained in native oxide when n-Si(100) wafers are immersed in ultrapure water containing 9ppm dissolved oxygen at room temperature. N_{Si} is the number of Si atoms in a unit area of wafer surface. Figure 3 indicates that a considerable amount of silicon is eluted from the wafer surface along with native oxide growth: the number of Si atoms eluted from the wafer surface is 10 times that contained in native oxide. This means native oxide growth and Si elution take place at the same time in ultrapure water and that these two phenomena are closely related. Figure 4 shows the results of a study to observe a wafer surface immersed in ultrapure water by the spotlight method. In this method, the light from an optical microscope is directed on the wafer surface to observe its diffuse reflection. No reflection is observed when the wafer suface is smooth, but some reflection is observed when the wafer surface is rough. As shown in Fig. 4, no reflection is observed for up to 3 days, but a clear reflection appears on day 7 of immersion and later.

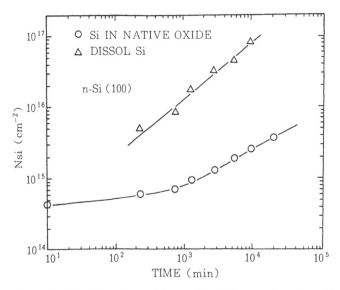

Figure 3   Time dependence of the number of Si atoms in native oxide and the number of dissolved Si atoms in ultrapure water for wafers immersed in ultrapure water. N_{Si} is the number of Si atoms per unit area on the wafer surface.

(1) 3 days immersed in ultrapure water with 9 ppm dissolved oxygen concentra-
tion.

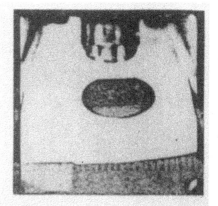

(2) 7 days immersed in ultrapure water with 9 ppm dissolved oxygen concentra-
tion.

(3) 14 days immersed in ultrapure water with 9 ppm dissolved oxygen concentra-
tion.

Figure 4 Observation of a water surface by the spotlight method.

Figures 5 through 7 show the measurement results: part 1 shows observation by three-dimensional indicate; part 2 shows the section profile. When wafers are dipped in ultrapure water, the roughness of the Si substrate is also deteriorated over time. The roughness level increases to as much as 100 Å in 20 days.

These experiments indicated that as wafers are immersed in ultrapure water containing dissolved oxygen, Si is eluted over time and the surface of the Si substrate as well as the surface of the native oxide is roughened.

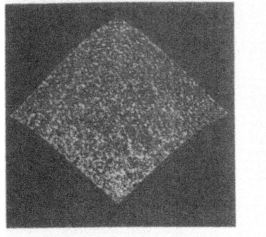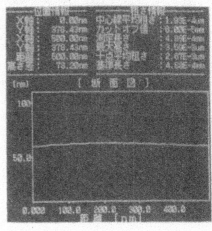

(1) Observation by 3D indicate   (2) A section profile

Figure 5   Surface condition of a wafer during cleaning with ultrapure water.

IV.   NATIVE OXIDE IN CLEAN ROOM AIR

In the LSI production line, wafers are exposed to the air in the clean room during drying and when they are transported from one process to another. Therefore the native oxide growing in the air should be closely examined.

Figure 8 shows the time dependence of the thickness of native oxide that grows when wafers are left in the air in the clean room. In clean room air, just as in ultrapure water, native oxide grows faster on n^+ wafers than on n wafers. Unlike the situation in ultrapure water, however, the native oxide grows step by step in clean room air. Table 2 shows the results of studying the effect of moisture in the air on native oxide growth. When wafers

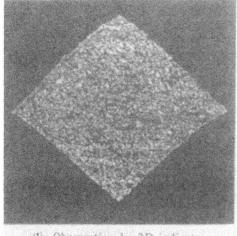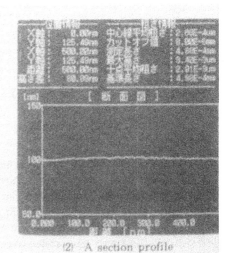

(1) Observation by 3D indicate   (2) A section profile

Figure 6   Surface conditions of a wafer immersed in ultrapure water for 6 days.

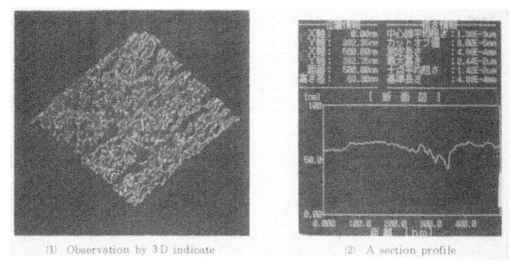

(1) Observation by 3 D indicate   (2) A section profile

Figure 7 Surface conditions of a wafer immersed in ultrapure water for 20 days.

are kept in N_2 with a low moisture concentration ($O_2/N_2 = 1:4$), no native oxide grows even in 7 days.

It was found that both oxygen and moisture are required for native oxide growth in clean room air. Wafers must therefore be transported and stored in an environment with a limited moisture concentration when oxygen is also present. One way to reduce the adsorbed impurities at low cost is to transport and store wafers in N_2, which contains limited impurities.

V. NATIVE OXIDE FORMED DURING VARIOUS TREATMENTS

Pretreated n-Si(100) wafers were treated by the following procedures: H_2SO_4-H_2O_2 cleaning (5 minutes), NH_4OH-H_2O_2 cleaning (10 minutes), and HCl-H_2O_2 cleaning (10

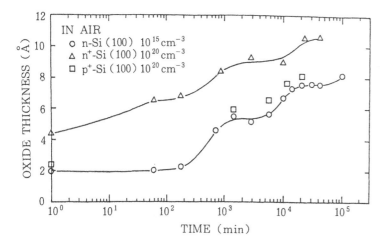

Figure 8 Oxide thickness as a function of exposure time of wafers to air at room temperature.

Table 2   Growth of Native Oxide with H₂O Concentration (7 Days)

(7 DAYS)

ambient	H₂O concentration	oxide thickness (Å)
air	~ 1.2 % (humidity : 42 % RH)	6.7
O₂ / N₂ (= 1 / 4)	< 0.1 ppm	1.7
N₂	< 0.1 ppm	1.9
immediately after hf cleaning	————	1.9

minutes). Figure 9 shows XPS spectra of native oxide formed in each cleaning process, and Fig. 10 shows XPS spectra of native oxide formed in ultrapure water (with 9 ppm dissolved oxygen) and in air.

Table 3 lists the results of studying wafers with native oxide formed in various treatments, including the contact angle when an ultrapure water droplet is dropped on the wafer and the results of wafer immersion in ultrapure water (with 9 ppm dissolved oxygen). The binding energy of the SiO_x ($x > 1$) peak of native oxide formed in ultrapure water is greater than that of the native oxide formed under other conditions. The native oxide formed in H_2SO_4-H_2O_2 cleaning is hydrophilic, with a low contact angle, and the native oxide formed in the air is hydrophobic. Little Si is eluted from the wafers with native oxide formed during H_2SO_4-H_2O_2 cleaning, NH_4OH-H_2O_2 cleaning, and 4 day air exposure when the wafers are immersed in ultrapure water.

It was found that the native oxide formed by different treatments exhibits different physical and chemical characteristics, including the binding detected by XPS, contact angles, and Si elution in ultrapure water.

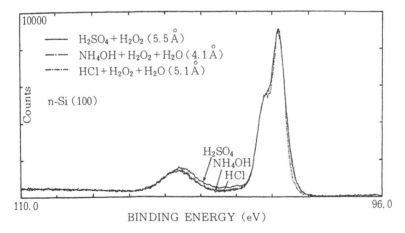

Figure 9 Si_{2p} XPS spectra of native oxides grown on *n*-Si by various chemical treatments.

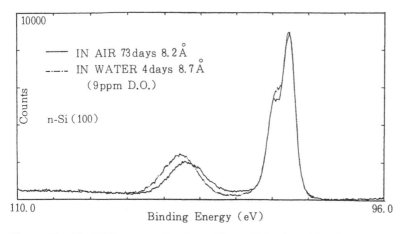

Figure 10 Si$_{2p}$ XPS spectra of native oxide on Si in air and in ultrapure water.

VI. NATIVE OXIDE FORMED IN ULTRAPURE WATER RINSING AND DRYING

As mentioned earlier, current wafer cleaning is conducted in a bath open to the air, and the native oxide formed in the ultrapure water rinsing and drying process must be noted. The effect of the ultrapure water rinsing time was studied using wafers whose native oxide was removed by DHF treatment. Figure 11 shows the results. Although the 2.6 Å native oxide formed in a 2 minute ultrapure water rinse, it did not grow remarkably after 2 minutes. It is therefore thought that the rapid growth of native oxide takes place when the wafers are transported from the DHF treatment bath to the ultrapure water rinsing bath or immediately after the ultrapure water rinse starts.

Table 4 shows the thickness of native oxide when wafers were treated with ultrapure water rinsing and drying in varying sequences. On both n and n^+ wafers, native oxide growth can be suppressed by reducing the concentration of dissolved oxygen. On n^+

Table 3 Characteristics of Native Oxides with Various Treatments

sample[†]	oxide thickness (Å)	ΔE[††] (eV)	contact angle (°)	dissolved Si[†††] weight ($\mu g / cm^2$)
in air 4 days	5.6	3.85	35.7	0.14 (9 days)
in ultrapure water (9 ppm D.O.) 2 days	5.6	4.08	11.4	4.03 (7 days)
H$_2$SO$_4$ + H$_2$O$_2$ cleaning	5.5	3.83	< 10	< 0.05 (7 days)
NH$_4$OH + H$_2$O$_2$ + H$_2$O cleaning	4.1	3.99	< 10	

†   n-Si (100)

†† $\Delta E = E - E$ (Si$^\circ_{2P3/2}$)

†††   into ultrapure water (9 ppm D.O.)

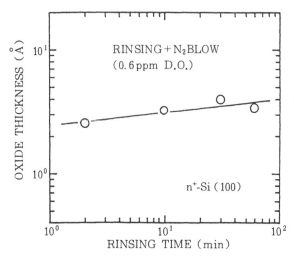

Figure 11   Average oxide thickness on n^+-Si(100) as a function of rinsing time by ultrapure water overflow. The dissolved oxygen concentration is 0.6 ppm.

wafers, the native oxide thickness differs by 1.2 Å when different drying methods are used. The thickness of native oxide formed during the 10 minute rinse with hot ultrapure water is almost the same as that formed in ultrapure water at room temperature (with 0.6 ppm dissolved oxygen).

Based on these findings, to suppress the native oxide in ultrapure water rinsing and drying, it is necessary to reduce the dissolved oxygen in the ultrapure water and to prevent oxygen in the air from dissolving in ultrapure water during rinsing and drying. The N_2 gas seal used in wet processing is effective.

Table 4   Growth of Native Oxides with Various Conditions of Cleaning and Drying

Condition	water temperature (°C)	Dissolved oxygen (ppm)	rinsing time (min)	oxide thickness (Å)	
				n-Si (100)	n^+-Si (100)
rinsing (0.04 ppm D.O.) +N_2 blow	room temperature	0.04	10	1.3	2.9
rinsing (0.04 ppm D.O.) +N_2 blow	″	0.6	10	1.9	3.2
rinsing (0.6 ppm D.O.) +hot water dipping (1 min)+hot Ar blow	″	0.6	10	1.9	4.4
hot water rinsing +N_2 blow	50.7	—	10	1.8	—
	68.1	—	10	2.0	—
	89.3	—	10	2.1	—

VII. SUMMARY

Native oxide growth and the roughness of the wafer surface will become more and more critical as ULSI is integrated further, with finer patterns. The findings presented in this chapter indicate the importance of reducing the dissolved oxygen in ultrapure water, the introduction of a N_2 gas seal to the cleaning and drying process, and the introduction of a N_2 gas-sealed wafer transportation and storage system. Further studies are required to determine the details of the mechanism of native oxide growth and Si elution.

ACKNOWLEDGMENTS

This study was conducted in the superclean room facility of the Electric Communication Laboratory at Tohoku University. The authors thank Seiko Instruments, Inc. for their cooperation in STM. We also appreciate the cooperation of Dr. Mizuho Morita and Mr. Eiji Hasegawa of Tohoku University.

REFERENCES

1. T. Ohmi, Ouyou Butsuri, *58*, 193 (1989).
2. M. Morita, T. Ohmi, E. Hasegawa, M. Kawakami, and K. Suma, Native Layer Free Oxidation for Very Thin Gate Oxides, VLSI Technology Symposium, Kyoto, May 1989.
3. M. Morita, T. Ohmi, E. Hasegawa, M. Kawakami, and K. Suma, Growth Kinetics of Native Oxide on Silicon Surface, Proceedings of 8th Symposium on ULSI Ultra Clean Technology, Tokyo, 1989, p. 169.

6
Moisture Adsorption on
Silicon Wafers

Norikuni Yabumoto
NTT, Kanagawa, Japan

Hiroyuki Harada
Mitsubishi Corporation, Tokyo, Japan

I. INTRODUCTON

The cleanliness of the Si wafer comes first in ultralarge-scale integrated (ULSI) circuit production. Since wafers are exposed to the air and many other process environments, however, particles, moisture, oxygen, hydrocarbons, and other impurities are adsorbed on the surface. In particular, moisture is expected to remain longest on the wafer surface because water is used during the final stage of cleaning. Moisture is also contained in process gas, penetrates the oxide made by chemical vapor deposition (CVD) forming silanol [1], and facilitates native oxide growth when the temperature is raised [2]. It is therefore very important to define moisture.

It is reported that when water molecules react with the Si surface in a high vacuum, they generate dissociative adsorption, forming Si-H and Si-OH bonds [3]. However, the state of water molecules on the cleaned surface are not clear. One analysis method that is sensitive to the surface condition is thermal desorption spectroscopy (TDS). TDS can determine the binding state of molecules and atoms by detecting the peak temperature of the desorption spectrum. Moreover, if the background can be removed, TDS is able to conduct a high-sensitivity analysis based on the characteristics of the quadrupole mass spectrometer. This chapter reports the results of a study that examined the cleaned wafer surface using a newly developed measurement method in which only the samples move through the TDS equipment. To remove the native oxide that forms during RCA cleaning, the wafers were treated with HF solution or HF gas, followed by final drying with inert gas or isopropyl alcohol (IPA).

II. THERMAL DESORPTION SPECTROSCOPY

The molecules and atoms adsorbed on the solid surface are desorbed as the temperature is raised. Molecules and atoms whose binding energy is weaker are desorbed earlier. TDS observes the adsorption state by detecting molecules and atoms that are desorbed from the solid surface at each temperature. In this experiment, the transportation mechanism in the TDS was improved so that samples can be transported in the TDS equipment, and the minute amount of adsorbed molecules desorbed from the surface of the samples can be discriminated from the background and detected [4].

Nondoped Si(100) wafers were used for the samples. Since the entire wafer is heated in TDS, both sides of the wafer were polished. The resistivity of the samples was 30 $\Omega \cdot$cm. Figure 1 shows the cleaning of wafers. First, wafers were treated by RCA cleaning: H_2SO_4-H_2O_2 cleaning, immersion in dilute HF (DHF), boiling and cleaning in NH_4OH-H_2O_2-H_2O, and boiling and cleaning in HCl-H_2O_2-H_2O.

Samples A and B were first immersed in 1% DHF and rinsed with DI water. Sample A was dried in Ar gas at about 80°C, sample B was dried in IPA vapor. Samples C and D were treated with a dry-cleaning method employing 1% HF gas balanced by N_2. Sample C was dried in N_2, and Sample D was dried in IPA vapor. The moisture concentration of Ar, N_2, and HF gas used in this experiment was 10 ppb or less.

III. ADSORBED MOLECULES ON THE CLEANED SI WAFER SURFACE

A. Major Adsorbed Molecules

Hydrogen has been reported as the major molecule adsorbed on a Si wafer cleaned with HF; hydroxyl groups, hydrocarbon, and fluorine are the minor molecules [5, 6]. The TDS analysis was conducted based on this finding. Figure 2 shows typical TDS spectra detected from sample A cleaned with HF solution and dried with hot Ar gas. The spectra at M/e = 2, 18, and 28 correspond to H_2, H_2O, and C_2H_4 or CO, respectively. Since M/e = 28 is symmetrical with its peak at about 500°C, just like M/e = 26 (C_2H_2) and 27 (C_2H_3), it is considered C_2H_4.

In Fig. 2, the solid line indicates the primary data, including information on the sample surface, and the dotted line indicates the background. More than 50 times as much H_2 was detected as H_2O and C_2H_4. H_2, H_2O, and C_2H_4 are the top three among mass numbers 1–100 that were detected. Considering that the transmission efficiency of a quadrupole mass spectrometer is H_2:H_2O:C_2H_4 = 1:5:6 [7] and considering the ioniza-

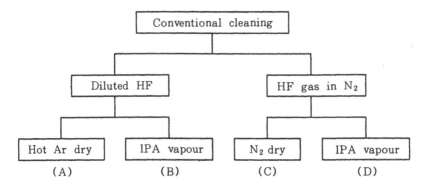

Figure 1 Wafer-cleaning methods.

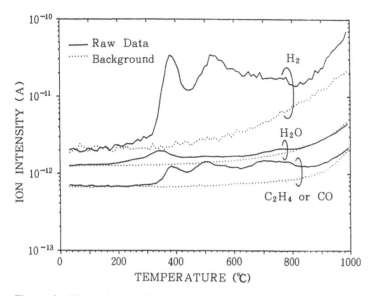

Figure 2   Thermal desorption spectra of H_2, H_2O, and C_2H_4 or CO from sample A.

tion efficiency at the ion source, more than 99% of the desorbed molecules must be hydrogen and most of the dangling bonds should be terminated by hydrogen.

In addition to the minor desorbed molecules CH_x ($x = 2, 3$), C_2H_y ($y = 2, 3, 5$), C_3H_z ($z = 2, 7$), C_4H_w ($w = 7, 9$), and CO_2 were also detected. All these, including hydrogen, moisture, and C_2H_4, were also detected from cleaned samples B, C, and D.

To examine oxidation after cleaning, AES analysis was carried out to compare a sample just taken from a N_2-sealed package and cleaned and a sample left in the air for 7 days after cleaning. Table 1 summarizes the results. The sample just taken from the package showed a few percent of oxygen on the surface, and the sample left in the air for 7 days exhibited about 10% oxygen. This is thought to be because the hydrogen that terminates the dangling bonds is effective in preventing oxidation. Because the oxygen detected in the sample taken from the sealed package actually included the oxygen in adsorbed moisture and hydroxyl groups as well as the oxygen in native oxide, the growth of native oxide was thought to be 2 Å or less on average [8].

Table 1   Oxygen Concentration on the Silicon Surface Immediately and 7 Days After Cleaning

	Immediately after cleaning	7 days after cleaning
A	5.8	9.0
B	3.7	8.2
C	5.3	13.3
D	3.8	8.4

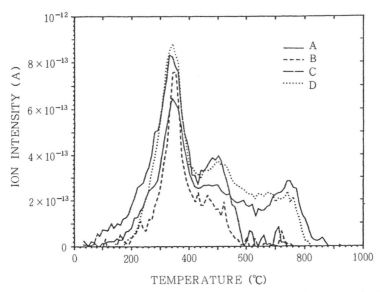

Figure 3   Thermal desorption spectra of H$_2$O for the four cleaning methods.

B.   Moisture Adsorption State

Figure 3 shows the TDS spectra of moisture in samples A, B, C, and D. The background was deducted from the primary data to equalize the values at room temperature and at 900°C. Therefore, Fig. 3 shows the moisture spectrum only from the sample surface. Each sample had a peak at 340°C in the range from 100 to 800°C. Some samples also had a peak at 500°C; others had a peak at 750°C. No difference was observed between wet cleaning and dry cleaning or between inert gas drying and IPA vapor drying. The difference in the spectra is thought to be caused by unknown factors, such surface irregularities and the presence of native oxide.

   The moisture desorption spectra are detected at high temperatures compared with its boiling point of 100°C. This indicates that moisture is chemically adsorbed on the Si surface by hydrogen bonds.

C.   Hydrogen Adsorption State

Figure 4 shows the TDS spectra of hydrogen in samples A, B, C, and D. The peaks at 380 and 520°C are thought to correspond to Si-H$_2$ and Si-H, respectively [9].

   The height of the Si-H peak at 520°C was common to all four samples regardless of differences in the cleaning method. In samples B and D, employing IPA drying, shoulder peaks were detected at 600°C or higher. They were thought to be released from native oxide in the form of islands. However, no relationship was observed between this phenomenon and moisture desorption (shoulder peaks were detected at 340°C or higher in samples A and D), although moisture desorption was also thought to depend on the presence of native oxide islands and Si surface irregularities.

   The shoulders at 600°C or higher detected in samples B and C were thought to be desorbed from native oxide islands that exist in small volume on the Si surface [9].

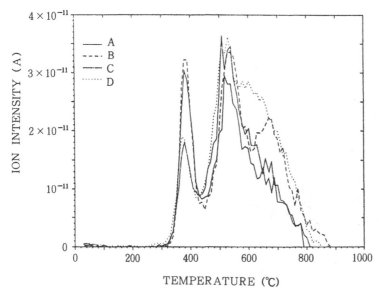

Figure 4 Thermal desorption spectra of H_2 for the four cleaning methods.

Although the formation of native oxide islands is thought to be close related to the cleaning method, no theoretical relation has been found.

D. Hydrocarbon Adsorption State

Many kinds of hydrocarbons were detected by TDS, mostly C_2H_y, followed by C_3H_2, CH_x, and hydrocarbons with more than four carbons. Figure 5 shows the relative desorption volumes of hydrocarbons by cleaning method. Each desorption volume is standardized based on the C_2H_4 volume desorbed from sample D. The hydrocarbon desorption volume was smaller in wet cleaning than in dry cleaning, which means the hydrocarbon adsorption volume is limited. In IPA drying, the volume of C_3H_7 increased. This is because of the adsorption of $(CH_3)_2CH$, which is the fragment of IPA $[(CH_3)_2CH(OH)]$. The difference in the hydrocarbon desorption volume among the different cleaning methods is not great, less than 2.

Figure 6 shows the TDS spectra of $CH_3CH(OH)$ ($M/e = 45$) for each cleaning method. In samples B and D, which were treated with IPA in the final stage, clear peaks appeared at 350°C. However, no such peaks were detected in samples A and C. The desorption of IPA molecules was not observed.

E. Trace Contaminants

Table 2 shows the results of secondary ion mass spectrometry (SIMS) of trace contaminants, such as F, S, and Cl. The F volume on the surface is greater after HF gas cleaning than HF solution cleaning. This is because, in wet cleaning, the Si-F bonds once generated in the solution change to Si-OH bonds by hydrolysis, resulting in a decrease

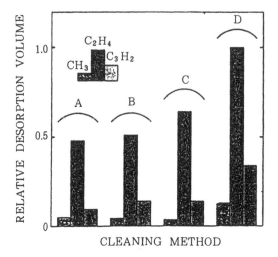

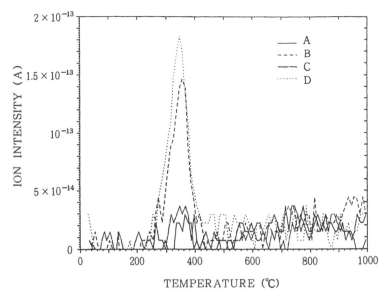

Figure 5   Relative desorption volumes of hydrocarbons for the four cleaning methods.

in F. IPA is found effective in removing F atoms. This is thought to be because Si-F is soluble in alcohol [10], and F is dissolved in IPA and removed from the surface.

S and Cl are present because H_2SO_4 and HCl are used in the cleaning solution. The volume of S is less than 2×10^{11} atoms cm^{-2}, and the volume of Cl is less than 1×10^{11} atoms cm^{-2}.

Figure 6   Thermal desorption spectra of $(CH_3)CH(OH)$ for the four cleaning methods.

Table 2 Contaminant Concentrations Determined by SIMS (Atoms cm^{-2})

	C	O	F	S	Cl
A	2.0×10^{13}	5.6×10^{13}	2.1×10^{11}	2.0×10^{11}	6.4×10^{10}
B	1.4×10^{13}	4.8×10^{13}	0.4×10^{11}	1.8×10^{11}	6.9×10^{10}
C	0.7×10^{13}	4.8×10^{13}	7.4×10^{11}	1.9×10^{11}	8.4×10^{10}
D	0.6×10^{13}	5.6×10^{13}	5.6×10^{11}	1.0×10^{11}	7.6×10^{10}

IV. SUMMARY

A new type of TDS to separate background noise was developed, and the adsorbed molecules on Si wafers treated with various cleaning methods were evaluated. As a result, the following were revealed:

1. The moisture adsorbed on the Si wafer surface was detected for the first time. The volume of adsorbed moisture was less than $1/100$ of that of adsorbed hydrogen. The desorption spectrum has a main peak at 340°C.
2. On the surface of wafers treated with HF solution, most of the broken bonds are terminated by hydrogen. This phenomenon was observed with both HF solution and HF gas.
3. Hydrocarbon exists in very small quantity, as small as moisture. The majority is C_2H_4. The desorption spectrum has a main peak at about 500°C.
4. IPA vapor drying decreases the volume of Si-F, but it increases IPA fragments, such as $(CH_3)_2CH(OH)$ and $(CH_3)_2CH$.

ACKNOWLEDGMENTS

The authors thank Professor Tadahiro Ohmi and Dr. Mizuho Morita of Tohoku University and Dr. Eisuke Arai, Dr. Kazuyuki Saito, and Mr. Kazushige Minegishi of the NTT LSI Laboratory for valuable advice and discussion. We also appreciate the cooperation of Mr. Eiji Hasegawa of Tohoku University and Mr. Yukio Komine of the NTT LSI Laboratory.

REFERENCES

1. E. Arai and Y. Terunuma, J. Electrochem. *Soc.*, *121*, 676 (1974).
2. T. Ohmi, Ouyo Butsuri, *58*, 193 (1989).
3. Y. J. Chabal, Phys. Rev. B, *29*, 3677 (1984).
4. N. Yabumoto, K. Minegishi, K. Saito, M. Morita, and T. Ohmi, Extended Abst. Electrochem. Soc. Fall Meeting, Hollywood, Florida, (1989), p. 592.
5. M. Grundner and H. Jacob, Appl. Phys. A, *39*, 73 (1986).
6. T. Takahagi, I. Nagai, A. Ishida, H. Kuroda, and Y. Nagasawa, J. Appl. Phys., *64*, 3516 (1988).

7. T. C. Ehlert, J. Phys. (E), *3*, 237 (1970).

8. M. Morita, T. Ohmi, E. Hasegawa, M. Kawakami, and K. Suma, 1989 Symp. VLSI Technol., Kyoto, (1989), p. 75.

9. N. Yabumoto, K. Minegishi, Y. Komine, and K. Saito, Jpn. J. Appl. Phys., *29*, L490 (1990).

10. Kagaku Binran Kiso Hen I 3rd ed., Nippon Kagaku Kai, Maruzen, 1988.

7

High-Temperature, High-Pressure Ultrapure Water Spray-Cleaning Technology

Yusuyuki Yagi
Hitachi Plant Engineering & Construction Co., Ltd., Chiba, Japan

Michiya Kawakami
Mitsubishi Gas Chemical Company, Inc., Tokyo, Japan

Kenichi Sato
Nihon Pall, Ltd., Tokyo, Japan

Tadahiro Ohmi
Tohoku University, Sendai, Japan

I. IMPORTANCE OF PRECISION CLEANING

The cleanliness of the Si wafer is vital to the semiconductor field. If contamination from dry process equipment is high, it is very difficult to keep the wafer surface clean. In other words, it is essential to maintain cleanliness on the inner surface of various chambers and components employed in the dry process in the ultralarge scale integrated (ULSI) circuit production line that might be affected by trace contaminants. In this sense, precision cleaning technology plays an important role in the semiconductor field.

When precision cleaning is not performed properly on stainless steel components for process chambers, they exhibit outgassing. Figure 1 shows the APIMS (atmospheric pressure ionization mass spectrometry) spectrum of outgassing from a stainless steel component for a process chamber that was not sufficiently cleaned after it was baked in argon at 400°C for 3 h. Figure 2 shows the results of the same measurement when the component was baked at 400°C for 29 h. It is obvious that because there was insufficient cleaning, the impurities at $M/Z = 21$ remained even after 29 h of baking at 400°C. Figure 3 is a photograph of the surface of this stainless steel component after it was oxidatively passivated [1] in high-purity oxygen. The photograph shows the deterioration of the portion that was not cleaned sufficiently, which renders oxidation passivation useless despite that this technology plays an important role in the high-purity gas delivery system. This means that, without sufficient cleaning, even the most outstanding surface processing technology cannot be applied effectively.

Precision cleaning technology is essential to promote the progress and developments in every cutting-edge technology, including the semiconductor field, and it will become increasingly important in the future. This chapter describes a precision cleaning technology, ultrapure water spray cleaning at high temperature and pressure.

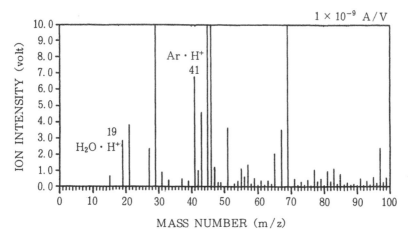

Figure 1   Outgassing from stainless steel parts after 3 h at 400°C.

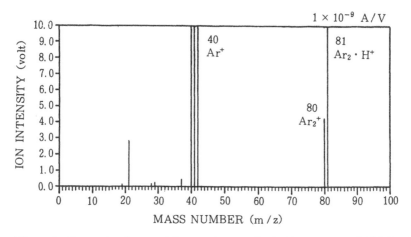

Figure 2   Outgassing from stainless steel parts after baking at up to 400°C for 29 h.

Figure 3   Surface of a stainless steel part after oxidative passivation.

II. PRECISION CLEANING

A. Outline

This precision cleaning combines the following different mechanisms. The cleaning systems are designed based on the characteristics of the impurities to be removed and the surface to be cleaned.

Solubility, diffusion: water, solvent (organic solvent)
Surface activation: surface-active agent
Chemical reaction: acid and alkali cleaning
Physical force: heat, ultrasound, pressure, surface flow, polish

Table 1 outlines present precision cleaning procedures. Basically precision cleaning is a process that captures impurities on the surface in chemicals. Therefore, it is important in precision cleaning both to separate impurities from the surface and remove the separated impurities from the system without having them readhere to the surface. It is also very important to keep the surface wet between each cleaning process because once the residual chemical has dried, it is hard to remove.

In the cleaning process, chemicals and water must have good wettability with the surface to be cleaned and it is necessary to investigate the properties of water and chemicals (described in Part I) to work out appropriate cleaning methods.

Figure 4 shows the temperature dependence of the resistivity of ultrapure water. When ultrapure water at room temperature is used for cleaning, electrostatic damage might be caused as a result of the high resistivity. As shown in Fig. 4, it is possible to overcome this problem by raising the temperature of the ultrapure water. As shown in Fig. 5, however, the surface tension of ultrapure water is higher than that of other liquids, and the surface tension is hardly reduced by the rise in temperature. The high surface tension of ultrapure water suppresses its wettability, resulting in poor efficiency of cleaning.

B. Stainless Steel

For the surface treatment of stainless material, electropolishing [1] and electrochemical buffing [2] are effective, both of which create a mirror-polished surface free of frag-

Table 1 Example of Precision Cleaning After Electrochemical Buffing

1. HIGH PRESS. CITY WATER CLEANING
2. HIGH PRESS. PURE WATER CLEANING (RESISTIVITY $1 M\Omega \cdot cm$)
3. ULTRAPURE WATER CLEANING
4. CHEMICAL CLEANING (NO. 1) ($HCl + H_2O_2 + H_2O$)
5. PURE WATER CLEANING
6. CHEMICAL CLEANING (NO. 2) ($NH_4OH + H_2O_2 + H_2O$)
7. PURE WATER CLEANING
8. ULTRAPURE WATER CLEANING
9. N_2 BLOW

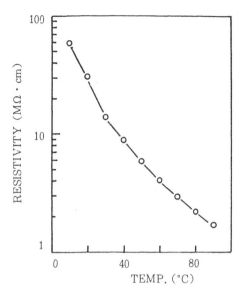

Figure 4 Temperature dependence of ultrapure water resistivity.

mented layers. This surface treatment is capable of suppressing particulation and outgassing to low levels because it creates an extremely smooth stainless steel surface, at the submicrometer level. Such polishing of the surface must be followed by precision cleaning. Figure 6 shows the result of measuring a mirrored surface treated by electrochemical buffing with IMA (Ion Micro Prove mass analyzer) [2]. The samples used in this measurement were cleaned briefly with neutral detergent. Impurities, such as Na (the residue of $NaNO_3$ used in the process) and Ca, are still observed in large volume on the polished surface.

Figure 7 shows the amount of Na contamination on the surface of a piping system in a

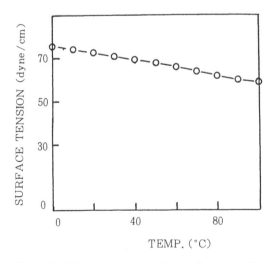

Figure 5 Temperature dependence of water surface tension.

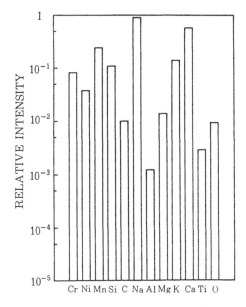

Figure 6   Impurity on electrochemically buffed surface (SUS 304) analyzed by IMA.

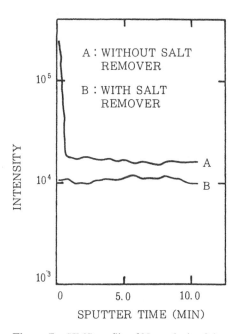

Figure 7   SIMS profile of Na analysis of the surface of inner stainless steel piping (electropolished SUS 316 L).

plant that manufactures electropolished SUS 316 L pipes before and after installing a salt remover in the clean room [3]. This plant is located close to the sea. As shown in Fig. 7, Na contamination was reduced by installing the salt remover. Thus various contaminants are deposited on the stainless steel surface during production. Therefore precision cleaning is vital to remove these contaminants.

As shown in Table 1, in precision cleaning, chemicals are generally used to clean chambers and components. However, much of the equipment and components used in the semiconductor industry are made of stainless material. Their sizes vary from small to large, such as a process chamber and a gas tank with a capacity of 20 m^3. For larger equipment, precision cleaning cannot always be used because large amounts of chemicals are required.

For ordinary SUS 316 L, HCl is deposited in many pinholes on the surface after HCl-H$_2$O$_2$ cleaning, which makes it necessary to neutralize the HCl by NH$_4$OH processing. Without this neutralizing, the HCl residue in the pinholes evaporates, eventually corroding the surface.

It is clear that precision cleaning with chemicals is a very effective and important cleaning technology if the cleaning operation is properly performed. However, the air pollution caused by chlorofluorocarbons and the water pollution caused by trichlene has become controversial. We must investigate a substitute cleaning technology.

The prerequisites for precision cleaning can be summarized as follows:
1. Completely eliminate the impurities on the surface.
2. Keep the impurities removed from the surface from readhering by solubilization and diffusion.
3. Control the cleaning environment.
4. Keep the surface wet between each cleaning process.
5. Make it possible to clean large-scale equipment.
6. Make it easy to process the waste chemicals.
7. Keep the cost low.

III.   EVALUATION OF HIGH-TEMPERATURE, HIGH-PRESSURE ULTRAPURE WATER SPRAY CLEANING

Spray cleaning with high-pressure ultrapure water is a precision cleaning method that satisfies the preceding prerequisites. Conventional high-pressure pumps, however, generate particles from the sliding parts, and impurities are eluted from the liquid-contacting surface, which made it impossible to apply the high-temperature, high-pressure ultrapure water spray cleaning. The development of a high-pressure pump for ultrapure water with a low level of dissolution and particle generation made this spray cleaning available. We describe the results of our study of high-temperature, high-pressure ultrapure water spray cleaning.

A.   Experimental Apparatus

The high-pressure pump for ultrapure water is produced by Nikoku Kikai Kogyo, with a maximum pressure of 30 kgf cm^{-2} and a three-step propeller. This is a spiral-flow pump controlled by an inverter. To lower particle generation and elution, this pump has the following features:

1. The particles generated from the mechanical seal, which is its only driving part, are drained away from the system as reject water.
2. SUS 316 L is used for the portion of the pump contacting the ultrapure water. The surface is electrochemically buffed to restrain particle generation and impurity dissolution.
3. The portion in contact with the ultrapure water is kept small by adopting a spiral-flow pump, which is compact and easily reaches high pressure.
4. The mechanical seal is outside to place the rotating ring, whose configuration is complicated, outside the portion in contact with the ultrapure water.
5. Dead space is eliminated by designing a passage that drains chemicals remaining in the joints between the casing or cover and the parting strips.

We compared this high-pressure ultrapure water pump with a conventional pump in terms of resistivity and particle generation immediately after supplying water. Figure 8 shows the resistivity at the pump exit immediately after supplying water. The resistivity of the high-pressure ultrapure water pump started up very quickly even during high-pressure operation at 30 kgf cm^{-2}, reaching a peak of 18.17 MΩ compared with the operation of a conventional pump at 6 kgf/cm^{-2}.

B. The Cleaning Unit

Figure 9 is a schematic diagram of the spray-cleaning unit. This unit is composed of ultrapure water distribution unit, high-pressure pump, heater, piping, and nozzle. Table 2 summarizes the major specifications.

SUS 316 L treated with by electrochemical buffing was employed for the piping to achieve excellent pressure resistance and limited elution.

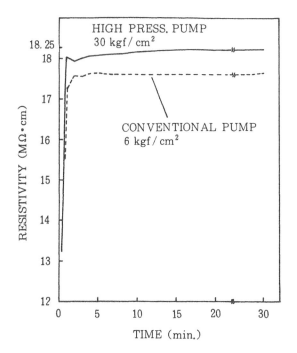

Figure 8 Time dependence of outlet resistivity for a high-pressure pump for ultrapure water.

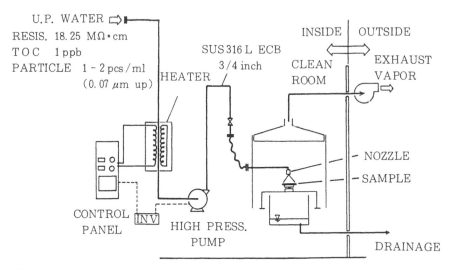

Figure 9   A spray-cleaning unit.

There are several types of nozzles, such as conic, fan, and bar, and the shapes of the water spray from slits are different, as shown in Fig. 10. In this experiment we used the fan nozzle (slit width of 0.3 mm), which sprays the water in a wide fan shape.

Table 3 lists the particulation from the pump and nozzle employed in this unit. The particles were counted with a particle counter (Klamic-100, Kurita Kogyo) after allowing the pump to rest for 2 h and after eliminating bubbles from the pump. The data in Table 3 show that the particle generation was cut to $1/10$ of that with the conventional type. It was observed, however, that the number of the particles in ultrapure water generated from the nozzle was increased three times. It is necessary to reduce particulation from the nozzle.

Table 2   Specifications for the Spray-Cleaning Unit

NAME	TYPE	MATERIAL
High Press Pump	Centrifugal Pump for Ultrapure Water (3 Stage Impeller) Max Press 30 kgf/cm²	SUS 316 L Electro-Chem. Buffing
Nozzle	Fan-Shaped High Press Spray Nozzle Slit Width 0.3 mm	Body SUS 304 Inner⎫ Part⎭Ceramics
Heating Equipment	PID Control 2 Step Heating (11 KW × 2)	SUS 316 L Electro-Chem. Buffing
Piping	MCG Joint 3/4 B	Same

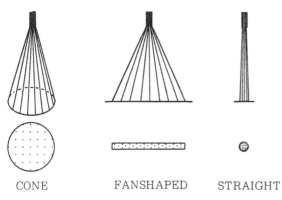

CONE   FANSHAPED   STRAIGHT

Figure 10   Types of nozzles.

C.   Evaluation of Performance

We studied the following three points: oil elimination, particle elimination, and metallic impurity elimination.

1.   *Evaluation of Oil Elimination*

In the test to evaluate the oil elimination, pieces of SUS 316 L ($35\Phi \times 0.5t$) were processed by electrochemical buffing, was coated with oil, and cleaned under different cleaning conditions. After cleaning, we dropped ultrapure water on the surfaces to measure the contact angle. To preclean the pieces, we used HCl-H_2O_2, NH_4OH-H_2O_2, and DHF.

Figure 11 shows the pressure dependence of oil elimination on the stainless surface when cleaning temperatures are fixed at 80°C. In this case, when the pressure was set at 30 kg cm^{-2}, the speed of ultrapure water spouted from the nozzle was about 110 m s^{-1}.

To compare these results with chemical cleaning, Fig. 11 also shows the result of isopropyl alcohol (IPA) ultrasound cleaning. The higher the pressure, the better is the oil elimination. When spray cleaning at 80°C and 30 kgf cm^{-2} is used, the result is very close to that after IPA ultrasound cleaning.

Figure 12 shows the temperature dependence of oil elimination on the stainless steel surface when the pressure is fixed at 30 kgf/cm^{-2}. After both 6 and 20 minutes of cleaning, higher cleaning temperatures give better results. The difference between the higher and lower temperatures is particularly significant when the cleaning time is short.

Table 3   Particles Generated by the Ultrapure Water Pump and Spray Nozzle

equipment size range	HIGH PRESSURE PUMP 30 kgf/cm₂	CONVENTIONAL PUMP 6 kgf/cm²	NOZZLE (FAN SHAPED) 30 kgf/cm²
$> 0.1\ \mu m$	147	1620	423
$> 0.2\ \mu m$	51	868	142
$> 0.5\ \mu m$	2	15	3

VALUES : AVERAGE OF TWO MEASUREMENTS

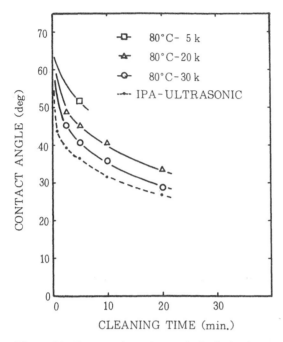

Figure 11   Pressure dependence of oil elimination on the stainless steel surface.

It is found when cleaning with ultrapure water that better oil elimination results were achieved when the pressure and temperature of the ultrapure water were set at higher levels. If this is done, we can expect good cleaning, close to that with IPA ultrasound.

2.   *Particle Elimination*

In this test, we first removed the native oxide from the surface of a 4 inch wafer [n-Si (100)], then dipped it in ultrapure water containing polystyrene latex (PSL), and dried it.

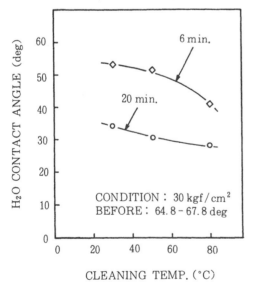

Figure 12   Temperature dependence of oil elimination on the stainless steel surface.

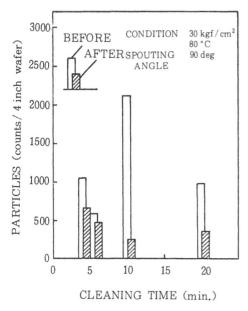

Figure 13 Time dependence of PSL particle elimination from the water surface.

We counted the number of particles before and after cleaning at 80°C and 30 kgf cm^{-2} using the WIS-150 (wafer surface defect scanner). Figure 13 shows the time dependence of PSL particle elimination from the wafer surface. After 10 minutes of cleaning, the number of particles decreased from 2124 per wafer (>0.5 μm) to 251 per wafer, and almost the same result was seen after 20 minutes of cleaning. This means that under these conditions the cleaning effect is limited, with about 250 particles left on the surface. Figure 14 shows the results of 10 minutes of cleaning at 80°C and 30 kgf cm^{-2}, using several different spray angles. As shown, there was no difference in the level of particle elimination even when the spray angles were changed.

Although we succeeded to some extent in particle elimination, the performance of this ultrapure water spray cleaning does not satisfy current standards in the semiconductor industry, and further study is required.

3. *Elimination of Metallic Impurities (Na and K)*

We treated pieces of electrochemically buffed SUS 316 L (35Φ × 0.5t) by acetone and ultrapure water cleaning and then tried various other cleaning methods. We then measured the residue by SIMS (secondary ion mass spectrometry). In these experiments, we used two different cleaning conditions: overflow rinsing using ultrapure water at 80°C and 30 kgf cm^{-2} and HCl-H$_2$O$_2$ + NH$_4$OH-H$_2$O$_2$ cleaning. Table 4 lists the Na and K residue on the surface. The SUS surface residue after spray cleaning at 80°C and 30 kgf cm^{-2} was twice as much as that after chemical cleaning, but far less than after overflow rinsing. Figures 15 and 16 show the depth profile of impurities using SUS. The elimination of Na residue after spray cleaning was greater than that after overflow rinsing up to a depth of 60 Å, but there was no difference between the two cleaning methods at 60 Å or deeper. K residue was eliminated up to a depth of 150 Å, but the elimination effect of spray cleaning was inferior to that of chemical cleaning.

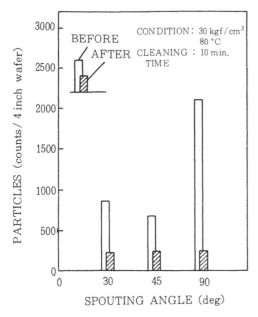

Figure 14   Spray angle dependence of PSL particle elimination on the wafer surface.

Ultrapure water spray cleaning shows a limited ability to eliminate metal impurities, not enough to satisfy the level required by the semiconductor industry.

4.   Effect of Cleaning on Oxidation-Passivated Film

We treated pieces of electrochemically buffed SUS 316 L ($35\Phi \times 0.5t$) by acetone and ultrapure water cleaning and then transported the pieces in a wet environment. The pieces were treated by various cleaning methods, followed by oxidation-passivated processing (400°C for 3 h). We tested three different cleaning conditions: 30°C and 5 kgf cm^{-2}, 80°C

Table 4   Na and K Residue on the Stainless Steel Surface After Various Cleaning Treatments

METHOD / METAL	OVER-FLOW 10 min	SPRAY 30 Kgcm^{-2} 80 °C 10 min	HCl-H$_2$O$_2$ NH$_4$OH-H$_2$O$_2$ 80 °C 10 min
Na INTENSITY (−)	5.0×10^5	1.2×10^5	5.0×10^4
Na CONCENTRATION (atoms/cm^2)	2.6×10^{13}	6.1×10^{12}	2.6×10^{12}
K INTENSITY (−)	3.5×10^5	5.3×10^4	2.5×10^4

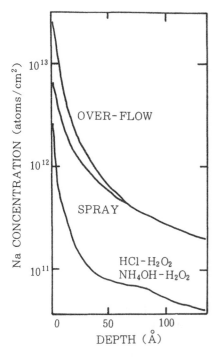

Figure 15   Na depth profile of the stainless steel surface after various cleaning treatments.

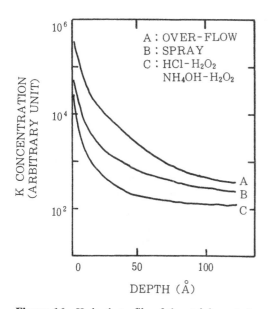

Figure 16   K depth profile of the stainless steel surface after various cleaning treatments.

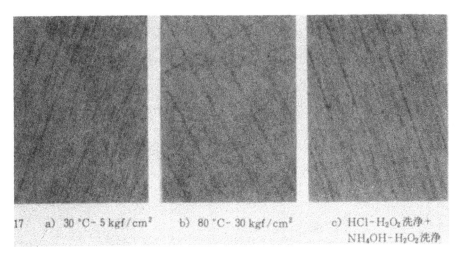

17   a) 30 °C - 5 kgf/cm²   b) 80 °C - 30 kgf/cm²   c) HCl - H₂O₂ 洗浄 +
                                                                        NH₄OH - H₂O₂ 洗浄

Figure 17   Stainless steel surface after various cleaning treatments (optical microscope, ×75).

and 30 kgf cm^{-2}, and HCl-H$_2$O$_2$ + NH$_4$OH-H$_2$O$_2$ cleaning. Optical microscope photographs of the stainless steel surface (magnification ×75) after oxidation are shown in Fig. 17. The surface treated by cleaning at 30°C and 5 kgf cm^{-2} has some white spots; the other two photographs show no spots. These spots are thought to be NaNO$_3$ residue from the polishing chemical that precipitated on the oxidized surface. This tells us that the lack of cleaning affects the oxidized surface.

D. Application of High-Temperature, High-Pressure Ultrapure Water Spray Cleaning

We describe a study in which we treated a process chamber with high-temperature, high-pressure ultrapure water spray cleaning. This is a cubic process chamber 30 cm on a side made of SUS 316 L. Table 5 lists the outline of the manufacturing process. Cutting and fabricating the stainless plate are followed by electrochemical buffing. After precision cleaning, the parts are assembled and welded, and then at the final stage, we spray cleaned with ultrapure water at 55°C and 20 kgf cm^{-2}. Figures 18 and 19 show the outgassing data when 350°C baking was used as a preprocessing before oxidative passivation.

   Figure 20 is a photograph of the inner surface of the oxidized chamber. The amount of outgassing is small, and the oxidized surface is even and free from damage. Figure 21 shows the change in vacuum when the unit was assembled and operated for the first time. As shown, the pressure ultimately reached 1 × 10^{-8} torr after 1.5 h of operation. When we treated the equipment by 29 h baking at 150°C, the pressure ultimately reached 5 × 10^{-11} torr. These results were obtained because the stainless steel was properly oxidatively passivated and the outgassing from the inner surface of the chamber was successfully suppressed to the lowest level.

IV. SUMMARY

The results observed during the various tests of high-temperature, high-pressure ultrapure water spray cleaning can be summarized as follows:

Table 5 Oxidation-Passivated Stainless Steel Chamber Manufacturing Process

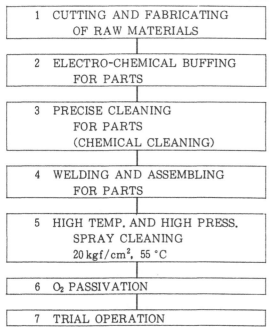

1 CUTTING AND FABRICATING
  OF RAW MATERIALS

2 ELECTRO-CHEMICAL BUFFING
  FOR PARTS

3 PRECISE CLEANING
  FOR PARTS
  (CHEMICAL CLEANING)

4 WELDING AND ASSEMBLING
  FOR PARTS

5 HIGH TEMP. AND HIGH PRESS.
  SPRAY CLEANING
  20 kgf/cm², 55 °C

6 O₂ PASSIVATION

7 TRIAL OPERATION

1. Cleaning at higher temperature and pressure yields better results for oil elimination.
2. This cleaning method has a limited effect on particle elimination, not sufficient for the semiconductor industry.
3. Although this cleaning method has some effect on metallic impurity elimination, further study is required to better understand the depth profile.
4. At the present stage of study, it is possible to apply this cleaning method to a final cleaning process. It can be also used when it is necessary to reduce the amount of chemicals.

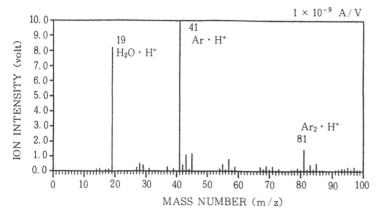

Figure 18 Outgassing from the process chamber immediately after baking at up to 350°C (APIMS).

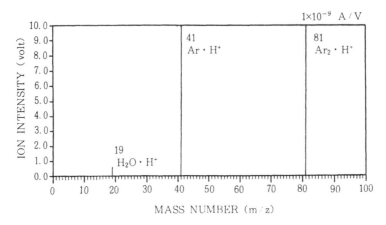

1×10⁻⁹ A / V

Figure 19   Outgassing from the process chamber after 15 h at 350°C (APIMS).

High-temperature, high-pressure ultrapure water spray cleaning technology is expected to be upgraded further when we investigate the nozzle shape and the use of pressures higher than 30 kgf cm^{-2}.

Because precision cleaning will play an increasingly important role in cutting-edge technology, such as semiconductors and biotechnology, we must study not only spray cleaning but also the wettability of ultrapure water and chemicals on the surface to be cleaned and the cleaning effect of high-temperature ultrapure water.

ACKNOWLEDGMENTS

All the results described in this chapter are from the superclean room in the Electric Communication Laboratory at Tohoku University and from the mini superclean room in the Engineering Department, Tohoku University. The authors express our thanks to all the people who helped us with the experiments, such as the staff of Prof. Ohmi's laboratory, Mr. Rokuheiji Sato of Nikuni Kikai Kogyo, and Mr. Kenji Sato and Mr. Yoshiyasu Baba of Nichizo Precision Polish.

Figure 20   Surface of the oxidation-passivated process chamber.

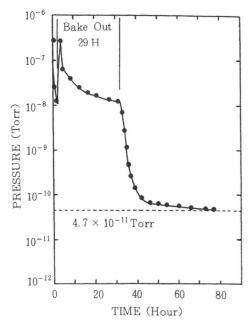

Figure 21 Pumping-down characteristics of a 30 cm cubic oxidation-passivated process chamber by a turbomolecular pump.

REFERENCES

1. K. Sugiyama et al., High Performance Technology Process III, Proceedings of Ultra Clean Technology Symposium No. 6, Ultra Clean Society, Tokyo, 1988, pp. 245–263.
2. H. Maehara et al., Ultra High-Purity Gas Supply System, Ultra Clean Society, Tokyo, 1986, pp. 332–369.
3. T. Ohmi et al., Ultra Clean Technology, Nikkei Micro Device, 1988, pp. 101–112.

8
The Future of Wet Processing

Tadahiro Ohmi
Tohoku University, Sendai, Japan

Ichiro Kawanabe and Masayuki Miyashita
Hashimoto Chemical Corporation, Sakai, Japan

Frederick W. Kern, Jr.
IBM Corporation, Essex Junction, Vermont

I. INTRODUCTION

The volume production of submicrometer ultralarge-scale integration (ULSI), such as the 4 Mbit DRAM (dynamic random access memory), has already been launched. Although dry processing has become the mainstream production process to deal with the extremely fine patterns, current dry processing, because of the insufficient performance of the equipment, causes many problems on the wafer surface, including metallic contamination and various defects. To remove the metallic contamination and defects, each dry processing step must currently be followed by wet processing.

The chemistry of aqueous solutions is a vast area. It can be applied to many phenomena within a limited range of heat energy, from room temperature to around 180°C. Selectivity is also very high. Wet processing is certain to continue as a very important technology in the future, if we can eliminate its single weak point, that it can conduct only isotropic etching in most cases.

The physics and chemistry of wet processing have hardly been described. This is because few evaluation technologies for the wet process have been developed. For wafer surface cleaning, the cleaning method developed by Kern of RCA in 1970 is still widely used [1]. At that time, the contamination level allowed on the wafer surface was 10^{14}–10^{16} atm cm^{-2}. At present, however, the allowed contamination level must be upgraded to 10^{9}–10^{10} atm cm^{-2}. Also, the critical dimension has changed from 10^{+} μm to the submicrometer level, and the junction depth is shallower, from a few micrometers to tens of nanometers. Therefore the RCA cleaning method, which in the past worked sufficiently, should be improved to cope with future devices with even finer patterns. This improvement will require a full understanding of the mechanism of wet processing. Simplification should be pursued in the course of this improvement, reducing the number of process steps and the consumption of chemicals.

It is important to recognize that wet processing is essential and to establish new wet processing techniques.

II.   PROCESS

Wet processing for semiconductor production can be categorized roughly as follows:

1.   Wafer surface cleaning
2.   Etching of each film
3.   Cleaning of process equipment and jigs

The RCA cleaning method is still employed, with modifications. For example, NH_4OH-H_2O_2-H_2O cleaning to remove adhered impurities has been modified by changing the composition ratio to remove adhered impurities without degrading the smoothness of the wafer surface [2]. In the original RCA cleaning method, the composition ratio was $NH_4OH/H_2O_2/H_2O = 1:1:5$. With this composition ratio, the surface smoothness is damaged although impurities can be removed. For a submicrometer device with shallow junctions, it is essential to maintain the surface smoothness during cleaning. It is effective to reduce the relative NH_4OH volume to maintain the surface condition while removing impurities.

Figure 1 shows an evaluation of cleaning efficiency when the NH_4OH concentration was changed. In this test, 3 inch wafers to which many polyethylene latex particle adhered were used. The NH_4OH concentration was changed from the original concentration of 1 to 0.5, 0.25, and 0.1. It was confirmed by SEM (scanning electron microscope) observation and by the spotlight method that the presence of haze reflects the surface condition of the wafer. It is obvious that NH_4OH-H_2O_2 cleaning at the original composition ratio degrades the surface [2] and causes serious damage to gate oxidation and storage capacitor oxidation, both of which are extremely thin films, and to shallow junction formation, such as source, drain, and emitter.

Figure 2 shows the particle removal efficiency. Even if the NH_4OH concentration is reduced to $1/10$ of the conventional concentration level, the particle removal efficiency remains sufficient. In practice, a composition ratio of $NH_4OH/H_2O_2/H_2O = 1:4:20$ is used in our laboratory at Tohoku University.

It was recently reported that HF-H_2O_2 cleaning is effective in removing heavy metals, particularly Cu [3].

It has been shown that the heavy metal contamination level should be suppressed to less than 1×10^{11} atm cm^{-2} (6×10^9 atm cm^{-2} if it is assumed the bulk is uniform) to suppress current leakage at the n^+p junction to less than 1×10^{-9} A cm^{-2} [4]. Figure 3 indicates the relationship between the contamination level of heavy metals (Cr and Ni) and the LOCOS defect generation ratio [5,6]. The heavy metal contamination level must be suppressed to less than 1×10^{10} atm cm^{-2} to limit defect generation in high-temperature processes such as oxidation, and to maintain current leakage of at the pn junction at low levels. Although no effective cleaning methods to remove heavy metals have been developed, HF-H_2O_2 cleaning is expected to work well. Figure 4 shows the heavy metal removal efficiency of various cleaning methods [3]. Note that the composition ratio should be properly controlled even with HF-H_2O_2 cleaning to maintain the surface smoothness.

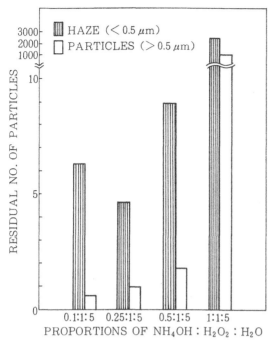

WAFER : CZ-N (100) 3-5 Ω • cm (3 in.)

POLYSTYRENE LATEX : 0.43 μm

INITIAL CONTAMINATION PER WAFER :
< 0.5 μm, 800 - 1800 ; 0.5 - 2.0 μm, 300 - 1000 ; 2 μm, 3 - 20.

REMOVAL OF POLYSTYRENE LATEX

Figure 1   Number of residual particles on wafers after cleaning with ammonia-hydrogen peroxide. (From Ref. 2.)

Thus various cleaning methods for Si wafers are being studied and improved. It is necessary to select the most suitable for each targeted contaminant considering the smoothness of the wafer surface and the generation of static electricity during cleaning, as well as efficiency of contaminant removal.

In semiconductor production, silicon and SiO_2 are the major materials to be etched. Despite the long history of wet etching, academic discussions of wet etching have just started, mainly because the wet etching reaction process was very much complicated by the large volumes of stabilizers added to chemicals whose purity was not sufficient. The residual impurities in chemicals have been markedly reduced as the process of producing and purifying chemicals has improved, the technology for filling bottles with chemicals has advanced, and high cleanliness levels have been achieved in the chemical distribution system. When impurities sharply decrease, most chemicals are stable because the causes of the decomposition reaction has disappeared. A typical example is H_2O_2. The purity level of H_2O_2 has risen so high that the original reaction process can be actualized. The Na concentration no longer rises even when H_2O_2 cleaning is repeated. Most of the Na is found in the added stabilizers.

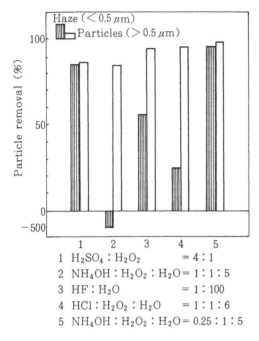

1 $H_2SO_4 : H_2O_2$ $= 4 : 1$
2 $NH_4OH : H_2O_2 : H_2O = 1 : 1 : 5$
3 $HF : H_2O$ $= 1 : 100$
4 $HCl : H_2O_2 : H_2O$ $= 1 : 1 : 6$
5 $NH_4OH : H_2O_2 : H_2O = 0.25 : 1 : 5$

Figure 2   Comparison of particle removal efficiencies of various cleaning methods when polystyrene latex spheres adhered to silicon wafers.

With silicon and SiO_2 as examples, points for the future improvement of wet processing are indicated here. With regard to wet processing, particularly etching, the following four points are important and essential: understanding (1) the reaction mechanism, (2) the saturation solubility of reaction products; (3) the stability of the etchant; and (4) the wettability of chemicals.

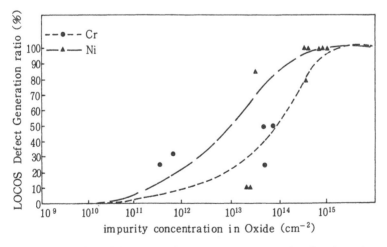

Figure 3   LOCOS defect generation ratio as a function of surface impurity concentration. (From Ref. 6.)

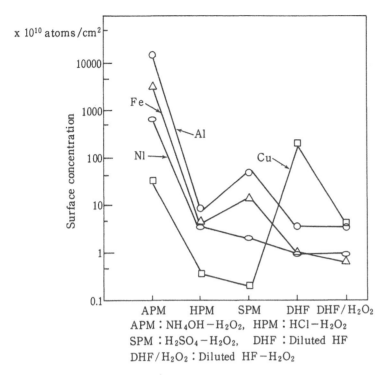

$x\ 10^{10}\ atoms/cm^2$

APM : $NH_4OH-H_2O_2$,   HPM : $HCl-H_2O_2$
SPM : $H_2SO_4-H_2O_2$,   DHF : Diluted HF
DHF/H_2O_2 : Diluted HF$-H_2O_2$

Figure 4   Surface impurity concentration for each cleaning solution. (From Ref. 3.)

1.   The Reaction Mechanism

First, the following questions must be answered with regard to etching: What kinds of ions exist in chemicals? Which ions govern the reaction? What kinds of reaction products are formed?

For example, the reaction between BHF and SiO_2 can be expressed as follows; the reaction product is ammonium hexafluorosilicate:

$$SiO_2 + 4HF + 2NH_4F \rightarrow (NH_4)_2SiF_6 + 2H_2O \tag{1}$$

The elementary process of Eq. (1) can be expressed as

$$SiO_2 + 3HF_2^- + H^+ \rightarrow SiF_6^{2-} + 2H_2O \qquad \text{surface reaction} \tag{2}$$

$$2NH_4 + SiF_6^{2-} \rightarrow (NH_4)_2SiF_6 \qquad \text{in the solution} \tag{3}$$

According to these equations, the major ions that contribute to etching are HF_2^- and H^+, and the reaction product is $(NH_4)_2SiF_6$ [7–9]. Because SiO_2 cannot be etched by NH_4F alone, which is a strong electrolyte, F^- does not directly contribute to the etching of SiO_2.

HF and NH_4F are ionized as follows:

$$HF \rightleftharpoons H^+ + F^- \tag{4}$$

$$NH_4F \rightleftharpoons NH_4^+ + F^- \tag{5}$$

$$HF + F \rightleftharpoons HF_2^- \tag{6}$$

Because HF is a weak electrolyte, HF_2^- is generated mainly from F^-, which is dissociated from NH_4F and neutral HF [7–9].

The relationship between the liquid composition and the HF_2^- can be measured by FT-IR [9]. The accurate empirical formula for the HF_2^- concentration, the H^+ concentration, and the etching rate of SiO_2 is also obtained [9]:

$$E_1 = 1282.8[HF_2^-] + 388.8[HF_2^-] \ \log \frac{[H^+]}{[HF_2^-]}$$

when

$$\frac{[H^+]}{[HF_2^-]} \geqq 2 \times 10^{-2} \tag{7}$$

$$E_2 = 757.9[HF_2^-] + 79.0[HF_2^-] \ \log \frac{[H^+]}{[HF_2^-]}$$

when

$$\frac{[H^+]}{[HF_2^-]} < 2 \times 10^{-2} \tag{8}$$

where: E_1, E_2 = etching rate, Å/minute

$[HF_2^-]$ = ion intensity of HF_2^-, mol · liter^{-1}

$[H^+]$ = ion intensity of H^+, mol · liter^{-1}

To etch accurately for submicrometer patterns, the concentration of HF_2^- and H^+ must be accurately controlled and the process must exhibit high accuracy in terms of time.

2.   Saturation Solubility of Reaction Products

In ideal etching, any reaction product should be immediately removed from the wafer surface. The reaction products must be identified and their properties and effects on etching fully grasped.

Figure 5 shows the relationship between the etching rate and the dissolved Si concentration in Si etching by HF-HNO_3 solution [10]. As the dissolved Si concentration increases, the etching rate decreases. The reaction can be expressed as

$$3Si + 4HNO_3 + 18HF \rightarrow 3H_2SiF_6 + 4NO + 8H_2O \tag{9}$$

Figure 5 indicates that the dissolved Si concentration that yields an etching rate of zero is around 9000 ppm. On the other hand, according to calculations based on Eq. (9), the dissolved Si concentration that yields an etching of rate zero is 18,000 ppm. This means that the drop in etching rate is actually twice as fast as in the theoretical case. This is thought to be because the reaction product of hydrogen hexafluorosilicate remains close to the Si surface and interferes with the reaction. Ideal etching cannot be realized if reaction products exist close to the wafer surface.

In the reaction between SiO_2 and BHF, occasionally the SiO_2 surface is not etched in uniformly in areas where the concentration of ammonium fluoride is high. This is another example of interference with the etching reaction by the insufficient saturation solubility of reaction products. It is possible to etch SiO_2 uniformly by lowering the concentration of ammonium fluoride, increasing the saturation solubility of ammonium hexafluorosilicate $[(NH_4)_2SiF_6]$ [9].

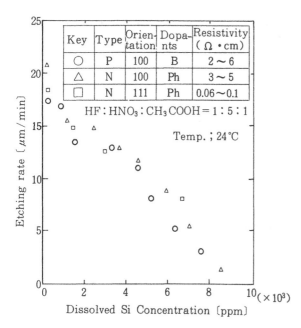

Figure 5 Relation between etching rate and dissolved Si concentration. (From Ref. 10.)

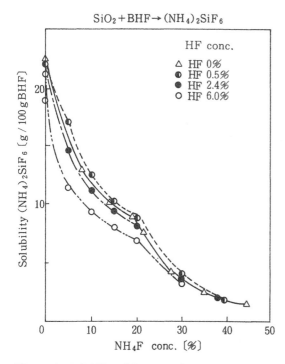

Figure 6 Solubility of the ammonium hexafluorosilicate [$(NH_4)_2SiF_6$] in NH_4F–HF–H_2O solution versus the NH_4F concentration at 25°C. (From Ref. 9.)

Figure 6 shows the saturation solubility of ammonium hexafluorosilicate in BHF. The solubility of ammonium hexafluorosilicate drops sharply as the concentration of ammonium fluoride increases. To ensure uniform etching of the wafer surface, higher solubility reaction products are preferred.

3.   Stability of Etchant

The chemicals should not change in quality, including liquid composition, as a result of temperature fluctuations during transportation and storage. Figure 7 shows the NH_4F concentration dependence of the crystal segregation temperature of BHF. The crystal segregation temperature changes over a wide range with the concentration of HF and NH_4F. Ice particles are segregated where the NH_4F concentration is low, and NH_4HF_2 is segregated where it is high. Once NH_4HF_2 is segregated, the liquid composition itself is changed because NH_4HF_2 is not fully dissolved even when the temperature returns to room temperature. Since the NH_4F concentration in conventional BHF is very high, 35–38%, the crystal segregation temperature is far higher. Such problems as the change in etching rate and particle deposition might be induced by NH_4HF_2 segregation unless the

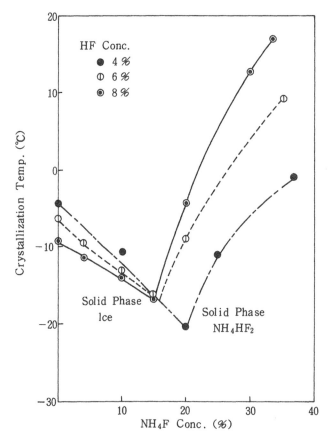

Figure 7   Solid-phase segregation temperature of NH_4F–HF–H_2O solution versus NH_4F concentration. (From Ref. 9.)

ambient temperature is properly controlled. To secure the reproducibility of wet processing, chemicals with as low a crystal segregation temperature as possible must be used.

4. *Wettability of Chemicals*

The sufficient wettability of chemicals on the targeted surface is essential to maintain a
uniform surface reaction, such as cleaning and etching, and to improve the reproducibility. In particular, a submicrometer device with a high aspect ratio requires good wettability of chemicals during cleaning and etching, so that the solid surface can become wet
enough, chemicals can penetrate the fine patterns, and the replacement of chemicals is
efficient. Also, the wettability contributes to removing reaction products from the wafer
surface in a short time.

In general, chemicals for semiconductors have low wettability on silicon wafers. The
wettability is evaluated based on the surface tension γ (dyn cm^{-1}) and the contact angle θ
(degrees) at the silicon surface. The wettability of major chemicals is as follows:

Ultrapure water: $\gamma = 72.5$ dyn cm^{-1}, $\theta = 85°$
H_2SO_4 (89%): $\gamma = 64$ dyn cm^{-1}, $\theta = 54.4°$
BHF (10:1): $\gamma = 87.8$ dyn cm^{-1}, $\theta = 70°$

The higher the value, the lower is the wettability. Therefore, ultrapure water and other
chemicals must be improved in various ways to demonstrate sufficient wettability on a
solid surface. The molecular structure can be changed, or surfactants can be added.
Sometimes the characteristics of the solid surface should be modified.

If the four parameters just discussed are well controlled, it becomes possible to
control the functions of chemicals and to accurately predict the results of wet processing.
This leads to the improvement in the production yield.

III. WET PROCESSING EQUIPMENT

At the present, the chemical, wafer-delay, and wafer transit system is widely employed in
wet processing for semiconductor production. There are some requirements for future wet
processing to deal with submicrometer patterns properly.

In the wafer transit system, in which wafers are transported and immersed in various
chemicals one after another, the wafers are more likely to be exposed to contaminants like
heavy metals and particles. Moreover, this system has a serious problem: many particles
on the surface of chemicals stick to the wafers when they are removed from the chemical
baths. Impurities also accumulate in etching and cleaning baths, which may increase the
possibility that impurities readhered to cleaned wafers and the etching is interfered with
by by-products formed in the baths. Circulating filtration is very effective in eliminating these problems. When impurities are present in the form of ions or colloids
and these adversely affect etching and cleaning, however, other countermeasures should
be taken.

Based on these considerations, the following concept is useful. Wafers are set in
place. Chemicals are supplied one after another. The wafers are always wet. These are
important points. In this process, chemicals are supplied only on the surface to be cleaned
or etched, which is cost effective. A new wafer-cleaning system based on this concept has
just become available on the market [11].

It has been confirmed that the electrical characteristics are dramatically improved on
a Si surface that is free of native oxide. For example, if native oxide is removed from the
wafer surface with HF and Al is deposited on this wafer at room temperature in an

environment in which native oxide cannot form (an environment in which oxygen and moisture do not coexist), a low-resistance ohmic contact of 0.4 $\mu\Omega\cdot$cm can be obtained [12]. This is why the suppression of native oxide growth during transportation constitutes a challenge.

It has also been reported that native oxide grows by several angstroms within a short period where oxygen and moisture coexist, such as during chemical cleaning, during ultrapure water rinsing, and in the air [13]. It is therefore how wafers are transported to the next process is very critical to suppressing native oxide formation.

The requirements for wet processing, including etching and cleaning, can be summarized as follows:

1.   Wafer-delay system for cleaning and etching
2.   Elimination of reaction products from the wafer surface
3.   Closed manufacturing

Finally, we discuss the chemical distribution system in wet processing. The concentration of heavy metals in chemicals for semiconductors is now shifting to the 0.1–0.01 ppb level. Yet this is not low enough. The cleanliness of chemicals should be further improved along with developments in analysis and evaluation technology.

The greatest problem with chemical distribution at present is the contamination caused by the piping system, including its components. Although fluorocarbon resins, including perfluoroalkoxyl, polytetrafluoroethylene, and polyvinylidene fluoride, polyether ether ketone, and clean polyvinyl chloride have been developed, none of these materials fully satisfy all the requirements, such as heat resistance, chemical resistance, low particulation, limited elution, and antielectrostatic properties. Semiconductor-grade fluorocarbon resin is now being developed by improving fluorocarbon resin materials and by controlling the environment of fluorocarbon resin production.

The technology to protect wafers from impurities is as important as the technology to remove impurities that adhere to wafers. With improvements in chemicals and other materials, the development of installation technology (the chemical distribution system, for example) must be addressed immediately.

IV.   SUMMARY

The following are now being studied in an attempt to realize more advanced wet processing:

Central control of chemicals
Automated chemical distribution system
Upgrading the purity and functionality of chemicals
Rapid sterilization of the ultrapure water delivery system
Suppression of total organic carbon

A drying process free of contamination is also essential. Drying methods using isopropyl alcohol vapor, spin drying with antielectrostatic measures, and high-temperature N_2 blow drying are now being studied.

A new cleaning and drying technology that is free of native oxide is being established. When moisture and oxygen coexist, the Si surface and the metal surface are oxidized even at room temperature. The wafer surface is oxidized most seriously when wafers are withdrawn from the final ultrapure water rinsing bath into the air. This is why a closed manufacturing system in which wafers are not exposed to the air is so important.

To develop new technologies, the evaluation technology to confirm the effect of certain new technologies is vital to conducting in situ quantitative observation of surface contamination with high accuracy.

REFERENCES

1. W. Kern and D. A. Puotinen, RCA Review, *31*, 187 (1970).
2. H. Mishima, T. Yasui, T. Mizuniwa, M. Abe, and T. Ohmi, IEEE Trans. Semiconductor Manufacturing, *2*(3), 69 (1989).
3. T. Shimono and M. Tsuji, Proceedings of 1st Workshop on Ultra Clean Technology, 1989, p. 50.
4. T. Nitta, N. Anzai, and J. Sugiura, IEICE Technical Report, 1989, p. 1.
5. Y. Matsushita, VLSI Symposium, 1989, p. 5.
6. H. Kamijo et al., Jpn. Appl. Phys. Spring Meeting, 1987.
7. H. Niesen and D. Hackleman, J. Electrochem. Soc., *30*(3), 708, (1987).
8. J. S. Judge, J. Electrochem. Soc., *118*(11), 1172 (1971).
9. H. Kikuyama, M. Miki, K. Saka, J. Takano, I. Kawanabe, M. Miyashita, and T. Ohmi, IEEE Trans. Semiconductor Manufacturing, *4*(1), 26(1991).
10. H. Mishima, N. Kowata, and T. Ohmi, Denshi Tsushin Gakkai, 29 (1986).
11. A. E. Walter and C. F. McConnell, Microcontamination, 35 (1990).
12. T. Ohmi and M. Miyawaki, Nikkei Microdevices, 112 (1990).
13. T. Ohmi, M. Morita, E. Hasegawa, M. Kawakami, and K. Suma, ULSI Science and Technology (1989).

IX

Future Tasks for Ultrapure Water

1
Future Challenges for Ultrapure Water

Hiroyuki Harada
Mitsubishi Corporation, Tokyo, Japan

Large-scale integration (LSI) has been integrated four times every 3 years since the development of the LSI device began with 1 kbit in 1970. The next generation 16 Mbit device is now in the final development stage, and the major development efforts are now shifting to the 64 Mbit device. Future LSI development will continue without slowing, 1 Gbit as well as 100 Mbit are now more likely to be realized.

Progress in LSI has been accompanied by progress in technology to improve the cleanliness level of gases, chemicals, and pure water, all of which form the very basis of LSI production. In the kilobit era, impurities were expressed as ppm. In the Mbit era, ppb is used to rate cleanliness and ppm order has come to be regarded as an index of dirtiness. In the coming Gbit era, ppt will probably supersede ppb as a measure of cleanliness. Thus the further integration of LSI requires the further development of technology to upgrade cleanliness in the basic technologies of LSI production.

The highest cleanliness level in the LSI production process is required by the ultrapure water used in the final rinsing stage of the wafer-cleaning process. The heavy metal and sodium concentration in chemicals used in semiconductor production have already been lowered below the ppb level. For the time being, ultrapure water must achieve the ppt level.

The first parameter to be noted in terms of evaluating ultrapure water quality is particulate contamination. For 64 and 256 kbit, 0.2 μm particles were expected to be removed from ultrapure water. As the LSI pattern is refined, the particle size to be controlled is also smaller. From the 1 Mbit era and on, we have tried to remove 0.1 μm particles. The particle count per milliliter was 50–100 in for 64 kbit. It was reduced to 10–50 for the 256 kbit device. For 1 Mbit, although the particle size to be controlled has decreased to 0.1 μm, the particle count per milliliter has reached even more stringent levels of 10–20. It is expected that 4 Mbit will allow a particle count of 5 or less and that

16 Mbit will allow less than 1. The particle size to be controlled will reach 0.05 μm in the 16 Mbit era.

Microorganisms are the major cause of particulate contamination in ultrapure water. As microorganisms propagate, the number of contaminants smaller than the microorganism itself increases. This is thought to be because microorganisms adhere to the inner surface of pipes to live, and they remain there even after they died. They gradually decompose and are peeled away. Therefore, once microorganisms propagate, particulation will not cease even after all of them are killed by sterilization.

The basic countermeasure to prevent microrganisms from propagating is to prevent them from penetrating the system. This is done using filters and ultraviolet (UV) sterilization. Furthermore, to prevent microorganisms from adhering to the inner surface of the system to propagate, it is very important to design entrapment-free pipes and sterilization methods for the entire system. Since microrganisms persistently adhere to the inner surface once they propagate, it is impossible to completely remove them to recover the original state.

For the piping technology to eliminate areas where liquid can be trapped and microorganisms multiply, butt welding is more effective than socket welding. In the past, however, the welded part had jigs and it was difficult to eliminate liquid entrapment completely. This problem has been addressed by developing better welding jigs, and it has become possible to obtain a smooth finish of the welded part. These advanced jigs are available on the market only for polyvinylidene fluoride resin at present, but they are expected to be modified and applied to other resins in the near future.

Sterilization of the ultrapure water system takes half a day to a full day when the system is filled with a very dilute hydrogen peroxide. Since the downtime caused by this sterilization method is as long as 1 day, the sterilization was conducted only a few times a year. This created long intervals between sterilizations, resulting in a high probability of microorganism propagation. Furthermore, if the hydrogen peroxide used in the sterilization is not pure enough, the system is contaminated and it takes a long time to return the system to normal conditions.

To tackle this problem, hot sterilization is mainly used now. This method supplies hot pure water to the system, and it is heated to above 80°C for 1 h. Since the time required for this method is less than 3 h even if the time for heating and cooling is included, the system can be sterilized frequently. Hot sterilization is highly reliable as a method to prevent microorganisms from entering the system.

If elution from pipes takes place because of hot water, the system is contaminated with the eluted materials, and it takes a long time to regain the original quality of ultrapure water after sterilization. Therefore piping materials with less elution on exposure to hot water should be selected. Suppression of elution from piping materials is also essential to upgrade the purity of ultrapure water. Polyether ether ketone (PEEK) is the best material in terms of suppressed elution. However, since the demand for PEEK is small, it is very expensive and not widely available. Recently more hot pure water systems employ PEEK, so the demand for PEEK is expected to grow.

PEEK is an excellent material not only in terms of its low elution but also in terms of its high heat resistance, light weight, small coefficient of thermal expansion, and high mechanical strength. It is therefore used for the cleaning jigs for semiconductors and liquid crystal, the carriers, and various wafer-handling jigs. For example, the PEEK wafer carrier is one-quarter lighter than the perfluoroalkoxyl (PFA) wafer carrier, and the processing accuracy of the PEEK wafer carrier is better than that of the PFA by one order of magnitude. Thus the PEEK resin will be used in LSI production and liquid crystal production, in particular in the automated process, so the demand for PEEK is likely to grow markedly.

Hot sterilization requires that the system be resistant to heat. The ion-exchange resin is poor in terms of heat resistance. An ion-exchange resin that withstands hot sterilization has not yet been developed. At present, hot sterilization is designed so that hot water does not pass the ion-exchange resins. Although they are not sterilized, this is not a great problem in terms of sterilization alone because the resins take up a very small amount of space in the ultrapure water circulating pipeline as a whole. Hot sterilization achieves a bacteria count of less than 1 liter^{-1}.

In the future, all the units connected to the ultrapure water production system that use ultrapure water, including the clean benches and the cleaning units, will need to be designed so that hot sterilization can be performed throughout the system, to the very end of the pipeline. Furthermore, in an attempt to suppress particulate contamination, piping components, such as valves and pumps, free of particulation and liquid entrapment, must be developed.

The second important problem for ultrapure water quality, following particulate contamination, is the total organic carbon (TOC) level. TOC elution from the piping system, whose surface area is the greatest in the pure water system, can be reduced by introducing PEEK and other materials with limited elution. When a TOC level of 1 ppb or less is the goal, elution from the ion-exchange resin should be addressed as well. Unfortunately, no plausible countermeasure to suppress elution from the ion-exchange resin has yet been found. Up to the ppb order, however, the TOC concentration has been lowered by combining the reverse osmosis (RO) membrane and UV oxidation.

Many LSI plants use an ultrapure water recycling system to conserve water and to process wastewater. The recycling system must remove low-molecular-weight organic materials, such as isopropyl alcohol (IPA). An RO membrane with an IPA removal efficiency of 90% or greater has been developed. By combining the RO membrane and UV oxidation, low-molecular-weight organic materials can be lowered to the ppb order. Water recycling technology has developed along with TOC removal technology. Some present LSI plants recycle 100% of the water and do not discharge wastewater at all. It is no longer true that clean water is indispensable to the semiconductor industry. Instead, people are indispensable, and healthy people are found where the water and air are clean. Future LSI ultrapure water systems should be designed based strictly on the water use program of the entire LSI plant.

Dissolved oxygen is another important parameter in evaluating water quality. In particular, the recent research of Ohmi et al. has revealed that dissolved oxygen triggers native oxide growth. Dissolved oxygen as well as microorganisms should be suppressed in the LSI production process. The suppression of native oxide growth is very important because native oxide formation leads to deviations in the contact resistance of through-holes and a rise in resistance. The data of Ohmi et al. indicate that dissolved oxygen should be lowered to the ppb order to completely suppress native oxide growth. The data also clearly indicate that the volume of oxygen dissolved from the air in ultrapure water is not negligible from the viewpoint of native oxide formation. This means a new wafer-cleaning unit based on the innovative concept of completely eliminating air should be introduced. It was mentioned earlier that hot sterilization requires integration of the entire ultrapure water system, including the clean benches and the cleaning units. Integration of the ultrapure water system, including every unit connected to it, is also required so that wafers can be cleaned without deteriorating the purity of water at the use points. This is the first priority for the future.

A native oxide-free cleaning process will be realized by introducing total system integration in which the purity of water at the use points is not degraded. This will enable some portion of the wastewater to be recycled to the middle stage of the ultrapure water production process to enhance the efficiency of the recycling operation; at present, all the wastewater is recycled back to the first stage of the ultrapure water production system, where source water is introduced.

Ultrapure water is used for cleaning because only a small volume is capable of dissolving anything. The cleaning efficiency of ultrapure water increases as it is heated. It is owing to this capability of ultrapure water that it can substituted for precision cleaning with Teflon. The demand for hot ultrapure water is expected to expand in the future, but the hot water circulating system has not yet been realized because of the poor heat resistance of the ion-exchange resin. A new heat-resistant ion-exchange resin should be developed along with piping materials that elute less. Moreover, piping technology that eliminates liquid entrapment, as well as new particulate-free and dead space-free valves, must be developed.

Perivaporization, which produces hot ultrapure water directly filtering steam with a molecular filter, has proved feasible in an experiment with a small pilot system. If a large pilot system (processing a few tons of water per hour) is constructed based on this result and it is put to practical use, the issue of hot pure water might be overcome.

The future progress of LSI integration is certain to be accompanied by the ultrahigh cleanliness of the basic technologies for LSI production, such as ultrapure water technology. To address this demand, analytic evaluation units with a higher sensitivity must be developed. The performance of the current TOC meter and particle counter should be improved. At the same time, a new evaluation method should be created. Macroscopic observation, a very basic evaluation approach, remains very important. A method in which particles ≤ 0.1 μm are retained in ultrapure water using a fine filter, for observation by scanning electron microscopy, should also be developed.

Although it has often been said that dry processing will completely take over wet processing in LSI production in a decade, this will not happen in the next 10 years. Advancements in ultrapure water technology will remain essential to future LSI developments.

2
Development of a Complete Ultrapure Water System

Takaaki Fukumoto
Mitsubishi Electronic Corporation, Kumamoto, Japan

I. INTRODUCTION

As semiconductor devices are increasingly integrated, the power, jigs, and equipment used in semiconductor production are required to feature higher levels of cleanliness. In particular, because ultrapure water directly contacts the silicon wafers and is used many times in each process, the quality of the ultrapure water greatly affects the yield and the reliability of the semiconductor devices. Therefore ultrapure water must be further improved to raise its quality. This chapter discusses the problems with current ultrapure water production systems and ultrapure water quality.

II. PROBLEMS

The current ultrapure water production system attempts to upgrade the purity of ultrapure water by employing the following:

1. Reverse osmosis membrane with low pressure and high demineralization efficiency and a high organic material removal efficiency
2. Ion-exchange resin with limited elution
3. Photooxidation to decompose and remove organic materials
4. Removal of dissolved oxygen with membranes and vacuum degasifier
5. Multilayer final purification filter, such as an ultrafiltration membrane

When these functions are introduced, the ultrapure water production system becomes very complicated, resulting in increased facility costs, maintenance costs, and installation space. The system is becoming more complicated because each unit fails to make the best use of its functions.

For example, the ion-exchange resin was originally installed for demineralization (ion removal). The regeneration cycle is determined by the ion concentration in the source

Table 1   Water Quality in a Two-Bed, Three-Tower Unit

	Inlet	Outlet	Rejection
Particle ($\geqq 0.1 \mu$m)	1.1×10^5	1.6×10^3	99 %
TOC (ppb)	1465	160	89 %
Colloidal iron (ppm)	4.9×10^{-4}	1.4×10^{-5}	97 %

water. However, the ion-exchange resin can remove particles, total organic carbon (TOC), and colloids, as well demineralized.

Table 1 lists the experimental results of measuring the particle count, the TOC concentration, and the colloidal iron content at the inlet and outlet of a two-bed, three-tower unit. It was found that particles, TOC, and colloidal irons are removed by 99, 89, and 97%, respectively, in the two-bed, three-tower unit. At present, however, the regeneration cycle and the method of ion exchange are not established by taking the other functions of the resin—removal of impurities other than ions—into consideration. Therefore the entire ultrapure water production system will be simplified if the functions of each unit not at present utilized are made the best use of and each unit is combined in an optimal manner.

Although the performance of each unit can be improved, the elution of organic materials is still detected from the reverse osmosis membrane, the ion-exchange resin, and the piping system, all of which are made using organic components. TOC elution from the reverse osmosis membrane and from the ion-exchange resin was measured. Figure 1 shows the initial TOC elution from the polyamide low-pressure reverse osmosis mem-

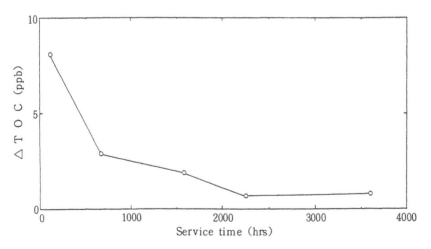

Figure 1   Relation between Δ TOC and service time in the RO unit.

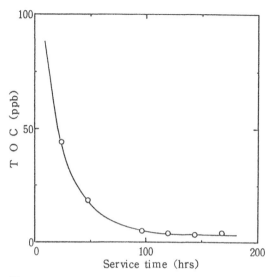

Figure 2   Relation between TOC and service time in the AP + CP unit.

brane as a function of time. The TOC elution immediately after the start of the operation was over 8 ppb and was still detected even after 2000 h. It takes more than 100 h to stabilize the TOC elution level.

Figure 2 shows the TOC concentration at the outlet of the anion polisher (AP) + cartridge polisher (CP) as a function of time. As with the reverse osmosis membrane, TOC elution is detected at the initial stage of the operation and over 100 h is required to stabilize the TOC elution level.

The experimental results shown in Figs. 1 and 2 indicate that the high TOC elution level from each component is observed at the initial stage and that a long period of rinsing is required to lower the initial elution. Moreover, the initial elution contaminates the pipes and units downstream. In other words, the initial elution causes self-contamination (internal contamination).

When the cartridge polisher is replaced, the differential pressure of the ultrafiltration unit, the final-stage filter in the ultrapure water production system, frequently rises. This phenomenon is caused when colloidal materials whose cutoff molecular weight is greater than that of the ultrafiltration membrane, 6000–10,000, elute from the ion-exchange resin and are adsorbed on the ultrafiltration membrane surface.

In an effort to overcome this problem, precleaning is conducted. Figure 3 compares a precleaned reverse osmosis membrane and a reverse osmosis membrane that was not precleaned by measuring the TOC concentration at operation start-up. The start-up time can be shortened by treating the reverse osmosis membrane by precleaning.

It is very important to take the following actions:

Suppress the elution from each component, such as the reverse osmosis membrane and the ion-exchange resin.
Install the component units in a clean manner.
Further improve the water quality by precleaning.
Start up the operation rapidly.
Reduce problems.

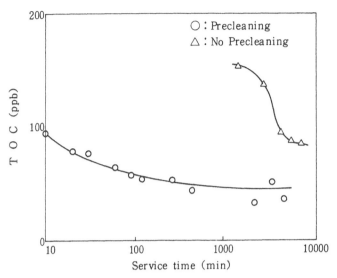

Figure 3   Relation between TOC and service time in the RO unit.

III.   COLLOID REMOVAL TECHNOLOGY

In the semiconductor production process, residue (stain) mainly composed of SiO_2 sometimes appears on the wafer surface after wet etching. Its causes are as follows:

1.   When wafers are rinsed with ultrapure water containing a few ppm dissolved oxygen, the silicon wafers dissolve in ultrapure water slightly, and the dissolved materials are occasionally left on the wafer surface as residue during the drying process.
2.   Colloids (colloidal silica, organic materials, and metals) in ultrapure water or chemicals remain on the wafer surface as residue during the drying process.

The colloids are not generally included at present in the checklist for controlling the ultrapure water quality. They are somewhat correlated with nonvolatile impurities since they remain on the wafer surface during drying. Some colloids can be removed during membrane separation, such as ion exchange, reverse osmosis, and ultrafiltration, but complete removal is not possible with present systems and methods.

To improve the colloid removal efficiency, the following are necessary:

1.   As mentioned earlier, colloids are removed in the two-bed, three-tower unit. Some colloids become particulates in the cation tower, where the pH is on the acid side. The particulate colloids are removed in the following anion tower. The two-bed, three-tower unit should be operated with the goal of colloid removal.
2.   Colloids are not completely removed even with a reverse osmosis membrane with a pore size of 5–10 Å. This is partly because colloids are very unstable in pure water and partly because there are pinholes on the reverse osmosis membrane. Therefore it is important to eliminate the pinholes and assemble the reverse osmosis membrane in a clean manner.
3.   A new technology to remove colloids efficiently should be developed and introduced.

Several new methods of colloid removal, including electromagnetic field treatment [1] and the oxidation treatment with oxygen bubbling [2], are described here.

A. Electromagnetic Field Treatment Technology

This method targets one of the characteristics of colloids in pure water, the ζ potential.

First the resistivity of pure water should be raised to 5–10 MΩ·cm or higher to eliminate ionic interference. Then, at an arbitrary stage, an electrical or magnetic field is applied to induce the colloids to adsorb to the electrodes or the magnetic poles. In this way, colloids that are not completely removed in present systems can be removed.

B. Oxidation Treatment with Oxygen Bubbling

Since colloids are very unstable in pure water, they cannot be completely removed with reverse osmosis. To improve the colloid removal efficiency of the reverse osmosis membrane, this method induces the unstable colloids to contact oxygen and be oxidized. [2]. The oxidized colloids become stable solid particles. This method is effective to easily remove colloids.

IV. CLOSED SYSTEM

Environmental issues are increasing in importance, including the reduction in chlorofluorocarbons to prevent ozone depletion and reduction in carcinogenic materials, such as trichloroethylene. It is therefore increasingly important for semiconductor plants to reduce wastewater drained to the outside by introducing a closed ultrapure water system to reclaim and reuse waste ultrapure water.

The major constituents in the wastewater drained from wet processing in the semiconductor production line are anions, such as HF, hydrochloric acid, and nitric acid. The reclamation system for this type of wastewater should include the following:

First, remove the trace surfactants and hydrogen peroxide from the wastewater with active carbon to protect the following ion-exchange resin unit from contamination and degradation.

Remove the anions, the major constituents, using a weakly basic anion-exchange resin unit.

Remove residual trace ions using a strongly acid cation-exchange resin unit and a strongly basic anion-exchange resin unit.

After this treatment, wastewater can be recycled back to the ultrapure water production system for reuse. Table 2 shows an example of the quality of wastewater treated in accordance with this agenda. It is clear that the recovered water can be reused in the ultrapure water production system.

V. REQUIREMENTS FOR ULTRAPURE WATER QUALITY

Ions, particles, TOC, bacteria, silica, and dissolved oxygen have been used to evaluate ultrapure water quality in terms of impurities. The resistivity is used to indicate the total volume of ions in pure water. The theoretical limit is 18.24 MΩ·cm, but the actual value in ultrapure water has been raised to 18 MΩ·cm or higher at 25°C. The actual values of impurities, such as particles, TOC, bacteria, silica, and dissolved oxygen, are also nearing the detection limits of present analysis methods. The specifications are usually set at the level of the detection limits. The principles of present high-sensitivity analyzers and their detection limits are described here.

Table 2   Water Quality in a Wastewater Recovery System

	Waste water	Treated water
Conductivity (μ S/cm)	460 ~ 848	0.22
pH (−)	2.4 ~ 2.7	7.4
TOC (ppb)	135	40

1. Resistivity Meter

The resistivity in water can be calculated by multiplying each ion concentration by the equivalent conductivity. However, the equivalent conductivity and the concentrations of H^+ and OH^- generated by the dissociation of water molecules are temperature dependent. Therefore, accurate temperature compensation is necessary. The resolution is 0.01 MΩ·cm; the measurement accuracy is within ±1% FS.

2. Particles

There are two methods of counting particles: direct microscopy and the particle counter. In direct microscopy, ultrapure water is filtered through a membrane filter, and the retained particles are measured by optical microscopy ($\geq$0.2 μm) or scanning electron microscopy (SEM, <0.2 μm).

The laser particle counter is currently popular. To improve its sensitivity, an Ar^+ laser is used. The minimum particle size is 0.07 μm; the measurement accuracy is 1 particle ml^{-1}.

A more advanced laser particle counter is now being developed. By improving the sensitivity of detecting scattered light, the measurement accuracy has been upgraded by three orders of magnitude to 1 particle per 1000 ml (minimum particle size of 0.2 μm).

3. TOC Meter

Potassium persulfate (oxidative reagent) and sulfuric acid for pH adjustment are added to a water sample. Inorganic carbon is then removed from the water sample by nitrogen gas. At high temperature and pressure, organic materials in the water sample are decomposed to carbon dioxide, which is measured by NDIR and converted to the TOC value. The detection limit is 1 ppb.

4. Bacteria

Sample water is filtered through a 0.45 μm membrane filter. Bacteria retained on the filter are incubated. The colonies that form are counted. The detection limit is 1 bacterium liter^{-1} (by increasing the volume of sample water to be filtered).

5. Silica Meter

The molybdate blue method uses ammonium molybdate and reducing agent. By improving the sensitivity of the absorbance detector, a very low concentration can be analyzed. The detection limit is 0.2 ppb.

6. Dissolved Oxygen Meter

With a membrane polarographic electrode, the dissolved oxygen concentration, even at low concentration levels, can be measured constantly. The detection limit is 0.2 ppb.

7. Nonvolatile Impurities

The sample water is atomized and dried, and the residual particles are measured with a particle counter. The resolution is 0.1 ppb.

VI. SUMMARY

When the required quality level of ultrapure water is considered, the impurity level must be determined, taking the effect of impurities in ultrapure water on semiconductor devices into consideration. To eliminate damage to semiconductor devices, the volume of residue on the wafer surface should be suppressed to less than 10^3 atoms cm^{-2} [4]. In other words, the nonvolatile impurities or colloids should be lowered to the ppt level. Current ultrapure water production technology is not able to lower the impurity level in ultrapure water to the ppt order.

Native oxide forms when moisture and oxygen coexist, and Si dissolution into ultrapure water depends on the dissolved oxygen concentration [5]. Based on these findings, it is obvious that native oxide growth can be suppressed by the following steps:
1. Reduce the moisture and oxygen on and around the wafer surface.
2. Reduce the dissolved oxygen concentration in the ultrapure water.

Moreover, the dissolved oxygen concentration in ultrapure water should also be lowered to suppress silicon dissolution into ultrapure water, which leads to the appearance of stain on the wafer surface.

The following challenges are expected to be overcome in the future:
1. Ultrapure water containing residues at the ppt level after evaporation
2. Ultrapure water with less dissolved oxygen (less than 1 ppb)
3. Measurement tools for ultrapure water for constant measurements with high sensitivity (ppt level)

VII. AFTERWORD

Along with the further integration of semiconductor devices, ultrapure water is expected to have higher quality. This trend makes the ultrapure water production system more and more complicated, resulting in increased facility costs, maintenance costs, and installation space. Another problem with current ultrapure water production systems is that they cannot completely remove colloids, which is one of the causes of stain on the wafer surface.

A complete ultrapure water production system that perfectly copes with the fluctuations in the quality of the source water is expected to be developed. It will be simple and inexpensive. The complete system should remove colloids effectively and adopt complete wastewater recycling.

REFERENCES

1. Japanese Patent Application 60-30714.
2. Japanese Patent Application 61-180601.

3. Japanese Patent No. 1385937.
4. T. Mizuniwa et al., ULSI Ultra Clean Technology Symposium No. 7, Ultra Clean Society, Realize, Inc., Tokyo, (1988), p. 31.
5. T. Ohmi et al., ULSI Ultra Clean Technology Symposium No. 8, Ultra Clean Society, Realize, Inc., Tokyo, (1989), p. 169.

Index

921

Printed and bound by CPI Group (UK) Ltd, Croydon, CR0 4YY

23/10/2024

01778259-0014